# Stereotaxic and Chemoarchitectural Atlas of the Brain of the Common Marmoset (*Callithrix jacchus*)

# Stereotaxic and Chemoarchitectural Atlas of the Brain of the Common Marmoset (*Callithrix jacchus*)

Craig D. Hardman
University of New South Wales
Sydney, Australia

Ken W. S. Ashwell
University of New South Wales
Sydney, Australia

CRC Press
Taylor & Francis Group
Boca Raton London New York

CRC Press is an imprint of the
Taylor & Francis Group, an **informa** business

CRC Press
Taylor & Francis Group
6000 Broken Sound Parkway NW, Suite 300
Boca Raton, FL 33487-2742



Printed in the United States of America on acid-free paper
Version Date: 20111213

International Standard Book Number: 978-1-4398-3778-8 (Hardback)



**Library of Congress Cataloging-in-Publication Data**

Hardman, Craig D.
Stereotaxic and chemoarchitectural atlas of the brain of the common marmoset (Callithrix jacchus) / authors, Craig D. Hardman, Ken W.S. Ashwell.
p. cm.
"A CRC title."
Includes bibliographical references and index.
ISBN 978-1-4398-3778-8 (hardcover : acid-free paper)
1. Marmosets--Anatomy--Atlases. 2. Brain--Anatomy--Atlases. 3. Stereotaxic techniques--Atlases. 4. Brain chemistry--Atlases. 5. Marmosets as laboratory animals--Atlases. I. Ashwell, Ken W. S. II. Title.

QL737.P925H37 2012
599.8'4--dc23 2011046926

**Visit the Taylor & Francis Web site at**
**http://www.taylorandfrancis.com**

**and the CRC Press Web site at**
**http://www.crcpress.com**

*To Rebecca, Xanthea, Louis, Juliette, Jennifer, Christine and Peter*

# Contents

Preface ............ ix

Acknowledgments ............ xi

Authors ............ xiii

**SECTION A Introduction** **1**

**SECTION B Named structures and abbreviations** **19**

**SECTION C Index of atlas plates** **31**

**SECTION D Atlas plates** **47**

**Coronal Plates** **49**

**Sagittal Plates** **449**

**Supplementary Resources Disclaimer**

Additional resources were previously made available for this title on CD. However, as CD has become a less accessible format, all resources have been moved to a more convenient online download option.

You can find these resources available here: www.routledge.com/9781439837788

Please note: Where this title mentions the associated disc, please use the downloadable resources instead.

# Preface

Developing better therapies for neurological conditions such as Parkinson's and Alzheimer's diseases remains an enduring problem for 21st century medicine. While the delivery of future treatments is likely to arise from molecular biological techniques, the testing of novel therapies will continue to require a robust experimental animal model. The marmoset is an ideal animal model for modern neurological research because of the species' convenient body size, ease of handling and the fundamental similarity of its neuroanatomy to the human brain. With this in mind, we have developed this comprehensive atlas, which combines modern imaging (MRI), with an accurate stereotaxic and chemoarchitectonic section atlas. We are confident that the work will prove to be a useful tool in research laboratories grappling with the problem of developing more effective therapies for disabling and debilitating neurological disorders.

# Acknowledgments

We would like to acknowledge the support and assistance of Prof. Malcolm Horne and Dr. David Finkelstein of the Department of Medicine, Monash University; and the encouragement of Prof. George Paxinos of Neuroscience Research Australia (NeuRA). We are also grateful to Dr. Maria Sarris of the Histology and Microscopy Unit of the School of Medical Sciences, The University of New South Wales (UNSW), for her assistance with slide scanning and to Xanthea Hardman for her assistance in editing the large number of illustrations. This work was supported in part by grants to Craig Hardman from the Australian Commonwealth National Health and Medical Research Council and the faculty of the UNSW Department of Medicine.

# Authors

**Craig Hardman, BSc(Hons), PhD, (UNSW),** has published extensively on the comparative neuroanatomy of motor pathways across species (rodents, non-human and human primates) and in disease states in humans. He also has more than 18 years of experience in teaching neurosciences and gross anatomy to medical and science students at the University of New South Wales, (UNSW).

**Ken Ashwell, BMedSc, MBBS, (UNSW), PhD, (USyd),** has published more than 100 papers on comparative and developmental neuroscience, as well as authoring five atlases of the mature and developing brains of rodents and marsupials. He also has more than 26 years of experience teaching in the fields of neuroscience, gross anatomy, embryology and anthropology to medical and science students at the UNSW and the University of Sydney (USyd).

Section A

# Introduction

Ecology, Behavior and Evolution of the Common Marmoset . . . . . . . . . . . . . . . . . . . . . . . . . . . . . . . . . . . . . . . 3
The Marmoset in Neuroscience Research . . . . . . . . . . . . . 3
Key Features of This Book . . . . . . . . . . . . . . . . . . . . . . . . 3
Animals and Ethical Considerations . . . . . . . . . . . . . . . . . 5
Methods: MRI, Perfusion, and Tissue Processing . . . . . . . . 5
Imaging, Principles of Delineation, and Abbreviation. . . . . 7
Quantitation of Volumes and Neuronal Populations of Motor Nuclei . . . . . . . . . . . . . . . . . . . . . . . . . . . . . . . . . . . 7
Bases for Delineation of Specific Regions and Nuclei . . . . 8
References . . . . . . . . . . . . . . . . . . . . . . . . . . . . . . . . . . . 14

## Ecology, Behavior and Evolution of the Common Marmoset

Understanding the natural environment, behavior and evolutionary history of an animal is essential to comprehending its brain structure and function and making intelligent use of it as a model of human disease.

The common marmoset (*Callithrix jacchus*) is a small New World primate belonging to the family Callitrichidae (marmosets and tamarins). In terms of size, some morphological features (e.g., having a mixture of claws and nails), and diet (insects and tree gum, as well as leaves and fruit), the marmosets are more like the so-called lower primates of the Old World (i.e., prosimians) than Old World monkeys. In fact, some authors consider that the marmosets occupy the same niche in the New World that tarsiers, lemurs and lorises occupy in the Old World (Klein 1999).

Marmosets are diurnal in habits and highly social (like cebids and cercopithecoids, the true monkeys of the New and Old World, respectively). The bottom teeth are adapted for gouging holes in bark to encourage gum to flow, but marmosets prefer to feed on fruit, nectar, flowers, invertebrates, and small vertebrates; turning to gum only when fruit is in short supply (Dunbar and Barrett 2000). As noted above, the digits have laterally compressed nails, except for the great toe, which has a true nail. In the wild, marmosets engage in scent-marking of gum holes to establish ownership. Marmosets prefer habitats with bushy vegetation and vine tangles, because these provide the safest sleeping sites and cover from the attention of monkey-eating birds of prey.

As diurnal primates, the marmosets are dependent mainly on vision, both for food acquisition and predator avoidance. This is reflected in the large size of visual pathway nuclei and primary visual and association cortical areas. The relatively smooth isocortex also has extensive somatosensory areas and a complex prefrontal cortex, making them ideal models for study of cortical function and degenerative disease. Nevertheless, researchers should bear in mind that the lineages of New and Old World primates separated 40 million years ago (Takahata and Satta 1997), so there has been considerable time for divergent isocortical evolution.

## The Marmoset in Neuroscience Research

Although marmosets are less widely used in neuroscience research than rodents, they are potentially very important as neurological models. Marmosets have several distinct advantages as a research species:

1. They are primates and therefore genetically and neurologically closer to humans than rodents.
2. They show more complex social behavior than non-primates and exhibit some features of primate neurological specialization (e.g., handedness; Rothe 1973) not seen readily in carnivores and rodents.
3. They are smaller and easier to handle than macaques. This represents potential savings to researchers and institutions in maintenance and breeding costs and avoids the hazards of working with larger primates.
4. They breed more rapidly than larger primates and have shorter generation and lifespan times. This facilitates long-term studies of neurodegenerative diseases such as Alzheimer's and Parkinson's disease

The number of neuroscientific studies of marmosets has been increasing steadily over the last two decades (Figure 1.1a), indicating that this species is finding its place as an important neurological research model.

## Key Features of This Book

Essential tools for studying the marmoset brain and making use of the species in neurological disease models are:

1. in vivo imaging of the brain in situ;
2. an accurate stereotaxic atlas
3. a chemoarchitectural atlas using modern high-contrast immunohistochemical techniques; and
4. reference quantitative data concerning neurologically important brain regions (e.g., cerebral cortex, basal ganglia and cerebellar nuclei). The present work brings all four elements together in a comprehensive research tool.

Stereotaxic atlases are essential for accurate placement of brain lesions and electrode insertion. To be of the maximal benefit to neuroscience researchers, such an atlas should preferably blend the most modern imaging techniques, e.g., MRI imaging, with classical stereotaxic techniques reproducible by laboratories throughout the world. This book provides the first accurate stereotaxic atlas of

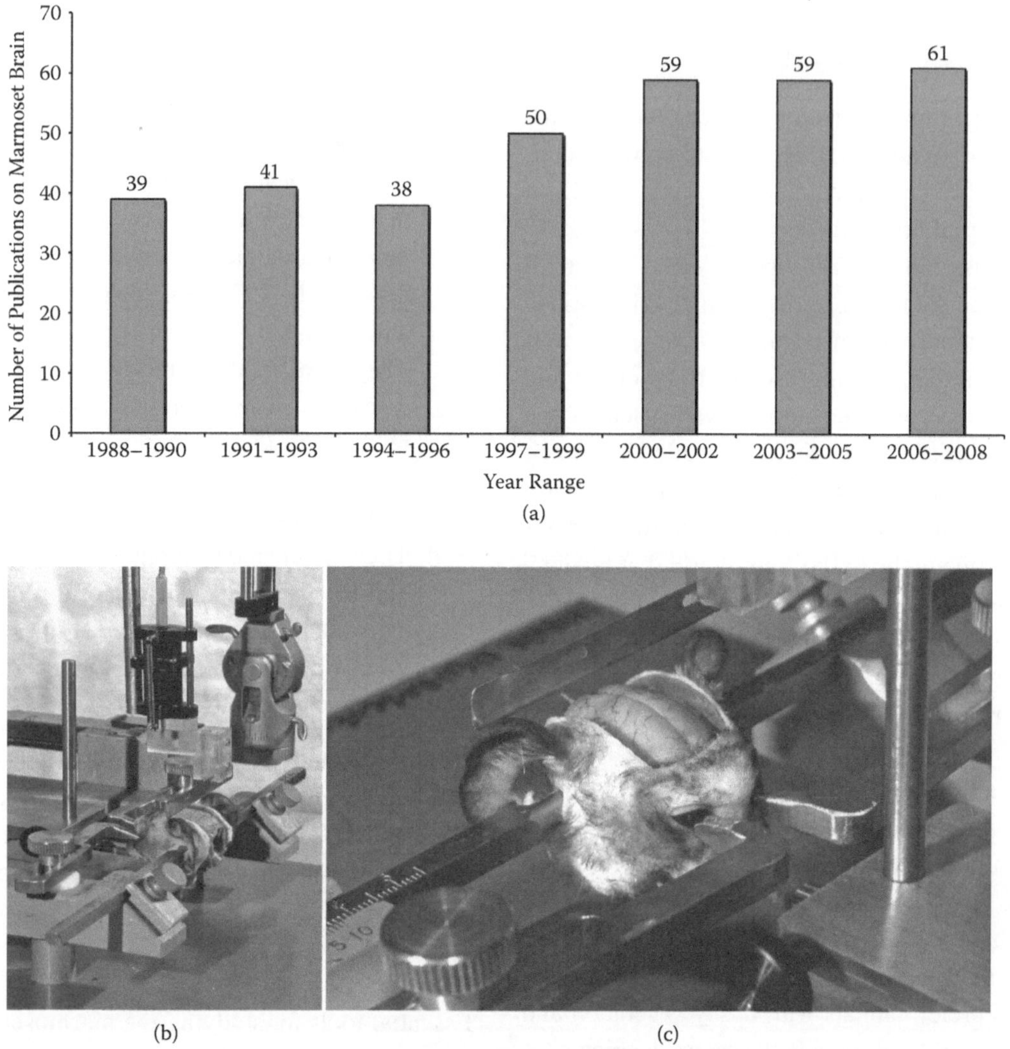


**FIGURE 1.1** **Figure 1.1a shows number of papers published on the marmoset brain during the years indicated. Figure 1.1b shows a left posterior view of a post-mortem marmoset head mounted in the stereotaxic apparatus used for insertion of fiducial needles and blocking of the brain. Figure 1.1c shows a right anterior view of a post-mortem marmoset head with the brain surface exposed prior to blocking in the coronal plane. Note the ear bars inserted into the external acoustic meatus and the eye bars in contact with the inferior margin of the orbit. These four points are in the same horizontal plane.**

all brain regions of the marmoset with comprehensive delineations using universally accepted nomenclature.

Chemoarchitectural atlases provide additional information on boundaries between brain regions, reinforcing the information contained in a stereotaxic atlas. In addition, chemoarchitecture with functionally important immunohistochemical and enzyme histochemical markers provides biochemically important information. In this book, both stereotaxic and immunohistochemical information have been represented using the same animal to allow the reader to match chemically identified regions with their stereotaxic coordinates.

For Nissl staining, slides were initially dehydrated, cleared of fat and rehydrated via immersion for 5 minutes in each of 70% ethanol, 95% ethanol, 3 changes of 100% ethanol, 3 changes of xylene, 3 changes of 100% ethanol, 95% ethanol, and 70% ethanol. They were then dipped in distilled water and stained for 15 to 20 minutes in 0.5% cresyl violet (pH 3.9). Sections were differentiated in water for 3 to 5 minutes and dehydrated with 70% ethanol, 95% ethanol and 3 changes of 100% ethanol before clearing in 3 changes of xylene and being coverslipped.

## Animals and Ethical Considerations

The animal ethics committees of both Monash University and the Howard Florey Institute in Melbourne approved all procedures described in this book. All activities were in accord with the National Health and Medical Research Council (Australia) guidelines for the care and use of experimental animals.

The findings described in this atlas have been based on experiments performed on 12 common marmosets (seven male and five female), although the plates depicted illustrate sectioned material from only two of these. The details of the two animals depicted are given in Table 1.

## Methods: MRI, Perfusion, and Tissue Processing

### *Magnetic Resonance Imaging*

The heads of two animals (marmosets 4 and 6) were imaged by T2 weighted MRI (TR = 1700 ms; TE = 18.5 ms) in the coronal and sagittal planes at the 4.7 Tesla (150 mm horizontal bore) Magnetic Resonance Imaging Facility of the Howard Florey Institute, Melbourne. To calm and immobilize the animals during transportation and imaging, the marmosets were administered an initial intramuscular 2.25 ml/kg body weight injection of an 8:1 mixture of Saffan (previously known as CT1341; 12 mg/ml) and atropine (0.6 mg/ml), followed by an intravenous 1ml/kg body weight injection of a 7:3 mixture of ketamine (100 mg/ml)

**Table 1 Details of marmosets depicted in the atlas plates**

| Identifying Number | Marmoset 7 | Marmoset 5 |
|---|---|---|
| Age (years) | 4.5 | 8 |
| Sex | male | female |
| Body mass (g) | 286 | 244 |
| Fixed brain mass (g) | 5.67 | 8.00 |
| Fixed brain volume (ml) | 5.50 | 7.65 |
| Fixed brain length (mm) | 31.0 | 31.5 |
| Fixed brain width (mm) | 21.0 | - |
| Section plane | Coronal | Sagittal |
| Stained series (in sequence) | SMI-32 Ab (neurofilament) | SMI-32 Ab (neurofilament) |
| | Parvalbumin | Acetylcholinesterase |
| | Calbindin | Cresyl Violet |
| | Calbindin | Cresyl Violet |
| | Cresyl Violet | Parvalbumin |
| | SMI-32 Ab (neurofilament) | Parvalbumin |
| | SMI-33 Ab (neurofilament) | Calbindin |
| | Acetylcholinesterase | SMI-32 Ab (neurofilament) |
| | Parvalbumin | Cresyl Violet |
| | Cresyl Violet | Calbindin |

and xylazine (20 mg/ml). A plastic framework was used to ensure that the head was in an orientation that matched the stereotaxic axes of the sectioned brains.

## Animal Perfusion, Stereotaxic Procedures and Tissue Preparation

All 12 animals were perfused with 0.9% saline in 0.1M phosphate buffer (pH 7.4), followed by 4% paraformaldehyde in 0.1 M phosphate buffer (pH 7.4). The skulls were positioned in a stereotaxic holder with the external acoustic meatus and the center of the inferior orbital margin oriented in the same horizontal plane. The description of methods is focussed on the two animals whose brains are depicted in the atlas plates, but all animals were processed in a similar fashion.

For marmoset 7, three 21-gauge needles were slowly advanced rostrocaudally through the left hemisphere under stereotaxic control to provide fiduciary marks. While the head was still in the stereotaxic frame (Kopf, Tujunga, CA, USA), the brain was cut into rostral and caudal portions by a coronal incision at the level of AP 0.0 (Figure 1.1b and Figure 1.1c). For marmoset 5, one 21-gauge needle was slowly advanced vertically through the right hemisphere of the brain at the level of the interaural line, such that the posterior margin of the hole lay at AP 0.0. Two 21-gauge needles were also advanced horizontally from the side in the coronal plane such that the anterior margins of the holes were at AP 0.0 and +7.1 mm, respectively. The dissected brains were photographed in lateral profile. After cryoprotection in 30% sucrose in 0.1M Tris (hydroxymethylmethylamine) buffered saline (pH = 7.4), left and right hemispheres were cut with the aid of a cryostat in the coronal plane (marmoset 7) or sagittal plane (marmoset 5) into 50 μm thick sections, each arranged into 10 series (see Table 1).

## Nissl and Acetylcholinesterase Staining

For the Nissl staining, the mounted sections were dehydrated, cleared of fat and rehydrated via immersion for 5 minutes in each of the following: 70% ethanol, 95% ethanol, 3 changes of 100% ethanol, 3 changes of xylene, 3 changes of 100% ethanol, 95% ethanol and 70% ethanol. They were then dipped in distilled water and stained for 15–20 minutes in 0.5% cresyl violet (pH = 3.9). Sections were then differentiated in water for 3 to 5 minutes, dehydrated via immersion in 70% ethanol, 95% ethanol, 3 changes of 100% ethanol and 3 changes of xylene for 5 minutes each and then coverslipped.

The method for acetylcholinesterase (AChE) followed the method of Koelle and Friedenwald (1949), and Lewis (1961) as adapted by Paxinos and Watson (1986).

## Immunohistochemistry

For the immunohistochemistry, free-floating sections were washed in 50% ethanol, a solution of 3% $H_2O_2$ in 50% ethanol, and 10% normal horse serum in 0.1% azide in 0.1 M Tris solution (pH 7.4) to expose target antigens and block endogenous peroxidase activity and non-specific binding. Following these treatments, sections for the atlas brains were incubated in either mouse monoclonal parvalbumin antibody (Parv, P3171, diluted 1:10,000, Sigma-Aldrich Corp. USA), SMI-32 mouse monoclonal (SMI-32, against non-phosphorylated neurofilament protein, diluted 1:10,000, Sternberger Monoclonals Inc.), SMI-33 mouse monoclonal (SMI-33, against non-phosphorylated neurofilament protein H and M, diluted 1:10,000, Sternberger Monoclonals Inc.), or mouse monoclonal calbindin antibody (Calb, C8666, diluted 1:1,000, Sigma-Aldrich Corp. USA). Sections from non-atlas brains used for quantification of basal ganglia were also incubated in antibodies for substance P (SP, rat monoclonal MAS055, diluted 1:1,000, Seralab UK), enkephalin (Enk, rat monoclonal MAS083, diluted 1:500, Seralab, UK) and tyrosine hydroxylase (TH, rabbit polyclonal, TE101, diluted 1:1,000, Eugene Tech, USA).

Sections were subsequently washed three times in 0.1M Tris for 15 minutes each wash and then incubated in solutions of biotinylated horse anti-mouse (Parv, Calb), rabbit anti-rat (SP and Enk) or goat anti-rabbit (TH) IgG (BA2000, BA4000 and BA1000, respectively, Vector Laboratories, CA, USA, diluted 1:200 in 0.1M Tris) and streptavidin-biotin-horseradish peroxidase complex (PK6100 Elite, Vector Laboratories, CA USA, diluted 1:500 in 0.1M Tris) for 1 hour each at room temperature with three 15-minute 0.1 M Tris washes in between and after these 1-hour incubations. Each series of sections was subsequently incubated in 5 ml of 0.6 mg/ml diaminobenzidine tetrahydrochloride (DAB) in 0.1 M Tris for 10 minutes, with addition of 5 ml of 10% $H_2O_2$ in 1 litre 0.1 M Tris for 5 minutes (final concentration of $H_2O_2$ 0.05%) for visualization of the tertiary complex. Following washes in 0.1M Tris solution, sections were mounted onto slides, allowed to dry overnight, dehydrated in graded ethanol and xylene, and coverslipped with DPX.

The specificity of the immunohistochemical reactions was tested by omitting the primary antisera as outlined previously (Hardman and Halliday 1999a; Hardman and Halliday 1999b; Hardman et al., 1997a, b). No peroxidase reaction was observed in these test sections.

## Imaging, Principles of Delineation, and Abbreviation

All series were digitally scanned at the Histology and Microscopy Unit of the School of Medical Sciences, The University of New South Wales (UNSW), with the aid of an Aperio ScanScope XT Slide Scanner. JPEG2000 images were extracted from Aperio SVS files, artefact around the pial surface digitally removed, and the images saved as Adobe Photoshop files. Incomplete sections at the interface between the two blocks were reconstructed in Photoshop to reflect the true appearance of the brain.

Images of the ten adjacent sections in each series were linked in Adobe Illustrator *CS2*. Structures were primarily outlined on cresyl violet stained sections, but linkage with nearby sections stained for AChE, Parv, Calb and SMI-32 allowed the matching of drawn outlines against chemoarchitecturally distinct boundaries in adjacent sections. Solid lines have been used for the boundaries between fiber tracts and adjacent structures, whereas dashed lines have been used for boundaries between nuclei. In some instances there is an overlap of fiber tracts and nuclei and this is reflected in overlapping of dashed and solid line outlines. Once the positions and boundaries of regions and nuclei were agreed upon in the diagram layer, these data were transferred to the imaged sections, with minor repositioning to allow for progressive positional changes in successive sections. Outlines have not been shown on the imaged sections, because we believe that (wherever possible) the contrast between regions in cyto- and chemoarchitecture should speak for itself.

We have used an abbreviation system for nuclei and regions based on that developed by Paxinos, Watson and colleagues for rodent, macaque and human brains (Paxinos, Huang, and Toga 2000; Paxinos et al., 2009), because of its wide acceptance. In most cases, the names of nuclei and tracts have been anglicized, except in those cases where the Latin name is in universal acceptance (e.g., fasciculus retroflexus, stria terminalis). Lower-case letters have been used for sulci and fiber tracts (e.g., af for amygdaloid fissure, f for fornix, st for stria terminalis) and upper-case letters for nuclei and regions.

The isocortex of the marmoset has a large number of physiologically and chemoarchitecturally identified regions with accepted abbreviations in the scientific literature. Some of these are commonly referred to as numbered Brodmann's areas, whereas others are named using an alphabetical system of nomenclature, often mixing upper- and lower-case letters. In some areas, both systems are in common use. We have tried to accommodate all of the common usages by providing alternative names for identified cortical regions. The accepted isocortical abbreviations occasionally overlap with those used by Paxinos and Watson for subcortical structures of rodent brain (e.g., PV, for parietoventral somatosensory cortex, might also be used for the paraventricular thalamic nucleus). In these cases we have preferentially used the accepted cortical abbreviation in the marmoset literature and modified the contraction for the subcortical structure (e.g., PVA, PVP). In the case of the many auditory core, belt and parabelt regions of temporal cortex (Aitkin, Kudo, and Irvine 1988), we have preceded the accepted abbreviation in the literature with "Au" to clearly indicate their auditory function.

## Quantitation of Volumes and Neuronal Populations of Motor Nuclei

This atlas also includes quantitative data on the number of defined neurons in the basal ganglia and motor thalamus, because primates like the common marmoset are frequently used in medically oriented basal ganglia research. The data set will be updated and extended as analysis proceeds and the results will be made available at our website, www.brainatlases.unsw.edu.au.

The boundaries of each nucleus were defined in the same manner as in our previous studies of the human, baboon, and rat basal ganglia (Hardman and Halliday 1999a, 1999b; Hardman et al., 1996; Hardman et al., 1997a, 1997b; McRitchie and Halliday 1995; McRitchie, Halliday, and Cartwright 1995; McRitchie, Hardman, Halliday 1996). The boundaries of the STh were defined in the series of sections stained for Parv, whereas the boundaries of the EGP were demarcated in the series of sections stained for Enk. The IGP and the entire SN boundaries were delineated in sections stained for substance P. In all cases, these nuclei could be readily separated from the surrounding structures by their intense neuropil labelling for the substances listed above.

For a select number of sections in the series used to define nuclear boundaries, the outline of the cross-sectional profile of the nucleus and the outline of the entire section were drawn by hand with the aid of a microscope-computer imaging system running the Magellan program (Department of Anatomy, School of Medical Sciences, UNSW). The number of sections used from each series depended on the anatomical dimensions of each nucleus. Drawings were made from an average of 7.0 ± 1.5 sections through each nucleus in each animal. For each of the drawings, the cross-sectional area of the nucleus was calculated and the volume of the entire nucleus estimated by multiplying the sum of these areas by the intersection distance (Cavalieri's basic estimator). Two investigators evaluated the reproducibility of boundary delineations in the STh. There was less than 15% variation in the inter-rater measurements for this structure (mean variation and standard deviation between researchers of 8 ± 4%). The final estimates for nuclear volume reported in this book were obtained by only one of the investigators (CH).

The drawn outlines were superimposed onto the adjacent Nissl-stained or parvalbumin-immunoreacted sections using the microscope-computer imaging system. All nucleolated Nissl-stained neurons and parvalbumin immunoreactive neuronal profiles located within the drawn boundaries were then plotted by hand. For the SN, the SNR and SNC components were delineated on the basis of immunoreactivity for tyrosine hydroxylase. All nucleolated Nissl-stained neurones and tyrosine hydroxylase immunoreactive (TH+) neuronal profiles within the cresyl violet counterstained SN were plotted. As above, the SNCD was identified as the compact group/s of dopaminergic neurons lying flush against the dorsal boundary of the SN. In rats, this region is called the SN pars compacta, whereas in humans it is called the dorsal tier of the SN pars compacta (Fearnley and Lees 1991; McRitchie, Hardman, and Halliday, 1996). In non-human primates this region is known as the β tier of the SN pars compacta (for review see McRitchie et al., 1998). In contrast, the SNCV was identified as the dopaminergic neuronal groups (often discontinuous) lying ventral to the SNCD. In rats, these cells are commonly known as the ventrally displaced dopaminergic neurons in the SN pars reticulata; whereas in humans these heavily populated compact cell groups are considered the ventral tier of the SN pars compacta (see Fearnley and Lees 1991; McRitchie, Hardman, and Halliday (1996). In non-human primates, these clusters are known as the α tier of the SN pars compacta (for review see McRitchie et al. 1998).

The percentages of neurons within the STh, EGP, IGP, and SNC that displayed immunoreactivity for Parv were calculated by dividing the total number of Parv+ neuronal profiles in each nucleus by the total number of neurones. TH+ neurones within the SN are not immunoreactive for Parv (Hontanilla, Parent, and Giménez-Amaya 1997; Hontanilla et al., 1998; McRitchie, Hardman, and Halliday 1996) and thus all Parv+ neurones plotted within the SN boundaries were considered as a subpopulation of SNR neurones.

The reproducibility of the quantitative method was tested by two investigators for the smallest structure (STh). There was <15% variation in the number of neurones plotted for this structure across all species (mean variation and standard deviation between researchers = 7±3%). Counts made by only one of the investigators (CH) are reported for all structures with little variation across animals of the same species in the numbers of neurones plotted for each nucleus (the range in the total number of neurones plotted for each nucleus in each species is given in Table 2). The coefficient of error was 0.045 for the marmoset. Estimates for the total number of neurones in each nucleus were made using the fractionator technique (Coggeshall 1992). Essentially, estimates were achieved by adding the numbers of neurones plotted for each section of a given nucleus and then multiplying this total number by the reciprocal of the fraction of sections in which neuronal plots were performed.

## Bases for Delineation of Specific Regions and Nuclei

### *Olfactory Bulb and Area*

Despite the reported use of odor signals in this species for territory marking and identification, the olfactory

**Table 2 Volume and neuronal population sizes for selected motor nuclei in the marmoset brain†**

| | Nucleus | Volume (mm3) (mean ± SD) | Neuronal Population (mean ± SD) | Parv + neurons (mean ± SD) | TH + neurons (mean ± SD) |
|---|---|---|---|---|---|
| Thalamus | VA | 1.82 ± 0.22 | 16783 ± 1737 | 12319 ± 1275 | – |
| | VLa | 2.58 ± 0.31 | 22238 ± 1584 | 19570 ± 1394 | – |
| | VLp | 4.60 ± 0.72 | 43796 ± 6107 | 40299 ± 5619 | – |
| Pallidum | EGP | 23.20 ± 1.63 | 124731 ± 8734 | 42509 ± 3275 | – |
| | IGP | 9.85 ± 1.14 | 39576 ± 5688 | 21176 ± 2132 | – |
| Subthalamus | STh | 2.75 ± 0.26 | 34638 ± 2692 | 18999 ± 2960 | – |
| Midbrain | A8 | – | – | – | 3934 ± 372 |
| | A9 | – | – | – | 29933 ± 2849 |
| | A10 | – | – | – | 16695 ± 1391 |
| | CLi | – | – | – | 1097 ± 148 |
| | IF | – | – | – | 803 ± 152 |
| | PaP | – | – | – | 151.5 ± 20.6 |
| | PBP | – | – | – | 9102 ± 1212 |
| | PN | – | – | – | 721 ± 170 |
| | RLi | – | – | – | 753 ± 132 |
| | RRF | – | – | – | 3934 ± 372 |
| | SNC | 12.65 ± 1.49* | 77288 ± 2427 | – | 29933 ± 2849 |
| | SNCD | – | – | – | 16932 ± 1101 |
| | SNCV | – | – | – | 13001 ± 2011 |
| | SNR | 12.65 ± 1.49* | 47355 ± 1400 | 19375 ± 4862 | – |
| | VTA | – | – | – | 4068 ± 440 |

† Data are based on analysis of 4 marmoset brains (3 females, 1 male). Average age was 8 years (range of 4 to 13.5 years). Average body mass (± SD) was 259 (± 42) g. Average brain mass (± SD) was 5.90 (± 0.08) g.

* Note that SNC and SNR are co-extensive and share the same volume. SNC is defined as dopaminergic cell groups within this space, whereas SNR is the non-dopaminergic cell groups within the same space.

bulb (OB) and anterior olfactory area (AO) are small and these structures occupy only a few sections of the atlas. Lamination of the OB and AO is similar to those in the macaque and rodent (Paxinos, Huang, and Toga 2000; Paxinos et al., 2009). See Liebetanz, et al. (2002) for connections in this species.

## *Frontal Cortex*

Delineation of the dorsolateral, medial and orbital cortical areas has been based on published maps (Burman et al. 2006, 2007; Burman and Rosa 2009; and Roberts et al., 2007). The reader is referred to those papers for details of cyto- and chemoarchitecture. We also delineated dorsal and ventral premotor areas (Burish, Stepniewska, and Kaas 2008). These areas have smaller layer 5 pyramidal neurons than area 4 (Burish, Stepniewska, and Kaas 2008; Burman et al., 2008). We also delineated the frontal eye field (FEF) within area 8Av (+16.5 to +12.0 mm AP) (Burman et al., 2006). Area M1 (area 4) is easily distinguished from the adjacent premotor and somatosensory areas because of its large Betz cells in layer 5 (seen in both cresyl violet and AChE reacted sections) and the loss of a Parv immunoreactive layer 4. Body representation within M1 cortex (area 4) has been based on microstimulation maps (Burish, Stepniewska, and Kaas 2008).

## Parietal Cortex

Topographic organization within areas 3a and 3b has been based on Krubitzer and Kaas (1990b); Huffman and Krubitzer (2001b); Fang, Jain, and Kaas (2002); and Iyengar et al. (2007). Somatotopic organization and topography of somatosensory areas PV, S2 and putative ventral somatosensory area (VSom?) have been based on maps in Qi et al., 2002. The reader is referred to the above papers as well as Garraghty, Pons, and Kaas (1990) and Zhang et al., (1996) for consideration of connections and functions. The caudal parietal lobe is occupied by dorsal and ventral posterior parietal regions (PPD, PPV, respectively) said to be separated by the intraparietal sulcus (Palmer and Rosa 2006), but we found that this sulcus often lays further caudally.

## Temporal Cortex

The most distinctive features of the temporal cortex are the auditory core, belt, and parabelt regions arranged along the crest of the temporal lobe (Aitkin, Kudo, and Irvine 1988). These features are characterized by strong Parv immunoreactivity (principally in layer 4, but also in supragranular layers) and have been delineated (including tonotopy where known) based on maps in Philibert et al. (2005); de la Mothe et al. (2006a, b); and Hackett et al. (2007).

Other temporal areas are arranged either ventral to the lateral parabelt along the lateral and ventral surface, or medial to medial parabelt in the lateral fissure. We have delineated the lateral regions (STL, ITA1 and 2, ITD1 and 2, ITV, TF, TH; Palmer and Rosa 2006) on the basis of cytoarchitecture and differential Parv immunoreactivity. A distinct perirhinal cortex intervenes between isocortical regions of the ITV and the EntL. The reader is referred to Ridley et al., 2004, 2005 for functional significance of these areas. The medial bank of the temporal lobe has an STM area contiguous with temporal prosiocortex.

The dorsocaudal extremity of the temporal lobe includes the middle temporal area (MT), a visual region concerned with processing information about motion in the visual field (Allman and Kaas 1971; Spatz 1977), and is surrounded ventrocaudally by a crescent-shaped MTc and dorsorostrally by the medial superior temporal area (MST)(Palmer and Rosa 2006). Please see Krubitzer and Kaas (1990a) for connections.

## Occipital Cortex

Primary visual cortex (V1) (Chappert-Piquemal et al., 2001; Soloman 2002) is clearly distinguishable by the striking cytoarchitecture of its layer 4 lamination and unique neurofilament protein immunoreactivity (Baldauf 2005). The region is centered around the calcarine fissure, but extends across the occipital pole and onto the caudolateral surface. Visuotopic organization has been indicated in selected plates based on the maps in Fritsches and Rosa 1996. Delineation of visual association areas V2, V3 (VLP), and V4 (VLA) and visuotopic organization, where appropriate, have been based on Cappe and Barone (2005); Rosa and Tweedale (2000); Rosa, Fritsches, and Elston (1997); and Rosa et al., (2005). Delineation of dorsal extrastriate areas (DA, DI, DM) is based on Rosa and Schmid (1995); and Rosa et al., (2005).

## Cingulate Cortex

The cingulate cortex is a crescent-shaped region arcing over the corpus callosum and is a central component in the limbic lobe. The fundamental division of the primate cingulate is into four regions (Vogt, Hof, and Vogt 2004): a rostral perigenual cingulate cortex (including areas 25 and rostral parts of 24 and 32) concerned with affect and visceromotor control; a midcingulate cortex (including areas 33 and caudal parts of 24 and 32) concerned with response selection and skeletomotor control; a caudal cingulate region (including areas 23 and 31) concerned with visuospatial processing; and a retrosplenial cortex (including areas 29 and 30) concerned with memory access (Vogt, Hof, and Vogt 2004). Our delineation of these regions is based on the above and includes an area 26 at the transition between induseum griseum and area 29.

## Hippocampus

The main components (DG and cornu Ammonis elements), transitional areas and fiber tracts are similar in all primates and delineation of these regions has been based on Paxinos, Huang, and Toga (2000); and Insausti and Amaral (2004). The existence of a distinct CA4, as often delineated in the primate cornu Ammonis (Paxinos, Huang, and Toga 2000), has been questioned by Insausti and Amaral (2004), so we have confined our subdivision of CA to three divisions only.

## *Insular Cortex and Claustrum*

By definition, primate insular cortex sits in the depths of the lateral fissure and/or lies external to the claustrum. The rostral parts of insular cortex extend onto the orbital surface of the frontal pole as far as +13 mm AP. Distinct medial and lateral rostral insular areas (Insrm, Insrl) are separated by a medial orbital sulcus. When the temporal lobe emerges at +10.5 mm AP, the insular cortex is divisible into a dorsolateral granular insular cortex (GI) with a distinct layer 4, and a ventrolateral agranular insular cortex (AI). These two insular regions have distinct patterns of Parv immunoreactivity and are contiguous caudally with the putative ventral somatic (VSom?) and retroinsular cortex (RIns) at about +6.5 mm AP.

The claustrum is easily demarcated by its fiber tract relations (external and extreme capsule), with the only ambiguity at the border with the dorsal endopiriform nucleus. A case could be made for dividing the claustrum into dorsal and ventral subunits, based on dissimilar levels of Parv immunoreactivity, but the difference is more clinal than precise, so we have not subdivided the claustrum in the atlas.

## *Piriform and Entorhinal Cortex, Endopiriform Nucleus*

The piriform cortex of the marmoset occupies the ventral surface of the frontal pole of the forebrain from +13.0 mm AP and expands onto the rostral surface of the temporal lobe at +10.5 mm AP. Lamination is best seen at the caudal levels.

The entorhinal cortex replaces the piriform cortex on the ventral surface of the temporal lobe at approximately +9.5 mm AP, whereas the piriform cortex is continuous with the amygdalopiriform cortex and ventral cortical amygdala on the medial surface at approximately the same level. We identified two distinct entorhinal cortical areas through the front half of the rostrocaudal extent of the region. EntL has a more strongly stained and discontinuous layer 2 in cresyl violet stained sections (+8.5 mm AP) and differs from EntM in the pattern of Parv immunoreactivity. At +6.5 mm AP, the mediolateral differences are lost and a caudal entorhinal cortex (EntC), with a more prominent lamina dissecans, emerges. EntC recedes medially to be replaced by isocortex (temporal area H) at approximately +2.5 mm AP. The extent, boundaries, and possible subdivisions of the endopiriform nucleus are difficult to define. At least part of the endopiriform nucleus should be within the piriform lobe, but there is a clear continuity of the region into the gap between the rostral temporal iso- and periallocortical pole and the amygdala. We have delineated separate dorsal and ventral endopiriform nuclei (DEn and VEn, respectively) on the basis of differential calcium-binding protein immunoreactivity. The DEn can be distinguished from the claustrum by the stronger Parv labelling of the latter and the DEn is more rostrally extensive than the VEn.

## *Striatum and Pallidum*

The striatum is divided into dorsal and ventral components. The dorsal striatum consists of caudate and putamen, whereas the ventral striatum is composed of nucleus accumbens and olfactory tubercle (Butler and Hodos 2005). Most of the boundaries of the dorsal striatum are clearly defined by fiber tracts and/or AChE activity and Parv immunoreactivity, but the fundus striatum (FStr) and amygdalostriatal transition areas (AStr) are more difficult to discriminate. The ventral striatal elements are readily delineated based on AChE activity and we have subdivided the accumbens (Acb) into shell and core components (AcbSh, AcbC, LAcbSh) after careful consideration of subtle gradations of differential AChE activity. While the superficial layers and mediolateral extent of the olfactory tubercle are clearly defined by AChE activity, defining the deeper boundaries can be difficult.

The pallidum also has dorsal and ventral components. The dorsal pallidum (external and internal segments of globus pallidus — EGP, IGP, respectively) are clearly defined by fiber tracts (medial and lateral medullary laminae) and chemoarchitecture (see AChE, Parv and Calb plates at approximately +7.0 mm AP). The rostral part of EGP extends ventral to the anterior commissure at +9.0 mm AP (note that this region has been mistakenly identified as ventral pallidum by Paxinos, Huang, and Toga (2000). The ventral pallidum is a more difficult region to delineate and consists of a shell around the ventral border of the rostral EGP.

## *Septal Region*

The septal region is usually defined anatomically by its position medial to the anterior horn of the lateral ventricle, but the region is more extensive than that simple definition would imply, and it has ventral extensions well away from the lateral ventricle (the diagonal band nuclei). The septal nuclear complex is often divided into four groups of nuclei (Risold 2004). These are the lateral group (LS, SFi, SHi), medial group (MS, VDB, HDB), posterior group (TS and bed nuclei of the anterior commissure and stria medullaris, although we did not find the last two in our analysis of the marmoset septum), and the BST (although the last may be considered part of the extended amygdala, see below). The lateral nucleus is complex and further subdivided into chemically distinct dorsal, intermediate and ventral subnuclei (LSD, LSI, LSV, see +10.5 mm AP). Subdivisions between LS subnuclei and within the lateral group are best seen in AChE reactive and Calb immunoreacted sections (see +10.0 mm AP). The medial group are clearly demarcated by bands of Parv immunoreactive neurons (and complementary low immunoreactivity for Calb) descending in a ventrolateral arc from the MS. The TS is confined to the posterior septum and best seen by its constituent Calb immunoreactive neurons (+9.25 mm AP). The BST subunits are discussed below.

## *Amygdala and Bed Nuclei of the Stria Terminalis*

Chemoarchitectonic parcellation of the amygdala follows natural transverse and vertical borders as revealed by immunoreactivity to calcium-binding proteins and AChE reactivity (Watson, Paxinos, and Tokuno 2010). The division into a dorsal tier (medial and central nuclei), as opposed to a ventral tier (basal and lateral elements), is best seen at +8.5 mm AP in Parv immunoreacted material. The central nuclear group can be divided into lateral, central, and medial subnuclei on the basis of differential AChE activity (see +9.0 mm AP). Subdivisions of the ventral tier elements by differential reactivity for AChE are best seen at +8.5 mm AP, with a strongly reactive centrally placed basolateral group (BLD, BLI, BLVL and BLVM) flanked by less reactive lateral nuclei (LaD, LaL, LaM) and basomedial nuclei (BMD, BMV, IMG and PLam). Further subdivision within basolateral and basomedial groups is largely based on cytoarchitecture.

Delineation of the cortical and transitional components of the amygdala is based on these regions in human and macaque brains (de Olmos, 2004; Paxinos, Huang, and Toga 2000). Transitional areas with allocortex rostral and caudal to the amygdala (i.e., amygdalopiriform and amygdalohippocampal areas, respectively) have been delineated, and with subdivision of the amygdalohippocampal region into magnocellular and parvicellular components.

The extended amygdala is a morphofunctional concept of an arc of gray matter arranged in a trajectory from the temporal amygdala to the midline (see de Olmos 2004 for discussion). This gray complex is divided into a medial subdivision, consisting of the medial bed nuclei of the stria terminalis (BSTMA, BSTMP, BSTMPI, BSTMPL, BSTMPM), medial sublenticular extended amygdala (EAM) and the rostral medial amygdala (Me); and a lateral subdivision consisting of the lateral bed nuclei of the stria terminalis (BSTLD, BSTLJ, BSTLP, BSTLV), the central sublenticular extended amygdala (EAC) and the central amygdaloid nucleus (CeC, CeL, CeM). The question of whether the primate extended amygdala has an interstitial nucleus of the posterior limb of the anterior commissure (IPAC, as seen in rodents, Paxinos et al., 2009) remains open (de Olmos 2004). We have identified a candidate IPAC contiguous with the Ce at +9.0 mm AP.

## *Prethalamus (Ventral Thalamus)*

The prethalamus includes the reticular nucleus, zona incerta, pregeniculate, and subthalamus. These structures are derived from prosomere 3 during development and are arranged as a shell around the lateral and ventral borders of the dorsal thalamus.

The reticular nucleus (Rt) is easily distinguished by strong AChE reactivity and some immunoreactivity to Parv (see +4.5 mm AP). Parts of Rt adjacent to the DLG (i.e., rostroventral) appear to be solely concerned with the DLG projection to primary visual cortex, whereas the dorsocaudal Rt is associated with higher order visual projections (Fitzgibbon, Szmajda, and Martin 2007).

The zona incerta (ZI) extends from +7.5 mm to +4.5 mm AP. Its most rostral division (ZIR) is not readily divisible, but throughout much of its extent (+6.5 to +4.5 mm AP) the ZI can be separated into dorsal (ZID) and ventral (ZIV) subunits on the basis of differential immunoreactivity to Parv and cytoarchitecture.

The pregeniculate (PrG) is homologous to the ventral lateral geniculate of non-primates (Butler and Hodos 2005), but the pregeniculate is probably a more appropriate term for the nucleus in all mammals, given that the PrG arises rostral to the thalamus in all vertebrates.

The subthalamus (STh) is distinctive in sections stained for a variety of markers illustrated in the atlas (Parv, Calb, AChE).

## *Dorsal Thalamus*

The motor thalamic nuclei can be seen at the rostral pole of the dorsal thalamus. The VA and VAMC occupy the most rostral position and the VAMC is traversed by the mammillothalamic tract. Distinguishing VA from VLa, and VLa from VLp, is quite difficult without reference to tract-tracing results and these boundaries are best distinguished by domains of terminals of the nigrothalamic (VA) as opposed to pallidal (VLa) and cerebellar (VLp) outflows (Hirai and Jones 1993). Our delineation consequently represents an extrapolation from tracing studies in the macaque, supplemented by differential immunoreactivity to Calb. We have used the abbreviations pal, cb and nigr in the plates to indicate the respective pallidal, cerebellar and nigral zones of input to these thalamic nuclei.

Midline and intralaminar nuclei are best distinguished in AChE, Parv and Calb material. The central medial thalamic nucleus (CM) is distinguished by strong AChE reactivity and poor Parv immunoreactivity. The strong AChE reactivity extends along the band of intralaminar nuclei into PC and CL, but the lateral intralaminar nuclei at that level have a lower level of Calb immunoreactivity. We have based our subdivisions of MD on differential staining for Parv and Calb. We did not follow the vertical subdivisions used by Paxinos, Huang, and Toga (2000) because the lines of separation between MDC, MDV, MDL and MDD are oriented as much horizontal as vertical. The small xiphoid nucleus in the midline exhibits stronger AChE reactivity than the surrounding VRe.

The ventral posterior nuclear complex has been extensively studied in this and related species. We have based our delineation of VP into five subdivisions (VPI, VPL, VPM, VPMPC and VPS) and the somatotopic organization of the somatosensory complex on maps and illustrations in Fang, Jain, and Kaas (2002); Huffman and Krubitzer (2001a); Iyengar et al., (2007); Krubitzer and Kaas (1992); Qi et al., (2002) and Wilson, Kitchener, and Snow (1999).

The dorsal lateral geniculate nucleus (DLG) and its lamination (Spatz 1978) is always clearly defined in cresyl violet, AChE, Parv, Calb and neurofilament immunoreacted sections (Kaas et al., 1978; Bourne and Rosa, 2003a). The reader is referred to White et al. (1998) for visuotopic maps of the DLG.

Our delineation of the medial geniculate nucleus is largely based on de la Mothe et al., (2006b) with identification of ventral, medial, anterodorsal, and anteroventral subnuclei (MGV, MGM, MGAD, MGAV, respectively) based on their illustrations. Delineation of the suprageniculate nucleus (SG) is also based on de la Mothe et al. (2006b).

Delineation of the pulvinar into anterior, inferior, lateral and medial subregions (PulA, PulI, PulL and PulM) is based on Paxinos, Huang, and Toga (2000) and Stepniewska and Kaas (1997).

## *Preoptic Area and Hypothalamus*

Differentiation of the preoptic area and hypothalamus is based mainly on Nissl cytoarchitecture and differential staining with AChE. Immunoreactivity for Parv is undiscriminating in this area and immunoreactivity for Calb labels altogether too much, although Cavalcante and colleagues have used Calb immunoreactivity in somata for defining the boundaries of SCh in this species (Cavalcante et al., 2008).

Fortunately, most of the nuclei in this region are conserved across mammals and there are recognizable homologies for many of the nuclei. We have subdivided the nuclei of the preoptic area along parallel lines to those applied to rodent and macaque brains because of the evident similarities (Paxinos, Huang, and Toga 2000). Similarly, the ventromedial hypothalamus shows clear division into medial and lateral subdivisions (VMHM, VMHL, respectively), as applied to the macaque (Paxinos, Huang, and Toga 2000).

## *Pretectal Region*

The pretectum is derived from prosomere 1 of the embryonic brain. Five pretectal nuclei are identified in all mammals as well as a pretectal (prosomere 1) reticular formation (p1Rt). In the marmoset, the pretectal nuclei lie between +3.0 and +1.5 mm AP and have a similar organization to that in other primates.

## Midbrain

Delineation of the substantia nigra is based on Beckstead and Frankfurter (1982). Additional subdivision is based on our published work for this region (Hardman et al., 1997b; McRitchie, Hardman, and Halliday 1996). The compact and reticularis zones of substantia nigra (SNC and SNR, respectively) are defined on the basis of tyrosine hydroxylase (TH) and substance P immunoreactivity. Those neurons that were TH+ and located within the substance P defined boundaries of the SN were termed SNC neurones, whereas those that were not were termed SNR neurones. SNC neurones were subdivided into dorsal (SNCD) and ventral (SNCV) groups or tiers based on descriptions from serial section cyto-and chemoarchitectural analyses of rat, baboon, and human midbrain (Gonzàlez-Hernàndez and Rodriguez 2000; McRitchie et al., 1998; McRitchie and Halliday, 1995). Briefly, the SNCD was identified as the compact group/s of dopaminergic neurons lying flush against the dorsal boundary of the SN, whereas the SNCV is the more ventral dopaminergic cell group extending into the SNR.

Lamination of the superior colliculus is quite conserved across mammals, and neurofilament expression in this region has been described in this species by Bourne and Rosa (2003b).

## Pontine Region

Pontine basal and tegmental nuclei in the common marmoset are similar in appearance and chemoarchitecture to those in macaques and humans and have been delineated accordingly (Paxinos and Huang 1995).

## Cerebellum and Precerebellar Nuclei

Delineation of the cerebellar folia is based on Larsell (1970) and Fujita et al. (2009). Deep cerebellar nuclei of the marmoset are cytoarchitectonically similar to those in the macaque and we have used similar delineations (Paxinos, Huang, and Toga 2000). Precerebellar nuclei (IO, LRt, Pn) are similar in organization to those seen in the macaque and we have adopted matching nomenclature (Paxinos, Huang, and Toga 2000).

## Medulla Oblongata

The medulla oblongata is the most evolutionarily conserved part of the cranial neuraxis and has a consistent structure across mammals. The underlying organization of the medulla is a division into radially arranged wedges with their apices located at the fourth ventricle surface. Since the cyto- and chemoarchitecture of the region is conserved, our delineations have been based largely on the Paxinos, Huang, and Toga (2000); and Paxinos et al. (2009) demarcations of the macaque and rodent brainstems, respectively. The exception to this is the identification of a median accessory nucleus (MnA) for processing of somatosensory information from the prehensile tail of the marmoset. The sagittal sections through the medulla exhibit some flexion of the caudal tip. We have made corrections for this in the line diagrams by depicting the position of caudal medullary nuclei as if no lateral flexion had occurred, while maintaining correct labelling in the photographs of the stained sections.

## Spinal Cord

Only the upper cervical levels of the spinal cord are depicted in this atlas, from the level of the pyramidal decussation at -5.5 mm AP to cervical segment 3 at –8.0 mm AP. Spinal trigeminal nucleus (pars caudalis) and spinal accessory nucleus extend into the upper spinal cord. We identified a central cervical nucleus at the upper labels on the basis of cytoarchitecture, AChE activity and Parv immunoreactivity; and a lateral spinal nucleus (LSp) ventral to the spinal trigeminal nucleus on the basis of cytoarchitecture. Rexed's laminae emerge from approximately –7.0 mm AP. Relative positions of major fiber tracts have been indicated in the white matter, but these are largely deduced from human and macaque data, because the relevant tracts have not been studied in detail in this species.

## References


Aitkin, L. M., Kudo, M., and D. R. Irvine. 1988. Connections of the primary auditory cortex in the common marmoset, *Callithrix jacchus jacchus*. *J. Comp. Neurol.* 269: 235–248.

Allman, J. M., and J. H. Kaas. 1971. A representation of the visual field in the caudal third of the middle temporal gyrus of the owl monkey (*Aotus trivirgatus*). *Brain Res.* 31: 85–105.

Baldauf, Z. S. 2005. SMI-32 parcellates the visual cortical areas of the marmoset. *Neurosci. Lett.* 383: 109–214.

Beckstead, R. M., and A. Frankfurter. 1982. The distribution and some morphological features of substantia nigra neurons that project to the thalamus, superior colliculus, and pedunculopontine nucleus in the monkey. *Neurosci.* 7: 2377–2388.

Bourne, J. A., and M. G. P. Rosa. 2003a. Neurofilament protein expression in the geniculostriate pathway of a New World monkey (*Callithrix jacchus*). *Exp. Brain Res.* 150: 19–24.

Bourne, J. A., and M. G. P. Rosa. 2003b. Laminar expression of neurofilament protein in the superior colliculus of the marmoset monkey. *Brain Res.* 973: 142–145.

Burish, M. J., Stepniewska, I., and J. H. Kaas. 2008. Microstimulation and architectonics of frontoparietal cortex in common marmosets (*Callithrix jacchus*). *J. Comp. Neurol.* 507: 1151–1168.

Burman, K. J., and M. G. P. Rosa. 2009. Architectural subdivisions of medial and orbital frontal cortices in the marmoset monkey (*Callithrix jacchus*). *J. Comp. Neurol.* 514: 11–29.

Burman, K. J., Lui, L. L., Rosa, M. G. P., and J. A. Bourne. 2007. Development of non-phosphorylated neurofilament protein expression in neurones of the New World monkey dorsolateral frontal cortex. *Eur. J. Neurosci.* 25: 1767–1779.

Burman, K. J., Palmer, S. M., Gamberini, M., and M. G. P. Rosa. 2006. Cytoarchitectonic subdivisions of the dorsolateral frontal cortex of the marmoset monkey (*Callithrix jacchus*), and their projections to dorsal visual areas. *J. Comp. Neurol.* 495: 149–172.

Burman, K. J., Palmer, S. M., Gamberini, M., Spitzer, M. W., and M. G. P. Rosa. 2008. Anatomical and physiological definition of the motor cortex of the marmoset monkey. *J. Comp. Neurol.* 506: 860–876.

Butler, A. B. and W. Hodos. 2005. *Comparative Vertebrate Neuroanatomy. Evolution and Adaptation.* 2nd ed.. New Jersey: Wiley.

Cappe, C., and P. Barone. 2005. Hetermodal connections supporting multisensory integration at low levels of cortical processing in the monkey. *Eur. J. Neurosci.* 22: 2886–2902.

Cavalcante, J. S., Britto, L. R. G., Toledo, C. A. B., Nascimento, E. S. Jr., Lima, R. R. M., Pontes, A. L. B., M. S. M. O Costa. 2008. Calcium-binding proteins in the circadian centers of the common marmoset (*Callithrix jacchus*) and the rock cavy (*Kerodon rupestris*) brains. *Brain Res. Bull.* 76: 354–360.

Chappert-Piquemal, C., Fonta, C., Malecaze, F., and M. Imbert. 2001. Ocular dominance columns in the adult New World monkey *Callithrix jacchus*. *Vis. Neurosci.* 18: 407–412.

Coggeshall, R. E. 1992. A consideration of neural counting methods. *Trends Neurosci.* 15: 9–13.

De la Mothe, L. A., Blumell, S., Kajikawa, Y., and T. A. Hackett. 2006a. Cortical connections of the auditory cortex in marmoset monkeys: Core and medial belt regions. *J. Comp. Neurol.* 496: 27–71.

De la Mothe, L. A., Blumell, S., Kajikawa, Y., and T. A. Hackett. 2006b. Thalamic connections of the auditory cortex in marmoset monkeys: Core and medial belt regions. *J. Comp. Neurol.* 496: 72–96.

De Olmos, J. S. 2004. Amygdala. In *The Human Nervous System*. 2nd ed., G. Paxinos, and J. K. Mai (eds.). San Diego: Elsevier.

Dunbar, R., and L. Barrett. 2000. *Cousins: Our Primate Relatives.* London: BBC Worldwide.

Fang, P-C., Jain, N., and J. H. Kaas. 2002. Few intrinsic connections cross the hand-face border of area 3b of New World monkeys. *J. Comp. Neurol.* 454: 310–319.

Fearnley, J. M., and A. J. Lees. 1991. Ageing and Parkinson's disease: Substantia nigra regional selectivity. *Brain* 114: 2283–2301.

Fitzgibbon, T., Szmajda, B. A., and P. R. Martin. 2007. First order connections of the visual sector of the thalamic reticular nucleus in marmoset monkeys (*Callithrix jacchus*). *Vis. Neurosci.* 24: 857–874.

Fritsches, K. A., and M. G. P. Rosa. 1996. Visuotopic organisation of striate cortex in the marmoset monkey (*Callithrix jacchus*). *J. Comp. Neurol.* 372: 264–282.

Fujita, H., Oh-Nishi, A., Obayashi, S., and I. Sugihara. 2009. Organization of the marmoset cerebellum in three-dimensional space: Lobulation, aldolase C compartmentalization and axonal projection. *J. Comp. Neurol.* 518: 1764–1791.

Garraghty, P. E., Pons, T. P., and J. H. Kaas. 1990. Ablation of areas 3b (SI proper) and 3a of somatosensory cortex in marmosets deactivate the second and parietal ventral somatosensory areas. *Somatosens Motor Res.* 7: 125–135.

Gonzàlez-Hernàndez, T., and M. Rodriguez. 2000. Compartmental organization and chemical profile of dopaminergic and GABAergic neurons in the substantia nigra of the rat. *J. Comp. Neurol.* 421: 107–135.

Hackett, T. A., Smiley, J. F., Ulbert, I., Karmos, G., Lakatos, P., de la Mothe, L. A., and C. E. Schroeder. 2007. Sources of somatosensory input to the caudal belt areas of auditory cortex. *Perception* 36: 1419–1430.

Hardman, C. D., and G. M. Halliday. 1999a. The external globus pallidus in patients with Parkinson's disease and progressive supranuclear palsy. *Mov. Disord.* 14: 626–633.

Hardman, C. D., and G. M. Halliday. 1999b. The internal globus pallidus is affected in progressive supranuclear palsy and Parkinson's disease. *Exp. Neurol.* 158: 135–142.

Hardman, C. D., McRitchie, D. A., Halliday, G. M., Cartwright, H. R., and J. G. L. Morris. 1996. Substantia nigra pars reticulata neurons in Parkinson's disease. *Neurodegeneration* 5:49–55.

Hardman, C. D., Halliday, G. M., McRitchie, D. A., and J. G. L. Morris. 1997a. The subthalamic nucleus in Parkinson's disease and progressive supranuclear palsy. *J. Neuropath. Expl. Neurol.* 56: 132–142.

Hardman, C. D., Halliday, G. M., McRitchie, D. A., Cartwright, H. R., and J. G. L. Morris. 1997b. Progressive supranuclear palsy affects both the substantia nigra pars compacta and reticulata. *Exp. Neurol.* 144: 183–192.

Hirai, T., and E. G. Jones. 1993. Comparative anatomical study of ventrolateral thalamic mass in humans and monkeys. *Sterotact. Funct. Neurosurg.* 60: 6–16.

Hontanilla, B., Parent, A., and J. M. Giménez-Amaya. 1997. Parvalbumin and calbindin D-28k in the entopeduncular nucleus, subthalamic nucleus, and substantia nigra of the rat as revealed by double-immunohistochemical methods. *Synapse* 25: 359–367.

Hontanilla, B., Parent, A., de las Heras, S., and J. M. Giménez-Amaya. 1998. Distribution of calbindin D-28k and parvalbumin neurons and fibers in the rat basal ganglia. *Brain Res. Bull.* 47: 107–116.

Huffman, K. J., and L. Krubitzer. 2001a. Thalamocortical connections of areas 3a and M1 in marmoset monkeys. *J. Comp. Neurol.* 435: 291–310.

Huffman, K. J., and L. Krubitzer. 2001b. Area 3a: topographic organization and cortical connections in marmoset monkeys. *Cerebral Cortex* 11: 849–867.

Insausti, R., and D. G. Amaral. 2004. Hippocampal formation. In *The Human Nervous System*. 2nd ed., G. Paxinos, and J. K. Mai (eds.). San Diego: Elsevier.

Iyengar, S., Qi, H-X., Jain, N., and J. H. Kaas. 2007. Cortical and thalamic connections of the representations of the teeth and tongue in somatosensory cortex of New World monkeys. *J. Comp. Neurol.* 501: 95–120.

Kaas, J. H., Huerta, M. F., Weber, J. T., and J. K. Harting. 1978. Patterns of retinal terminations and laminar organization of the lateral geniculate nucleus of primates. *J. Comp. Neurol.* 182: 517–553.

Klein, R. G. 1999. *The Human Career. Human Biological and Cultural Origins.* 2nd ed.. Chicago: University of Chicago Press.

Koelle, G. G., and J. S. Friedenwald. 1949. A histochemical method for localizing cholinesterase activity. *Proc. Soc. Exp. Biol. Med.* 70: 617–622.

Krubitzer, L. A., and J. H. Kaas. 1990a. Cortical connections of MT in four species of primates: Areal, modular and retinotopic patterns. *Vis. Neurosci.* 5: 165–204.

Krubitzer, L. A., and J. H. Kaas. 1990b. The organization and connections of somatosensory cortex in marmosets. *J. Neurosci.* 10: 952–974.

Krubitzer, L. A., and J. H. Kaas. 1992. The somatosensory thalamus of monkeys: Cortical connections and a redefinition of nuclei in marmosets. *J. Comp. Neurol.* 319: 123–140.

Larsell, O. 1970. *The Comparative Anatomy and Histology of the Cerebellum from Monotremes through Apes.* Minneapolis: University of Minnesota Press.

Lewis, P. R. 1961. The effect of varying the conditions in the Koelle method. *Bibliotheca Anat.* 2: 11–20.

Liebetanz, D., Nistche, M. A., Fromm, C., and C. K. Reyher. 2002. Central olfactory connections in the microsmatic marmoset monkey (*Callithrix jacchus*). *Cells Tissue Organs* 172: 53–69.

McRitchie, D. A., and G. M. Halliday. 1995. Calbindin D28k containing neurons are restricted to the medial substantia nigra in humans. *Neuroscience* 65: 87–91.

McRitchie, D. A., Halliday, G. M., and H. Cartwright. 1995. Quantitative analysis of the variability of substantia nigra pigmented cell clusters in the humans. *Neuroscience* 65: 87–91.

McRitchie, D. A., Hardman, C. D., and G. M. Halliday. 1996. Cytoarchitectural distribution of calcium binding proteins in midbrain dopaminergic regions of rats and humans. *J. Comp. Neurol.* 364: 121–150.

McRitchie, D. A., Cartwright, H., Pond, S. M., van der Schyf, C. J., Castagnoli, N. J., van der Nest, D. G., and G. M. Halliday. 1998. The midbrain dopaminergic cell groups in the baboon *Papio ursinus*. *Brain Res. Bull.* 47: 611–623.

Palmer, S. M., and M. G. P. Rosa. 2006. Quantitative analysis of the corticocortical projections to the middle temporal area in the marmoset monkey: Evolutionary and functional implications. *Cerebral Cortex* 16: 1361–1375.

Paxinos, G., and X-F. Huang. 1995. *Atlas of the Human Brainstem.* San Diego: Academic.

Paxinos, G., and C. R. R. Watson. 1986. *The Rat Brain in Stereotaxic Co-ordinates.* 2nd ed. San Diego: Academic.

Paxinos, G., Huang, X-F., and A. W. Toga. 2000. *The Rhesus Monkey Brain in Stereotaxic Coordinates.* San Diego: Academic.

Paxinos, G., Watson, C., Carrive, P., Kirkcaldie, M., and K. W. S. Ashwell. 2009. *Chemoarchitectonic Atlas of the Rat Brain.* 2nd ed. San Diego: Academic.

Philibert, B., Beitel, R. E., Nagarajan, S.S., Bonham, B. H., Schreiner, C. E., and S.W. Cheung. 2005. Functional organization and hemispheric comparison of primary auditory cortex in the common marmoset (*Callithrix jacchus*). *J. Comp. Neurol.* 487: 391–406.

Qi, H-X., Lyon, D. C., and J. H. Kaas. 2002. Cortical and thalamic connections of the parietal ventral somatosensory area in marmoset monkeys (*Callithrix jacchus*). *J. Comp. Neurol.* 443: 168–182.

Ridley, R. M., Baker, H. F., Cummings, R. M., Green, M. E., and A. Leow-Dyke. 2005. Mild topographical memory impairment following crossed unilateral lesions of the mediodorsal thalamic nucleus and the inferotemporal cortex. *Behav. Neurosci.* 119: 518–525.

Ridley, R. M., Baker, H. F., Mills, D. A., Green, M. E., and R. M. Cummings. 2004. Topographical memory impairments after unilateral lesions of the anterior thalamus and contralateral inferotemporal cortex. *Neuropsychologia* 42: 1178–1191.

Risold, P. Y. 2004. The septal region. In *The Rat Nervous System* 3rd ed.. G. Paxinos (ed.).. San Diego: Elsevier.

Roberts, A. C., Tomic, D. L., Parkinson, C. H., Roeling, T. A., Cutter, D. J., Robbins, T. W., and B. J. Everitt. 2007. Forebrain connectivity of the prefrontal cortex in the marmoset monkey (*Callithrix jacchus*): An anterograde and retrograde tract-tracing study. *J. Comp. Neurol.* 502: 86–112.

Rosa, M. G. P., and L. M. Schmid. 1995. Visual areas in the dorsal and medial extrastriate cortices of the marmoset. *J. Comp. Neurol.* 359: 272–299.

Rosa, M. G. P., and R. Tweedale. 2000. Visual areas in lateral and ventral extrastriate cortices of the marmoset monkey. *J. Comp. Neurol.* 422: 621–651.

Rosa, M. G. P., Fritsches, K. A., and G. N. Elston. 1997. The second visual area in the marmoset monkey: Visuotopic organisation, magnification factors, architectonical boundaries, and modularity. *J. Comp. Neurol.* 387: 547–567.

Rosa, M. G. P., Palmer, S. M., Gamberini, M., Tweedale, R., Piñon, M. C., and J. A. Bourne. 2005. Resolving the organization of the New World monkey third visual complex: The dorsal extrastriate cortex of the marmoset (*Callithrix jacchus*). *J. Comp. Neurol.* 483: 164–191.

Rothe, H. 1973. Handedness in the common marmoset (*Callithrix jacchus*). *Am. J. Phys. Anthropol.* 38: 561–565.

Soloman, S. G. 2002. Striate cortex in dichromatic and trichromatic marmosets: Neurochemical compartmentalization and geniculate input. *J. Comp. Neurol.* 450: 366–381.

Spatz, W. B. 1977. Topographically organized reciprocal connections between areas 17 and MT (visual area of superior temporal sulcus) in the marmoset *Callithrix jacchus*. *Exp. Brain Res.* 27: 559–572.

Spatz, W. B. 1978. The retinogeniculocortical pathway in *Callithrix*. I. Intraspecific variations in the lamination pattern of the lateral geniculate nucleus. *Exp. Brain Res.* 33: 551–563.

Stepniewska, I., and J. H. Kaas. 1997. Architectonic subdivisions of the inferior pulvinar in New World and Old World monkeys. *Vis. Neurosci.* 14: 1043-1060.

Takahata, N., and Y. Satta. 1997. Evolution of the primate lineage leading to modern humans: Phylogenetic and demographic inferences from DNA sequences. *Proc. Natl. Acad. Sci. USA* 94: 4811–4815.

Vogt, B. A., Hof, P. R., and L. J. Vogt. 2004. Cingulate gyrus. In *The Human Nervous System.* 2nd ed. G. Paxinos, and J. K. Mai (eds.). San Diego: Elsevier.

Watson, C. R., Paxinos, G., and H. Tokuno. 2010. Using a panel of immunomarkers to define homologies in mammalian brains. *Front Hum. Neurosci.* 4: 13–17.

White, A. J. R., Wilder, H. D., Goodchild, A. K., Sefton, A., and P. R. Martin. 1998. Segregation of receptive field properties in the lateral geniculate nucleus of a New World monkey, the marmoset *Callithrix jacchus*. *J. Neurophysiol.* 80: 2063–2076.

Wilson, P., Kitchener, P. D., and P. J. Snow. 1999. Cutaneous receptive field organization in the ventral posterior nucleus of the thalamus in the common marmoset. *J. Neurophysiol.* 82: 1865–1875.

Zhang, H. Q., Murray, G. M., Turman, A. B., Mackie, P. D., Coleman, G. T., and M. J. Rowe. 1996. Parallel processing in cerebral cortex of the marmoset monkey: Effect of reversible SI inactivation on tactile responses in SII. *J. Neurophysiol.* 76: 3633–3655.

Section B

# Named structures and abbreviations

## List of named structures and abbreviations

A1 noradrenaline cells, A1
A5 noradrenaline cells, A5
A8 dopaminergic cells of midbrain, A8
A9 dopaminergic cells of midbrain, A9
A10 dopaminergic cells of midbrain, A10
abducens nerve, 6n
abducens nucleus, 6N
accessory facial nucleus, acs7
accessory motor nucleus, 11N
accessory neurosecretory nucleus, ANS
accumbens nucleus, Acb
accumbens nucleus core, AcbC
accumbens nucleus shell, AcbSh
accumbens nucleus shell, lateral part, LAcbSh
agranular insular cortex, AI
alar nucleus, Al
alveus of hippocampus, alv
ambiguus nucleus, Amb
ambiguus nucleus, compact part, AmbC
amygdalohippocampal area, magnocellular part, AHiMC
amygdalohippocampal area, parvicellular part, AHiPC
amygdalohippocampal area, unspecified, AHi
amygdaloid fissure, af
amygdalopiriform transition area, APir
amygdalostriatal transition zone, AStr
ansa lenticularis, al
anterior amygdaloid area, dorsal, AAD
anterior amygdaloid area, lateral, AAL
anterior amygdaloid area, unspecified, AA
anterior amygdaloid area, ventral, AAV
anterior commissure, ac
anterior commissure, anterior limb, aca
anterior commissure, posterior limb, acp
anterior hypothalamic area, anterior part, AHA
anterior hypothalamic area, posterior part, AHP
anterior lobe of pituitary gland, APit
anterior olfactory area, unspecified, AO
anterior pretectal nucleus, APT
anterior tegmental nucleus, ATg
anterodorsal thalamic nucleus, AD
anteromedial thalamic nucleus, AM
anteroventral periventricular hypothalamic nucleus, AVPe
anteroventral thalamic nucleus, AV
arcuate hypothalamic nucleus, dorsal part, ArcD
arcuate hypothalamic nucleus, lateral part, ArcL
arcuate hypothalamic nucleus, medial part, ArcM
arcuate sulcus, arc
areas 1/2 of cerebral cortex, 1 and 2
area 3a of cerebral cortex, 3a
area 3b of cerebral cortex, 3b
area 4 of cerebral cortex, 4
area 6 (dorsal, caudal) of cerebral cortex, 6Dc
area 6 (dorsal, rostral) of cerebral cortex, 6Dr
area 6 (medial) of cerebral cortex, 6M
area 6 (ventral) of cerebral cortex, 6V
area 8 (anterodorsal) of cerebral cortex, 8Ad
area 8 (anteroventral) of cerebral cortex, 8Av
area 8B of cerebral cortex, 8B
area 9 of cerebral cortex, 9
area 10 (mediocaudal) of cerebral cortex, 10mc
area 10 of cerebral cortex, 10
area 11 (lateral) of cerebral cortex, 11l
area 11 (medial) of cerebral cortex, 11m
area 12 (lateral) of cerebral cortex, 12l
area 12 (medial) of cerebral cortex, 12m
area 12 (orbital) of cerebral cortex, 12o
area 13a of cerebral cortex, 13a
area 13b of cerebral cortex, 13b
area 13l of cerebral cortex, 13l
area 13m of cerebral cortex, 13m
area 14 (caudal) of cerebral cortex, 14c
area 14 (rostral) of cerebral cortex, 14r
area 19 (medial) of cerebral cortex, 19m
area 23 of cerebral cortex, 23
area 23a of cerebral cortex, 23a
area 23b of cerebral cortex, 23b
area 23v of cerebral cortex, 23v
area 24 of cerebral cortex, 24
area 24a of cerebral cortex, 24a
area 24b of cerebral cortex, 24b
area 25 of cerebral cortex, 25
area 26 of cerebral cortex, 26
area 29 of cerebral cortex, 29
area 29a to c of cerebral cortex, 29a-c
area 29d of cerebral cortex, 29d

area 30 of cerebral cortex, 30
area 31 of cerebral cortex, 31
area 32 of cerebral cortex, 32
area 36 of cerebral cortex, 36
area 45 of cerebral cortex, 45
area 46 of cerebral cortex, 46
area postrema, AP
ascending fibers of facial nerve, asc7
auditory belt cortex, anterolateral, AuAL
auditory belt cortex, caudolateral, AuCL
auditory belt cortex, caudomedial, AuCM
auditory belt cortex, middle lateral, AuML
auditory belt cortex, rostromedial, AuRM
auditory belt cortex, rostrotemporolateral, AuRTL
auditory belt cortex, rostrotemporomedial, AuRTM
auditory core cortex, primary, Au1
auditory core cortex, primary, lateral, Au1L
auditory core cortex, primary, medial, Au1M
auditory core cortex, rostral, AuR
auditory core cortex, rostrotemporal, AuRT
auditory parabelt cortex, caudal, AuCPB
auditory parabelt cortex, rostral, AuRPB
Barrington's nucleus, Bar
basal interstitial nucleus, BI
basal nucleus of Meynert, B
basilar artery, bas
basolateral amygdaloid nucleus, dorsal, BLD
basolateral amygdaloid nucleus, dorsolateral, BLDL
basolateral amygdaloid nucleus, intermediate, BLI
basolateral amygdaloid nucleus, ventrolateral, BLVL
basolateral amygdaloid nucleus, ventromedial, BLVM
basomedial amygdaloid nucleus, anterior, BMA
basomedial amygdaloid nucleus, dorsal, BMD
basomedial amygdaloid nucleus, ventral, BMV
bed nucleus of stria terminalis, intraamygdalar division, BSTIA
bed nucleus of stria terminalis, lateral anterior division, BSTLA
bed nucleus of stria terminalis, lateral dorsal division, BSTLD
bed nucleus of stria terminalis, lateral juxtacapsular division, BSTLJ
bed nucleus of stria terminalis, lateral posterior division, BSTLP
bed nucleus of stria terminalis, lateral ventral division, BSTLV
bed nucleus of stria terminalis, medial anterior division, BSTMA
bed nucleus of stria terminalis, medial posterior division, intermediate, BSTMPI
bed nucleus of stria terminalis, medial posterior division, lateral, BSTMPL
bed nucleus of stria terminalis, medial posterior division, medial, BSTMPM
bed nucleus of stria terminalis, medial posterior division, BSTMP
blood vessel, bv
brachium of inferior colliculus, bic
brachium of superior colliculus, bsc
C1 adrenaline cells, C1
C2 adrenaline cells, C2
C3 adrenaline cells, C3
calcarine sulcus, ca
callosomarginal sulcus of cortex, cm
caudal linear nucleus, CLi
caudate nucleus, Cd
cell bridges of ventral striatum, CB
central amygdaloid nucleus, central part, CeC
central amygdaloid nucleus, lateral part, CeL
central amygdaloid nucleus, medial part, CeM
central canal of spinal cord, CC
central cervical nucleus of spinal cord, CeCv
central gray of pons, CGPn
central medial thalamic nucleus, CM
central nucleus of inferior colliculus, CIC
central tegmental tract, ctg
centrolateral thalamic nucleus, CL
centrolateral thalamic nucleus, anterior division, CLa
centrolateral thalamic nucleus, posterior division, CLp
centromedian thalamic nucleus, CMn
cerebellar commissure, cbc
cerebellar input zone of motor thalamus (VLp), cb
cerebellar lobule 1, 1Cb
cerebellar lobule 2, 2Cb
cerebellar lobule 3, 3Cb
cerebellar lobule 4, 4Cb
cerebellar lobule 5, 5Cb
cerebellar lobule 6, 6Cb
cerebellar lobule 7, 7Cb
cerebellar lobule 8, 8Cb
cerebellar lobule 9, 9Cb
cerebellar lobule 10, 10Cb

cerebral aqueduct, Aq
cerebral peduncle, cp
cingulate gyrus of cortex, CG
cingulate sulcus of cortex, CGS
cingulum, cg
circular sulcus, ci
claustrum, Cl
cochlear division of vestibulocochlear nerve, cn
commissure of inferior colliculus, cic
commissure of superior colliculus, csc
copula of cerebellar pyramis, Cop
cornu ammonis 1, CA1
cornu ammonis 2, CA2
cornu ammonis 3, CA3
corpus callosum, cc
corpus callosum rostrum, ccr
cortex–amygdala transition zone, CxA
crus 1 of ansiform lobule of cerebellum, Crus1
crus 2 of ansiform lobule of cerebellum, Crus2
cuneate fasciculus, cu
cuneate nucleus, Cu
cuneate nucleus, rotunda part, CuR
cuneate nucleus, triangular part, CuT
cuneiform nucleus, CnF
deep cortical white matter, dcw
deep gray of superior colliculus, DpG
deep white of superior colliculus, DpWh
dentate gyrus, DG
dorsal (motor) nucleus of vagus nerve, 10N
dorsal (motor) nucleus of vagus nerve, caudal intermediate part, 10CaI
dorsal (motor) nucleus of vagus nerve, central intermediate part, 10CeI
dorsal (motor) nucleus of vagus nerve, dorsal intermediate part, 10DI
dorsal (motor) nucleus of vagus, ventrorostral subnucleus, VeR
dorsal acoustic stria, das
dorsal cochlear nucleus, DC
dorsal cochlear nucleus, molecular layer, DCMo
dorsal cortex of inferior colliculus, DCIC
dorsal endopiriform nucleus, DEn
dorsal funiculus of spinal cord, dfu
dorsal hippocampal commissure, dhc
dorsal horn of spinal cord, DH
dorsal nucleus of lateral lemniscus, DLL
dorsal paragigantocellular nucleus, DPGi
dorsal raphe nucleus, caudal part, DRC
dorsal raphe nucleus, dorsal part, DRD
dorsal raphe nucleus, interfascicular part, DRI
dorsal raphe nucleus, ventral part, DRV
dorsal raphe nucleus, ventrolateral part, DRVL
dorsal spinocerebellar tract, dsc
dorsal tegmental decussation, dtgd
dorsal tegmental nucleus, central part, DTgC
dorsal tegmental nucleus, pericentral part, DTgP
dorsal terminal nucleus, DTnl
dorsoanterior extrastriate visual cortex, DA
dorsointermediate extrastriate visual cortex, DI
dorsolateral periaqueductal gray, DLPAG
dorsolateral tract of Lissauer, dlL
dorsomedial extrastriate visual cortex (V3d), DM
dorsomedial hypothalamic nucleus, compact, DMC
dorsomedial hypothalamic nucleus, dorsal, DMD
dorsomedial hypothalamic nucleus, ventral, DMV
dorsomedial periaqueductal gray, DMPAG
dorsomedial spinal trigeminal nucleus, DMSp5
dorsomedial tegmental area, DMTg
Edinger Westphal nucleus, EW
entorhinal cortex, caudal, EntC
entorhinal cortex, lateral, EntM
entorhinal cortex, medial, EntL
entorhinal cortex, rostral, EntR
ependymal lining of ventricle, E
extended amygdala, EA
extended amygdala, central part, EAC
extended amygdala, medial part, EAM
external capsule, ec
external cortex of inferior colliculus, ECIC
external cuneate nucleus, ECu
external medullary lamina, eml
external plexiform layer, olfactory bulb, EPl
external segment of globus pallidus, EGP
extreme capsule, ex
facial (motor) nucleus, 7N
facial nerve, 7n
fasciculus retroflexus, fr
fimbria of hippocampus, fi
fimbriodentate sulcus, fd
flocculus of cerebellum, Fl
forceps major of corpus callosum, fma
forceps minor of corpus callosum, fmi

fornix, f
fourth ventricle, 4V
frontal eye field of cortex, FEF
fundus of striatum, FStr
fundus of superior temporal cortex, FST
gelatinosus layer of caudal spinal trigeminal nucleus, Ge5
geniculocalcarine tract, gc
genu of facial nerve, g7
gigantocellular reticular nucleus, Gi
gigantocellular reticular nucleus, ventral part, GiV
glomerular layer, olfactory bulb, Gl
gracile fasciculus, gr
gracile nucleus, Gr
granular insular cortex, GI
granule cell cluster, GCC
granule cell cluster, magna, GCCM
granule cell layer of dentate gyrus, GrDG
granule cell layer, olfactory bulb, GrO
gustatory area of cortex, Gu
H field of Forel, prerubral field, ansa lenticularis, H (al)
H1 field of Forel, thalamic fasciculus, H1 (thf)
H2 field of Forel, lenticular fasciculus, H2 (lentf)
habenular commissure, hbc
hippocampal fissure, hi
horizontal limb of diagonal band nucleus, HDB
hypoglossal nerve, 12n
hypoglossal nucleus, 12N
hypoglossal nucleus, ventrolateral part, 12VL
induseum griseum, IG
inferior cerebellar peduncle, icp
inferior colliculus, unspecified subregion, IC
inferior medullary velum, imv
inferior olivary nucleus, cap of Kooy subnucleus, IOK
inferior olivary nucleus, dorsal subnucleus, IOD
inferior olivary nucleus, dorsomedial subnucleus, IODM
inferior olivary nucleus, medial subnucleus, IOM
inferior olivary nucleus, principal subnucleus, IOPr
inferior olivary nucleus, subnucleus A of medial nucleus, IOA
inferior olivary nucleus, subnucleus β, IOβ
inferior olivary nucleus, subnucleus B of medial nucleus, IOB
inferior olivary nucleus, subnucleus C of medial nucleus, IOC
inferior salivatory nucleus, IS
inferior temporal cortex, anterior area, ITA
inferior temporal cortex, anterior area 1, ITA1
inferior temporal cortex, anterior area 2, ITA2
inferior temporal cortex, dorsal area, ITD
inferior temporal cortex, dorsal area 1, ITD1
inferior temporal cortex, dorsal area 2, ITD2
inferior temporal cortex, ventral area, ITV
inferior temporal sulcus, it
infundibular stalk, InfS
insular cortex, rostrolateral, Insrl
insular cortex, rostromedial, Insrm
insular proisocortex, IPro
interanteromedial thalamic nucleus, IAM
intercalated nuclei of amygdala, I
intercalated nucleus of medulla, In
intercollicular nucleus, InCo
interfascicular nucleus, IF
intermediate (anterior) deep cerebellar nucleus, IntA
intermediate (posterior) deep cerebellar nucleus, IntP
intermediate gray of superior colliculus, InG
intermediate lobe of pituitary gland, IPit
intermediate nucleus of lateral lemniscus, ILL
intermediate reticular formation, IRt
intermediate temporal sulcus of cortex, int
intermediate white of superior colliculus, InWh
intermediodorsal thalamic nucleus, IMD
internal arcuate fiber, ia
internal basal nucleus, IB
internal capsule, ic
internal medullary lamina, iml
internal segment of globus pallidus, IGP
interpeduncular fossa, IPF
interpeduncular nucleus, caudal subnucleus, IPC
interpeduncular nucleus, dorsal subnucleus, IPD
interpeduncular nucleus, dorsolateral subnucleus, IPDL
interpeduncular nucleus, dorsomedial subnucleus, IPDM
interpeduncular nucleus, interfascicular subnucleus, IPI
interpeduncular nucleus, lateral subnucleus, IPL
interpeduncular nucleus, rostral subnucleus, IPR
interpositus nucleus, IPo
interstitial nucleus of Cajal, InC
interstitial nucleus of Cajal, shell part, InCSh
interstitial nucleus of the posterior limb of ac, IPAC
interstitial nucleus of vestibular division, 8th nerve, I8Ve

intertrigeminal nucleus, I5
interventricular foramen, IVF
intramedullary gray of amygdala, IMG
intraparietal sulcus of cortex, ip
isthmic reticular formation, isRt
Kölliker-Fuse nucleus, KF
lamina terminalis, LTer
lateral amygdaloid nucleus, dorsal part, LaD
lateral amygdaloid nucleus, lateral part, LaL
lateral amygdaloid nucleus, medial part, LaM
lateral corticospinal tract, lcs
lateral deep cerebellar nucleus, Lat
lateral fissure, lf
lateral funiculus of spinal cord, lfu
lateral geniculate thalamic nucleus, internal interlaminar, LGii
lateral geniculate thalamic nucleus, external interlaminar, LGei
lateral geniculate thalamic nucleus, magnocell. ext., LGme
lateral geniculate thalamic nucleus, magnocell. int., LGmi
lateral geniculate thalamic nucleus, parvicell. ext., LGpe
lateral geniculate thalamic nucleus, parvicell. int., LGpi
lateral geniculate thalamic nucleus, superficial, LGs
lateral geniculate thalamic nucleus, LG
lateral habenular nucleus, LHb
lateral habenular nucleus, dorsal part, LHbD
lateral habenular nucleus, ventral part, LHbV
lateral hypothalamic area, LH
lateral lemniscus, ll
lateral mamillary nucleus, LM
lateral medullary lamina, lml
lateral olfactory tract, lo
lateral parabrachial nucleus, central part, LPBC
lateral parabrachial nucleus, dorsal part, LPBD
lateral parabrachial nucleus, external part, LPBE
lateral parabrachial nucleus, ventral part, LPBV
lateral paragiganticellular nucleus, LPGi
lateral periaqueductal gray, LPAG
lateral pericuneate nucleus, LPCu
lateral periolivary nucleus, LaPO
lateral preoptic area, LPO
lateral recess of fourth ventricle, LR4V
lateral reticular nucleus, LRt
lateral reticular nucleus, parvicellular part, LRtPC
lateral reticular nucleus, subtrigeminal part, LRtS5
lateral septal nucleus, dorsal, LSD
lateral septal nucleus, intermediate, LSI
lateral septal nucleus, ventral part, LSV
lateral spinal nucleus, LSp
lateral superior olivary nucleus, LSO
lateral terminal nucleus, LTnl
lateral tuberal hypothalamic nucleus, LTu
lateral ventricle, LV
lateral vestibular nucleus, LVe
lateral vestibulospinal tract, lvs
lateroanterior hypothalamic nucleus, LA
laterodorsal tegmental nucleus, LDTg
laterodorsal tegmental nucleus, dorsal part, LDTgD
laterodorsal tegmental nucleus, ventral part, LDTgV
laterodorsal thalamic nucleus, LD
lateroposterior thalamic nucleus, LP
lateroventral periolivary nucleus, LVPO
layer 1 of cerebral cortex, 1Cx
layer 2 of cerebral cortex, 2Cx
layer 3 of cerebral cortex, 3Cx
layer 4 of cerebral cortex, 4Cx
layer 4a of cerebral cortex, 4aCx
layer 4b of cerebral cortex, 4bCx
layer 4c of cerebral cortex, 4cCx
layer 5 of cerebral cortex, 5Cx
layer 6 of cerebral cortex, 6Cx
lenticular fasciculus, lentf
limitans nucleus, Lim
locus coeruleus, LC
longitudinal fibers of pons, lfp
lunate sulcus, lu
magnocellular nucleus of lateral hypothalamus, MCLH
magnocellular nucleus of posterior commissure, MCPC
magnocellular preoptic nucleus, MCPO
mamillary peduncle, mp
mamillary recess of third ventricle, MRe
mamillotegmental tract, mtg
mamillothalamic tract, mt
matrix sensory area, Mx
medial accessory oculomotor nucleus, MA3
medial amygdaloid nucleus, Me
medial deep cerebellar nucleus, Med
medial forebrain bundle, mfb

medial geniculate nucleus, anterodorsal subnucleus, MGAD
medial geniculate nucleus, marginal zone, MGMZ
medial geniculate nucleus, medial subnucleus, MGM
medial geniculate nucleus, posterodorsal subnucleus, MGPD
medial geniculate nucleus, ventral subnucleus, MGV
medial habenular nucleus, MHb
medial lemniscus, ml
medial longitudinal fasciculus, mlf
medial mamillary nucleus, lateral part, ML
medial mamillary nucleus, medial part, MM
medial medullary lamina, mml
medial parabrachial nucleus, MPB
medial parabrachial nucleus, external part, MPBE
medial periolivary nucleus, MePO
medial preoptic area, MPA
medial preoptic nucleus, compact part, MPOC
medial preoptic nucleus, lateral part, MPOL
medial preoptic nucleus, medial part, MPOM
medial pretectal nucleus, MPT
medial septal nucleus, MS
medial superior olivary nucleus, MSO
medial superior temporal area of cortex, MST
medial temporal sulcus of cortex, mte
medial terminal nucleus, MTnl
medial tuberal hypothalamic nucleus, MTu
medial vestibular nucleus, MVe
medial vestibular nucleus, magnocellular part, MVeMC
medial vestibular nucleus, parvicellular part, MVePC
median accessory nucleus of medulla, MnA
median eminence of hypothalamus, ME
median eminence, external layer, MEE
median eminence, internal layer, MEI
median preoptic nucleus, MnPO
median raphe nucleus, MnR
mediodorsal thalamic nucleus, MD
mediodorsal thalamic nucleus, central part, MDC
mediodorsal thalamic nucleus, dorsal part, MDD
mediodorsal thalamic nucleus, lateral part, MDL
mediodorsal thalamic nucleus, medial part, MDM
medioventral periolivary nucleus, MVPO
medullary reticular nucleus, dorsal part, MdD
medullary reticular nucleus, ventral part, MdV
mesencephalic reticular formation, mRt
mesencephalic trigeminal nucleus, Me5
mesencephalic trigeminal tract, me5
microcellular tegmental nucleus, MiTg
middle cerebellar peduncle, mcp
middle temporal area of cortex, MT
middle temporal crescent area of cortex, MTc
molecular layer of dentate gyrus, MoDG
motor root of trigeminal nerve, m5
motor trigeminal nucleus, Mo5
nanocellular hypothalamic nucleus, NC
neurohypophyseal tract, nehy
nigral input zone of motor thalamus (VA), nigr
nigrostriatal tract, ns
nucleus of optic tract, OT
nucleus of brachium of inferior colliculus, BIC
nucleus of Darkschewitsch, Dk
nucleus of lateral olfactory tract, LOT
nucleus of Roller, Ro
nucleus of solitary tract, Sol
nucleus of solitary tract, central subnucleus, SolCe
nucleus of solitary tract, commissural subnucleus, SolC
nucleus of solitary tract, dorsolateral subnucleus, SolDL
nucleus of solitary tract, dorsomedial subnucleus, SolDM
nucleus of solitary tract, gelatinosus subnucleus, SolG
nucleus of solitary tract, intermediate subnucleus, SolIM
nucleus of solitary tract, intrafascicular subnucleus, SolI
nucleus of solitary tract, lateral subnucleus, SolL
nucleus of solitary tract, medial subnucleus, SolM
nucleus of solitary tract, ventral subnucleus, SolV
nucleus of solitary tract, ventrolateral subnucleus, SolVL
nucleus of the stria medullaris thalami, SM
nucleus of trapezoid body, Tz
nucleus X, X
nucleus Y, Y
oculomotor nerve, 3n
oculomotor nucleus, 3N
oculomotor nucleus, parvicellular part, 3PC
olfactory bulb, OB
olfactory nerve layer, olfactory bulb, ON
olfactory tubercle, Tu
olfactory ventricle, OV
olivary pretectal nucleus, OPT
olivocerebellar tract, oc

optic chiasm, och
optic nerve, 2n
optic tract, opt
orbital periallocortex, OPAl
orbital sulcus (lateral) of cortex, orbl
orbital sulcus (medial) of cortex, orbm
orbital sulcus of cortex, orb
pallidal input zone of motor thalamus (VLa), pal
parabducens nucleus, Pa6
parabigeminal nucleus, PBG
parabrachial pigmented nucleus, PBP
paracentral thalamic nucleus, PC
paracollicular tegmentum, PCTg
paracommissural nucleus, PCom
parafascicular thalamic nucleus, Pf
paraflocculus, PFl
paraflocculus posterior, PFlP
paraflocculus ventral, PFlV
paralambdoid septal zone, PLd
paralaminar amygdaloid nucleus, PLam
paralaminar amygdaloid nucleus, posterior part, PLamP
paramedian lobule of cerebellum, PM
paramedian lobule of cerebellum, anterior, PMa
paramedian lobule of cerebellum, posterior, PMp
paramedian raphe nucleus, PMnR
paranigral nucleus of VTA, PN
parapeduncular nucleus, PaP
pararubral area, PaR
parasolitary nucleus, PSol
parastriatal nucleus, PS
parasubiculum, PaS
parasubthalamic nucleus, PSTh
paratenial thalamic nucleus, PT
paraventricular hypothalamic nucleus, anterior parvicellular part, PaAP
paraventricular hypothalamic nucleus, magnocellular part, PaM
paraventricular hypothalamic nucleus, magnocellular, dorsal part, PaMD
paraventricular hypothalamic nucleus, parvicellular, dorsal part, PaPD
paraventricular hypothalamic nucleus, posterior part, PaPo
paraventricular hypothalamic nucleus, ventral part, PaV
paraventricular thalamic nucleus, anterior part, PVA
paraventricular thalamic nucleus, posterior part, PVP
paravestibular nucleus, PaVe
paraxiphoid thalamic nucleus, PaXi
parietal field E of cortex, PE
parietal field E of cortex, caudal subdivision, PEc
parietal field F of cortex, PF
parietal field G of cortex, PG
parietal rostral somatosensory area of cortex, PRSom
parietoventral somatosensory cortex, PV
parvicellular oculomotor nucleus, PC3
parvicellular reticular nucleus, PCRt
parvicellular reticular nucleus, alpha part, PCRtα
parvicellular trigeminal motor nucleus, PC5
peduncular part of lateral hypothalamus, PLH
pedunculopontine tegmental nucleus, PPTg
pedunculopontine tegmental nucleus, compact part, PPTgC
pedunculopontine tegmental nucleus, diffuse part, PPTgD
pericuneate matrix, PCuMx
perifacial nucleus, P7
perifornical hypothalamic nucleus, PeF
perifornical hypothalamic nucleus, lateral part, PeFL
perifornical hypothalamic nucleus, medial part, PeFM
peripeduncular nucleus, PP
perirhinal cortex, PRh
peritrigeminal nucleus, P5
periventricular gray, PVG
periventricular hypothalamic nucleus, Pe
pineal gland, Pi
piriform cortex, Pir
piriform cortex, layer 1, Pir1
piriform cortex, layer 2, Pir2
piriform cortex, layer 3, Pir3
polymorphic layer of dentate gyrus, PoDG
pontine caudal nucleus, PnC
pontine nucleus, Pn
pontine oral nucleus, PnO
pontine raphe nucleus, PnR
posterior commissure, pc
posterior hypothalamic area, PH
posterior lobe of pituitary gland, PPit
posterior parietal cortex, dorsal area 1, PPD1
posterior parietal cortex, dorsal area 2, PPD2
posterior parietal cortex, medial area, PPM

posterior parietal cortex, ventral area, PPV
posterior pretectum, PPT
posterior superior fissure of cerebellum, psf
posterior temporal sulcus of cortex, pt
posterior thalamic nucleus, Po
posterolateral fissure of cerebellum, plf
precentral fissure of cerebellum, prcf
precentral opercular cortex, PrCO
precommissural nucleus, PrC
preculminate fissure of cerebellum, pcf
prefrontal sulcus of cortex, pr
pregeniculate nucleus, PrG
premamillary hypothalamic nucleus, dorsal part, PMaD
premamillary hypothalamic nucleus, ventral part, PMaV
premotor cortex, dorsal part, PMD
premotor cortex, ventral part, PMV
prepositus hypoglossi nucleus, Pr
prepyramidal fissure of cerebellum, ppf
prerubral field, PRF
presubiculum, PrS
primary fissure of cerebellum, prf
primary visual (striate) cortex, V1
principal trigeminal sensory nucleus, Pr5
prosomere 1 reticular formation, p1Rt
prostriate cortex, ProSt
prosubiculum, ProS
pulvinar nucleus, anterior part, PulA
pulvinar nucleus, inferior part, PulI
pulvinar nucleus, lateral part, PulL
pulvinar nucleus, medial part, PulM
putamen, Pu
pyramidal tract, py
pyramidal tract decussation, pyd
raphe magnus nucleus, RMg
raphe obscurus nucleus, ROb
raphe pallidus nucleus, RPa
recess of inferior colliculus, ReIC
red nucleus, magnocellular part, RMC
red nucleus, parvicellular part, RPC
reticular (thalamic) nucleus, Rt
reticulotegmental nucleus, RtTg
retroambiguus nucleus, RAmb
retrochiasmatic nucleus, lateral part, RChL
retrochiasmatic nucleus, medial part, RChM
retroinsular area of cortex, RIns
retromamillary decussation, rmd
retromamillary nucleus, lateral part, RML
retromamillary nucleus, medial part, RMM
retrorubral field, RRF
reuniens thalamic nucleus, Re
Rexed's lamina 1 of spinal cord, R1
Rexed's lamina 2 of spinal cord, R2
Rexed's lamina 3 of spinal cord, R3
Rexed's lamina 4 of spinal cord, R4
Rexed's lamina 5 of spinal cord, R5
Rexed's lamina 7 of spinal cord, R7
Rexed's lamina 8 of spinal cord, R8
Rexed's lamina 9 of spinal cord, R9
rhabdoid nucleus, Rbd
rhinal fissure, rf
rhomboid thalamic nucleus, Rh
rostral interstitial nucleus, RI
rostral linear nucleus (midbrain), RLi
rostral periolivary nucleus, RPO
rostral ventrolateral area of medulla, RVL
rubrospinal tract, rs
sagulum nucleus, Sag
secondary fissure of cerebellum, sf
secondary somatosensory area of cortex, S2
secondary visual area (lateral) of cortex, V2l
secondary visual area (medial) of cortex, V2m
secondary visual area (ventral) of cortex, V2v
secondary visual area of cortex, V2
sensory root of trigeminal nerve, s5
septofimbrial nucleus, SFi
septohippocampal nucleus, SHi
septohypothalamic nucleus, SHy
simplex lobule of cerebellum, Sim
simplex lobule of cerebellum, anterior, Sima
simplex lobule of cerebellum, posterior, Simp
solitary tract, sol
spinal trigeminal nucleus, caudalis part, Sp5C
spinal trigeminal nucleus, interpolaris part, Sp5I
spinal trigeminal nucleus, oralis part, Sp5O
spinal trigeminal tract, sp5
spinal vestibular nucleus, SpVe
spinothalamic tract, spth
stratum lacunosum moleculare of cornu ammonis, LMol
stratum opticum of superior colliculus, Op

stratum oriens of cornu ammonis, Or
stratum pyramidale of cornu ammonis, Py
stratum radiatum of cornu ammonis, Rad
stratum zonale of superior colliculus, Zo
stria medullaris thalami, sm
stria terminalis, st
striatohypothalamic nucleus, StHy
striatonigral fibers, sn
subcoeruleus nucleus, dorsal part, SubCD
subcoeruleus nucleus, ventral part, SubCV
subcommissural organ, SCO
subfornical organ, SFO
subiculum, S
subincertal nucleus, SubI
subparafascicular thalamic nucleus, SPf
subparafascicular thalamic nucleus, parvicellular part, SPfPC
subpeduncular tegmental nucleus, SPTg
substantia innominata, SI
substantia nigra, compact part, SNC
substantia nigra, compact part, dorsal tier, SNCD
substantia nigra, compact part, medial tier, SNM
substantia nigra, compact part, ventral tier, SNCV
substantia nigra, compact part, ventrolateral tier, SNVL
substantia nigra, compact part, ventromedial tier, SNVM
substantia nigra, lateral part, SNL
substantia nigra, reticular part, SNR
subthalamic nucleus, STh
superficial gray of superior colliculus, SuG
superior cerebellar peduncle, scp
superior cerebellar peduncle decussation, scpd
superior colliculus, SC
superior medullary velum, smv
superior salivatory nucleus, SS
superior temporal cortex, lateral area, STL
superior temporal cortex, medial area, STM
superior temporal sulcus, sts
superior vestibular nucleus, SuVe
supplementary motor area of cortex, SMA
suprachiasmatic hypothalamic nucleus, dorsal, SChD
suprachiasmatic hypothalamic nucleus, ventral, SChV
suprageniculate thalamic nucleus, SG
supraoculomotor cap, Su3C
supraoculomotor periaqueductal gray, Su3
supraoptic decussation, sod
supraoptic hypothalamic nucleus, SO
supraoptic hypothalamic nucleus, retrochiasmatic part, SOR
supratrigeminal nucleus, Su5
tectospinal tract, ts
temporal field F of cortex, TF
temporal field H of cortex, TH
temporal parietotemporal area of Cx, TPt
temporal proisocortex, TPro
terete nucleus of hypothalamus, Te
tertiary visual area (dorsal) of cortex, V3d
third ventricle, 3V
transverse cerebral fissure, TCF
transverse fibers of pons, tfp
trapezoid body, tz
trapezoid decussation, tzd
triangular septal nucleus, TS
trigeminal motor nucleus, 5N
trigeminal motor nucleus, masseter region, 5Ma
trigeminal motor nucleus, pterygoid region, 5Pt
trigeminal motor nucleus, temporalis region, 5Te
trigeminal nerve, 5n
trigeminosolitary transition zone, 5Sol
trigeminothalamic tract, tth
trochlear nerve, 4n
trochlear nerve decussation, 4nd
trochlear nucleus, 4N
tuberal area of lateral hypothalamus, TuLH
tuberomamillary hypothalamic area, TM
vascular organ of lamina terminalis, VOLT
ventral anterior cortical amygdala, VACo
ventral anterior thalamic nucleus, VA
ventral anterior thalamic nucleus, lateral part, VAL
ventral anterior thalamic nucleus, medial part, VAM
ventral anterior thalamic nucleus, mediocentral part, VAMC
ventral cochlear nucleus, anterior part, VCA
ventral cochlear nucleus, posterior part, VCP
ventral corticospinal tract, vcs
ventral endopiriform nucleus, VEn
ventral funiculus of spinal cord, vfu
ventral hippocampal commissure, vhc
ventral horn of spinal cord, VH
ventral inferior cortical amygdala, VCoI
ventral intermediate cortical amygdala, VCoIn
ventral median fissure of spinal cord, vmf

ventral nucleus of lateral lemniscus, VLL
ventral pallidum, VP
ventral periventricular zone of hypothalamus, VPe
ventral pontine nucleus, PnV
ventral posterior thalamic nucleus, inferior part, VPI
ventral posterior thalamic nucleus, lateral part, VPL
ventral posterior thalamic nucleus, medial part, VPM
ventral posterior thalamic nucleus, medial parvicellular part, VPMPC
ventral posterior thalamic nucleus, superior part, VPS
ventral reuniens thalamic nucleus, VRe
ventral somatosensory area (putative) of cortex, VSom?
ventral spinocerebellar tract, vsc
ventral superior cortical amygdala, VCoS
ventral tegmental area, VTA
ventral tegmental decussation, vtgd
ventral tuberomamillary area, VTM
ventrolateral anterior extrastriate cortex (V4), VLA (V4)
ventrolateral hypothalamic area, VLH
ventrolateral periaqueductal gray, VLPAG
ventrolateral posterior extrastriate cortex (V3v), VLP (V3v)
ventrolateral preoptic nucleus, VLPO
ventrolateral thalamic nucleus, anterior part, VLA
ventrolateral thalamic nucleus, medial part, VLM
ventrolateral thalamic nucleus, posterior part, VLP
ventromedial hypothalamic nucleus, unspecified, VMH
ventromedial hypothalamic nucleus, lateral part, VMHL
ventromedial hypothalamic nucleus, medial part, VMHM
ventromedial preoptic nucleus, VMPO
vertical limb of diagonal band nuclei, VDB
vestibular division of vestibulocochlear nerve, 8vn
vestibulocerebellar nucleus, VeCb
vestibulocochlear nerve, 8n
xiphoid thalamic nucleus, Xi
zona incerta, dorsal part, ZID
zona incerta, rostral part, ZIR
zona incerta, ventral part, ZIV

Section C

# Index of atlas plates

| Abbreviation | Long Name | Plates |
|---|---|---|
| 1&2 | areas 1/2 of cerebral cortex | 26–29, 62, 63, 70–75 |
| 1Cb | cerebellar lobule 1 | 41, 63, 64 |
| 2Cb | cerebellar lobule 2 | 38–41, 62–69 |
| 2n | optic nerve | 15 |
| 3a | area 3a of cerebral cortex | 17–24, 66–75 |
| 3b | area 3b of cerebral cortex | 17–27, 62–75 |
| 3Cb | cerebellar lobule 3 | 39–43, 62–70 |
| 3n | oculomotor nerve | 28–31, 62–64 |
| 3N | oculomotor nucleus | 30–34, 62 |
| 3PC | oculomotor nucleus, parvicellular part | 33, 34, 62, 63 |
| 3V | third ventricle | 16–32 |
| 4 | area 4 of cerebral cortex | 19–25, 62, 63, 70–74, 75 |
| 4Cb | cerebellar lobule 4 | 40–44, 62–75 |
| 4n | trochlear nerve | 35–38, 63–66 |
| 4N | trochlear nucleus | 35, 62, 63 |
| 4nd | trochlear nerve decussation | 38, 62–64 |
| 4V | fourth ventricle | 38–47, 62–71 |
| 5Cb | cerebellar lobule 5 | 40–51, 62–73 |
| 5Ma | trigeminal motor nucleus, masseter region | 37, 67 |
| 5n | trigeminal nerve | 36–38 |
| 5Pt | trigeminal motor nucleus, pterygoid region | 37, 66 |
| 5Sol | trigeminosolitary transition zone | 40–44, 66 |
| 5Te | trigeminal motor nucleus, temporalis region | 37, 67 |
| 6Cb | cerebellar lobule 6 | 46–55, 62–69 |
| 6Dc | area 6 (dorsal, caudal) of cerebral cortex | 12–18, 67–74 |
| 6Dr | area 6 (dorsal, rostral) of cerebral cortex | 10, 11, 65–70 |
| 6M | area 6 (medial) of cerebral cortex | 12–20, 62–66 |
| 6n | abducens nerve | 63–65 |
| 6N | abducens nucleus | 38, 39, 63, 64 |
| 6V | area 6 (ventral) of cerebral cortex | 12–18, 73–75 |
| 7Cb | cerebellar lobule 7 | 49–58, 62–69 |
| 7n | facial nerve | 38, 65, 67, 68 |
| 7N | facial (motor) nucleus | 38, 39, 66, 67 |
| 8Ad | area 8 (anterodorsal) of cerebral cortex | 2–11, 63–73 |

| Abbreviation | Long Name | Plates |
|---|---|---|
| 8Av | area 8 (anteroventral) of cerebral cortex | 4–14, 67–75 |
| 8B | area 8B of cerebral cortex | 5–11, 62–66 |
| 8Cb | cerebellar lobule 8 | 48–58, 62–69 |
| 8n | vestibulocochlear nerve | 38, 40, 70–72 |
| 8vn | vestibular division of vestibulocochlear nerve | 39, 43, 67–69, 71 |
| 9 | area 9 of cerebral cortex | 2–4, 62, 63 |
| 9Cb | cerebellar lobule 9 | 46–53, 62–67 |
| 10 | area 10 of cerebral cortex | 1–4, 62, 63 |
| 10CaI | dorsal (motor) nucleus of vagus nerve, caudal intermediate part | 44, 45, 63 |
| 10Cb | cerebellar lobule 10 | 45–48, 62–68 |
| 10CeI | dorsal (motor) nucleus of vagus nerve, central intermediate part | 43–45, 63 |
| 10DI | dorsal (motor) nucleus of vagus nerve, dorsal intermediate part | 43–45 |
| 10mc | area 10 (mediocaudal) of cerebral cortex | 5–9, 62–64 |
| 10N | dorsal (motor) nucleus of vagus nerve | 46, 64, 65 |
| 11l | area 11 (lateral) of cerebral cortex | 2–10, 66–72 |
| 11m | area 11 (medial) of cerebral cortex | 2–10, 65–69 |
| 11N | accessory motor nucleus | 46–48, 64 |
| 12l | area 12 (lateral) of cerebral cortex | 2–11, 67–75 |
| 12m | area 12 (medial) of cerebral cortex | 3–11, 68–74 |
| 12n | hypoglossal nerve | 43, 44 |
| 12N | hypoglossal nucleus | 43–45, 62 |
| 12o | area 12 (orbital) of cerebral cortex | 11–14, 74, 75 |
| 12VL | hypoglossal nucleus, ventrolateral part | 45 |
| 13a | area 13a of cerebral cortex | 10, 11, 65–67 |
| 13b | area 13b of cerebral cortex | 2–9, 63–66 |
| 13l | area 13l of cerebral cortex | 11, 70–72 |
| 13m | area 13m of cerebral cortex | 5–12, 67–70 |
| 14c | area 14 (caudal) of cerebral cortex | 10, 11, 62–64 |
| 14r | area 14 (rostral) of cerebral cortex | 2–9, 62–64 |

| Abbreviation | Long Name | Plates |
|---|---|---|
| 19m | area 19 (medial) of cerebral cortex | 36–41, 62–64 |
| 23 | area 23 of cerebral cortex | 26–32, 37, 38, 63 |
| 23a | area 23a of cerebral cortex | 33–36, 62–64 |
| 23b | area 23b of cerebral cortex | 33–36, 62–64 |
| 23v | area 23v of cerebral cortex | 38–41, 62–69 |
| 24 | area 24 of cerebral cortex | 12 |
| 24a | area 24a of cerebral cortex | 10–25, 62–64 |
| 24b | area 24b of cerebral cortex | 10–25, 62–64 |
| 25 | area 25 of cerebral cortex | 10–14, 62, 63 |
| 26 | area 26 of cerebral cortex | 23–36 |
| 29 | area 29 of cerebral cortex | 23–29, 35–38 |
| 29a-c | area 29 of cerebral cortex | 30–34, 68 |
| 29d | area 29d of cerebral cortex | 30–34, 68, 69 |
| 30 | area 30 of cerebral cortex | 23–38, 63 |
| 31 | area 31 of cerebral cortex | 26–29, 62, 63 |
| 32 | area 32 of cerebral cortex | 2–9, 62, 63 |
| 36 | area 30 of cerebral cortex | 20–27, 74 |
| 45 | area 45 of cerebral cortex | 11, 74, 75 |
| 46 | area 46 of cerebral cortex | 2–10, 70–75 |
| A1 | A1 noradrenaline cells | 45, 46 |
| A5 | A5 noradrenaline cells | 37, 38, 66, 68 |
| AA | anterior amygdaloid area, unspecified | 20–23, 68–71 |
| AAD | anterior amygdaloid area, dorsal | 17–19, 68, 70–73 |
| AAL | anterior amygdaloid area, . lateral | 18, 19, 72, 73 |
| AAV | anterior amygdaloid area, ventral | 17, 18, 70–73 |
| ac | anterior commissure | 19, 20, 62–71 |
| aca | anterior commissure, anterior limb | 15–18, 65–68 |
| Acb | accumbens nucleus, unspecified | 19 |
| AcbC | accumbens nucleus core | 13–18, 63–67 |
| AcbSh | accumbens nucleus shell | 13–17, 63–67 |
| acp | anterior commissure, posterior limb | 19–24, 72–75 |
| acs7 | accessory facial nucleus | 66 |
| AD | anterodorsal thalamic nucleus | 23–27, 63–65 |
| af | amygdaloid fissure | 18–21, 67, 68 |
| AHA | anterior hypothalamic area, anterior part | 18–21, 62–64 |

| Abbreviation | Long Name | Plates |
|---|---|---|
| AHi | amygdalohippocampal area, unspecified | 22, 23, 67 |
| AHiMC | amygdalohippocampal area, magnocellular part | 20, 21, 68 |
| AHiPC | amygdalohippocampal area, parvicellular part | 20, 21, 67–69 |
| AHP | anterior hypothalamic area, posterior part | 22 |
| AI | agranular insular cortex | 16–25, 75 |
| al | ansa lenticularis | 21–23, 66–69 |
| AI | agranular insular cortex | 18 |
| Al | alar nucleus | 62 |
| alv | alveus of hippocampus | 25–37, 69, 72–75 |
| AM | anteromedial thalamic nucleus | 23–26, 62, 63 |
| Amb | ambiguus nucleus | 44, 45, 65–68 |
| AmbC | ambiguus nucleus, compact part | 40–43 |
| ANS | accessory neurosecretory nucleus | 23, 24, 62, 63 |
| AO | anterior olfactory area, unspecified | 1 |
| AP | area postrema | 43, 44, 47, 62, 63 |
| APir | amygdalopiriform transition area | 18–22, 67–69 |
| APit | anterior lobe of pituitary gland | 21–25 |
| APT | anterior pretectal nucleus | 31–33, 65–67 |
| Aq | cerebral aqueduct | 33–37, 62 |
| arc | arcuate sulcus | 16, 66–68 |
| ArcD | arcuate hypothalamic nucleus, dorsal part | 22, 23, 62 |
| ArcL | arcuate hypothalamic nucleus, lateral part | 21–24, 62, 63 |
| ArcM | arcuate hypothalamic nucleus, medial part | 21–24, 62 |
| asc7 | ascending fibers of facial nerve | 39, 63–65 |
| AStr | amygdalostriatal transition zone | 22, 23, 73–75 |
| ATg | anterior tegmental nucleus | 35, 36, 62 |
| Au1 | auditory core cortex, primary | 27, 28, 30 |
| Au1L | auditory core cortex, primary, lateral | 29 |
| Au1M | auditory core cortex, primary, medial | 29 |

| Abbreviation | Long Name | Plates |
|---|---|---|
| AuAL | auditory belt cortex, anterolateral | 27 |
| AuCL | auditory belt cortex, caudolateral | 29 |
| AuCM | auditory belt cortex, caudomedial | 27–30 |
| AuCPB | auditory parabelt cortex, caudal | 28–30 |
| AuML | auditory belt cortex, middle lateral | 28–30 |
| AuR | auditory core cortex, rostral | 24–27 |
| AuRM | auditory belt cortex, rostromedial | 24–26 |
| AuRPB | auditory parabelt cortex, rostral | 21–27 |
| AuRT | auditory core cortex, rostrotemporal | 21–23 |
| AuRTL | auditory belt cortex, rostrotemporolateral | 18–27 |
| AuRTM | auditory belt cortex, rostrotemporomedial | 18–23 |
| AV | anteroventral thalamic nucleus | 23–27, 63–66 |
| AVPe | anteroventral periventricular hypothalamic nucleus | 17–19 |
| B | basal nucleus of Meynert | 18–26, 64–75 |
| Bar | Barrington's nucleus | 38 |
| bas | basilar artery | 62 |
| BI | basal interstitial nucleus | 43–46, 68–71 |
| bic | brachium of inferior colliculus | 32–38, 68–70 |
| BIC | nucleus of brachium of inferior colliculus | 33–38, 67–69 |
| BLD | basolateral amygdaloid nucleus, dorsal | 19–23, 72–74 |
| BLDL | basolateral amygdaloid nucleus, dorsolateral | 18–20, 72–74 |
| BLI | basolateral amygdaloid nucleus, intermediate | 21–23, 74 |
| BLV | basolateral amygdaloid nucleus, ventral | 72–75 |
| BLVL | basolateral amygdaloid nucleus, ventrolateral | 18–21, 72–74 |
| BLVM | basolateral amygdaloid nucleus, ventromedial | 18–23, 71, 72 |
| BMA | basomedial amygdaloid nucleus, anterior | 18–20, 68–71 |
| BMD | basomedial amygdaloid nucleus, dorsal | 20–23, 70, 71, 73 |
| BMV | basomedial amygdaloid nucleus, ventral | 21–23, 69–71, 73 |
| bsc | brachium of superior colliculus | 32–35, 63–74 |
| BSTIA | bed nucleus of stria terminalis, intraamygdalar. division | 24, 25, 74, 75 |
| BSTLA | bed nucleus of stria terminalis, lateral anterior division | 64 |
| BSTLD | bed nucleus of stria terminalis, lateral dorsal division | 19–21, 64, 65 |
| BSTLJ | bed nucleus of stria terminalis, lateral juxtacapsular division | 19–21, 65 |
| BSTLP | bed nucleus of stria terminalis, lateral posterior division | 19–21, 64, 65 |
| BSTLV | bed nucleus of stria terminalis, lateral ventral division | 19, 20, 64, 65 |
| BSTMA | bed nucleus of stria terminalis, medial anterior division | 18–20, 62, 63 |
| BSTMP | bed nucleus of stria terminalis, medial posterior division | 19, 20, 23, 62, 63, 66 |
| BSTMPI | bed nucleus of stria terminalis, medial posterior division, intermediate | 21, 22, 62–64 |
| BSTMPL | bed nucleus of stria terminalis, medial posterior division, lateral | 21, 22, 64, 65 |
| BSTMPM | bed nucleus of stria terminalis, medial posterior division, medial | 21, 63 |
| bv | blood vessel | 16, 27, 41 |
| C1 | C1 adrenaline cells | 38–44, 67, 68 |
| C2 | C2 adrenaline cells | 40 |
| C3 | C3 adrenaline cells | 40–42 |
| ca | calcarine sulcus | 38–75 |
| CA1 | cornu ammonis 1 | 22–39, 67–75 |
| CA2 | cornu ammonis 2 | 22–37, 68–75 |
| CA3 | cornu ammonis 3 | 22–37, 66–75 |
| cbc | cerebellar commissure | 42, 43, 62 |

| Abbreviation | Long Name | Plates |
|---|---|---|
| cc | corpus callosum | 12–38, 62–68 |
| CC | central canal of spinal cord | 46–53, 62 |
| ccr | corpus callosum rostrum | 12–14, 62–65 |
| Cd | caudate nucleus | 12–36, 64–75 |
| CeC | central amygdaloid nucleus, central part | 20–23, 73, 74 |
| CeCv | central cervical nucleus of spinal cord | 47–53, 62 |
| CeL | central amygdaloid nucleus, lateral part | 20–24, 72–75 |
| CeM | central amygdaloid nucleus, medial part | 20–23, 70–74 |
| cg | cingulum | 8–39, 65, 67, 68 |
| CG | cingulate gyrus of cortex | 27, 28, 32–38, 39, 62, 63, 68 |
| CGPn | central gray of pons | 38, 64, 65 |
| cgs | cingulate sulcus of cortex | 8–19, 24, 27, 28 |
| ci | circular sulcus | 16–24 |
| cic | commissure of inferior colliculus | 38, 39, 62, 63 |
| CIC | central nucleus of inferior colliculus | 37–39, 63–68 |
| Cl | claustrum | 14–26, 71–75 |
| CL | centrolateral thalamic nucleus | 31 |
| CLa | centrolateral thalamic nucleus, anterior division | 25–28, 63–66 |
| CLi | caudal linear nucleus | 29–32, 62 |
| CLp | centrolateral thalamic nucleus, posterior division | 27–30, 64–66 |
| cm | callosomarginal sulcus of cortex | 32–42 |
| CM | central medial thalamic nucleus | 24–28, 62 |
| CMn | centromedian thalamic nucleus | 28–31, 63–66 |
| CnF | cuneiform nucleus | 35–38, 64–67 |
| Cop | copula of cerebellar pyramis | 46–53, 68–74 |
| cp | cerebral peduncle | 25–32, 64–71 |
| Crus1 | crus 1 of ansiform lobule of cerebellum | 44–53, 70–75 |
| Crus2 | crus 2 of ansiform lobule of cerebellum | 46–52, 70–75 |

| Abbreviation | Long Name | Plates |
|---|---|---|
| csc | commissure of superior colliculus | 33–35, 62, 63 |
| ctg | central tegmental tract | 29–41, 62–65 |
| cu | cuneate fasciculus | 46–53, 64–67 |
| Cu | cuneate nucleus | 45, 49, 63, 64, 67 |
| CuR | cuneate nucleus, rotunda part | 46–48, 65, 66 |
| CuT | cuneate nucleus, triangular part | 46–48, 65, 66 |
| CxA | cortex–amygdala transition zone | 18 |
| DA | dorsoanterior extrastriate visual cortex | 36–40, 67–74 |
| das | dorsal acoustic stria | 68, 69 |
| DC | dorsal cochlear nucleus | 38–40, 71, 72 |
| DCIC | dorsal cortex of inferior colliculus | 37–39, 62–66 |
| dcw | deep cortical white matter | 11–14, 64–66 |
| DEn | dorsal endopiriform nucleus | 14–20, 68, 70–75 |
| dfu | dorsal funiculus of spinal cord | 62, 63 |
| DG | dentate gyrus | 22, 23, 66–75 |
| DH | dorsal horn of spinal cord | 63, 64 |
| dhc | dorsal hippocampal commissure | 23, 24 |
| DI | dorsointermediate extrastriate visual cortex | 41, 74, 75 |
| Dk | nucleus of Darkschewitsch | 29–32, 62 |
| dlL | dorsolateral tract of Lissauer | 50–53 |
| DLL | dorsal nucleus of lateral lemniscus | 34–38, 68, 69 |
| DLPAG | dorsolateral periaqueductal gray | 33–38, 62, 63 |
| DM | dorsomedial extrastriate visual cortex (V3d) | 36–49, 62–75 |
| DMC | dorsomedial hypothalamic nucleus, compact | 21–23, 62, 63 |
| DMD | dorsomedial hypothalamic nucleus, dorsal | 21–24, 62, 63 |
| DMPAG | dorsomedial periaqueductal gray | 34–38, 62 |
| DMSp5 | dorsomedial spinal trigeminal nucleus | 39–43, 67, 68 |
| DMTg | dorsomedial tegmental area | 37, 38, 62–64 |

| Abbreviation | Long Name | Plates |
|---|---|---|
| DMV | dorsomedial hypothalamic nucleus, ventral | 24 |
| DpG | deep gray of superior colliculus | 34–38, 62–66 |
| DPGi | dorsal paragigantocellular nucleus | 38–44, 62, 63 |
| DpWh | deep white of superior colliculus | 34–38, 62–66 |
| DR | dorsal raphe nucleus | 28–33 |
| DRC | dorsal raphe nucleus, caudal part | 37 |
| DRD | dorsal raphe nucleus, dorsal part | 34–36 |
| DRI | dorsal raphe nucleus, interfascicular part | 33–36, 62 |
| DRV | dorsal raphe nucleus, ventral part | 35, 36, 62 |
| DRVL | dorsal raphe nucleus, ventrolateral part | 35, 36, 62, 63 |
| dsc | dorsal spinocerebellar tract | 43–53, 66, 67 |
| DTgC | dorsal tegmental nucleus, central part | 37, 63 |
| dtgd | dorsal tegmental decussation | 32 |
| DTgP | dorsal tegmental nucleus, pericentral part | 37, 62 |
| DTnl | dorsal terminal nucleus | 34–36, 68, 69 |
| E | ependymal lining of ventricle | 17, 22, 23, 63, 65, 67, 69 |
| EA | extended amygdala | 69 |
| EAC | extended amygdala, central part | 20, 21, 64–68 |
| EAM | extended amygdala, medial part | 20–22, 64–68 |
| ec | external capsule | 12–27, 64–75 |
| ECIC | external cortex of inferior colliculus | 36–39, 63–69 |
| ECu | external cuneate nucleus | 44–48, 66–69 |
| EGP | external segment of globus pallidus | 19–26, 67–75 |
| eml | external medullary lamina | 24–34, 65–75 |
| EntC | entorhinal cortex, caudal | 25–31, 69–75 |
| EntL | entorhinal cortex, medial | 20–24, 71–74 |
| EntM | entorhinal cortex, lateral | 20–24, 68–72 |
| EntR | entorhinal cortex, rostral | 19, 70–74 |
| EPl | external plexiform layer, olfactory bulb | 1 |
| EW | Edinger Westphal nucleus | 30–35 |
| ex | extreme capsule | 14–26, 72–75 |
| f | fornix | 20–35, 62–75 |
| fd | fimbriodentate sulcus | 26–33, 36, 37, 67–72 |
| FEF | frontal eye field of cortex | 5–14, 71–75 |
| fi | fimbria of hippocampus | 26–37, 71–75 |
| Fl | flocculus of cerebellum | 41–45, 68–74 |
| fmi | forceps minor of corpus callosum | 4–12, 66–68 |
| fr | fasiculus retroflexus | 27–31, 62–64 |
| FST | fundus of superior temporal cortex | 29, 30 |
| FStr | fundus of striatum | 14–19, 68–73 |
| g7 | genu of facial nerve | 38, 39, 63, 64 |
| gc | geniculocalcarine tract | 36–59, 68, 72–75 |
| GCC | granule cell cluster | 14–17, 63–70 |
| GCCM | granule cell cluster, magna | 15, 16 |
| Ge5 | gelatinosus layer of caudal spinal trigeminal nucleus | 46–49, 65–67 |
| Gi | gigantocellular reticular nucleus | 38–44, 62–65 |
| GI | granular insular cortex | 16–24 |
| GiV | gigantocellular reticular nucleus, ventral part | 38–41, 62–65 |
| Gl | glomerular layer, olfactory bulb | 1 |
| gr | gracile fasciculus | 48–53, 62, 64, 65 |
| Gr | gracile nucleus | 46–50, 62–65 |
| GrDG | granule cell layer of dentate gyrus | 24–37, 72 |
| GrO | granule cell layer, olfactory bulb | 1 |
| Gu | gustatory area of cortex | 12–16 |
| H | H field of Forel, prerubral field, lenticular fasciculus | 25 |
| H1 (al) | H field of Forel, prerubral field, ansa lenticularis | 64 |
| H1 (thf) | H1 field of Forel, thalamic fasciculus | 25, 26, 66–69 |
| H2 (lentf) | H2 field of Forel, lenticular fasciculus | 26, 67–69 |
| hbc | habenular commissure | 33, 62 |
| HDB | horizontal limb of diagonal band nucleus | 15–20, 62–68 |
| hi | hippocampal fissure | 24–33, 68–73 |

| Abbreviation | Long Name | Plates |
|---|---|---|
| I | intercalated nuclei of amygdala | 18–20, 69, 70 |
| I5 | intertrigeminal nucleus | 37, 38 |
| I8Ve | interstitial nucleus of vestibular division, 8th nerve | 39, 70–72 |
| ia | internal arcuate fibers | 46, 47, 63 |
| IAM | interanteromedial thalamic nucleus | 24, 25, 62 |
| IB | internal basal nucleus | 48, 49, 51–53 |
| ic | internal capsule | 13–27, 64–75 |
| IC | inferior colliculus (unspecified region) | 40 |
| icp | inferior cerebellar peduncle | 40–46, 69–71 |
| IF | interfascicular nucleus | 28 |
| IG | induseum griseum | 12–24, 31–38, 62, 64–68 |
| IGP | internal segment of globus pallidus | 22–25, 66–72 |
| ILL | intermediate nucleus of lateral lemniscus | 35, 36, 67–69 |
| IMD | intermediodorsal thalamic nucleus | 26–30, 62 |
| IMG | intramedullary gray of amygdala | 18, 20–22, 71 |
| iml | internal medullary lamina | 25–27, 29–31, 63, 64, 66 |
| imv | inferior medullary velum | 42 |
| In | intercalated nucleus of medulla | 43–45 |
| InC | interstitial nucleus of Cajal | 30–32, 62, 63 |
| InCo | intercollicular nucleus | 35, 36, 38, 64–66 |
| InCSh | interstitial nucleus of Cajal, shell part | 30–32, 62, 63 |
| InfS | infundibular stalk | 21, 22, 62 |
| InG | intermediate gray of superior colliculus | 35–38, 62–67 |
| Insrl | insular cortex, rostrolateral | 12–15, 70–75 |
| Insrm | insular cortex, rostromedial | 13–15, 70–75 |
| int | intermediate temporal sulcus of cortex | 27, 28, 30, 31 |
| IntA | intermediate (anterior) deep cerebellar nucleus | 42–45, 66–68 |
| IntP | intermediate (posterior) deep cerebellar nucleus | 43–47, 64–67 |
| InWh | intermediate white of superior colliculus | 35–38, 62–66 |

| Abbreviation | Long Name | Plates |
|---|---|---|
| IOA | inferior olivary nucleus, subnucleus A of medial nucleus | 41–44, 64, 65 |
| IOβ | inferior olivary nucleus, subnucleus beta | 41–44, 62 |
| IOB | inferior olivary nucleus, subnucleus B of medial nucleus | 41–44, 63 |
| IOC | inferior olivary nucleus, subnucleus C of medial nucleus | 41–44 |
| IOD | inferior olivary nucleus, dorsal subnucleus | 38–44, 63–65 |
| IODM | inferior olivary nucleus, dorsomedial subnucleus | 38–40, 62 |
| IOK | inferior olivary nucleus, cap of Kooy subnucleus | 43 |
| IOPr | inferior olivary nucleus, principal subnucleus | 38–41, 63–65 |
| ip | intraparietal sulcus of cortex | 28–41, 70–74 |
| IPAC | interstitial nucleus of the posterior limb of ac | 20–23, 75 |
| IPC | interpeduncular nucleus, caudal subnucleus | 29–32 |
| IPD | interpeduncular nucleus, dorsal subnucleus | 29, 30 |
| IPF | interpeduncular fossa | 27, 28, 62, 64 |
| IPI | interpeduncular nucleus, interfascicular subnucleus | 29, 30, 62 |
| IPit | intermediate lobe of pituitary gland | 23, 24 |
| IPL | interpeduncular nucleus, lateral subnucleus | 29, 30, 62 |
| IPo | interpositus nucleus | 40–42, 63 |
| IPR | interpeduncular nucleus, rostral subnucleus | 29, 30 |
| IPro | insular proisocortex | 18, 19, 75 |
| IRt | intermediate reticular formation | 39–48, 50, 63–67 |
| IS | inferior salivatory nucleus | 40, 65 |
| isRt | isthmic reticular formation | 36, 67, 68 |
| it | inferior temporal sulcus | 25–42 |
| ITA | inferior temporal cortex, anterior area | 17–21, 75 |
| ITA1 | inferior temporal cortex, anterior area 1 | 22–26 |
| ITA2 | inferior temporal cortex, anterior area 2 | 22–26 |

| Abbreviation | Long Name | Plates |
|---|---|---|
| ITD | inferior temporal cortex, dorsal area | 36 |
| ITD1 | inferior temporal cortex, dorsal area 1 | 27–35 |
| ITD2 | inferior temporal cortex, dorsal area 2 | 32–35 |
| ITV | inferior temporal cortex, ventral area | 27–31 |
| IVF | interventricular foramen | 21, 22, 26, 62, 64 |
| KF | Kölliker-Fuse nucleus | 37, 38, 68 |
| LA | lateroanterior hypothalamic nucleus | 20, 62 |
| LAcbSh | accumbens nucleus shell, lateral part | 15–18, 67–71 |
| LaD | lateral amygdaloid nucleus, dorsal part | 21–23, 75 |
| LaL | lateral amygdaloid nucleus, lateral part | 20–24, 75 |
| LaM | lateral amygdaloid nucleus, medial part | 20–24, 75 |
| LaPO | lateral periolivary nucleus | 36–38, 66, 67 |
| Lat | lateral deep cerebellar nucleus | 42–46, 67–71 |
| LC | locus coeruleus | 38, 39, 65 |
| lcs | lateral corticospinal tract | 49–53, 63–65 |
| Ld | lambdoid nucleus of septum | 16 |
| LD | laterodorsal thalamic nucleus | 26–30, 64–67 |
| LDTgD | laterodorsal tegmental nucleus, dorsal part | 36, 37, 62, 63 |
| LDTgV | laterodorsal tegmental nucleus, ventral part | 37, 64 |
| lentf | H2 field of Forel, lenticular fasciculus | 23, 25, 65, 66 |
| lf | lateral fissure | 16–36, 44, 73–75 |
| lfp | longitudinal fibers of pons | 31–33, 62–66 |
| lfu | lateral funiculus of spinal cord | 51 |
| LGei | lateral geniculate thalamic nucleus, external interlaminar | 28–32, 74, 75 |
| LGii | lateral geniculate thalamic nucleus, internal interlaminar | 29, 75 |
| LGme | lateral geniculate thalamic nucleus, magnocellular exterior | 28–31, 72–75 |
| LGmi | lateral geniculate thalamic nucleus, magnocellular interior | 28–32, 72–75 |

| Abbreviation | Long Name | Plates |
|---|---|---|
| LGpe | lateral geniculate thalamic nucleus, parvicellular exterior | 27–32, 72–75 |
| LGpi | lateral geniculate thalamic nucleus, parvicellular interior | 27–32, 72–75 |
| LGs | lateral geniculate thalamic nucleus, superficial | 29, 30, 72–74 |
| LH | lateral hypothalamic area | 21–25, 62–66 |
| LHb | lateral habenular nucleus | 30–32, 62–64 |
| LHbD | lateral habenular nucleus, dorsal part | 29–31 |
| LHbV | lateral habenular nucleus, ventral part | 30, 31 |
| Lim | limitans nucleus | 32–34, 65–67 |
| ll | lateral lemniscus | 34–37, 67–69 |
| LM | lateral mamillary nucleus | 25, 26, 63 |
| lml | lateral medullary lamina | 18–26, 68–74 |
| LMol | stratum lacunosum moleculare of cornu ammonis | 24, 26–33, 36–38, 72–75 |
| lo | lateral olfactory tract | 12, 13, 17, 18, 63–66 |
| LOT | nucleus of lateral olfactory tract | 18, 69 |
| LPAG | lateral periaqueductal gray | 32–38, 62–64 |
| LPBC | lateral parabrachial nucleus, central part | 37, 38, 66, 67 |
| LPBD | lateral parabrachial nucleus, dorsal part | 37, 38, 64–67 |
| LPBE | lateral parabrachial nucleus, external part | 37, 38, 68 |
| LPBV | lateral parabrachial nucleus, ventral part | 38, 66 |
| LPCu | lateral pericuneate nucleus | 45, 68, 69 |
| LPGi | lateral paragiganticellular nucleus | 38–41, 64–66 |
| LPO | lateral preoptic area | 17–20, 62–65 |
| LR4V | lateral recess of fourth ventricle | 44–47, 68 |
| LRt | lateral reticular nucleus | 42–46, 64–68 |
| LRtPC | lateral reticular nucleus, parvicellular part | 43, 44 |
| LRtS5 | lateral reticular nucleus, subtrigeminal part | 42–45, 67, 68 |
| LSD | lateral septal nucleus, dorsal | 15–21, 62, 63 |
| LSI | lateral septal nucleus, intermediate | 15–20, 62, 63 |

| Abbreviation | Long Name | Plates |
|---|---|---|
| LSO | lateral superior olivary nucleus | 36–38, 67 |
| LSp | lateral spinal nucleus | 47–53 |
| LSV | lateral septal nucleus, ventral part | 15–19, 62, 63 |
| LTer | lamina terminalis of third ventricle | 16, 62 |
| LTnl | lateral terminal nucleus | 30–32, 68–70 |
| LTu | lateral tuberal hypothalamic nucleus | 23, 65, 66 |
| lu | lunate sulcus | 39–41 |
| LV | lateral ventricle | 10–41, 62–75 |
| LVe | lateral vestibular nucleus | 39–42, 66–70 |
| LVPO | lateroventral periolivary nucleus | 36–38, 67 |
| lvs | lateral vestibulospinal tract | 48–53 |
| m5 | motor root of trigeminal nerve | 35–38, 68 |
| MA3 | medial accessory oculomotor nucleus | 31 |
| MCLH | magnocellular nucleus of lateral hypothalamus | 23, 24, 65, 66 |
| mcp | middle cerebellar peduncle | 29–43, 65–72 |
| MCPC | magnocellular nucleus of posterior commissure | 33, 63, 64 |
| MCPO | magnocellular preoptic nucleus | 18, 19, 67–71 |
| MD | mediodorsal thalamic nucleus | 32 |
| MDC | mediodorsal thalamic nucleus, central part | 27–31, 62 |
| MdD | medullary reticular nucleus, dorsal part | 45–49, 64–66 |
| MDD | mediodorsal thalamic nucleus, dorsal part | 26–30, 62–65 |
| MDL | mediodorsal thalamic nucleus, lateral part | 26–31, 63–65 |
| MDM | mediodorsal thalamic nucleus, medial part | 26–30, 62–64 |
| MdV | medullary reticular nucleus, ventral part | 45–49, 62–64 |
| Me | medial amygdaloid nucleus | 18–23, 68–72 |
| ME | median eminence of hypothalamus | 24 |
| me5 | mesencephalic trigeminal tract | 36, 64 |
| Me5 | mesencephalic trigeminal nucleus | 34–38, 63, 64 |
| Med | medial deep cerebellar nucleus | 41–47, 62–65 |
| MEE | median eminence, external layer | 20–23 |
| MEI | median eminence, internal layer | 20, 21, 23 |
| MePO | medial periolivary nucleus | 36–38, 65 |
| mfb | medial forebrain bundle | 23–27, 62, 64 |
| MGAD | medial geniculate nucleus, anterodorsal subnucleus | 30–32, 70, 71 |
| MGM | medial geniculate nucleus, medial subnucleus | 30–33, 69, 70 |
| MGPD | medial geniculate nucleus, posterodorsal subnucleus | 31–33, 70, 71 |
| MGV | medial geniculate nucleus, ventral subnucleus | 30–33, 70–73 |
| MHb | medial habenular nucleus | 30–33, 62, 63 |
| MiTg | microcellular tegmental nucleus | 34–37, 67–69 |
| ml | medial lemniscus | 29–45, 62–68 |
| ML | medial mamillary nucleus, lateral part | 25, 26, 62, 63 |
| mlf | medial longitudinal fasciculus | 31–53, 62, 63 |
| MM | medial mamillary nucleus, medial part | 25, 26, 62 |
| mml | medial medullary lamina | 22–25, 67–71, 73 |
| MnA | median accessory nucleus of medulla | 48, 49, 62 |
| MnPO | median preoptic nucleus | 17–19 |
| MnR | median raphe nucleus | 31–38 |
| Mo5 | motor trigeminal nucleus | 38, 68, 69 |
| MoDG | molecular layer of dentate gyrus | 26–38, 72, 73 |
| mp | mamillary peduncle | 25–27, 64 |
| MPA | medial preoptic area | 17, 18, 20, 62, 63 |
| MPB | medial parabrachial nucleus | 37–39, 65, 66 |
| MPBE | medial parabrachial nucleus, external part | 37, 38, 67, 68 |
| MPOC | medial preoptic nucleus, compact part | 19, 20, 63 |
| MPOL | medial preoptic nucleus, lateral part | 18–20, 62, 63 |

| Abbreviation | Long Name | Plates |
|---|---|---|
| MPOM | medial preoptic nucleus, medial part | 18–20, 62, 63 |
| MPT | medial pretectal nucleus | 33–36, 62–64 |
| MRe | mamillary recess of third ventricle | 25, 62 |
| mRt | mesencephalic reticular formation | 28, 30–34, 62–70 |
| MS | medial septal nucleus | 15–19 |
| MSO | medial superior olivary nucleus | 36–38, 65, 66 |
| MST | medial superior temporal area of cortex | 30 |
| mt | mamillothalamic tract | 24–26, 62–64 |
| MT | middle temporal area of cortex | 30–34 |
| MTc | middle temporal crescent area of cortex | 30–36 |
| mte | medial temporal sulcus of cortex | 32–35, 38–42, 70, 71, 73 |
| mtg | mamillotegmental tract | 63 |
| MTnl | medial terminal nucleus | 25–27, 64, 65 |
| MTu | medial tuberal hypothalamic nucleus | 24 |
| MVe | medial vestibular nucleus | 38, 44, 65, 66 |
| MVeMC | medial vestibular nucleus, magnocellular part | 39–43, 65, 66 |
| MVePC | medial vestibular nucleus, parvicellular part | 39–43, 63–65 |
| MVPO | medioventral periolivary nucleus | 36–38, 66 |
| Mx | matrix sensory area | 39–44, 67, 68 |
| NC | nanocellular hypothalamic nucleus | 23 |
| nehy | neurohypophyseal tract | 21, 62 |
| ns | nigrostriatal tract | 23, 25 |
| OB | olfactory bulb | 63 |
| oc | olivocerebellar tract | 40–43, 66–69 |
| och | optic chiasm | 16–19, 62, 63 |
| ON | olfactory nerve layer, olfactory bulb | 1, 63 |
| Op | stratum opticum of superior colliculus | 35–38, 62–68 |
| OPAl | orbital periallocortex | 13, 62, 63 |
| opt | optic tract | 16–28, 30, 64–74 |
| OPT | olivary pretectal nucleus | 32, 33, 64 |
| Or | stratum oriens of cornu ammonis | 24–33, 36–38, 74 |
| orb | orbital sulcus of cortex | 1–7 |
| orbl | orbital sulcus (lateral) of cortex | 8–15, 68–75 |
| orbm | orbital sulcus (medial) of cortex | 8–14, 68–72 |
| OT | nucleus of optic tract | 34, 35, 65–68 |
| OV | olfactory ventricle | 1 |
| p1Rt | prosomere 1 reticular formation | 28–33, 63–68 |
| P5 | peritrigeminal nucleus | 37, 38, 66–68 |
| P7 | perifacial nucleus | 38, 39, 65–67 |
| Pa6 | parabducens nucleus | 38, 39, 63–65 |
| PaAP | paraventricular hypothalamic nucleus, anterior parvicellular part | 21, 62 |
| PaM | paraventricular hypothalamic nucleus, magnocellular | 20, 22 |
| PaMD | paraventricular hypothalamic nucleus, magnocellular, dorsal part | 20–22, 62, 63 |
| PaP | parapeduncular nucleus | 28, 63 |
| PaPD | paraventricular hypothalamic nucleus, parvicellular, dorsal part | 21, 22 |
| PaPo | paraventricular hypothalamic nucleus, posterior part | 23, 62, 63 |
| PaR | pararubral area | 32, 33 |
| PaS | parasubiculum | 26–37, 70–73 |
| PaV | paraventricular hypothalamic nucleus, ventral part | 21, 22, 62 |
| PaVe | paravestibular nucleus | 44, 45, 66, 67 |
| PaXi | paraxiphoid thalamic nucleus | 23, 62 |
| PBG | parabigeminal nucleus | 33–36, 38, 68, 69 |
| PBP | parabrachial pigmented nucleus | 27–33, 64–68 |
| pc | posterior commissure | 32, 62 |
| PC | paracentral thalamic nucleus | 25–29, 62, 63 |
| PC5 | parvicellular trigeminal motor nucleus | 37 |
| pcf | preculminate fissure of cerebellum | 41–43, 62–69 |
| PCom | paracommissural nucleus | 31–33, 62–64 |
| PCRt | parvicellular reticular nucleus | 39–44, 64–67 |

| Abbreviation | Long Name | Plates |
|---|---|---|
| PCRtα | parvicellular reticular nucleus, alpha part | 38, 65, 66 |
| PCTg | paracollicular tegmentum | 38, 65–67 |
| PCuMx | pericuneate matrix | 45–48, 64–67 |
| Pe | periventricular hypothalamic nucleus | 19–24 |
| PE | parietal field E of cortex | 28–32, 62–73 |
| PEc | parietal field E of cortex, caudal subdivision | 33–35, 62–72 |
| PeF | perifornical hypothalamic nucleus | 24, 64 |
| PeFL | perifornical hypothalamic nucleus, lateral part | 23 |
| PeFM | perifornical hypothalamic nucleus, medial part | 23 |
| Pf | parafascicular thalamic nucleus | 27–31, 62–64 |
| PF | parietal field F of cortex | 28–30, 74, 75 |
| PFl | paraflocculus | 36–41, 71–75 |
| PFlP | paraflocculus posterior | 42–45 |
| PFlV | paraflocculus ventral | 37, 38 |
| PG | parietal field G of cortex | 31–35, 73–75 |
| PH | posterior hypothalamic area | 24–28, 62, 63 |
| Pi | pineal gland | 34, 35 |
| Pir | piriform cortex | 63–66, 68, 70, 75 |
| Pir1 | piriform cortex, layer 1 | 12–18, 67, 69–74 |
| Pir2 | piriform cortex, layer 2 | 12–18, 67, 69–74 |
| Pir3 | piriform cortex, layer 3 | 12–18, 67, 69–74 |
| PLam | paralaminar amygdaloid nucleus | 18–23, 69–74 |
| PLamP | paralaminar amygdaloid nucleus, posterior part | 24 |
| PLd | paralambdoid septal zone | 16–19, 62 |
| plf | posterolateral fissure of cerebellum | 41, 43, 46, 47, 62–70, 72, 73 |
| PLH | peduncular part of lateral hypothalamus | 26, 27, 62–66 |
| PM | paramedian lobule of cerebellum | 48–54, 70 |
| PMa | paramedian lobule of cerebellum, anterior | 70–75 |
| PMaD | premamillary hypothalamic nucleus, dorsal part | 24, 63 |

| Abbreviation | Long Name | Plates |
|---|---|---|
| PMaV | premamillary hypothalamic nucleus, ventral part | 24 |
| PMD | premotor cortex, dorsal part | 16, 67–73 |
| PMnR | paramedian raphe nucleus | 31, 32, 36–38, 62 |
| PMp | paramedian lobule of cerebellum, posterior | 68–71, 74, 75 |
| PMV | premotor cortex, ventral part | 16 |
| Pn | pontine nucleus | 28, 30–35, 62, 63, 65–69 |
| PN | paranigral nucleus of VTA | 29, 30, 62 |
| PnC | pontine caudal nucleus | 35–38, 63–66 |
| PnO | pontine oral nucleus | 33–36, 63–68 |
| PnR | pontine raphe nucleus | 38, 39 |
| Po | posterior thalamic nucleus | 31–33, 66–70 |
| PoDG | polymorphic layer of dentate gyrus | 26–30, 36 |
| PP | peripeduncular nucleus | 31, 32, 68, 69 |
| PPD1 | posterior parietal cortex, dorsal area 1 | 28–35, 65–73 |
| PPD2 | posterior parietal cortex, dorsal area 2 | 30–35, 62–66 |
| ppf | prepyramidal fissure of cerebellum | 49–57, 62–69 |
| PPit | posterior lobe of pituitary gland | 23–26 |
| PPM | posterior parietal cortex, medial area | 32–37, 62–67 |
| PPT | posterior pretectum | 33, 34, 63–66 |
| PPTg | pedunculopontine tegmental nucleus | 33 |
| PPTgC | pedunculopontine tegmental nucleus, compact part | 34–37, 63–67 |
| PPTgD | pedunculopontine tegmental nucleus, diffuse part | 34–36, 65–68 |
| PPV | posterior parietal cortex, ventral area | 28–37, 72–75 |
| pr | prefrontal sulcus of cortex | 8–11, 16, 72–74 |
| Pr | prepositus hypoglossi nucleus | 40–42, 62 |
| Pr5 | principal trigeminal sensory nucleus | 37, 69, 70 |
| PrC | precommissural nucleus | 32, 62, 63 |
| prcf | precentral fissure of cerebellum | 39, 62–64 |
| PrCO | precentral opercular cortex | 12–16 |

| Abbreviation | Long Name | Plates |
|---|---|---|
| prf | primary fissure of cerebellum | 43–48, 50, 51, 62–73, 75 |
| PRF | prerubral field | 26, 27, 63–65 |
| PrG | pregeniculate nucleus | 27, 28, 72–74 |
| PRh | perirhinal cortex | 17–30, 74, 75 |
| ProS | prosubiculum | 23–38, 69–75 |
| ProSt | prostriate cortex | 38–40, 72–75 |
| PrS | presubiculum | 23–38, 68–75 |
| PRSom | parietal rostral somatosensory area of cortex | 17 |
| PS | parastriatal nucleus | 19, 20 |
| psf | posterior superior fissure of cerebellum | 64–69 |
| PSol | parasolitary nucleus | 44, 45 |
| PSTh | parasubthalamic nucleus | 26, 64, 65 |
| pt | posterior temporal sulcus of cortex | 35–42 |
| PT | paratenial thalamic nucleus | 23–25, 62 |
| Pu | putamen | 13–28, 68–75 |
| PulA | pulvinar nucleus, anterior part | 29–31, 67–73 |
| PulI | pulvinar nucleus, inferior part | 31–34, 69–74 |
| PulL | pulvinar nucleus, lateral part | 32–35, 69–75 |
| PulM | pulvinar nucleus, medial part | 32–35, 65–69 |
| PV | parietoventral somatosensory cortex | 17–24 |
| PVA | paraventricular thalamic nucleus, anterior part | 22–29, 62 |
| PVG | periventricular gray | 29–31, 62, 63 |
| PVP | paraventricular thalamic nucleus, posterior part | 29, 30, 62 |
| py | pyramidal tract | 34–47, 62–65 |
| Py | stratum pyramidale of cornu ammonis | 36 |
| pyd | pyramidal tract decussation | 45–48, 62, 63 |
| R1 | Rexed's lamina 1 of spinal cord | 50–52 |
| R2 | Rexed's lamina 2 of spinal cord | 50–53 |
| R3 | Rexed's lamina 3 of spinal cord | 50–53 |
| R4 | Rexed's lamina 4 of spinal cord | 50–53, 63 |

| Abbreviation | Long Name | Plates |
|---|---|---|
| R5 | Rexed's lamina 5 of spinal cord | 51–53, 62, 63 |
| R7 | Rexed's lamina 7 of spinal cord | 51–53, 63 |
| R8 | Rexed's lamina 8 of spinal cord | 49–53, 63 |
| R9 | Rexed's lamina 9 of spinal cord | 49–53, 63 |
| Rad | stratum radiatum of cornu ammonis | 24–33, 36–38, 72–75 |
| RAmb | retroambiguus nucleus | 46, 66 |
| Rbd | rhabdoid nucleus | 33, 35, 36, 62 |
| RChL | retrochiasmatic nucleus, lateral part | 20, 62 |
| RChM | retrochiasmatic nucleus, medial part | 20 |
| Re | reuniens thalamic nucleus | 23–27, 62 |
| ReIC | recess of inferior colliculus | 38 |
| rf | rhinal fissure | 1–8, 14–30, 72–75 |
| RI | rostral interstitial nucleus | 28, 62, 63 |
| RIns | retroinsular area of cortex | 26–29, 75 |
| RLi | rostral linear nucleus (midbrain) | 28, 29 |
| RMC | red nucleus, magnocellular part | 29–32, 63, 64, 66 |
| rmd | retromamillary decussation | 25 |
| RMg | raphe magnus nucleus | 38–40, 62 |
| RML | retromamillary nucleus, lateral part | 25, 26, 62, 63 |
| RMM | retromamillary nucleus, medial part | 25–27 |
| Ro | nucleus of Roller | 43 |
| ROb | raphe obscurus nucleus | 39–45 |
| RPa | raphe pallidus nucleus | 38–43 |
| RPC | red nucleus, parvicellular part | 28–31, 63–66 |
| RPO | rostral periolivary nucleus | 36 |
| RRF | retrorubral field | 30–34, 63–67 |
| rs | rubrospinal tract | 36–53, 64, 68 |
| Rt | reticular (thalamic) nucleus | 22–34, 64–75 |
| RtTg | reticulotegmental nucleus | 32–36, 62–65 |
| RVL | rostral ventrolateral area of medulla | 38, 39, 68 |
| S | subiculum | 23–38, 68–75 |

| Abbreviation | Long Name | Plates |
|---|---|---|
| S2 | secondary somatosensory of cortex | 23–27, 75 |
| s5 | sensory root of trigeminal nerve | 33–38, 70, 71 |
| Sag | sagulum nucleus | 36, 37, 69 |
| SC | superior colliculus | 39 |
| SChD | suprachiasmatic hypothalamic nucleus, dorsal | 18, 19, 62, 63 |
| SChV | suprachiasmatic hypothalamic nucleus, ventral | 18, 19, 62, 63 |
| SCO | subcommissural organ | 32 |
| scp | superior cerebellar peduncle | 33–43, 64–69 |
| scpd | superior cerebellar peduncle decussation | 32–35, 62, 63 |
| sf | secondary fissure of cerebellum | 50–53, 62–69 |
| SFi | septofimbrial nucleus | 18–21, 62 |
| SFO | subfornical organ | 20, 21, 62 |
| SG | suprageniculate thalamic nucleus | 30–34, 67–70 |
| SHi | septohippocampal nucleus | 15–21 |
| SHy | septohypothalamic nucleus | 18–20 |
| SI | substantia innominata | 14–21, 62–72, 74 |
| Sim | simplex lobule of cerebellum | 41–54 |
| Sima | simplex lobule of cerebellum, anterior | 43, 70–75 |
| Simp | simplex lobule of cerebellum, posterior | 43, 70–74 |
| sm | stria medullaris thalami | 21–30, 62–64 |
| SM | nucleus of stria medullaris thalami | 22, 64 |
| SMA | supplementary motor area of cortex | 22–24, 62–64 |
| smv | superior medullary velum | 38, 40, 62 |
| sn | striatonigral fibers | 24–26, 66, 67, 70, 71 |
| SNC | substantia nigra, compact part | 27, 28, 64–67, 69 |
| SNCD | substantia nigra, compact part, dorsal tier | 29–31, 68 |
| SNCV | substantia nigra, compact part, ventral tier | 29–32, 68 |
| SNR | substantia nigra, reticular part | 26–32, 64–70 |

| Abbreviation | Long Name | Plates |
|---|---|---|
| SO | supraoptic hypothalamic nucleus | 17–21, 64–67 |
| sod | supraoptic decussation | 19–28, 62–72 |
| sol | solitary tract | 43–47, 63, 65, 66 |
| Sol | nucleus of solitary tract | 39–41, 63–65 |
| SolC | nucleus of solitary tract, commissural subnucleus | 46–48, 62 |
| SolCe | nucleus of solitary tract, central subnucleus | 42–45, 65 |
| SolDL | nucleus of solitary tract, dorsolateral subnucleus | 44–47, 65 |
| SolDM | nucleus of solitary tract, dorsomedial subnucleus | 42, 43 |
| SolG | nucleus of solitary tract, gelatinosus subnucleus | 44–46, 63, 64 |
| SolI | nucleus of solitary tract, intrafascicular subnucleus | 41, 42 |
| SolIM | nucleus of solitary tract, intermediate subnucleus | 42, 43 |
| SolL | nucleus of solitary tract, lateral subnucleus | 42, 43, 66 |
| SolM | nucleus of solitary tract, intermediate subnucleus | 44–49 |
| SolM | nucleus of solitary tract, medial subnucleus | 62 |
| SolV | nucleus of solitary tract, ventral subnucleus | 42–47, 65 |
| SolVL | nucleus of solitary tract, ventrolateral subnucleus | 42–48, 65, 66 |
| SOR | supraoptic hypothalamic nucleus, retrochiasmatic part | 20–22, 62–64 |
| sp5 | spinal trigeminal tract | 38–49, 65–70 |
| Sp5C | spinal trigeminal nucleus, caudalis part | 45–49, 65–68 |
| Sp5I | spinal trigeminal nucleus, interpolaris part | 38–45, 67–69 |
| Sp5O | spinal trigeminal nucleus, oralis part | 38, 68, 69 |
| SPf | subparafascicular thalamic nucleus | 26, 27, 63 |
| SPTg | subpeduncular tegmental nucleus | 35, 36, 63–65 |
| spth | spinothalamic tract | 44, 46–53, 66 |
| SpVe | spinal vestibular nucleus | 41–45, 65–69 |
| SS | superior salivatory nucleus | 66 |
| st | stria terminalis | 19–35, 64–75 |

| Abbreviation | Long Name | Plates |
|---|---|---|
| STh | subthalamic nucleus | 26–28, 66–70 |
| StHy | striatohypothalamic nucleus | 21–23, 63 |
| STL | superior temporal cortex, lateral area | 16–21 |
| STM | superior temporal cortex, medial area | 16–22 |
| sts | superior temporal sulcus | 16–34 |
| Su3 | supraoculomotor periaqueductal gray | 33–35, 62, 63 |
| Su3C | supraoculomotor cap | 33–35, 62, 63 |
| Su5 | supratrigeminal nucleus | 37, 38, 67, 68 |
| SubCD | subcoeruleus nucleus, dorsal part | 37, 38, 65–67 |
| SubCV | subcoeruleus nucleus, ventral part | 37, 38, 65–68 |
| SubI | subincertal nucleus | 23, 24, 64, 65 |
| SuG | superficial gray of superior colliculus | 35–38, 62–68 |
| SuVe | superior vestibular nucleus | 38–42, 65–70 |
| TCF | transverse cerebral fissure | 22–37, 63 |
| Te | terete nucleus of hypothalamus | 23, 24 |
| TF | temporal field F of cortex | 27–36, 75 |
| tfp | transverse fibers of pons | 29, 32–35, 62–64 |
| TH | temporal field H of cortex | 32–35, 72–75 |
| TM | tuberomamillary hypothalamic area | 23, 64, 65 |
| TPro | temporal proisocortex | 18–26 |
| TPt | temporal parietotemporal area of Cx | 30, 75 |
| ts | tectospinal tract | 36–53, 62, 63 |
| TS | triangular septal nucleus | 20, 21, 62 |
| tth | trigeminothalamic tract | 34, 64 |
| Tu | olfactory tubercle | 13–17, 64–71 |
| TuLH | tuberal area of lateral hypothalamus | 20, 21, 63 |
| tz | trapezoid body | 36–38, 63–67 |
| Tz | nucleus of trapezoid body | 36–38, 63, 65 |
| tzd | trapezoid decussation | 62 |
| V1 | primary visual (striate) cortex | 41–75 |
| V2 | secondary visual area of cortex | 53, 54 |
| V2l | secondary visual area (lateral) of cortex | 50–53, 71–75 |

| Abbreviation | Long Name | Plates |
|---|---|---|
| V2m | secondary visual area (medial) of cortex | 39–52, 62–75 |
| V2v | secondary visual area (ventral) of cortex | 35–51, 64–75 |
| V3d | tertiary visual area (dorsal) of cortex | 41 |
| VA | ventral anterior thalamic nucleus | 24–27, 63–66 |
| VACo | ventral anterior cortical amygdala | 18 |
| VAMC | ventral anterior thalamic nucleus, mediocentral part | 24, 25, 63, 64 |
| VCA | ventral cochlear nucleus, anterior part | 38, 39, 71, 72 |
| VCoI | ventral inferior cortical amygdala | 18–20, 67, 68 |
| VCoIn | ventral intermediate cortical amygdala | 18, 19, 67, 68 |
| VCoS | ventral superior cortical amygdala | 18–20, 67–69 |
| VCP | ventral cochlear nucleus, posterior part | 40–43, 70–72 |
| vcs | ventral corticospinal tract | 49–53, 62 |
| VDB | vertical limb of diagonal band nuclei | 14–17, 62, 63 |
| VeCb | vestibulocerebellar nucleus | 40–42, 65, 66 |
| VEn | ventral endopiriform nucleus | 16–19, 71–75 |
| VeR | dorsal motor nucleus of vagus, ventrorostral subnucleus | 42 |
| vfu | ventral funiculus of spinal cord | 51, 62–64 |
| VH | ventral horn of spinal cord | 47–53, 62, 63, 65 |
| vhc | ventral hippocampal commissure | 22–24, 28–36, 62 |
| VLa | ventrolateral thalamic nucleus, anterior part | 25–28, 65–68 |
| VLA (V4) | ventrolateral anterior extrastriate cortex (V4) | 35–41, 75 |
| VLH | ventrolateral hypothalamic area | 21, 22, 64, 65 |
| VLL | ventral nucleus of lateral lemniscus | 34, 35, 67, 68 |
| VLp | ventrolateral thalamic nucleus, posterior part | 26–30, 66, 67, 69, 71 |
| VLP (V3v) | ventrolateral posterior extrastriate cortex (V3v) | 34–49, 68–75 |

| Abbreviation | Long Name | Plates |
|---|---|---|
| VLPAG | ventrolateral periaqueductal gray | 34–38, 62–65 |
| VLPO | ventrolateral preoptic nucleus | 17, 63 |
| vmf | ventral median fissure of spinal cord | 48, 49, 51–53 |
| VMH | ventromedial hypothalamic nucleus | 20, 21, 24 |
| VMHL | ventromedial hypothalamic nucleus, lateral part | 22, 23, 62, 63 |
| VMHM | ventromedial hypothalamic nucleus, medial part | 22, 23, 62 |
| VMPO | ventromedial preoptic nucleus | 17, 62 |
| VOLT | vascular organ of lamina terminalis | 16 |
| VP | ventral pallidum | 18, 19, 66–70 |
| VPe | ventral periventricular zone of hypothalamus | 62 |
| VPI | ventral posterior thalamic nucleus, inferior part | 28–30, 67–70 |
| VPL | ventral posterior thalamic nucleus, lateral part | 28–31, 68–73 |
| VPM | ventral posterior thalamic nucleus, medial part | 28–31, 65–68 |

| Abbreviation | Long Name | Plates |
|---|---|---|
| VPMPC | ventral posterior thalamic nucleus, medial parvicellular part | 28, 29, 64–66 |
| VPS | ventral posterior thalamic nucleus, superior part | 28–30, 66–69 |
| VRe | ventral reuniens thalamic nucleus | 23–27, 62, 63 |
| vsc | ventral spinocerebellar tract | 37–41, 44, 46, 48–69 |
| VSom? | ventral somatosensory area (putative) of cortex | 25–28, 75 |
| VTA | ventral tegmental area | 28–31, 62–64 |
| vtgd | ventral tegmental decussation | 31, 62, 64 |
| VTM | ventral tuberomamillary area | 24, 25, 64, 65 |
| X | nucleus X | 43, 69 |
| Xi | xiphoid thalamic nucleus | 23–25 |
| Y | nucleus Y | 43, 68 |
| ZID | zona incerta, dorsal part | 25–29, 64–71 |
| ZIR | zona incerta, rostral part | 23, 24, 62–65 |
| ZIV | zona incerta, ventral part | 25–29, 64–71 |
| Zo | stratum zonale of superior colliculus | 37–39, 64–68 |

Section D

# Atlas plates

# Coronal Plates

# Plate 1

Interaural +19.05 mm

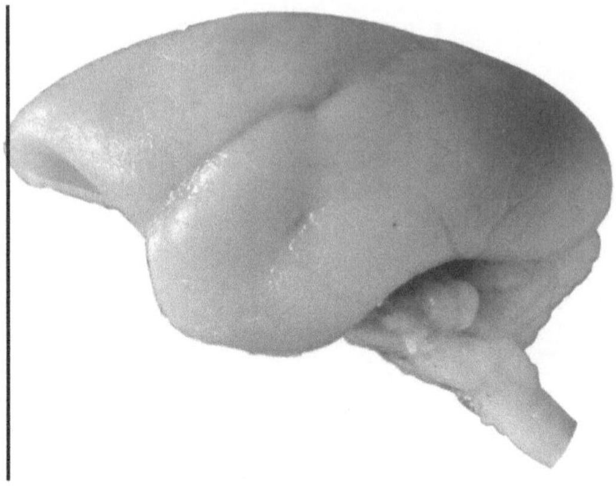

# Plate 1a

**Calbindin**

Interaural +19.35 mm

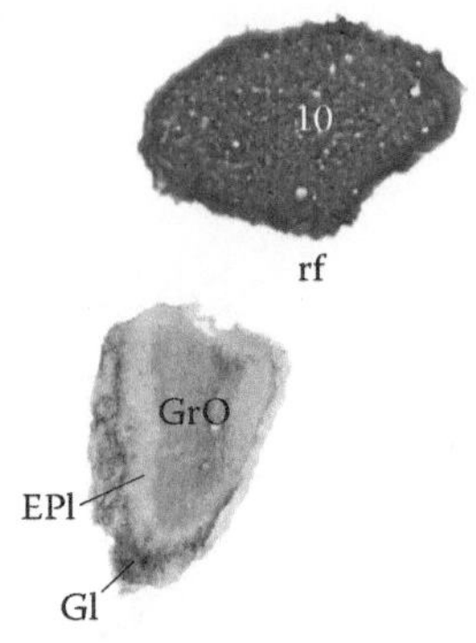


# Plate 1b

**Neurofilament**

Interaural +19.20 mm

# Plate 1c

**Parvalbumin**

Interaural +19.10 mm

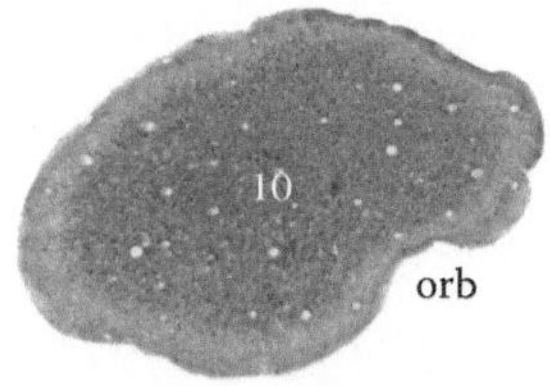

# Plate 1d

**Cresyl Violet**

Interaural +19.05 mm

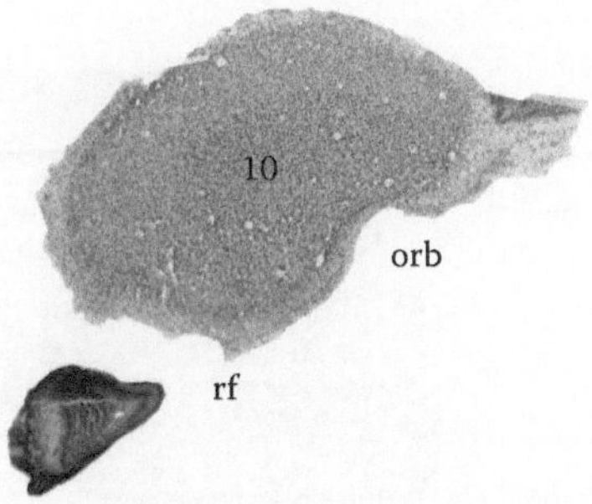


# Plate 1

Interaural +19.05 mm

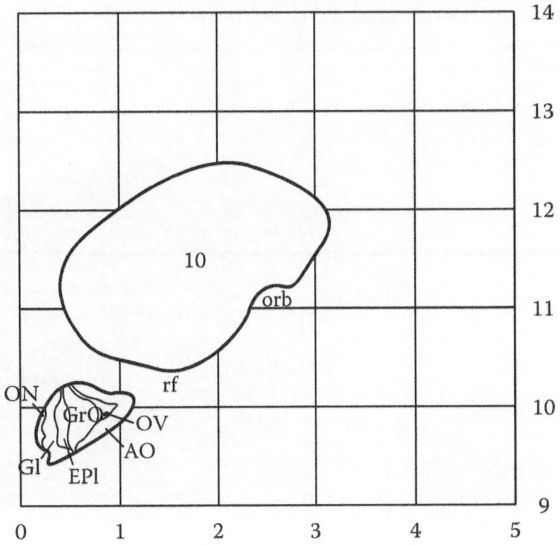


| | | | |
|---|---|---|---|
| 10 | area 10 Cx | ON | olf. n. layer, olf. bulb |
| AO | ant. olf. area | orb | orbital sulcus |
| EPl | ext. plex. layer, olf. bulb | OV | olf. ventricle |
| Gl | glom. layer, olf. bulb | rf | rhinal fissure |
| GrO | gran. cell layer, olf. bulb | | |

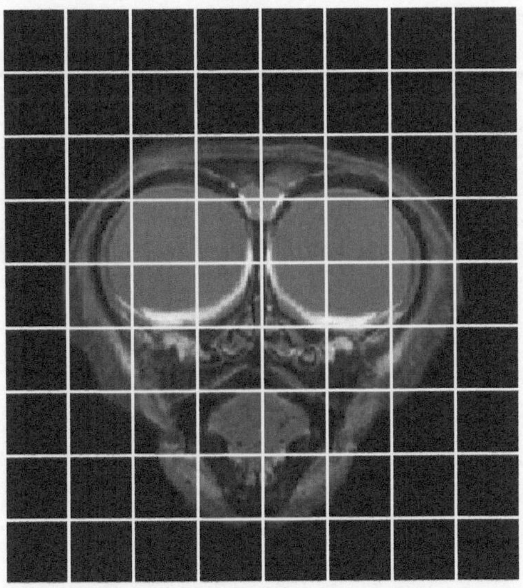

Interaural +19.15 mm

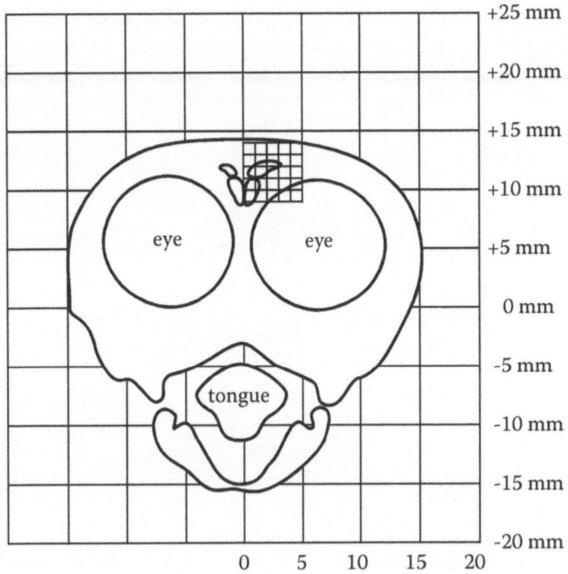

# Plate 2

Interaural +18.50 mm

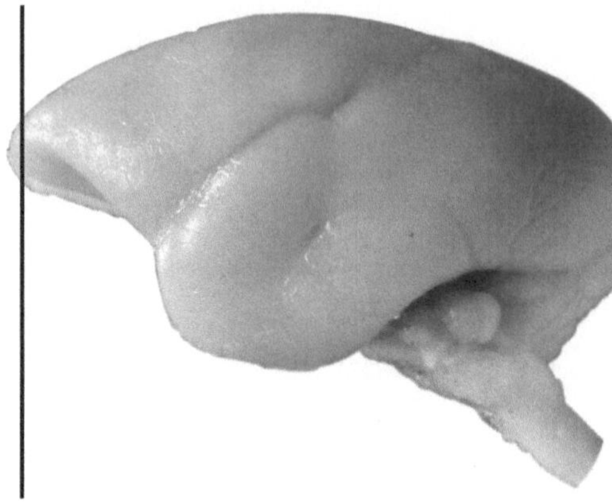

# Plate 2a

**Calbindin**

Interaural +18.80 mm

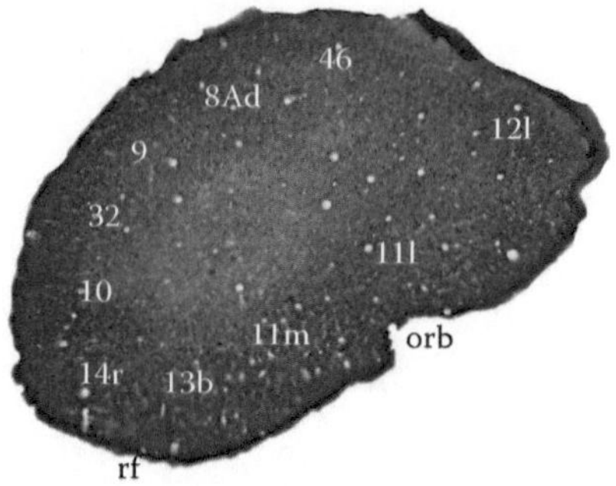


# Plate 2b

**Acetylcholinesterase**

Interaural +18.60 mm

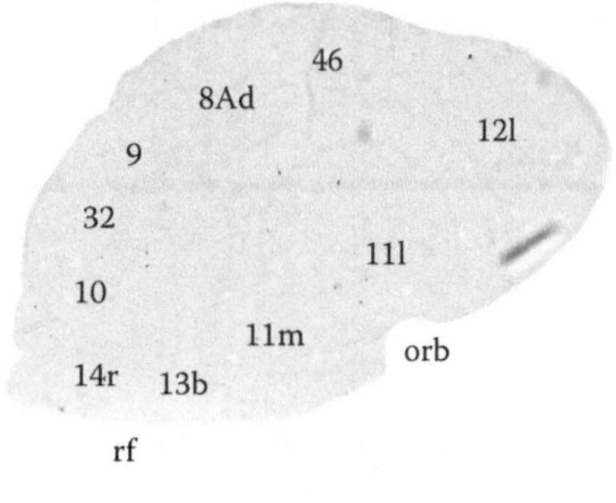


# Plate 2c

**Parvalbumin**

Interaural +18.55 mm

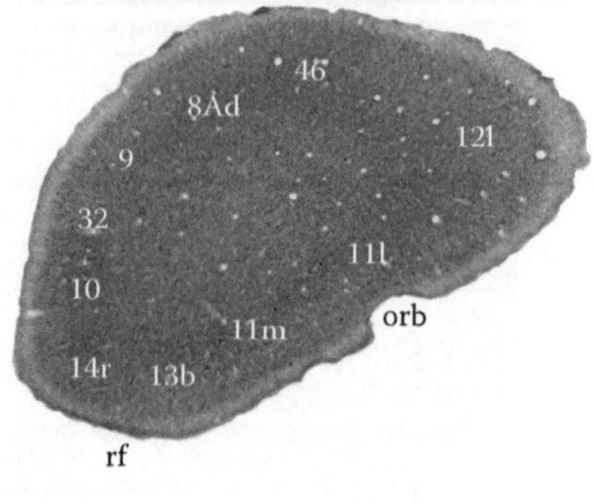

# Plate 2d

**Cresyl Violet**

Interaural +18.50 mm

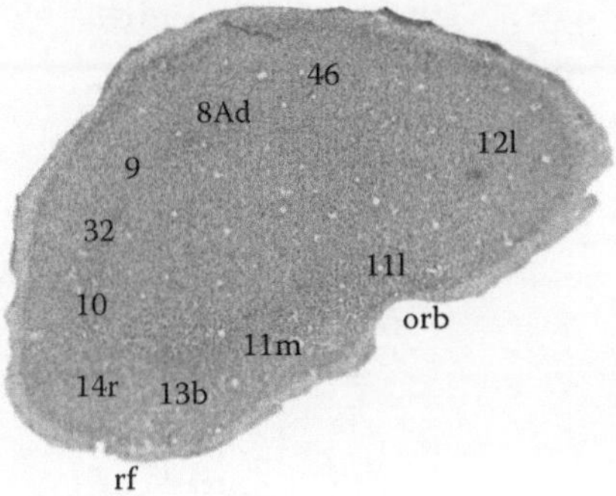


# Plate 2

Interaural +18.50 mm

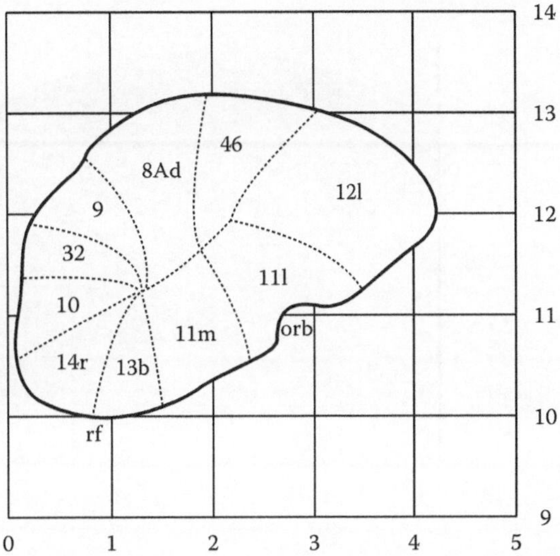


| | |
|---|---|
| 8Ad | area 8Ad Cx |
| 9 | area 9 Cx |
| 10 | area 10 Cx |
| 11l | area 11l Cx |
| 11m | area 11m Cx |
| 12l | area 12l Cx |
| 13b | area 13b Cx |
| 14r | area 14r Cx |
| 32 | area 32 Cx |
| 46 | area 46 Cx |
| orb | orbital sulcus |
| rf | rhinal fissure |

# Plate 3

Interaural +18.00 mm

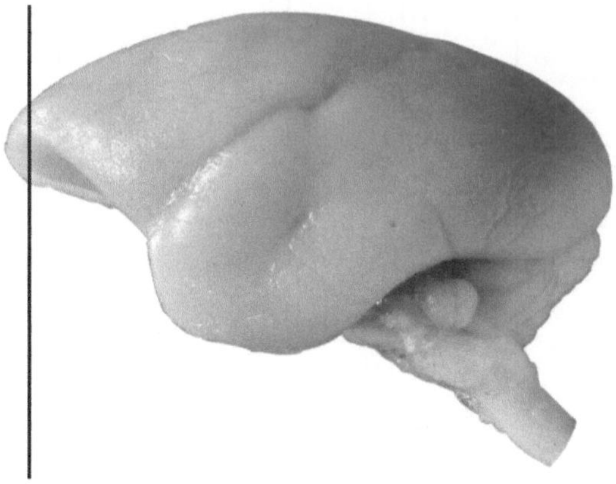

| | |
|---|---|
| 8Ad | area 8Ad Cx |
| 9 | area 9 Cx |
| 10 | area 10 Cx |
| 11l | area 11l Cx |
| 11m | area 11m Cx |
| 12l | area 12l Cx |
| 12m | area 12m Cx |
| 13b | area 13b Cx |
| 14r | area 14r Cx |
| 32 | area 32 Cx |
| 46 | area 46 Cx |
| orb | orbital sulcus |

# Plate 3a

**Calbindin**

Interaural +18.30 mm

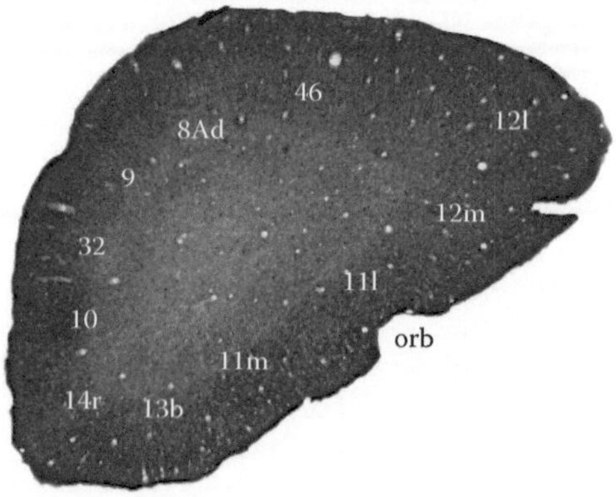

# Plate 3b

**Neurofilament**
Interaural +18.15 mm

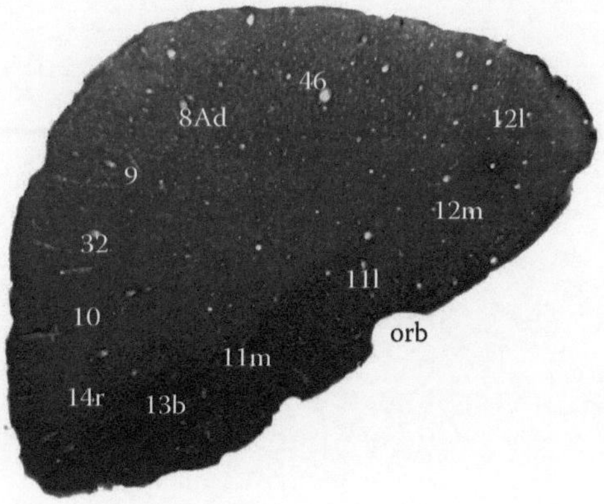


# Plate 3c

**Acetylcholinesterase**
Interaural +18.10 mm

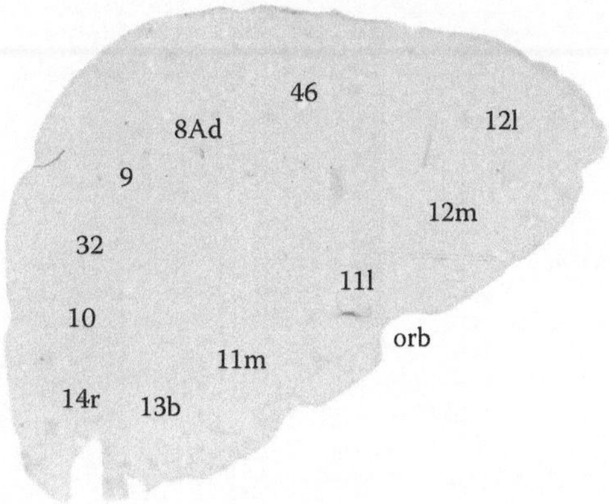

# Plate 3d

**Parvalbumin**

Interaural +18.05 mm

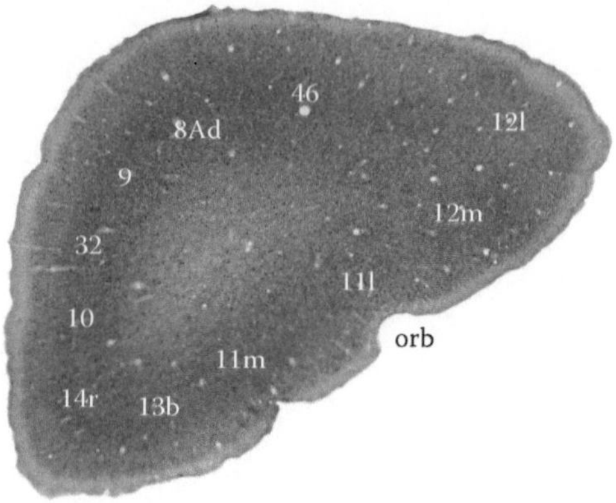


# Plate 3e

**Cresyl Violet**

Interaural +18.00 mm

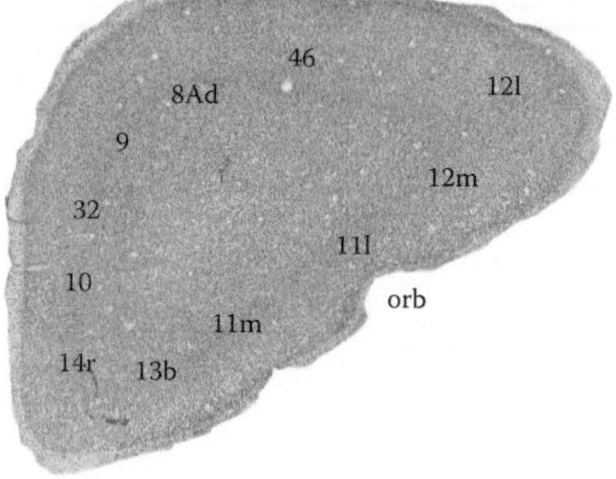

# Plate 3

## Interaural +18.00 mm

| | |
|---|---|
| 8Ad | area 8Ad Cx |
| 9 | area 9 Cx |
| 10 | area 10 Cx |
| 11l | area 11l Cx |
| 11m | area 11m Cx |
| 12l | area 12l Cx |
| 12m | area 12m Cx |
| 13b | area 13b Cx |
| 14r | area 14r Cx |
| 32 | area 32 Cx |
| 46 | area 46 Cx |
| orb | orbital sulcus |

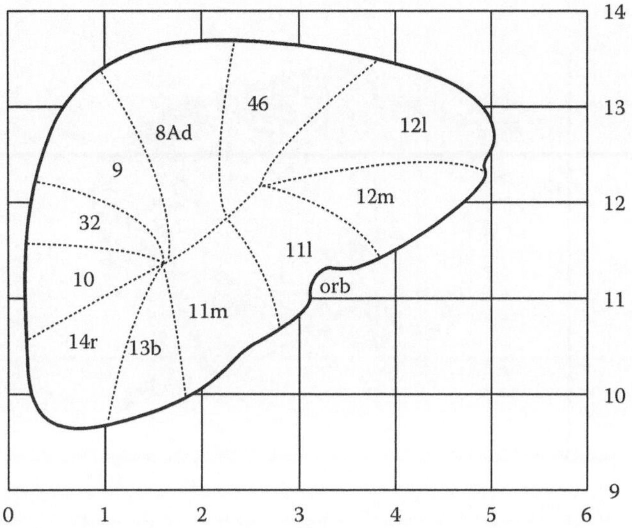


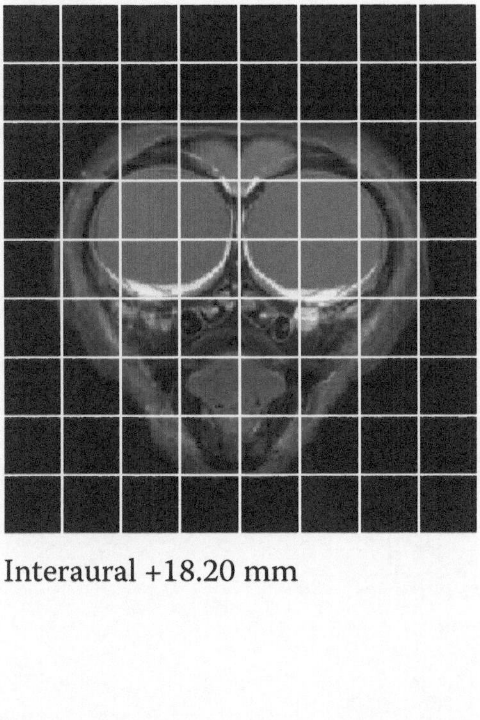

Interaural +18.20 mm

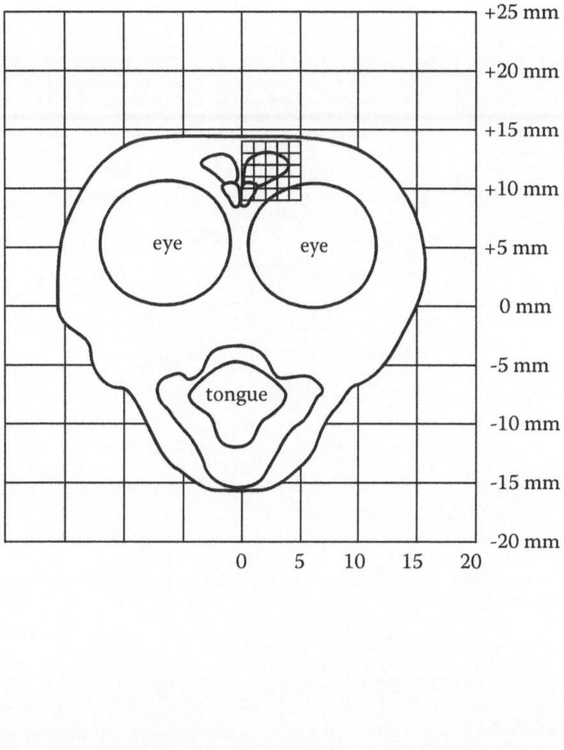

# Plate 4

Interaural +17.50 mm

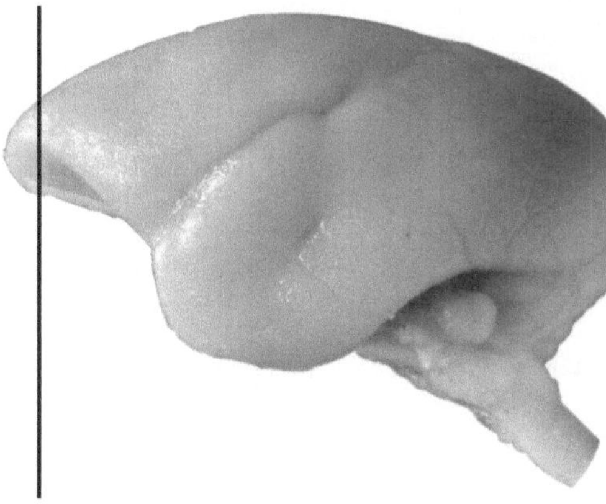

| | |
|---|---|
| 8Ad | area 8Ad Cx |
| 8Av | area 8Av Cx |
| 9 | area 9 Cx |
| 10 | area 10 Cx |
| 11l | area 11l Cx |
| 11m | area 11m Cx |
| 12l | area 12l Cx |
| 12m | area 12m Cx |
| 13b | area 13b Cx |
| 14r | area 14r Cx |
| 32 | area 32 Cx |
| 46 | area 45 Cx |
| fmi | forceps minor |
| orb | orbital sulcus |
| rf | rhinal fissure |

# Plate 4a

**Calbindin**

Interaural +17.80 mm

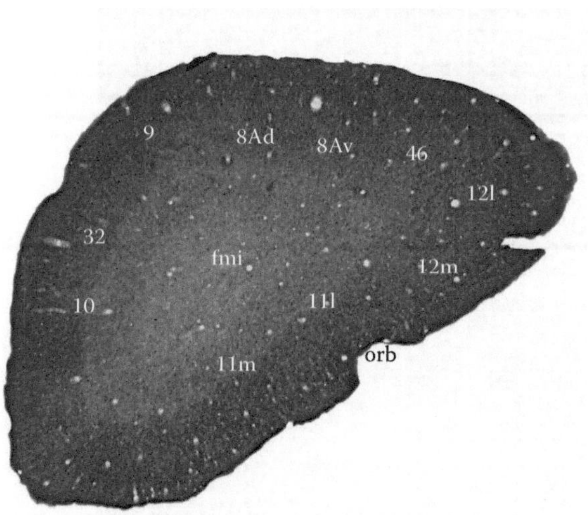

# Plate 4b

**Neurofilament**

Interaural +17.65 mm

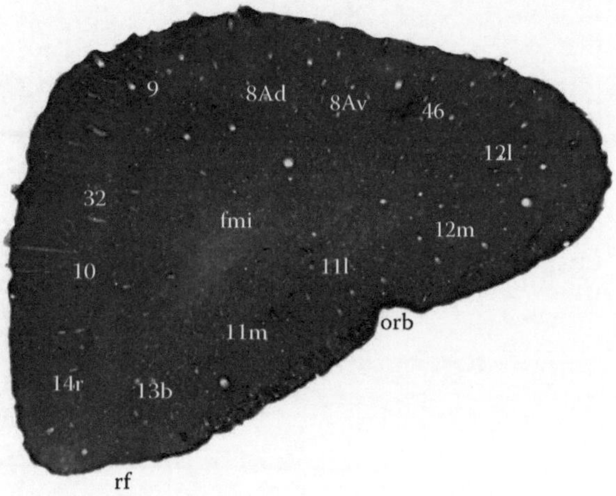


# Plate 4c

**Acetylcholinesterase**

Interaural +17.60 mm

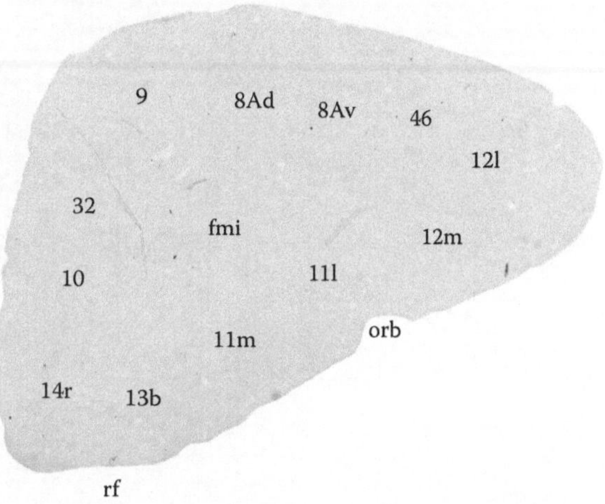

# Plate 4d

**Parvalbumin**

Interaural +17.55 mm

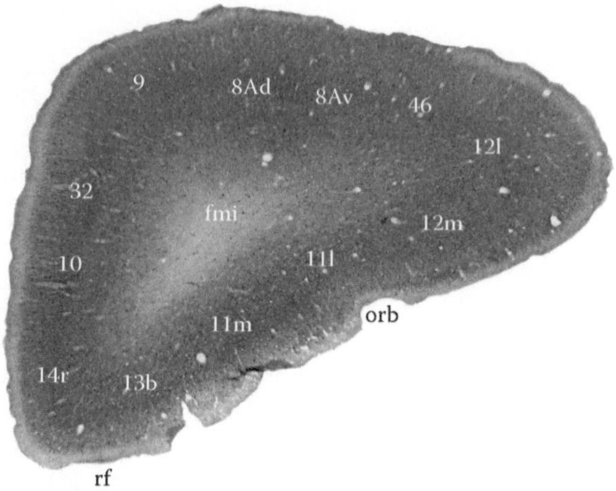


# Plate 4e

**Cresyl Violet**

Interaural +17.50 mm

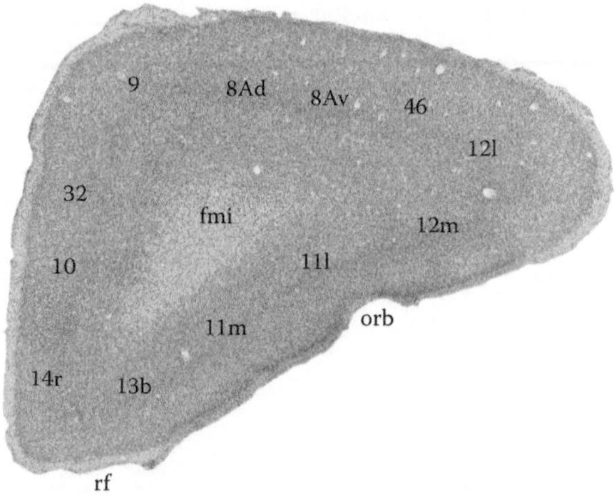

# Plate 4

## Interaural +17.50 mm

| | |
|---|---|
| 8Ad | area 8Ad Cx |
| 8Av | area 8Av Cx |
| 9 | area 9 Cx |
| 10 | area 10 Cx |
| 11l | area 11l Cx |
| 11m | area 11m Cx |
| 12l | area 12l Cx |
| 12m | area 12m Cx |
| 13b | area 13b Cx |
| 14r | area 14r Cx |
| 32 | area 32 Cx |
| 46 | area 45 Cx |
| fmi | forceps minor |
| orb | orbital sulcus |
| rf | rhinal fissure |

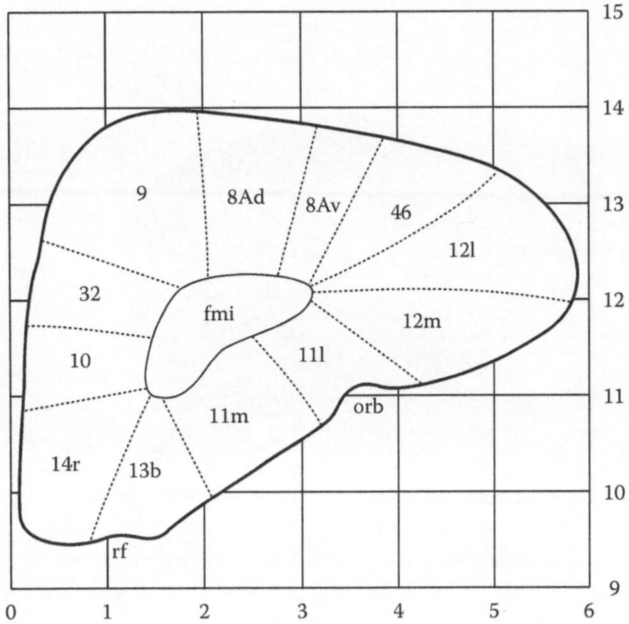


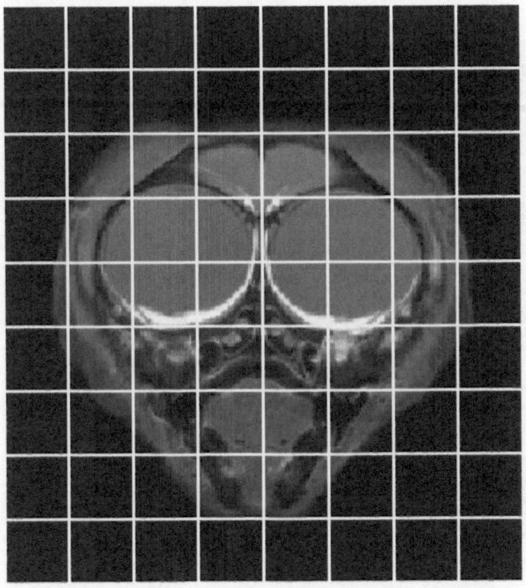

Interaural +17.25 mm

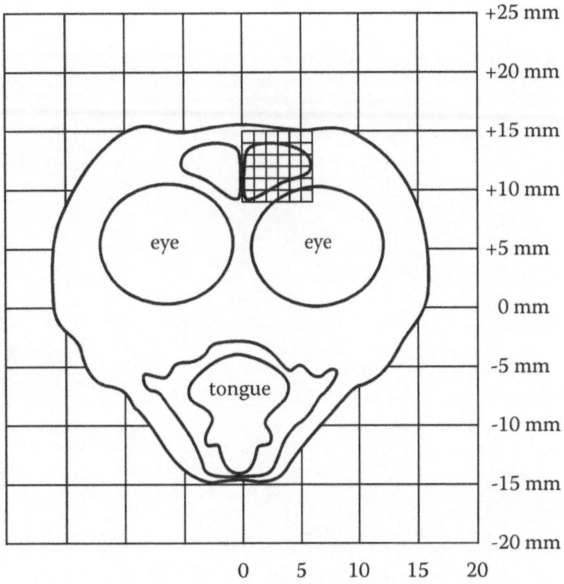

# Plate 5

Interaural +16.95 mm

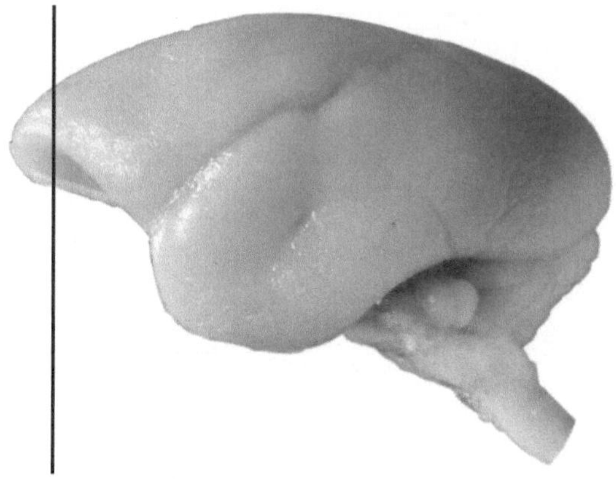

| | |
|---|---|
| 8Ad | area 8Ad Cx |
| 8Av | area 8Av Cx |
| 8B | area 8B Cx |
| 10mc | area 10mc Cx |
| 11l | area 11l Cx |
| 11m | area 11m Cx |
| 12l | area 12l Cx |
| 12m | area 12m Cx |
| 13b | area 13b Cx |
| 13m | area 13m Cx |
| 14r | area 14r Cx |
| 32 | area 32 Cx |
| 46 | area 45 Cx |
| FEF | frontal eye field Cx |
| fmi | forceps minor |
| orb | orbital sulcus |
| rf | rhinal fissure |

# Plate 5a

**Calbindin**

Interaural +17.25 mm

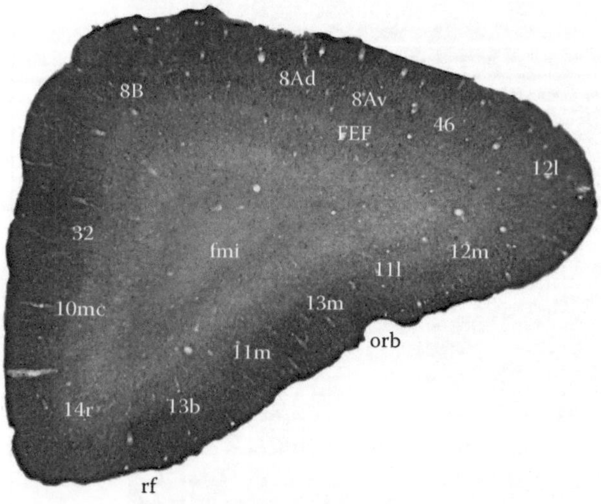

# Plate 5b

**Neurofilament**
Interaural +17.10 mm

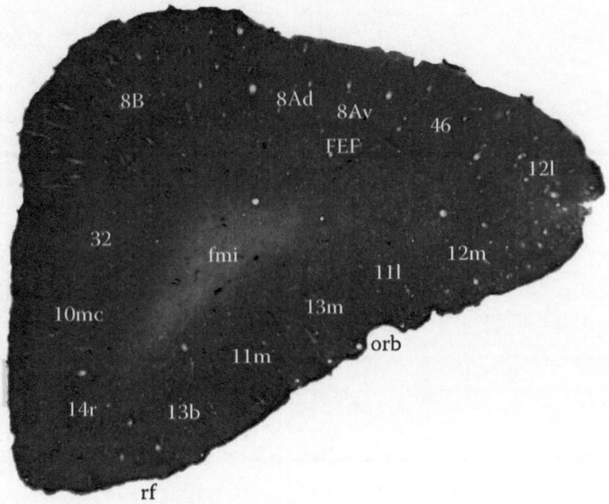


# Plate 5c

**Acetylcholinesterase**
Interaural +17.05 mm

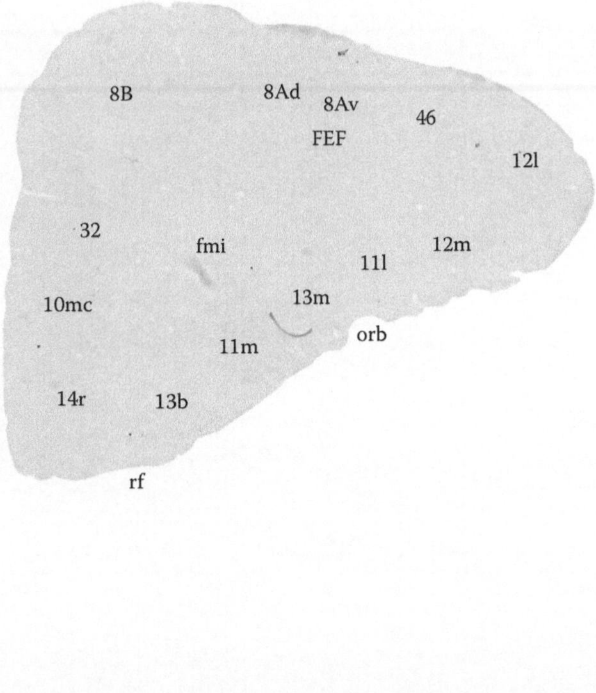

# Plate 5d

**Parvalbumin**

Interaural +17.00 mm

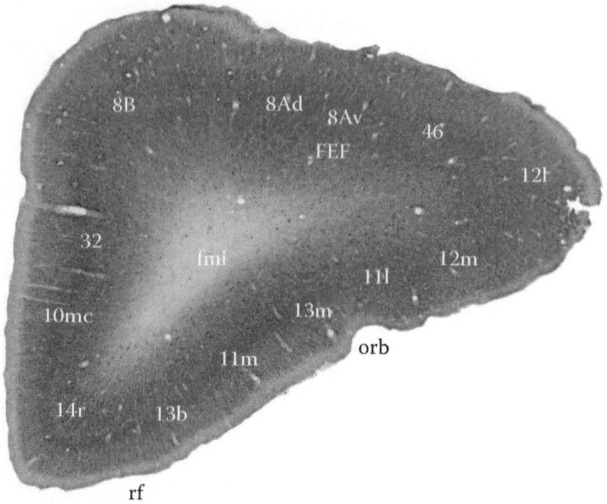


# Plate 5e

**Cresyl Violet**

Interaural +16.95 mm

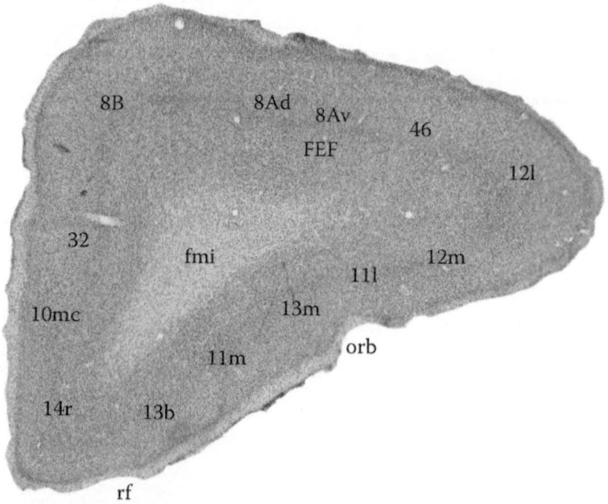

# Plate 5

## Interaural +16.95 mm

| | |
|---|---|
| 8Ad | area 8Ad Cx |
| 8Av | area 8Av Cx |
| 8B | area 8B Cx |
| 10mc | area 10mc Cx |
| 11l | area 11l Cx |
| 11m | area 11m Cx |
| 12l | area 12l Cx |
| 12m | area 12m Cx |
| 13b | area 13b Cx |
| 13m | area 13m Cx |
| 14r | area 14r Cx |
| 32 | area 32 Cx |
| 46 | area 45 Cx |
| FEF | frontal eye field Cx |
| fmi | forceps minor |
| orb | orbital sulcus |
| rf | rhinal fissure |

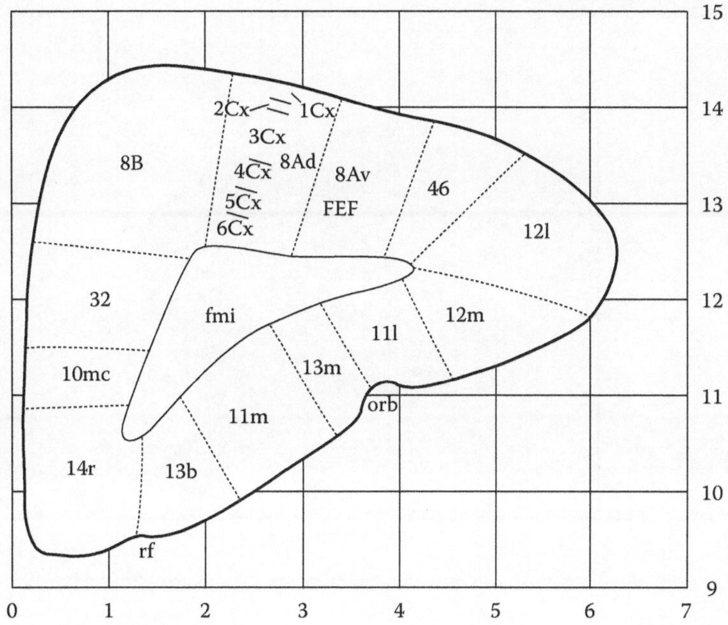

# Plate 6

Interaural +16.45 mm

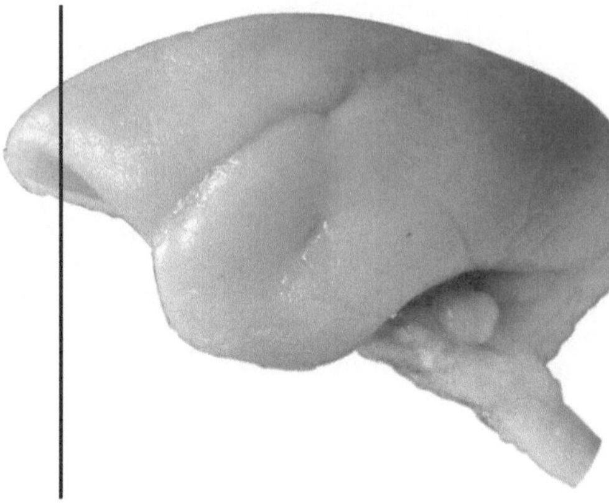

| | |
|---|---|
| 8Ad | area 8Ad Cx |
| 8Av | area 8Av Cx |
| 8B | area 8B Cx |
| 10mc | area 10mc Cx |
| 11l | area 11l Cx |
| 11m | area 11m Cx |
| 12l | area 12l Cx |
| 12m | area 12m Cx |
| 13b | area 13b Cx |
| 13m | area 13m Cx |
| 14r | area 14r Cx |
| 32 | area 32 Cx |
| 46 | area 45 Cx |
| FEF | frontal eye field Cx |
| fmi | forceps minor |
| orb | orbital sulcus |
| rf | rhinal fissure |

# Plate 6a

**Calbindin**

Interaural +16.70 mm

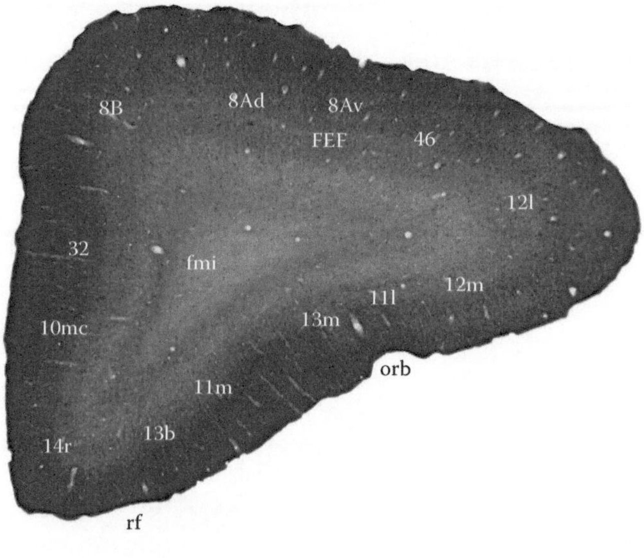

# Plate 6b

**Neurofilament**
Interaural +16.60 mm

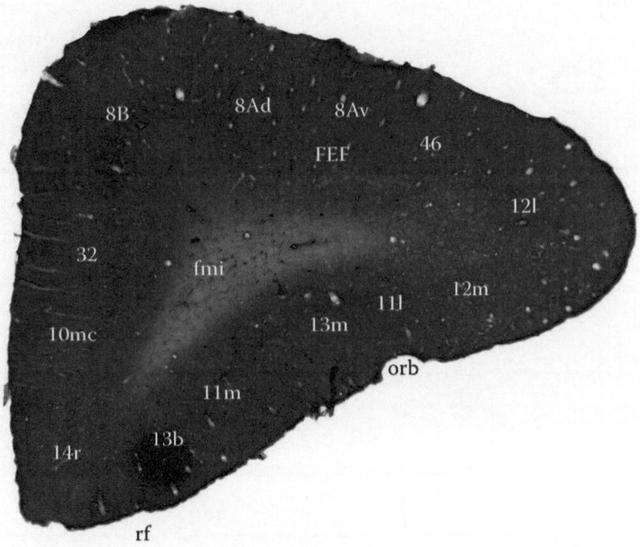


# Plate 6c

**Acetylcholinesterase**
Interaural +16.55 mm

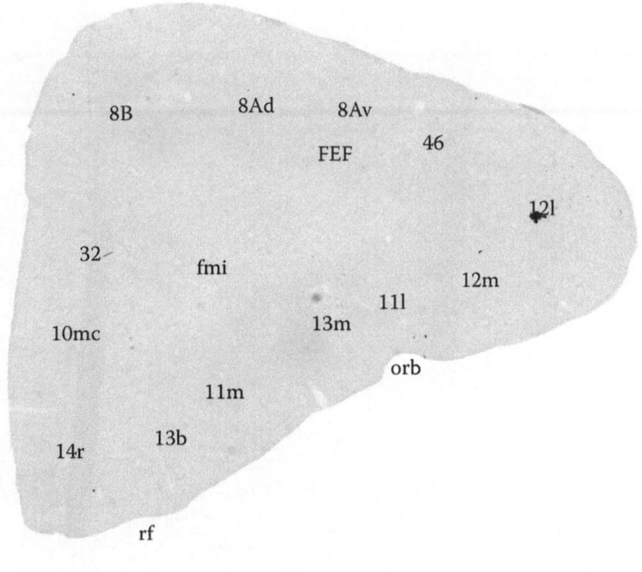

# Plate 6d

**Parvalbumin**

Interaural +16.50 mm

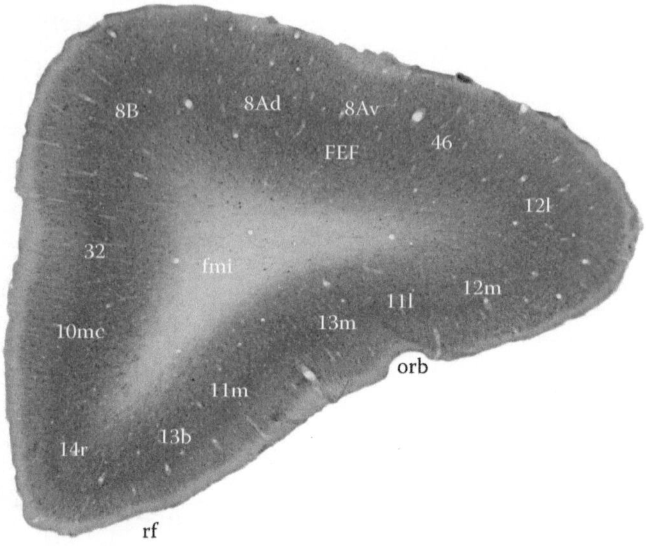


# Plate 6e

**Cresyl Violet**

Interaural +16.45 mm

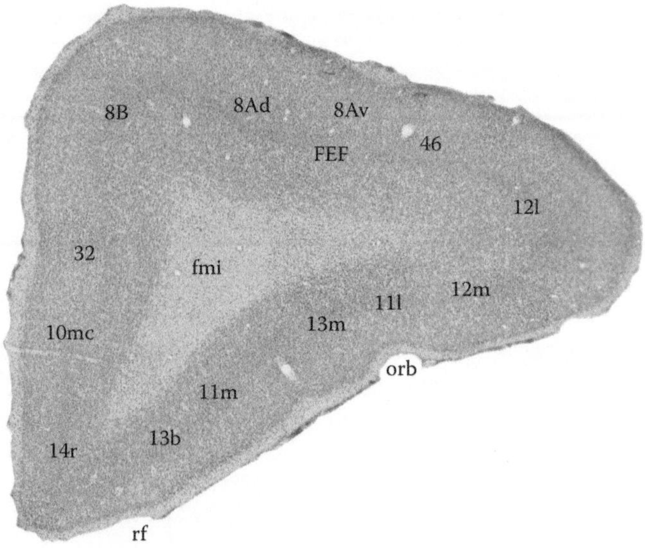

# Plate 6

## Interaural +16.45 mm

| | |
|---|---|
| 8Ad | area 8Ad Cx |
| 8Av | area 8Av Cx |
| 8B | area 8B Cx |
| 10mc | area 10mc Cx |
| 11l | area 11l Cx |
| 11m | area 11m Cx |
| 12l | area 12l Cx |
| 12m | area 12m Cx |
| 13b | area 13b Cx |
| 13m | area 13m Cx |
| 14r | area 14r Cx |
| 32 | area 32 Cx |
| 46 | area 45 Cx |
| FEF | frontal eye field Cx |
| fmi | forceps minor |
| orb | orbital sulcus |
| rf | rhinal fissure |

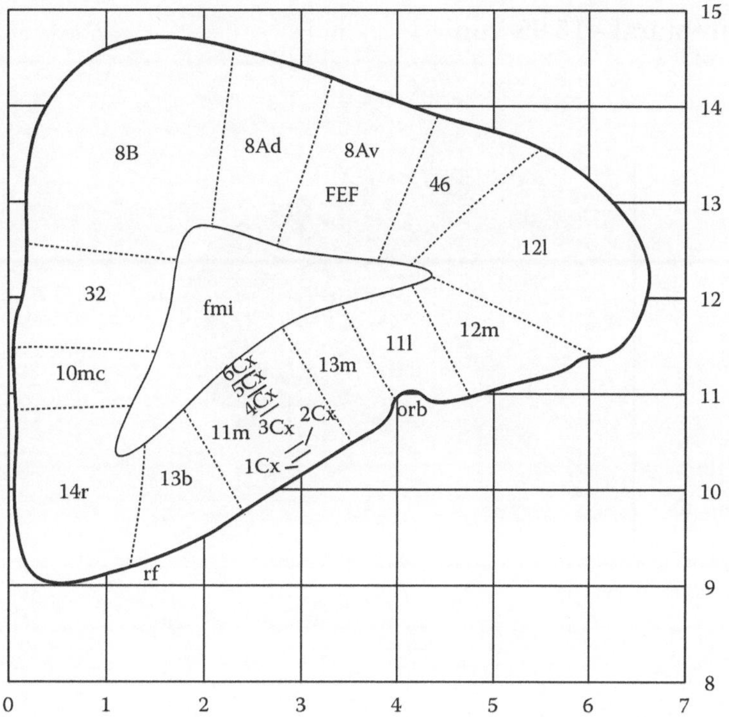


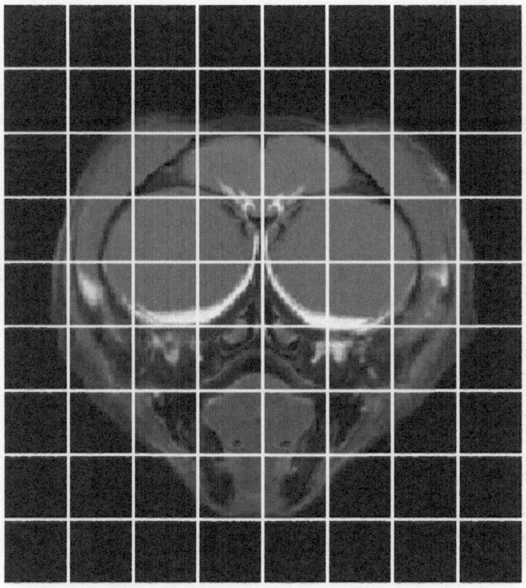

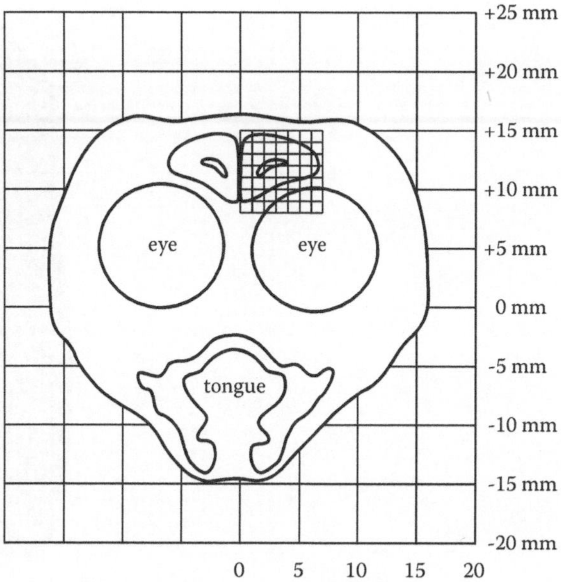


Interaural +16.30 mm

# Plate 7

Interaural +15.95 mm

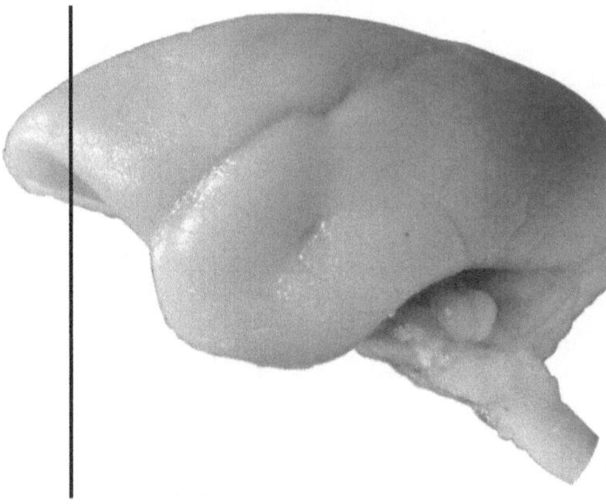

| | |
|---|---|
| 8Ad | area 8Ad Cx |
| 8Av | area 8Av Cx |
| 8B | area 8B Cx |
| 10mc | area 10mc Cx |
| 11l | area 11l Cx |
| 11m | area 11m Cx |
| 12l | area 12l Cx |
| 12m | area 12m Cx |
| 13b | area 13b Cx |
| 13m | area 13m Cx |
| 14r | area 14r Cx |
| 32 | area 32 Cx |
| 46 | area 45 Cx |
| FEF | frontal eye field Cx |
| fmi | forceps minor |
| orb | orbital sulcus |
| rf | rhinal fissure |

# Plate 7a

**Calbindin**

Interaural +16.25 mm

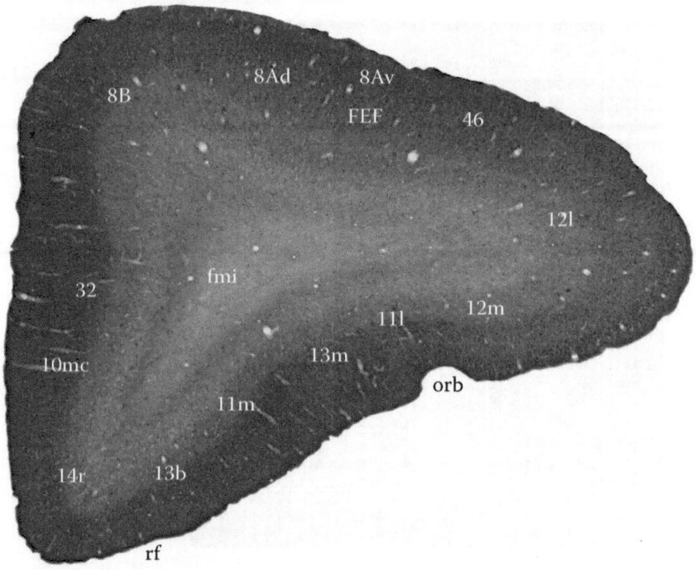

# Plate 7b

**Neurofilament**

Interaural +16.10 mm

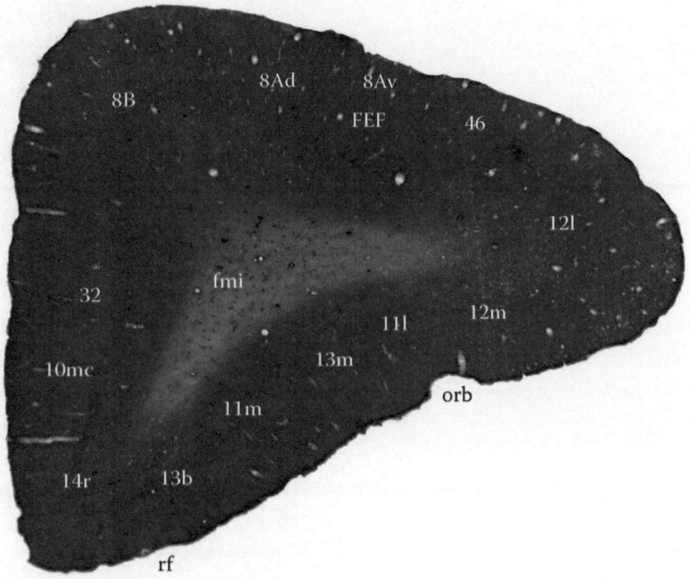


# Plate 7c

**Acetylcholinesterase**

Interaural +16.05 mm

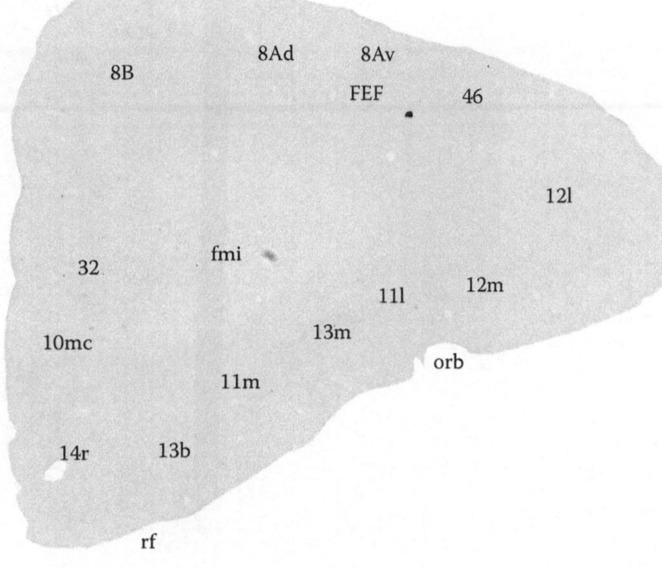

# Plate 7d

**Parvalbumin**
Interaural +16.00 mm

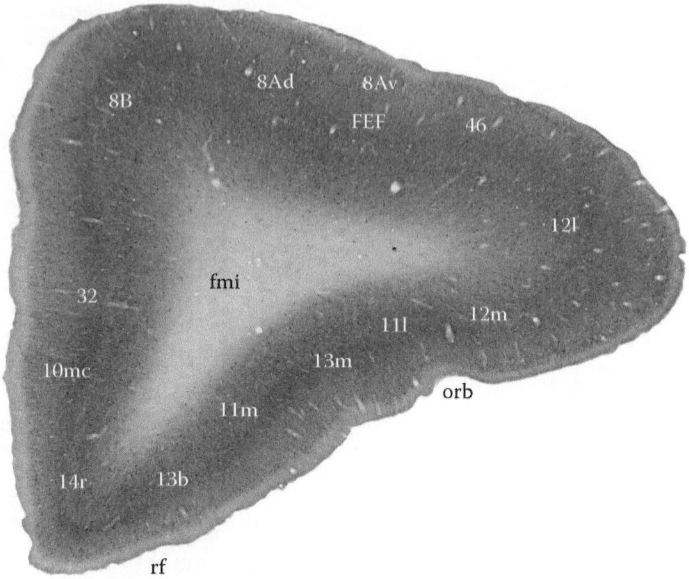


# Plate 7e

**Cresyl Violet**
Interaural +15.95 mm

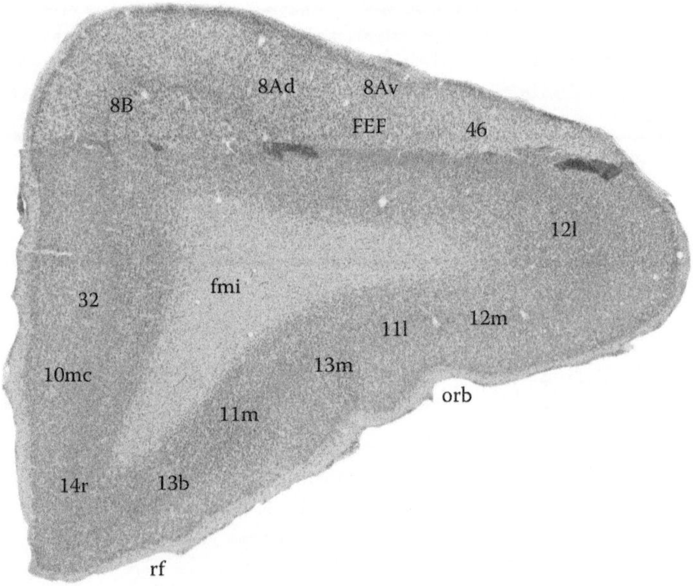

# Plate 7

## Interaural +15.95 mm

| | |
|---|---|
| 8Ad | area 8Ad Cx |
| 8Av | area 8Av Cx |
| 8B | area 8B Cx |
| 10mc | area 10mc Cx |
| 11l | area 11l Cx |
| 11m | area 11m Cx |
| 12l | area 12l Cx |
| 12m | area 12m Cx |
| 13b | area 13b Cx |
| 13m | area 13m Cx |
| 14r | area 14r Cx |
| 32 | area 32 Cx |
| 46 | area 45 Cx |
| FEF | frontal eye field Cx |
| fmi | forceps minor |
| orb | orbital sulcus |
| rf | rhinal fissure |

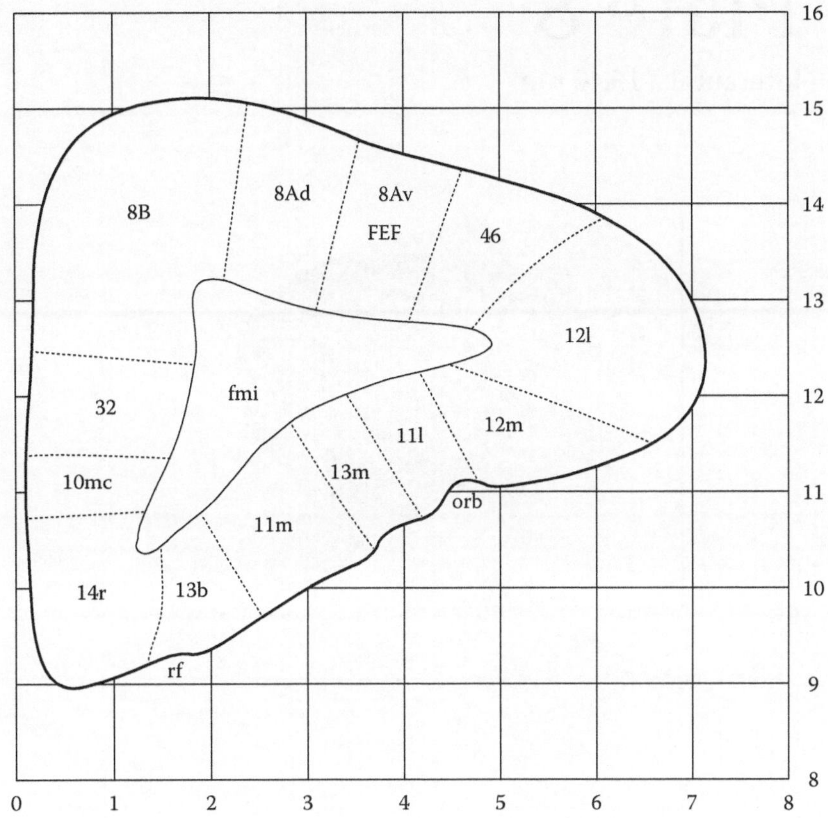

# Plate 8

Interaural +15.40 mm

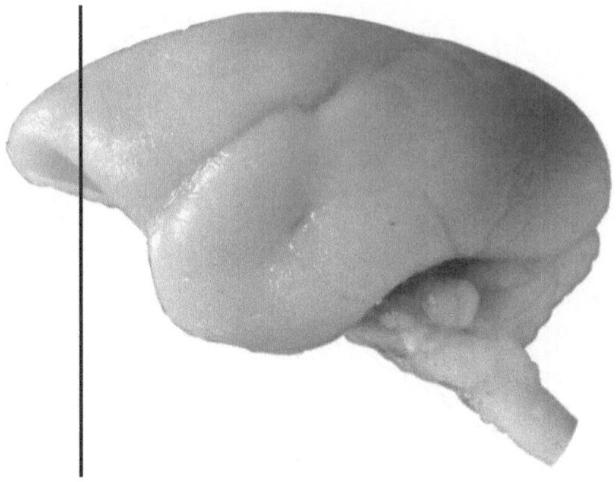

| | |
|---|---|
| 8Ad | area 8Ad Cx |
| 8Av | area 8Av Cx |
| 8B | area 8B Cx |
| 10mc | area 10mc Cx |
| 11l | area 11l Cx |
| 11m | area 11m Cx |
| 12l | area 12l Cx |
| 12m | area 12m Cx |
| 13b | area 13b Cx |
| 13m | area 13m Cx |
| 14r | area 14r Cx |
| 32 | area 32 Cx |
| 46 | area 45 Cx |
| cg | cingulum |
| cgs | cingulate sulc. |
| FEF | frontal eye field Cx |
| fmi | forceps minor |
| orbl | orbital sulc. lat. |
| orbm | orbital sulc. med. |
| pr | prefrontal sulcus |
| rf | rhinal fissure |

# Plate 8a

**Calbindin**

Interaural +15.70 mm

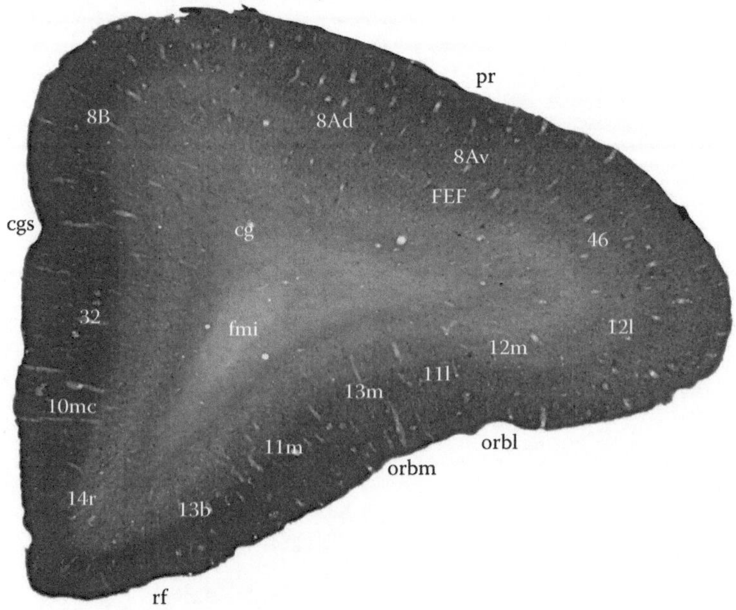

# Plate 8b

**Neurofilament**
Interaural +15.55 mm

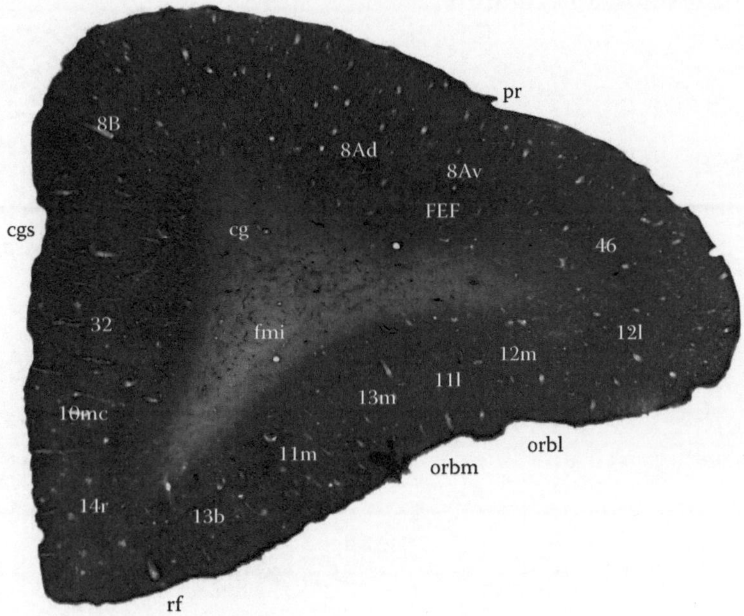


# Plate 8c

**Acetylcholinesterase**
Interaural +15.50 mm

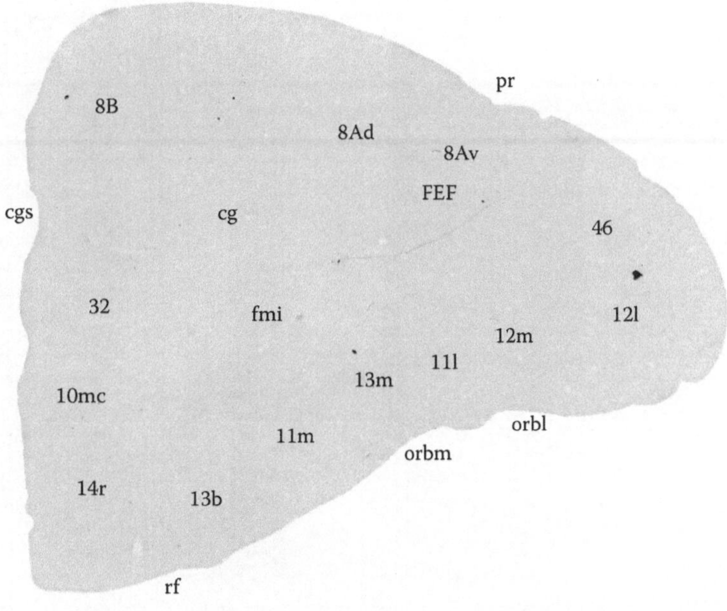

# Plate 8d

**Parvalbumin**

Interaural +15.45 mm

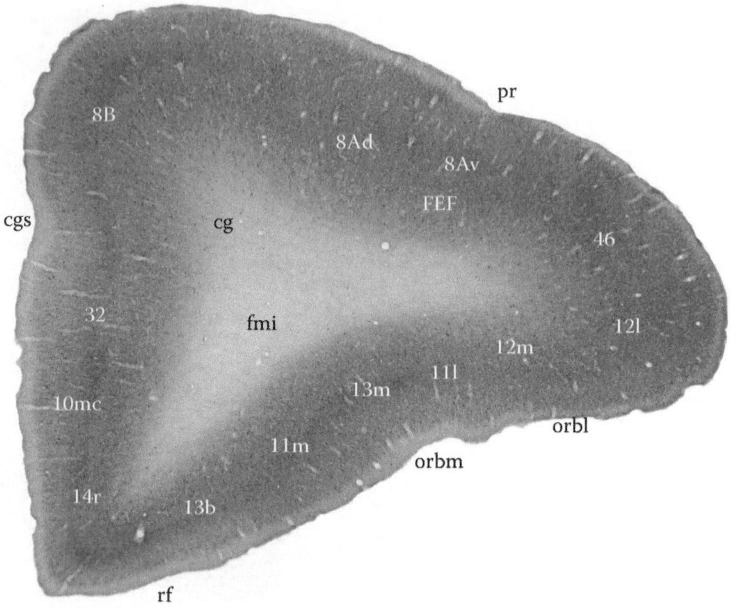


# Plate 8e

**Cresyl Violet**

Interaural +15.40 mm

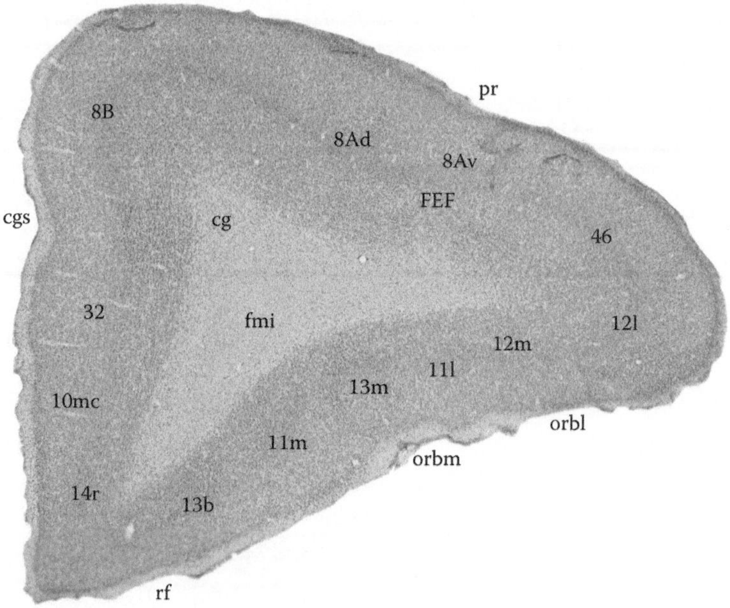

# Plate 8

## Interaural +15.40 mm

| | |
|---|---|
| 8Ad | area 8Ad Cx |
| 8Av | area 8Av Cx |
| 8B | area 8B Cx |
| 10mc | area 10mc Cx |
| 11l | area 11l Cx |
| 11m | area 11m Cx |
| 12l | area 12l Cx |
| 12m | area 12m Cx |
| 13b | area 13b Cx |
| 13m | area 13m Cx |
| 14r | area 14r Cx |
| 32 | area 32 Cx |
| 46 | area 45 Cx |
| cg | cingulum |
| cgs | cingulate sulc. |
| FEF | frontal eye field Cx |
| fmi | forceps minor |
| orbl | orbital sulc. lat. |
| orbm | orbital sulc. med. |
| pr | prefrontal sulcus |
| rf | rhinal fissure |

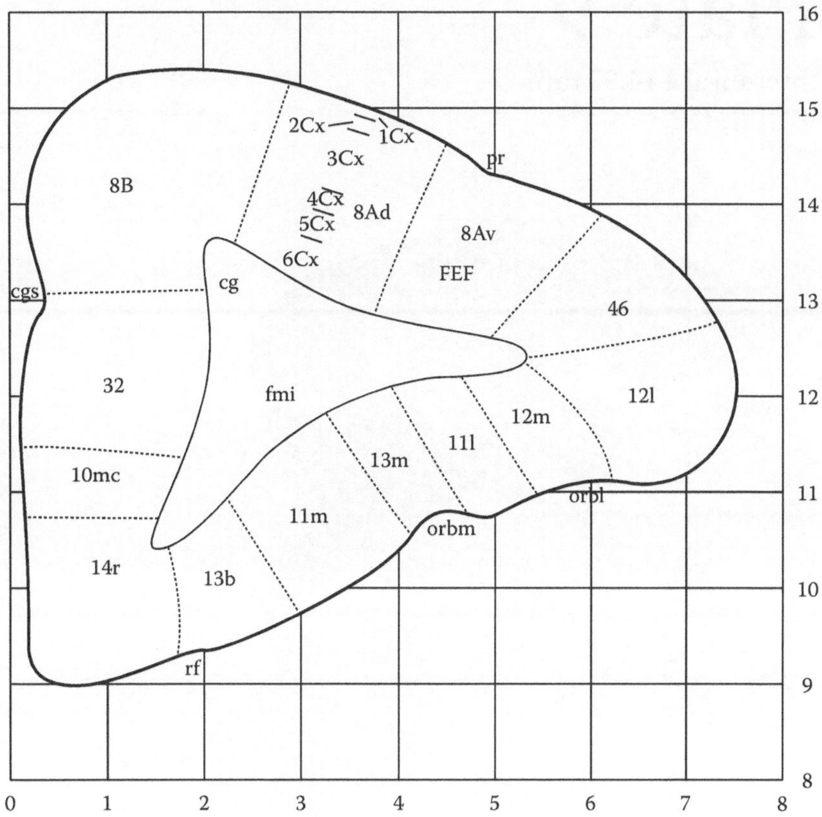


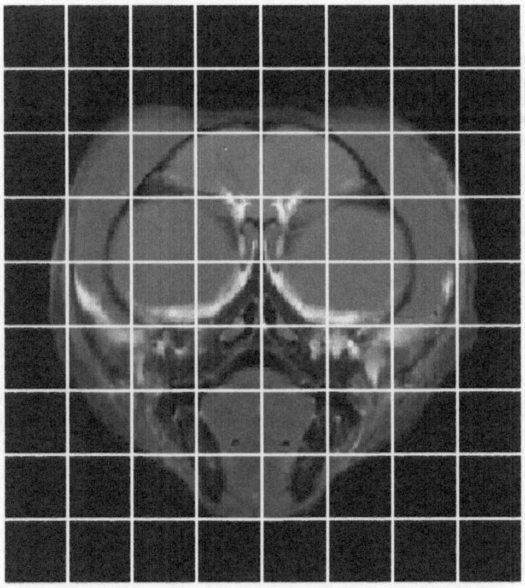

Interaural +15.35 mm

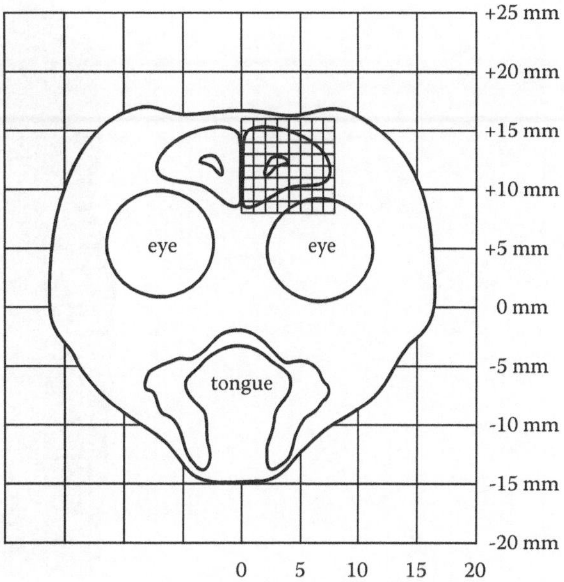

# Plate 9

Interaural +14.90 mm

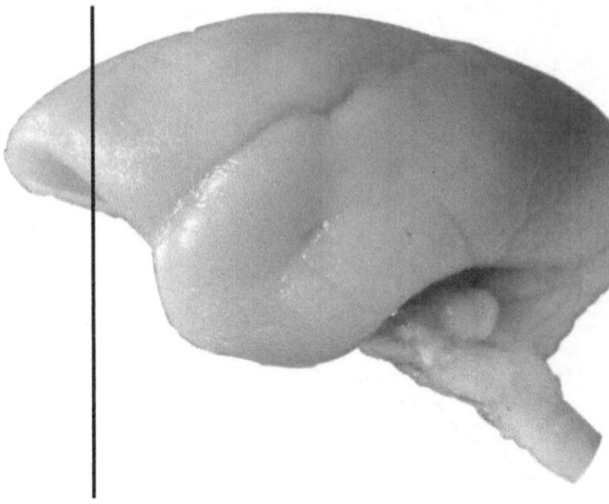

| | |
|---|---|
| 8Ad | area 8Ad Cx |
| 8Av | area 8Av Cx |
| 8B | area 8B Cx |
| 10mc | area 10mc Cx |
| 11l | area 11l Cx |
| 11m | area 11m Cx |
| 12l | area 12l Cx |
| 12m | area 12m Cx |
| 13b | area 13b Cx |
| 13m | area 13m Cx |
| 14r | area 14r Cx |
| 32 | area 32 Cx |
| 46 | area 45 Cx |
| cg | cingulum |
| cgs | cingulate sulc. |
| FEF | frontal eye field Cx |
| fmi | forceps minor |
| orbl | orbital sulc. lat. |
| orbm | orbital sulc. med. |
| pr | prefrontal sulcus |

# Plate 9a

**Calbindin**

Interaural +15.20 mm

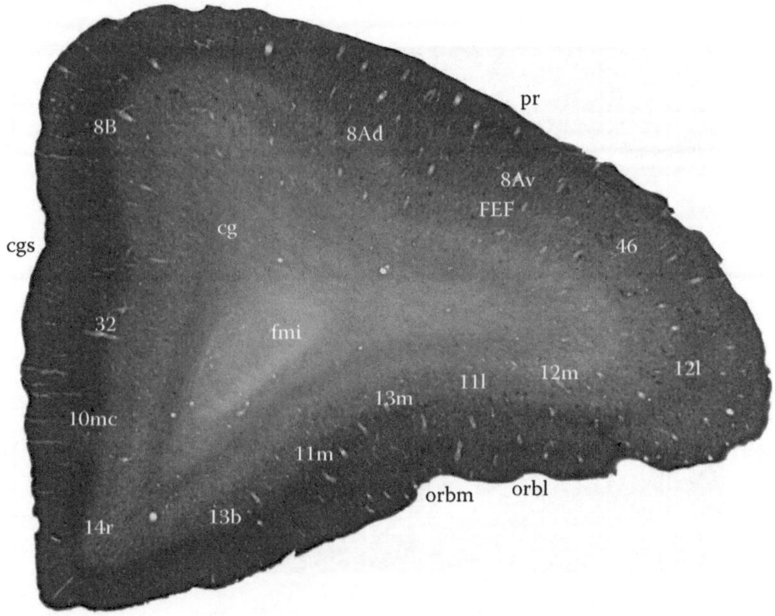

# Plate 9b

**Neurofilament**
Interaural +15.05 mm

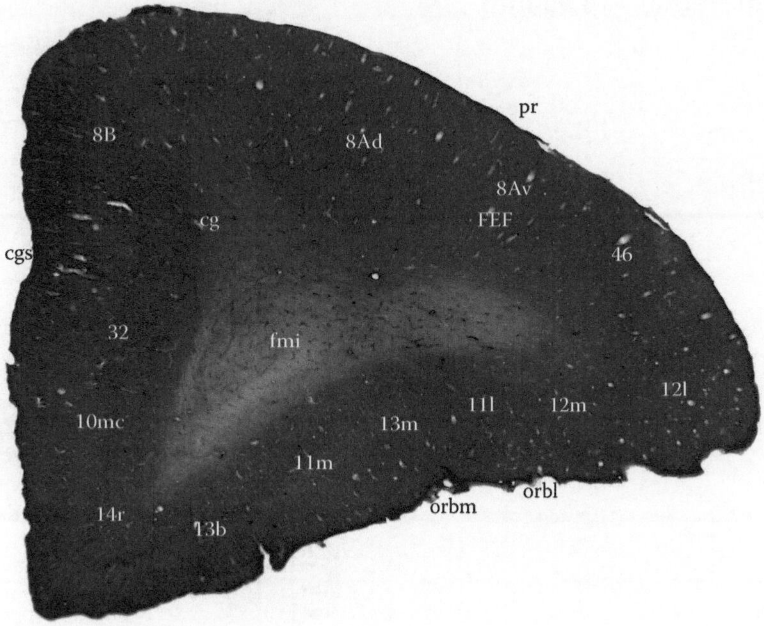


# Plate 9c

**Acetylcholinesterase**
Interaural +15.00 mm

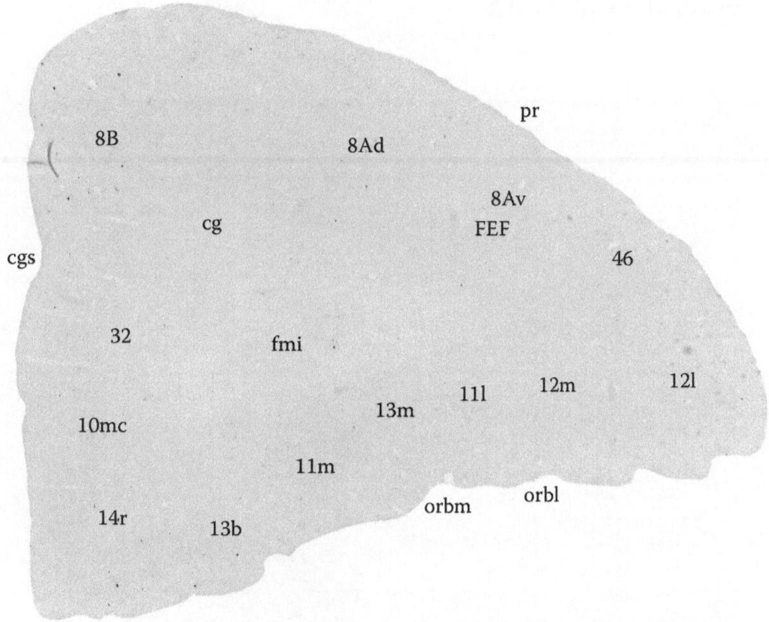

# Plate 9d

**Parvalbumin**
Interaural +14.95 mm

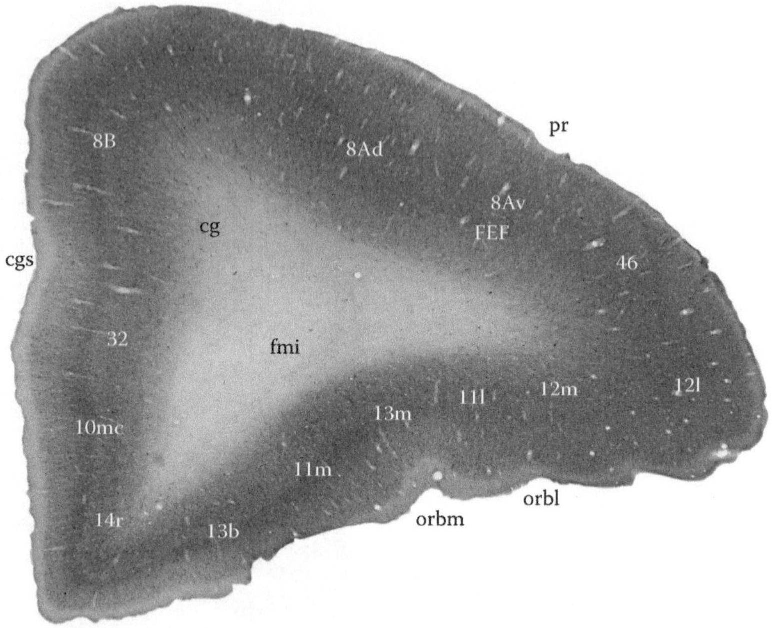


# Plate 9e

**Cresyl Violet**
Interaural +14.90 mm

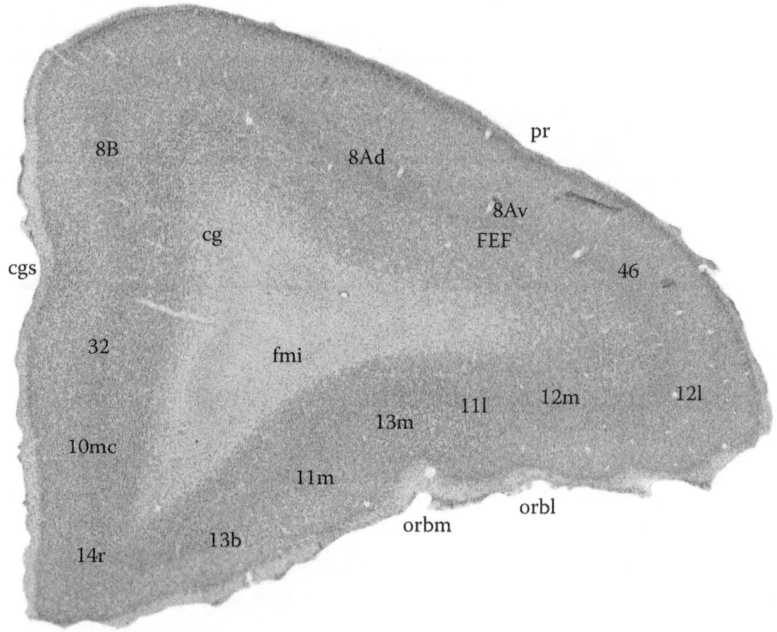

# Plate 9

## Interaural +14.90 mm

| | |
|---|---|
| 8Ad | area 8Ad Cx |
| 8Av | area 8Av Cx |
| 8B | area 8B Cx |
| 10mc | area 10mc Cx |
| 11l | area 11l Cx |
| 11m | area 11m Cx |
| 12l | area 12l Cx |
| 12m | area 12m Cx |
| 13b | area 13b Cx |
| 13m | area 13m Cx |
| 14r | area 14r Cx |
| 32 | area 32 Cx |
| 46 | area 45 Cx |
| cg | cingulum |
| cgs | cingulate sulc. |
| FEF | frontal eye field Cx |
| fmi | forceps minor |
| orbl | orbital sulc. lat. |
| orbm | orbital sulc. med. |
| pr | prefrontal sulcus |

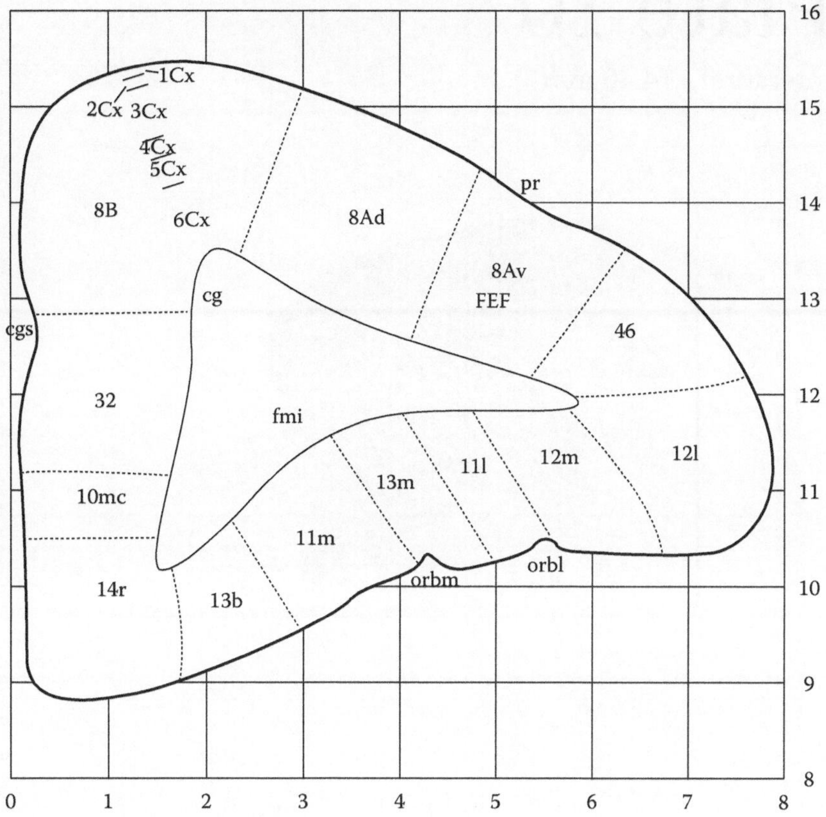

# Plate 10

Interaural +14.40 mm

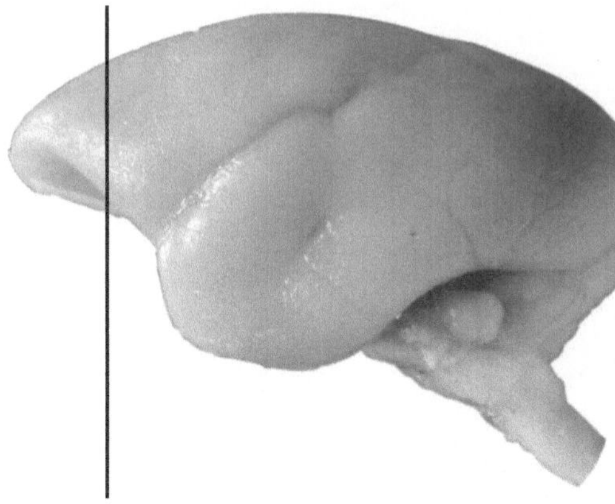

| | |
|---|---|
| 6Dr | area 6Dr Cx |
| 8Ad | area 8Ad Cx |
| 8Av | area 8Av Cx |
| 8B | area 8B Cx |
| 11l | area 11l Cx |
| 11m | area 11m Cx |
| 12l | area 12l Cx |
| 12m | area 12m Cx |
| 13a | area 13a Cx |
| 13m | area 13m Cx |
| 14c | area 14c Cx |
| 24a | area 24a Cx |
| 24b | area 24b Cx |
| 25 | area 25 Cx |
| 46 | area 45 Cx |
| cg | cingulum |
| cgs | cingulate sulc. |
| FEF | frontal eye field Cx |
| fmi | forceps minor |
| LV | lateral ventricle |
| orbl | orbital sulc. lat. |
| orbm | orbital sulc. med. |
| pr | prefrontal sulcus |

# Plate 10a

**Calbindin**

Interaural +14.70 mm

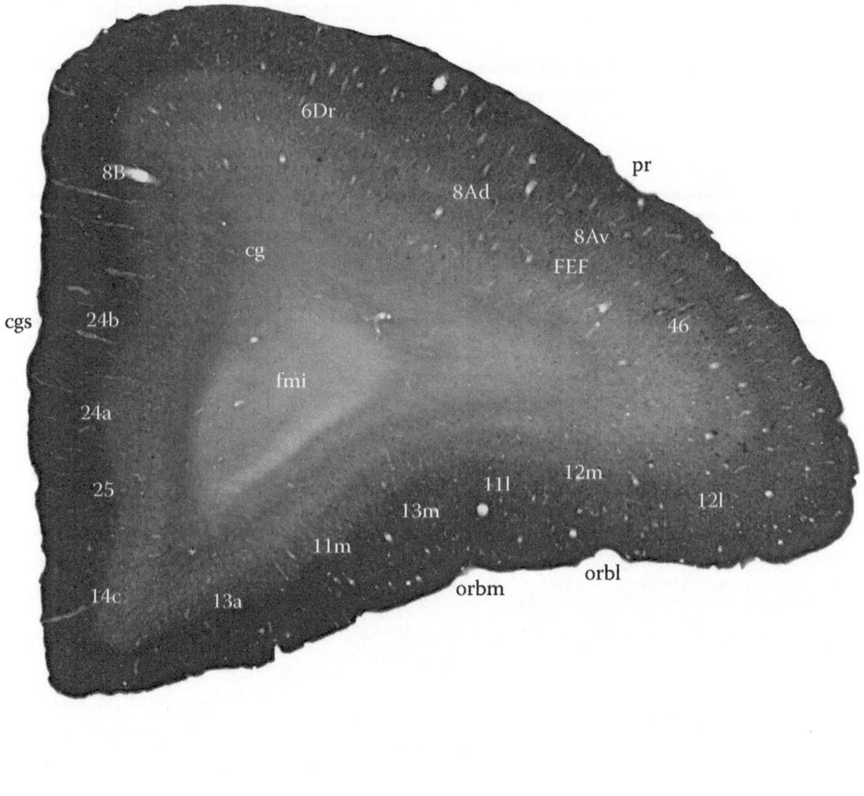

# Plate 10b

**Neurofilament**

Interaural +14.55 mm

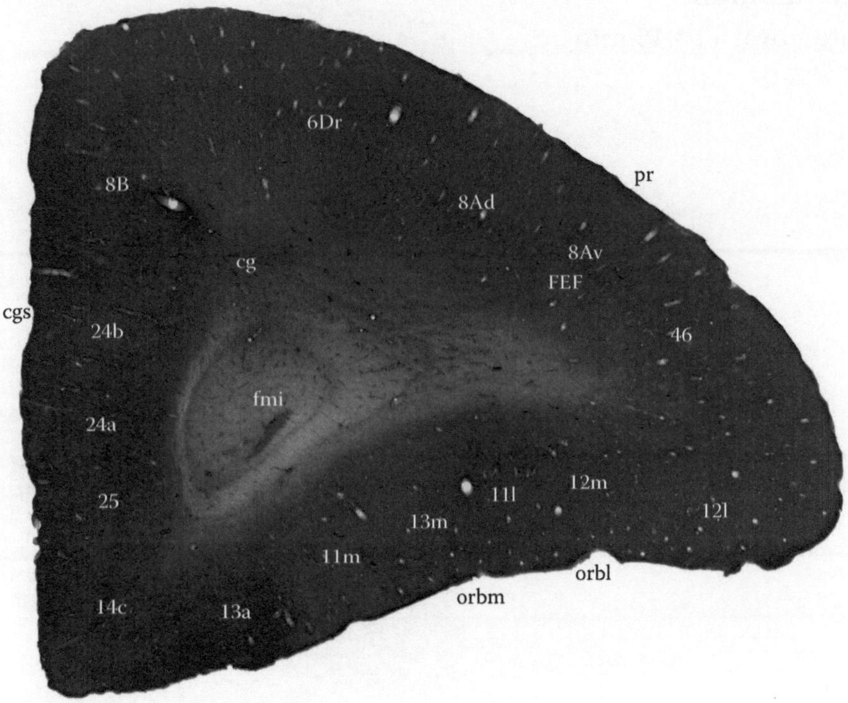


# Plate 10c

**Acetylcholinesterase**

Interaural +14.50 mm

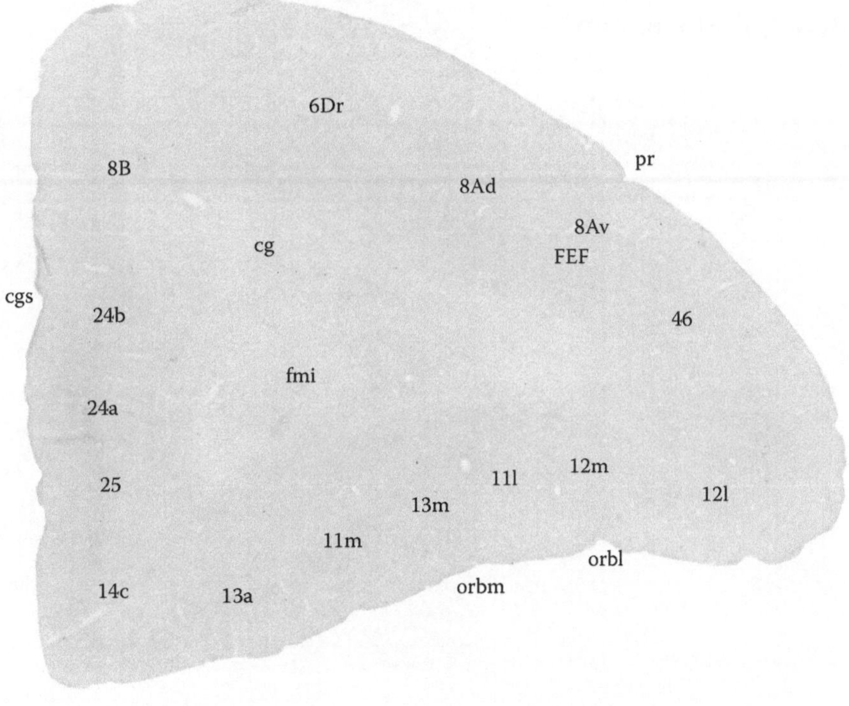

# Plate 10d

**Parvalbumin**
Interaural +14.45 mm

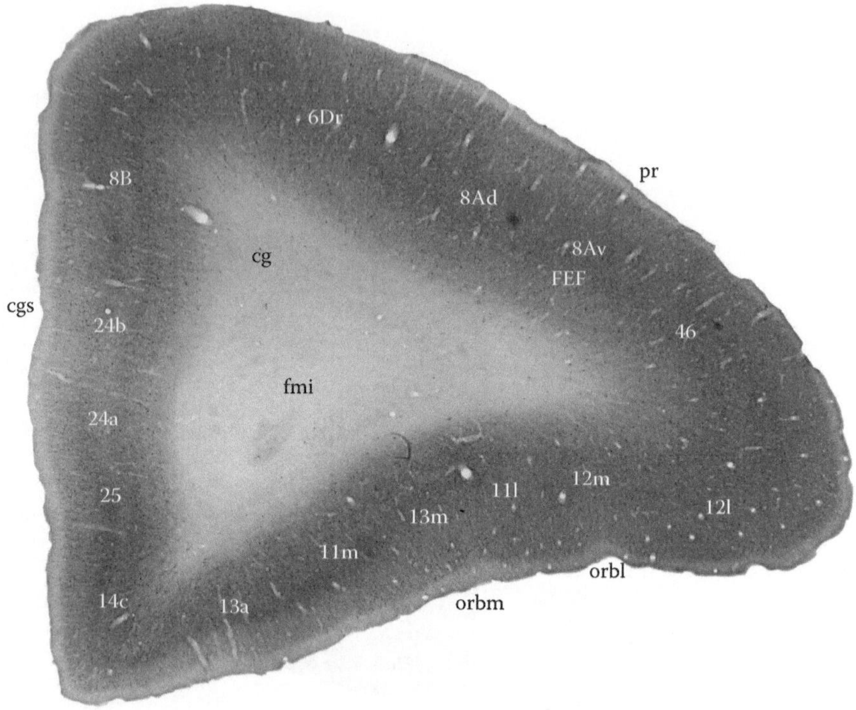


# Plate 10e

**Cresyl Violet**
Interaural +14.40 mm

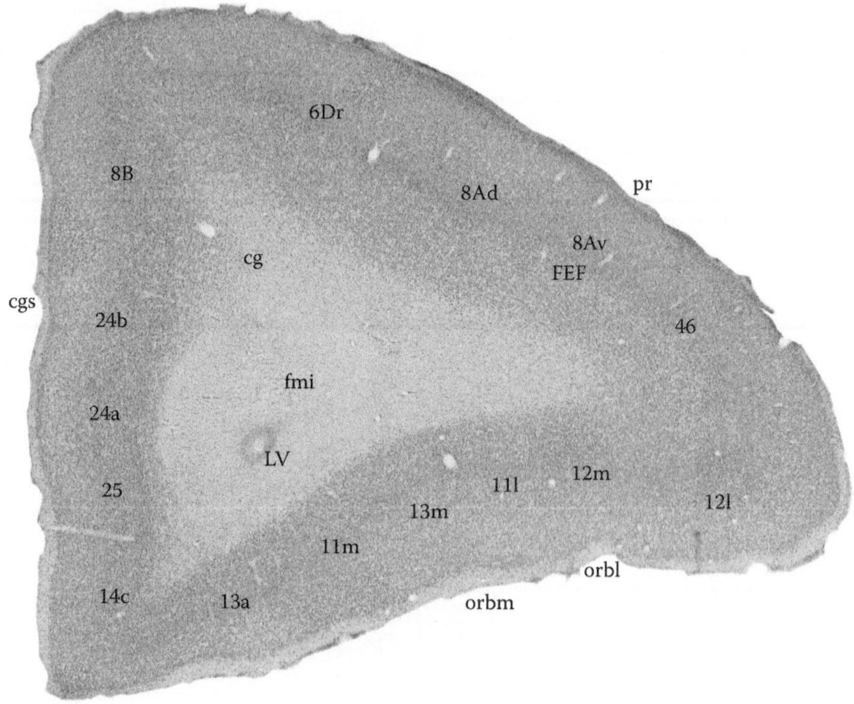

# Plate 10

## Interaural +14.40 mm

| | |
|---|---|
| 6Dr | area 6Dr Cx |
| 8Ad | area 8Ad Cx |
| 8Av | area 8Av Cx |
| 8B | area 8B Cx |
| 11l | area 11l Cx |
| 11m | area 11m Cx |
| 12l | area 12l Cx |
| 12m | area 12m Cx |
| 13a | area 13a Cx |
| 13m | area 13m Cx |
| 14c | area 14c Cx |
| 24a | area 24a Cx |
| 24b | area 24b Cx |
| 25 | area 25 Cx |
| 46 | area 45 Cx |
| cg | cingulum |
| cgs | cingulate sulc. |
| FEF | frontal eye field Cx |
| fmi | forceps minor |
| LV | lateral ventricle |
| orbl | orbital sulc. lat. |
| orbm | orbital sulc. med. |
| pr | prefrontal sulcus |

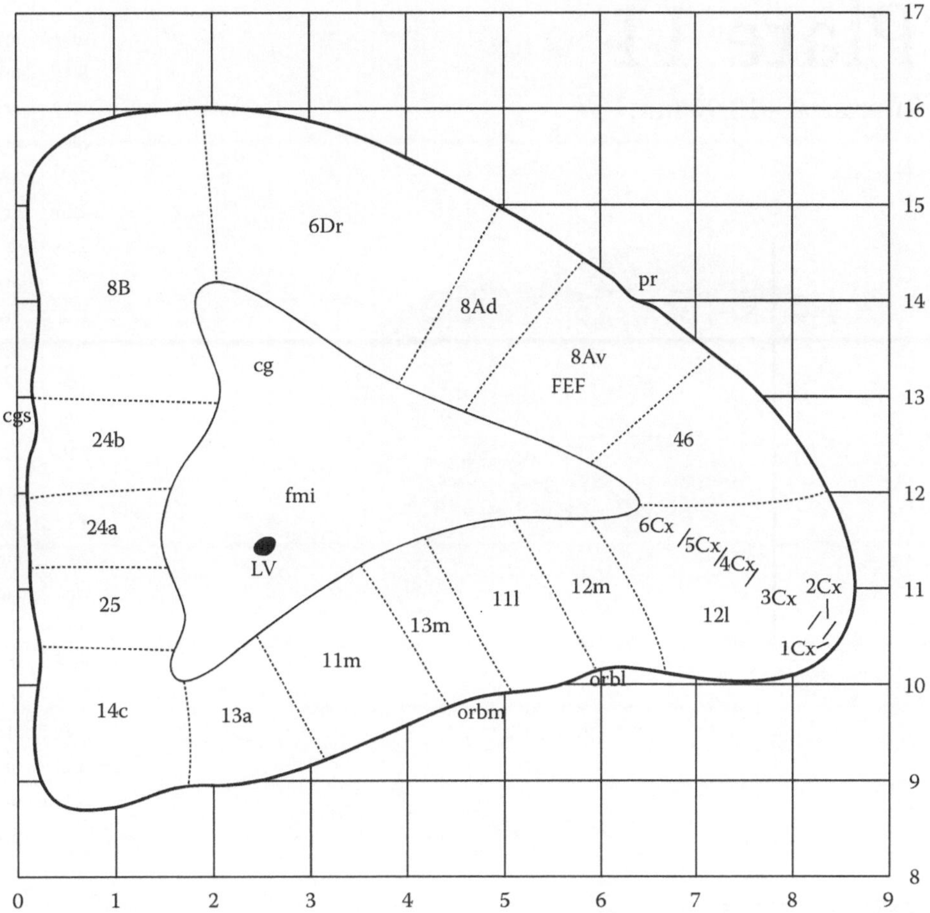


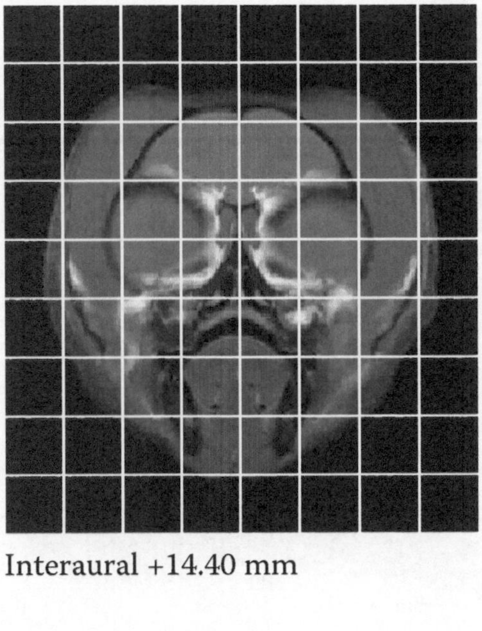

Interaural +14.40 mm

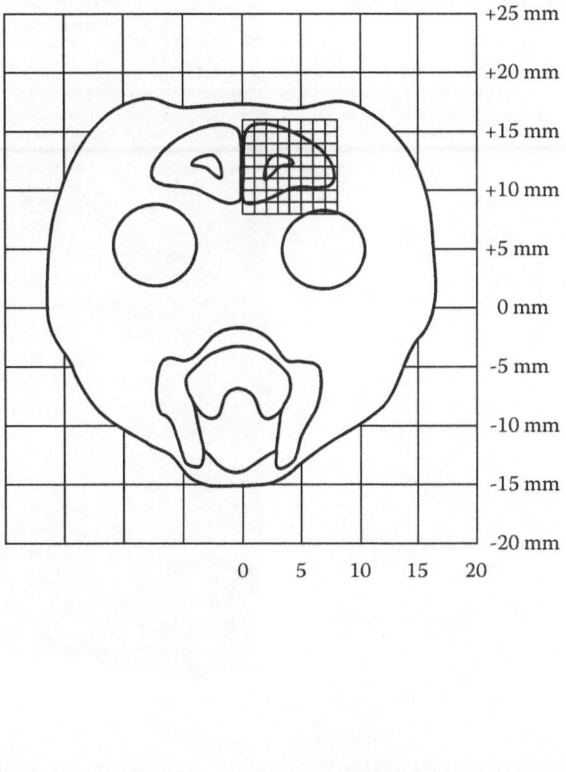

# Plate 11

Interaural +13.90 mm

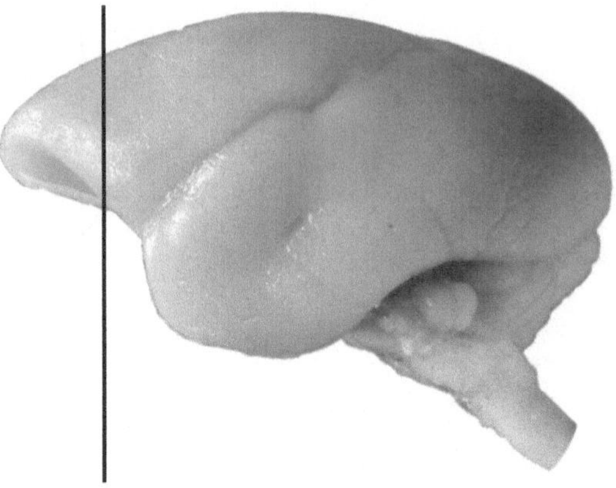

| | |
|---|---|
| 6Dr | area 6Dr Cx |
| 8Ad | area 8Ad Cx |
| 8Av | area 8Av Cx |
| 8B | area 8B Cx |
| 12l | area 12l Cx |
| 12m | area 12m Cx |
| 12o | area 12o Cx |
| 13a | area 13a Cx |
| 13l | area 13l Cx |
| 13m | area 13m Cx |
| 14c | area 14c Cx |
| 24a | area 24a Cx |
| 24b | area 24b Cx |
| 25 | area 25 Cx |
| 45 | area 45 Cx |
| cg | cingulum |
| cgs | cingulate sulc. |
| dcw | deep cort. white |
| FEF | frontal eye field Cx |
| fmi | forceps minor |
| LV | lateral ventricle |
| orbl | orbital sulc. lat. |
| orbm | orbital sulc. med. |
| pr | prefrontal sulcus |

# Plate 11a

**Calbindin**

Interaural +14.20 mm

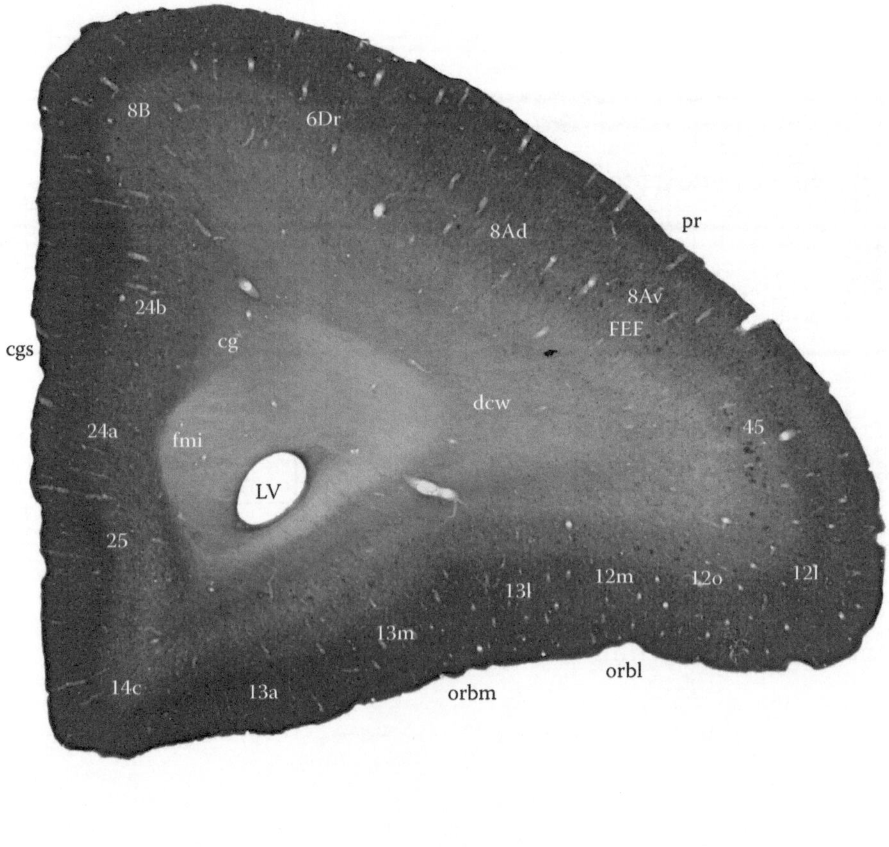

# Plate 11b

**Neurofilament**
Interaural +14.05 mm

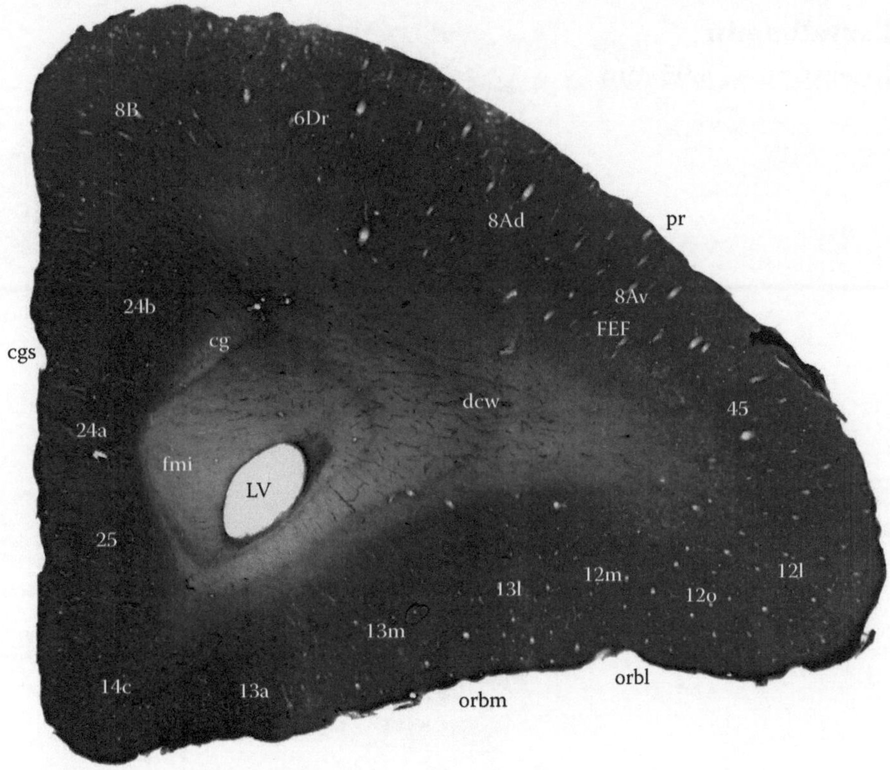


# Plate 11c

**Acetylcholinesterase**
Interaural +14.00 mm

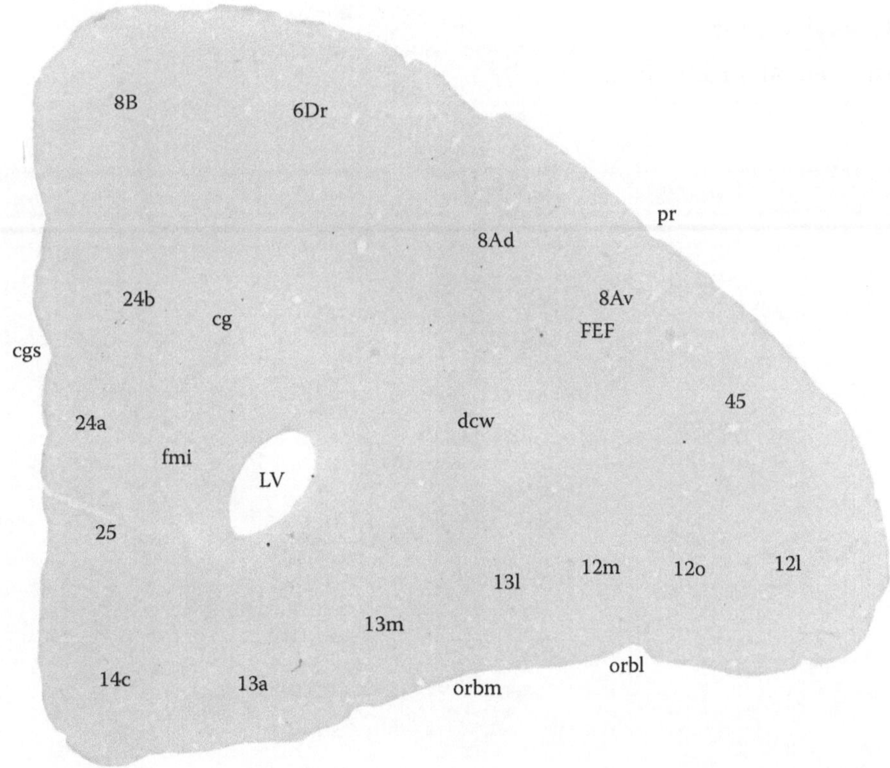

# Plate 11d

**Parvalbumin**
Interaural +13.95 mm

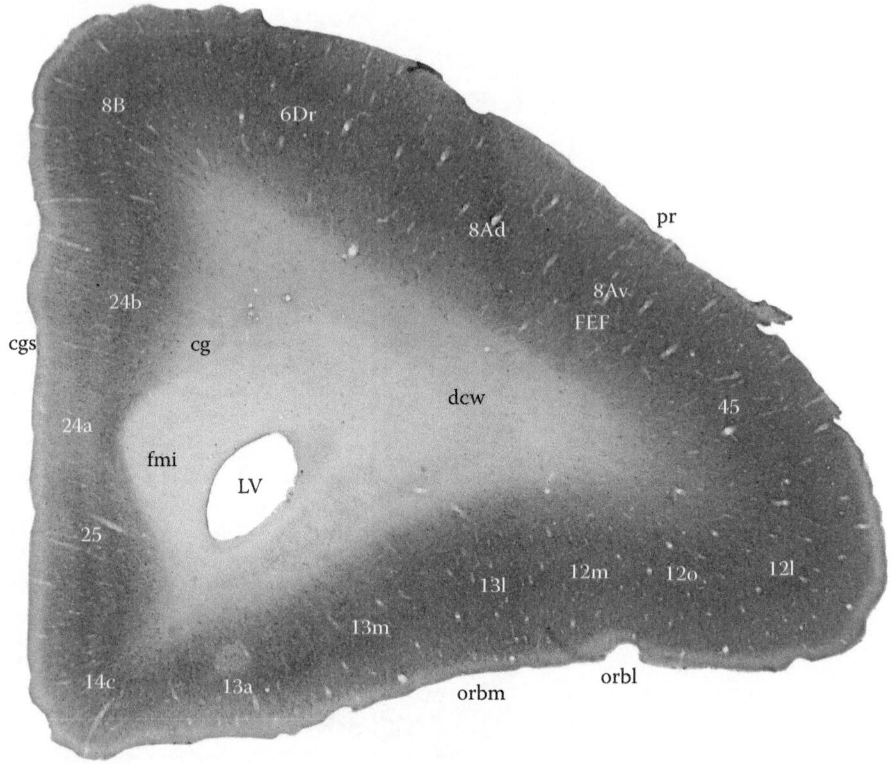


# Plate 11e

**Cresyl Violet**
Interaural +13.90 mm

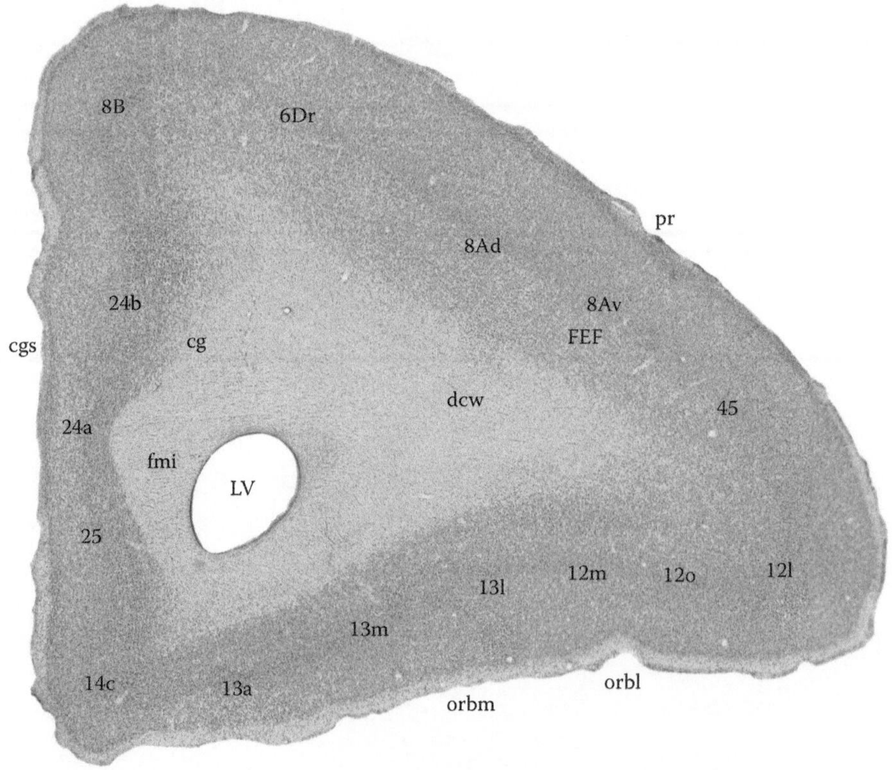

# Plate 11

## Interaural +13.90 mm

| | |
|---|---|
| 6Dr | area 6Dr Cx |
| 8Ad | area 8Ad Cx |
| 8Av | area 8Av Cx |
| 8B | area 8B Cx |
| 12l | area 12l Cx |
| 12m | area 12m Cx |
| 12o | area 12o Cx |
| 13a | area 13a Cx |
| 13l | area 13l Cx |
| 13m | area 13m Cx |
| 14c | area 14c Cx |
| 24a | area 24a Cx |
| 24b | area 24b Cx |
| 25 | area 25 Cx |
| 45 | area 45 Cx |
| cg | cingulum |
| cgs | cingulate sulc. |
| dcw | deep cort. white |
| FEF | frontal eye field Cx |
| fmi | forceps minor |
| LV | lateral ventricle |
| orbl | orbital sulc. lat. |
| orbm | orbital sulc. med. |
| pr | prefrontal sulcus |

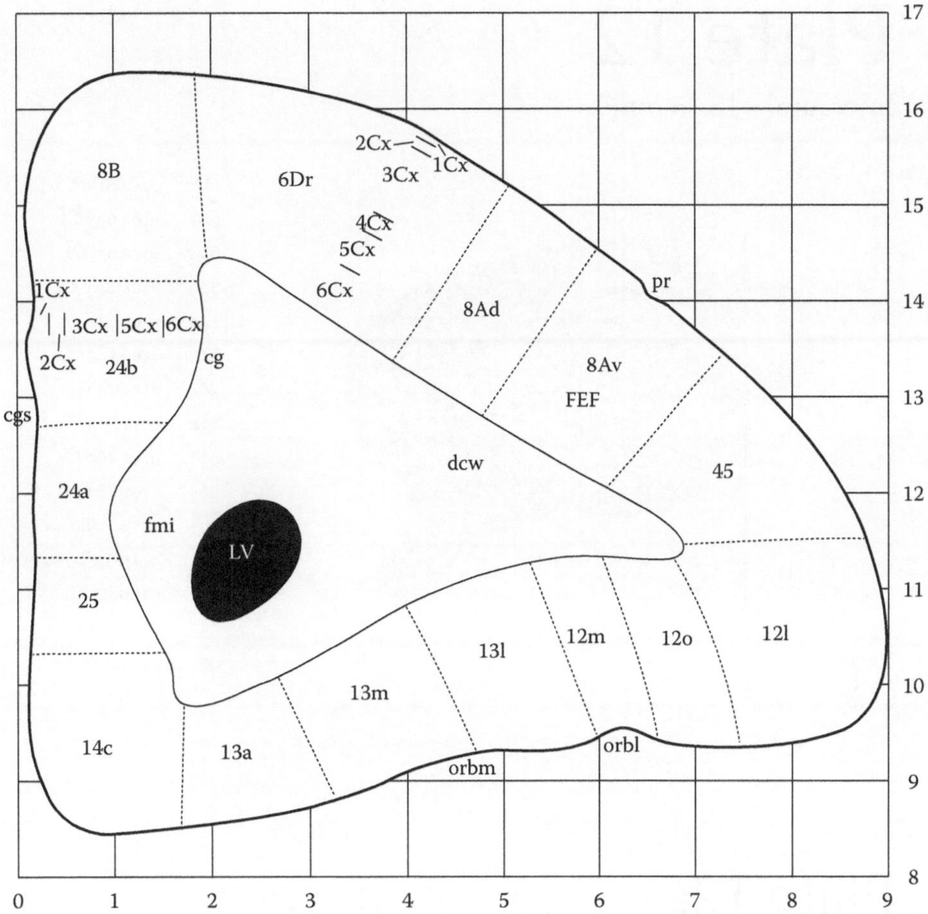

# Plate 12

Interaural +13.35 mm

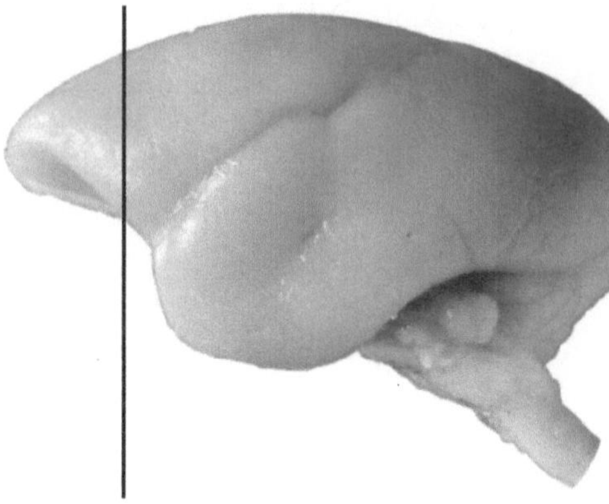

| | | | |
|---|---|---|---|
| 6Dc | area 6Dc Cx | dcw | deep cort. white |
| 6M | area 6M Cx | ec | external capsule |
| 6V | area 6V Cx | FEF | frontal eye field Cx |
| 8Av | area 8Av Cx | fmi | forceps minor |
| 12o | area 12o Cx | Gu | gustatory Cx |
| 13m | area 13m Cx | IG | intern. caps. |
| 24 | area 24 Cx | Insrl | insular Cx, rost. lat. |
| 24a | area 24a Cx | lo | lat. olf. tract |
| 24b | area 24b Cx | LV | lateral ventricle |
| 25 | area 25 Cx | orbl | orbital sulc. lat. |
| cc | corp. callosum | orbm | orbital sulc. med. |
| ccr | corp. call. rostrum | Pir1 | piriform Cx 1 |
| Cd | caudate nu. | Pir2 | piriform Cx 2 |
| cg | cingulum | Pir3 | piriform Cx 3 |
| cgs | cingulate sulc. | PrCO | precent. operc. Cx |

# Plate 12a

**Calbindin**

Interaural +13.65 mm

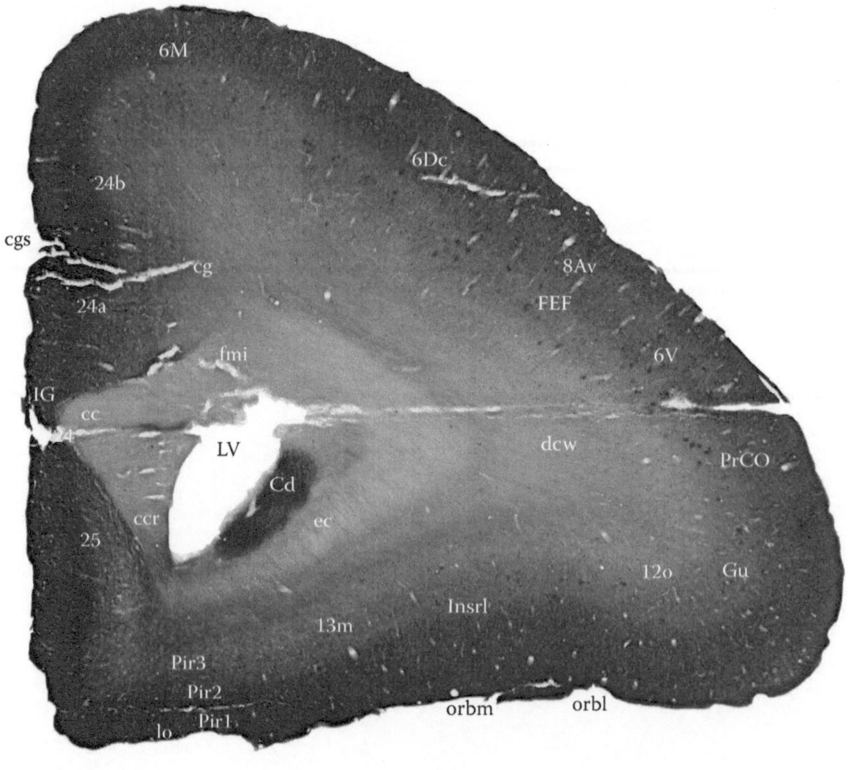

# Plate 12b

**Neurofilament**
Interaural +13.50 mm

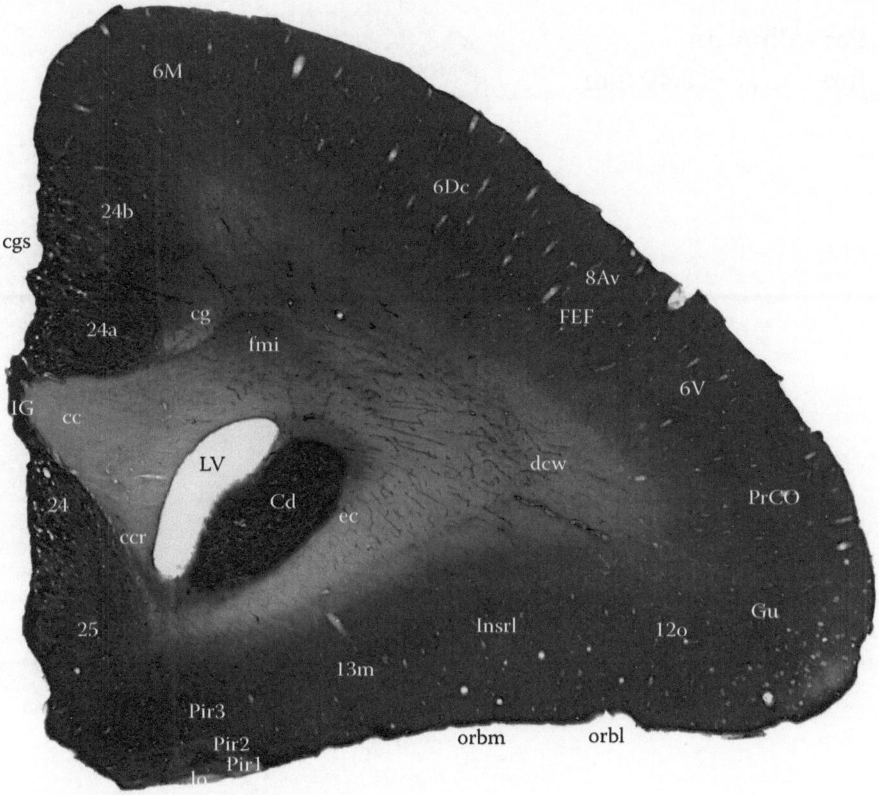


# Plate 12c

**Acetylcholinesterase**
Interaural +13.45 mm

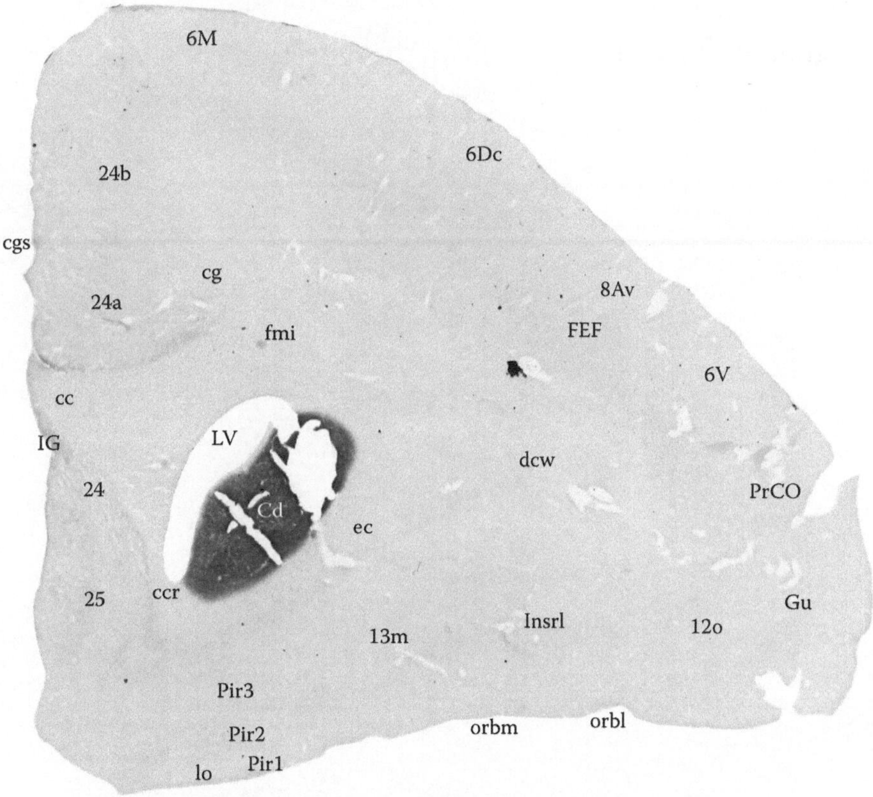

# Plate 12d

**Parvalbumin**
Interaural +13.40 mm

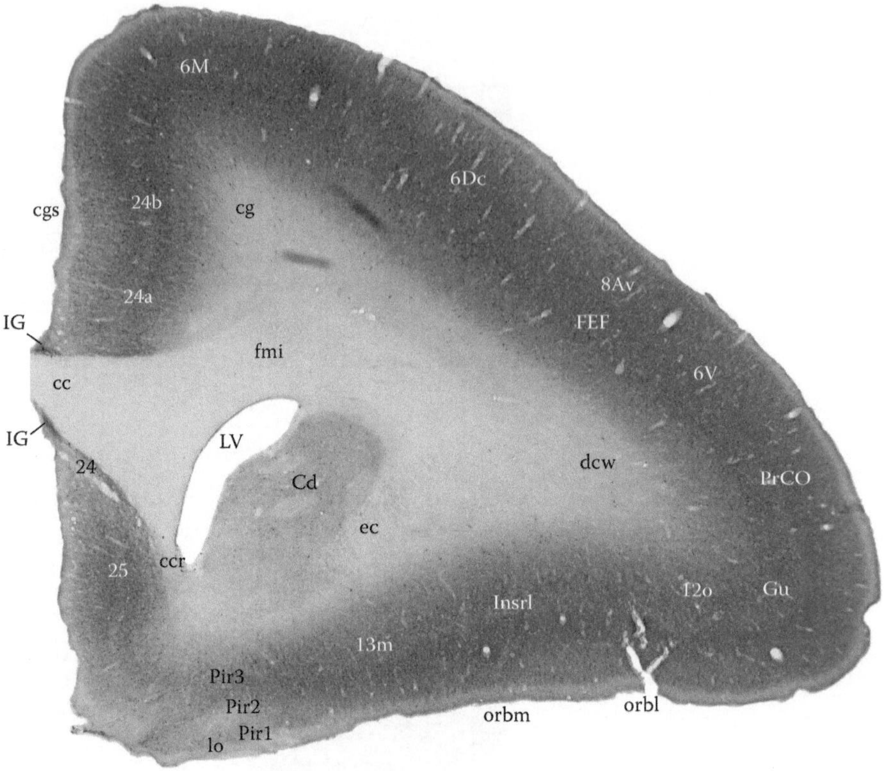


# Plate 12e

**Cresyl Violet**
Interaural +13.35 mm

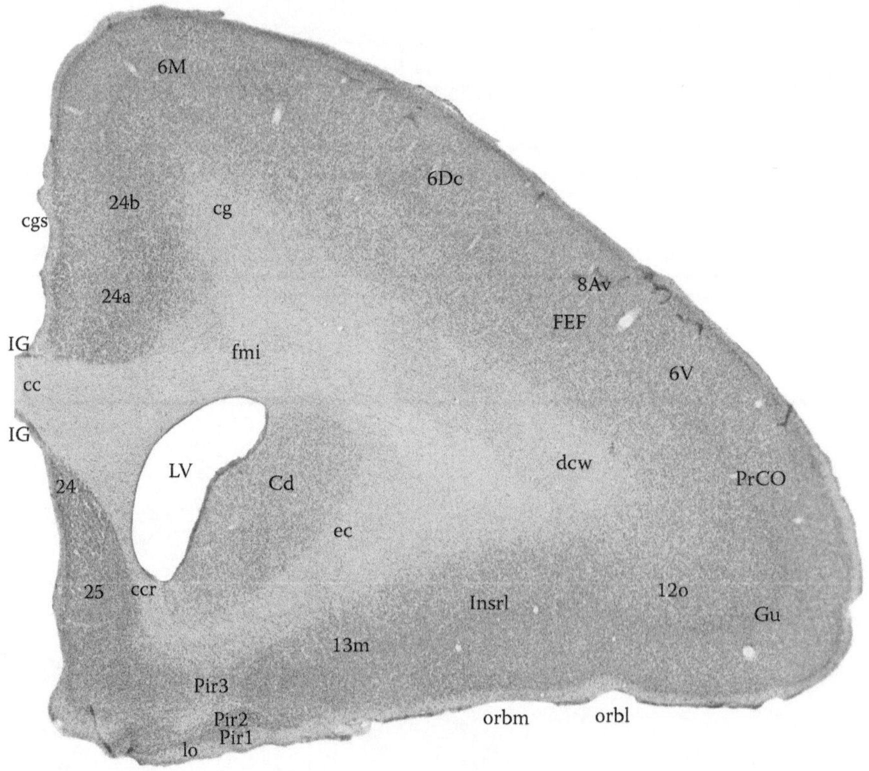

# Plate 12

6Dc area 6Dc Cx
6M area 6M Cx
6V area 6V Cx
8Av area 8Av Cx
12o area 12o Cx
13m area 13m Cx
24 area 24 Cx
24a area 24a Cx
24b area 24b Cx
25 area 25 Cx
cc corp. callosum
ccr corp. call. rostrum
Cd caudate nu.
cg cingulum
cgs cingulate sulc.
dcw deep cort. white
ec external capsule
FEF frontal eye field Cx
fmi forceps minor
Gu gustatory Cx
IG intern. caps.
Insrl insular Cx, rost. lat.
lo lat. olf. tract
LV lateral ventricle
orbl orbital sulc. lat.
orbm orbital sulc. med.
Pir1 piriform Cx 1
Pir2 piriform Cx 2
Pir3 piriform Cx 3
PrCO precent. operc. Cx

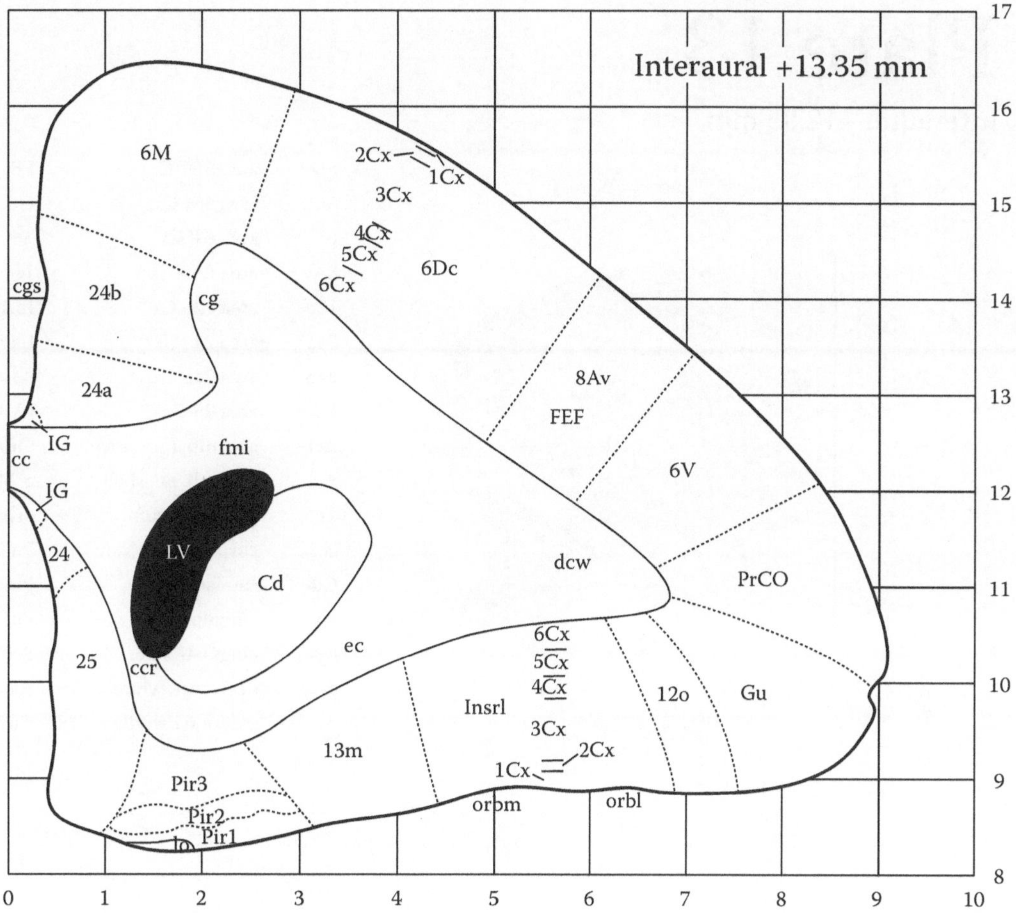


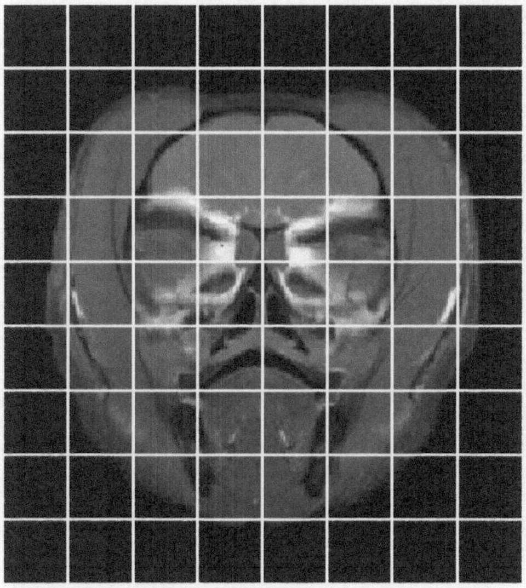

Interaural +13.45 mm

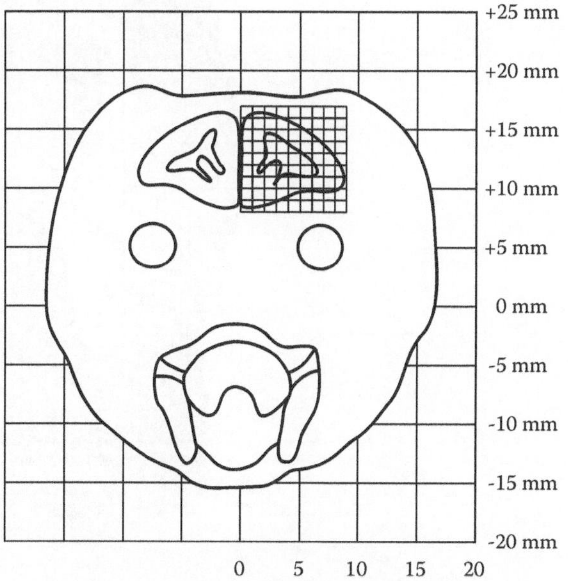

# Plate 13

Interaural +12.85 mm

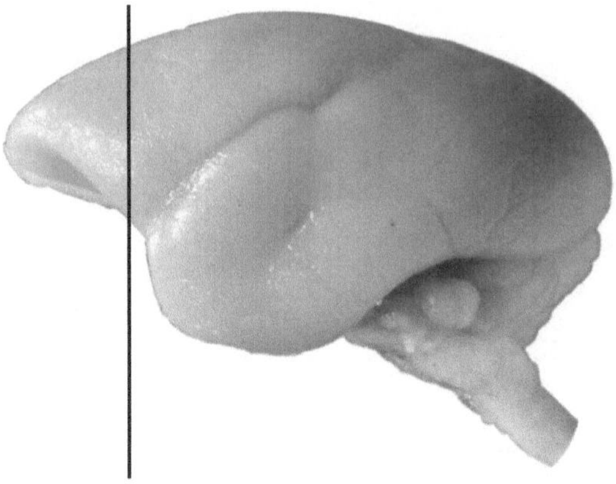

| | | | |
|---|---|---|---|
| 6Dc | area 6Dc Cx | FEF | frontal eye field Cx |
| 6M | area 6M Cx | Gu | gustatory Cx |
| 6V | area 6V Cx | ic | intern. caps. |
| 8Av | area 8Av Cx | IG | induseum gris. |
| 12o | area 12o Cx | Insrl | insular Cx, rost. lat. |
| 24a | area 24a Cx | Insrm | insular Cx, rost. med. |
| 24b | area 24b Cx | lo | lat. olf. tract |
| 25 | area 25 Cx | LV | lateral ventricle |
| AcbC | accumb. nu. core | OPAl | orbital periallocortex |
| AcbSh | accumb. nu. shell | orbl | orbital sulc. lat. |
| cc | corp. callosum | orbm | orbital sulc. med. |
| ccr | corp. call. rostrum | Pir1 | piriform Cx 1 |
| Cd | caudate nu. | Pir2 | piriform Cx 2 |
| cg | cingulum | Pir3 | piriform Cx 3 |
| cgs | cingulate sulc. | PrCO | precent. operc. Cx |
| dcw | deep cort. white | Pu | putamen |
| ec | external capsule | Tu | olf. tubercle |

## Plate 13a

**Calbindin**

Interaural +13.15 mm

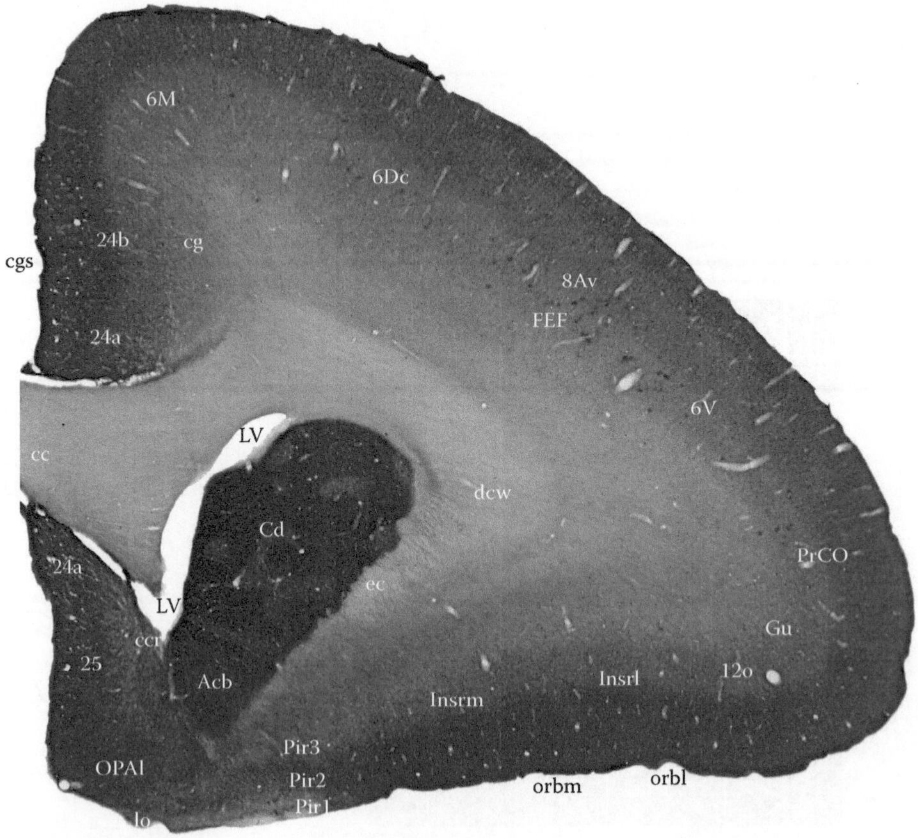

# Plate 13b

**Neurofilament**
Interaural +13.00 mm

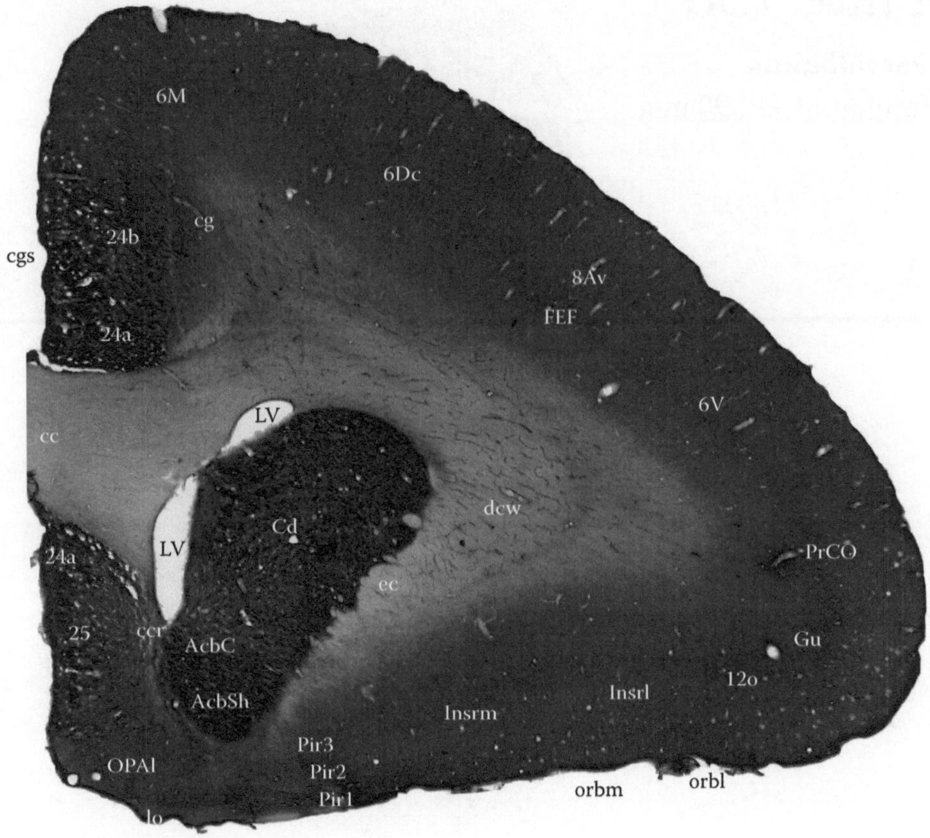


# Plate 13c

**Acetylcholinesterase**
Interaural +12.95 mm

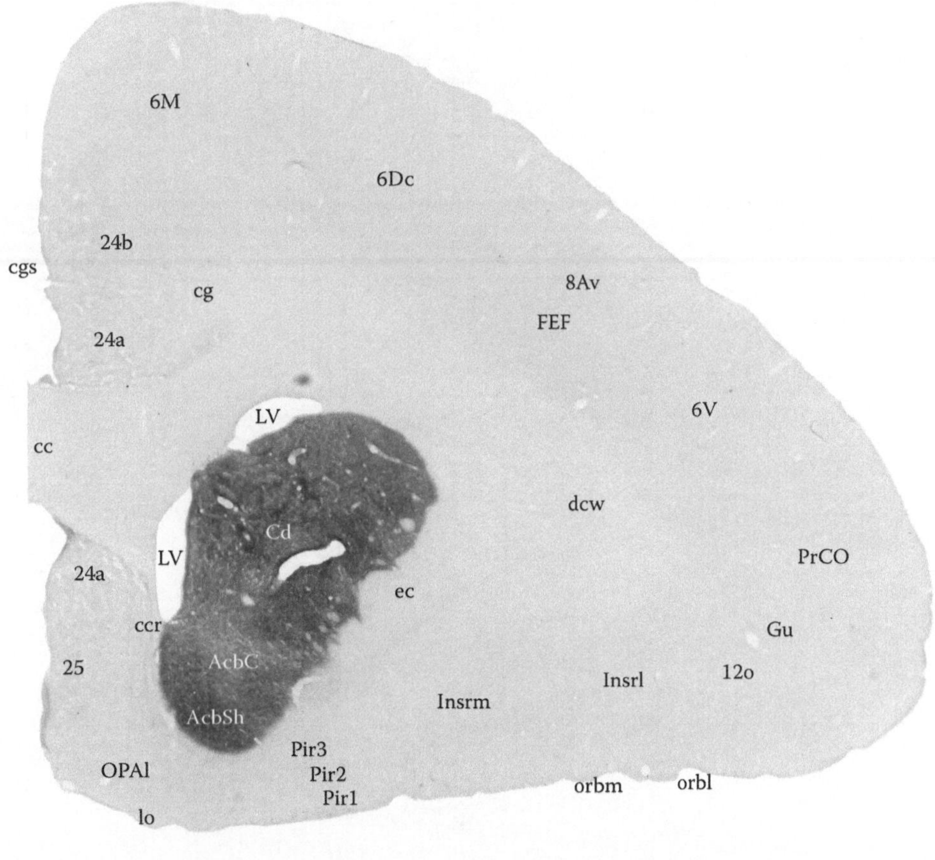

# Plate 13d

**Parvalbumin**
Interaural +12.90 mm

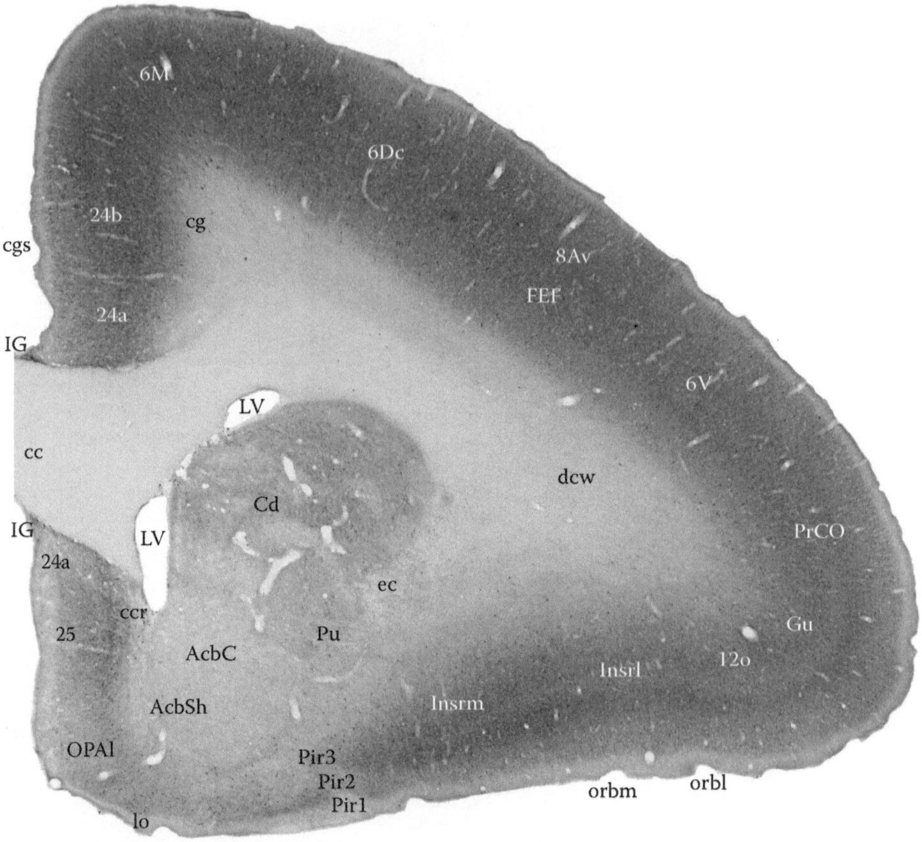


# Plate 13e

**Cresyl Violet**
Interaural +12.85 mm

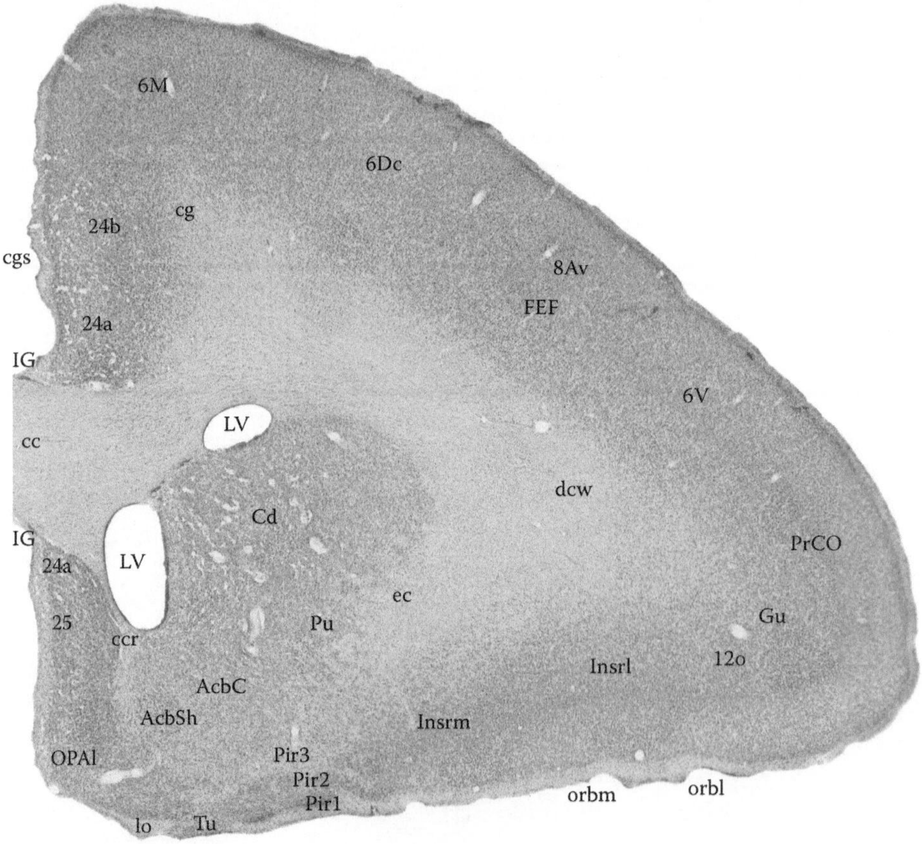

# Plate 13

| | |
|---|---|
| 6Dc | area 6Dc Cx |
| 6M | area 6M Cx |
| 6V | area 6V Cx |
| 8Av | area 8Av Cx |
| 12o | area 12o Cx |
| 24a | area 24a Cx |
| 24b | area 24b Cx |
| 25 | area 25 Cx |
| AcbC | accumb. nu. core |
| AcbSh | accumb. nu. shell |
| cc | corp. callosum |
| ccr | corp. call. rostrum |
| Cd | caudate nu. |
| cg | cingulum |
| cgs | cingulate sulc. |
| dcw | deep cort. white |
| ec | external capsule |
| FEF | frontal eye field Cx |
| Gu | gustatory Cx |
| ic | intern. caps. |
| IG | induseum gris. |
| Insrl | insular Cx, rost. lat. |
| Insrm | insular Cx, rost. med. |
| lo | lat. olf. tract |
| LV | lateral ventricle |
| OPAl | orbital periallocortex |
| orbl | orbital sulc. lat. |
| orbm | orbital sulc. med. |
| Pir1 | piriform Cx 1 |
| Pir2 | piriform Cx 2 |
| Pir3 | piriform Cx 3 |
| PrCO | precent. operc. Cx |
| Pu | putamen |
| Tu | olf. tubercle |

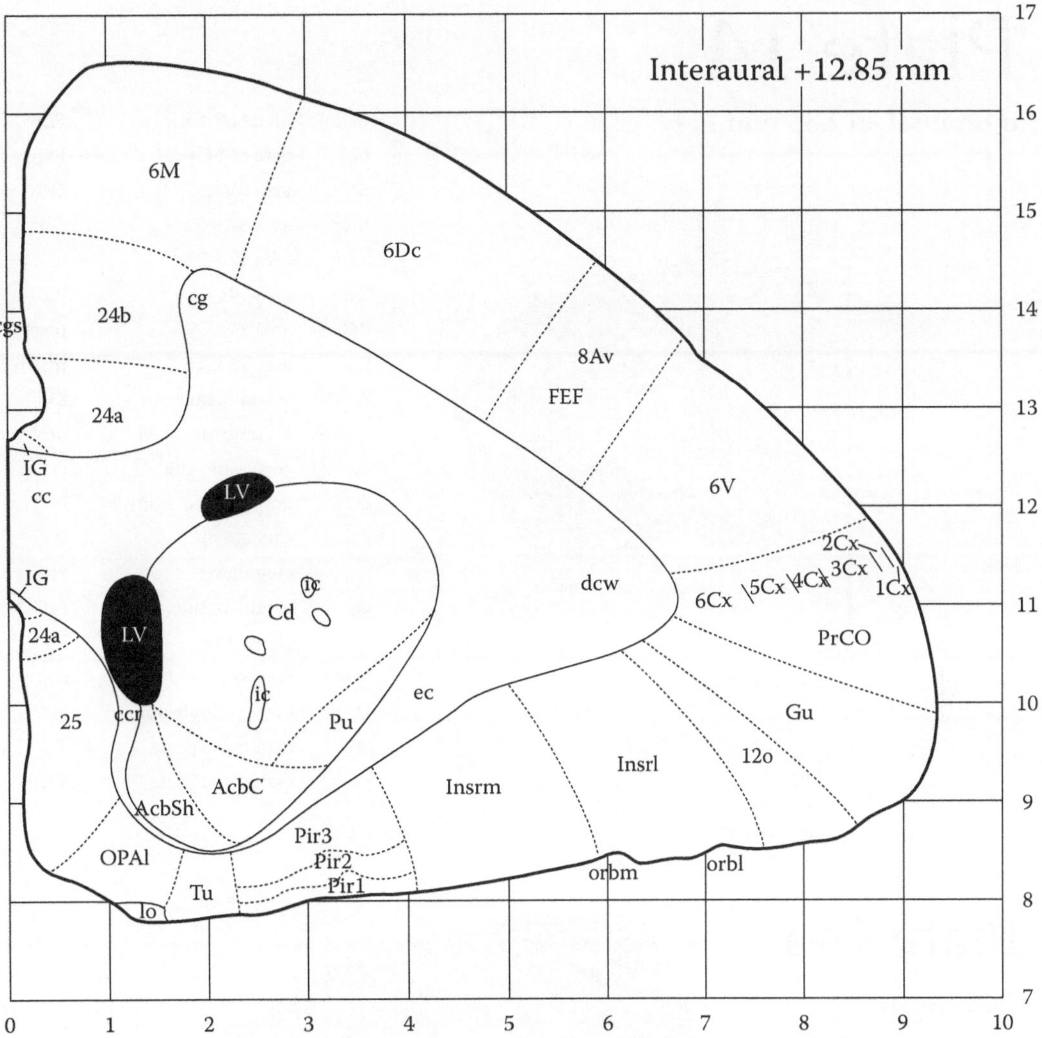

# Plate 14

Interaural +12.35 mm

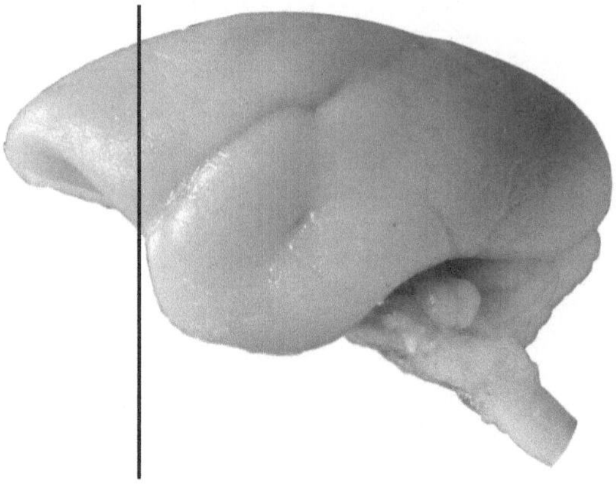

| | | | |
|---|---|---|---|
| 6Dc | area 6Dc Cx | FEF | frontal eye field Cx |
| 6M | area 6M Cx | FStr | fundus striatum |
| 6V | area 6V Cx | GCC | gran. cell clust. |
| 8Av | area 8Av Cx | Gu | gustatory Cx |
| 12o | area 12o Cx | ic | intern. caps. |
| 24a | area 24a Cx | IG | induseum gris. |
| 24b | area 24b Cx | Insrl | insular Cx, rost. lat. |
| 25 | area 25 Cx | Insrm | insular Cx, rost. med. |
| AcbC | accumb. nu. core | LV | lateral ventricle |
| AcbSh | accumb. nu. shell | orbl | orbital sulc. lat. |
| cc | corp. callosum | orbm | orbital sulc. med. |
| ccr | corp. call. rostrum | Pir1 | piriform Cx 1 |
| Cd | caudate nu. | Pir2 | piriform Cx 2 |
| cg | cingulum | Pir3 | piriform Cx 3 |
| cgs | cingulate sulc. | PrCO | precent. operc. Cx |
| Cl | claustrum | Pu | putamen |
| dcw | deep cort. white | rf | rhinal fiss. |
| DEn | dors. endopir. nu. | SI | subst. innominata |
| ec | external capsule | Tu | olf. tubercle |
| ex | extreme capsule | VDB | vert. limb diag. band. nu. |

# Plate 14a

**Calbindin**

Interaural +12.65 mm

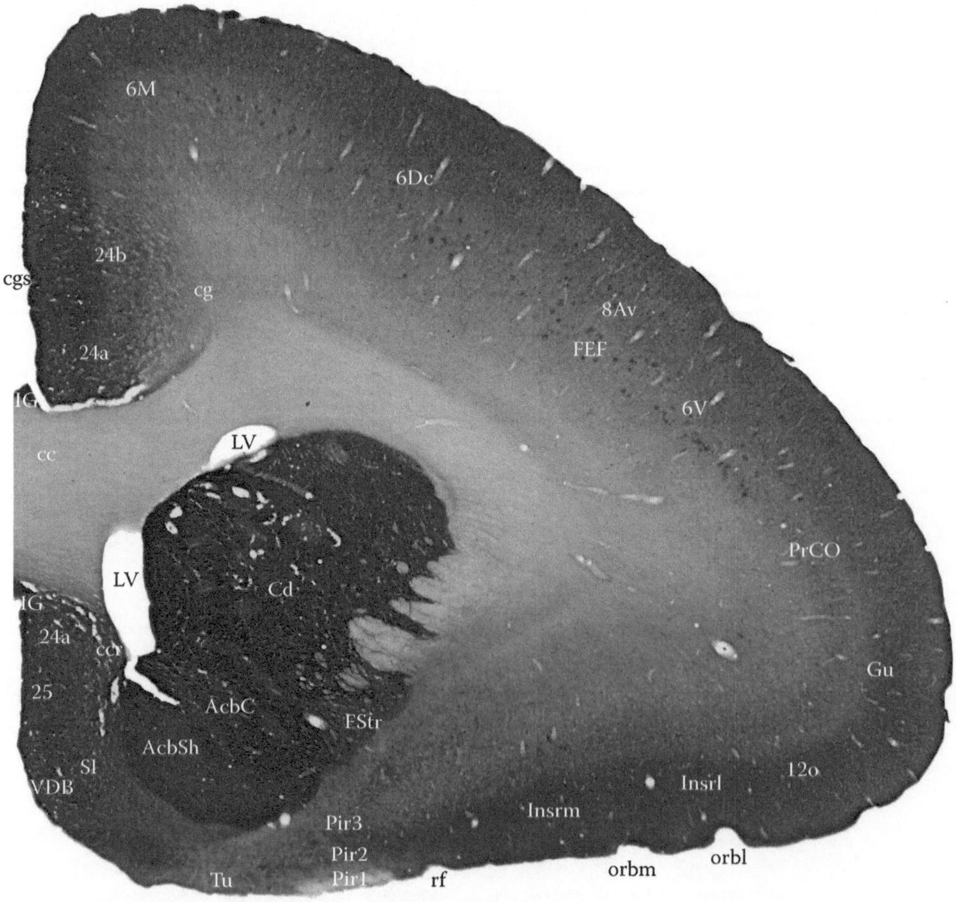

# Plate 14b

**Neurofilament**
Interaural +12.50 mm

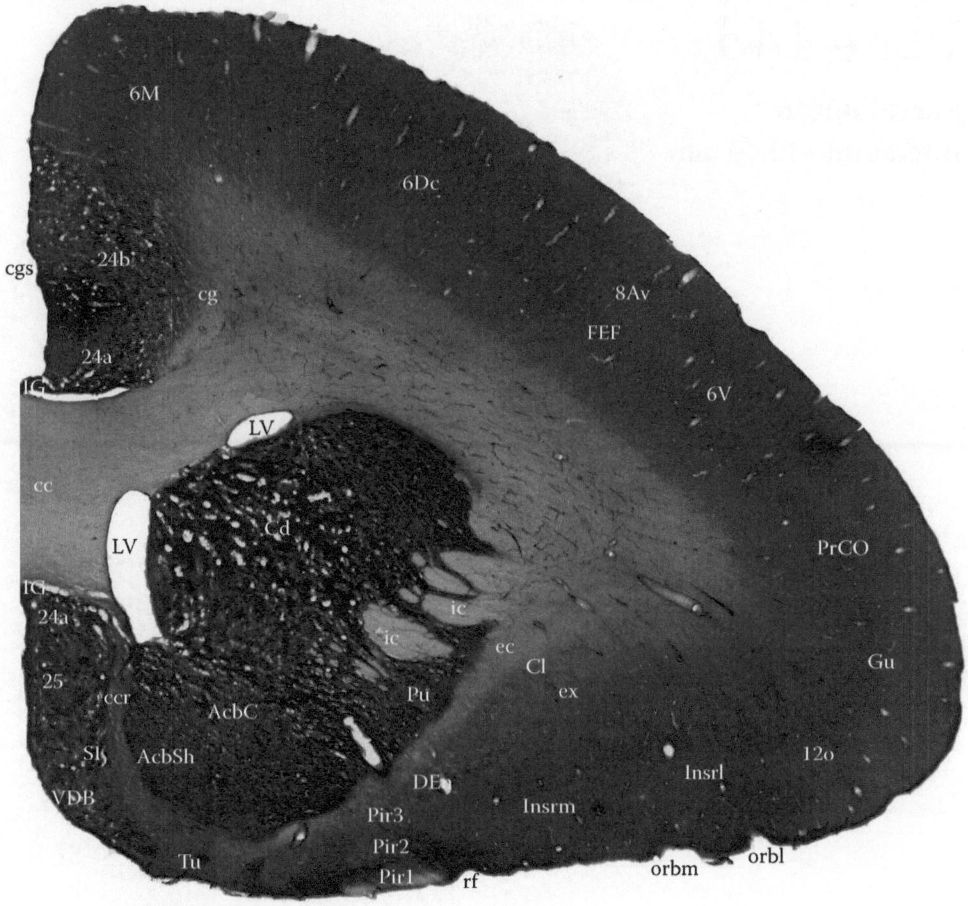


# Plate 14c

**Acetylcholinesterase**
Interaural +12.45 mm

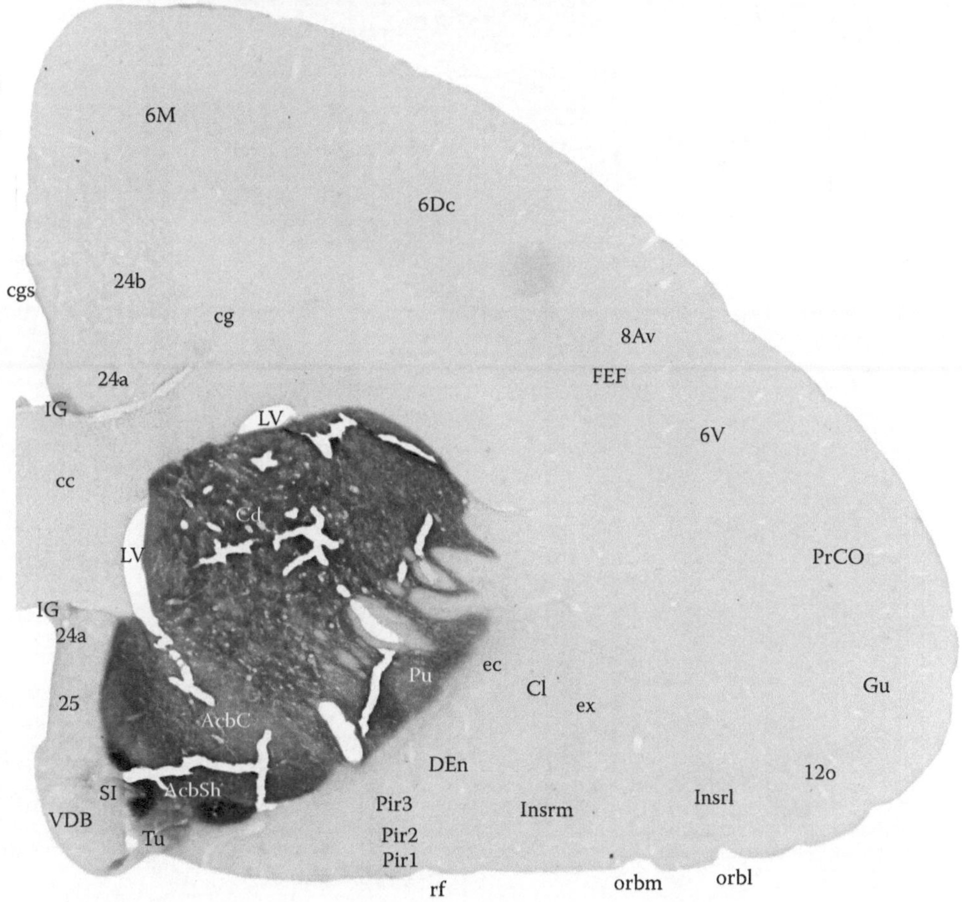

# Plate 14d

**Parvalbumin**
Interaural +12.40 mm

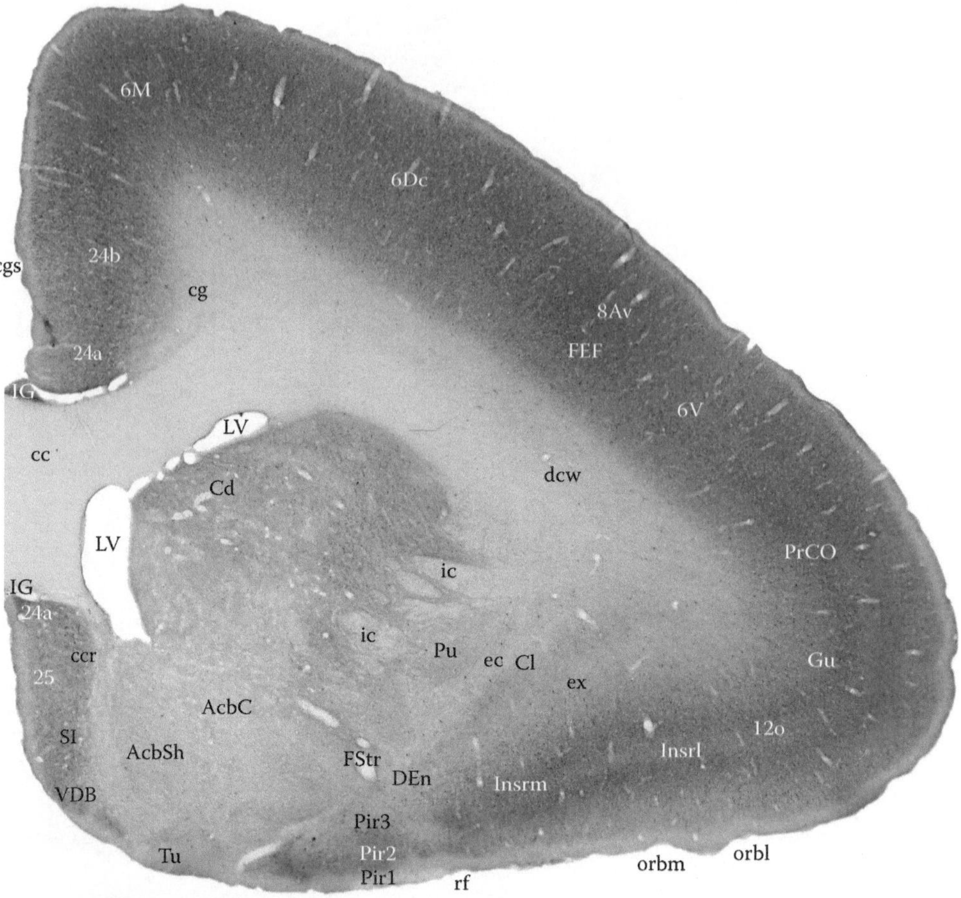


# Plate 14e

**Cresyl Violet**
Interaural +12.35 mm

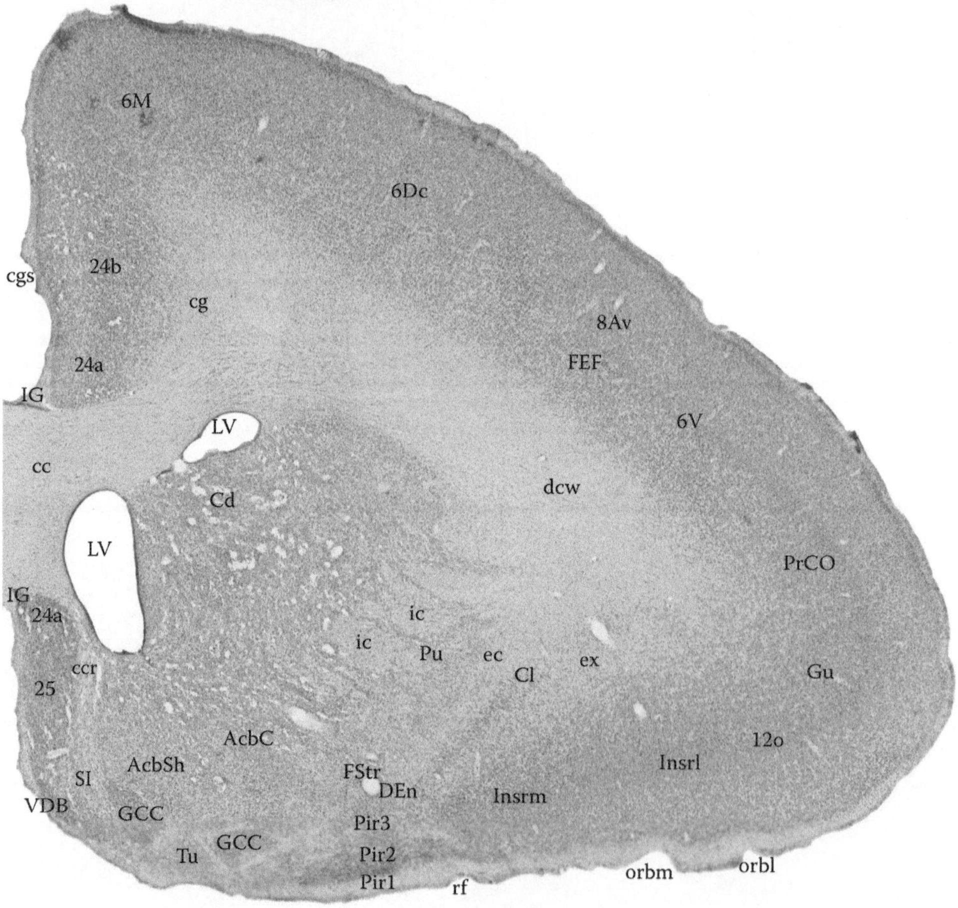

# Plate 14

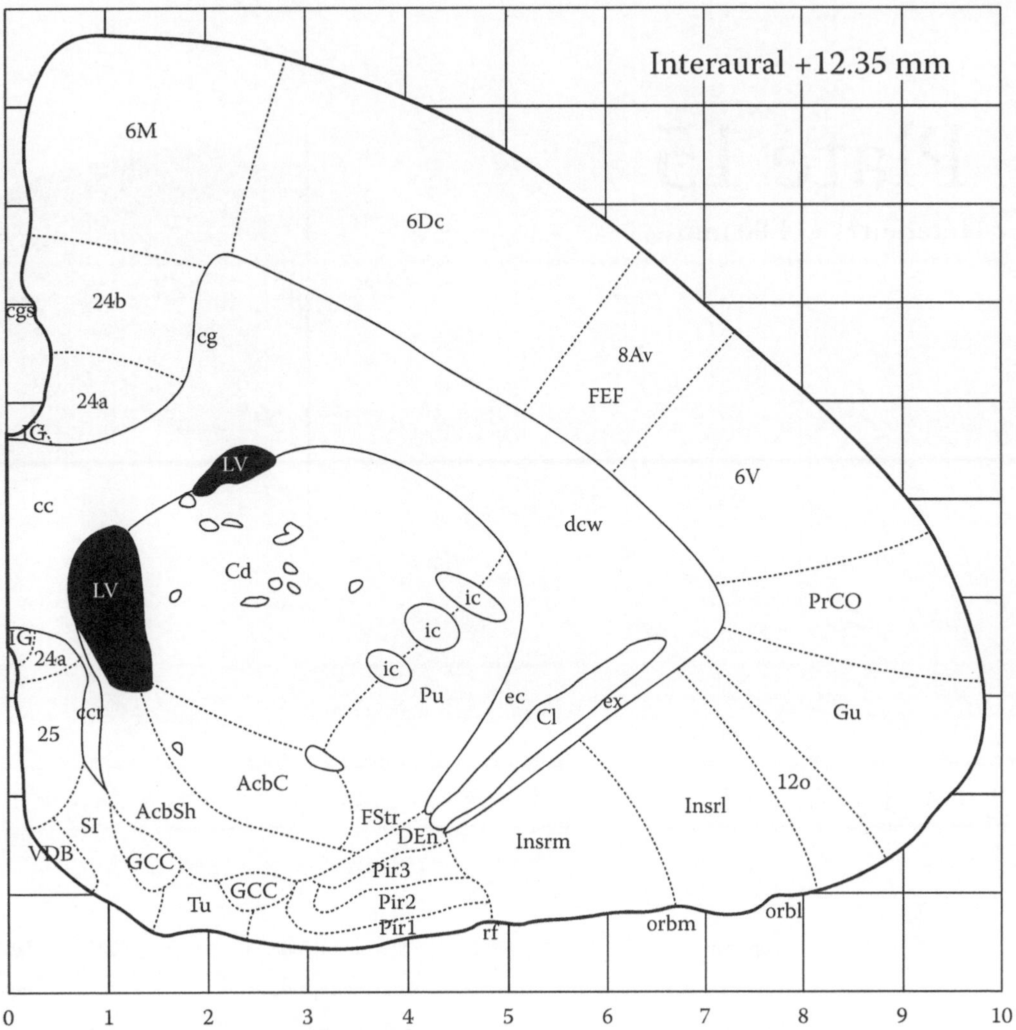


| | |
|---|---|
| 6Dc | area 6Dc Cx |
| 6M | area 6M Cx |
| 6V | area 6V Cx |
| 8Av | area 8Av Cx |
| 12o | area 12o Cx |
| 24a | area 24a Cx |
| 24b | area 24b Cx |
| 25 | area 25 Cx |
| AcbC | accumb. nu. core |
| AcbSh | accumb. nu. shell |
| cc | corp. callosum |
| ccr | corp. call. rostrum |
| Cd | caudate nu. |
| cg | cingulum |
| cgs | cingulate sulc. |
| Cl | claustrum |
| dcw | deep cort. white |
| DEn | dors. endopir. nu. |
| ec | external capsule |
| ex | extreme capsule |
| FEF | frontal eye field Cx |
| FStr | fundus striatum |
| GCC | gran. cell clust. |
| Gu | gustatory Cx |
| ic | intern. caps. |
| IG | induseum gris. |
| Insrl | insular Cx, rost. lat. |
| Insrm | insular Cx, rost. med. |
| LV | lateral ventricle |
| orbl | orbital sulc. lat. |
| orbm | orbital sulc. med. |
| Pir1 | piriform Cx 1 |
| Pir2 | piriform Cx 2 |
| Pir3 | piriform Cx 3 |
| PrCO | precent. operc. Cx |
| Pu | putamen |
| rf | rhinal fiss. |
| SI | subst. innominata |
| Tu | olf. tubercle |
| VDB | vert. limb diag. band. nu. |

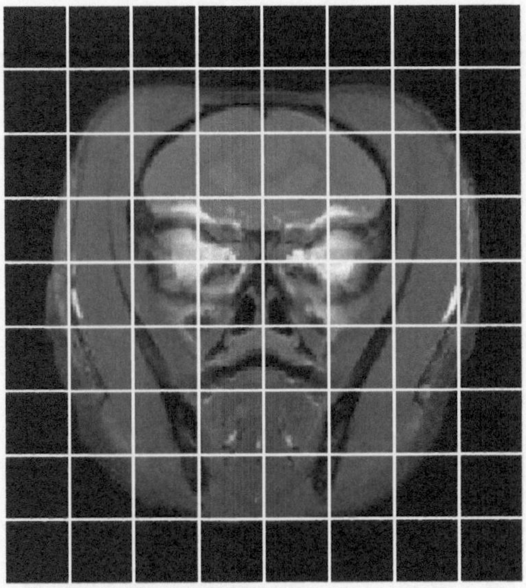

Interaural +12.50 mm

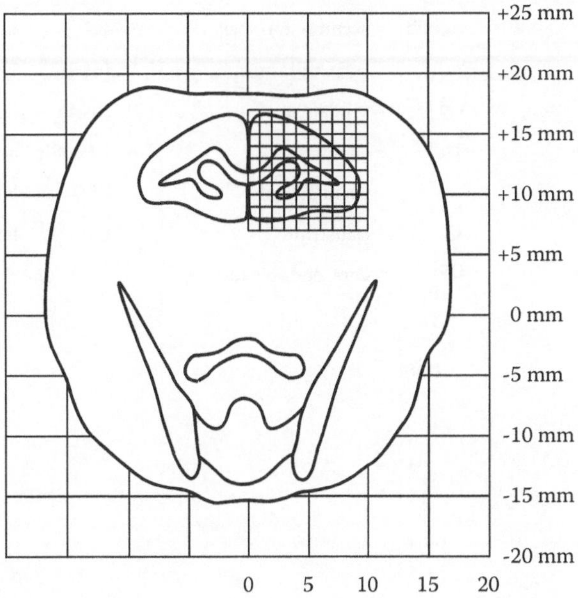

# Plate 15

Interaural +11.80 mm

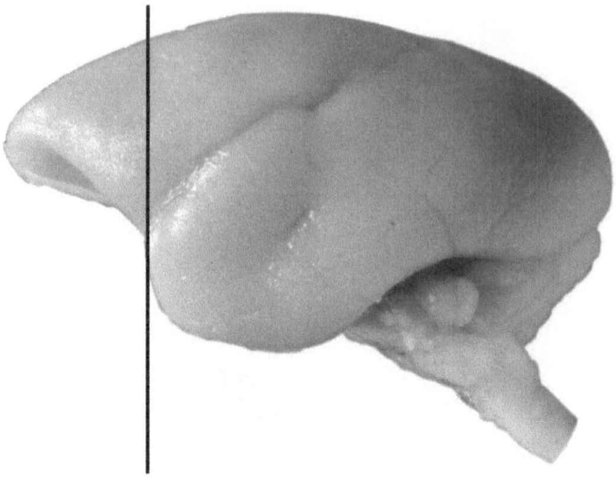

| | | | | | |
|---|---|---|---|---|---|
| 2n | optic nerve | ec | external capsule | LSV | lat. sept. nu., vent. |
| 6Dc | area 6Dc Cx | ex | extreme capsule | LV | lateral ventricle |
| 6M | area 6M Cx | FStr | fundus striatum | MS | med. sept. nu. |
| 6V | area 6V Cx | GCC | gran. cell clust. | orbl | orbital sulc. lat. |
| 24a | area 24a Cx | GCCM | gran. cell clust. magn. | Pir1 | piriform Cx 1 |
| 24b | area 24b Cx | Gu | gustatory Cx | Pir2 | piriform Cx 2 |
| aca | ant. comm. ant. | HDB | horiz. limb diag. | Pir3 | piriform Cx 3 |
| AcbC | accumb. nu. core | ic | intern. caps. | PrCO | precent. operc. Cx |
| AcbSh | accumb. nu. shell | IG | induseum gris. | Pu | putamen |
| cc | corp. callosum | Insrl | insular Cx, rost. lat. | rf | rhinal fiss. |
| Cd | caudate nu. | Insrm | insular Cx, rost. med. | SHi | septohippo. nu. |
| cg | cingulum | LAcbSh | accumb. shell lat. | SI | subst. innominata |
| cgs | cingulate sulc. | LSD | lat. sept. nu., dors. | Tu | olf. tubercle |
| Cl | claustrum | LSI | lat. sept. nu., int. | VDB | vert. limb diag. band. nu. |
| DEn | dors. endopir. nu. | | | | |

# Plate 15a

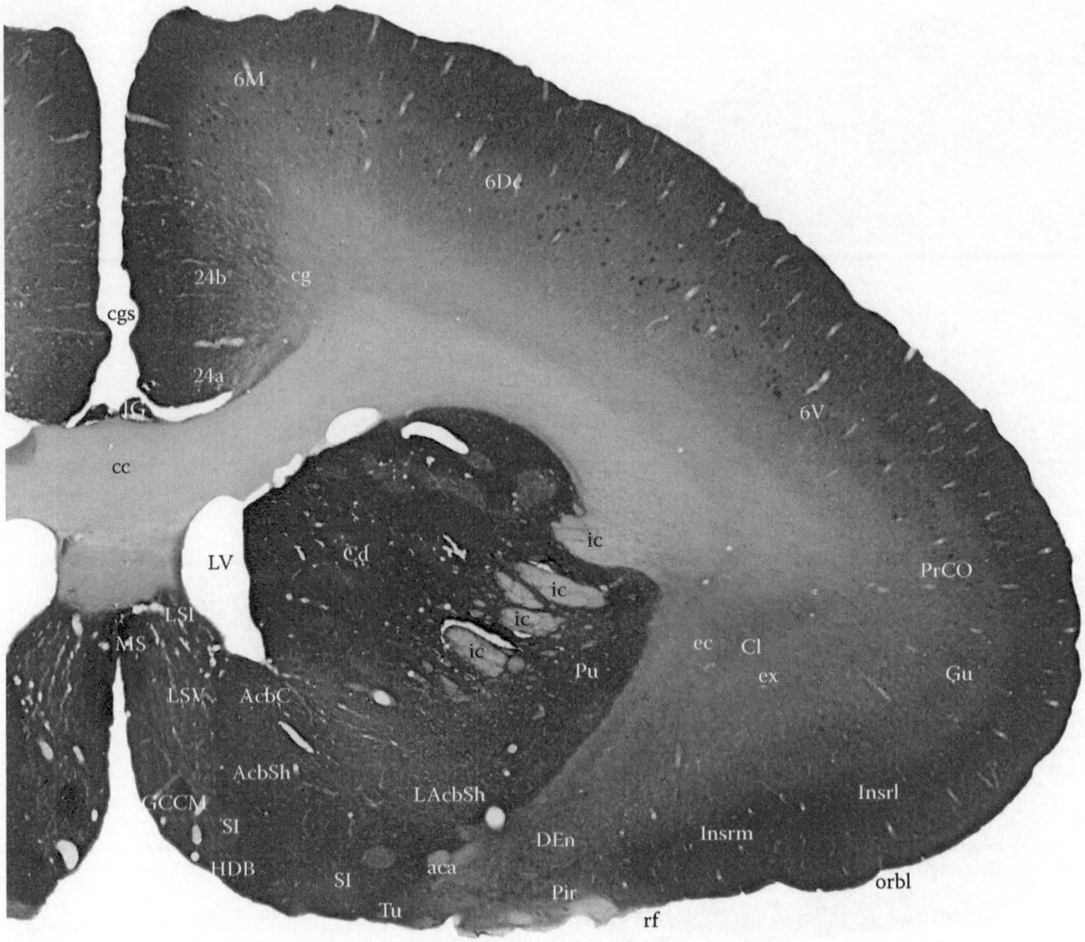


**Calbindin**
Interaural +12.10 mm

# Plate 15b

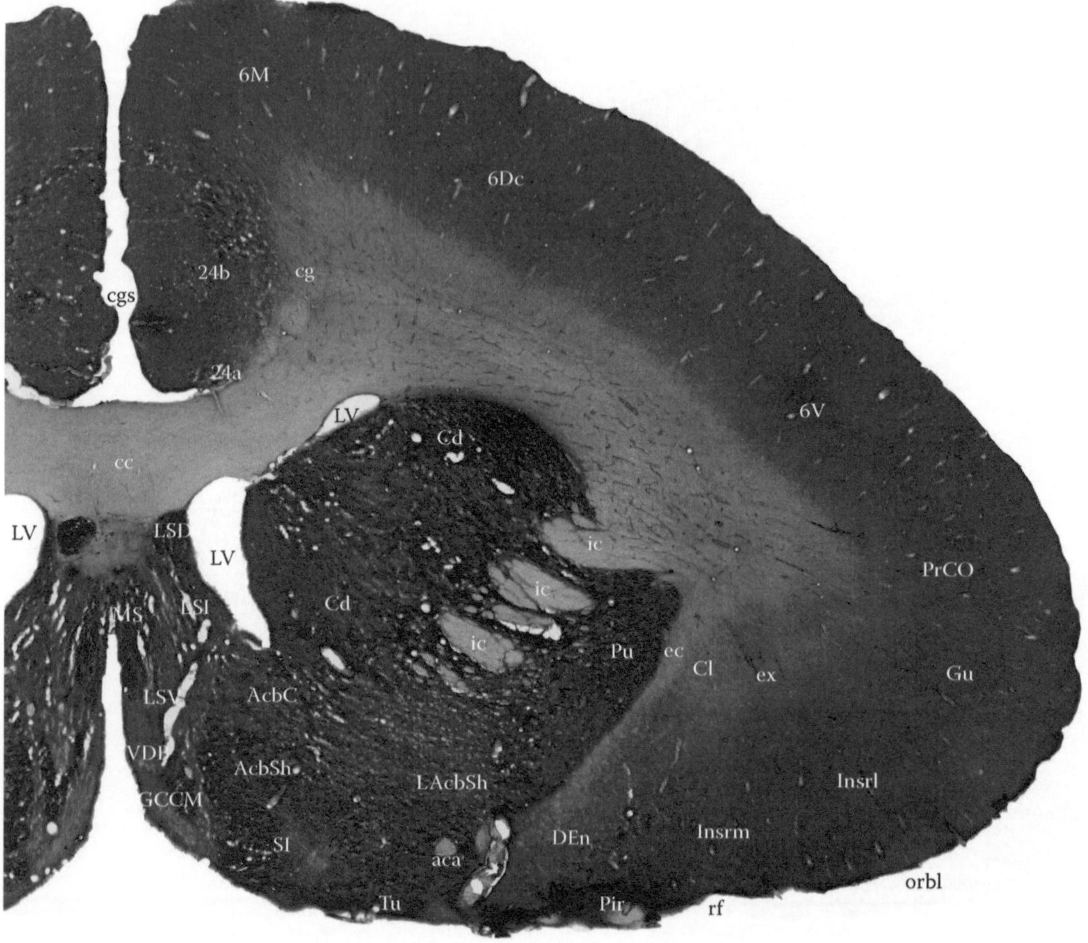


**Neurofilament**
Interaural +11.95 mm

# Plate 15c

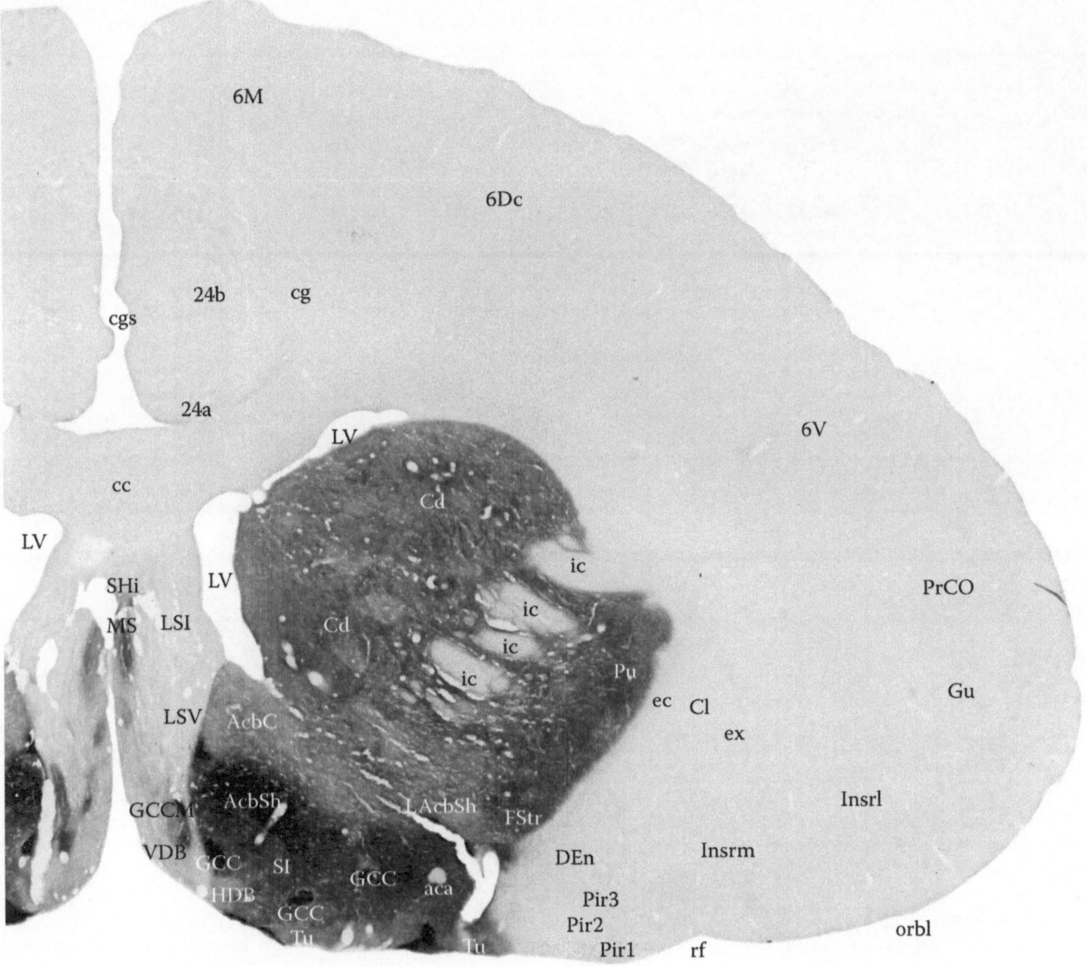


**Acetylcholinesterase**
Interaural +11.90 mm

# Plate 15d

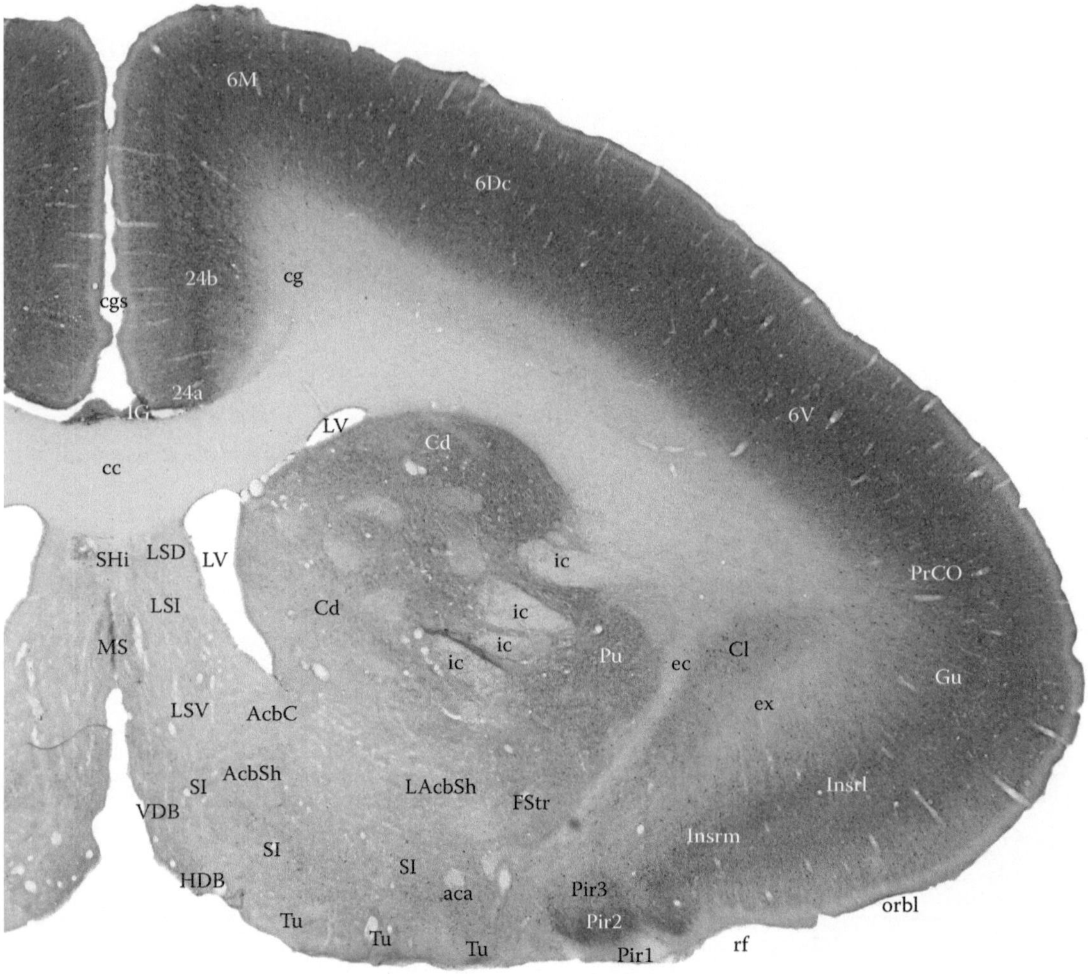


**Parvalbumin**
Interaural +11.85 mm

# Plate 15e

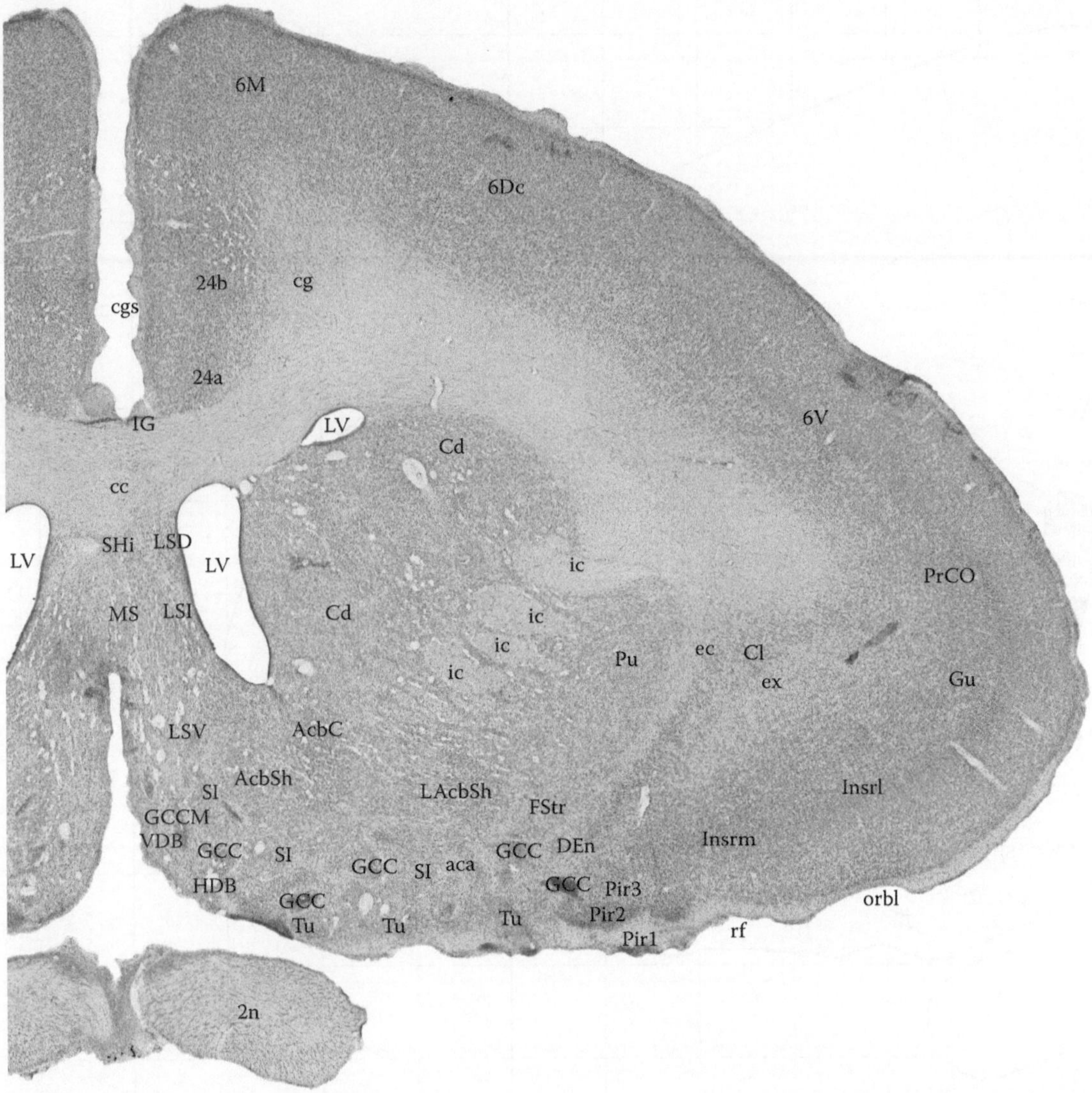


**Cresyl Violet**
Interaural +11.80 mm

# Plate 15

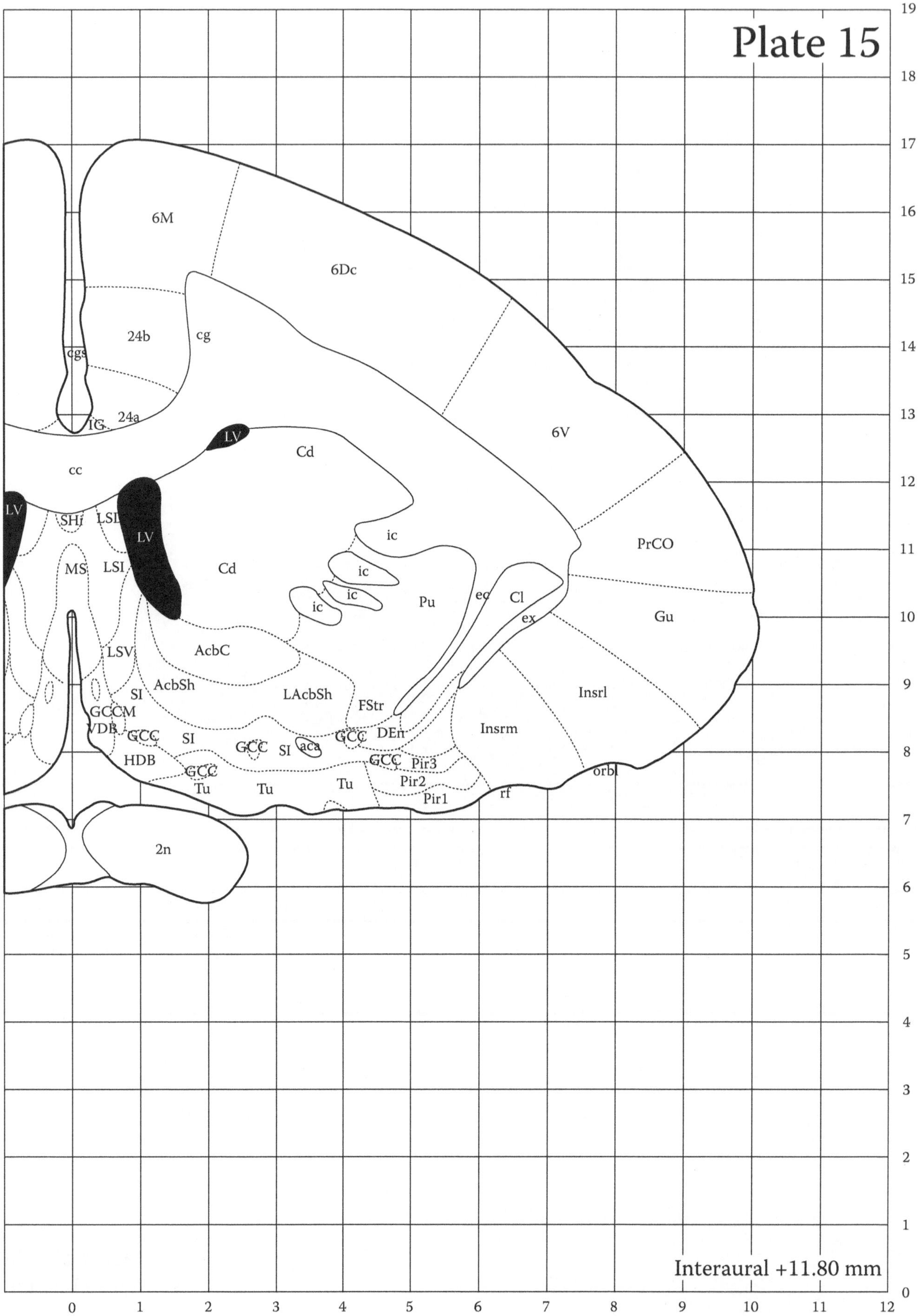


Interaural +11.80 mm

| | |
|---|---|
| 2n | optic nerve |
| 6Dc | area 6Dc Cx |
| 6M | area 6M Cx |
| 6V | area 6V Cx |
| 24a | area 24a Cx |
| 24b | area 24b Cx |
| aca | ant. comm. ant. |
| AcbC | accumb. nu. core |
| AcbSh | accumb. nu. shell |
| cc | corp. callosum |
| Cd | caudate nu. |
| cg | cingulum |
| cgs | cingulate sulc. |
| Cl | claustrum |
| DEn | dors. endopir. nu. |
| ec | external capsule |
| ex | extreme capsule |
| FStr | fundus striatum |
| GCC | gran. cell clust. |
| GCCM | gran. cell clust. magn. |
| Gu | gustatory Cx |
| HDB | horiz. limb diag. |
| ic | intern. caps. |
| IG | induseum gris. |
| Insrl | insular Cx, rost. lat. |
| Insrm | insular Cx, rost. med. |
| LAcbSh | accumb. shell lat. |
| LSD | lat. sept. nu., dors. |
| LSI | lat. sept. nu., int. |
| LSV | lat. sept. nu., vent. |
| LV | lateral ventricle |
| MS | med. sept. nu. |
| orbl | orbital sulc. lat. |
| Pir1 | piriform Cx 1 |
| Pir2 | piriform Cx 2 |
| Pir3 | piriform Cx 3 |
| PrCO | precent. operc. Cx |
| Pu | putamen |
| rf | rhinal fiss. |
| SHi | septohippo. nu. |
| SI | subst. innominata |
| Tu | olf. tubercle |
| VDB | vert. limb diag. band. nu. |

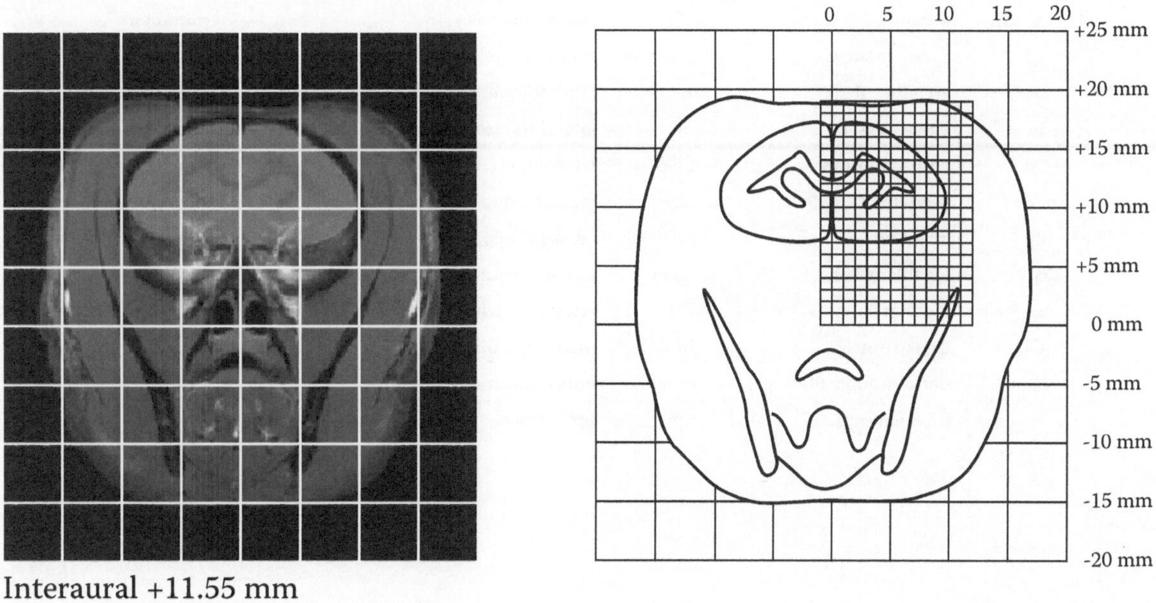


Interaural +11.55 mm

# Plate 16

Interaural +11.30 mm

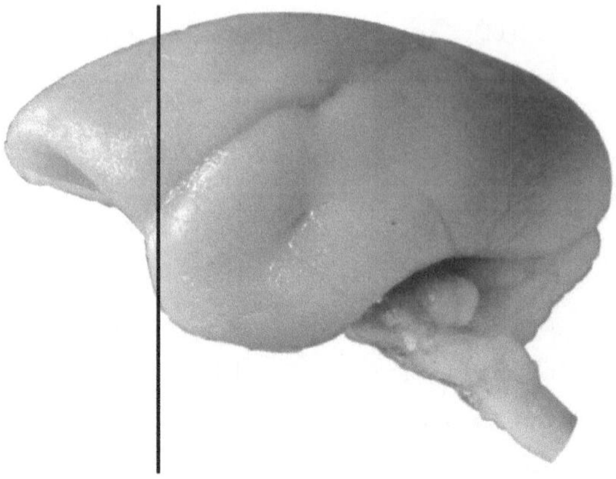

| | | | | | |
|---|---|---|---|---|---|
| 3V | third ventricle | ex | extreme capsule | Pir1 | piriform Cx 1 |
| 6Dc | area 6Dc Cx | FStr | fundus striatum | Pir2 | piriform Cx 2 |
| 6M | area 6M Cx | GCC | gran. cell clust. | Pir3 | piriform Cx 3 |
| 6V | area 6V Cx | GCCM | gran. cell clust. magn. | PLd | paralambdoid septum |
| 24a | area 24a Cx | GI | granular insul. Cx | PMD | premotor Cx dors. |
| 24b | area 24b Cx | Gu | gustatory Cx | PMV | premotor Cx vent. |
| aca | ant. comm. ant. | HDB | horiz. limb diag. band nu. | pr | prefrontal sulcus |
| AcbC | accumb. nu. core | ic | intern. caps. | PrCO | precent. operc. Cx |
| AcbSh | accumb. nu. shell | IG | induseum gris. | Pu | putamen |
| AI | agran. insul. Cx | LAcbSh | accumb. shell lat. | rf | rhinal fiss. |
| arc | arcuate sulcus | Ld | lambdoid nu. | SHi | septohippo. nu. |
| bv | blood vessel | lf | lateral fissure | SI | subst. innominata |
| cc | corp. callosum | LSD | lat. sept. nu., dors. | STL | sup. temp. Cx, lat. |
| Cd | caudate nu. | LSI | lat. sept. nu., int. | STM | sup. temp. Cx, med. |
| cg | cingulum | LSV | lat. sept. nu., vent. | sts | sup. temp. sulcus |
| cgs | cingulate sulc. | LTer | lamina terminalis | Tu | olf. tubercle |
| ci | circular sulcus | LV | lateral ventricle | VDB | vert. limb diag. band. nu. |
| Cl | claustrum | MS | med. sept. nu. | VEn | vent. endopir. nu. |
| DEn | dors. endopir. nu. | och | optic chiasm | VOLT | vasc. organ LTer |
| ec | external capsule | opt | optic tract | | |

# Plate 16a

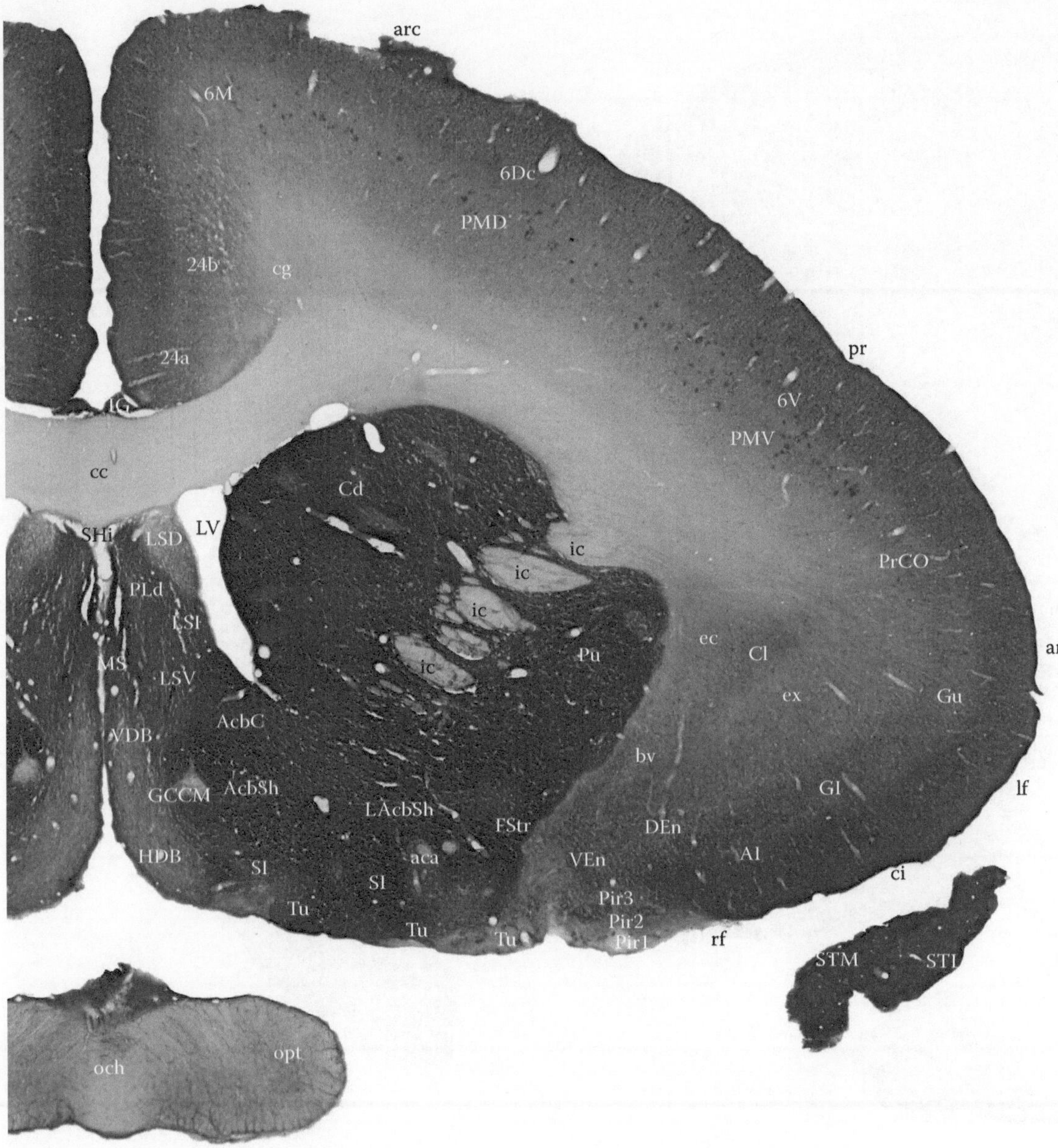


**Calbindin**
Interaural +11.60 mm

# Plate 16b

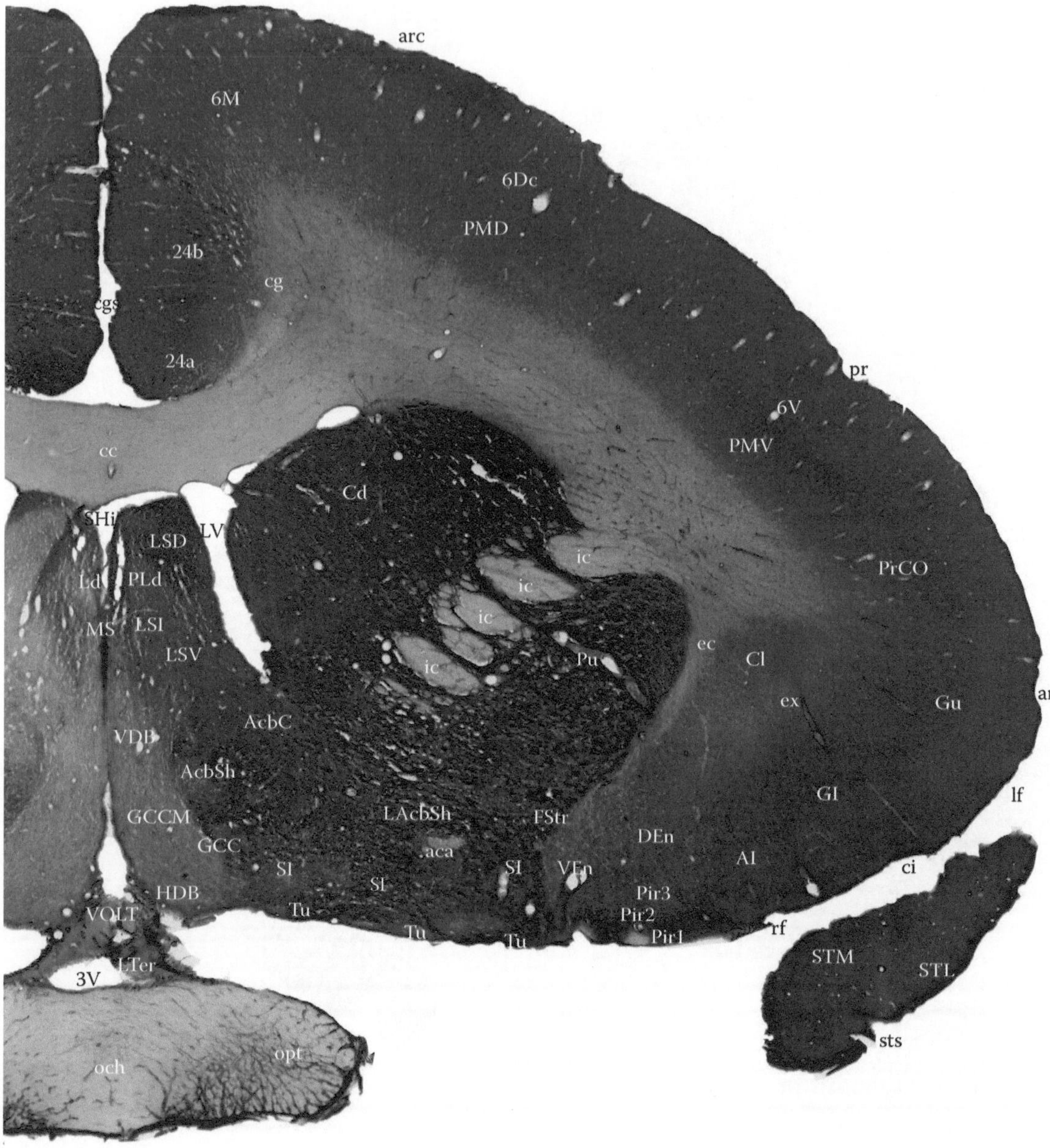


**Neurofilament**
Interaural +11.45 mm

# Plate 16c

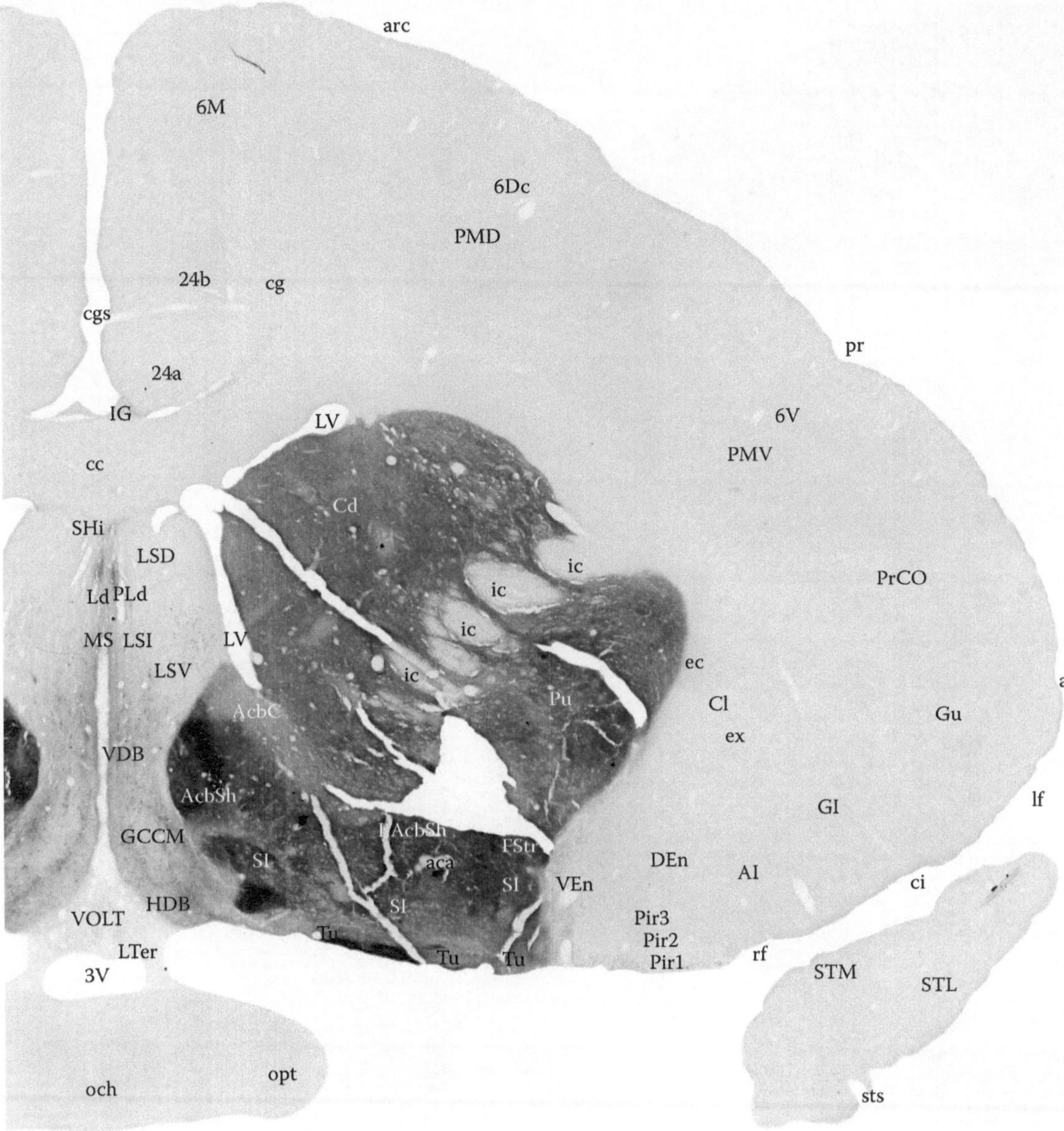


**Acetylcholinesterase**
Interaural +11.40 mm

# Plate 16d

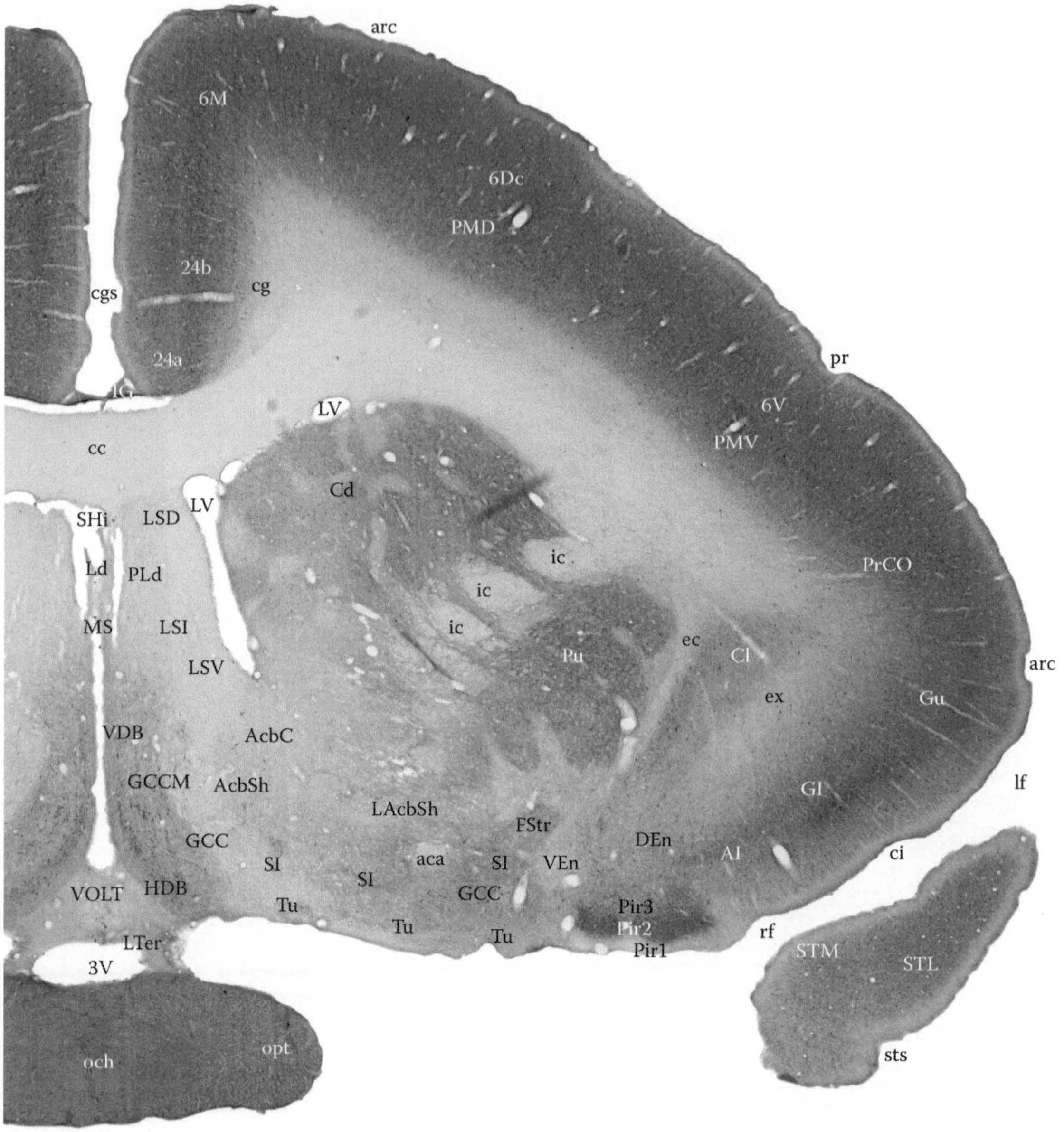


**Parvalbumin**
Interaural +11.35 mm

# Plate 16e

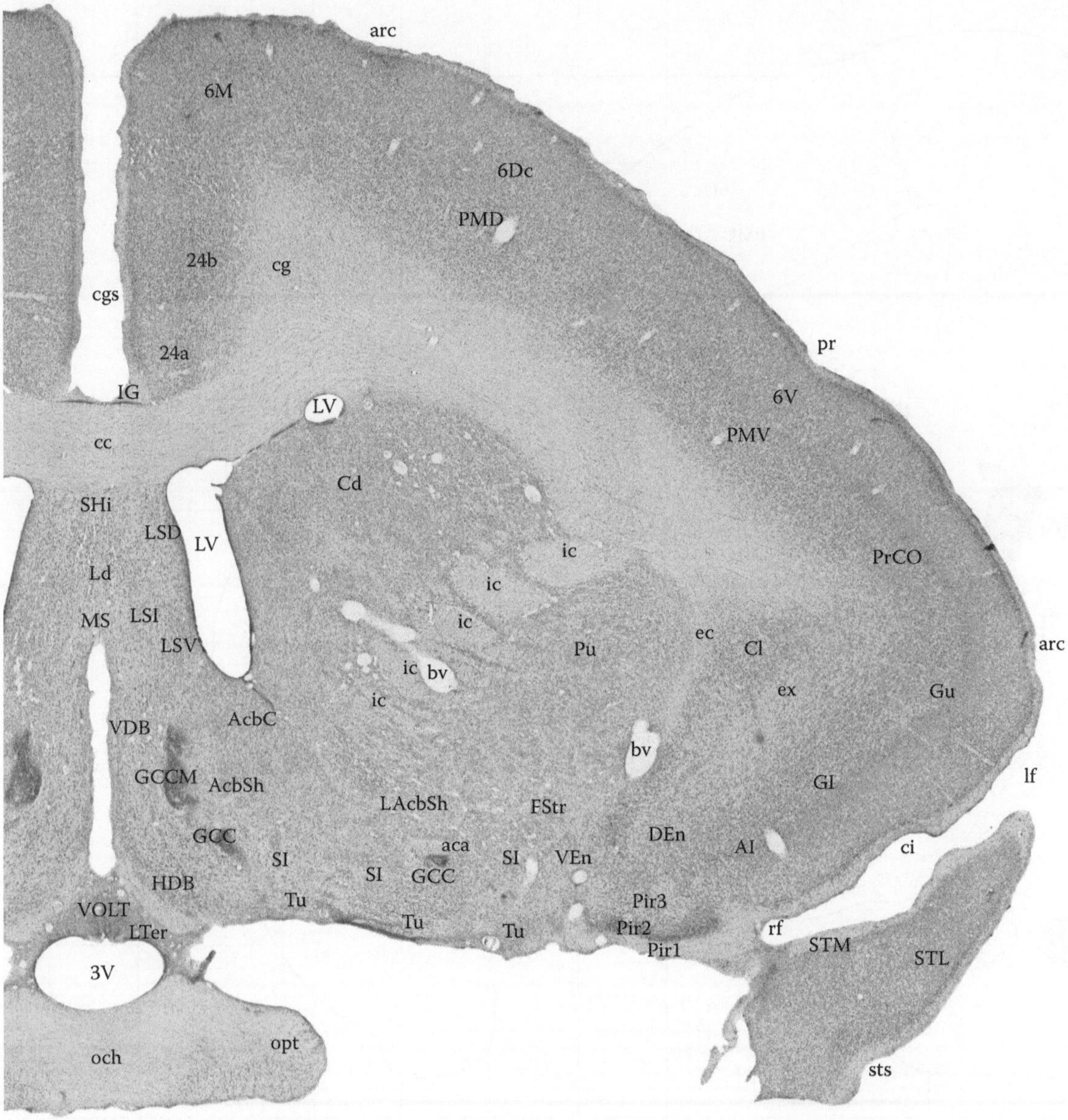


**Cresyl Violet**
Interaural +11.30 mm

# Plate 16

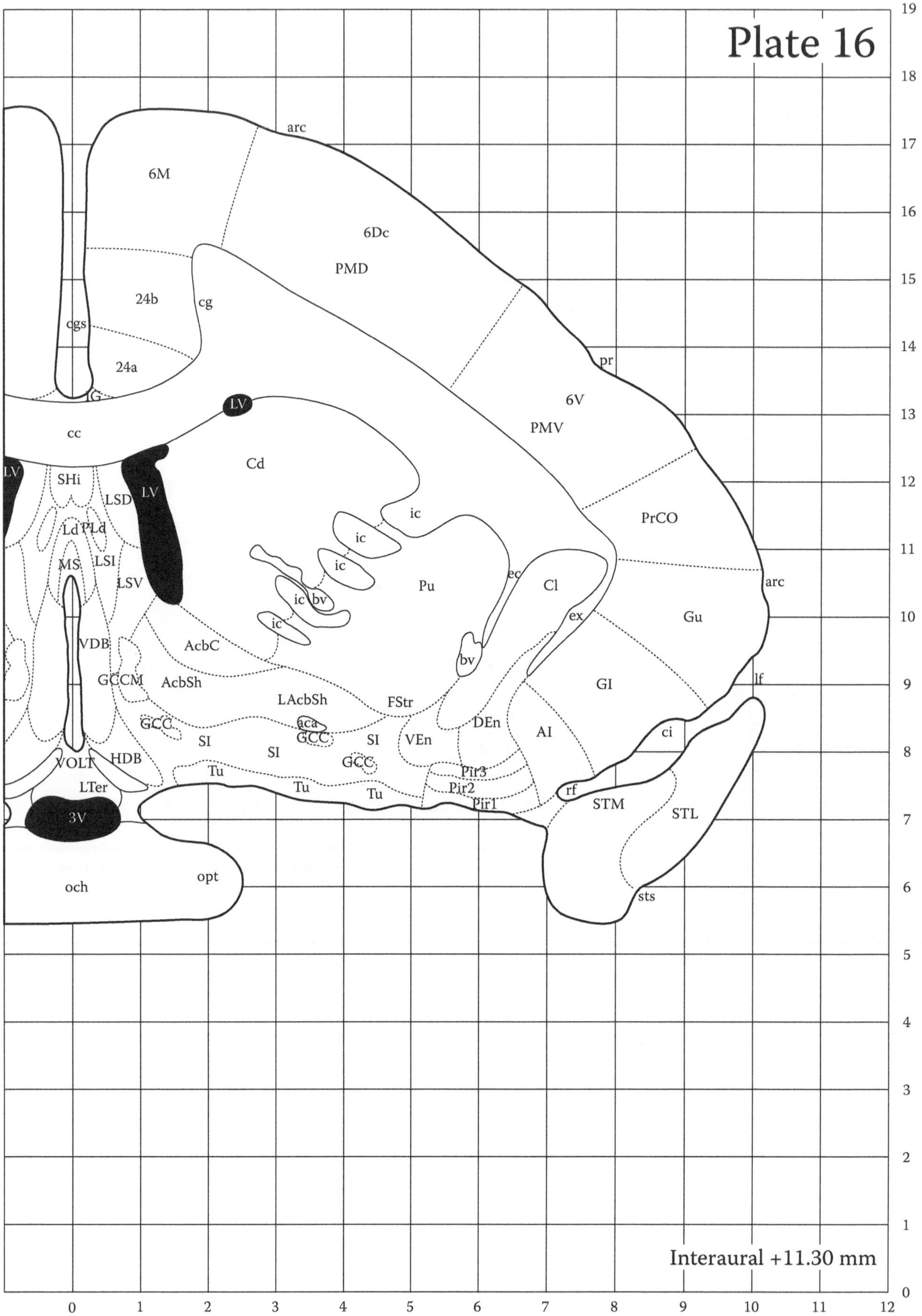
arc
6M
6Dc
PMD
24b
cg
cgs
24a
IG
pr
6V
PMV
LV
cc
Cd
LV
SHi
LSD
LV
Ld
PLd
ic
PrCO
ic
MS
LSI
ic
Pu
ec
Cl
arc
LSV
ic
bv
ex
Gu
ic
VDB
AcbC
bv
GCCM
AcbSh
GI
lf
LAcbSh
FStr
GCC
aca
DEn
AI
ci
GCC
SI
SI
SI
VEn
VOLT
HDB
GCC
Pir3
Tu
rf
LTer
Tu
Pir2
Tu
Pir1
STM
STL
3V
opt
och
sts
Interaural +11.30 mm
0
1
2
3
4
5
6
7
8
9
10
11
12
13
14
15
16
17
18
19

| | |
|---|---|
| 3V | third ventricle |
| 6Dc | area 6Dc Cx |
| 6M | area 6M Cx |
| 6V | area 6V Cx |
| 24a | area 24a Cx |
| 24b | area 24b Cx |
| aca | ant. comm. ant. |
| AcbC | accumb. nu. core |
| AcbSh | accumb. nu. shell |
| AI | agran. insul. Cx |
| arc | arcuate sulcus |
| bv | blood vessel |
| cc | corp. callosum |
| Cd | caudate nu. |
| cg | cingulum |
| cgs | cingulate sulc. |
| ci | circular sulcus |
| Cl | claustrum |
| DEn | dors. endopir. nu. |
| ec | external capsule |
| ex | extreme capsule |
| FStr | fundus striatum |
| GCC | gran. cell clust. |
| GCCM | gran. cell clust. magn. |
| GI | granular insul. Cx |
| Gu | gustatory Cx |
| HDB | horiz. limb diag. band nu. |
| ic | intern. caps. |
| IG | induseum gris. |
| LAcbSh | accumb. shell lat. |
| Ld | lambdoid nu. |
| lf | lateral fissure |
| LSD | lat. sept. nu., dors. |
| LSI | lat. sept. nu., int. |
| LSV | lat. sept. nu., vent. |
| LTer | lamina terminalis |
| LV | lateral ventricle |
| MS | med. sept. nu. |
| och | optic chiasm |
| opt | optic tract |
| Pir1 | piriform Cx 1 |
| Pir2 | piriform Cx 2 |
| Pir3 | piriform Cx 3 |
| PLd | paralambdoid septum |
| PMD | premotor Cx dors. |
| PMV | premotor Cx vent. |
| pr | prefrontal sulcus |
| PrCO | precent. operc. Cx |
| Pu | putamen |
| rf | rhinal fiss. |
| SHi | septohippo. nu. |
| SI | subst. innominata |
| STL | sup. temp. Cx, lat. |
| STM | sup. temp. Cx, med. |
| sts | sup. temp. sulcus |
| Tu | olf. tubercle |
| VDB | vert. limb diag. band. nu. |
| VEn | vent. endopir. nu. |
| VOLT | vasc. organ LTer |

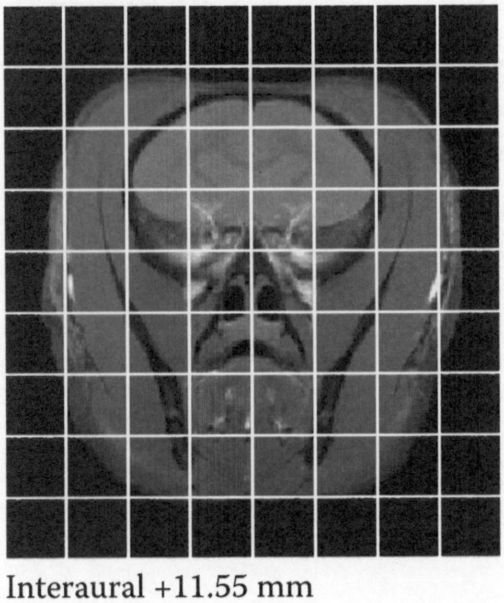

Interaural +11.55 mm

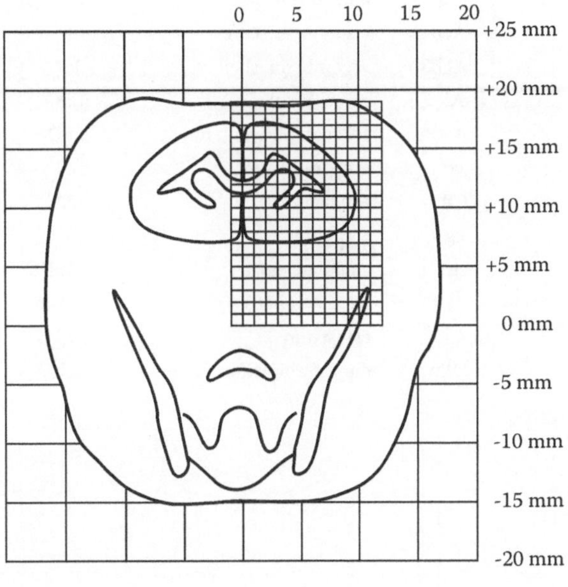

# Plate 17

Interaural +10.80 mm

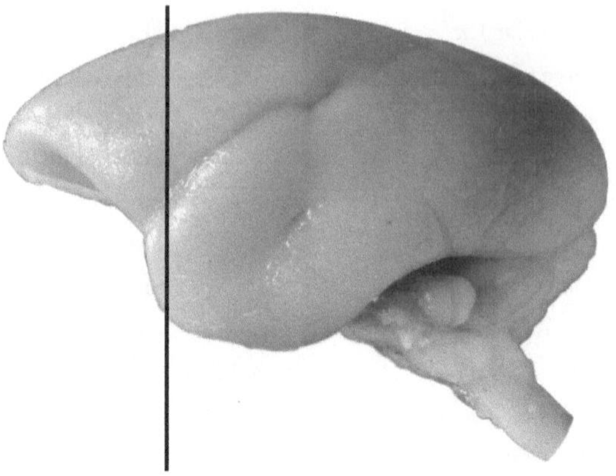

| | | | | | |
|---|---|---|---|---|---|
| 3a | area 3a Cx | E | ependyma | opt | optic tract |
| 3b | area 3b Cx | ec | external capsule | Pir1 | piriform Cx 1 |
| 3V | third ventricle | ex | extreme capsule | Pir2 | piriform Cx 2 |
| 6Dc | area 6Dc Cx | FStr | fundus striatum | Pir3 | piriform Cx 3 |
| 6M | area 6M Cx | GCC | gran. cell clust. | PLd | paralambdoid septum |
| 6V | area 6V Cx | GI | granular insul. Cx | PRh | perirhinal Cx |
| 24a | area 24a Cx | HDB | horiz. limb diag. band nu. | PRSom | par. rost. som. area Cx |
| 24b | area 24b Cx | ic | intern. caps. | Pu | putamen |
| AAD | ant. amyg. dors. | IG | induseum gris. | PV | parietovent. somato. Cx |
| AAV | ant. amyg. vent. | ITA | inf. temp. ant. Cx | rf | rhinal fiss. |
| aca | ant. comm. ant. | LAcbSh | accumb. shell lat. | SHi | septohippo. nu. |
| AcbC | accumb. nu. core | lf | lateral fissure | SI | subst. innominata |
| AcbSh | accumb. nu. shell | lo | lat. olf. tract | SO | supraoptic nu. |
| AI | agran. insul. Cx | LPO | lat. preopt. area | STL | sup. temp. Cx, lat. |
| AVPe | antvent. perivent. nu. | LSD | lat. sept. nu., dors. | STM | sup. temp. Cx, med. |
| cc | corp. callosum | LSI | lat. sept. nu., int. | sts | sup. temp. sulcus |
| Cd | caudate nu. | LSV | lat. sept. nu., vent. | Tu | olf. tubercle |
| cg | cingulum | LV | lateral ventricle | VDB | vert. limb diag. band. nu. |
| cgs | cingulate sulc. | MnPO | median preopt. nu. | VEn | vent. endopir. nu. |
| ci | circular sulcus | MPA | medial preopt. area | VLPO | ventrolat. preopt. nu. |
| Cl | claustrum | MS | med. sept. nu. | VMPO | ventromed. preopt. nu. |
| DEn | dors. endopir. nu. | och | optic chiasm | | |

# Plate 17a

**Calbindin**
Interaural +11.10 mm

# Plate 17b

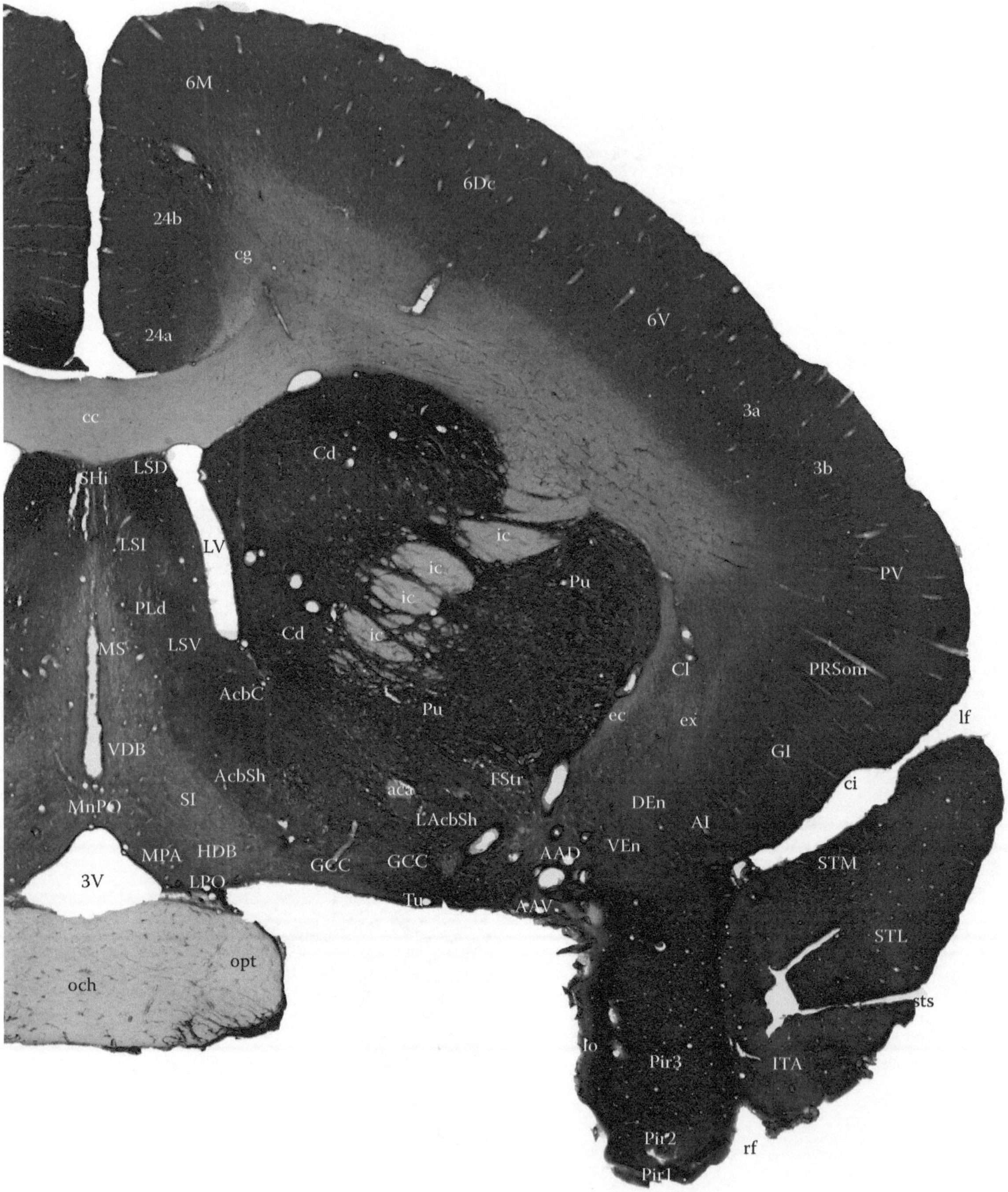


**Neurofilament**
Interaural +10.95 mm

# Plate 17c

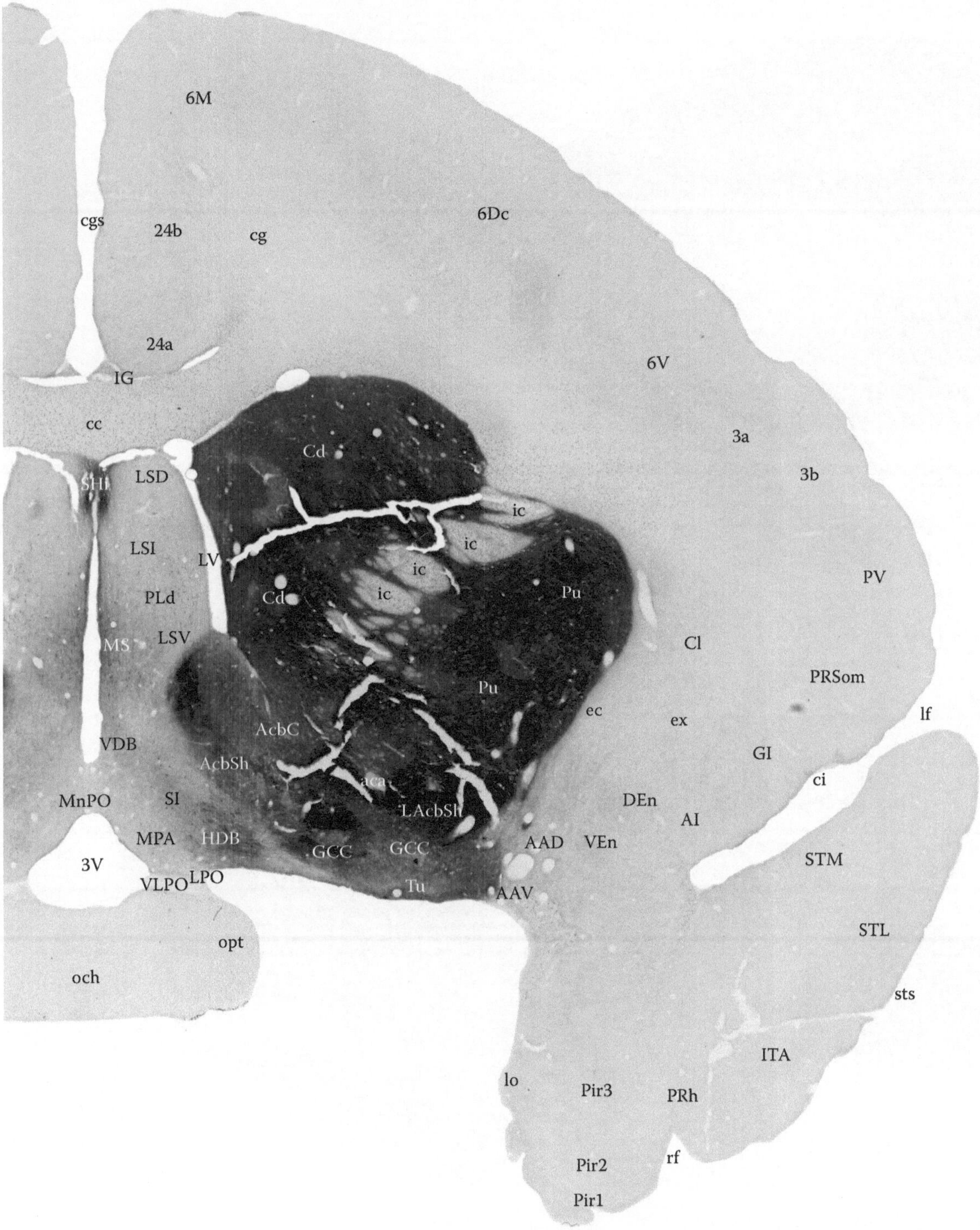


**Acetylcholinesterase**
Interaural +10.90 mm

# Plate 17d

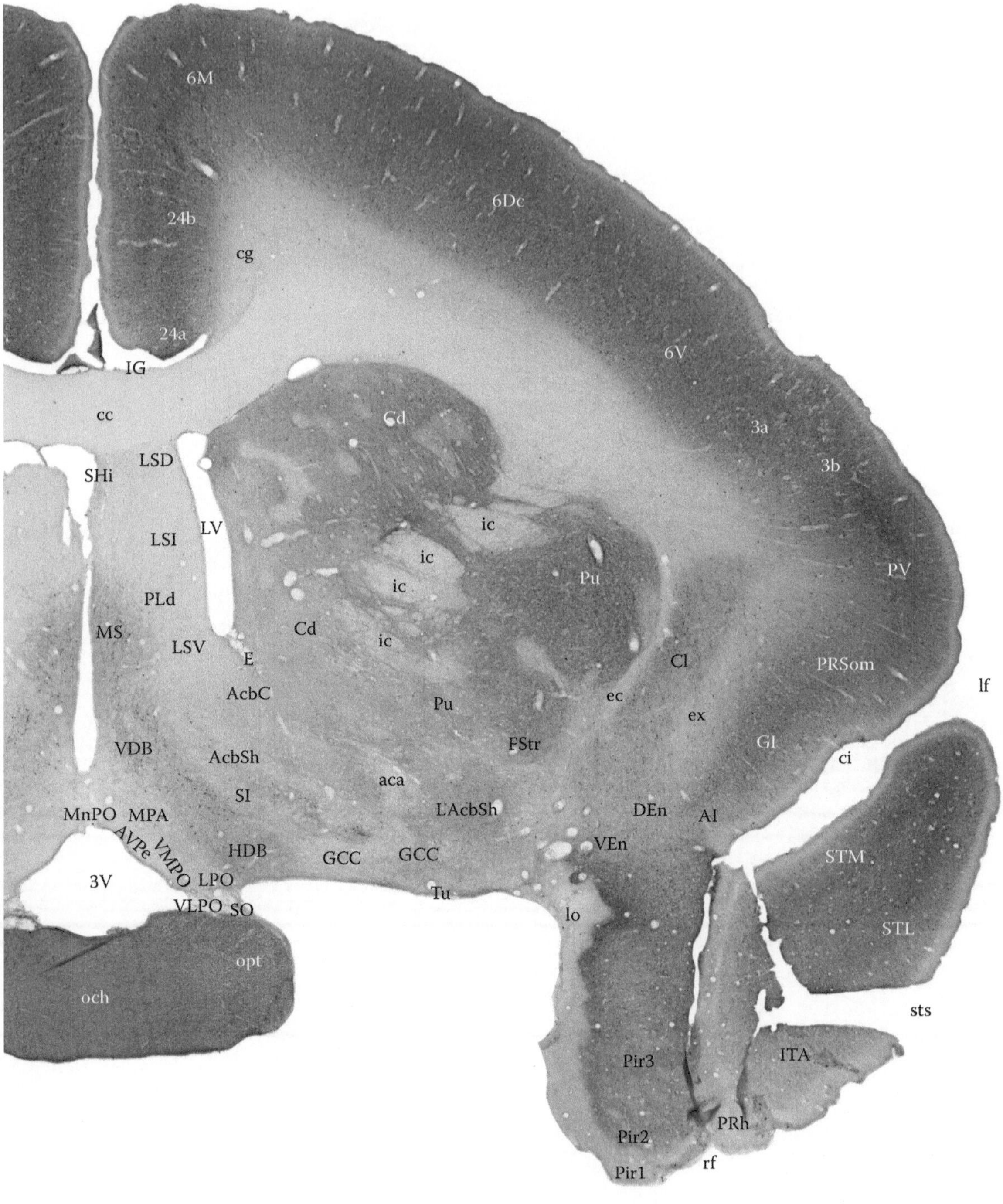


**Parvalbumin**
Interaural +10.85 mm

# Plate 17e

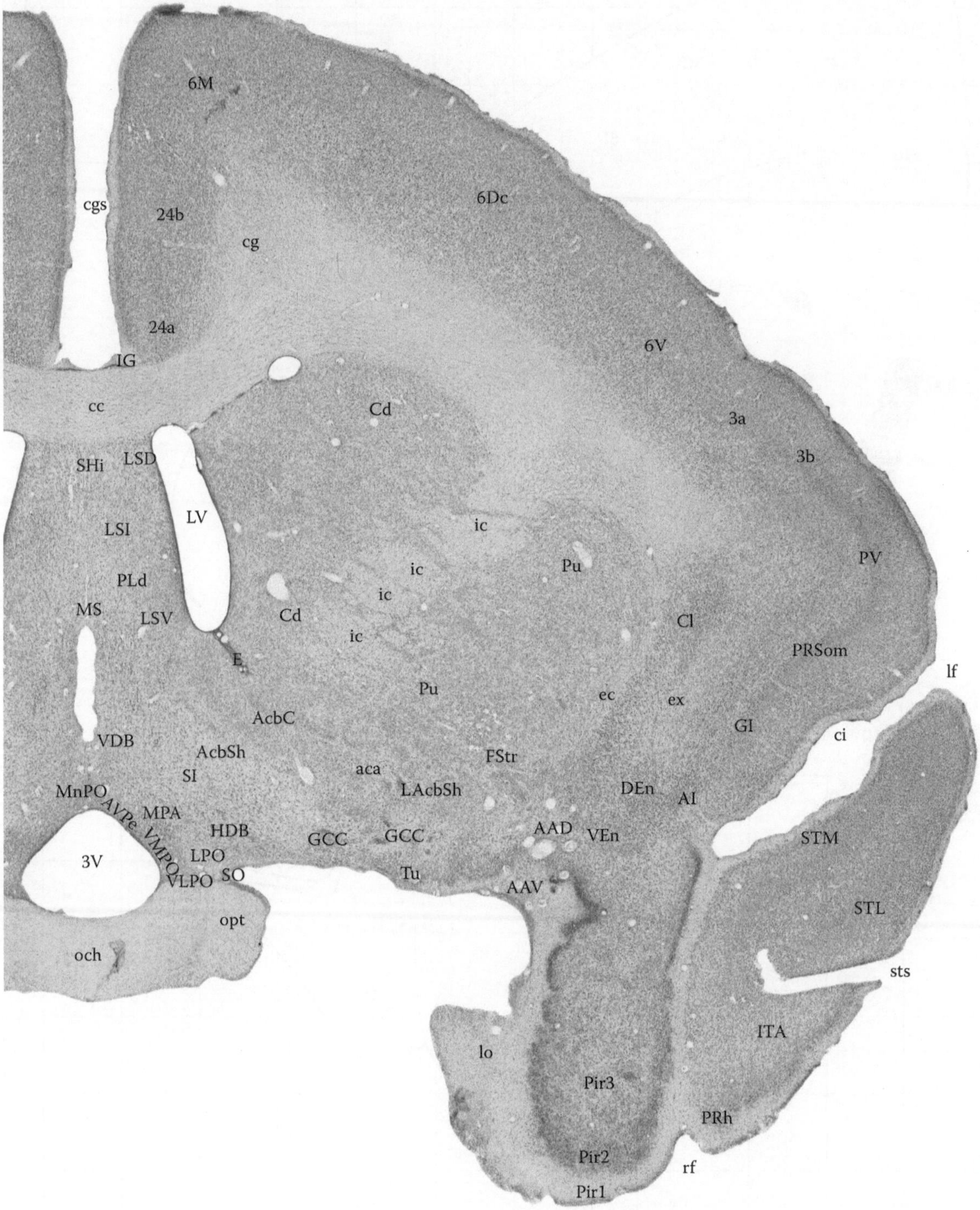


**Cresyl Violet**
Interaural +10.80 mm

# Plate 17

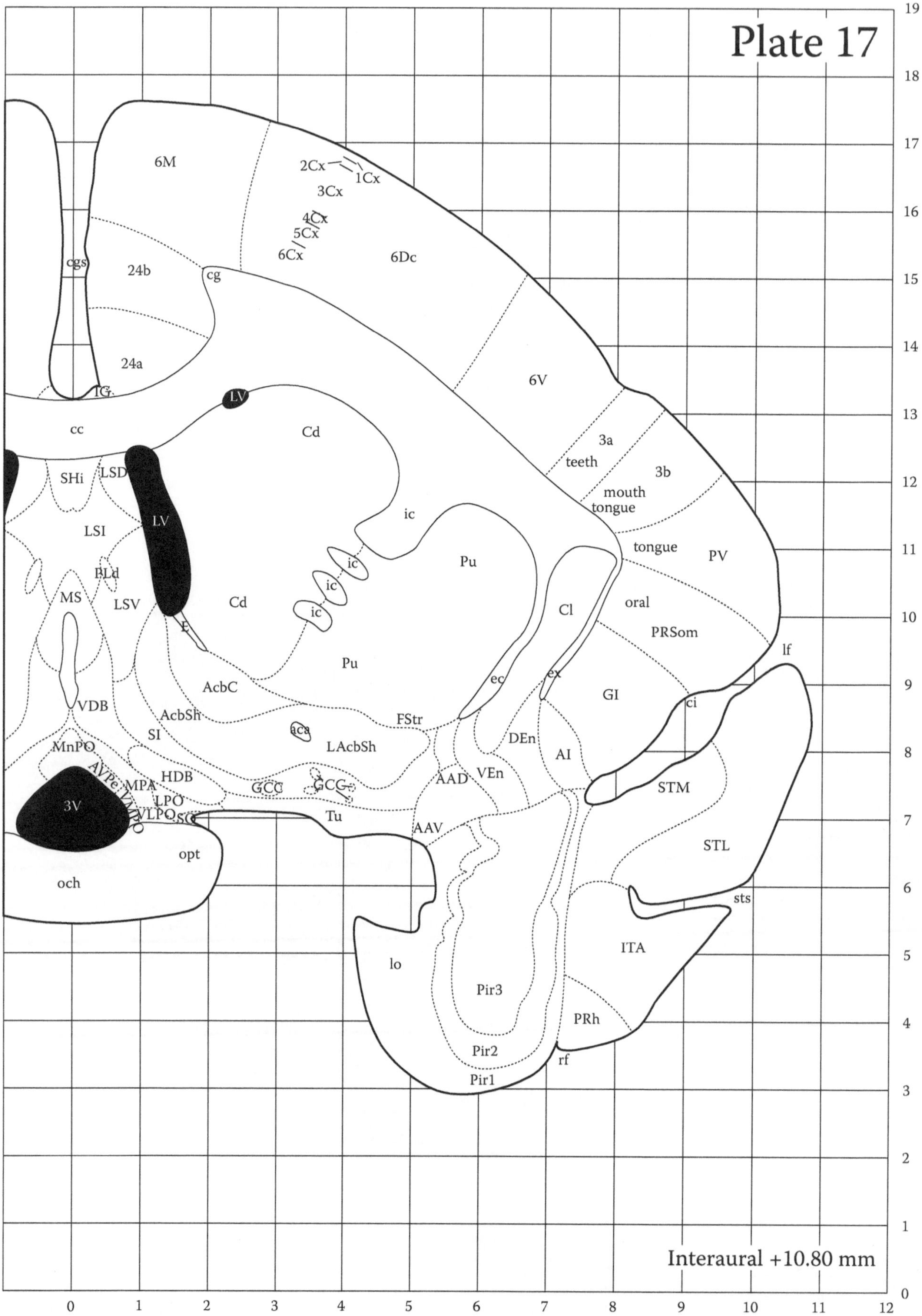

6M
2Cx
1Cx
3Cx
4Cx
5Cx
6Cx
6Dc
cgs
24b
cg
24a
6V
IG
LV
cc
Cd
3a
teeth
3b
mouth
tongue
SHi
LSD
ic
LV
LSI
tongue
PV
Pu
ic
LSV
ic
Cl
oral
MS
Cd
ic
PRSom
E
Pu
lf
ec
ex
AcbC
GI
VDB
ci
AcbSh
FStr
SI
aca
DEn
AI
MnPO
LAcbSh
AVPe
VEn
MPA
HDB
AAD
STM
GCC
GCC
3V
VMPO
LPO
VLPO
SO
Tu
AAV
STL
opt
och
sts
ITA
lo
Pir3
PRh
Pir2
rf
Pir1
Interaural +10.80 mm

| | |
|---|---|
| 3a | area 3a Cx |
| 3b | area 3b Cx |
| 3V | third ventricle |
| 6Dc | area 6Dc Cx |
| 6M | area 6M Cx |
| 6V | area 6V Cx |
| 24a | area 24a Cx |
| 24b | area 24b Cx |
| AAD | ant. amyg. dors. |
| AAV | ant. amyg. vent. |
| aca | ant. comm. ant. |
| AcbC | accumb. nu. core |
| AcbSh | accumb. nu. shell |
| AI | agran. insul. Cx |
| AVPe | antvent. perivent. nu. |
| cc | corp. callosum |
| Cd | caudate nu. |
| cg | cingulum |
| cgs | cingulate sulc. |
| ci | circular sulcus |
| Cl | claustrum |
| DEn | dors. endopir. nu. |
| E | ependyma |
| ec | external capsule |
| ex | extreme capsule |
| FStr | fundus striatum |
| GCC | gran. cell clust. |
| GI | granular insul. Cx |
| HDB | horiz. limb diag. band nu. |
| ic | intern. caps. |
| IG | induseum gris. |
| ITA | inf. temp. ant. Cx |
| LAcbSh | accumb. shell lat. |
| lf | lateral fissure |
| lo | lat. olf. tract |
| LPO | lat. preopt. area |
| LSD | lat. sept. nu., dors. |
| LSI | lat. sept. nu., int. |
| LSV | lat. sept. nu., vent. |
| LV | lateral ventricle |
| MnPO | median preopt. nu. |
| MPA | medial preopt. area |
| MS | med. sept. nu. |
| och | optic chiasm |
| opt | optic tract |
| Pir1 | piriform Cx 1 |
| Pir2 | piriform Cx 2 |
| Pir3 | piriform Cx 3 |
| PLd | paralambdoid septum |
| PRh | perirhinal Cx |
| PRSom | par. rost. som. area Cx |
| Pu | putamen |
| PV | parietovent. somato. Cx |
| rf | rhinal fiss. |
| SHi | septohippo. nu. |
| SI | subst. innominata |
| SO | supraoptic nu. |
| STL | sup. temp. Cx, lat. |
| STM | sup. temp. Cx, med. |
| sts | sup. temp. sulcus |
| Tu | olf. tubercle |
| VDB | vert. limb diag. band. nu. |
| VEn | vent. endopir. nu. |
| VLPO | ventrolat. preopt. nu. |
| VMPO | ventromed. preopt. nu. |

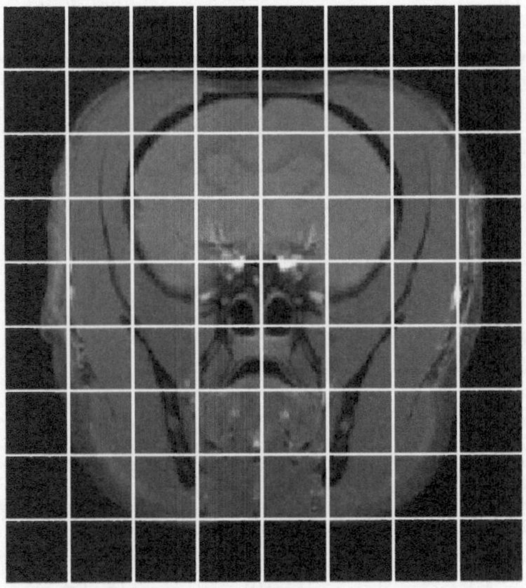

Interaural +10.60 mm

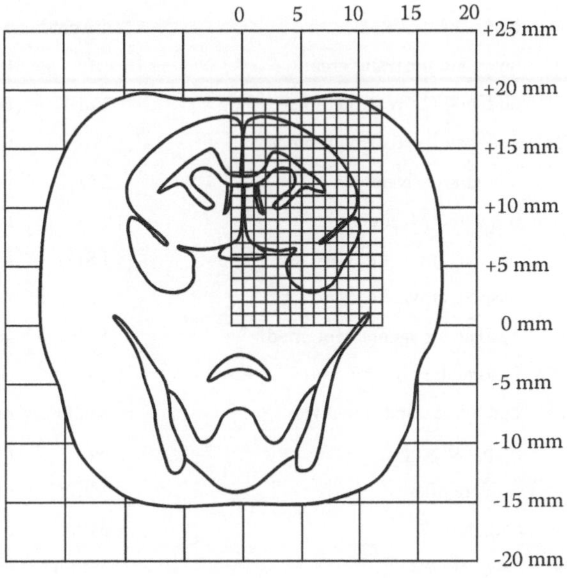

# Plate 18

Interaural +10.25 mm

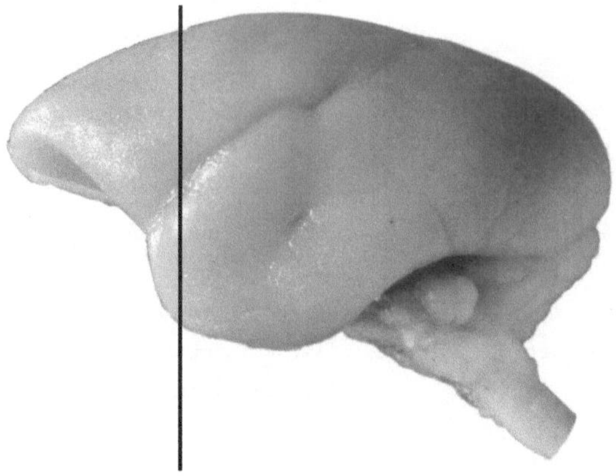

| | |
|---|---|
| 3a | area 3a Cx |
| 3b | area 3b Cx |
| 3V | third ventricle |
| 6Dc | area 6Dc Cx |
| 6M | area 6M Cx |
| 6V | area 6V Cx |
| 24a | area 24a Cx |
| 24b | area 24b Cx |
| AAD | ant. amyg. dors. |
| AAL | ant. amyg. lat. |
| AAV | ant. amyg. vent. |
| aca | ant. comm. ant. |
| AcbC | accumb. nu. core |
| af | amygdaloid fiss. |
| AHA | ant. hypoth. area, ant. |
| AI | agran. insul. Cx |
| APir | amygdalopir. trans. area |
| AuRTL | aud. belt Cx, rostrotemplat. |
| AuRTM | aud. belt Cx, rostrotempmed. |
| AVPe | antvent. perivent. nu. |
| B | basal nu. (Meynert) |
| BLDL | basolat. amyg. nu., dors. lat. |
| BLVL | basolat. amyg. nu., vent. lat. |
| BLVM | basolat. amyg. nu., vent. med. |
| BMA | basomed. amyg. nu., ant. |
| BSTMA | bed nu. st, med. ant. divn |
| cc | corp. callosum |
| Cd | caudate nu. |
| cg | cingulum |
| cgs | cingulate sulc. |
| ci | circular sulcus |
| Cl | claustrum |
| CxA | cortex-amygdala |
| DEn | dors. endopir. nu. |
| ec | external capsule |
| ex | extreme capsule |
| FStr | fundus striatum |
| GI | granular insul. Cx |
| HDB | horiz. limb diag. band nu. |
| I | intercal. nu. amyg. |
| ic | intern. caps. |
| IG | induseum gris. |
| IMG | intramedull. gr. amyg. |
| IPro | insular proisocortex |
| ITA | inf. temp. ant. Cx |
| LAcbSh | accumb. shell lat. |
| lf | lateral fissure |
| lml | lat. medull. lamina |
| lo | lat. olf. tract |
| LOT | nu. of lat. olf. tract |
| LPO | lat. preopt. area |
| LSD | lat. sept. nu., dors. |
| LSI | lat. sept. nu., int. |
| LSV | lat. sept. nu., vent. |
| LV | lateral ventricle |
| MCPO | magnocell. preopt. nu. |
| Me | medial amyg. nu. |
| MnPO | median preopt. nu. |
| MPA | medial preopt. area |
| MPOL | med. preopt. nu., lat. |
| MPOM | med. preopt. nu., med. |
| MS | med. sept. nu. |
| och | optic chiasm |
| opt | optic tract |
| Pir1 | piriform Cx 1 |
| Pir2 | piriform Cx 2 |
| Pir3 | piriform Cx 3 |
| PLam | paralam. amyg. nu. |
| PLd | paralambdoid septum |
| PRh | perirhinal Cx |
| Pu | putamen |
| PV | parietovent. somato. Cx |
| rf | rhinal fiss. |
| SChD | suprachiasm. nu., dors. |
| SChV | suprachiasm. nu., vent. |
| SFi | septofimbrial nu. |
| SHi | septohippo. nu. |
| SHy | septohypothal. nu. |
| SI | subst. innominata |
| SO | supraoptic nu. |
| STL | sup. temp. Cx, lat. |
| STM | sup. temp. Cx, med. |
| sts | sup. temp. sulcus |
| TPro | temp. proisocortex |
| VACo | vent. ant. cort. amyg. |
| VCoI | vent. inf. cort. amyg. |
| VCoIn | vent. inter. cort. amyg. |
| VCoS | vent. sup. cort. amyg. |
| VEn | vent. endopir. nu. |
| VP | vent. pallidum |

# Plate 18a

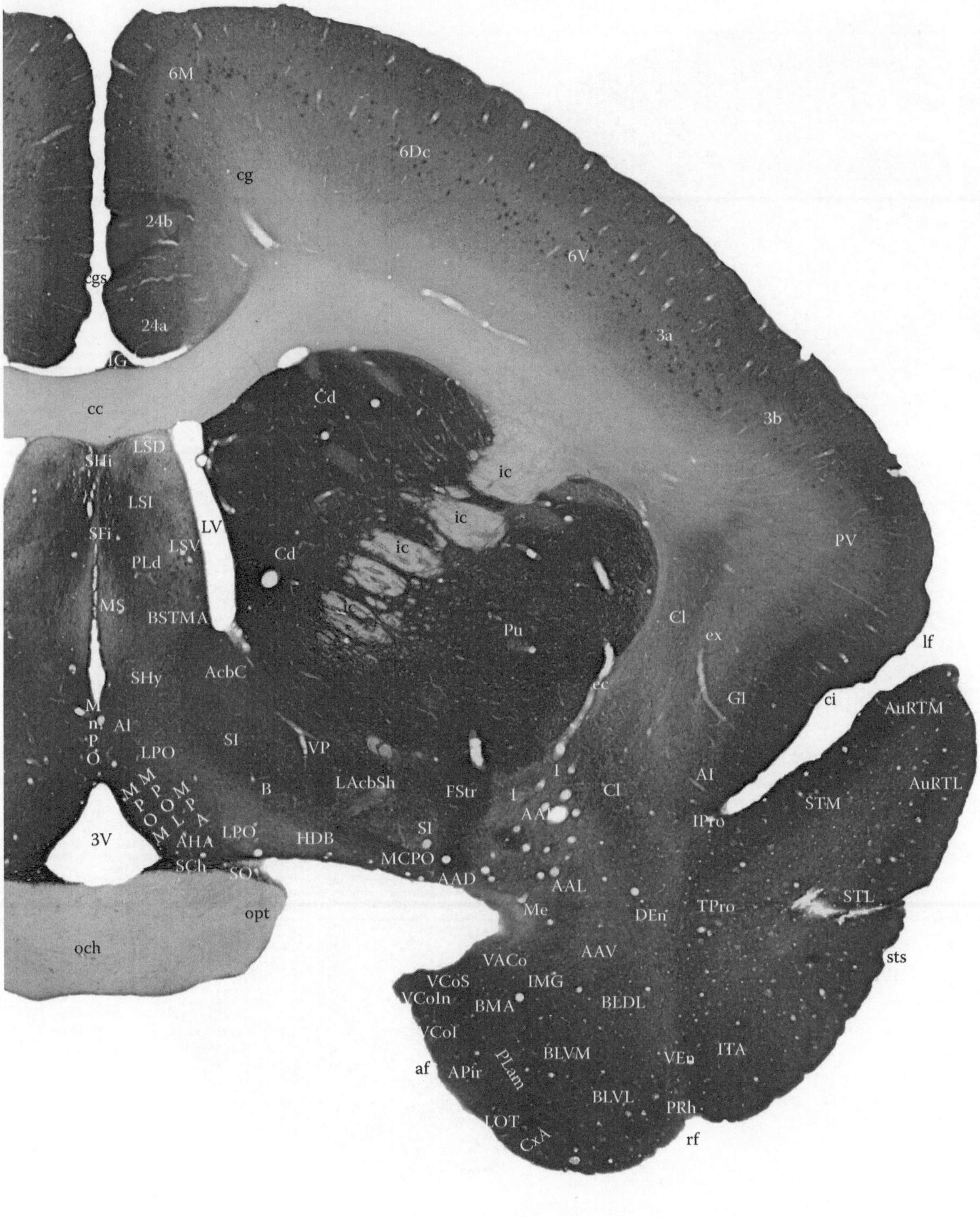


**Calbindin**
Interaural +10.55 mm

# Plate 18b

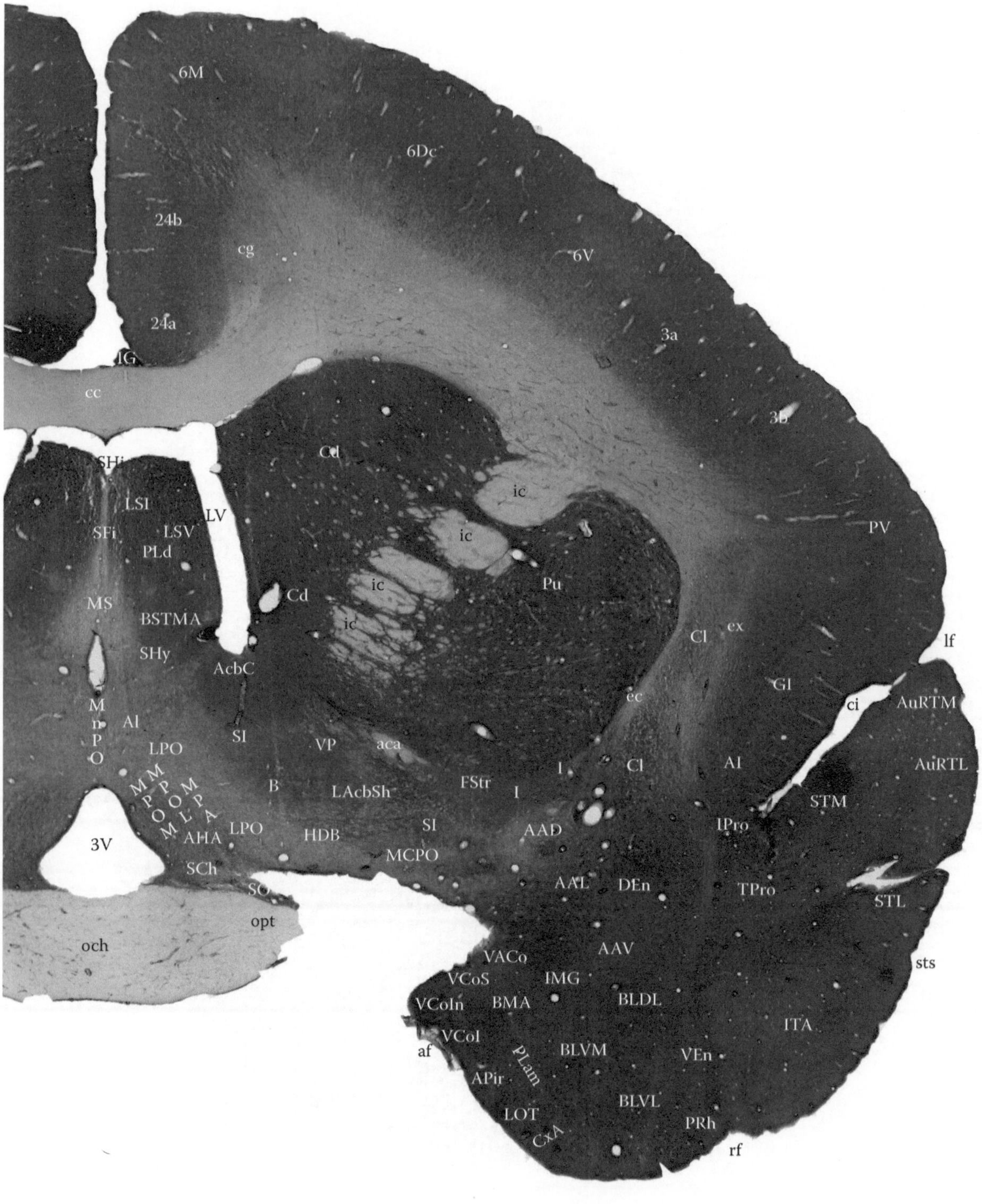


**Neurofilament**
Interaural +10.40 mm

# Plate 18c

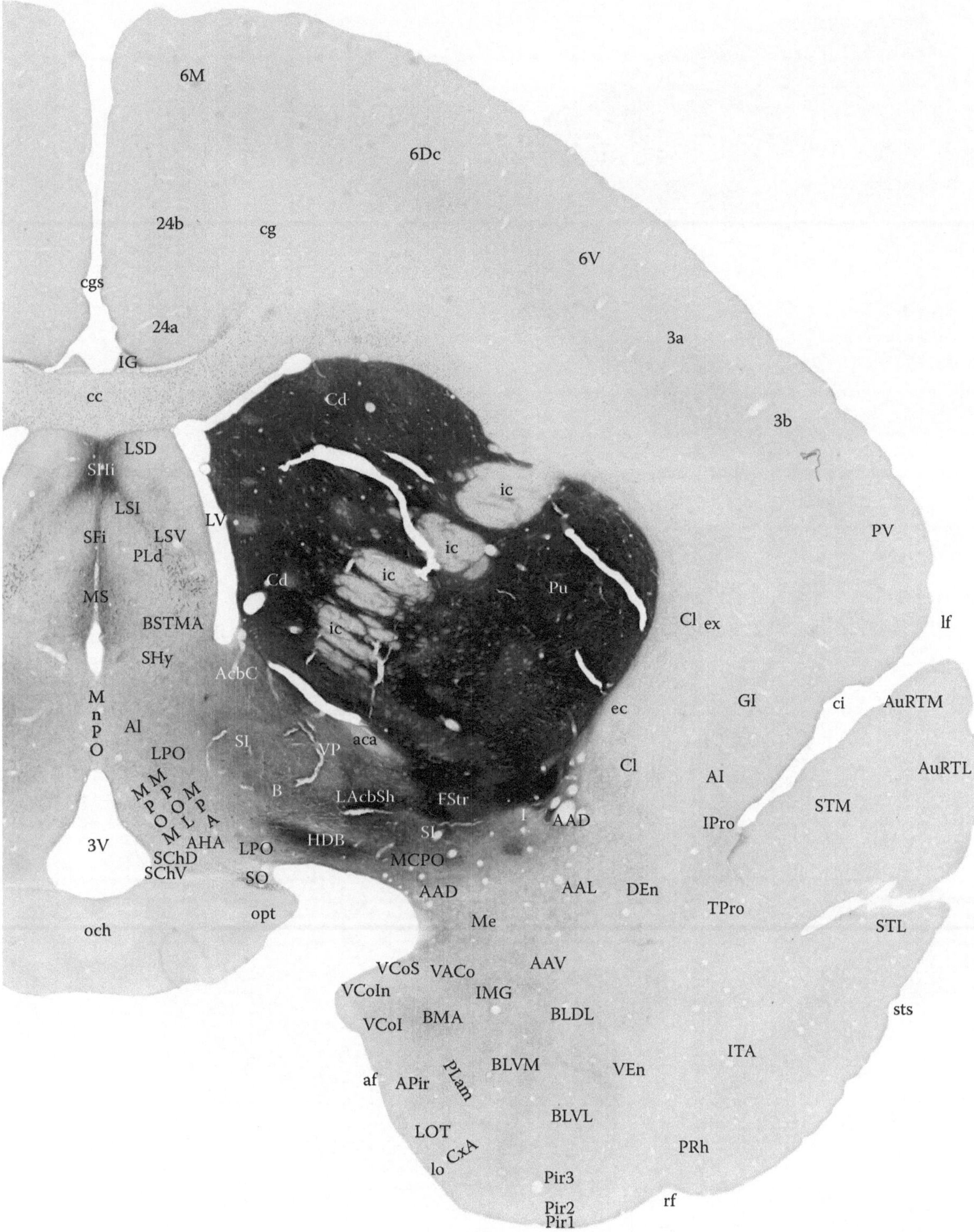


**Acetylcholinesterase**
Interaural +10.35 mm

# Plate 18d

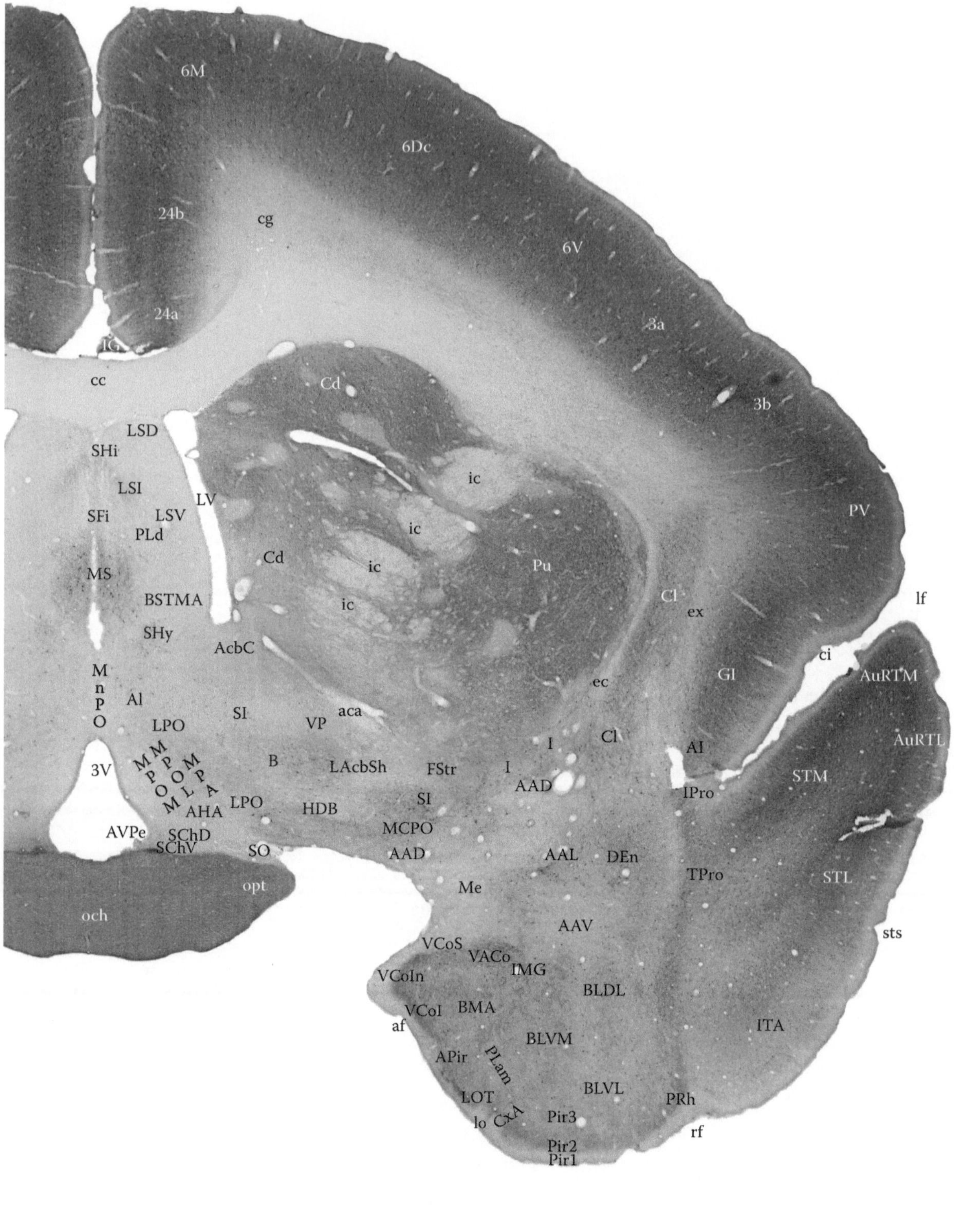


**Parvalbumin**
Interaural +10.30 mm

# Plate 18e

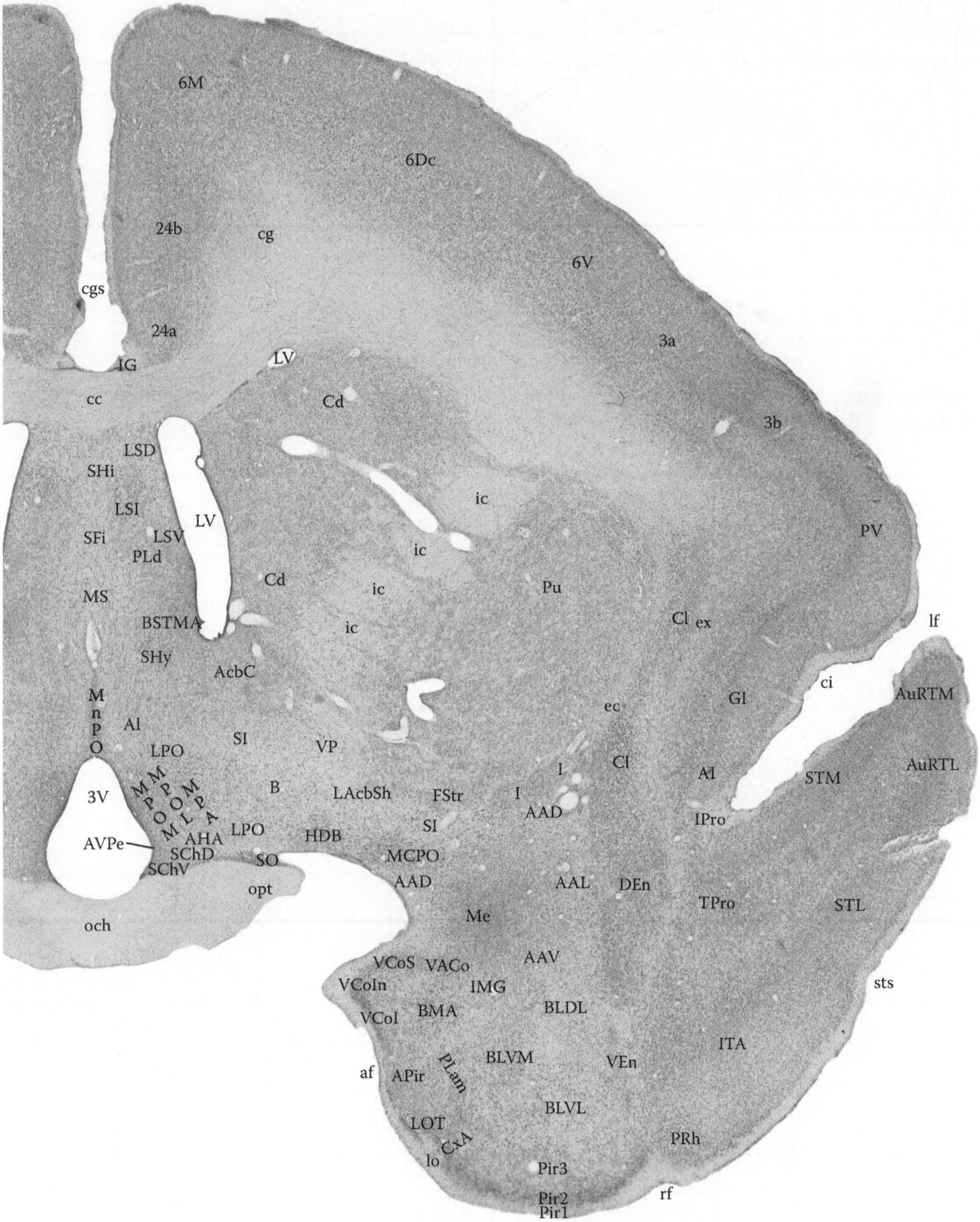


**Cresyl Violet**
Interaural +10.25 mm

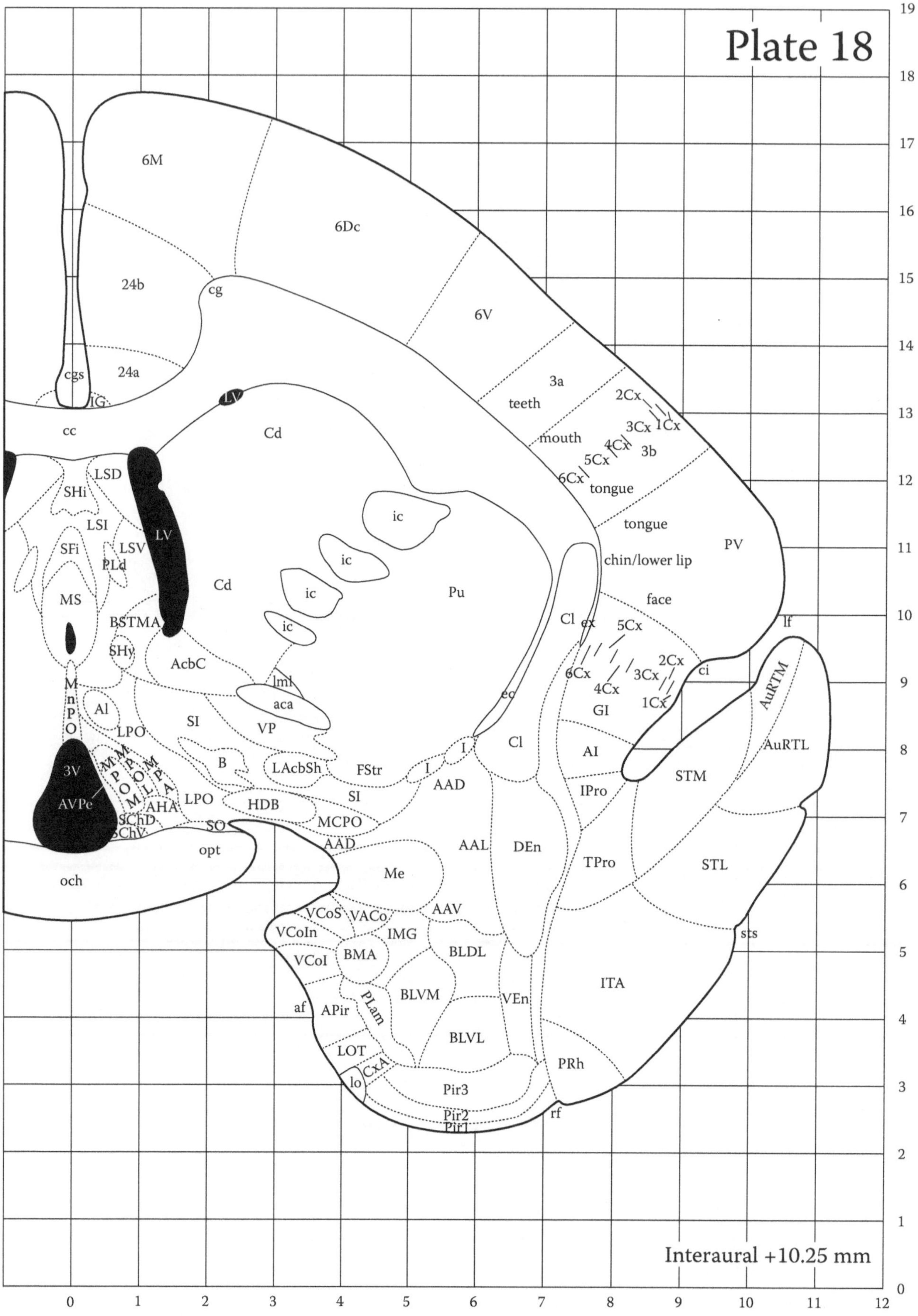

Plate 18
6M
6Dc
24b
cg
6V
24a
cgs
IG
3a
teeth
2Cx
3Cx
1Cx
mouth
4Cx
3b
5Cx
6Cx
tongue
LV
cc
Cd
LSD
SHi
ic
LSI
LV
tongue
SFi
LSV
PLd
ic
chin/lower lip
PV
Cd
ic
Pu
MS
face
BSTMA
Cl
ex
5Cx
lf
ic
SHy
AcbC
2Cx
6Cx
4Cx
3Cx
ci
lml
AuRTM
MnPO
aca
ec
1Cx
Al
SI
GI
LPO
VP
AuRTL
I
Cl
AI
3V
MPO
MPOM
LPO
MPA
B
LAcbSh
FStr
I
STM
AAD
IPro
AVPe
AHA
SChD
LPO
HDB
SI
SChV
SO
MCPO
opt
AAD
AAL
DEn
TPro
STL
och
Me
AAV
VCoS
VACo
VCoIn
IMG
sts
BMA
BLDL
VCoI
ITA
BLVM
PLam
VEn
af
APir
BLVL
LOT
PRh
CxA
lo
Pir3
rf
Pir2
Pir1
Interaural +10.25 mm

| | |
|---|---|
| 3a | area 3a Cx |
| 3b | area 3b Cx |
| 3V | third ventricle |
| 6Dc | area 6Dc Cx |
| 6M | area 6M Cx |
| 6V | area 6V Cx |
| 24a | area 24a Cx |
| 24b | area 24b Cx |
| AAD | ant. amyg. dors. |
| AAL | ant. amyg. lat. |
| AAV | ant. amyg. vent. |
| aca | ant. comm. ant. |
| AcbC | accumb. nu. core |
| af | amygdaloid fiss. |
| AHA | ant. hypoth. area, ant. |
| Al | agran. insul. Cx |
| APir | amygdalopir. trans. area |
| AuRTL | aud. belt Cx, rostrotemplat. |
| AuRTM | aud. belt Cx, rostrotempmed. |
| AVPe | antvent. perivent. nu. |
| B | basal nu. (Meynert) |
| BLDL | basolat. amyg. nu., dors. lat. |
| BLVL | basolat. amyg. nu., vent. lat. |
| BLVM | basolat. amyg. nu., vent. med. |
| BMA | basomed. amyg. nu., ant. |
| BSTMA | bed nu. st, med. ant. divn |
| cc | corp. callosum |
| Cd | caudate nu. |
| cg | cingulum |
| cgs | cingulate sulc. |
| ci | circular sulcus |
| Cl | claustrum |
| CxA | cortex-amygdala |
| DEn | dors. endopir. nu. |
| ec | external capsule |
| ex | extreme capsule |
| FStr | fundus striatum |
| GI | granular insul. Cx |
| HDB | horiz. limb diag. band nu. |
| I | intercal. nu. amyg. |
| ic | intern. caps. |
| IG | induseum gris. |
| IMG | intramedull. gr. amyg. |
| IPro | insular proisocortex |
| ITA | inf. temp. ant. Cx |
| LAcbSh | accumb. shell lat. |
| lf | lateral fissure |
| lml | lat. medull. lamina |
| lo | lat. olf. tract |
| LOT | nu. of lat. olf. tract |
| LPO | lat. preopt. area |
| LSD | lat. sept. nu., dors. |
| LSI | lat. sept. nu., int. |
| LSV | lat. sept. nu., vent. |
| LV | lateral ventricle |
| MCPO | magnocell. preopt. nu. |
| Me | medial amyg. nu. |
| MnPO | median preopt. nu. |
| MPA | medial preopt. area |
| MPOL | med. preopt. nu., lat. |
| MPOM | med. preopt. nu., med. |
| MS | med. sept. nu. |
| och | optic chiasm |
| opt | optic tract |
| Pir1 | piriform Cx 1 |
| Pir2 | piriform Cx 2 |
| Pir3 | piriform Cx 3 |
| PLam | paralam. amyg. nu. |
| PLd | paralambdoid septum |
| PRh | perirhinal Cx |
| Pu | putamen |
| PV | parietovent. somato. Cx |
| rf | rhinal fiss. |
| SChD | suprachiasm. nu., dors. |
| SChV | suprachiasm. nu., vent. |
| SFi | septofimbrial nu. |
| SHi | septohippo. nu. |
| SHy | septohypothal. nu. |
| SI | subst. innominata |
| SO | supraoptic nu. |
| STL | sup. temp. Cx, lat. |
| STM | sup. temp. Cx, med. |
| sts | sup. temp. sulcus |
| TPro | temp. proisocortex |
| VACo | vent. ant. cort. amyg. |
| VCoI | vent. inf. cort. amyg. |
| VCoIn | vent. inter. cort. amyg. |
| VCoS | vent. sup. cort. amyg. |
| VEn | vent. endopir. nu. |
| VP | vent. pallidum |

# Plate 19

Interaural +09.75 mm

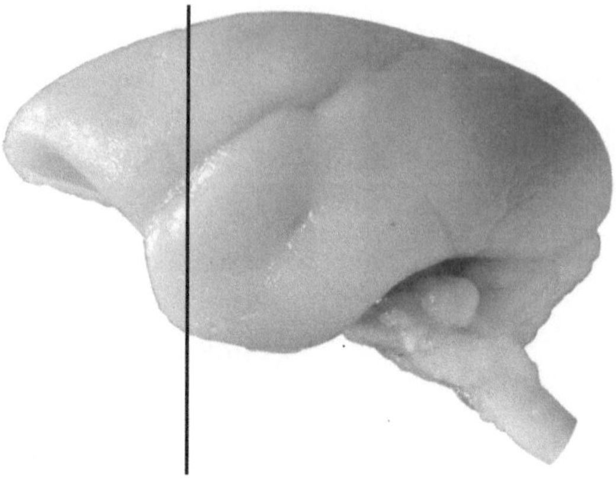

| | |
|---|---|
| 3a | area 3a Cx |
| 3b | area 3b Cx |
| 3V | third ventricle |
| 4 | area 4 Cx |
| 6M | area 6M Cx |
| 24a | area 24a Cx |
| 24b | area 24b Cx |
| AAD | ant. amyg. dors. |
| AAL | ant. amyg. lat. |
| ac | ant. comm. |
| Acb | accumb. nu. |
| acp | ant. comm. post. |
| af | amygdaloid fiss. |
| AHA | ant. hypoth. area, ant. |
| AI | agran. insul. Cx |
| APir | amygdalopir. trans. area |
| AuRTL | aud. belt Cx, rostrotemplat. |
| AuRTM | aud. belt Cx, rostrotempmed. |
| AVPe | antvent. perivent. nu. |
| B | basal nu. (Meynert) |
| BLD | basolat. amyg. nu., dors. |
| BLDL | basolat. amyg. nu., dors. lat. |
| BLVL | basolat. amyg. nu., vent. lat. |
| BLVM | basolat. amyg. nu., vent. med. |
| BMA | basomed. amyg. nu., ant. |
| BSTLD | bed nu. st, lat. dors. divn |
| BSTLJ | bed nu. st, lat. juxtacap. divn |
| BSTLP | bed nu. st, lat. post. divn |
| BSTLV | bed nu. st, lat. vent. divn |
| BSTMA | bed nu. st, med. ant. divn |
| BSTMP | bed nu. st, med. post. divn |
| cc | corp. callosum |
| Cd | caudate nu. |
| cg | cingulum |
| cgs | cingulate sulc. |
| ci | circular sulcus |
| Cl | claustrum |
| DEn | dors. endopir. nu. |
| ec | external capsule |
| EGP | ext. globus pallidus |
| EntR | entorhin. Cx, rost. |
| ex | extreme capsule |
| FStr | fundus striatum |
| GI | granular insul. Cx |
| HDB | horiz. limb diag. band nu. |
| I | intercal. nu. amyg. |
| ic | intern. caps. |
| IG | induseum gris. |
| IPro | insular proisocortex |
| ITA | inf. temp. ant. Cx |
| lf | lateral fissure |
| lml | lat. medull. lamina |
| LPO | lat. preopt. area |
| LSD | lat. sept. nu., dors. |
| LSI | lat. sept. nu., int. |
| LSV | lat. sept. nu., vent. |
| LV | lateral ventricle |
| MCPO | magnocell. preopt. nu. |
| Me | medial amyg. nu. |
| MnPO | median preopt. nu. |
| MPOC | med. preopt. nu., comp. |
| MPOL | med. preopt. nu., lat. |
| MPOM | med. preopt. nu., med. |
| MS | med. sept. nu. |
| och | optic chiasm |
| opt | optic tract |
| Pe | perivent. hypoth. nu. |
| PLam | paralam. amyg. nu. |
| PLd | paralambdoid septum |
| PRh | perirhinal Cx |
| PS | parastriatal nu. |
| Pu | putamen |
| PV | parietovent. somato. Cx |
| rf | rhinal fiss. |
| SChD | suprachiasm. nu., dors. |
| SChV | suprachiasm. nu., vent. |
| SFi | septofimbrial nu. |
| SHi | septohippo. nu. |
| SHy | septohypothal. nu. |
| SI | subst. innominata |
| SO | supraoptic nu. |
| sod | suproptic decn |
| st | stria terminalis |
| STL | sup. temp. Cx, lat. |
| STM | sup. temp. Cx, med. |
| sts | sup. temp. sulcus |
| TPro | temp. proisocortex |
| VCoI | vent. inf. cort. amyg. |
| VCoIn | vent. inter. cort. amyg. |
| VCoS | vent. sup. cort. amyg. |
| VEn | vent. endopir. nu. |
| VP | vent. pallidum |

# Plate 19a

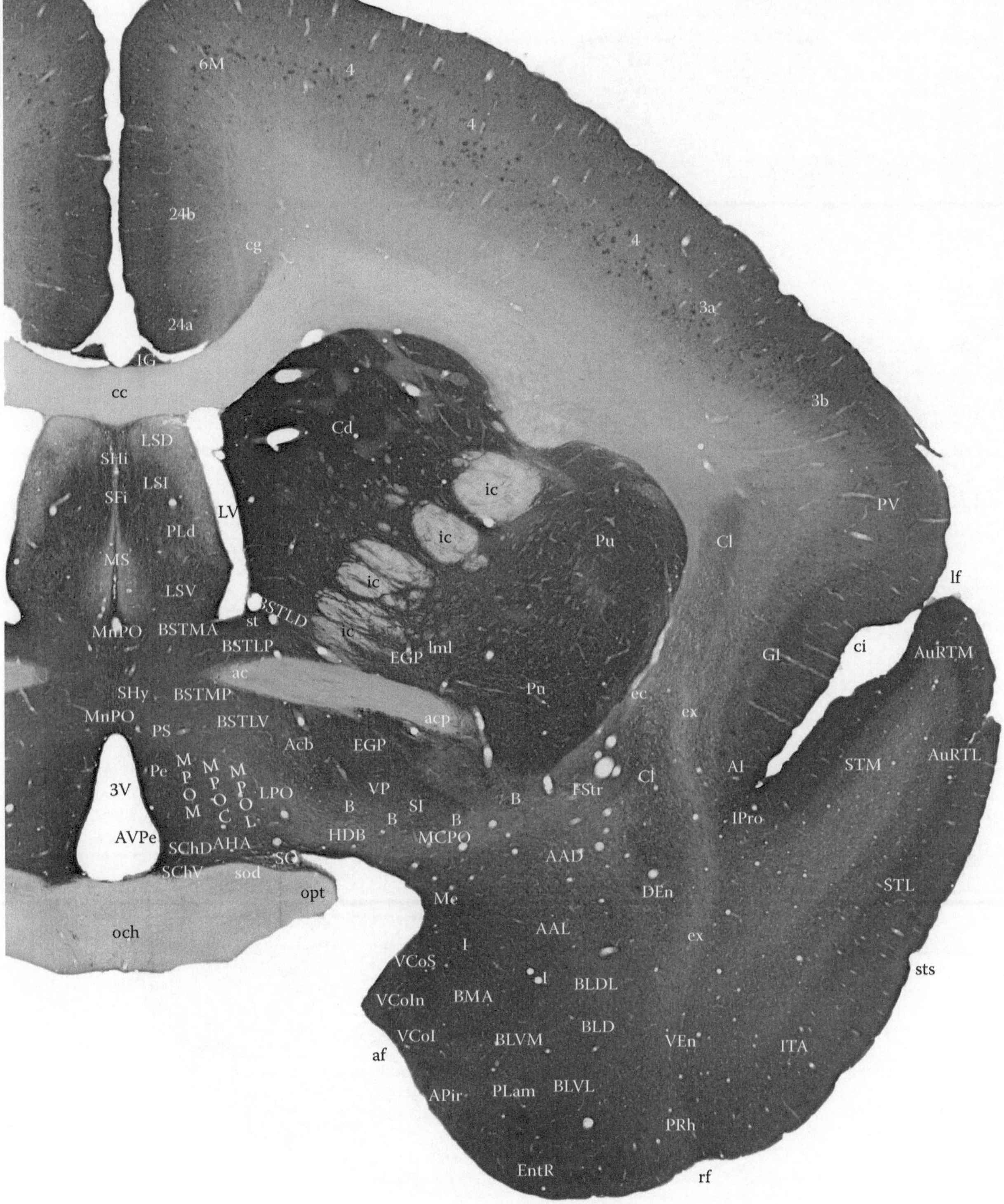


**Calbindin**
Interaural +10.05 mm

# Plate 19b

**Neurofilament**
Interaural +09.90 mm

# Plate 19c

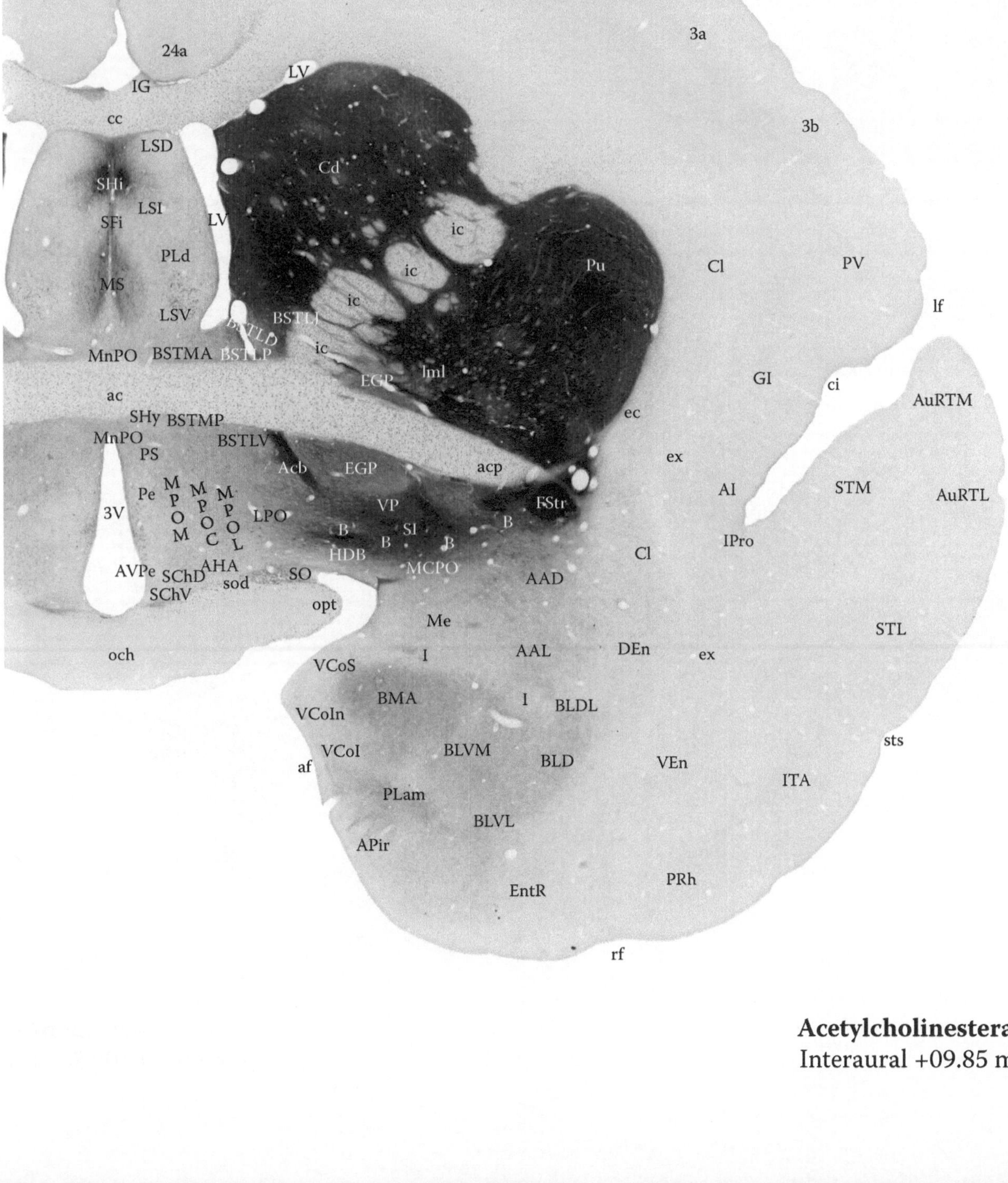


**Acetylcholinesterase**
Interaural +09.85 mm

# Plate 19d

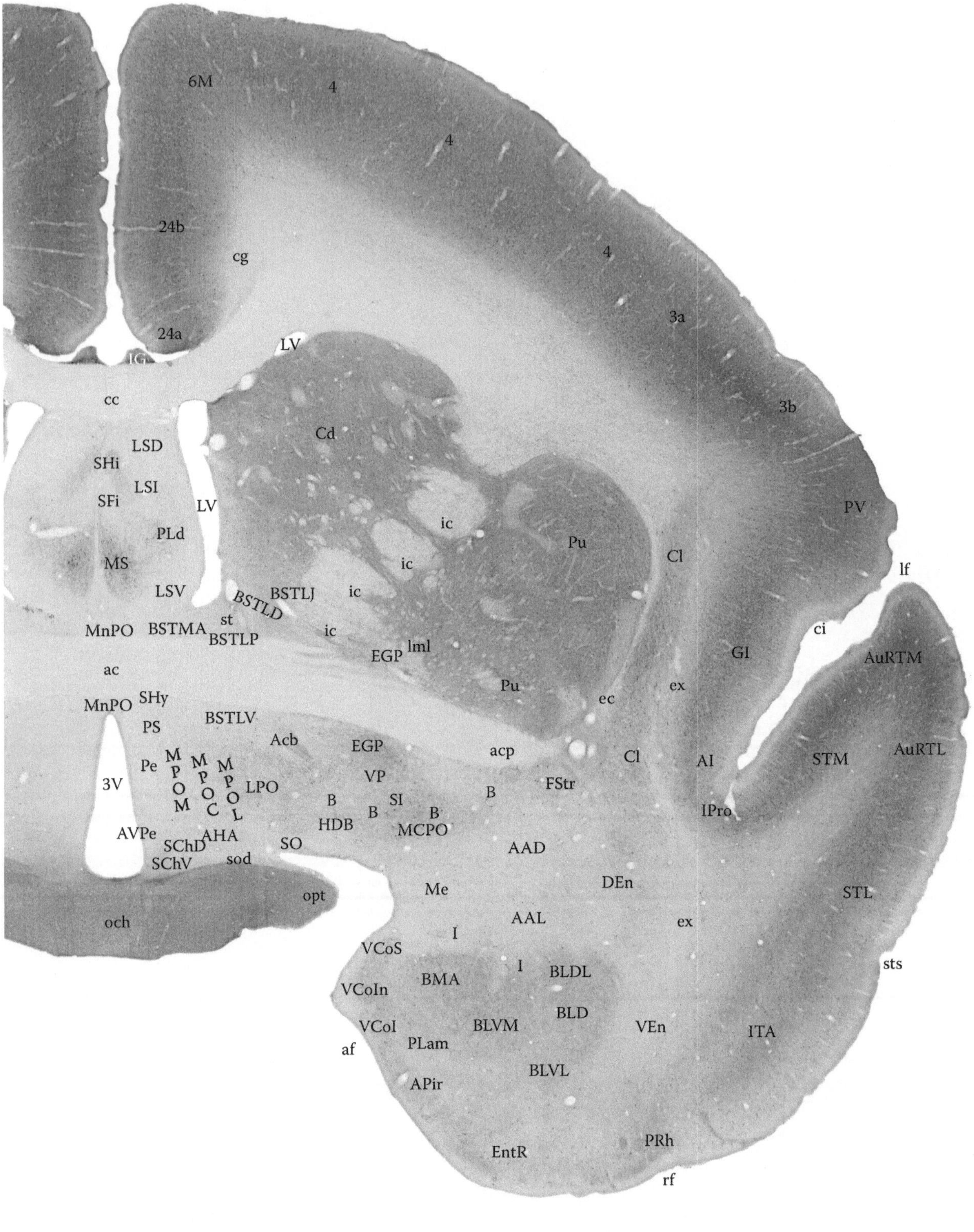


**Parvalbumin**
Interaural +09.80 mm

# Plate 19e

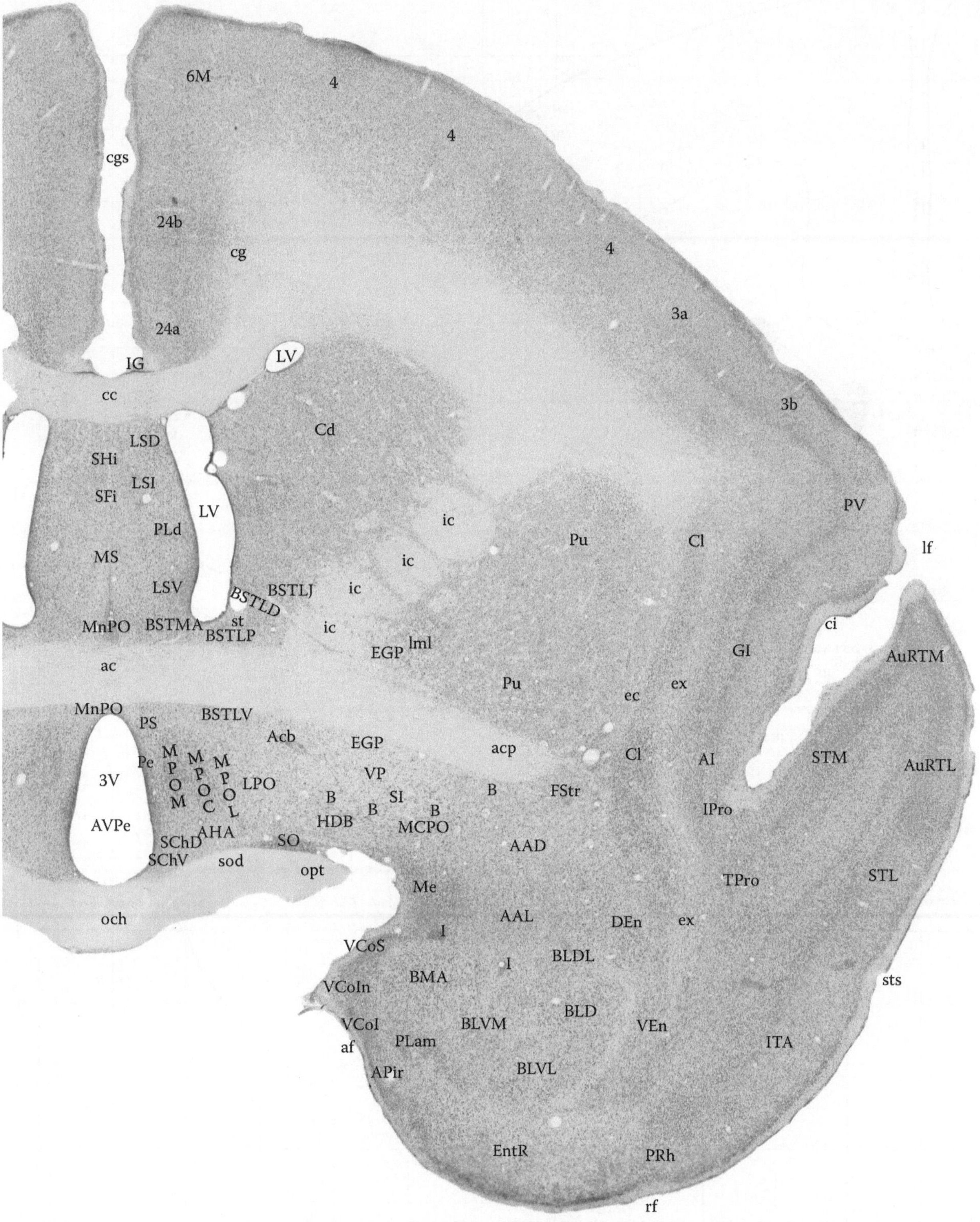


**Cresyl Violet**
Interaural +09.75 mm

# Plate 19

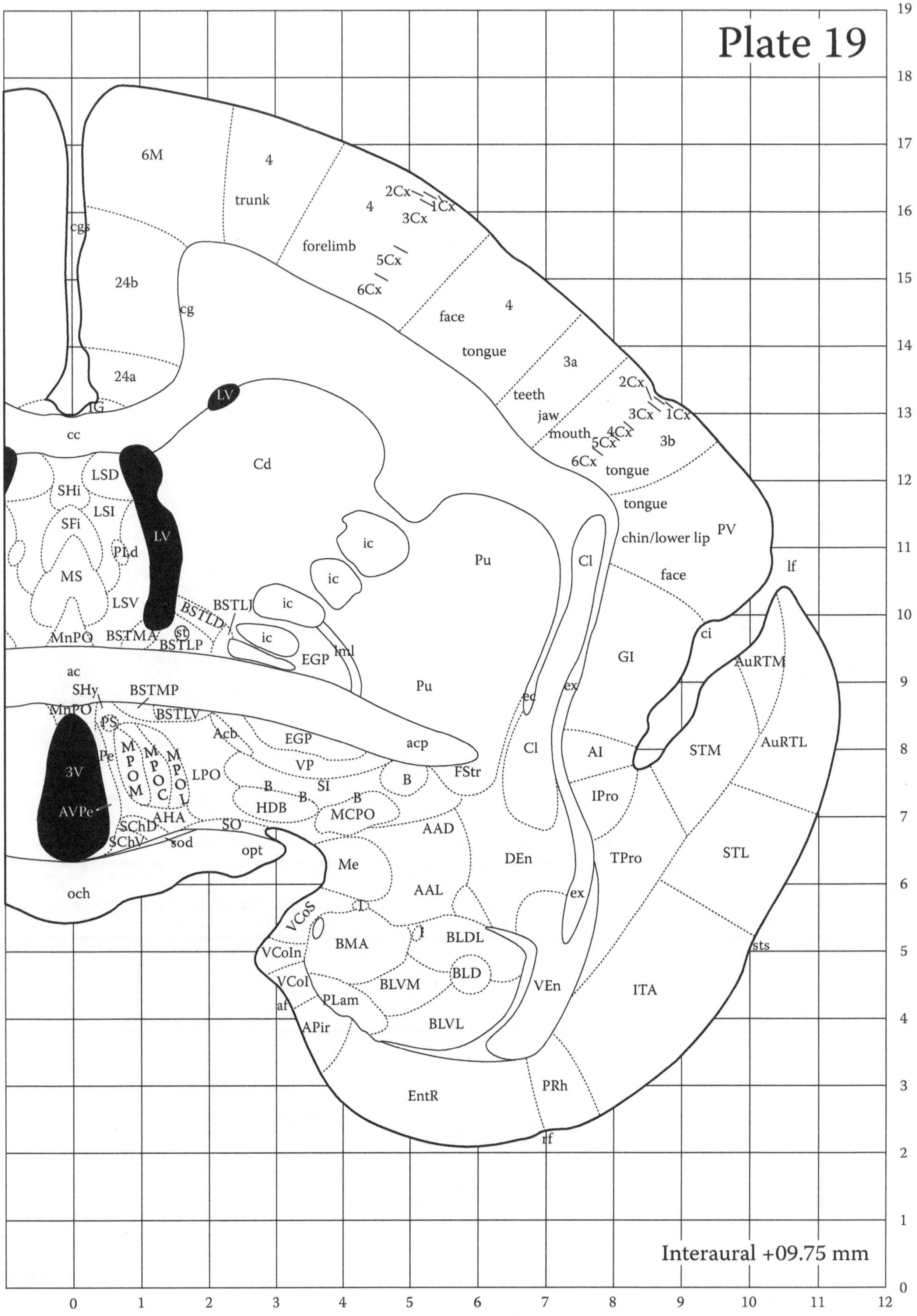
6M
4
trunk
2Cx
1Cx
4
3Cx
forelimb
5Cx
6Cx
cgs
24b
cg
face
4
tongue
3a
24a
LV
2Cx
teeth
jaw
3Cx
1Cx
mouth
4Cx
5Cx
3b
IG
cc
Cd
6Cx
tongue
LSD
SHi
LSI
SFi
LV
PLd
MS
LSV
tongue
ic
ic
Pu
Cl
chin/lower lip
PV
face
lf
ic
BSTLJ
BSTLD
MnPO
BSTMA
st
BSTLP
ic
EGP
lml
ci
GI
AuRTM
ac
SHy
BSTMP
Pu
ec
ex
MnPO
BSTLV
PS
Acb
EGP
acp
Cl
AI
STM
AuRTL
Pe
MPOM
MPOC
MPOL
LPO
VP
FStr
3V
SI
B
B
B
B
IPro
HDB
MCPO
AVPe
AHA
AAD
SChD
SO
SChV
sod
opt
DEn
TPro
STL
Me
och
AAL
ex
VCoS
I
BMA
I
BLDL
sts
VCoIn
BLD
VCoI
BLVM
VEn
ITA
af
PLam
APir
BLVL
EntR
PRh
rf
Interaural +09.75 mm
0
1
2
3
4
5
6
7
8
9
10
11
12
0
1
2
3
4
5
6
7
8
9
10
11
12
13
14
15
16
17
18
19

| | |
|---|---|
| 3a | area 3a Cx |
| 3b | area 3b Cx |
| 3V | third ventricle |
| 4 | area 4 Cx |
| 6M | area 6M Cx |
| 24a | area 24a Cx |
| 24b | area 24b Cx |
| AAD | ant. amyg. dors. |
| AAL | ant. amyg. lat. |
| ac | ant. comm. |
| Acb | accumb. nu. |
| acp | ant. comm. post. |
| af | amygdaloid fiss. |
| AHA | ant. hypoth. area, ant. |
| AI | agran. insul. Cx |
| APir | amygdalopir. trans. area |
| AuRTL | aud. belt Cx, rostrotemplat. |
| AuRTM | aud. belt Cx, rostrotempmed. |
| AVPe | antvent. perivent. nu. |
| B | basal nu. (Meynert) |
| BLD | basolat. amyg. nu., dors. |
| BLDL | basolat. amyg. nu., dors. lat. |
| BLVL | basolat. amyg. nu., vent. lat. |
| BLVM | basolat. amyg. nu., vent. med. |
| BMA | basomed. amyg. nu., ant. |
| BSTLD | bed nu. st, lat. dors. divn |
| BSTLJ | bed nu. st, lat. juxtacap. divn |
| BSTLP | bed nu. st, lat. post. divn |
| BSTLV | bed nu. st, lat. vent. divn |
| BSTMA | bed nu. st, med. ant. divn |
| BSTMP | bed nu. st, med. post. divn |
| cc | corp. callosum |
| Cd | caudate nu. |
| cg | cingulum |
| cgs | cingulate sulc. |
| ci | circular sulcus |
| Cl | claustrum |
| DEn | dors. endopir. nu. |
| ec | external capsule |
| EGP | ext. globus pallidus |
| EntR | entorhin. Cx, rost. |
| ex | extreme capsule |
| FStr | fundus striatum |
| GI | granular insul. Cx |
| HDB | horiz. limb diag. band nu. |
| I | intercal. nu. amyg. |
| ic | intern. caps. |
| IG | induseum gris. |
| IPro | insular proisocortex |
| ITA | inf. temp. ant. Cx |
| lf | lateral fissure |
| lml | lat. medull. lamina |
| LPO | lat. preopt. area |
| LSD | lat. sept. nu., dors. |
| LSI | lat. sept. nu., int. |
| LSV | lat. sept. nu., vent. |
| LV | lateral ventricle |
| MCPO | magnocell. preopt. nu. |
| Me | medial amyg. nu. |
| MnPO | median preopt. nu. |
| MPOC | med. preopt. nu., comp. |
| MPOL | med. preopt. nu., lat. |
| MPOM | med. preopt. nu., med. |
| MS | med. sept. nu. |
| och | optic chiasm |
| opt | optic tract |
| Pe | perivent. hypoth. nu. |
| PLam | paralam. amyg. nu. |
| PLd | paralambdoid septum |
| PRh | perirhinal Cx |
| PS | parastriatal nu. |
| Pu | putamen |
| PV | parietovent. somato. Cx |
| rf | rhinal fiss. |
| SChD | suprachiasm. nu., dors. |
| SChV | suprachiasm. nu., vent. |
| SFi | septofimbrial nu. |
| SHi | septohippo. nu. |
| SHy | septohypothal. nu. |
| SI | subst. innominata |
| SO | supraoptic nu. |
| sod | suproptic decn |
| st | stria terminalis |
| STL | sup. temp. Cx, lat. |
| STM | sup. temp. Cx, med. |
| sts | sup. temp. sulcus |
| TPro | temp. proisocortex |
| VCoI | vent. inf. cort. amyg. |
| VCoIn | vent. inter. cort. amyg. |
| VCoS | vent. sup. cort. amyg. |
| VEn | vent. endopir. nu. |
| VP | vent. pallidum |

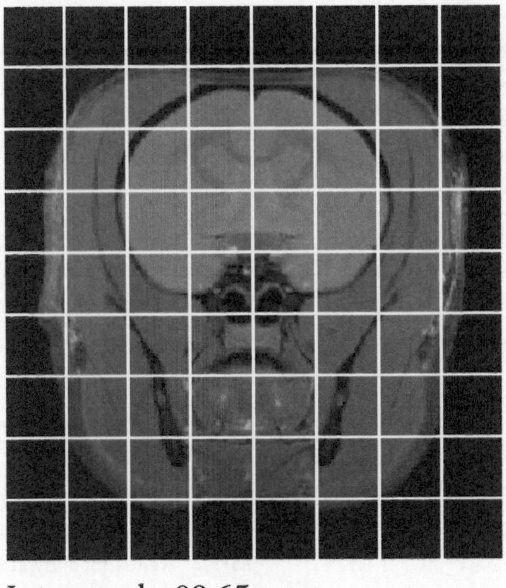

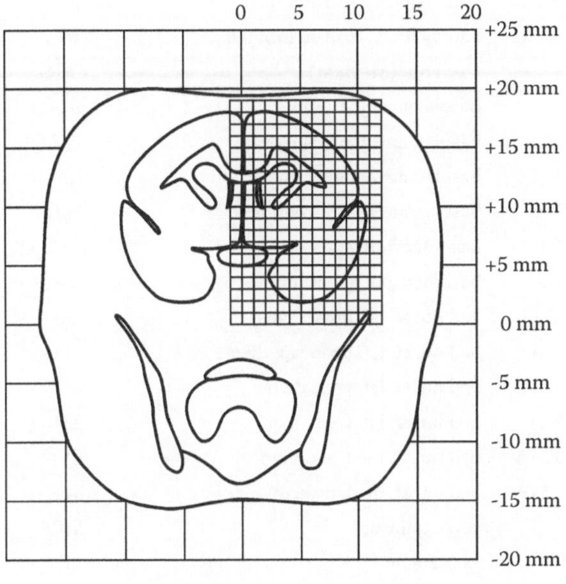


Interaural +09.65 mm

# Plate 20

Interaural +09.25 mm

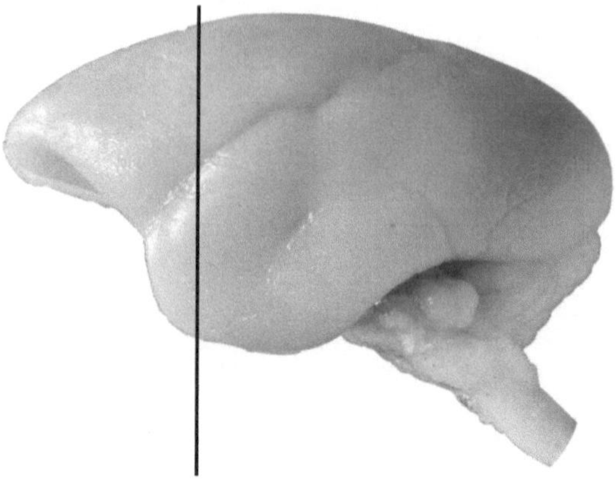

| | |
|---|---|
| 3a | area 3a Cx |
| 3b | area 3b Cx |
| 3V | third ventricle |
| 4 | area 4 Cx |
| 6M | area 6M Cx |
| 24a | area 24a Cx |
| 24b | area 24b Cx |
| 36 | area 30 Cx |
| AA | ant. amyg. area |
| ac | ant. comm. |
| acp | ant. comm. post. |
| af | amygdaloid fiss. |
| AHA | ant. hypoth. area, ant. |
| AHiMC | amygdalohipp., parv. |
| AHiPC | amygdalohipp., magn. |
| AI | agran. insul. Cx |
| APir | amygdalopir. trans. area |
| AuRTL | aud. belt Cx, rostrotemplat. |
| AuRTM | aud. belt Cx, rostrotempmed. |
| B | basal nu. (Meynert) |
| BLD | basolat. amyg. nu., dors. |
| BLDL | basolat. amyg. nu., dors. lat. |
| BLVL | basolat. amyg. nu., vent. lat. |
| BLVM | basolat. amyg. nu., vent. med. |
| BMA | basomed. amyg. nu., ant. |
| BMD | basomed. amyg. nu., dors. |
| BSTLD | bed nu. st, lat. dors. divn |
| BSTLJ | bed nu. st, lat. juxtacap. divn |
| BSTLP | bed nu. st, lat. post. divn |
| BSTLV | bed nu. st, lat. vent. divn |
| BSTMA | bed nu. st, med. ant. divn |
| BSTMP | bed nu. st, med. post. divn |
| cc | corp. callosum |
| Cd | caudate nu. |
| CeC | cent. amyg. nu., cent. |
| CeL | cent. amyg. nu., lat. |
| CeM | cent. amyg. nu., med. |
| cg | cingulum |
| ci | circular sulcus |
| Cl | claustrum |
| DEn | dors. endopir. nu. |
| EAC | extend amyg. central |
| EAM | extend amyg. medial |
| ec | external capsule |
| EGP | ext. globus pallidus |
| EntL | entorhin. Cx, lat. |
| EntM | entorhin. Cx, med. |
| ex | extreme capsule |
| f | fornix |
| GI | granular insul. Cx |
| HDB | horiz. limb diag. band nu. |
| I | intercal. nu. amyg. |
| ic | intern. caps. |
| IG | induseum gris. |
| IMG | intramedull. gr. amyg. |
| IPAC | interst. nu. acp |
| ITA | inf. temp. ant. Cx |
| LA | lateroant. hypoth. nu. |
| LaL | lat. amyg. nu., lat. |
| LaM | lat. amyg. nu., med. |
| lf | lateral fissure |
| lml | lat. medull. lamina |
| LPO | lat. preopt. area |
| LSD | lat. sept. nu., dors. |
| LSI | lat. sept. nu., int. |
| LV | lateral ventricle |
| Me | medial amyg. nu. |
| MEE | median eminence, ext. |
| MEI | median eminence, int. |
| MPA | medial preopt. area |
| MPOC | med. preopt. nu., comp. |
| MPOL | med. preopt. nu., lat. |
| MPOM | med. preopt. nu., med. |
| opt | optic tract |
| PaM | paravent. nu., magno. |
| PaMD | paravent. nu., magno. dors. |
| Pe | perivent. hypoth. nu. |
| PLam | paralam. amyg. nu. |
| PRh | perirhinal Cx |
| PS | parastriatal nu. |
| Pu | putamen |
| PV | parietovent. Cx |
| RChL | retrochiasm. nu, lat. |
| RChM | retrochiasm. nu, med. |
| rf | rhinal fiss. |
| SFi | septofimbrial nu. |
| SFO | subfornical organ |
| SHi | septohippo. nu. |
| SHy | septohypothal. nu. |
| SI | subst. innominata |
| SO | supraoptic nu. |
| sod | suproptic decn |
| SOR | supraoptic nu, retroch. |
| st | stria terminalis |
| STL | sup. temp. Cx, lat. |
| STM | sup. temp. Cx, med. |
| sts | sup. temp. sulcus |
| TPro | temp. proisocortex |
| TS | triang. sept. nu. |
| TuLH | tub. lat. hypothal. |
| VCoI | vent. inf. cort. amyg. |
| VCoS | vent. sup. cort. amyg. |
| VMH | ventromed. hypoth. nu. |

# Plate 20a

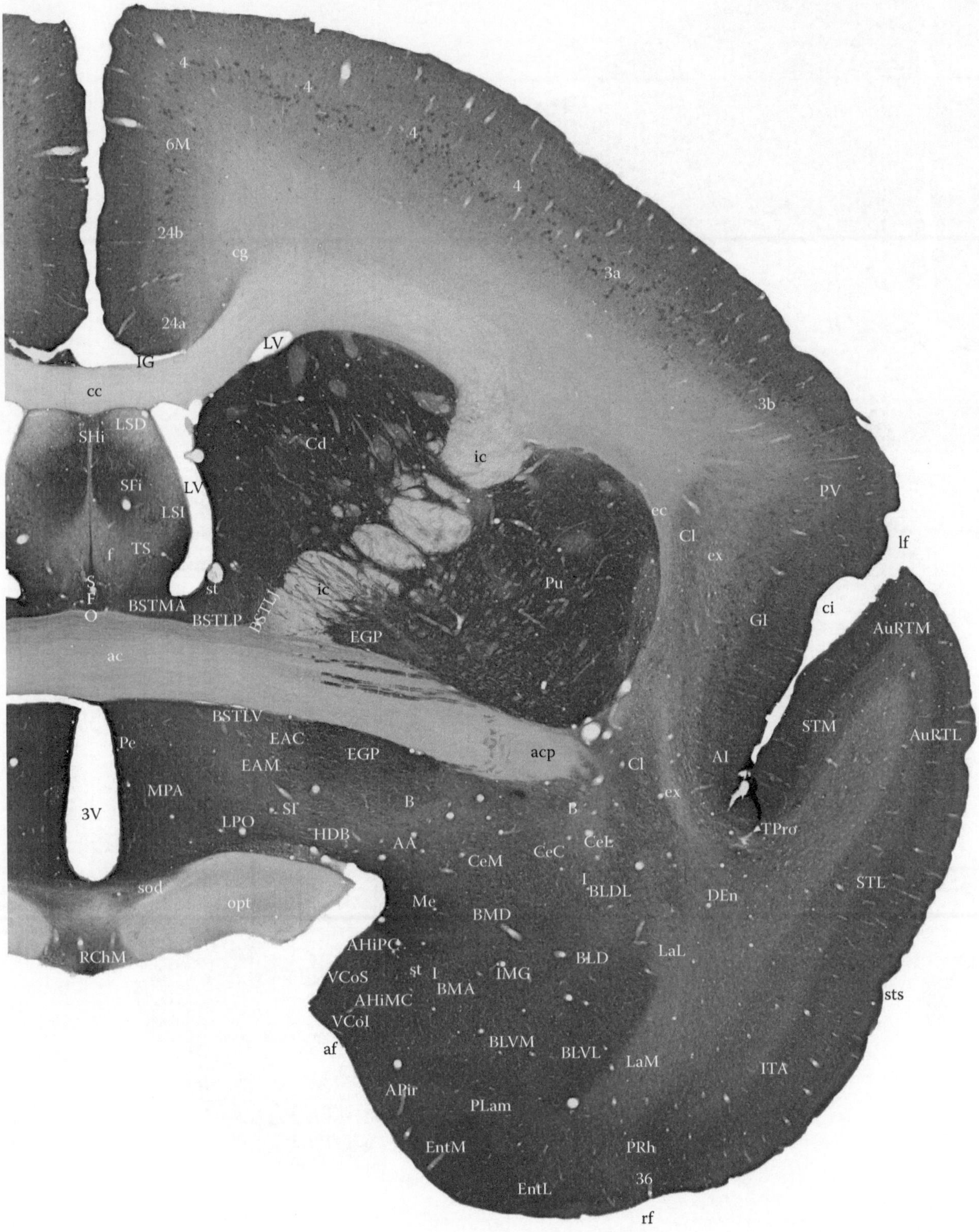


**Calbindin**
Interaural +09.55 mm

# Plate 20b

**Neurofilament**
Interaural +09.40 mm

# Plate 20c

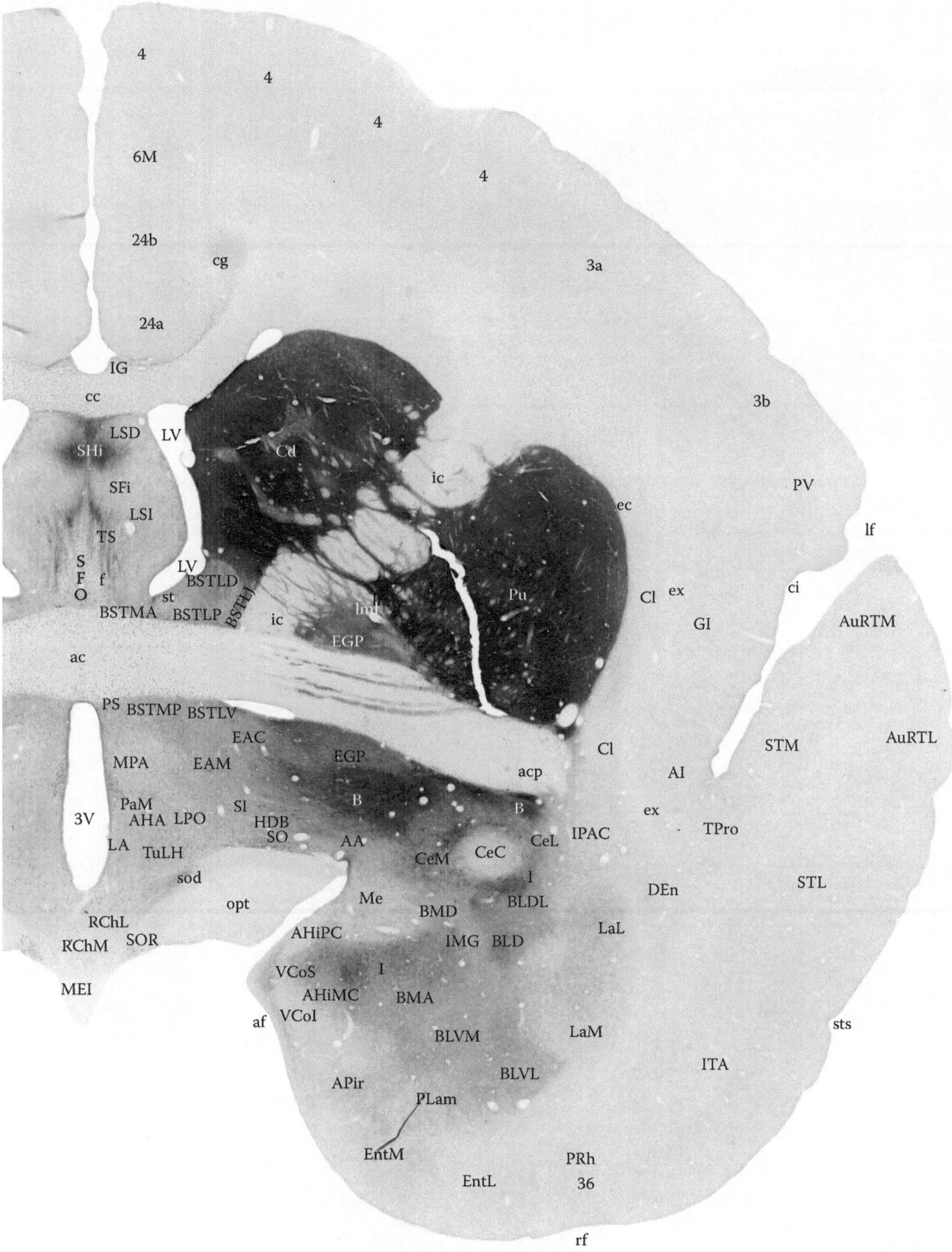


**Acetylcholinesterase**
Interaural +09.35 mm

# Plate 20d

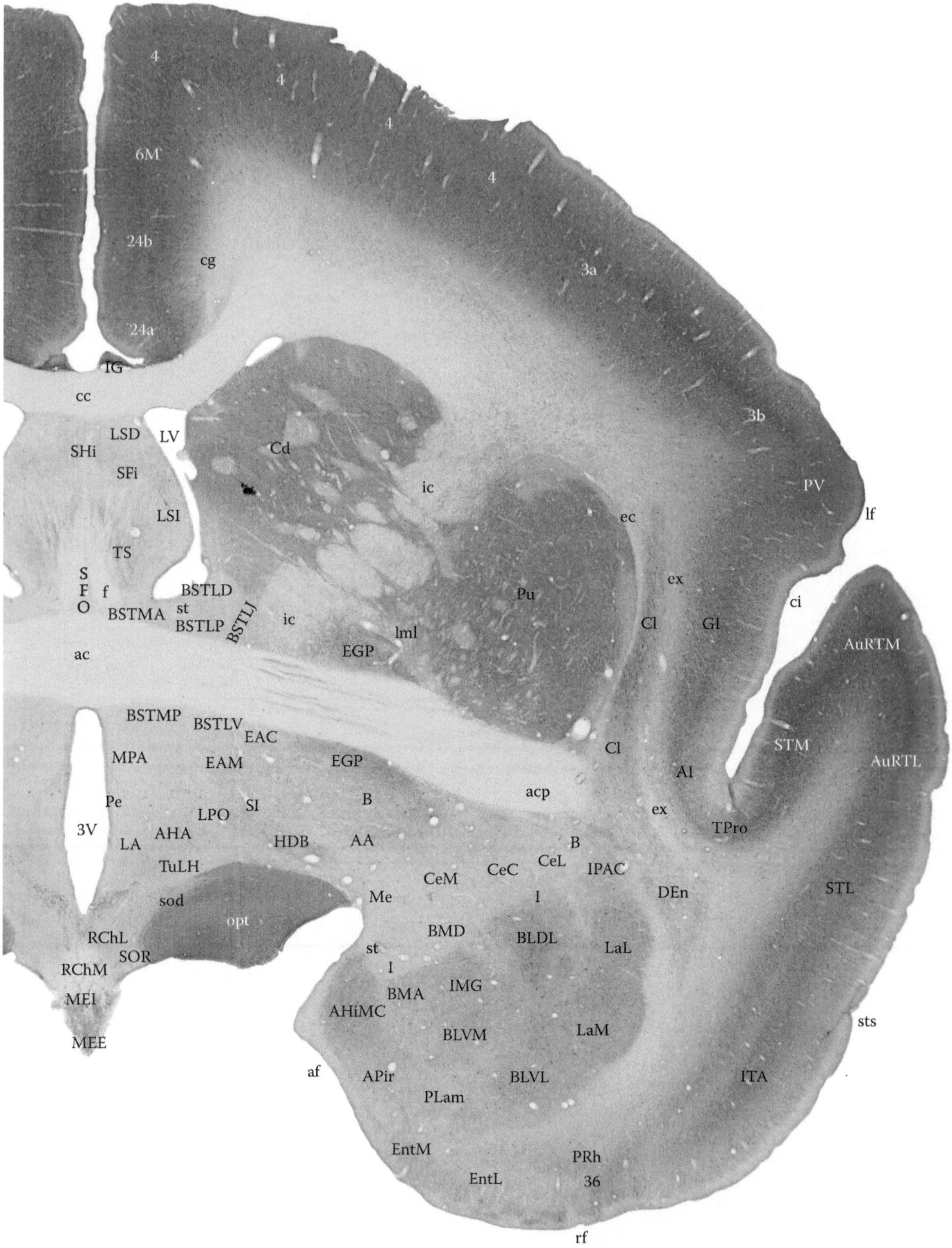


**Parvalbumin**
Interaural +09.30 mm

# Plate 20e

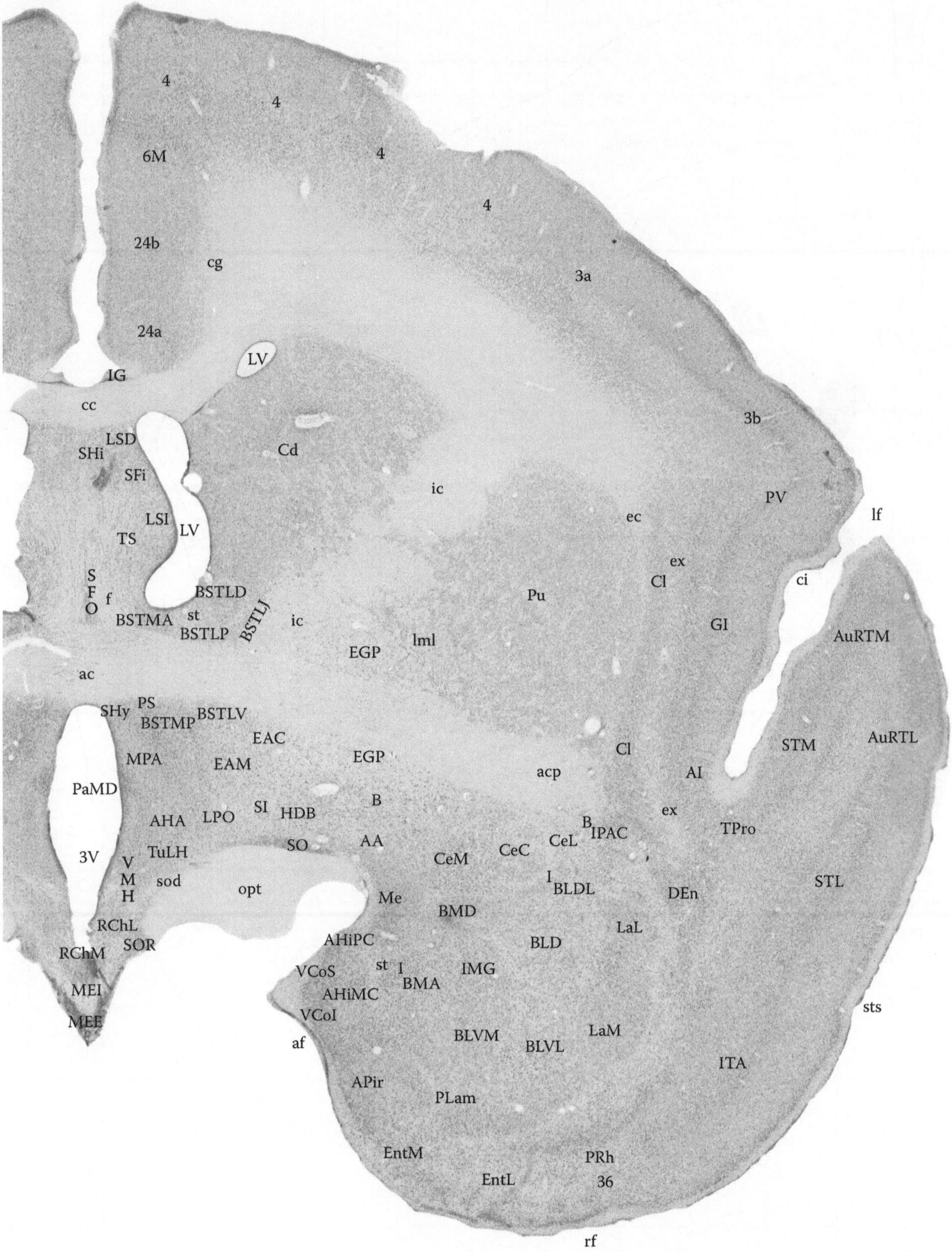


**Cresyl Violet**
Interaural +09.25 mm

# Plate 20

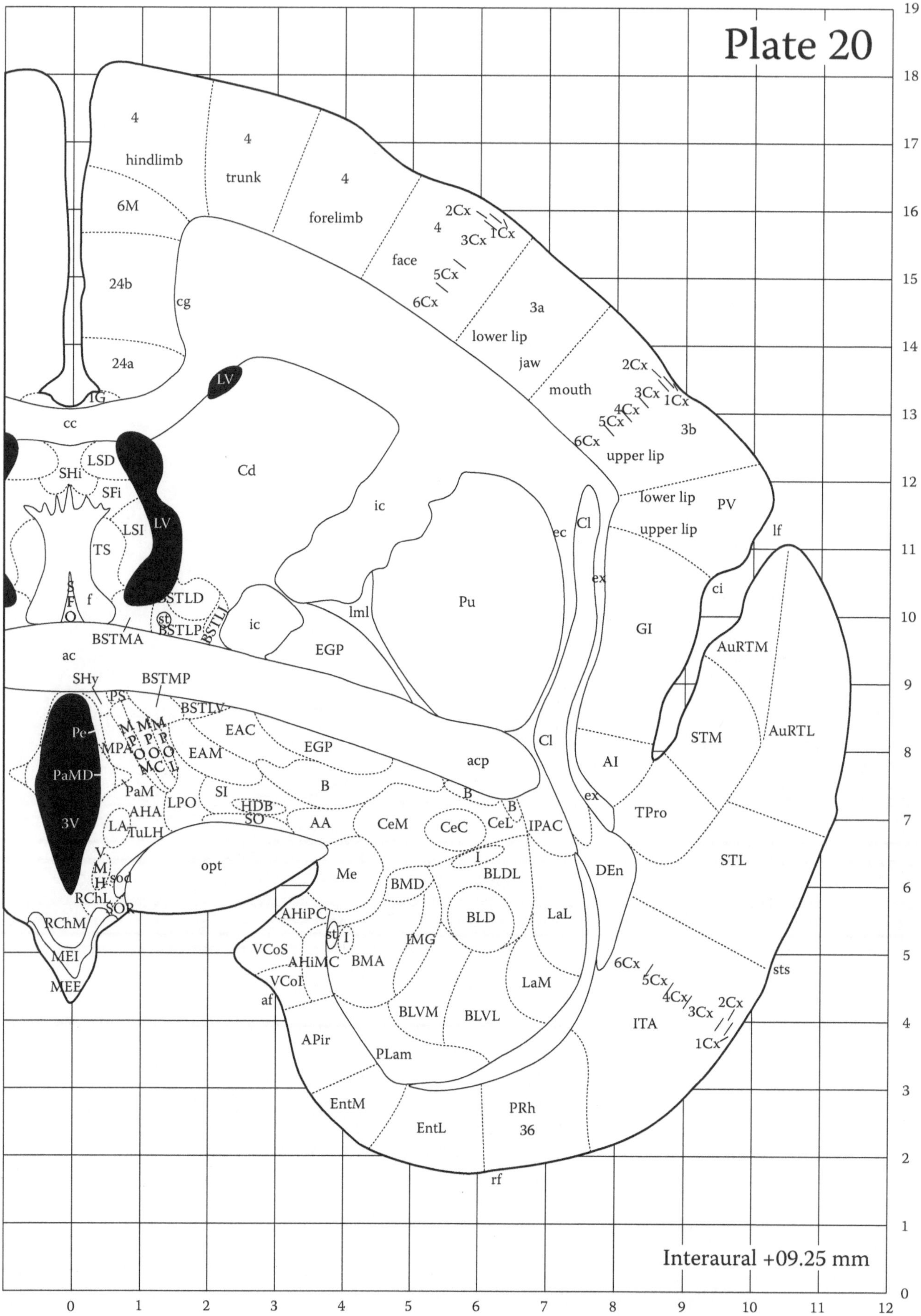


Interaural +09.25 mm

| | |
|---|---|
| 3a | area 3a Cx |
| 3b | area 3b Cx |
| 3V | third ventricle |
| 4 | area 4 Cx |
| 6M | area 6M Cx |
| 24a | area 24a Cx |
| 24b | area 24b Cx |
| 36 | area 30 Cx |
| AA | ant. amyg. area |
| ac | ant. comm. |
| acp | ant. comm. post. |
| af | amygdaloid fiss. |
| AHA | ant. hypoth. area, ant. |
| AHiMC | amygdalohipp., parv. |
| AHiPC | amygdalohipp., magn. |
| AI | agran. insul. Cx |
| APir | amygdalopir. trans. area |
| AuRTL | aud. belt Cx, rostrotemplat. |
| AuRTM | aud. belt Cx, rostrotempmed. |
| B | basal nu. (Meynert) |
| BLD | basolat. amyg. nu., dors. |
| BLDL | basolat. amyg. nu., dors. lat. |
| BLVL | basolat. amyg. nu., vent. lat. |
| BLVM | basolat. amyg. nu., vent. med. |
| BMA | basomed. amyg. nu., ant. |
| BMD | basomed. amyg. nu., dors. |
| BSTLD | bed nu. st, lat. dors. divn |
| BSTLJ | bed nu. st, lat. juxtacap. divn |
| BSTLP | bed nu. st, lat. post. divn |
| BSTLV | bed nu. st, lat. vent. divn |
| BSTMA | bed nu. st, med. ant. divn |
| BSTMP | bed nu. st, med. post. divn |
| cc | corp. callosum |
| Cd | caudate nu. |
| CeC | cent. amyg. nu., cent. |
| CeL | cent. amyg. nu., lat. |
| CeM | cent. amyg. nu., med. |
| cg | cingulum |
| ci | circular sulcus |
| Cl | claustrum |
| DEn | dors. endopir. nu. |
| EAC | extend amyg. central |
| EAM | extend amyg. medial |
| ec | external capsule |
| EGP | ext. globus pallidus |
| EntL | entorhin. Cx, lat. |
| EntM | entorhin. Cx, med. |
| ex | extreme capsule |
| f | fornix |
| GI | granular insul. Cx |
| HDB | horiz. limb diag. band nu. |
| I | intercal. nu. amyg. |
| ic | intern. caps. |
| IG | induseum gris. |
| IMG | intramedull. gr. amyg. |
| IPAC | interst. nu. acp |
| ITA | inf. temp. ant. Cx |
| LA | lateroant. hypoth. nu. |
| LaL | lat. amyg. nu., lat. |
| LaM | lat. amyg. nu., med. |
| lf | lateral fissure |
| lml | lat. medull. lamina |
| LPO | lat. preopt. area |
| LSD | lat. sept. nu., dors. |
| LSI | lat. sept. nu., int. |
| LV | lateral ventricle |
| Me | medial amyg. nu. |
| MEE | median eminence, ext. |
| MEI | median eminence, int. |
| MPA | medial preopt. area |
| MPOC | med. preopt. nu., comp. |
| MPOL | med. preopt. nu., lat. |
| MPOM | med. preopt. nu., med. |
| opt | optic tract |
| PaM | paravent. nu., magno. |
| PaMD | paravent. nu., magno. dors. |
| Pe | perivent. hypoth. nu. |
| PLam | paralam. amyg. nu. |
| PRh | perirhinal Cx |
| PS | parastriatal nu. |
| Pu | putamen |
| PV | parietovent. Cx |
| RChL | retrochiasm. nu, lat. |
| RChM | retrochiasm. nu, med. |
| rf | rhinal fiss. |
| SFi | septofimbrial nu. |
| SFO | subfornical organ |
| SHi | septohippo. nu. |
| SHy | septohypothal. nu. |
| SI | subst. innominata |
| SO | supraoptic nu. |
| sod | suproptic decn |
| SOR | supraoptic nu, retroch. |
| st | stria terminalis |
| STL | sup. temp. Cx, lat. |
| STM | sup. temp. Cx, med. |
| sts | sup. temp. sulcus |
| TPro | temp. proisocortex |
| TS | triang. sept. nu. |
| TuLH | tub. lat. hypothal. |
| VCoI | vent. inf. cort. amyg. |
| VCoS | vent. sup. cort. amyg. |
| VMH | ventromed. hypoth. nu. |

# Plate 21

Interaural +08.70 mm

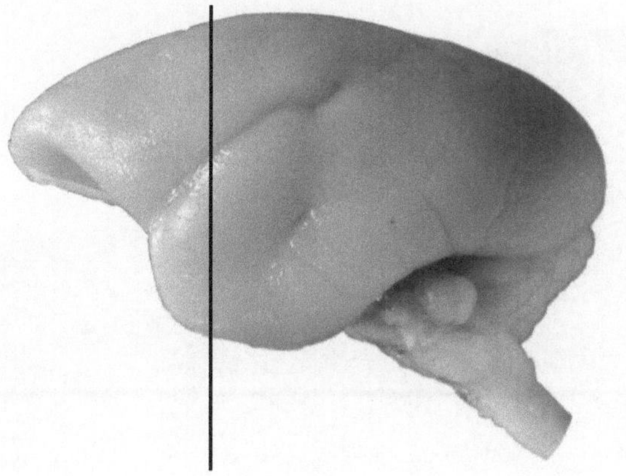

| | | | | | |
|---|---|---|---|---|---|
| 3a | area 3a Cx | BSTMPM | bed nu. st, med. post. med. | LV | lateral ventricle |
| 3b | area 3b Cx | cc | corp. callosum | Me | medial amyg. nu. |
| 3V | third ventricle | Cd | caudate nu. | MEE | median eminence, ext. |
| 4 | area 4 Cx | CeC | cent. amyg. nu., cent. | MEI | median eminence, int. |
| 24a | area 24a Cx | CeL | cent. amyg. nu., lat. | nehy | neurohypophyseal tr. |
| 24b | area 24b Cx | CeM | cent. amyg. nu., med. | opt | optic tract |
| 36 | area 30 Cx | cg | cingulum | PaAP | paravent. nu., ant. parv. |
| AA | ant. amyg. area | ci | circular sulcus | PaMD | paravent. nu., magno. dors. |
| acp | ant. comm. post. | Cl | claustrum | PaPD | paravent. nu., parv. dors. |
| af | amygdaloid fiss. | DMC | dorsmed. hypoth. nu., comp. | PaV | paravent. nu., vent. |
| AHA | ant. hypoth. area, ant. | DMD | dorsmed. hypoth. nu., dors. | Pe | perivent. hypoth. nu. |
| AHiMC | amygdalohipp., parv. | EAC | extend amyg. central | PLam | paralam. amyg. nu. |
| AHiPC | amygdalohipp., magn. | EAM | extend amyg. medial | PRh | perirhinal Cx |
| AI | agran. insul. Cx | ec | external capsule | Pu | putamen |
| al | ansa lenticularis | EGP | ext. globus pallidus | PV | PV somato. Cx |
| APir | amygdalopir. trans. area | EntL | entorhin. Cx, lat. | rf | rhinal fiss. |
| APit | ant. pituit. gland | EntM | entorhin. Cx, med. | SFi | septofimbrial nu. |
| ArcL | arc. hypoth. nu., lat. | ex | extreme capsule | SFO | subfornical organ |
| ArcM | arc. hypoth. nu., med. | f | fornix | SHi | septohippo. nu. |
| AuRPB | aud. parabelt Cx, rost. | GI | granular insul. Cx | SI | subst. innominata |
| AuRT | aud. core Cx, rosttemp. | ic | intern. caps. | sm | stria medullaris |
| AuRTL | aud. belt Cx, rostrotemplat. | IG | induseum gris. | SO | supraoptic nu. |
| AuRTM | aud. belt Cx, rostrotempmed. | IMG | intramedull. gr. amyg. | sod | suproptic decn |
| B | basal nu. (Meynert) | InfS | infundib. stalk | SOR | supraoptic nu, retroch. |
| BLD | basolat. amyg. nu., dors. | IPAC | interst. nu. acp | st | stria terminalis |
| BLI | basolat. amyg. nu., int. | ITA | inf. temp. ant. Cx | StHy | striatohypothal. nu. |
| BLVL | basolat. amyg. nu., vent. lat. | IVF | intervent. foramen | STL | sup. temp. Cx, lat. |
| BLVM | basolat. amyg. nu., vent. med. | LaD | lat. amyg. nu., dors. | STM | sup. temp. Cx, med. |
| BMD | basomed. amyg. nu., dors. | LaL | lat. amyg. nu., lat. | sts | sup. temp. sulcus |
| BMV | basomed. amyg. nu., vent. | LaM | lat. amyg. nu., med. | TPro | temp. proisocortex |
| BSTLD | bed nu. st, lat. dors. divn | lf | lateral fissure | TS | triang. sept. nu. |
| BSTLJ | bed nu. st, lat. juxtacap. divn | LH | lat. hypothal. area | TuLH | tub. lat. hypothal. |
| BSTLP | bed nu. st, lat. post. divn | lml | lat. medull. lamina | VLH | vent. lat. hypothal. |
| BSTMPI | bed nu. st, med. post. int. | LSD | lat. sept. nu., dors. | VMH | ventromed. hypoth. nu. |
| BSTMPL | bed nu. st, med. post. lat | | | | |

# Plate 21a

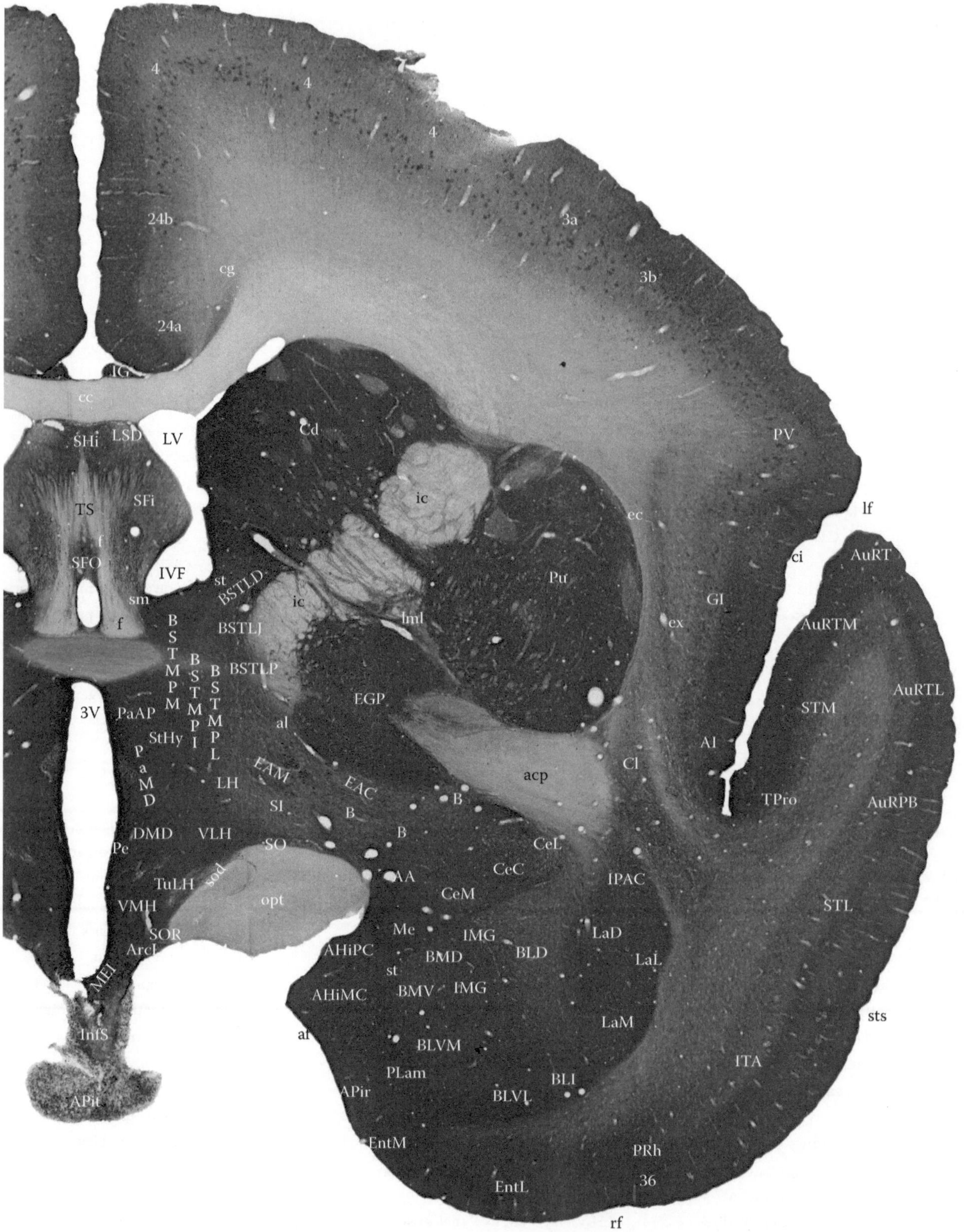


**Calbindin**
Interaural +09.00 mm

# Plate 21b

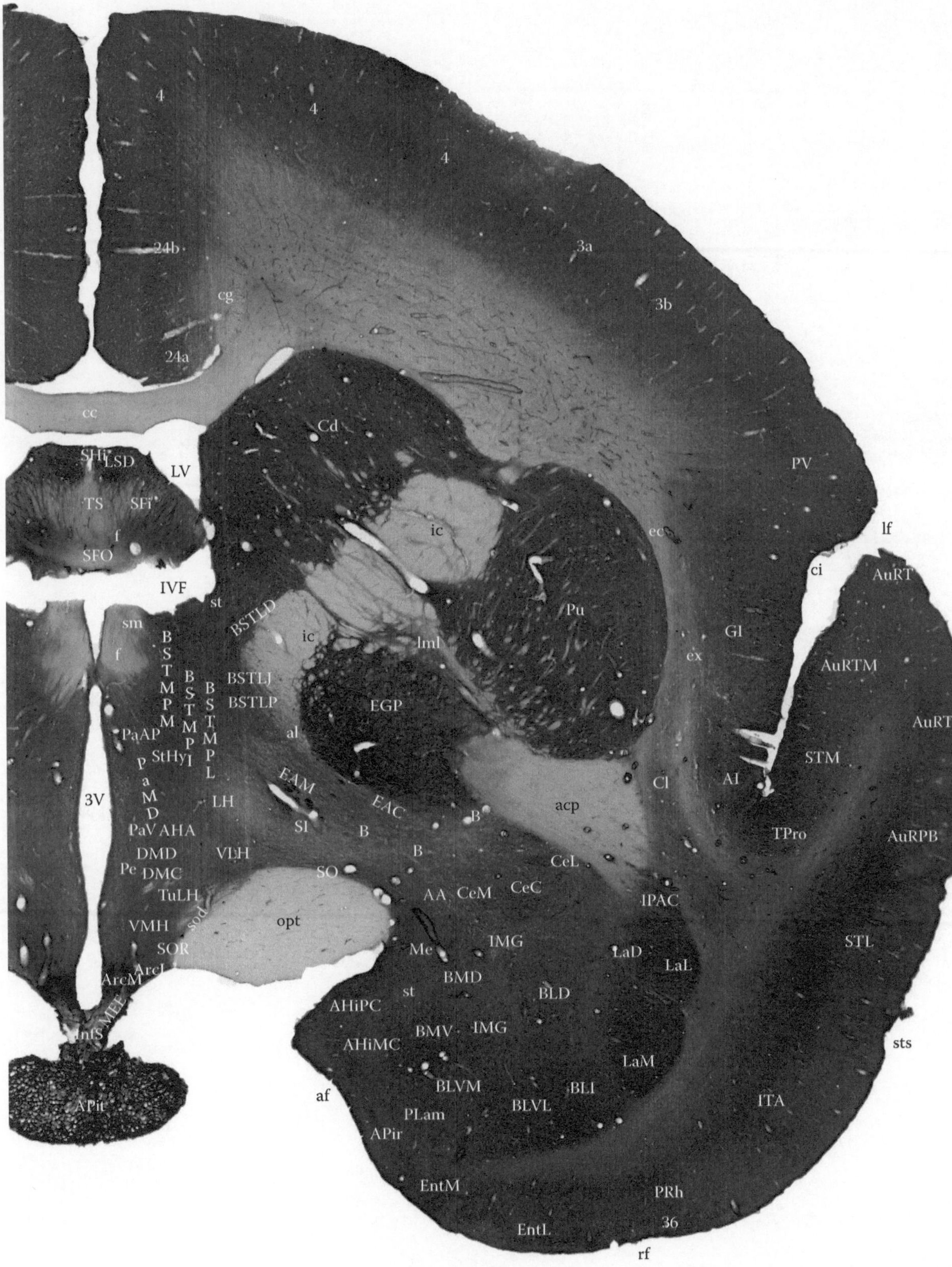


**Neurofilament**
Interaural +08.85 mm

# Plate 21c

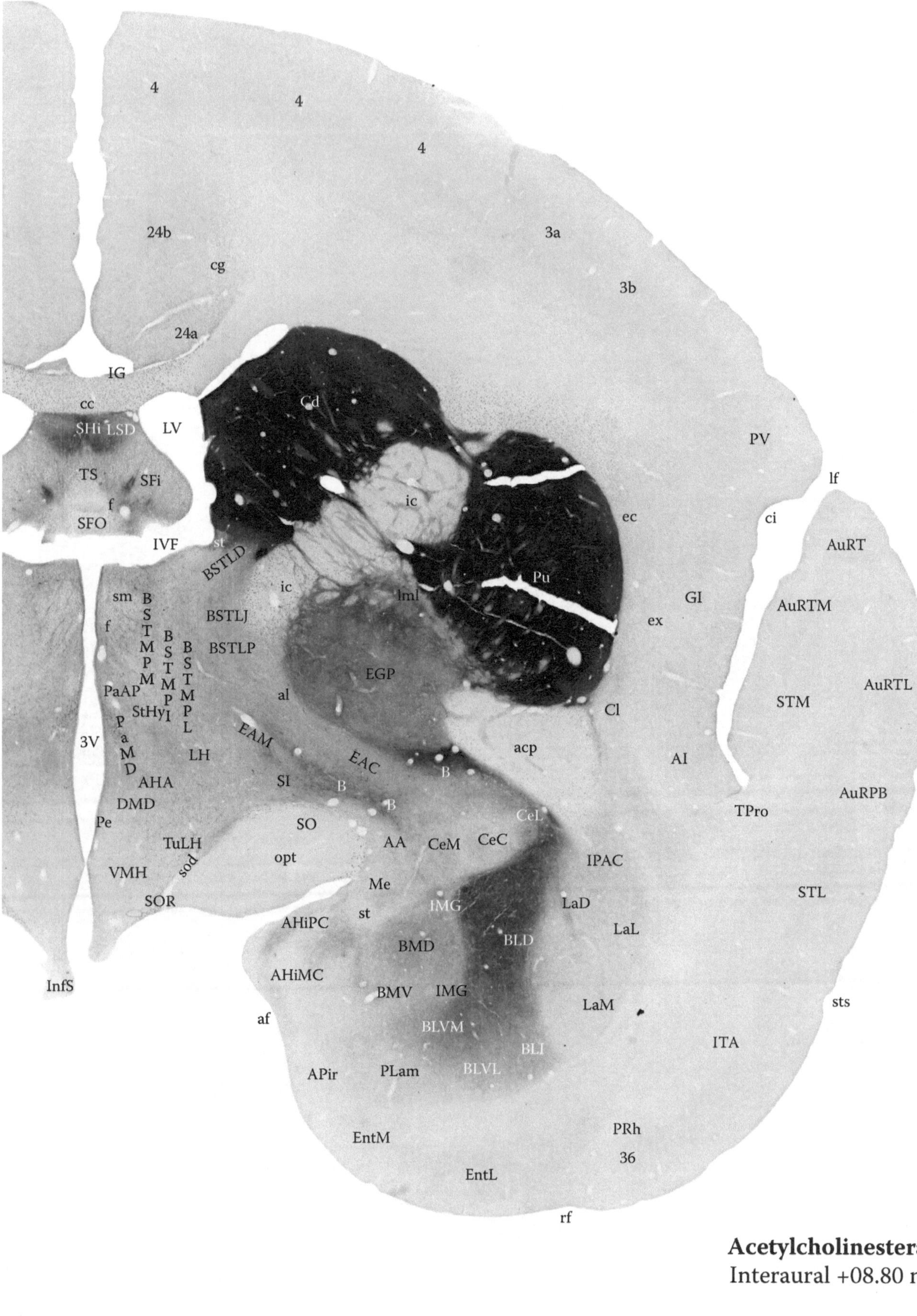


**Acetylcholinesterase**
Interaural +08.80 mm

# Plate 21d

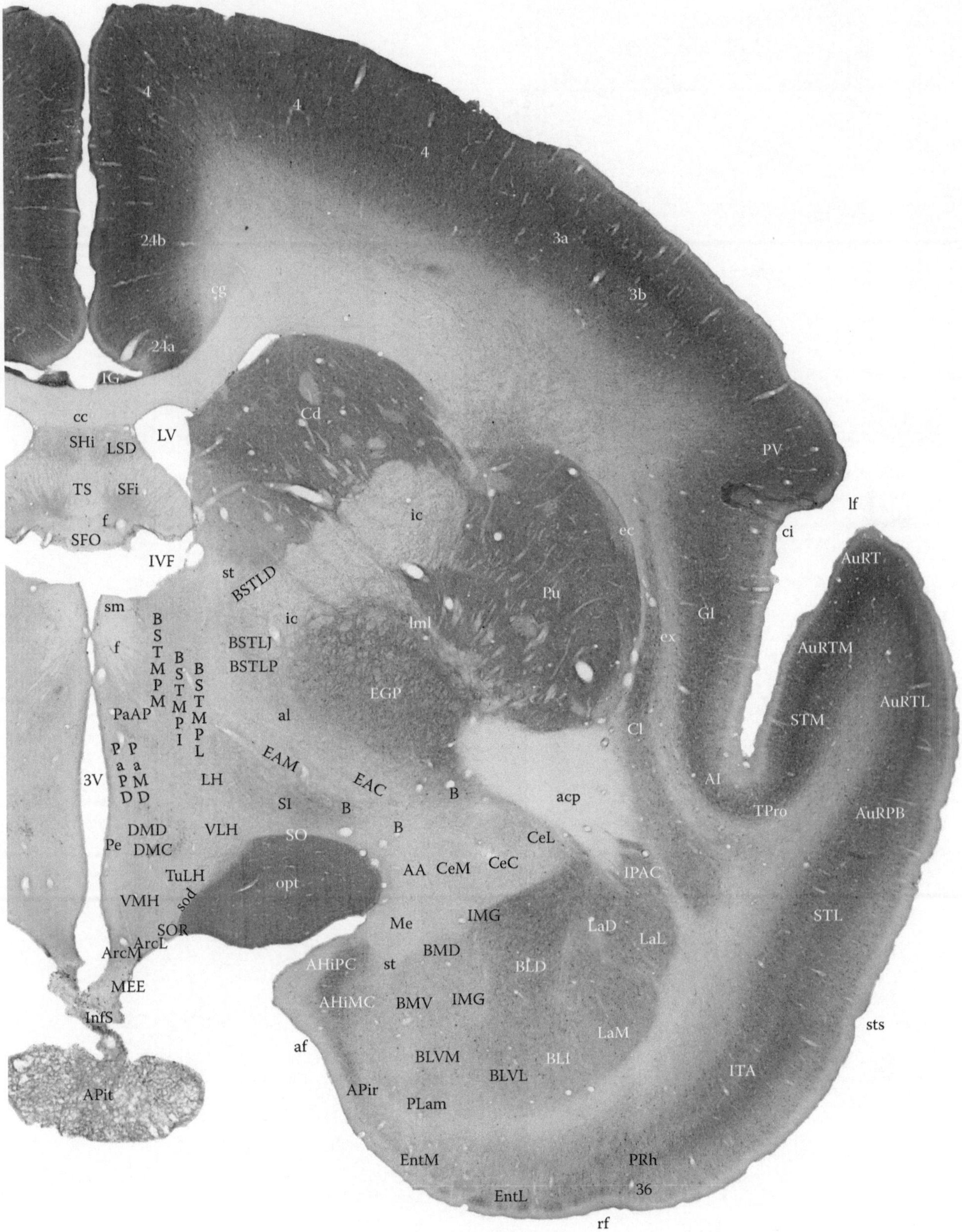


**Parvalbumin**
Interaural +08.75 mm

# Plate 21e

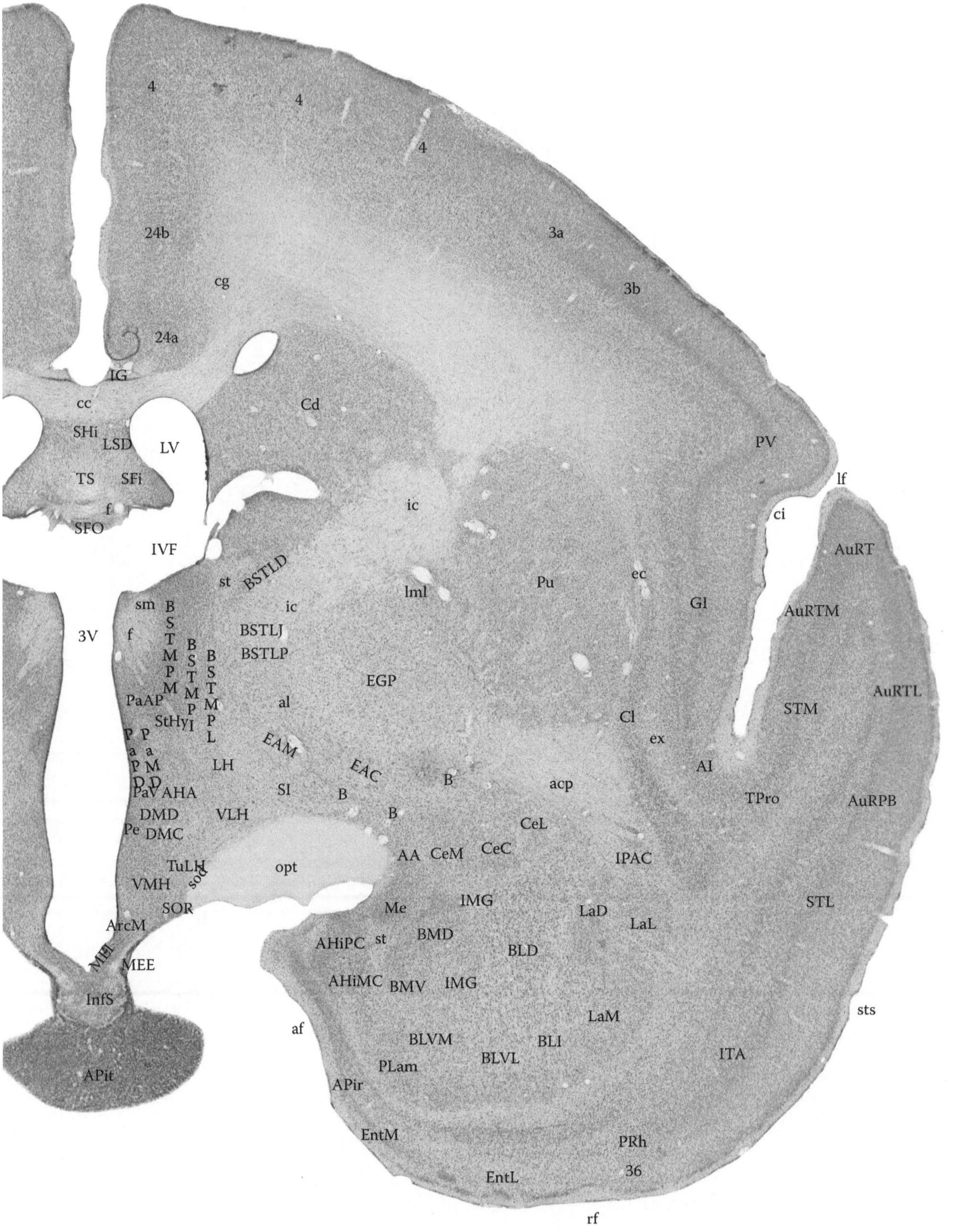


**Cresyl Violet**
Interaural +08.70 mm

# Plate 21

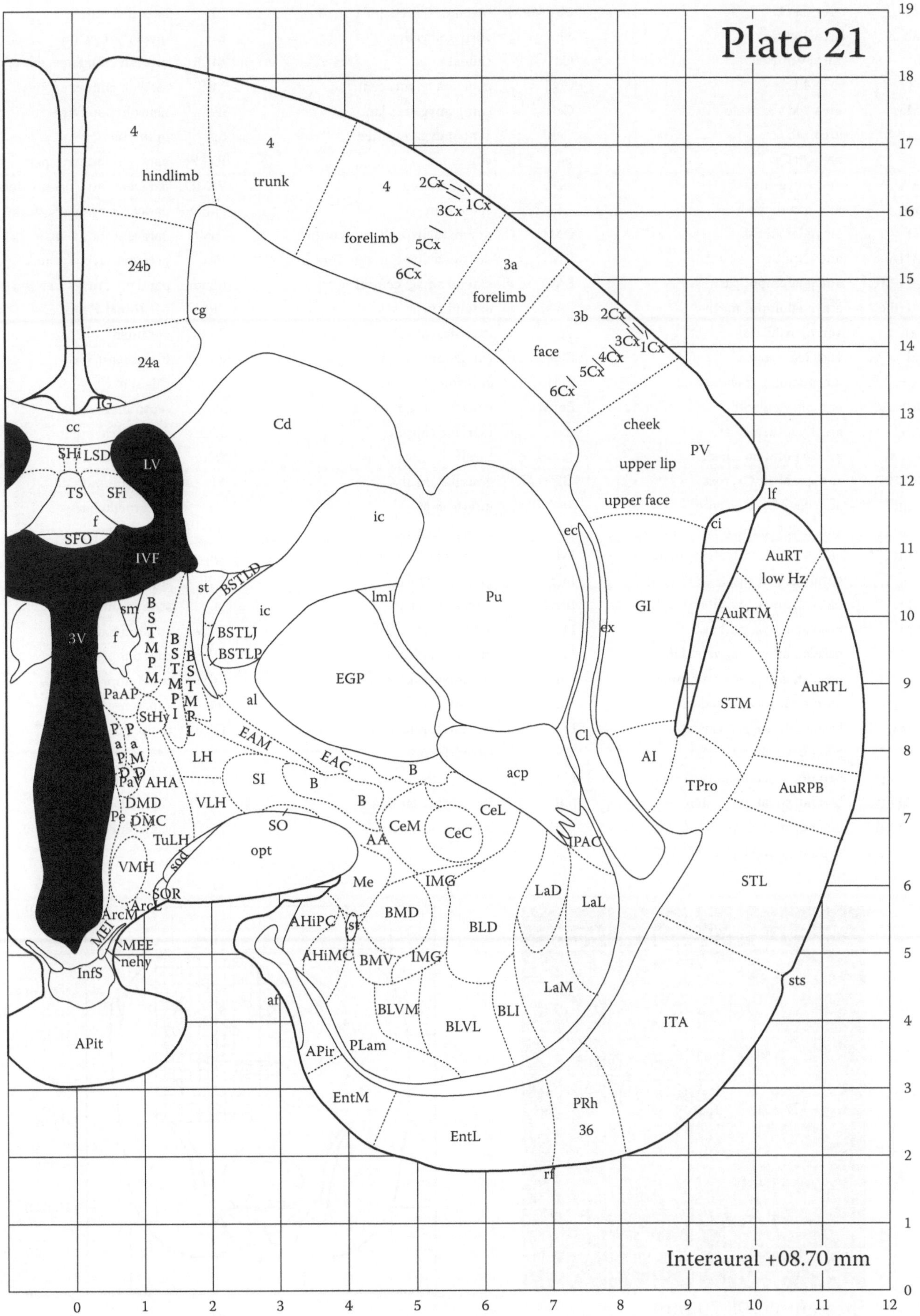


Interaural +08.70 mm

| | |
|---|---|
| 3a | area 3a Cx |
| 3b | area 3b Cx |
| 3V | third ventricle |
| 4 | area 4 Cx |
| 24a | area 24a Cx |
| 24b | area 24b Cx |
| 36 | area 30 Cx |
| AA | ant. amyg. area |
| acp | ant. comm. post. |
| af | amygdaloid fiss. |
| AHA | ant. hypoth. area, ant. |
| AHiMC | amygdalohipp., parv. |
| AHiPC | amygdalohipp., magn. |
| AI | agran. insul. Cx |
| al | ansa lenticularis |
| APir | amygdalopir. trans. area |
| APit | ant. pituit. gland |
| ArcL | arc. hypoth. nu., lat. |
| ArcM | arc. hypoth. nu., med. |
| AuRPB | aud. parabelt Cx, rost. |
| AuRT | aud. core Cx, rosttemp. |
| AuRTL | aud. belt Cx, rostrotemplat. |
| AuRTM | aud. belt Cx, rostrotempmed. |
| B | basal nu. (Meynert) |
| BLD | basolat. amyg. nu., dors. |
| BLI | basolat. amyg. nu., int. |
| BLVL | basolat. amyg. nu., vent. lat. |
| BLVM | basolat. amyg. nu., vent. med. |
| BMD | basomed. amyg. nu., dors. |
| BMV | basomed. amyg. nu., vent. |
| BSTLD | bed nu. st, lat. dors. divn |
| BSTLJ | bed nu. st, lat. juxtacap. divn |
| BSTLP | bed nu. st, lat. post. divn |
| BSTMPI | bed nu. st, med. post. int. |
| BSTMPL | bed nu. st, med. post. lat |
| BSTMPM | bed nu. st, med. post. med. |
| cc | corp. callosum |
| Cd | caudate nu. |
| CeC | cent. amyg. nu., cent. |
| CeL | cent. amyg. nu., lat. |
| CeM | cent. amyg. nu., med. |
| cg | cingulum |
| ci | circular sulcus |
| Cl | claustrum |
| DMC | dorsmed. hypoth. nu., comp. |
| DMD | dorsmed. hypoth. nu., dors. |
| EAC | extend amyg. central |
| EAM | extend amyg. medial |
| ec | external capsule |
| EGP | ext. globus pallidus |
| EntL | entorhin. Cx, lat. |
| EntM | entorhin. Cx, med. |
| ex | extreme capsule |
| f | fornix |
| GI | granular insul. Cx |
| ic | intern. caps. |
| IG | induseum gris. |
| IMG | intramedull. gr. amyg. |
| InfS | infundib. stalk |
| IPAC | interst. nu. acp |
| ITA | inf. temp. ant. Cx |
| IVF | intervent. foramen |
| LaD | lat. amyg. nu., dors. |
| LaL | lat. amyg. nu., lat. |
| LaM | lat. amyg. nu., med. |
| lf | lateral fissure |
| LH | lat. hypothal. area |
| lml | lat. medull. lamina |
| LSD | lat. sept. nu., dors. |
| LV | lateral ventricle |
| Me | medial amyg. nu. |
| MEE | median eminence, ext. |
| MEI | median eminence, int. |
| nehy | neurohypophyseal tr. |
| opt | optic tract |
| PaAP | paravent. nu., ant. parv. |
| PaMD | paravent. nu., magno. dors. |
| PaPD | paravent. nu., parv. dors. |
| PaV | paravent. nu., vent. |
| Pe | perivent. hypoth. nu. |
| PLam | paralam. amyg. nu. |
| PRh | perirhinal Cx |
| Pu | putamen |
| PV | PV somato. Cx |
| rf | rhinal fiss. |
| SFi | septofimbrial nu. |
| SFO | subfornical organ |
| SHi | septohippo. nu. |
| SI | subst. innominata |
| sm | stria medullaris |
| SO | supraoptic nu. |
| sod | suproptic decn |
| SOR | supraoptic nu, retroch. |
| st | stria terminalis |
| StHy | striatohypothal. nu. |
| STL | sup. temp. Cx, lat. |
| STM | sup. temp. Cx, med. |
| sts | sup. temp. sulcus |
| TPro | temp. proisocortex |
| TS | triang. sept. nu. |
| TuLH | tub. lat. hypothal. |
| VLH | vent. lat. hypothal. |
| VMH | ventromed. hypoth. nu. |

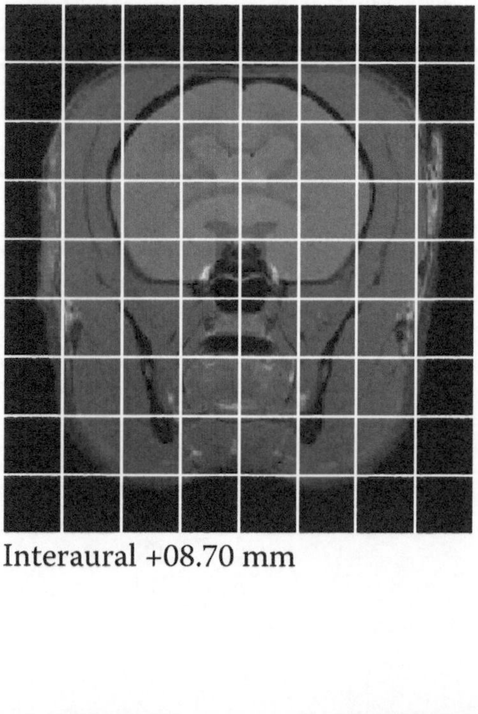

Interaural +08.70 mm

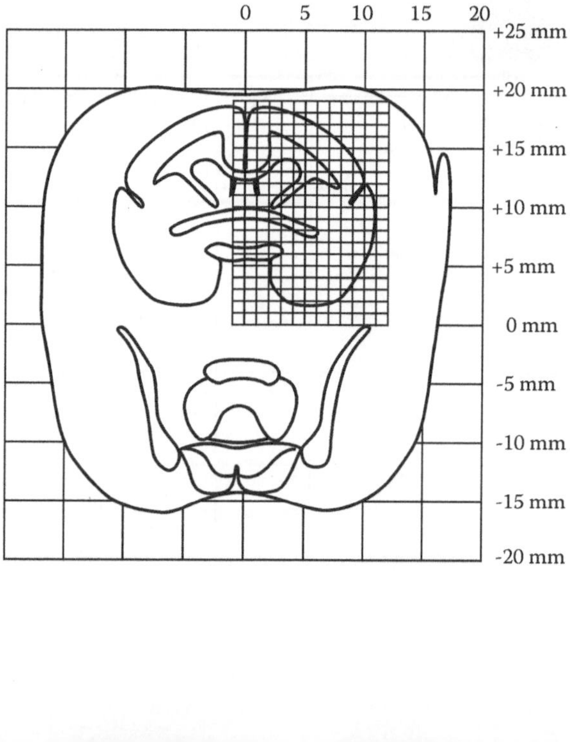

# Plate 22

Interaural +08.20 mm

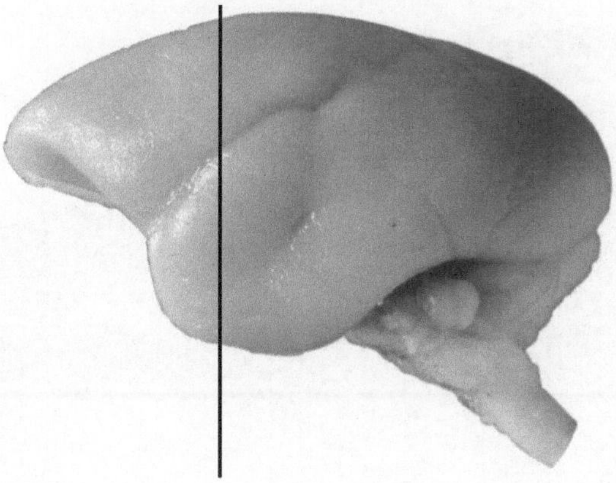

| | |
|---|---|
| 3a | area 3a Cx |
| 3b | area 3b Cx |
| 3V | third ventricle |
| 4 | area 4 Cx |
| 24a | area 24a Cx |
| 24b | area 24b Cx |
| 36 | area 30 Cx |
| AA | ant. amyg. area |
| acp | ant. comm. post. |
| AHi | amygdalohipp. area |
| AHP | ant. hypoth. area, post. |
| AI | agran. insul. Cx |
| al | ansa lenticularis |
| APir | amygdalopir. trans. area |
| APit | ant. pituit. gland |
| ArcD | arc. hypoth. nu., dors. |
| ArcL | arc. hypoth. nu., lat. |
| ArcM | arc. hypoth. nu., med. |
| AStr | amygdalostr. trans. zone |
| AuRPB | aud. parabelt Cx, rost. |
| AuRT | aud. core Cx, rosttemp. |
| AuRTL | aud. belt Cx, rostrotemplat. |
| AuRTM | aud. belt Cx, rostrotempmed. |
| B | basal nu. (Meynert) |
| BLD | basolat. amyg. nu., dors. |
| BLI | basolat. amyg. nu., int. |
| BLVM | basolat. amyg. nu., vent. med. |
| BMD | basomed. amyg. nu., dors. |
| BMV | basomed. amyg. nu., vent. |
| BSTMPI | bed nu. st, med. post. int. |
| BSTMPL | bed nu. st, med. post. lat |
| CA1 | cornu ammonis 1 |
| CA2 | cornu ammonis 2 |
| CA3 | cornu ammonis 3 |
| cc | corp. callosum |
| Cd | caudate nu. |
| CeC | cent. amyg. nu., cent. |
| CeL | cent. amyg. nu., lat. |
| CeM | cent. amyg. nu., med. |
| cg | cingulum |
| ci | circular sulcus |
| Cl | claustrum |
| DG | dentate gyrus |
| DMC | dorsmed. hypoth. nu., comp. |
| DMD | dorsmed. hypoth. nu., dors. |
| E | ependyma |
| EAM | extend amyg. medial |
| ec | external capsule |
| EGP | ext. globus pallidus |
| EntL | entorhin. Cx, lat. |
| EntM | entorhin. Cx, med. |
| ex | extreme capsule |
| f | fornix |
| GI | granular insul. Cx |
| ic | intern. caps. |
| IG | induseum gris. |
| IGP | int. globus pallidus |
| IMG | intramedull. gr. amyg. |
| InfS | infundib. stalk |
| IPAC | interst. nu. acp |
| ITA1 | inf. temp. ant. Cx, 1 |
| ITA2 | inf. temp. ant. Cx, 2 |
| IVF | intervent. foramen |
| LaD | lat. amyg. nu., dors. |
| LaL | lat. amyg. nu., lat. |
| LaM | lat. amyg. nu., med. |
| lf | lateral fissure |
| LH | lat. hypothal. area |
| lml | lat. medull. lamina |
| LV | lateral ventricle |
| Me | medial amyg. nu. |
| MEE | median eminence, ext. |
| mml | med. medull. lamina |
| opt | optic tract |
| PaM | paravent. nu., magno. |
| PaMD | paravent. nu., magno. dors. |
| PaPD | paravent. nu., parv. dors. |
| PaV | paravent. nu., vent. |
| Pe | perivent. hypoth. nu. |
| PLam | paralam. amyg. nu. |
| PRh | perirhinal Cx |
| Pu | putamen |
| PV | parietovent. Cx |
| PVA | paravent. thal. nu., ant. |
| rf | rhinal fiss. |
| Rt | reticular nu. |
| sm | stria medullaris |
| SM | stria medullaris nu. |
| SMA | supp. motor area Cx |
| sod | suproptic decn |
| SOR | supraoptic nu., retroch. |
| st | stria terminalis |
| StHy | striatohypothal. nu. |
| STM | sup. temp. Cx, med. |
| sts | sup. temp. sulcus |
| TCF | transverse cerebral fiss. |
| TPro | temp. proisocortex |
| vhc | vent. hipp. comm. |
| VLH | vent. lat. hypothal. |
| VMHL | ventromed. hypoth. nu., lat. |
| VMHM | ventromed. hypoth. nu., med. |

# Plate 22a

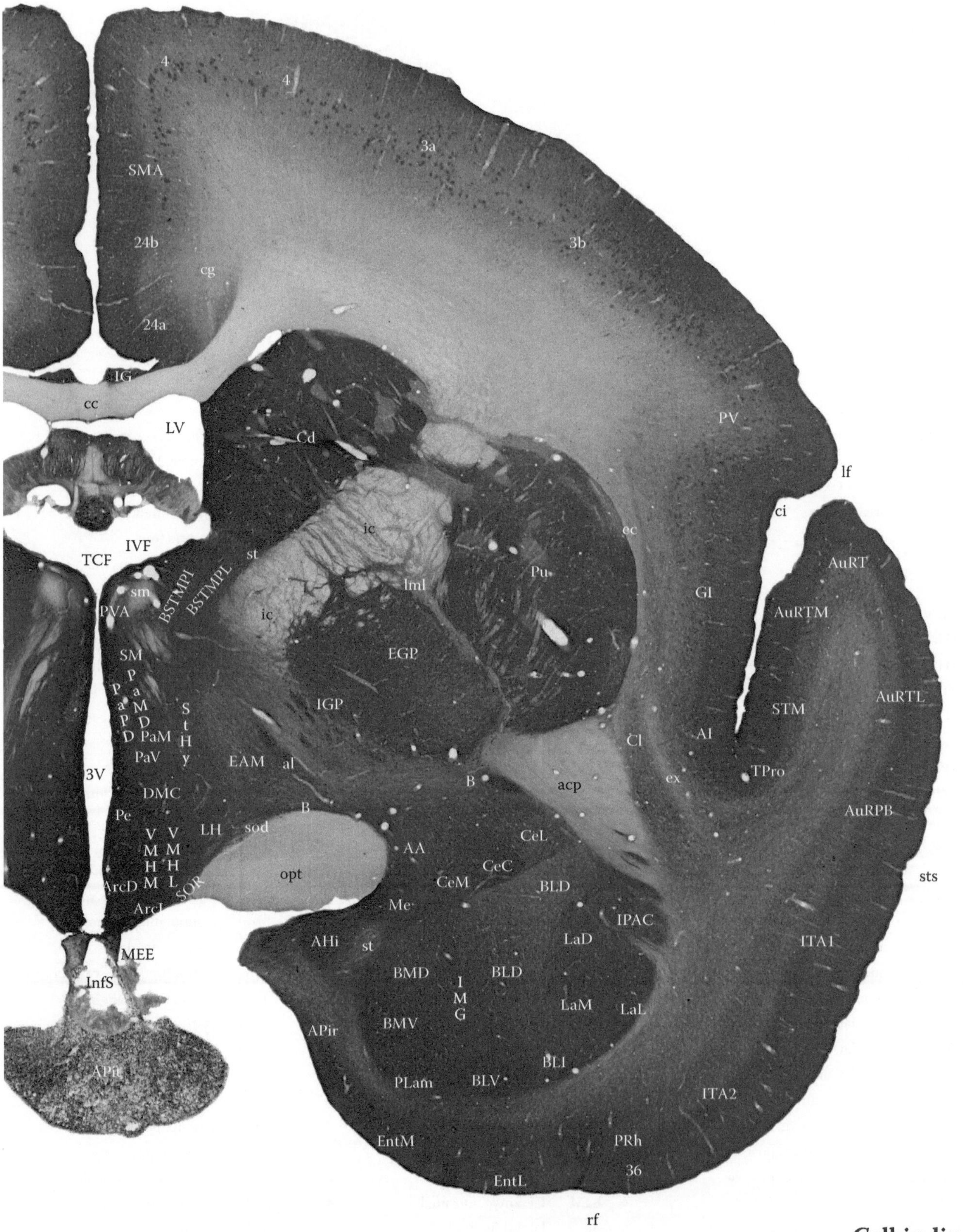


**Calbindin**
Interaural +08.50 mm

# Plate 22b

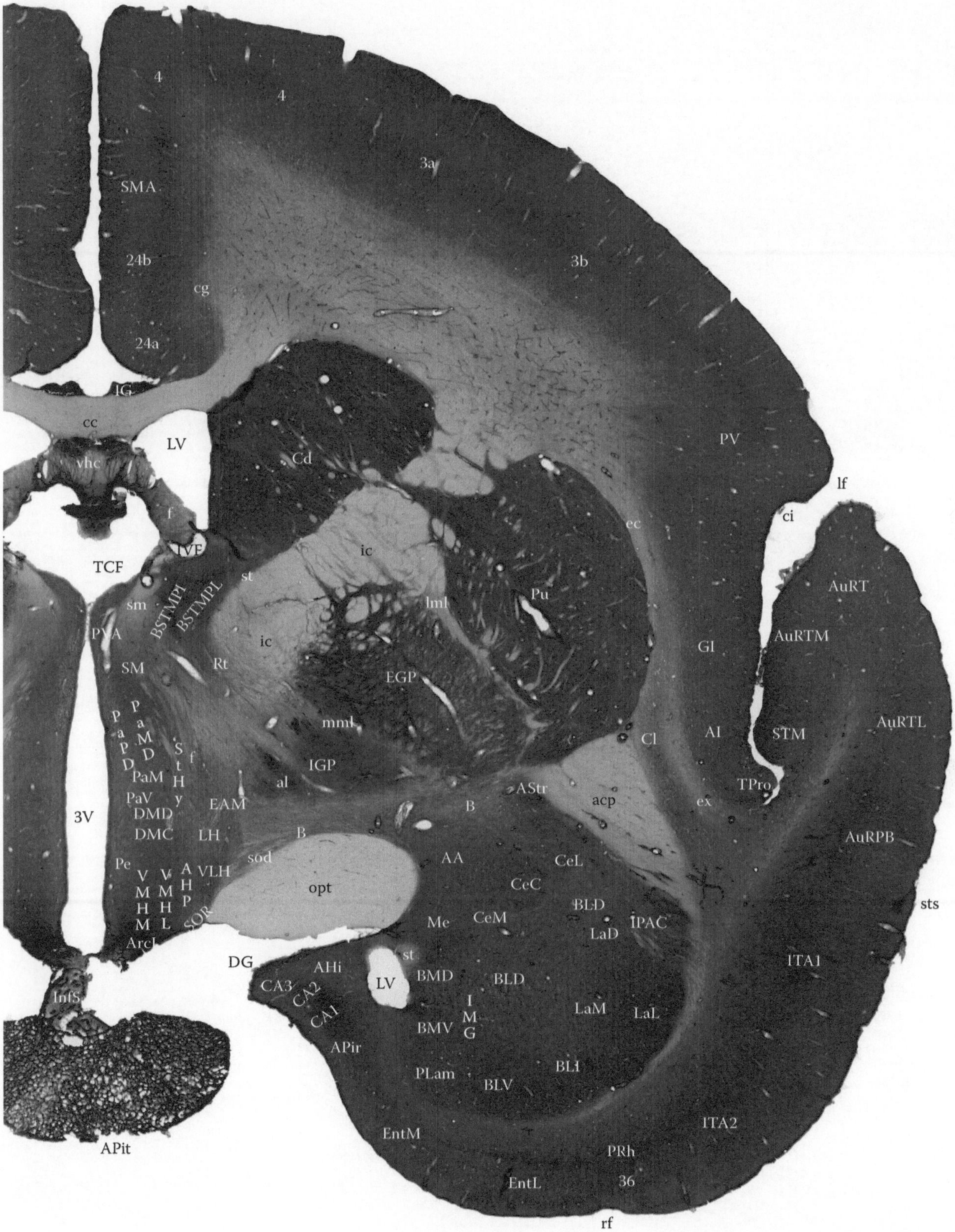


**Neurofilament**
Interaural +08.35 mm

# Plate 22c

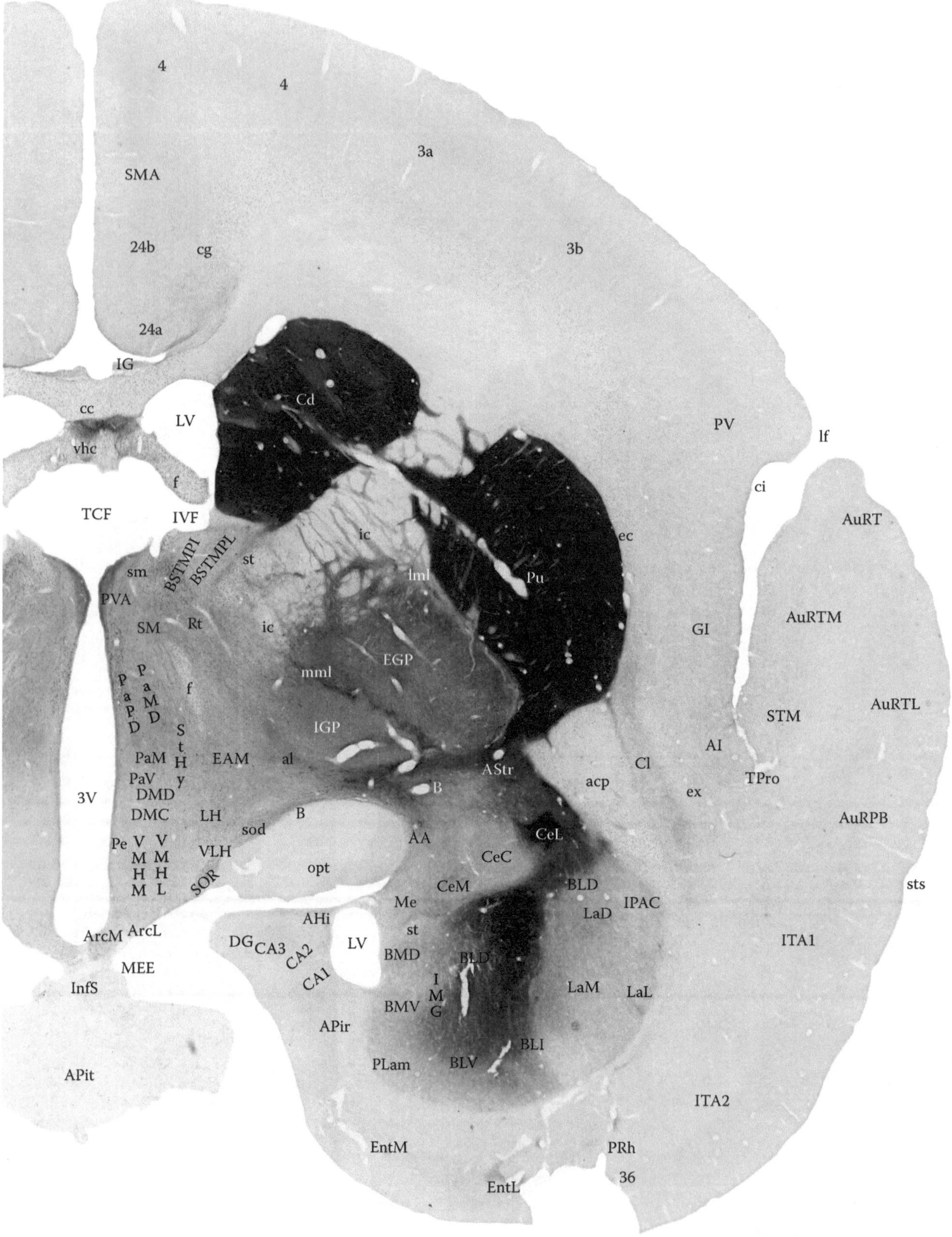


**Acetylcholinesterase**
Interaural +08.30 mm

# Plate 22d

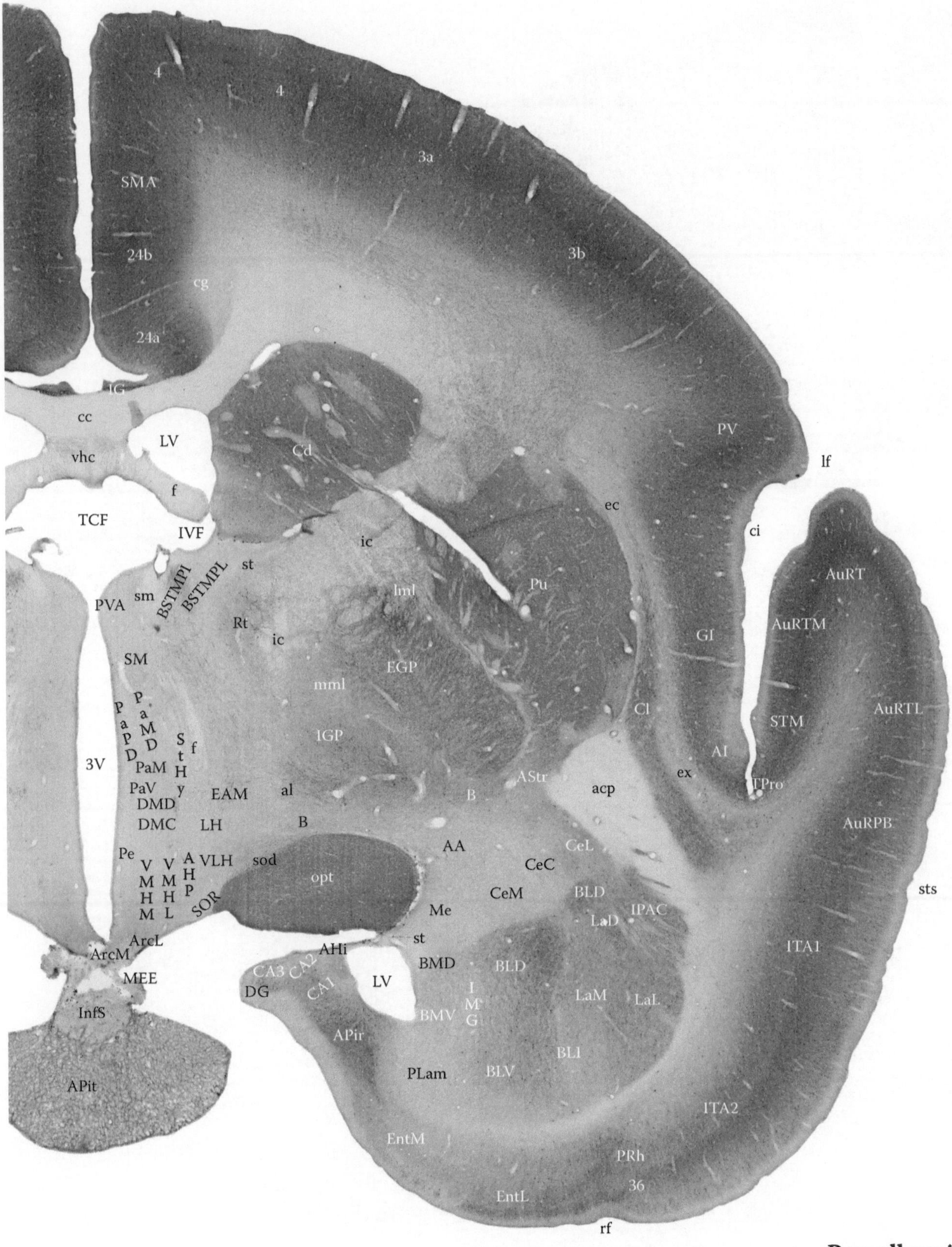


**Parvalbumin**
Interaural +08.25 mm

# Plate 22e

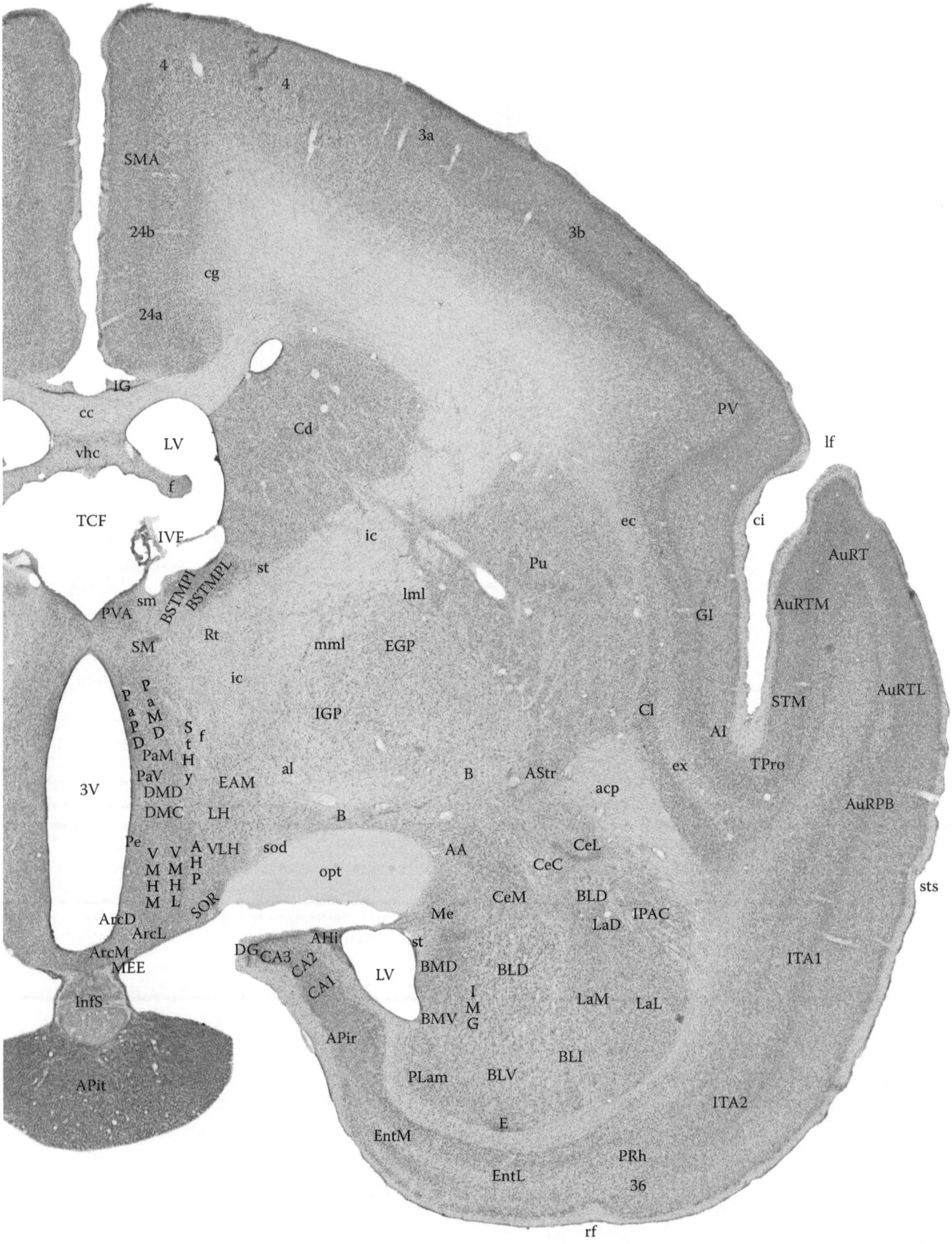


**Cresyl Violet**
Interaural +08.20 mm

# Plate 22

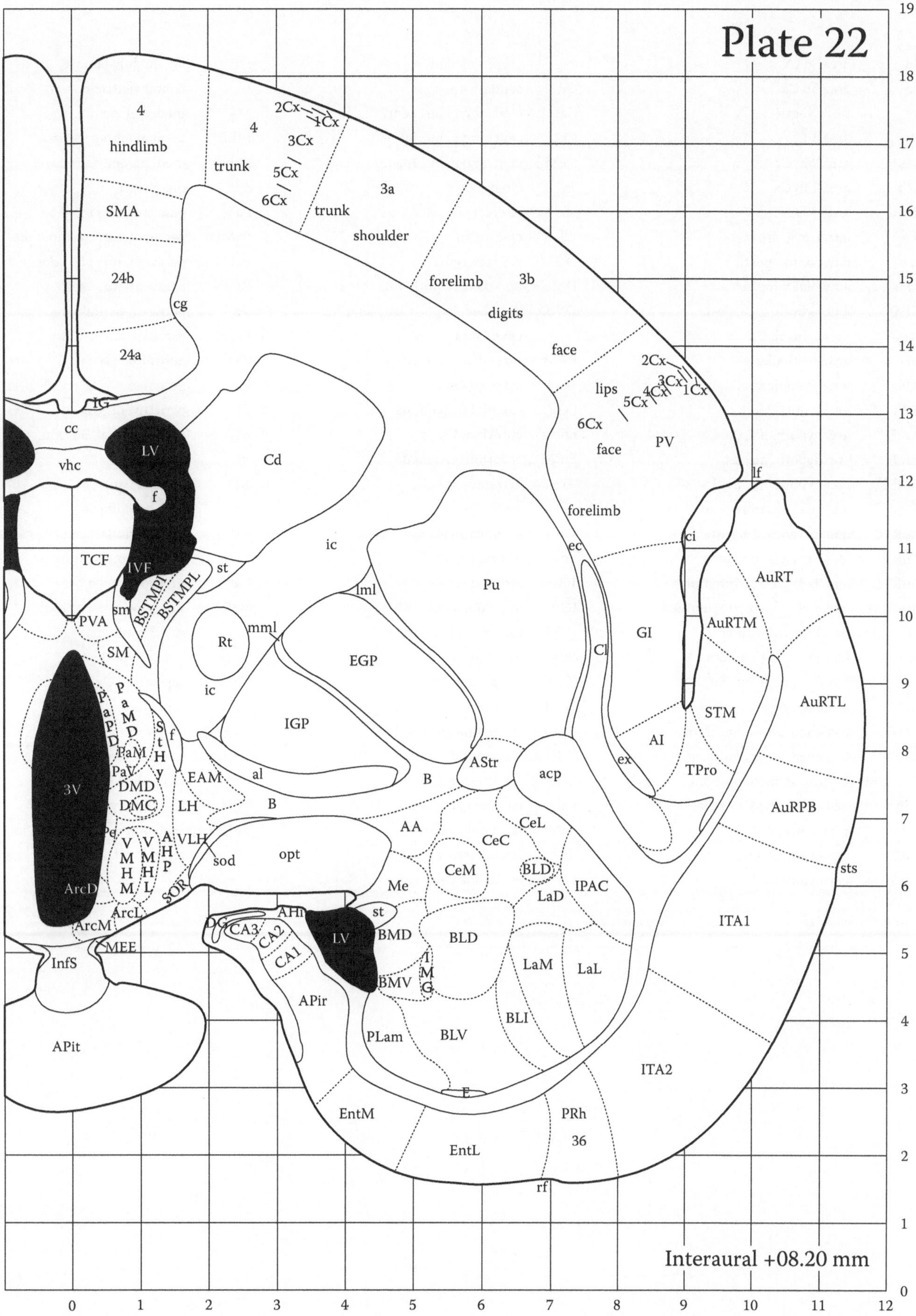


Interaural +08.20 mm

| | |
|---|---|
| 3a | area 3a Cx |
| 3b | area 3b Cx |
| 3V | third ventricle |
| 4 | area 4 Cx |
| 24a | area 24a Cx |
| 24b | area 24b Cx |
| 36 | area 30 Cx |
| AA | ant. amyg. area |
| acp | ant. comm. post. |
| AHi | amygdalohipp. area |
| AHP | ant. hypoth. area, post. |
| AI | agran. insul. Cx |
| al | ansa lenticularis |
| APir | amygdalopir. trans. area |
| APit | ant. pituit. gland |
| ArcD | arc. hypoth. nu., dors. |
| ArcL | arc. hypoth. nu., lat. |
| ArcM | arc. hypoth. nu., med. |
| AStr | amygdalostr. trans. zone |
| AuRPB | aud. parabelt Cx, rost. |
| AuRT | aud. core Cx, rosttemp. |
| AuRTL | aud. belt Cx, rostrotemplat. |
| AuRTM | aud. belt Cx, rostrotempmed. |
| B | basal nu. (Meynert) |
| BLD | basolat. amyg. nu., dors. |
| BLI | basolat. amyg. nu., int. |
| BLVM | basolat. amyg. nu., vent. med. |
| BMD | basomed. amyg. nu., dors. |
| BMV | basomed. amyg. nu., vent. |
| BSTMPI | bed nu. st, med. post. int. |
| BSTMPL | bed nu. st, med. post. lat |
| CA1 | cornu ammonis 1 |
| CA2 | cornu ammonis 2 |
| CA3 | cornu ammonis 3 |
| cc | corp. callosum |
| Cd | caudate nu. |
| CeC | cent. amyg. nu., cent. |
| CeL | cent. amyg. nu., lat. |
| CeM | cent. amyg. nu., med. |
| cg | cingulum |
| ci | circular sulcus |
| Cl | claustrum |
| DG | dentate gyrus |
| DMC | dorsmed. hypoth. nu., comp. |
| DMD | dorsmed. hypoth. nu., dors. |
| E | ependyma |
| EAM | extend amyg. medial |
| ec | external capsule |
| EGP | ext. globus pallidus |
| EntL | entorhin. Cx, lat. |
| EntM | entorhin. Cx, med. |
| ex | extreme capsule |
| f | fornix |
| GI | granular insul. Cx |
| ic | intern. caps. |
| IG | induseum gris. |
| IGP | int. globus pallidus |
| IMG | intramedull. gr. amyg. |
| InfS | infundib. stalk |
| IPAC | interst. nu. acp |
| ITA1 | inf. temp. ant. Cx, 1 |
| ITA2 | inf. temp. ant. Cx, 2 |
| IVF | intervent. foramen |
| LaD | lat. amyg. nu., dors. |
| LaL | lat. amyg. nu., lat. |
| LaM | lat. amyg. nu., med. |
| lf | lateral fissure |
| LH | lat. hypothal. area |
| lml | lat. medull. lamina |
| LV | lateral ventricle |
| Me | medial amyg. nu. |
| MEE | median eminence, ext. |
| mml | med. medull. lamina |
| opt | optic tract |
| PaM | paravent. nu., magno. |
| PaMD | paravent. nu., magno. dors. |
| PaPD | paravent. nu., parv. dors. |
| PaV | paravent. nu., vent. |
| Pe | perivent. hypoth. nu. |
| PLam | paralam. amyg. nu. |
| PRh | perirhinal Cx |
| Pu | putamen |
| PV | parietovent. Cx |
| PVA | paravent. thal. nu., ant. |
| rf | rhinal fiss. |
| Rt | reticular nu. |
| sm | stria medullaris |
| SM | stria medullaris nu. |
| SMA | supp. motor area Cx |
| sod | suproptic decn |
| SOR | supraoptic nu., retroch. |
| st | stria terminalis |
| StHy | striatohypothal. nu. |
| STM | sup. temp. Cx, med. |
| sts | sup. temp. sulcus |
| TCF | transverse cerebral fiss. |
| TPro | temp. proisocortex |
| vhc | vent. hipp. comm. |
| VLH | vent. lat. hypothal. |
| VMHL | ventromed. hypoth. nu., lat. |
| VMHM | ventromed. hypoth. nu., med. |

# Plate 23

Interaural +07.70 mm

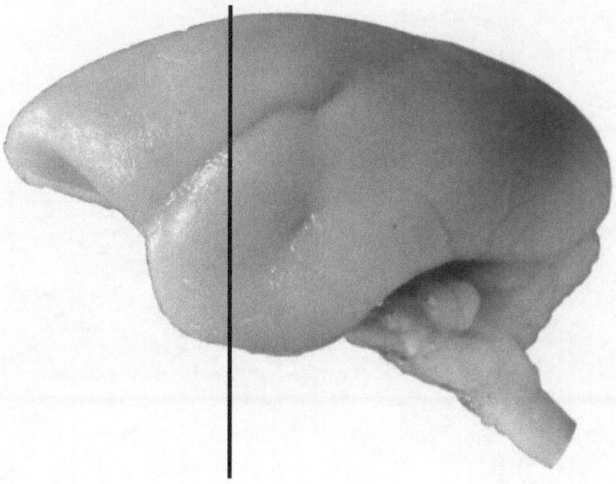

| | |
|---|---|
| 3a | area 3a Cx |
| 3b | area 3b Cx |
| 3V | third ventricle |
| 4 | area 4 Cx |
| 24a | area 24a Cx |
| 24b | area 24b Cx |
| 26 | area 26 Cx |
| 29 | area 29 Cx |
| 30 | area 30 Cx |
| 36 | area 36 Cx |
| AA | ant. amyg. area |
| acp | ant. comm. post. |
| AD | anterodors. thal. nu. |
| AHi | amygdalohipp. area |
| AI | agran. insul. Cx |
| al | ansa lenticularis |
| AM | anteromed. thal. nu. |
| ANS | access. neurosec. nu. |
| APit | ant. pituit. gland |
| ArcD | arc. hypoth. nu., dors. |
| ArcL | arc. hypoth. nu., lat. |
| ArcM | arc. hypoth. nu., med. |
| AStr | amygdalostr. trans. zone |
| AuRPB | aud. parabelt Cx, rost. |
| AuRT | aud. core Cx, rosttemp. |
| AuRTL | aud. belt Cx, rostrotemplat. |
| AuRTM | aud. belt Cx, rostrotempmed. |
| AV | anterovent. thal. nu. |
| B | basal nu. (Meynert) |
| BLD | basolat. amyg. nu., dors. |
| BLI | basolat. amyg. nu., int. |
| BLVM | basolat. amyg. nu., vent. med. |
| BMD | basomed. amyg. nu., dors. |
| BMV | basomed. amyg. nu., vent. |
| BSTMP | bed nu. st, med. post. divn |
| CA1 | cornu ammonis 1 |
| CA2 | cornu ammonis 2 |
| CA3 | cornu ammonis 3 |
| cc | corp. callosum |
| Cd | caudate nu. |
| CeC | cent. amyg. nu., cent. |
| CeL | cent. amyg. nu., lat. |
| CeM | cent. amyg. nu., med. |
| cg | cingulum |
| ci | circular sulcus |
| Cl | claustrum |
| DG | dentate gyrus |
| dhc | dors. hipp. comm. |
| DMC | dorsmed. hypoth. nu., comp. |
| DMD | dorsmed. hypoth. nu., dors. |
| E | ependyma |
| ec | external capsule |
| EGP | ext. globus pallidus |
| EntL | entorhin. Cx, lat. |
| EntM | entorhin. Cx, med. |
| ex | extreme capsule |
| f | fornix |
| GI | granular insul. Cx |
| ic | intern. caps. |
| IG | induseum gris. |
| IGP | int. globus pallidus |
| IPAC | interst. nu. acp |
| IPit | intermed. pituit. gland |
| ITA1 | inf. temp. ant. Cx, 1 |
| ITA2 | inf. temp. ant. Cx, 2 |
| LaD | lat. amyg. nu., dors. |
| LaL | lat. amyg. nu., lat. |
| LaM | lat. amyg. nu., med. |
| lentf | lenticular fasc. |
| lf | lateral fissure |
| LH | lat. hypothal. area |
| lml | lat. medull. lamina |
| LTu | lat. tuberal nu. |
| LV | lateral ventricle |
| MCLH | magno. nu. lat. hypothal. |
| Me | medial amyg. nu. |
| MEE | median eminence, ext. |
| MEI | median eminence, int. |
| mfb | med. forebrain bundle |
| mml | med. medull. lamina |
| NC | nanocell. hypoth. nu. |
| ns | nigrostriatal tract |
| opt | optic tract |
| PaPo | paravent. nu., post. |
| PaXi | paraxiphoid nu. |
| Pe | perivent. hypoth. nu. |
| PeFL | perifornical nu., lat. |
| PeFM | perifornical nu., med. |
| PLam | paralam. amyg. nu. |
| PPit | post. pituit. gland |
| PRh | perirhinal Cx |
| ProS | prosubiculum |
| PrS | presubiculum |
| PT | paratenial nu. |
| Pu | putamen |
| PV | PV somatosens. Cx |
| PVA | paravent. thal. nu., ant. |
| Re | reuniens nu. |
| rf | rhinal fiss. |
| Rt | reticular nu. |
| S | subiculum |
| S2 | second. somat. area Cx |
| sm | stria medullaris |
| SMA | supp. motor area Cx |
| sod | suproptic decn |
| st | stria terminalis |
| StHy | striatohypothal. nu. |
| sts | sup. temp. sulcus |
| SubI | subincertal nu. |
| TCF | transverse cerebral fiss. |
| Te | terete hypothal. nu. |
| TM | tuberomamill. area |
| TPro | temp. proisocortex |
| vhc | vent. hipp. comm. |
| VMHL | ventromed. hypo. nu., lat. |
| VMHM | ventromed. hypo. nu., med. |
| VRe | vent. reuniens nu. |
| Xi | xiphoid nu. |
| ZIR | zona incerta, rost. |

# Plate 23a

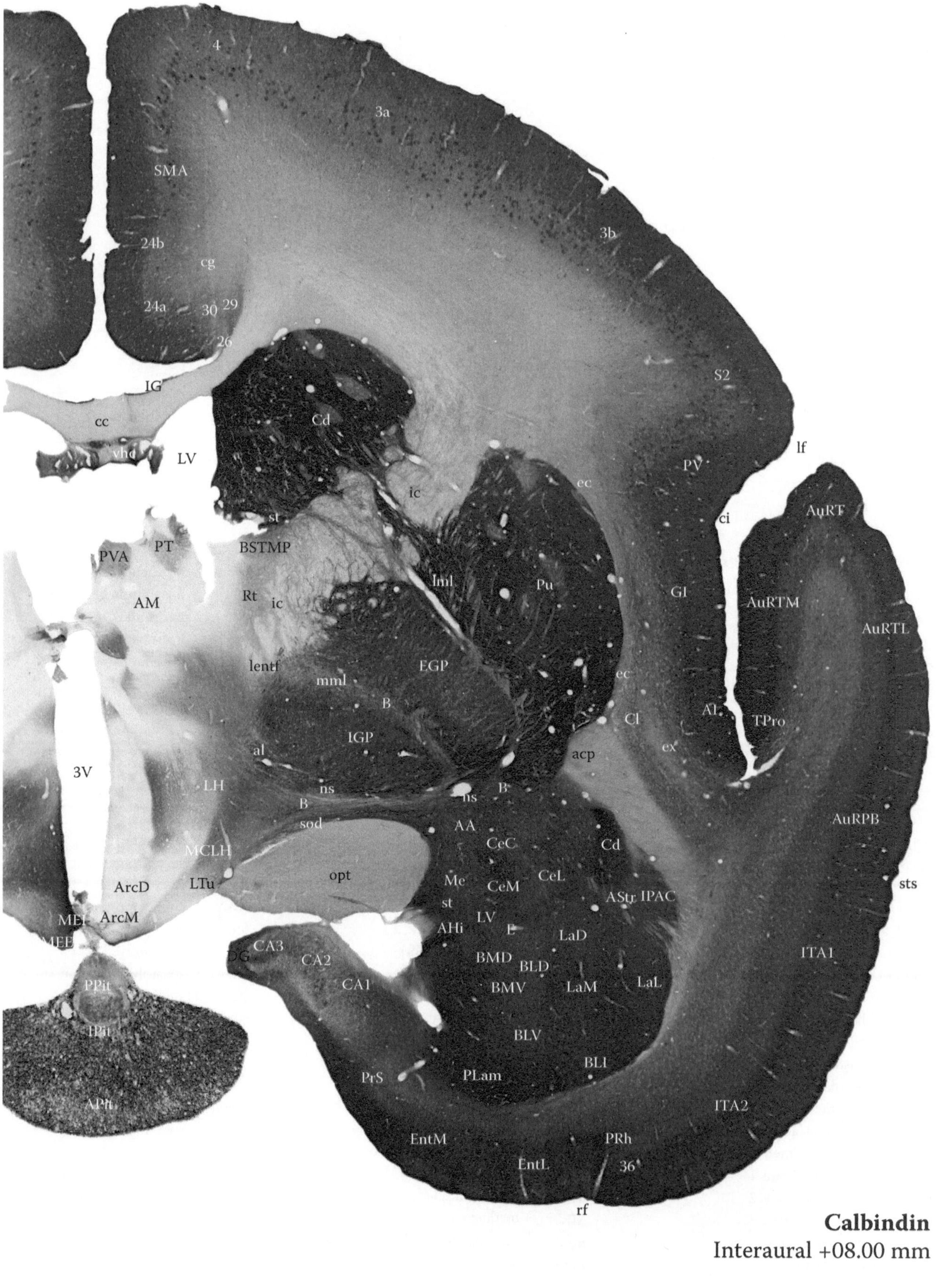


**Calbindin**
Interaural +08.00 mm

# Plate 23b

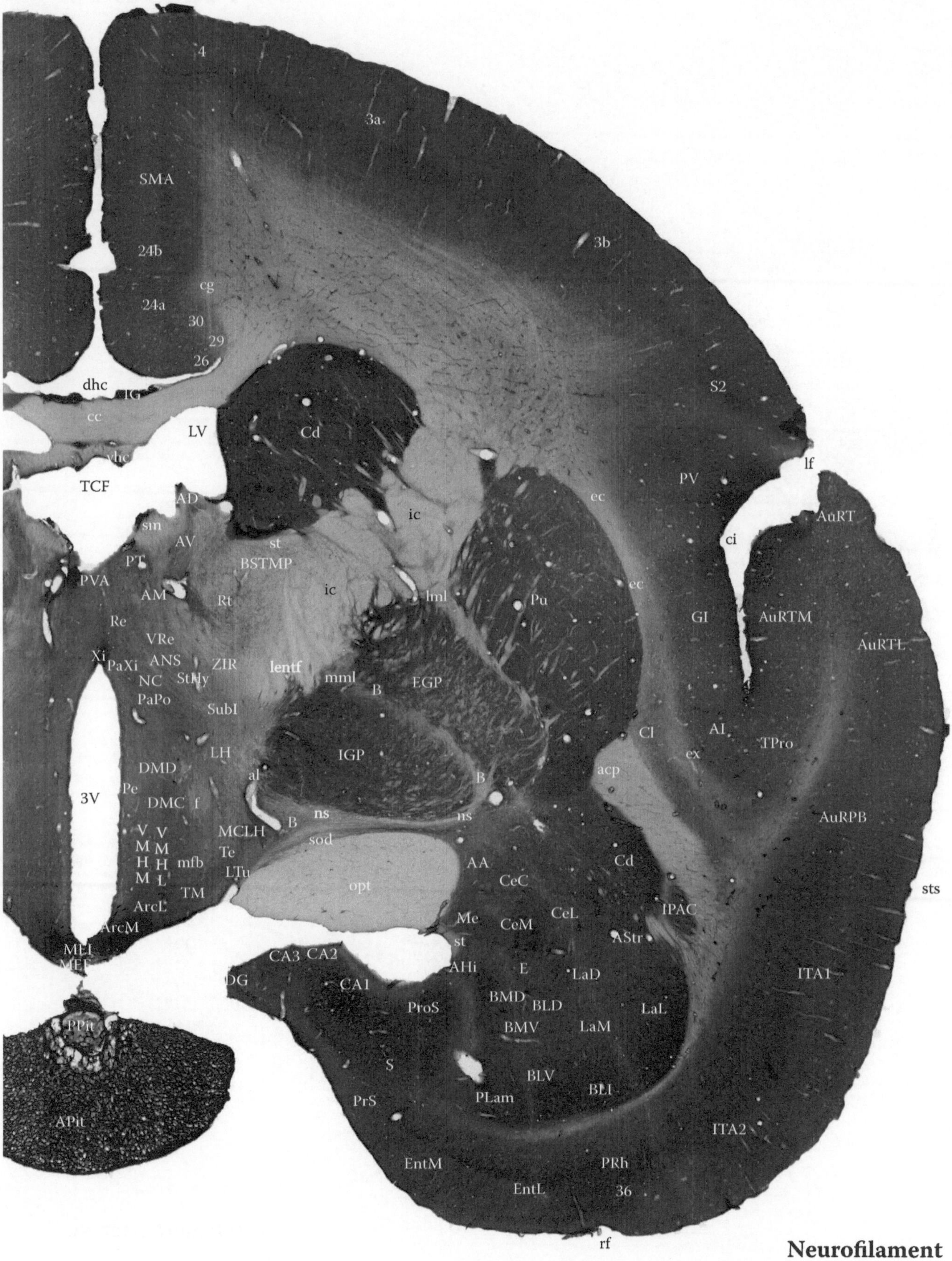


**Neurofilament**
Interaural +07.85 mm

# Plate 23c

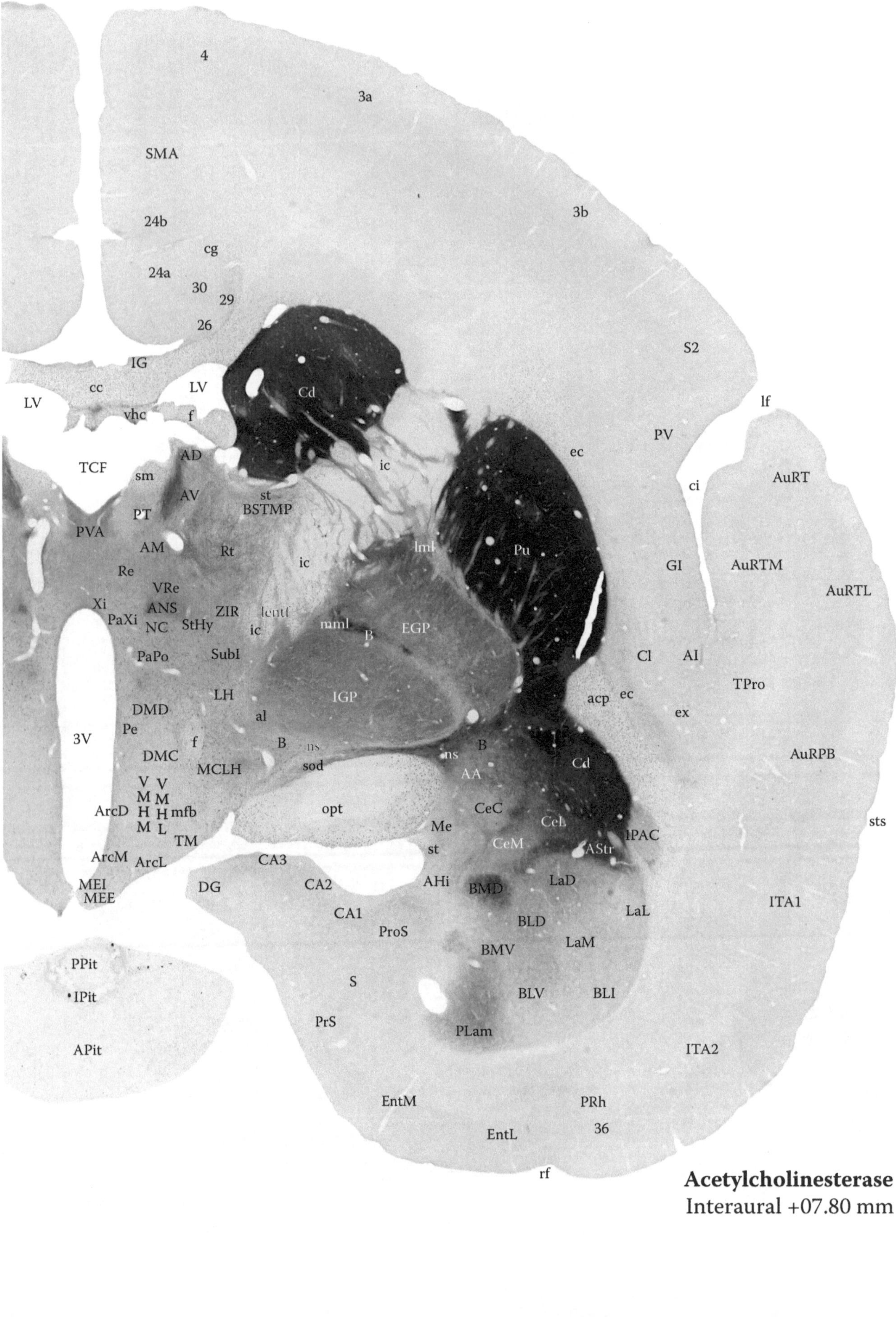


**Acetylcholinesterase**
Interaural +07.80 mm

# Plate 23d

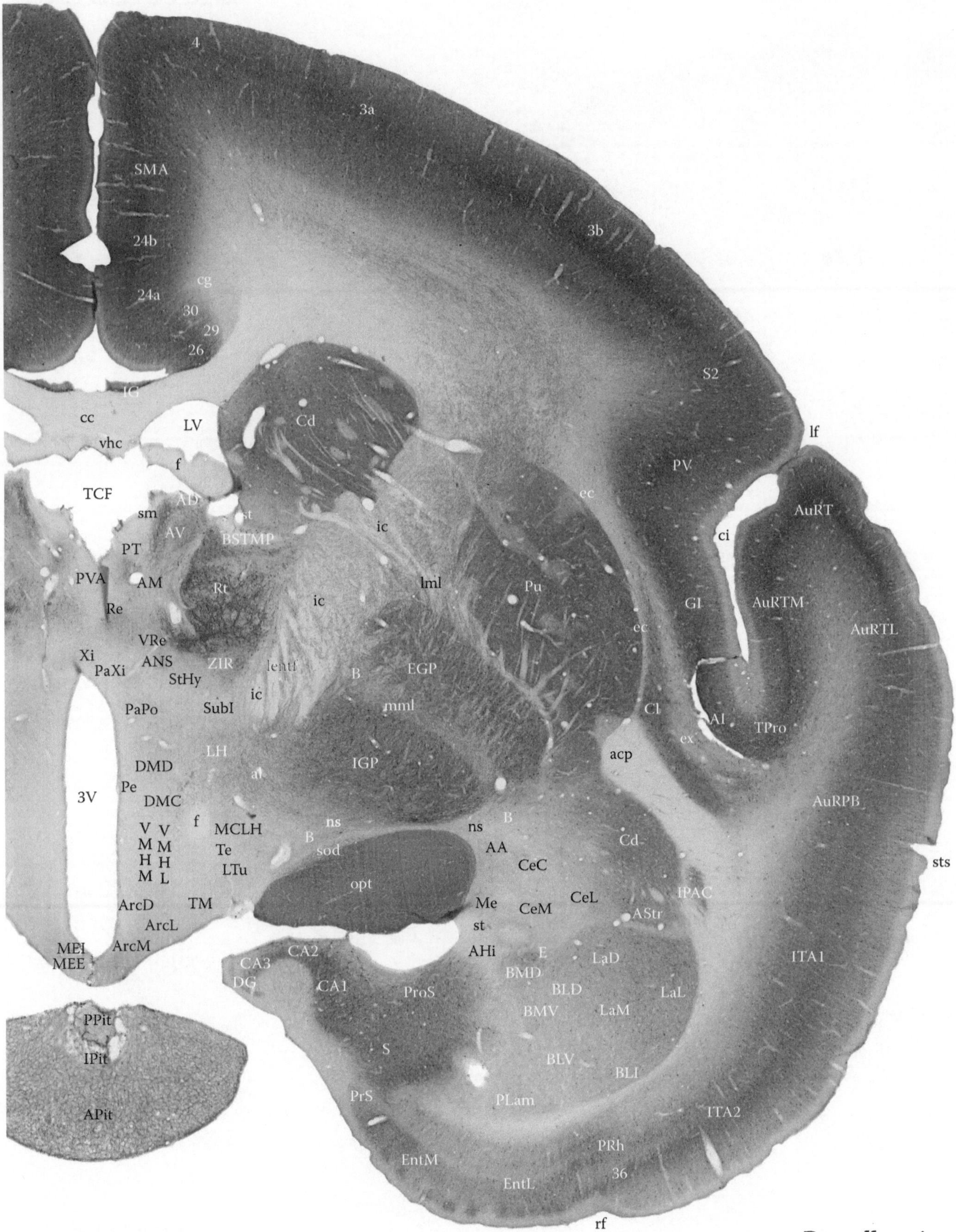


**Parvalbumin**
Interaural +07.75 mm

# Plate 23e

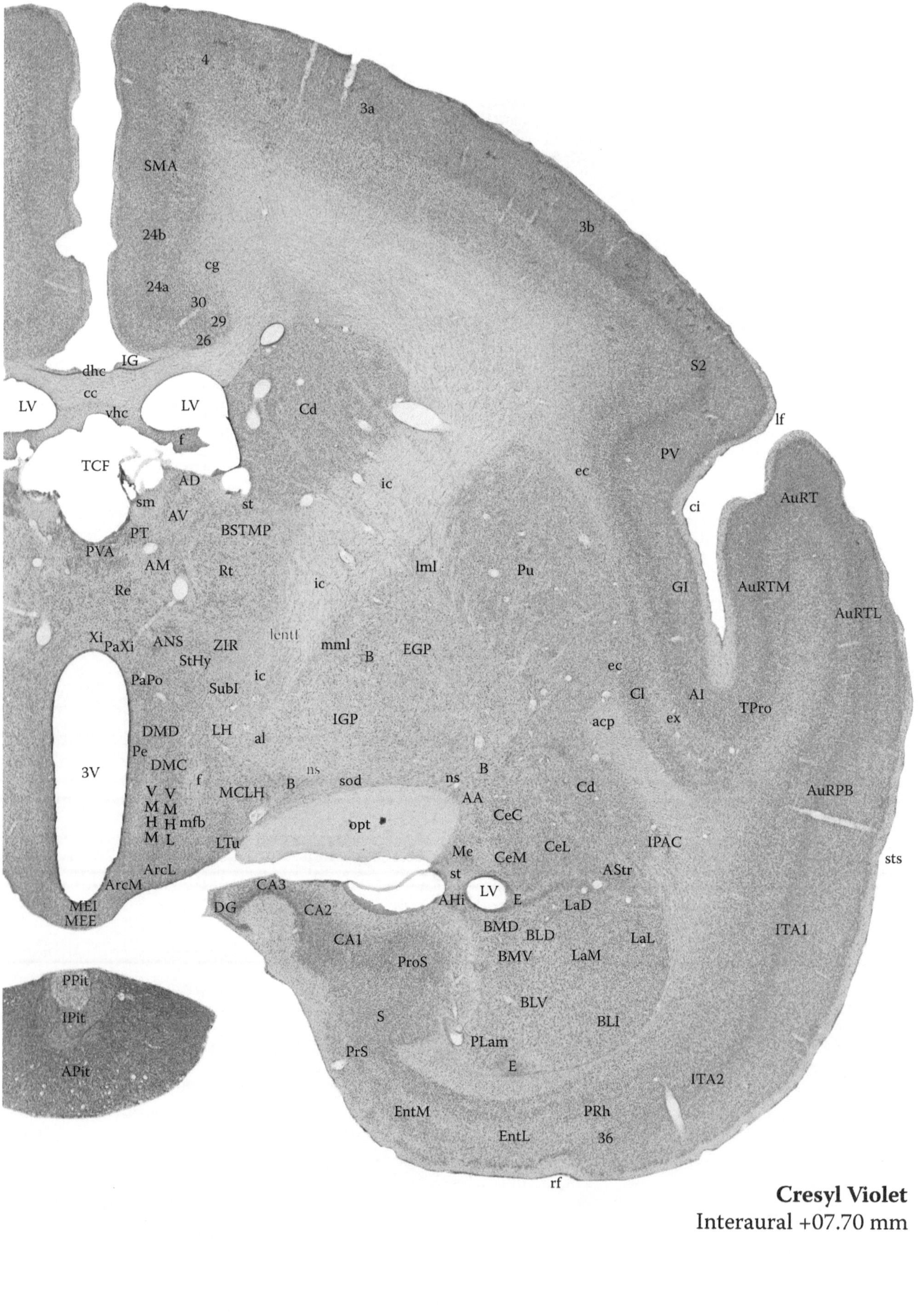


**Cresyl Violet**
Interaural +07.70 mm

# Plate 23

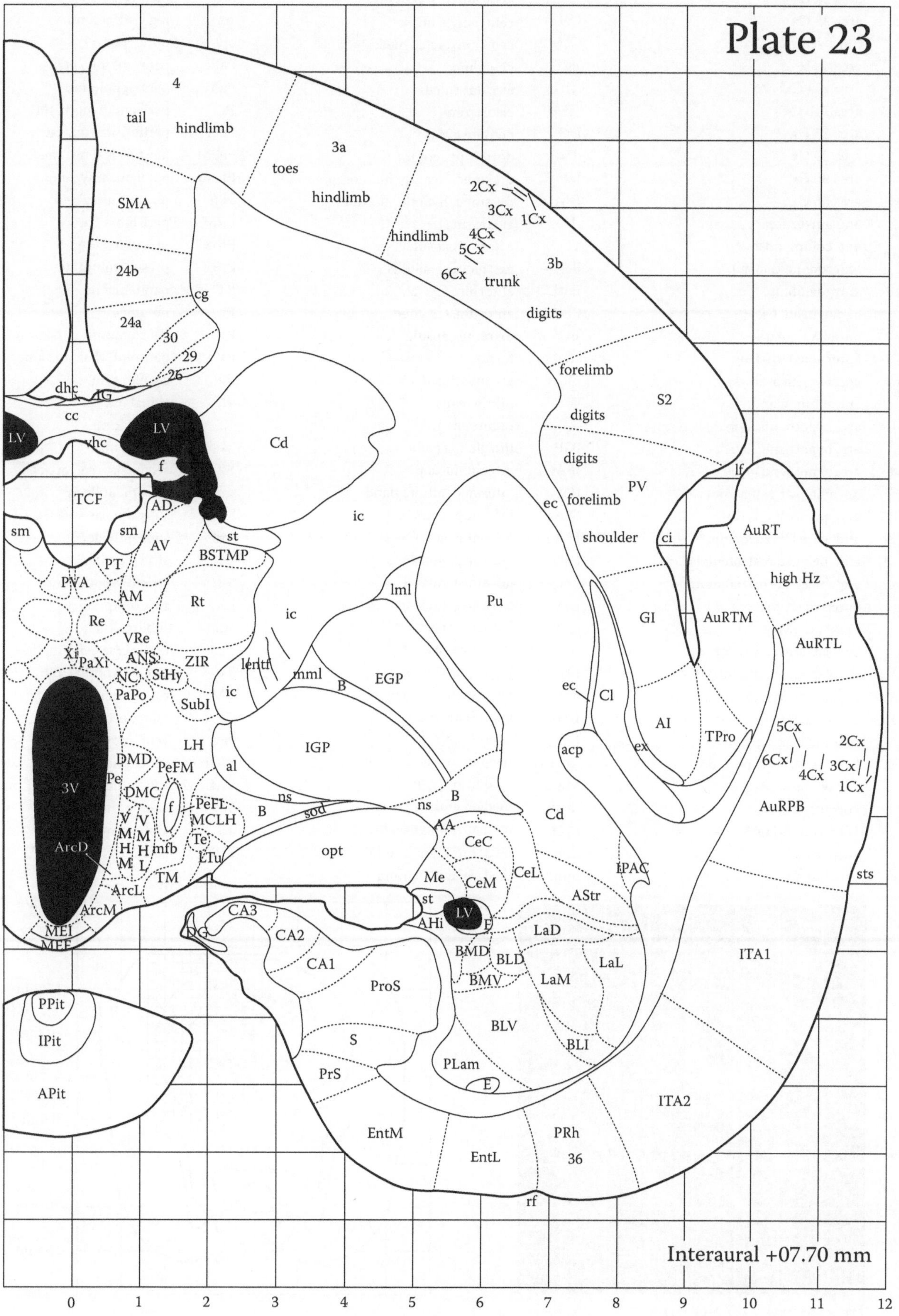


Interaural +07.70 mm

| | |
|---|---|
| 3a | area 3a Cx |
| 3b | area 3b Cx |
| 3V | third ventricle |
| 4 | area 4 Cx |
| 24a | area 24a Cx |
| 24b | area 24b Cx |
| 26 | area 26 Cx |
| 29 | area 29 Cx |
| 30 | area 30 Cx |
| 36 | area 36 Cx |
| AA | ant. amyg. area |
| acp | ant. comm. post. |
| AD | anterodors. thal. nu. |
| AHi | amygdalohipp. area |
| AI | agran. insul. Cx |
| al | ansa lenticularis |
| AM | anteromed. thal. nu. |
| ANS | access. neurosec. nu. |
| APit | ant. pituit. gland |
| ArcD | arc. hypoth. nu., dors. |
| ArcL | arc. hypoth. nu., lat. |
| ArcM | arc. hypoth. nu., med. |
| AStr | amygdalostr. trans. zone |
| AuRPB | aud. parabelt Cx, rost. |
| AuRT | aud. core Cx, rosttemp. |
| AuRTL | aud. belt Cx, rostrotemplat. |
| AuRTM | aud. belt Cx, rostrotempmed. |
| AV | anterovent. thal. nu. |
| B | basal nu. (Meynert) |
| BLD | basolat. amyg. nu., dors. |
| BLI | basolat. amyg. nu., int. |
| BLVM | basolat. amyg. nu., vent. med. |
| BMD | basomed. amyg. nu., dors. |
| BMV | basomed. amyg. nu., vent. |
| BSTMP | bed nu. st, med. post. divn |
| CA1 | cornu ammonis 1 |
| CA2 | cornu ammonis 2 |
| CA3 | cornu ammonis 3 |
| cc | corp. callosum |
| Cd | caudate nu. |
| CeC | cent. amyg. nu., cent. |
| CeL | cent. amyg. nu., lat. |
| CeM | cent. amyg. nu., med. |
| cg | cingulum |
| ci | circular sulcus |
| Cl | claustrum |
| DG | dentate gyrus |
| dhc | dors. hipp. comm. |
| DMC | dorsmed. hypoth. nu., comp. |
| DMD | dorsmed. hypoth. nu., dors. |
| E | ependyma |
| ec | external capsule |
| EGP | ext. globus pallidus |
| EntL | entorhin. Cx, lat. |
| EntM | entorhin. Cx, med. |
| ex | extreme capsule |
| f | fornix |
| GI | granular insul. Cx |
| ic | intern. caps. |
| IG | induseum gris. |
| IGP | int. globus pallidus |
| IPAC | interst. nu. acp |
| IPit | intermed. pituit. gland |
| ITA1 | inf. temp. ant. Cx, 1 |
| ITA2 | inf. temp. ant. Cx, 2 |
| LaD | lat. amyg. nu., dors. |
| LaL | lat. amyg. nu., lat. |
| LaM | lat. amyg. nu., med. |
| lentf | lenticular fasc. |
| lf | lateral fissure |
| LH | lat. hypothal. area |
| lml | lat. medull. lamina |
| LTu | lat. tuberal nu. |
| LV | lateral ventricle |
| MCLH | magno. nu. lat. hypothal. |
| Me | medial amyg. nu. |
| MEE | median eminence, ext. |
| MEI | median eminence, int. |
| mfb | med. forebrain bundle |
| mml | med. medull. lamina |
| NC | nanocell. hypoth. nu. |
| ns | nigrostriatal tract |
| opt | optic tract |
| PaPo | paravent. nu., post. |
| PaXi | paraxiphoid nu. |
| Pe | perivent. hypoth. nu. |
| PeFL | perifornical nu., lat. |
| PeFM | perifornical nu., med. |
| PLam | paralam. amyg. nu. |
| PPit | post. pituit. gland |
| PRh | perirhinal Cx |
| ProS | prosubiculum |
| PrS | presubiculum |
| PT | paratenial nu. |
| Pu | putamen |
| PV | PV somatosens. Cx |
| PVA | paravent. thal. nu., ant. |
| Re | reuniens nu. |
| rf | rhinal fiss. |
| Rt | reticular nu. |
| S | subiculum |
| S2 | second. somat. area Cx |
| sm | stria medullaris |
| SMA | supp. motor area Cx |
| sod | suproptic decn |
| st | stria terminalis |
| StHy | striatohypothal. nu. |
| sts | sup. temp. sulcus |
| SubI | subincertal nu. |
| TCF | transverse cerebral fiss. |
| Te | terete hypothal. nu. |
| TM | tuberomamill. area |
| TPro | temp. proisocortex |
| vhc | vent. hipp. comm. |
| VMHL | ventromed. hypo. nu., lat. |
| VMHM | ventromed. hypo. nu., med. |
| VRe | vent. reuniens nu. |
| Xi | xiphoid nu. |
| ZIR | zona incerta, rost. |

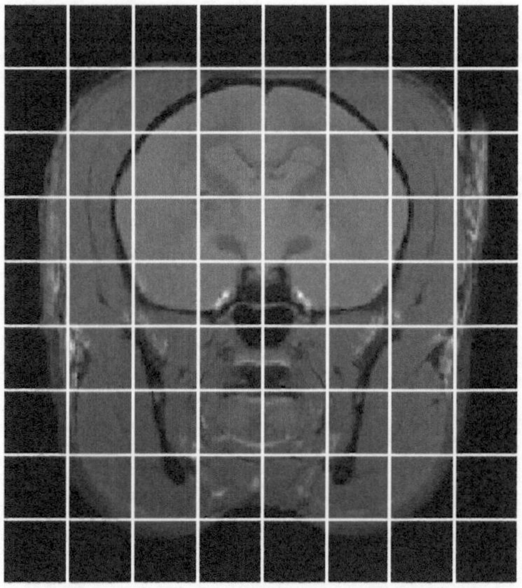

Interaural +07.80 mm

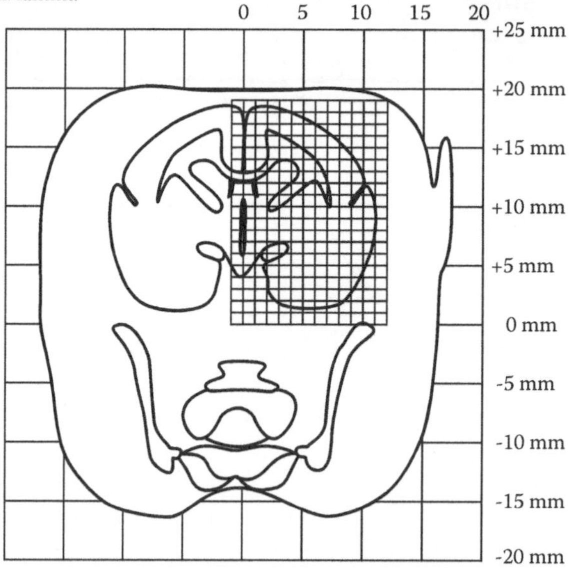

# Plate 24

Interaural +07.20 mm

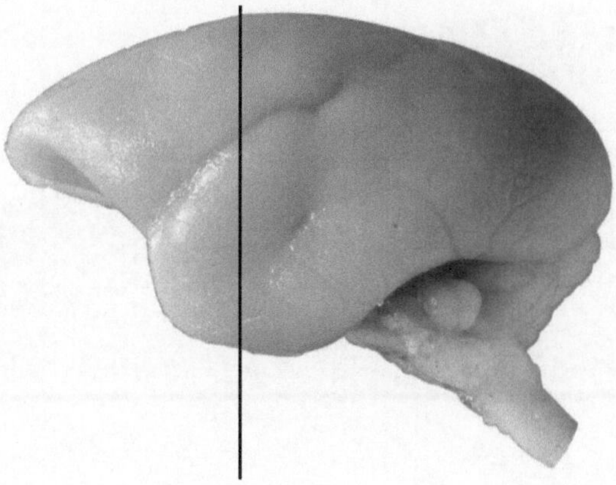

| | |
|---|---|
| 3a | area 3a Cx |
| 3b | area 3b Cx |
| 3V | third ventricle |
| 4 | area 4 Cx |
| 24a | area 24a Cx |
| 24b | area 24b Cx |
| 26 | area 26 Cx |
| 29 | area 29 Cx |
| 30 | area 30 Cx |
| 36 | area 36 Cx |
| acp | ant. comm. post. |
| AD | anterodors. thal. nu. |
| AI | agran. insul. Cx |
| AM | anteromed. thal. nu. |
| ANS | access. neurosec. nu. |
| APit | ant. pituit. gland |
| ArcL | arc. hypoth. nu., lat. |
| ArcM | arc. hypoth. nu., med. |
| AuR | aud. core Cx, rost. |
| AuRM | aud. belt Cx, rostmed. |
| AuRPB | aud. parabelt Cx, rost. |
| AuRTL | aud. belt Cx, rostrotemplat. |
| AV | anterovent. thal. nu. |
| B | basal nu. (Meynert) |
| BSTIA | bed nu. st, intraamyg. |
| CA1 | cornu ammonis 1 |
| CA2 | cornu ammonis 2 |
| CA3 | cornu ammonis 3 |
| cc | corp. callosum |
| Cd | caudate nu. |
| CeL | cent. amyg. nu., lat. |
| cg | cingulum |
| cgs | cingulate sulc. |
| ci | circular sulcus |
| Cl | claustrum |
| CM | cent. medial thal. nu. |
| dhc | dors. hipp. comm. |
| DMD | dorsmed. hypoth. nu., dors. |
| DMV | dorsmed. hypoth. nu., vent. |
| ec | external capsule |
| EGP | ext. globus pallidus |
| eml | extern. medull. lamina |
| EntL | entorhin. Cx, lat. |
| EntM | entorhin. Cx, med. |
| ex | extreme capsule |
| f | fornix |
| GI | granular insul. Cx |
| GrDG | gran. cell layer DG |
| hi | hippocamp. fiss. |
| IAM | interantmed. nu. |
| ic | intern. caps. |
| IG | induseum gris. |
| IGP | int. globus pallidus |
| IPit | intermed. pituit. gland |
| ITA1 | inf. temp. ant. Cx, 1 |
| ITA2 | inf. temp. ant. Cx, 2 |
| LaL | lat. amyg. nu., lat. |
| LaM | lat. amyg. nu., med. |
| lf | lateral fissure |
| LH | lat. hypothal. area |
| lml | lat. medull. lamina |
| LMol | strat. lacun. mol. CA |
| LV | lateral ventricle |
| MCLH | magno. nu. lat. hypothal. |
| ME | median eminence |
| mfb | med. forebrain bundle |
| mml | med. medull. lamina |
| mt | mamillothal. tr. |
| MTu | med. tuberal nu. |
| opt | optic tract |
| Or | strat. oriens CA |
| Pe | perivent. hypoth. nu. |
| PeF | perifornical nu. |
| PH | post. hypothal. area |
| PLamP | paralam. amyg. nu., post. |
| PMaD | premamill. nu., dorsal |
| PMaV | premamill. nu., ventral |
| PPit | post. pituit. gland |
| PRh | perirhinal Cx |
| ProS | prosubiculum |
| PrS | presubiculum |
| PT | paratenial nu. |
| Pu | putamen |
| PV | PV somatosens. Cx |
| PVA | paravent. thal. nu., ant. |
| Rad | strat. radiat. CA |
| Re | reuniens nu. |
| rf | rhinal fiss. |
| Rt | reticular nu. |
| S | subiculum |
| S2 | second. somat. area Cx |
| sm | stria medullaris |
| SMA | supp. motor area Cx |
| sn | striatonigral fibers |
| sod | suproptic decn |
| st | stria terminalis |
| sts | sup. temp. sulcus |
| SubI | subincertal nu. |
| TCF | transverse cerebral fiss. |
| Te | terete hypothal. nu. |
| TPro | temp. proisocortex |
| VA | vent. ant. thal. nu. |
| VAMC | vent. ant. thal. nu., mediocent. |
| vhc | vent. hipp. comm. |
| VMH | ventromed. hypoth. nu. |
| VRe | vent. reuniens nu. |
| VTM | vent. tuberomamill. area |
| Xi | xiphoid nu. |
| ZIR | zona incerta, rost. |

# Plate 24a

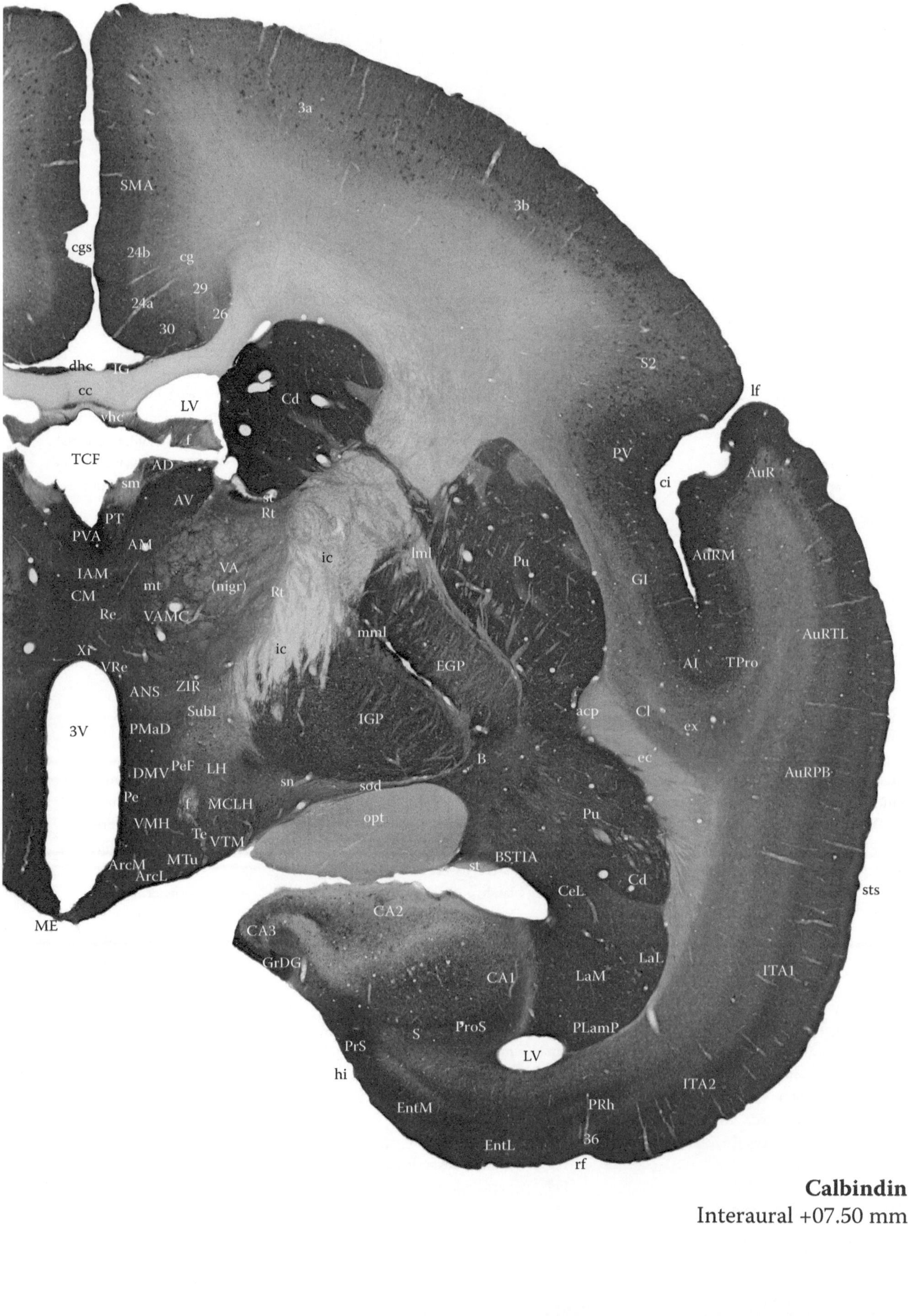


**Calbindin**
Interaural +07.50 mm

# Plate 24b

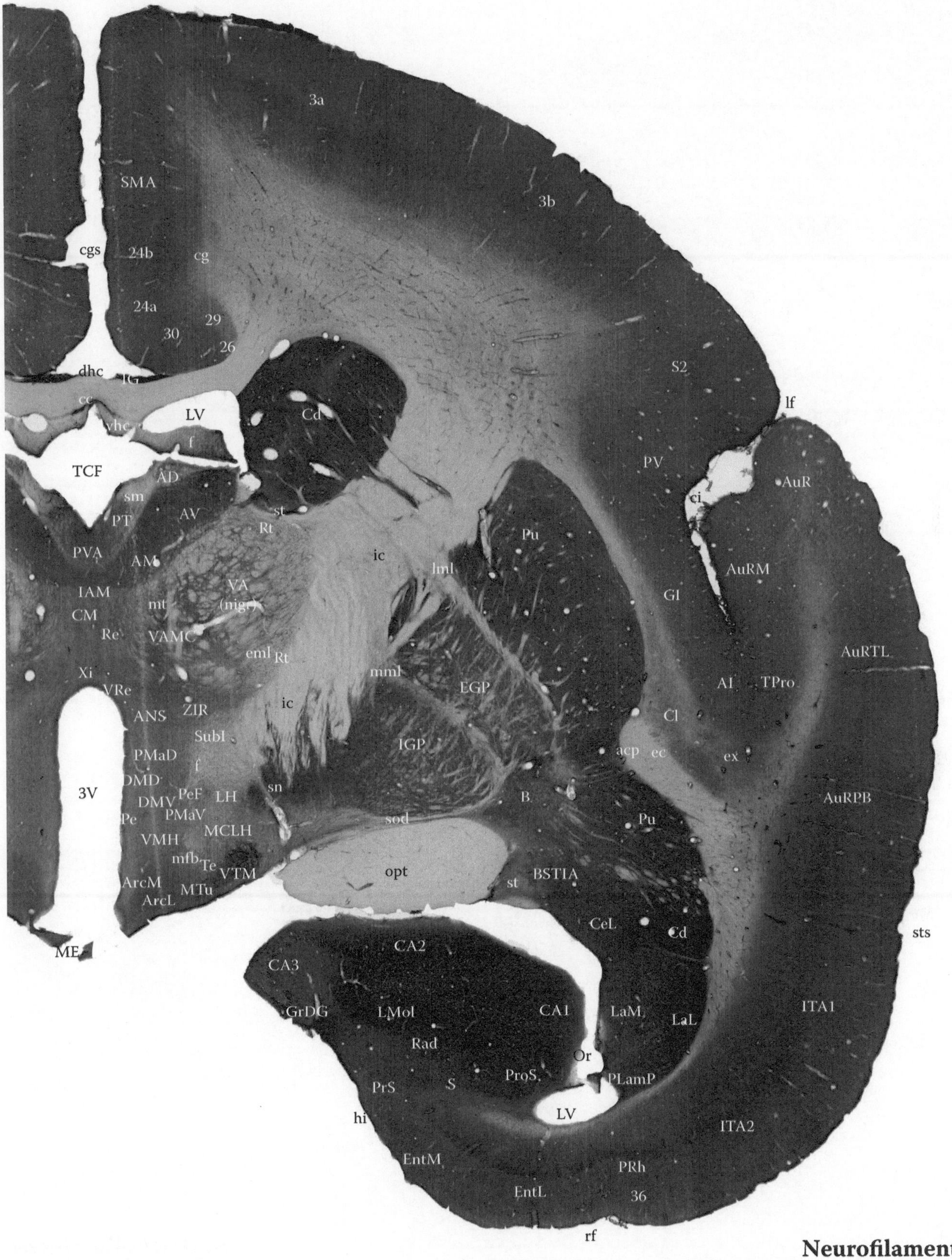


**Neurofilament**
Interaural +07.35 mm

# Plate 24c

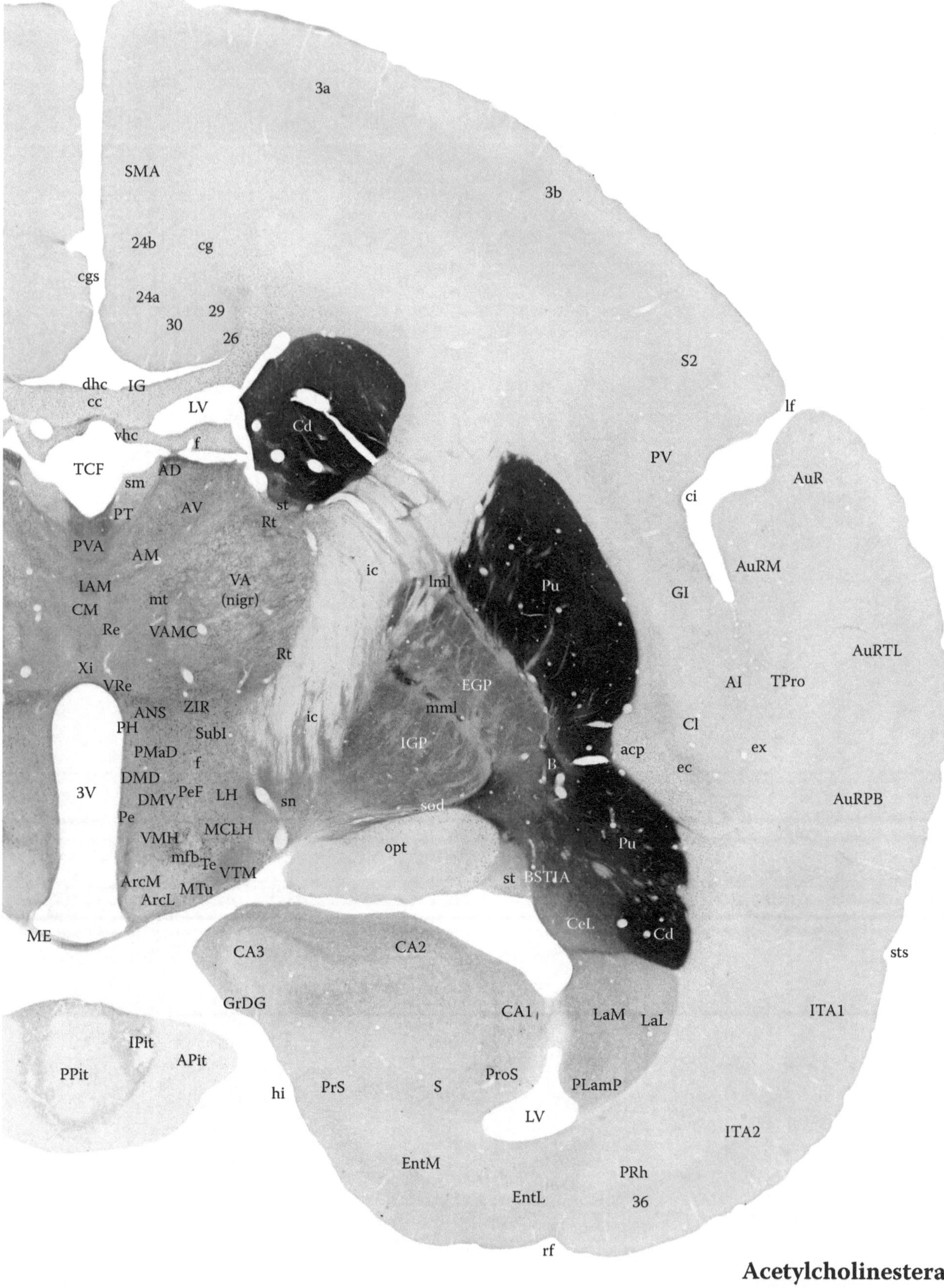


**Acetylcholinesterase**
Interaural +07.30 mm

# Plate 24d

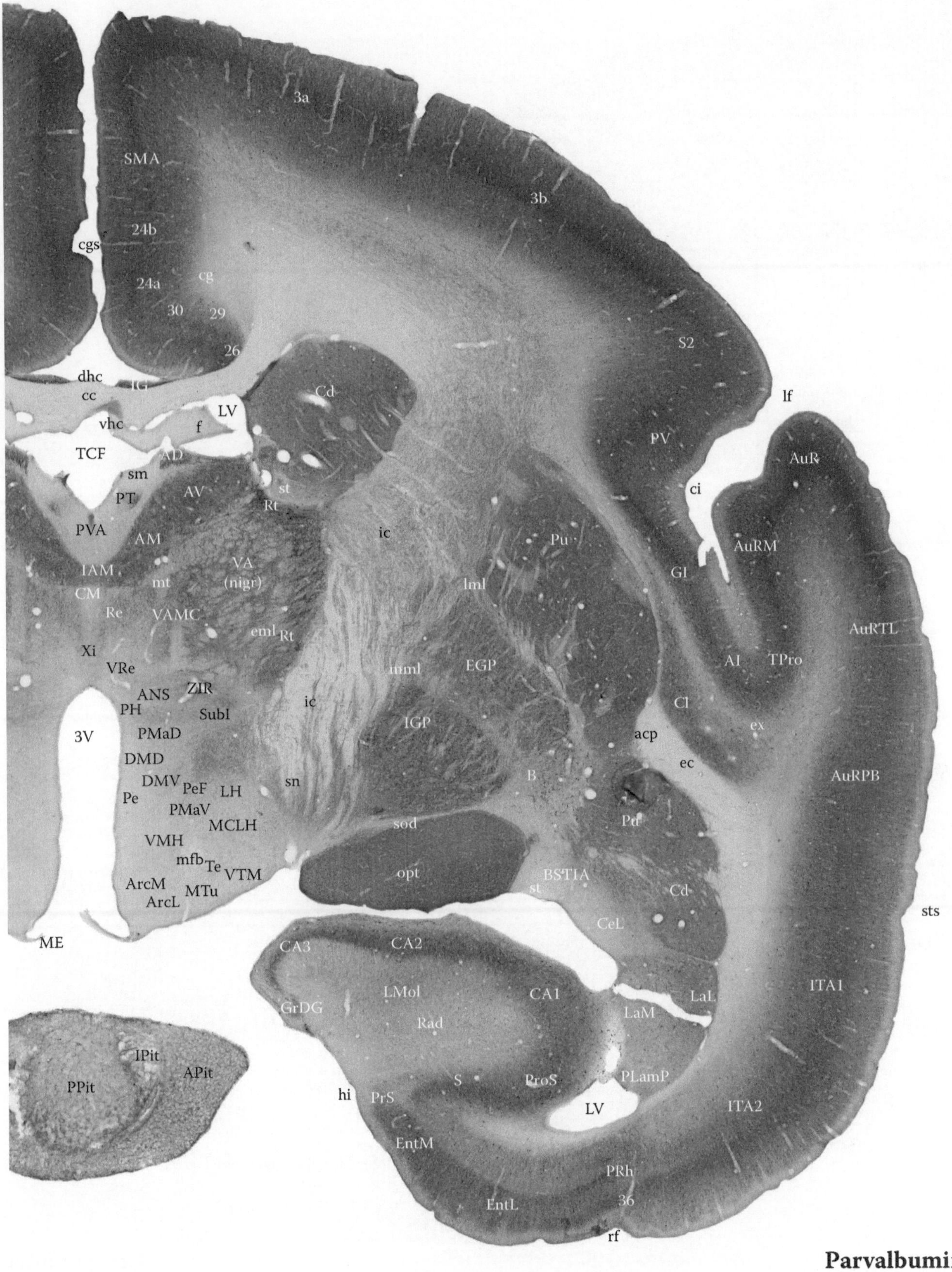


**Parvalbumin**
Interaural +07.25 mm

# Plate 24e

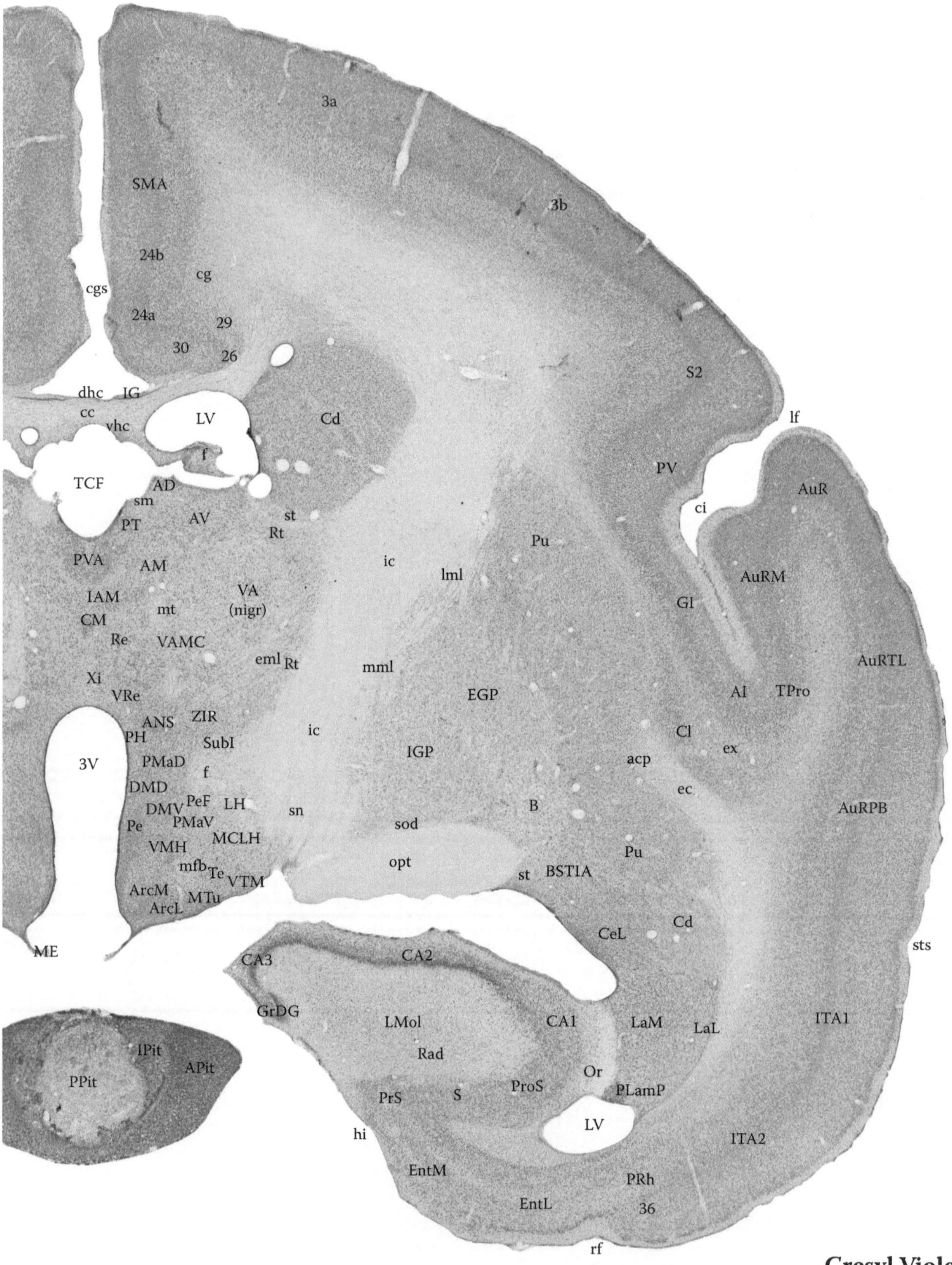


**Cresyl Violet**
Interaural +07.20 mm

# Plate 24

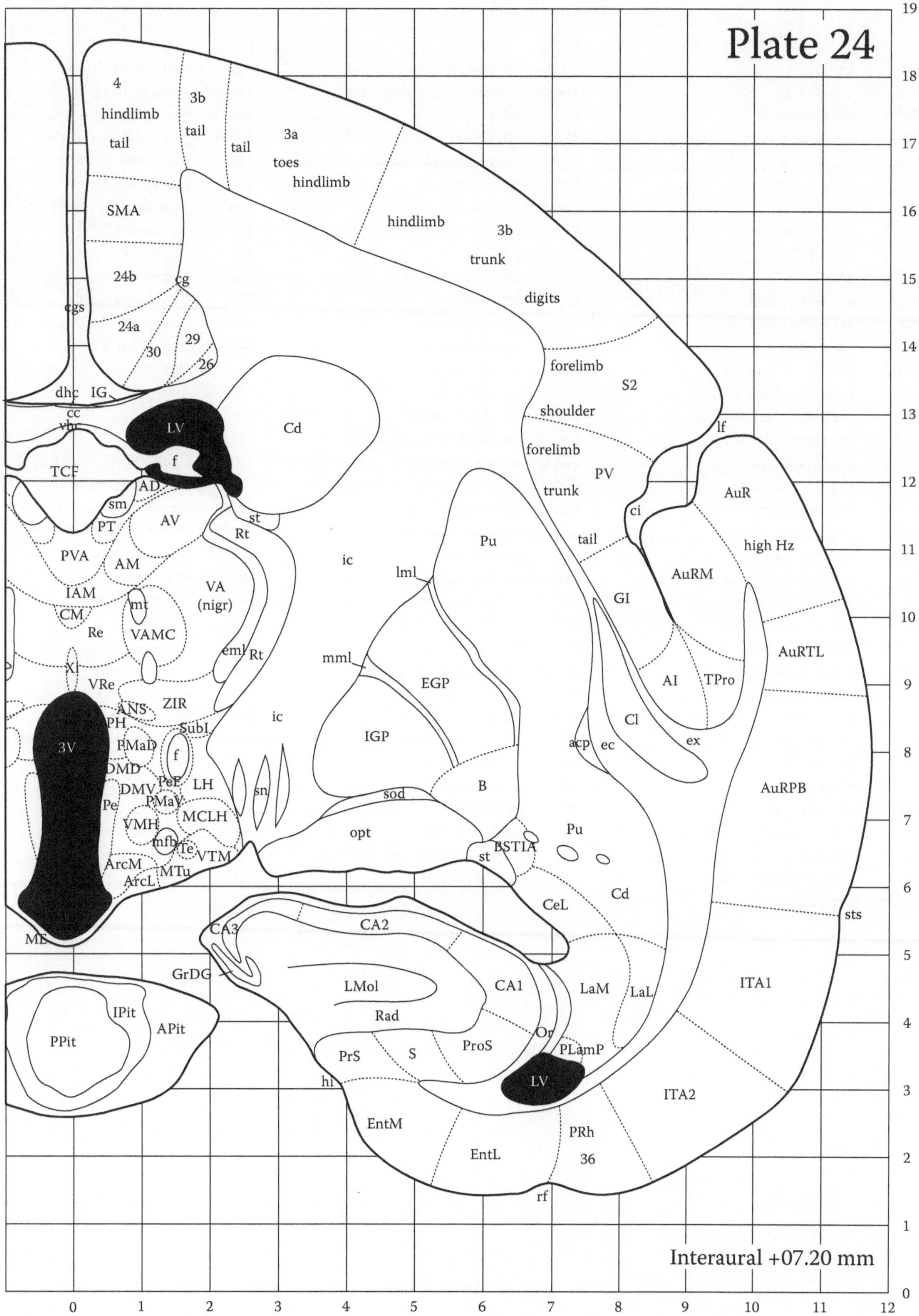


Interaural +07.20 mm

| | |
|---|---|
| 3a | area 3a Cx |
| 3b | area 3b Cx |
| 3V | third ventricle |
| 4 | area 4 Cx |
| 24a | area 24a Cx |
| 24b | area 24b Cx |
| 26 | area 26 Cx |
| 29 | area 29 Cx |
| 30 | area 30 Cx |
| 36 | area 36 Cx |
| acp | ant. comm. post. |
| AD | anterodors. thal. nu. |
| AI | agran. insul. Cx |
| AM | anteromed. thal. nu. |
| ANS | access. neurosec. nu. |
| APit | ant. pituit. gland |
| ArcL | arc. hypoth. nu., lat. |
| ArcM | arc. hypoth. nu., med. |
| AuR | aud. core Cx, rost. |
| AuRM | aud. belt Cx, rostmed. |
| AuRPB | aud. parabelt Cx, rost. |
| AuRTL | aud. belt Cx, rostrotemplat. |
| AV | anterovent. thal. nu. |
| B | basal nu. (Meynert) |
| BSTIA | bed nu. st, intraamyg. |
| CA1 | cornu ammonis 1 |
| CA2 | cornu ammonis 2 |
| CA3 | cornu ammonis 3 |
| cc | corp. callosum |
| Cd | caudate nu. |
| CeL | cent. amyg. nu., lat. |
| cg | cingulum |
| cgs | cingulate sulc. |
| ci | circular sulcus |
| Cl | claustrum |
| CM | cent. medial thal. nu. |
| dhc | dors. hipp. comm. |
| DMD | dorsmed. hypoth. nu., dors. |
| DMV | dorsmed. hypoth. nu., vent. |
| ec | external capsule |
| EGP | ext. globus pallidus |
| eml | extern. medull. lamina |
| EntL | entorhin. Cx, lat. |
| EntM | entorhin. Cx, med. |
| ex | extreme capsule |
| f | fornix |
| GI | granular insul. Cx |
| GrDG | gran. cell layer DG |
| hi | hippocamp. fiss. |
| IAM | interantmed. nu. |
| ic | intern. caps. |
| IG | induseum gris. |
| IGP | int. globus pallidus |
| IPit | intermed. pituit. gland |
| ITA1 | inf. temp. ant. Cx, 1 |
| ITA2 | inf. temp. ant. Cx, 2 |
| LaL | lat. amyg. nu., lat. |
| LaM | lat. amyg. nu., med. |
| lf | lateral fissure |
| LH | lat. hypothal. area |
| lml | lat. medull. lamina |
| LMol | strat. lacun. mol. CA |
| LV | lateral ventricle |
| MCLH | magno. nu. lat. hypothal. |
| ME | median eminence |
| mfb | med. forebrain bundle |
| mml | med. medull. lamina |
| mt | mamillothal. tr. |
| MTu | med. tuberal nu. |
| opt | optic tract |
| Or | strat. oriens CA |
| Pe | perivent. hypoth. nu. |
| PeF | perifornical nu. |
| PH | post. hypothal. area |
| PLamP | paralam. amyg. nu., post. |
| PMaD | premamill. nu., dorsal |
| PMaV | premamill. nu., ventral |
| PPit | post. pituit. gland |
| PRh | perirhinal Cx |
| ProS | prosubiculum |
| PrS | presubiculum |
| PT | paratenial nu. |
| Pu | putamen |
| PV | PV somatosens. Cx |
| PVA | paravent. thal. nu., ant. |
| Rad | strat. radiat. CA |
| Re | reuniens nu. |
| rf | rhinal fiss. |
| Rt | reticular nu. |
| S | subiculum |
| S2 | second. somat. area Cx |
| sm | stria medullaris |
| SMA | supp. motor area Cx |
| sn | striatonigral fibers |
| sod | suproptic decn |
| st | stria terminalis |
| sts | sup. temp. sulcus |
| SubI | subincertal nu. |
| TCF | transverse cerebral fiss. |
| Te | terete hypothal. nu. |
| TPro | temp. proisocortex |
| VA | vent. ant. thal. nu. |
| VAMC | vent. ant. thal. nu., mediocent. |
| vhc | vent. hipp. comm. |
| VMH | ventromed. hypoth. nu. |
| VRe | vent. reuniens nu. |
| VTM | vent. tuberomamill. area |
| Xi | xiphoid nu. |
| ZIR | zona incerta, rost. |

# Plate 25

Interaural +06.65 mm

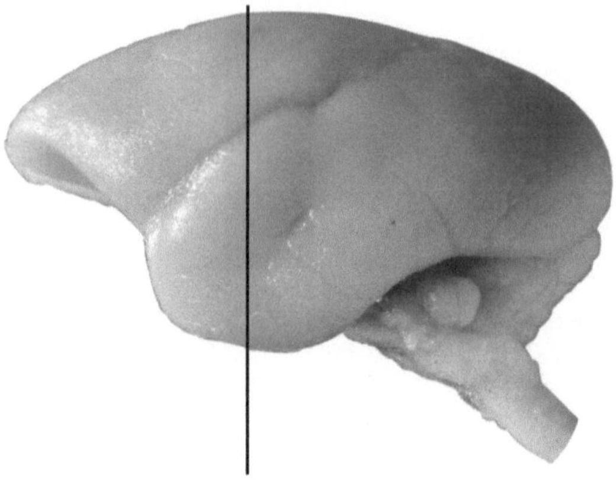

| | |
|---|---|
| 3b | area 3b Cx |
| 3V | third ventricle |
| 4 | area 4 Cx |
| 24a | area 24a Cx |
| 24b | area 24b Cx |
| 26 | area 26 Cx |
| 29 | area 29 Cx |
| 30 | area 30 Cx |
| 36 | area 36 Cx |
| AD | anterodors. thal. nu. |
| AI | agran. insul. Cx |
| alv | alveus |
| AM | anteromed. thal. nu. |
| APit | ant. pituit. gland |
| AuR | aud. core Cx, rost. |
| AuRM | aud. belt Cx, rostmed. |
| AuRPB | aud. parabelt Cx, rost. |
| AuRTL | aud. belt Cx, rostrotemplat. |
| AV | anterovent. thal. nu. |
| B | basal nu. (Meynert) |
| BSTIA | bed nu. st, intraamyg. |
| CA1 | cornu ammonis 1 |
| CA2 | cornu ammonis 2 |
| CA3 | cornu ammonis 3 |
| cc | corp. callosum |
| Cd | caudate nu. |
| cg | cingulum |
| Cl | claustrum |
| CLa | centrolat. thal. nu., ant. |
| CM | cent. medial thal. nu. |
| cp | cerebral peduncle |
| ec | external capsule |
| EGP | ext. globus pallidus |
| eml | extern. medull. lamina |
| EntC | entorhin. Cx, caud. |
| ex | extreme capsule |
| f | fornix |
| GrDG | gran. cell layer DG |
| H1 (thf) | H1 thalamic fasc. |
| H (al) | H ansa lentic. |
| hi | hippocamp. fiss. |
| IAM | interantmed. nu. |
| ic | intern. caps. |
| IGP | int. globus pallidus |
| iml | intern. medull. lamina |
| it | inf. temporal sulcus |
| ITA1 | inf. temp. ant. Cx, 1 |
| ITA2 | inf. temp. ant. Cx, 2 |
| lentf | lenticular fasc. |
| lf | lateral fissure |
| LH | lat. hypothal. area |
| LM | lat. mamill. nu. |
| lml | lat. medull. lamina |
| LV | lateral ventricle |
| mfb | med. forebrain bundle |
| ML | med. mamill. nu., lat. |
| MM | med. mamill. nu., med. |
| mml | med. medull. lamina |
| mp | mamill. peduncle |
| MRe | mamill. recess |
| mt | mamillothal. tr. |
| MTnl | med. terminal nu. |
| ns | nigrostriatal tract |
| opt | optic tract |
| PC | paracent. thal. nu. |
| PH | post. hypothal. area |
| PPit | post. pituit. gland |
| PRh | perirhinal Cx |
| ProS | prosubiculum |
| PrS | presubiculum |
| PT | paratenial nu. |
| Pu | putamen |
| PVA | paravent. thal. nu., ant. |
| Re | reuniens nu. |
| rf | rhinal fiss. |
| rmd | retromamill. decn |
| RML | retromamill. nu., lat. |
| RMM | retromamill. nu., med. |
| Rt | reticular nu. |
| S | subiculum |
| S2 | second. somat. area Cx |
| sm | stria medullaris |
| sn | striatonigral fibers |
| sod | suproptic decn |
| st | stria terminalis |
| TCF | transverse cerebral fiss. |
| TPro | temp. proisocortex |
| VA | vent. ant. thal. nu. |
| VAMC | vent. ant. thal. nu., mediocent. |
| VLa | ventrolat. thal. nu., ant. |
| VRe | vent. reuniens nu. |
| VSom? | vent. somatosens. Cx |
| VTM | vent. tuberomamill. area |
| Xi | xiphoid nu. |
| ZID | zona incerta, dors. |
| ZIV | zona incerta, vent. |

# Plate 25a

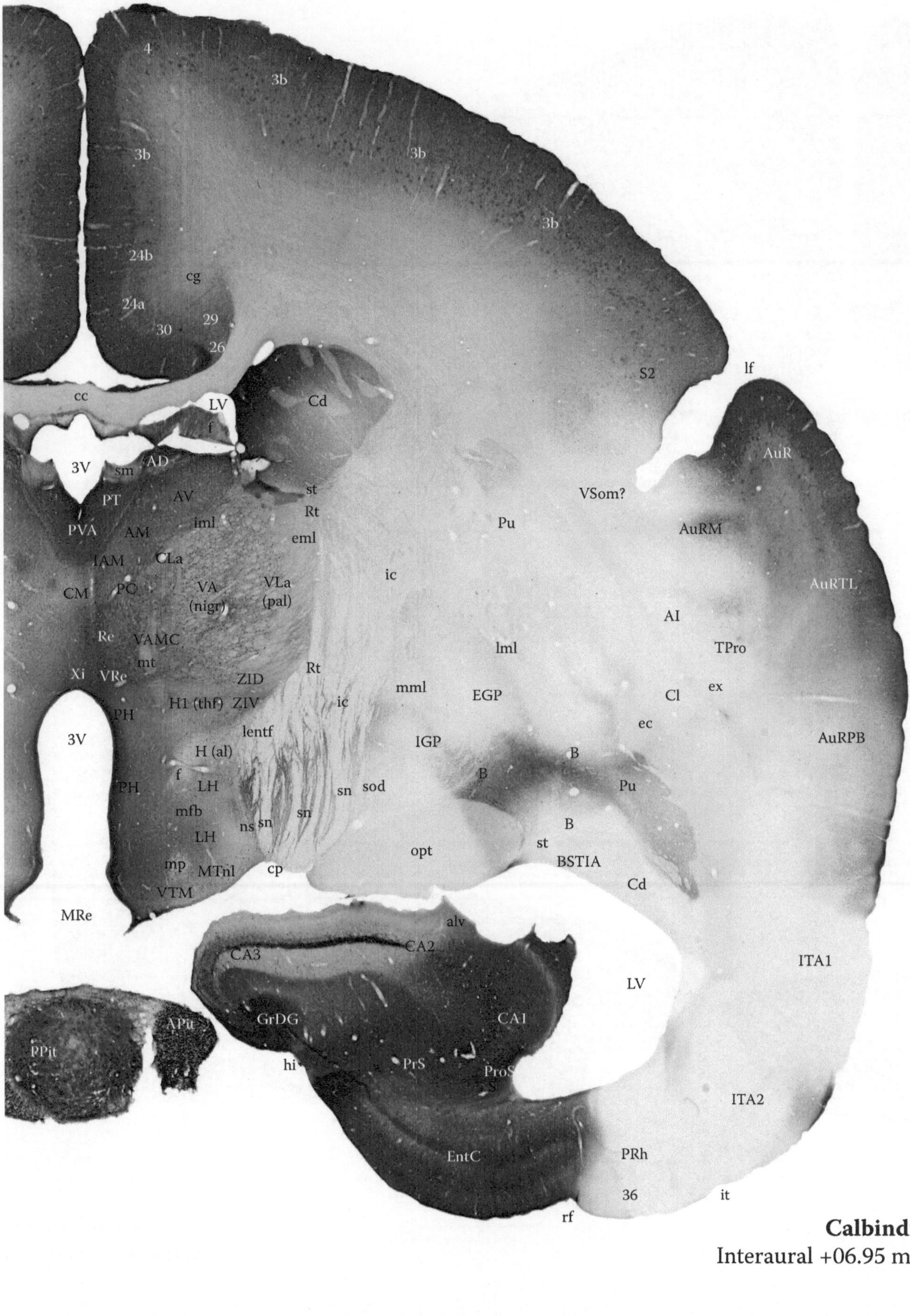


**Calbindin**
Interaural +06.95 mm

# Plate 25b

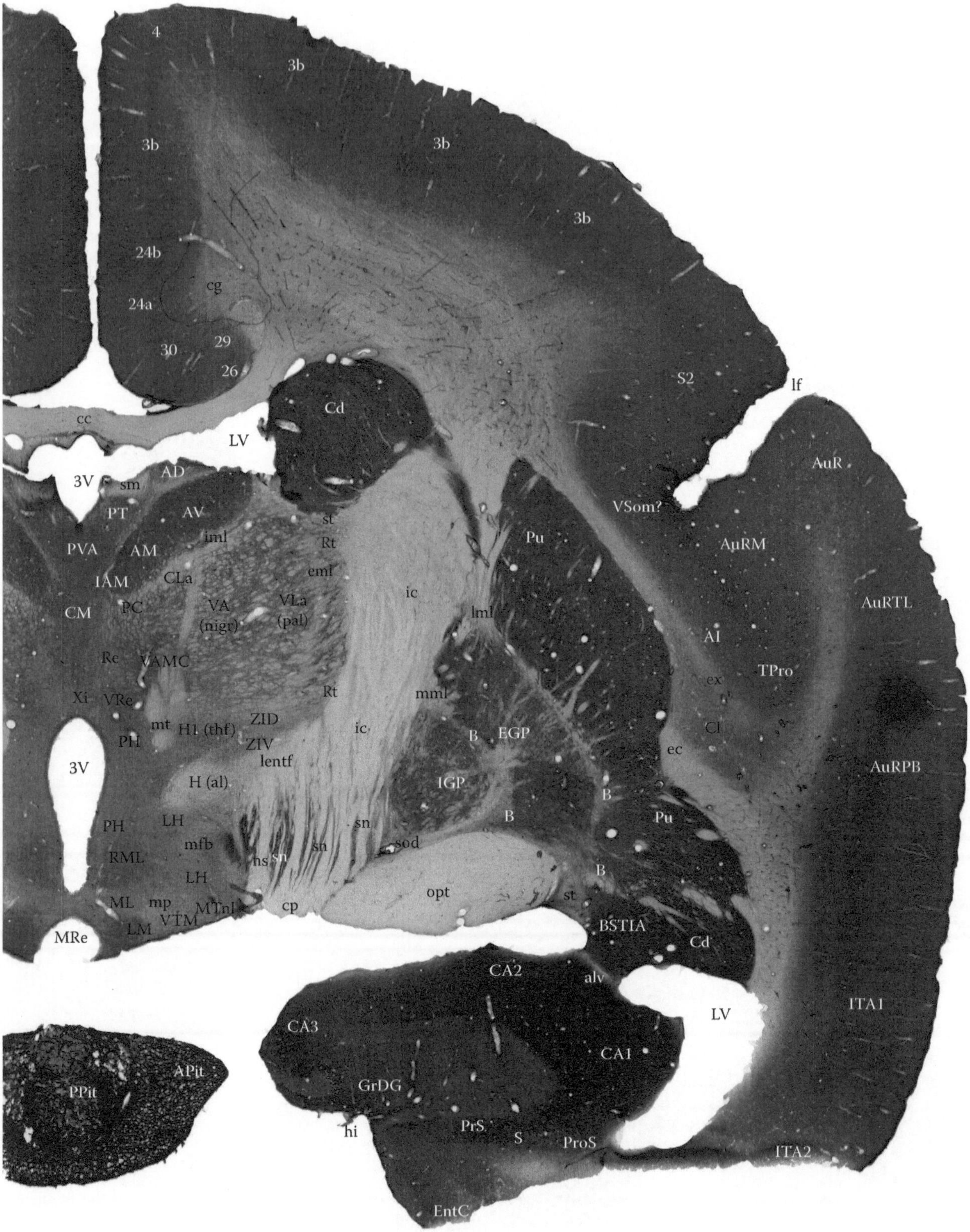


**Neurofilament**
Interaural +06.80 mm

# Plate 25c

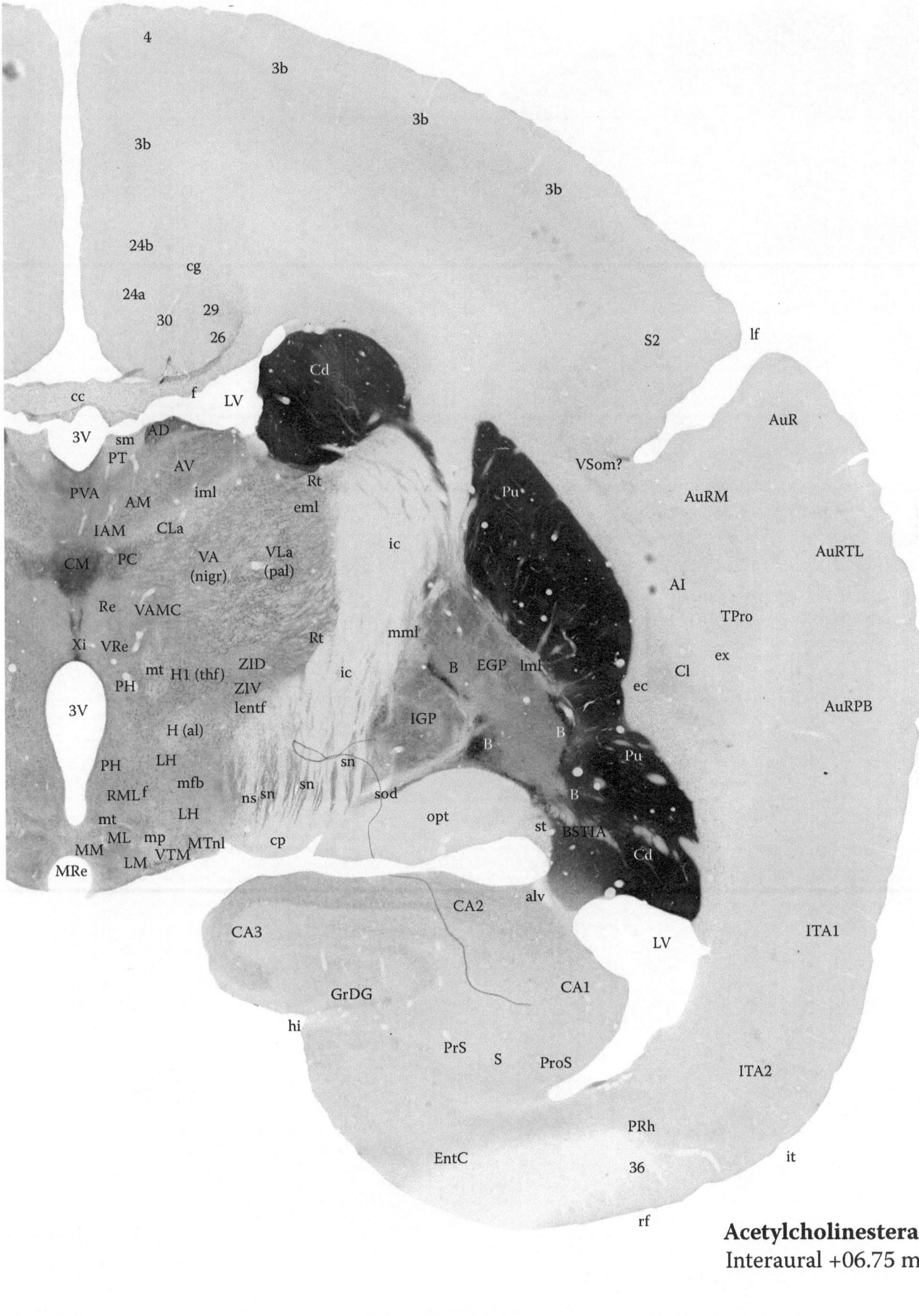


**Acetylcholinesterase**
Interaural +06.75 mm

# Plate 25d

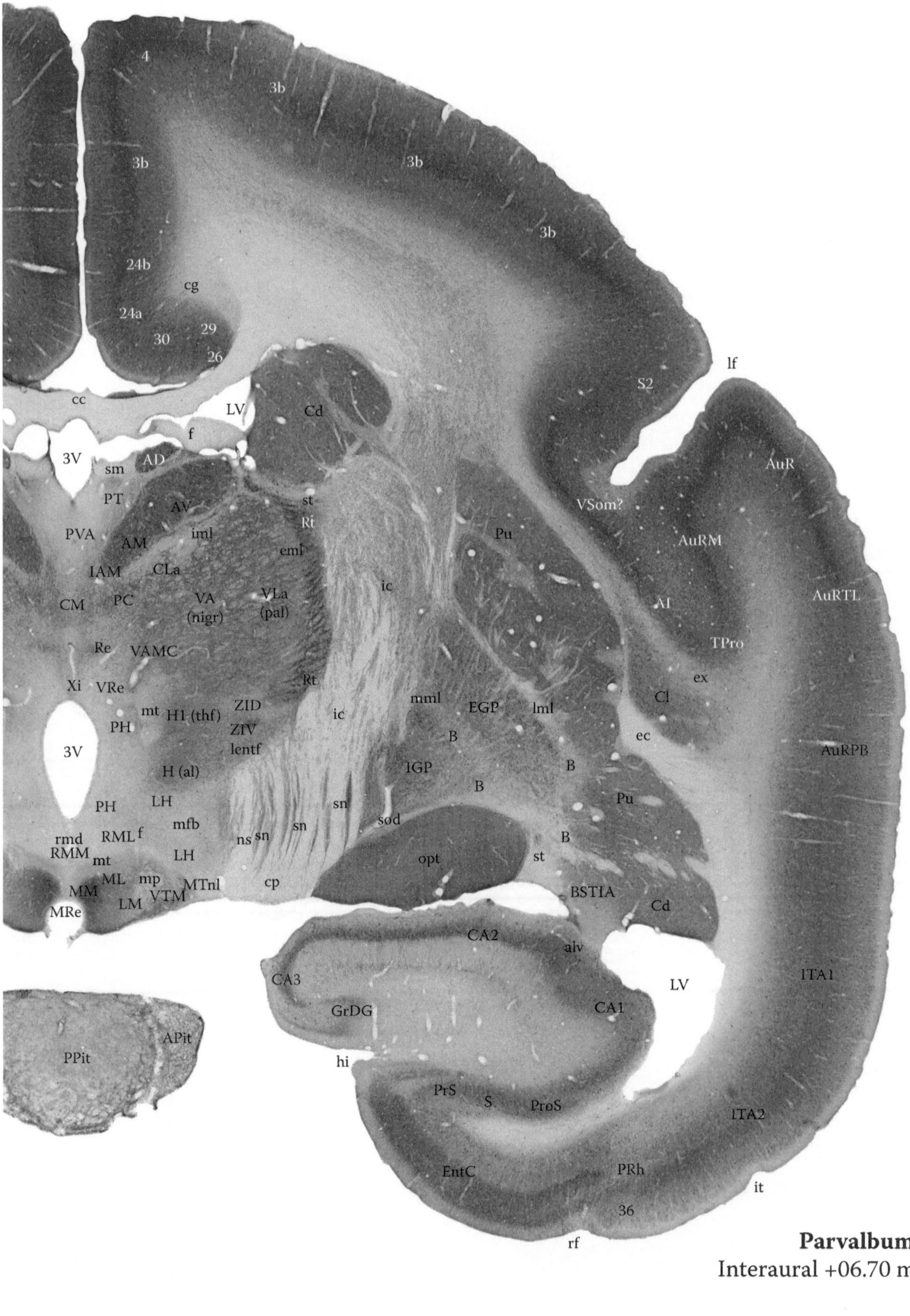


**Parvalbumin**
Interaural +06.70 mm

# Plate 25e

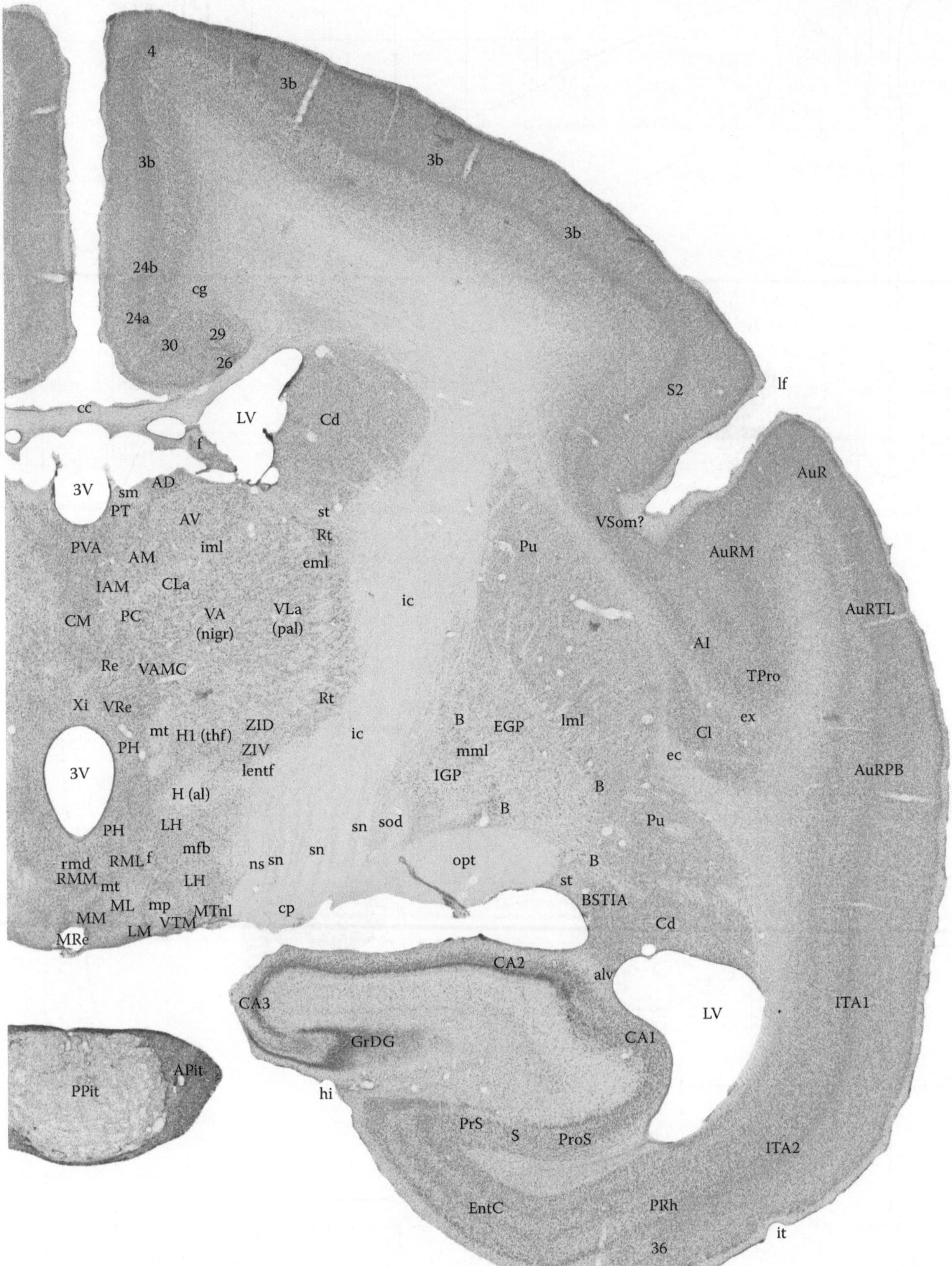


**Cresyl Violet**
Interaural +06.65 mm

# Plate 25

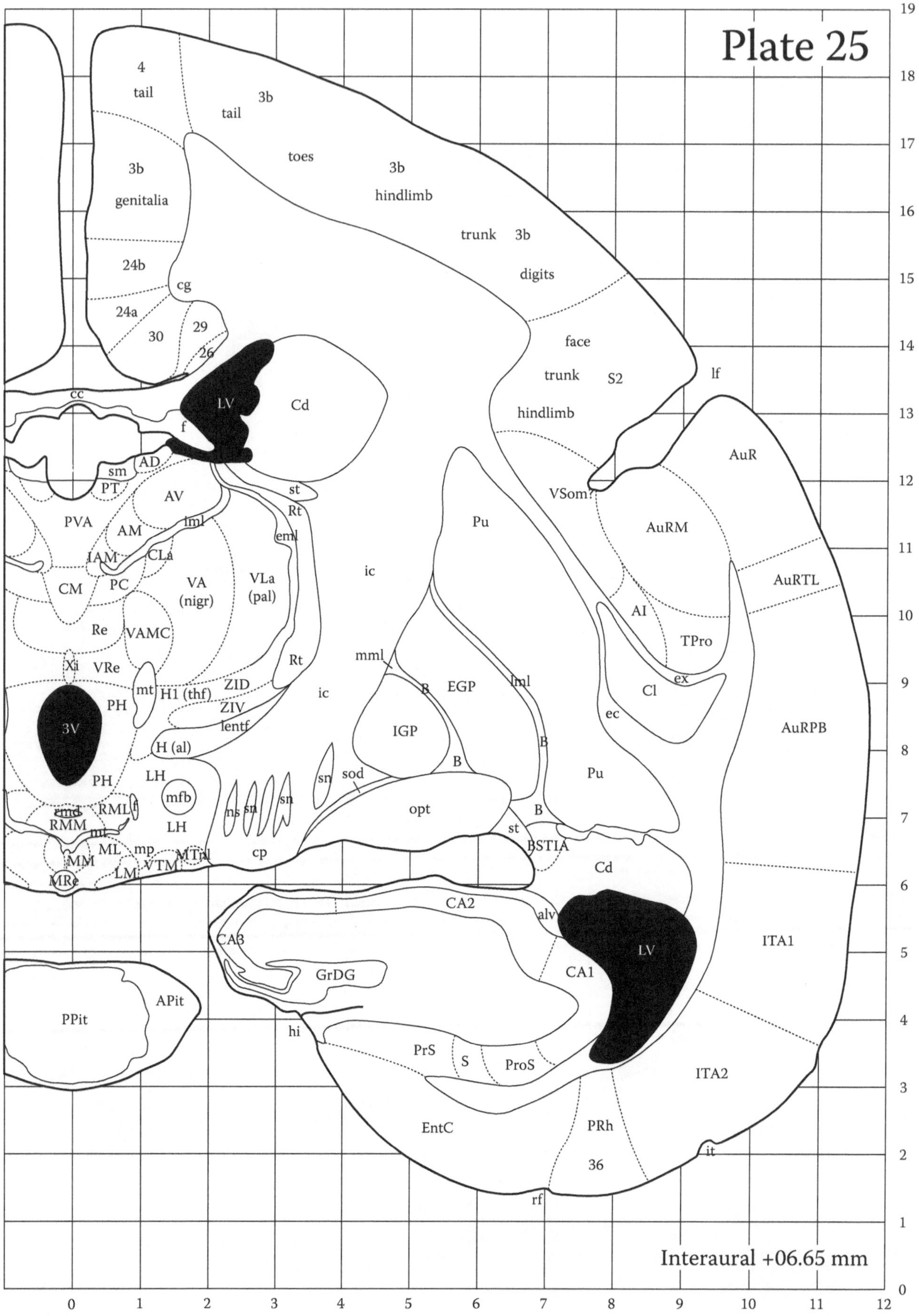
4
tail
3b
tail
toes
3b
genitalia
3b
hindlimb
trunk
3b
24b
digits
cg
24a
30
29
26
face
trunk
S2
lf
cc
LV
Cd
hindlimb
f
AuR
sm
AD
PT
st
VSom?
AV
Rt
Pu
PVA
AM
lml
eml
AuRM
IAM
CLa
CM
PC
VA
(nigr)
VLa
(pal)
ic
AuRTL
AI
Re
VAMC
TPro
mml
Rt
Xi
VRe
ex
ZID
mt
H1 (thf)
B
EGP
lml
Cl
PH
ic
ZIV
ec
3V
lentf
IGP
AuRPB
H (al)
B
PH
LH
sn
sod
B
Pu
mfb
sn
RML
f
rmd
ns
sn
opt
B
RMM
mt
LH
st
ML
mp
MTu
cp
BSTIA
MM
VTM
LM
Cd
MRe
CA2
alv
CA3
LV
ITA1
CA1
GrDG
APit
PPit
hi
PrS
S
ProS
ITA2
EntC
PRh
it
36
rf
Interaural +06.65 mm
0
1
2
3
4
5
6
7
8
9
10
11
12
0
1
2
3
4
5
6
7
8
9
10
11
12
13
14
15
16
17
18
19

| | | | | | |
|---|---|---|---|---|---|
| 3b | area 3b Cx | EGP | ext. globus pallidus | PC | paracent. thal. nu. |
| 3V | third ventricle | eml | extern. medull. lamina | PH | post. hypothal. area |
| 4 | area 4 Cx | EntC | entorhin. Cx, caud. | PPit | post. pituit. gland |
| 24a | area 24a Cx | ex | extreme capsule | PRh | perirhinal Cx |
| 24b | area 24b Cx | f | fornix | ProS | prosubiculum |
| 26 | area 26 Cx | GrDG | gran. cell layer DG | PrS | presubiculum |
| 29 | area 29 Cx | H1 (thf) | H1 thalamic fasc. | PT | paratenial nu. |
| 30 | area 30 Cx | H (al) | H ansa lentic. | Pu | putamen |
| 36 | area 36 Cx | hi | hippocamp. fiss. | PVA | paravent. thal. nu., ant. |
| AD | anterodors. thal. nu. | IAM | interantmed. nu. | Re | reuniens nu. |
| AI | agran. insul. Cx | ic | intern. caps. | rf | rhinal fiss. |
| alv | alveus | IGP | int. globus pallidus | rmd | retromamill. decn |
| AM | anteromed. thal. nu. | iml | intern. medull. lamina | RML | retromamill. nu., lat. |
| APit | ant. pituit. gland | it | inf. temporal sulcus | RMM | retromamill. nu., med. |
| AuR | aud. core Cx, rost. | ITA1 | inf. temp. ant. Cx, 1 | Rt | reticular nu. |
| AuRM | aud. belt Cx, rostmed. | ITA2 | inf. temp. ant. Cx, 2 | S | subiculum |
| AuRPB | aud. parabelt Cx, rost. | lentf | lenticular fasc. | S2 | second. somat. area Cx |
| AuRTL | aud. belt Cx, rostrotemplat. | lf | lateral fissure | sm | stria medullaris |
| AV | anterovent. thal. nu. | LH | lat. hypothal. area | sn | striatonigral fibers |
| B | basal nu. (Meynert) | LM | lat. mamill. nu. | sod | suproptic decn |
| BSTIA | bed nu. st, intraamyg. | lml | lat. medull. lamina | st | stria terminalis |
| CA1 | cornu ammonis 1 | LV | lateral ventricle | TCF | transverse cerebral fiss. |
| CA2 | cornu ammonis 2 | mfb | med. forebrain bundle | TPro | temp. proisocortex |
| CA3 | cornu ammonis 3 | ML | med. mamill. nu., lat. | VA | vent. ant. thal. nu. |
| cc | corp. callosum | MM | med. mamill. nu., med. | VAMC | vent. ant. thal. nu., mediocent. |
| Cd | caudate nu. | mml | med. medull. lamina | VLa | ventrolat. thal. nu., ant. |
| cg | cingulum | mp | mamill. peduncle | VRe | vent. reuniens nu. |
| Cl | claustrum | MRe | mamill. recess | VSom? | vent. somatosens. Cx |
| CLa | centrolat. thal. nu., ant. | mt | mamillothal. tr. | VTM | vent. tuberomamill. area |
| CM | cent. medial thal. nu. | MTnl | med. terminal nu. | Xi | xiphoid nu. |
| cp | cerebral peduncle | ns | nigrostriatal tract | ZID | zona incerta, dors. |
| ec | external capsule | opt | optic tract | ZIV | zona incerta, vent. |

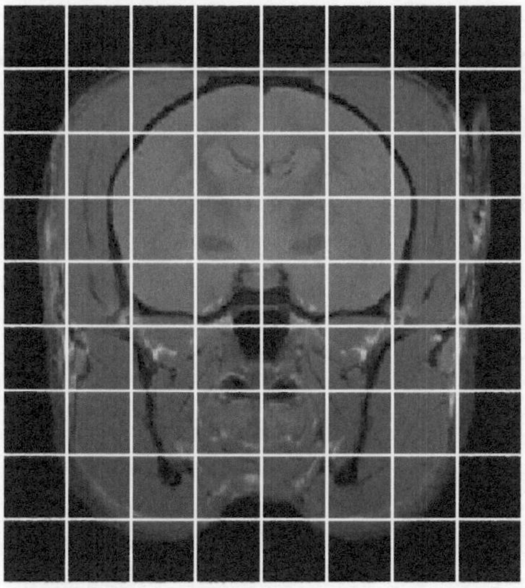

Interaural +06.80 mm

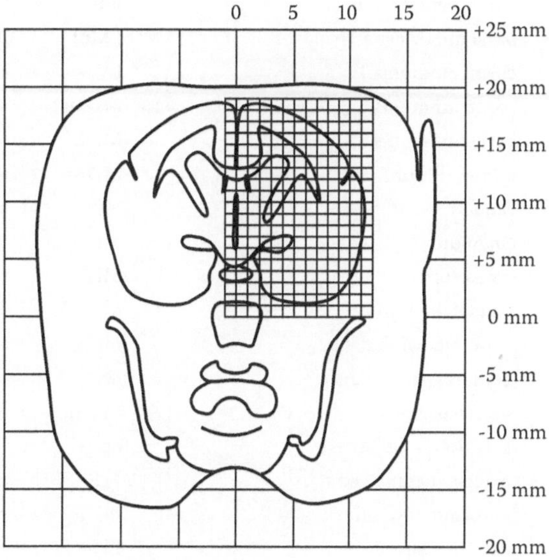

# Plate 26

Interaural +06.15 mm

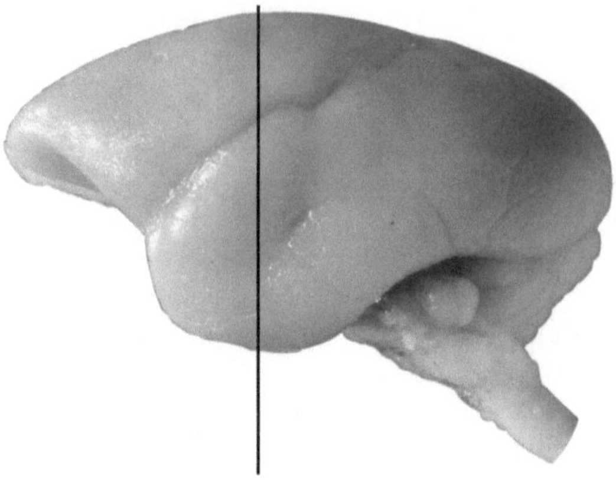

| | |
|---|---|
| 1&2 | areas 1/2 Cx |
| 3b | area 3b Cx |
| 3V | third ventricle |
| 23 | area 23 Cx |
| 26 | area 26 Cx |
| 29 | area 29 Cx |
| 30 | area 30 Cx |
| 31 | area 31 Cx |
| 36 | area 36 Cx |
| AD | anterodors. thal. nu. |
| alv | alveus |
| AM | anteromed. thal. nu. |
| AuR | aud. core Cx, rost. |
| AuRM | aud. belt Cx, rostmed. |
| AuRPB | aud. parabelt Cx, rost. |
| AuRTL | aud. belt Cx, rostrotemplat. |
| AV | anterovent. thal. nu. |
| B | basal nu. (Meynert) |
| CA1 | cornu ammonis 1 |
| CA2 | cornu ammonis 2 |
| CA3 | cornu ammonis 3 |
| cc | corp. callosum |
| Cd | caudate nu. |
| cg | cingulum |
| Cl | claustrum |
| CLa | caudal lin. nu. |
| CM | cent. medial thal. nu. |
| cp | cerebral peduncle |
| ec | external capsule |
| EGP | ext. globus pallidus |
| eml | extern. medull. lamina |
| EntC | entorhin. Cx, caud. |
| ex | extreme capsule |
| f | fornix |
| fd | fimbriodent. sulcus |
| fi | fimbria |
| GrDG | gran. cell layer DG |
| H1 (thf) | H1 thalamic fasc. |
| H1 (lentf) | H2 lenticular fasc. |
| hi | hippocamp. fiss. |
| ic | intern. caps. |
| IMD | intermediodors. thal. nu. |
| iml | intern. medull. lamina |
| it | inf. temporal sulcus |
| ITA1 | inf. temp. ant. Cx, 1 |
| ITA2 | inf. temp. ant. Cx, 2 |
| IVF | intervent. foramen |
| LD | laterodors. thal. nu. |
| lf | lateral fissure |
| LM | lat. mamill. nu. |
| lml | lat. medull. lamina |
| LMol | strat. lacun. mol. CA |
| LV | lateral ventricle |
| MDD | mediodors. thal. nu., dors. |
| MDL | mediodors. thal. nu., lat. |
| MDM | mediodors. thal. nu., med. |
| mfb | med. forebrain bundle |
| ML | med. mamill. nu., lat. |
| MM | med. mamill. nu., med. |
| MoDG | molecular layer DG |
| mp | mamill. peduncle |
| mt | mamillothal. tr. |
| MTnl | med. terminal nu. |
| opt | optic tract |
| PaS | parasubiculum |
| PC | paracent. thal. nu. |
| PH | post. hypothal. area |
| PLH | pedunc. lat. hypothal. |
| PoDG | polymorphic layer DG |
| PPit | post. pituit. gland |
| PRF | prerubral field |
| PRh | perirhinal Cx |
| ProS | prosubiculum |
| PrS | presubiculum |
| PSTh | parasubthal. nu. |
| Pu | putamen |
| PVA | paravent. thal. nu., ant. |
| Rad | strat. radiat. CA |
| Re | reuniens nu. |
| rf | rhinal fiss. |
| RIns | retroinsular Cx |
| RML | retromamill. nu., lat. |
| RMM | retromamill. nu., med. |
| Rt | reticular nu. |
| S | subiculum |
| S2 | second. somat. area Cx |
| sm | stria medullaris |
| sn | striatonigral fibers |
| SNR | subst. nigra, retic. |
| sod | suproptic decn |
| SPf | subparafasc. thal. nu. |
| st | stria terminalis |
| STh | subthalamic nu. |
| sts | sup. temp. sulcus |
| TPro | temp. proisocortex |
| VA | vent. ant. thal. nu. |
| VLa | ventrolat. thal. nu., ant. |
| VLp | ventrolat. thal. nu., post. |
| VRe | vent. reuniens nu. |
| VSom? | vent. somatosens. Cx |
| ZID | zona incerta, dors. |
| ZIV | zona incerta, vent. |

# Plate 26a

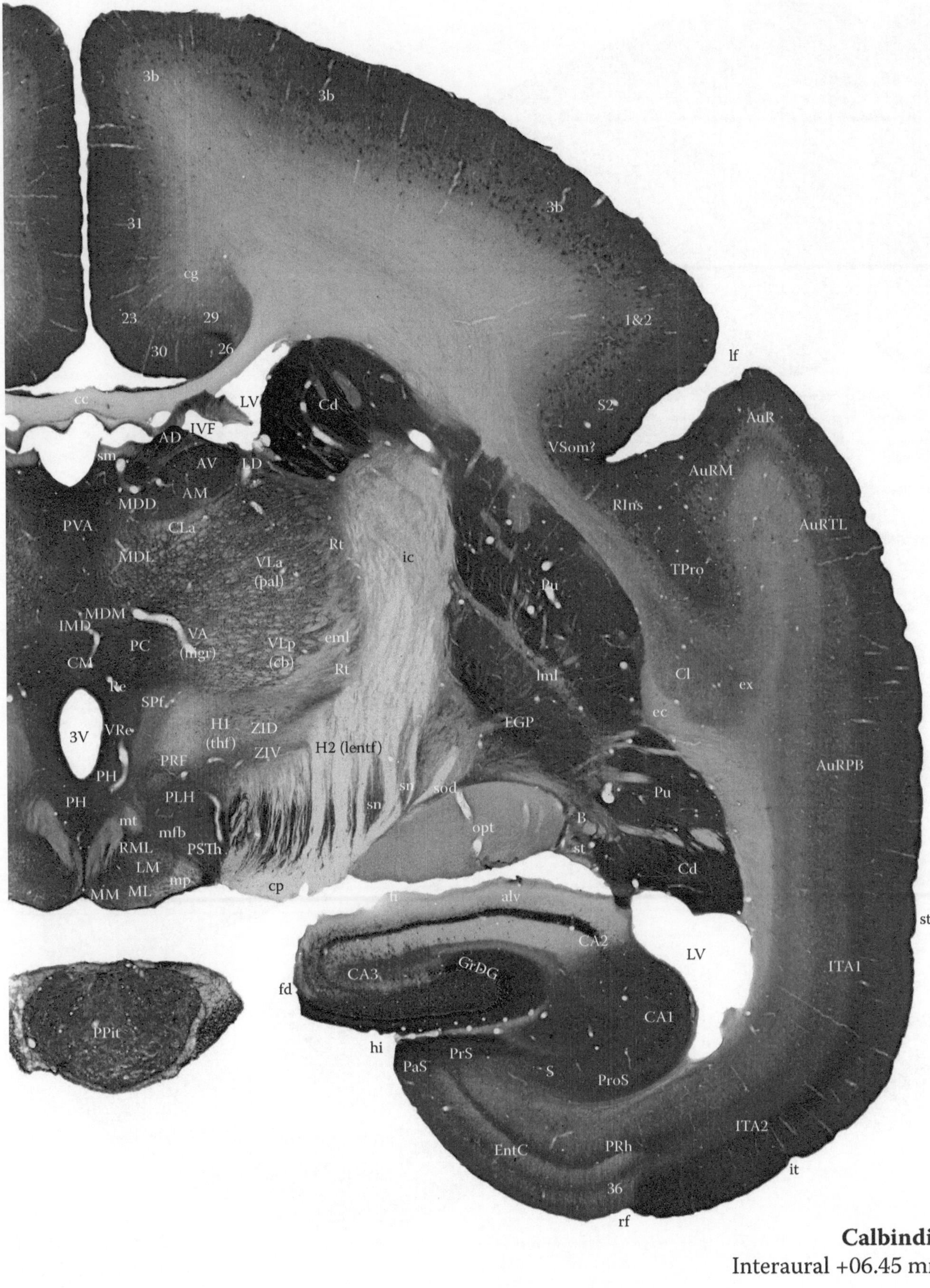


**Calbindin**
Interaural +06.45 mm

# Plate 26b

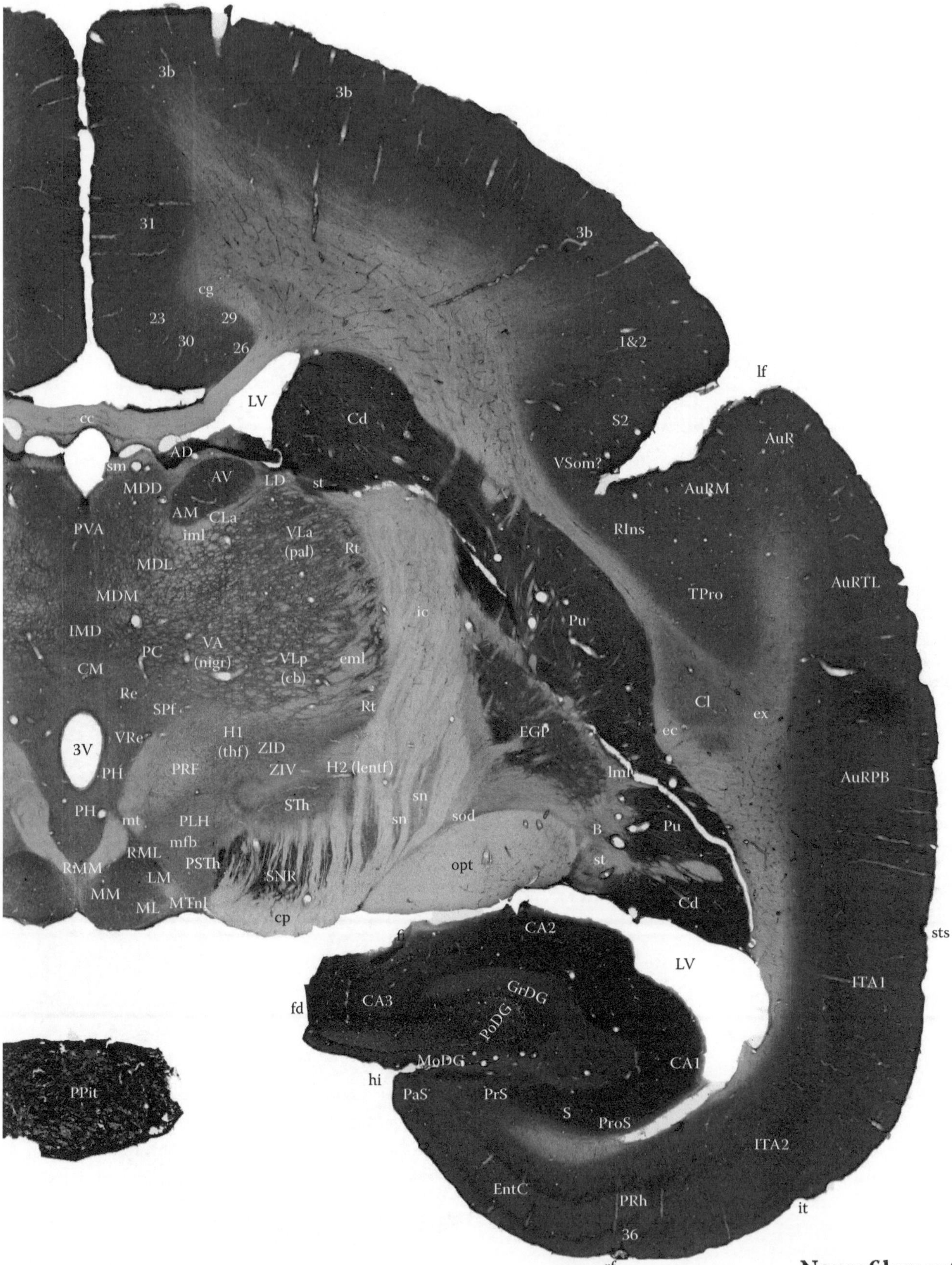


**Neurofilament**
Interaural +06.30 mm

# Plate 26c

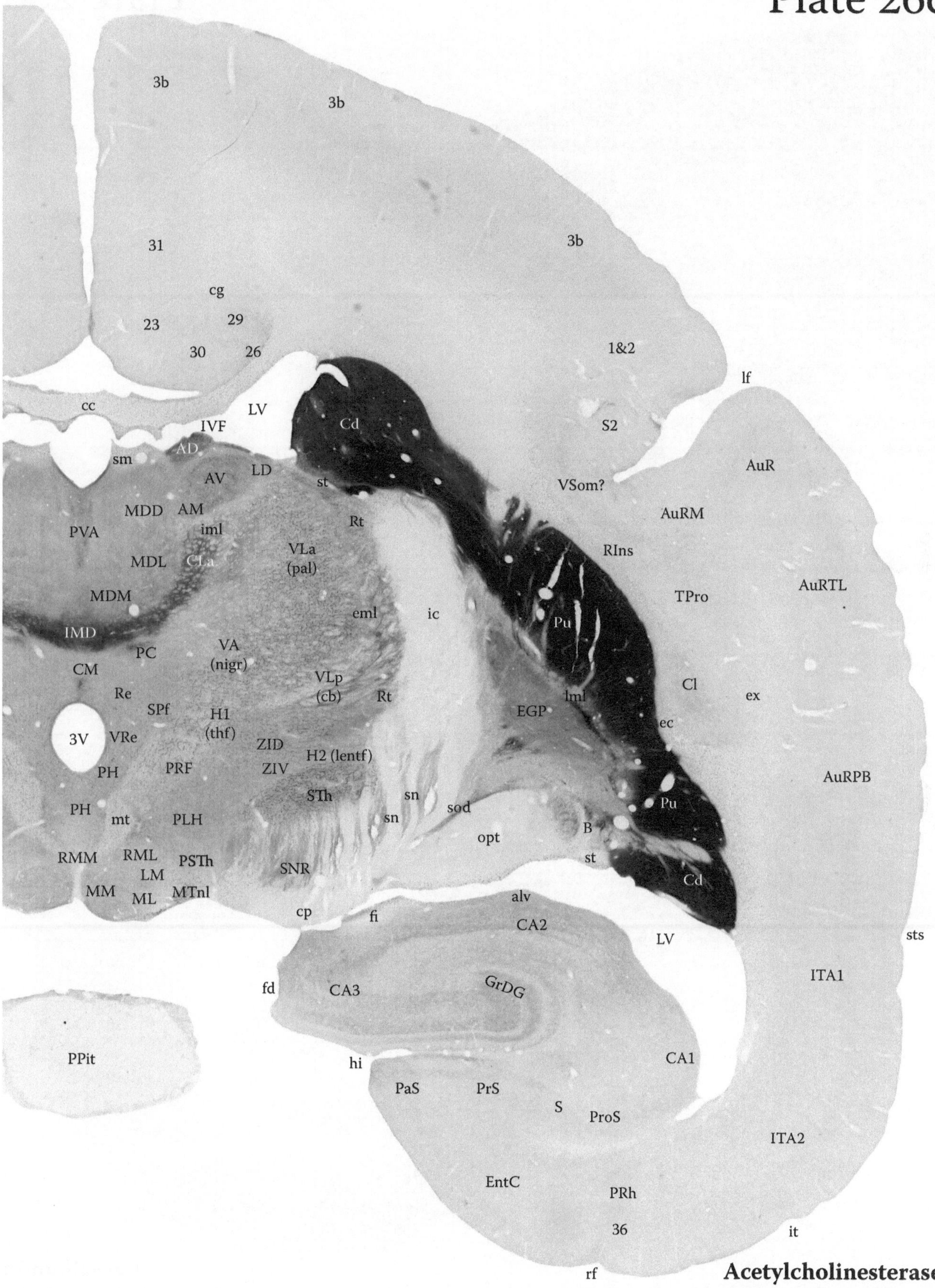


**Acetylcholinesterase**
Interaural +06.25 mm

# Plate 26d

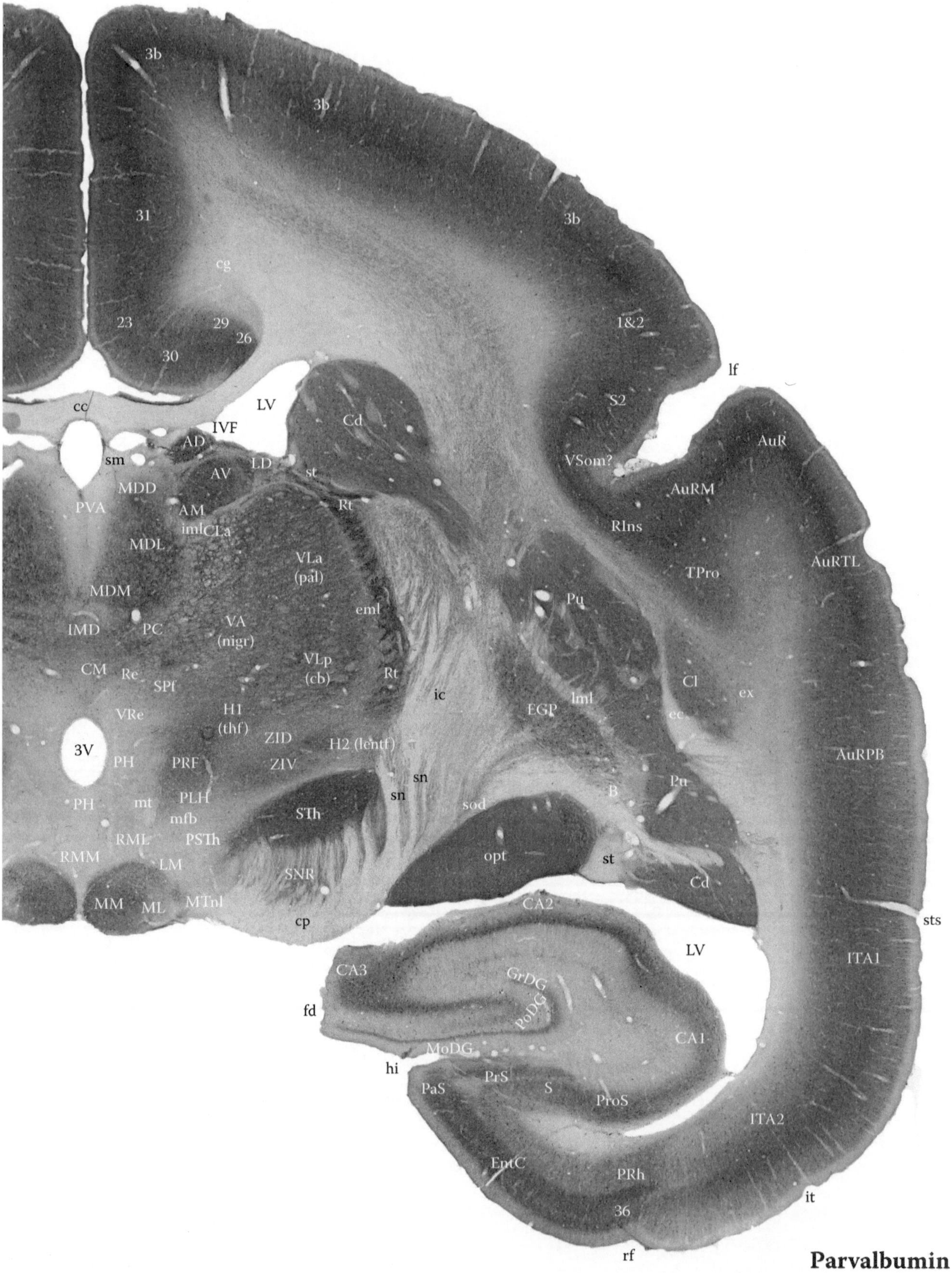


**Parvalbumin**
Interaural +06.20 mm

# Plate 26e

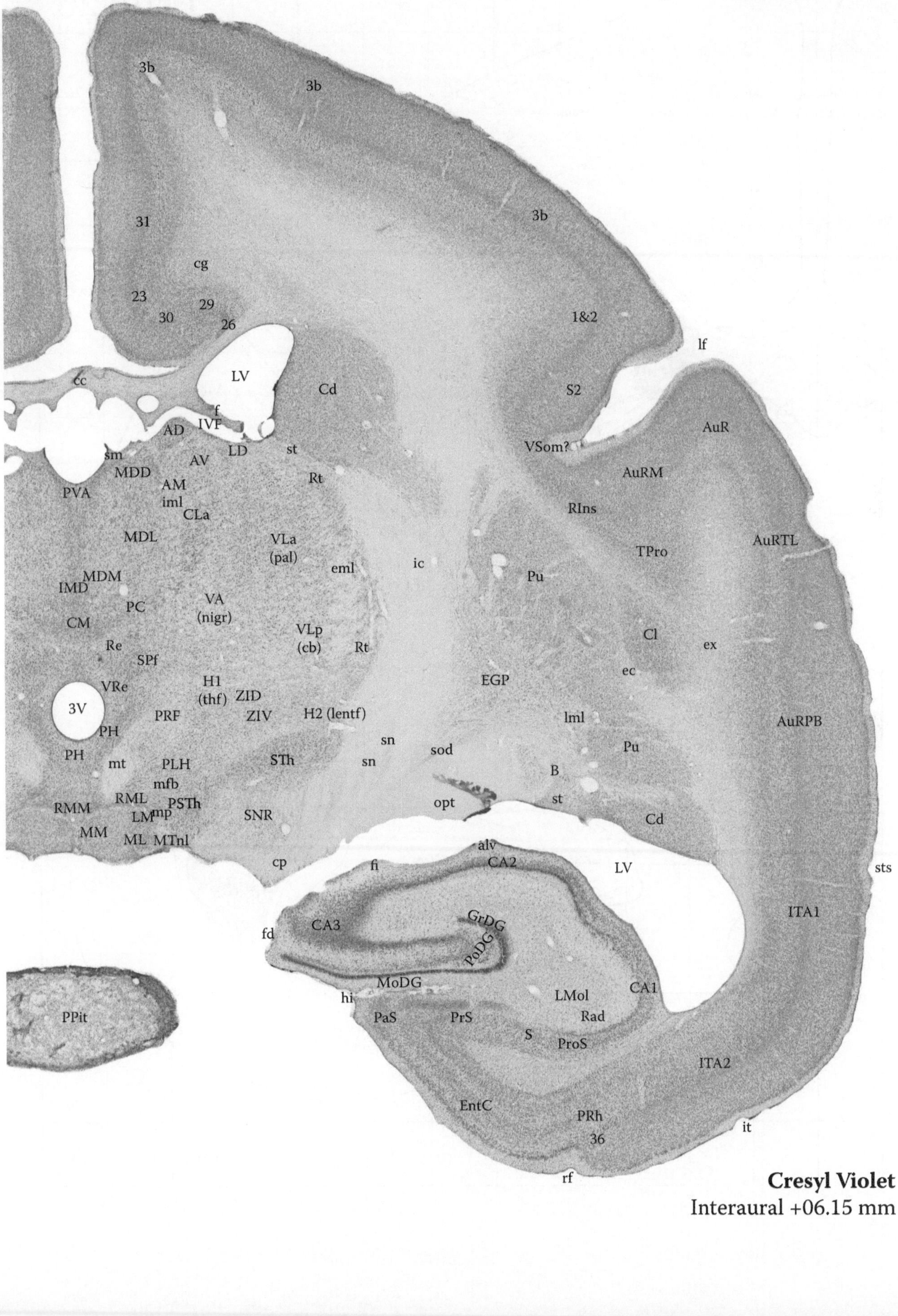


**Cresyl Violet**
Interaural +06.15 mm

# Plate 26

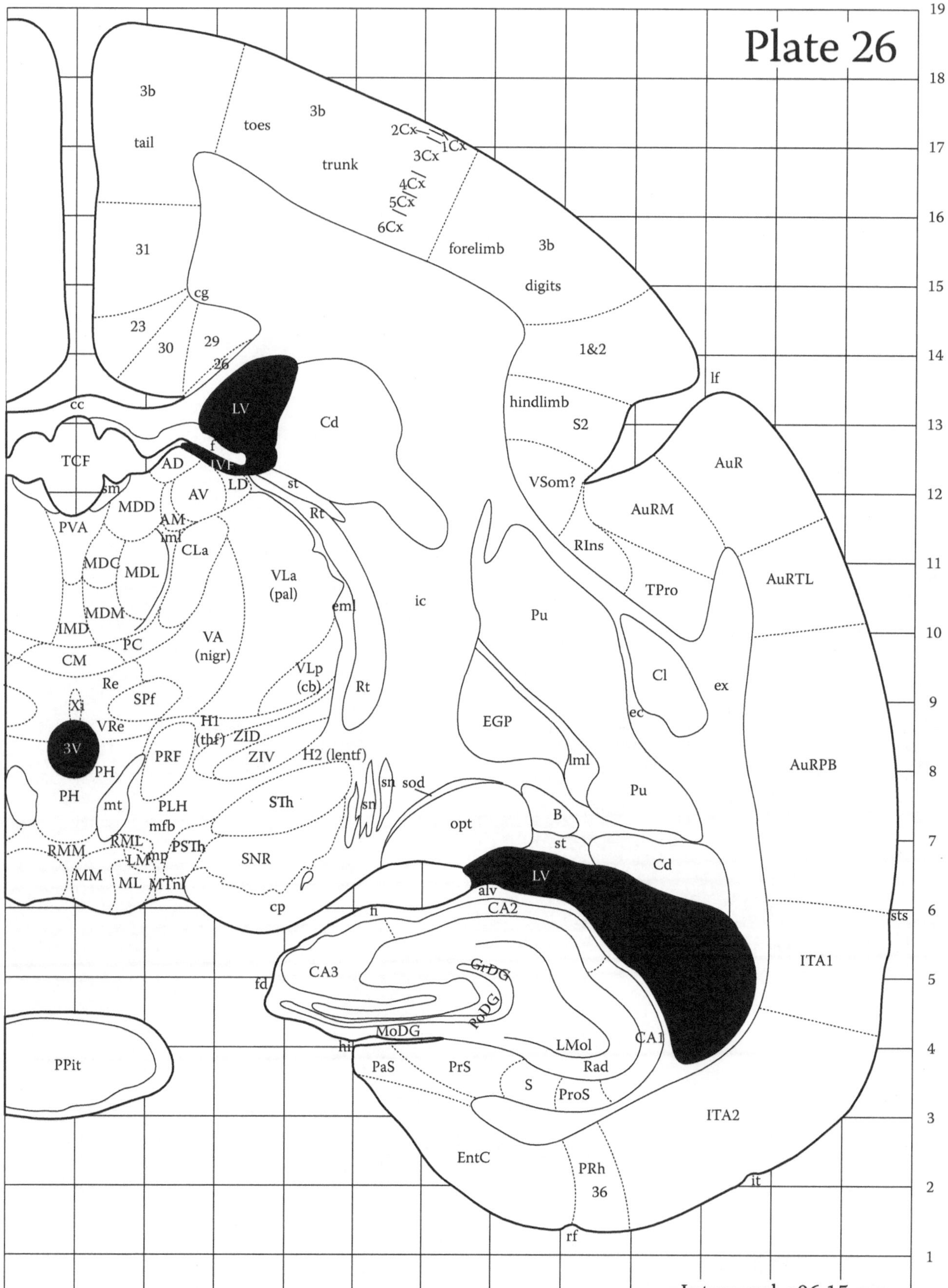
3b
tail
toes
3b
trunk
2Cx
1Cx
3Cx
4Cx
5Cx
6Cx
31
forelimb
3b
digits
cg
23
30
29
26
1&2
lf
cc
LV
Cd
hindlimb
S2
AuR
TCF
f
IVF
AD
LD
st
sm
AV
VSom?
MDD
AM
Rt
AuRM
PVA
iml
CLa
RIns
MDC
MDL
VLa
(pal)
AuRTL
TPro
eml
ic
Pu
MDM
IMD
PC
VA
(nigr)
CM
VLp
(cb)
Rt
Cl
ex
Re
SPf
Xi
VRe
H1
(thf)
ZID
EGP
ec
3V
PRF
ZIV
H2 (lentf)
lml
AuRPB
PH
PH
sn
sod
Pu
mt
PLH
STh
sn
opt
B
mfb
RML
st
RMM
LMmp
PSTh
SNR
Cd
MM
ML
MTnl
LV
alv
cp
fi
CA2
sts
ITA1
CA3
GrDG
fd
PoDG
MoDG
CA1
hi
LMol
PPit
PaS
PrS
Rad
S
ProS
ITA2
EntC
PRh
36
it
rf
Interaural +06.15 mm

| | |
|---|---|
| 1&2 | areas 1/2 Cx |
| 3b | area 3b Cx |
| 3V | third ventricle |
| 23 | area 23 Cx |
| 26 | area 26 Cx |
| 29 | area 29 Cx |
| 30 | area 30 Cx |
| 31 | area 31 Cx |
| 36 | area 36 Cx |
| AD | anterodors. thal. nu. |
| alv | alveus |
| AM | anteromed. thal. nu. |
| AuR | aud. core Cx, rost. |
| AuRM | aud. belt Cx, rostmed. |
| AuRPB | aud. parabelt Cx, rost. |
| AuRTL | aud. belt Cx, rostrotemplat. |
| AV | anterovent. thal. nu. |
| B | basal nu. (Meynert) |
| CA1 | cornu ammonis 1 |
| CA2 | cornu ammonis 2 |
| CA3 | cornu ammonis 3 |
| cc | corp. callosum |
| Cd | caudate nu. |
| cg | cingulum |
| Cl | claustrum |
| CLa | caudal lin. nu. |
| CM | cent. medial thal. nu. |
| cp | cerebral peduncle |
| ec | external capsule |
| EGP | ext. globus pallidus |
| eml | extern. medull. lamina |
| EntC | entorhin. Cx, caud. |
| ex | extreme capsule |
| f | fornix |
| fd | fimbriodent. sulcus |
| fi | fimbria |
| GrDG | gran. cell layer DG |
| H1 (thf) | H1 thalamic fasc. |
| H1 (lentf) | H2 lenticular fasc. |
| hi | hippocamp. fiss. |
| ic | intern. caps. |
| IMD | intermediodors. thal. nu. |
| iml | intern. medull. lamina |
| it | inf. temporal sulcus |
| ITA1 | inf. temp. ant. Cx, 1 |
| ITA2 | inf. temp. ant. Cx, 2 |
| IVF | intervent. foramen |
| LD | laterodors. thal. nu. |
| lf | lateral fissure |
| LM | lat. mamill. nu. |
| lml | lat. medull. lamina |
| LMol | strat. lacun. mol. CA |
| LV | lateral ventricle |
| MDD | mediodors. thal. nu., dors. |
| MDL | mediodors. thal. nu., lat. |
| MDM | mediodors. thal. nu., med. |
| mfb | med. forebrain bundle |
| ML | med. mamill. nu., lat. |
| MM | med. mamill. nu., med. |
| MoDG | molecular layer DG |
| mp | mamill. peduncle |
| mt | mamillothal. tr. |
| MTnl | med. terminal nu. |
| opt | optic tract |
| PaS | parasubiculum |
| PC | paracent. thal. nu. |
| PH | post. hypothal. area |
| PLH | pedunc. lat. hypothal. |
| PoDG | polymorphic layer DG |
| PPit | post. pituit. gland |
| PRF | prerubral field |
| PRh | perirhinal Cx |
| ProS | prosubiculum |
| PrS | presubiculum |
| PSTh | parasubthal. nu. |
| Pu | putamen |
| PVA | paravent. thal. nu., ant. |
| Rad | strat. radiat. CA |
| Re | reuniens nu. |
| rf | rhinal fiss. |
| RIns | retroinsular Cx |
| RML | retromamill. nu., lat. |
| RMM | retromamill. nu., med. |
| Rt | reticular nu. |
| S | subiculum |
| S2 | second. somat. area Cx |
| sm | stria medullaris |
| sn | striatonigral fibers |
| SNR | subst. nigra, retic. |
| sod | suproptic decn |
| SPf | subparafasc. thal. nu. |
| st | stria terminalis |
| STh | subthalamic nu. |
| sts | sup. temp. sulcus |
| TPro | temp. proisocortex |
| VA | vent. ant. thal. nu. |
| VLa | ventrolat. thal. nu., ant. |
| VLp | ventrolat. thal. nu., post. |
| VRe | vent. reuniens nu. |
| VSom? | vent. somatosens. Cx |
| ZID | zona incerta, dors. |
| ZIV | zona incerta, vent. |

# Plate 27

Interaural +05.60 mm

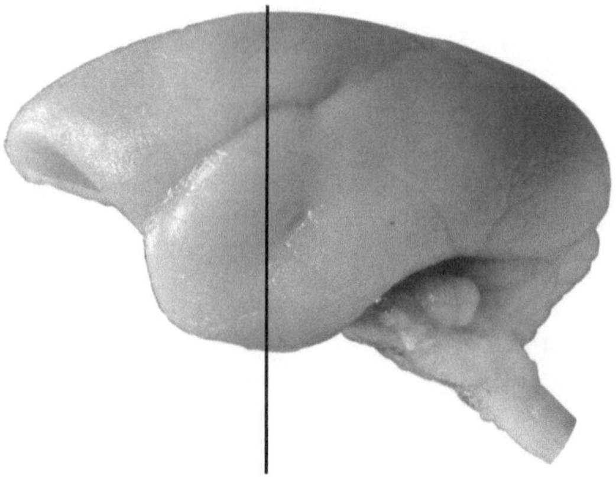

| | |
|---|---|
| 1&2 | areas 1/2 Cx |
| 3b | area 3b Cx |
| 3V | third ventricle |
| 23 | area 23 Cx |
| 26 | area 26 Cx |
| 29 | area 29 Cx |
| 30 | area 30 Cx |
| 31 | area 31 Cx |
| 36 | area 36 Cx |
| AD | anterodors. thal. nu. |
| alv | alveus |
| Au1 | aud. core Cx, primary |
| AuAL | aud. belt Cx, anterolat. |
| AuCM | aud. belt Cx, caudomed. |
| AuR | aud. core Cx, rost. |
| AuRPB | aud. parabelt Cx, rost. |
| AuRTL | aud. belt Cx, rostrotemplat. |
| AV | anterovent. thal. nu. |
| bv | blood vessel |
| CA1 | cornu ammonis 1 |
| CA2 | cornu ammonis 2 |
| CA3 | cornu ammonis 3 |
| cc | corp. callosum |
| Cd | caudate nu. |
| cg | cingulum |
| CG | cingulate gyrus Cx |
| cgs | cingulate sulc. |
| CLa | centrolat. thal. nu., ant. |
| CLp | centrolat. thal. nu., post. |
| CM | cent. medial thal. nu. |
| cp | cerebral peduncle |
| ec | external capsule |
| eml | extern. medull. lamina |
| EntC | entorhin. Cx, caud. |
| f | fornix |
| fd | fimbriodent. sulcus |
| fi | fimbria |
| fr | fasciculus retroflexus |
| GrDG | gran. cell layer DG |
| hi | hippocamp. fiss. |
| ic | intern. caps. |
| IMD | intermediodors. thal. nu. |
| iml | intern. medull. lamina |
| int | intermed. temp. sulc. |
| IPF | interpedunc. fossa |
| it | inf. temporal sulcus |
| ITD1 | inf. temp. dors. Cx 1 |
| ITV | inf. temp. vent. Cx |
| LD | laterodors. thal. nu. |
| lf | lateral fissure |
| LGpe | LG, parvicell. external |
| LGpi | LG, parvicell. internal |
| LMol | strat. lacun. mol. CA |
| LV | lateral ventricle |
| MDC | mediodors. thal. nu., cent. |
| MDD | mediodors. thal. nu., dors. |
| MDL | mediodors. thal. nu., lat. |
| MDM | mediodors. thal. nu., med. |
| mfb | med. forebrain bundle |
| MoDG | molecular layer DG |
| mp | mamill. peduncle |
| MTnl | med. terminal nu. |
| opt | optic tract |
| PaS | parasubiculum |
| PBP | parabrach. pigm. nu. |
| PC | paracent. thal. nu. |
| Pf | parafascicular nu. |
| PH | post. hypothal. area |
| PLH | pedunc. lat. hypothal. |
| PoDG | polymorphic layer DG |
| PRF | prerubral field |
| PrG | pregeniculate nu. |
| PRh | perirhinal Cx |
| ProS | prosubiculum |
| PrS | presubiculum |
| Pu | putamen |
| PVA | paravent. thal. nu., ant. |
| Rad | strat. radiat. CA |
| Re | reuniens nu. |
| rf | rhinal fiss. |
| RIns | retroinsular Cx |
| RMM | retromamill. nu., med. |
| Rt | reticular nu. |
| S | subiculum |
| S2 | second. somat. area Cx |
| sm | stria medullaris |
| SNC | subst. nigra, comp. |
| SNR | subst. nigra, retic. |
| sod | suproptic decn |
| SPf | subparafasc. thal. nu. |
| st | stria terminalis |
| STh | subthalamic nu. |
| sts | sup. temp. sulcus |
| TCF | transverse cerebral fiss. |
| TF | temp. field F Cx |
| VA | vent. ant. thal. nu. |
| VLa | ventrolat. thal. nu., ant. |
| VLp | ventrolat. thal. nu., post. |
| VRe | vent. reuniens nu. |
| VSom? | vent. somatosens. Cx |
| ZID | zona incerta, dors. |
| ZIV | zona incerta, vent. |

# Plate 27a

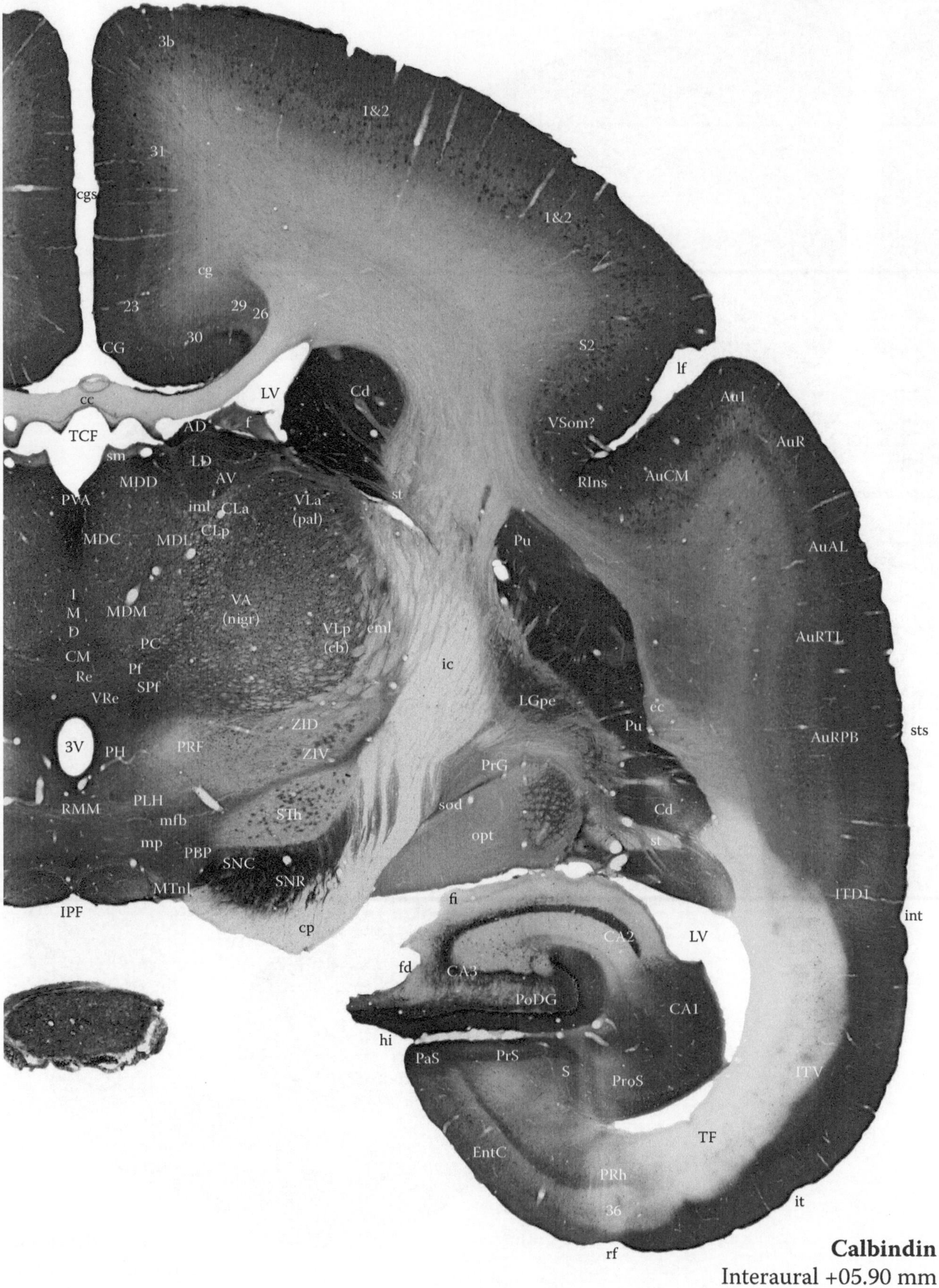


**Calbindin**

Interaural +05.90 mm

# Plate 27b

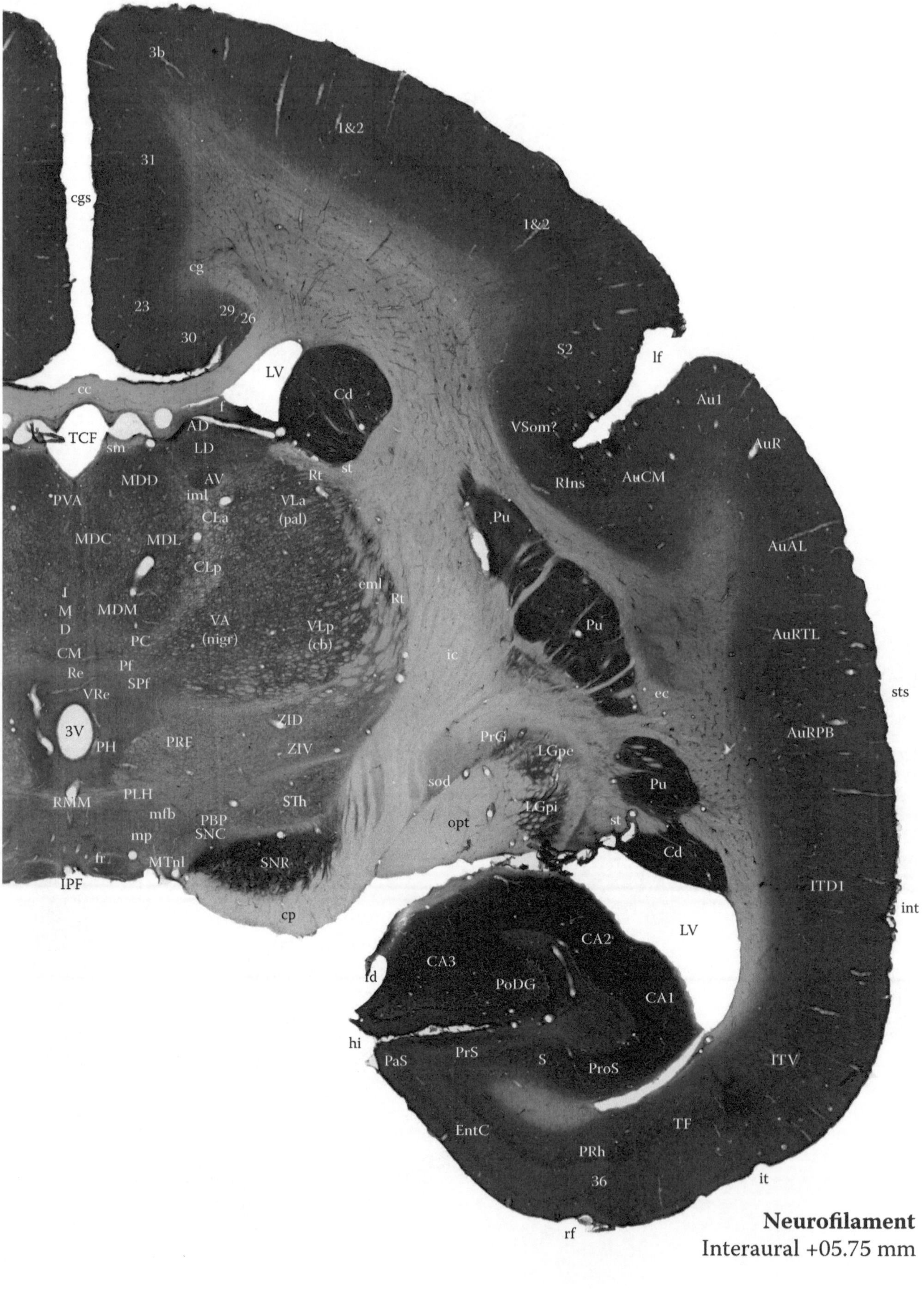


**Neurofilament**
Interaural +05.75 mm

# Plate 27c

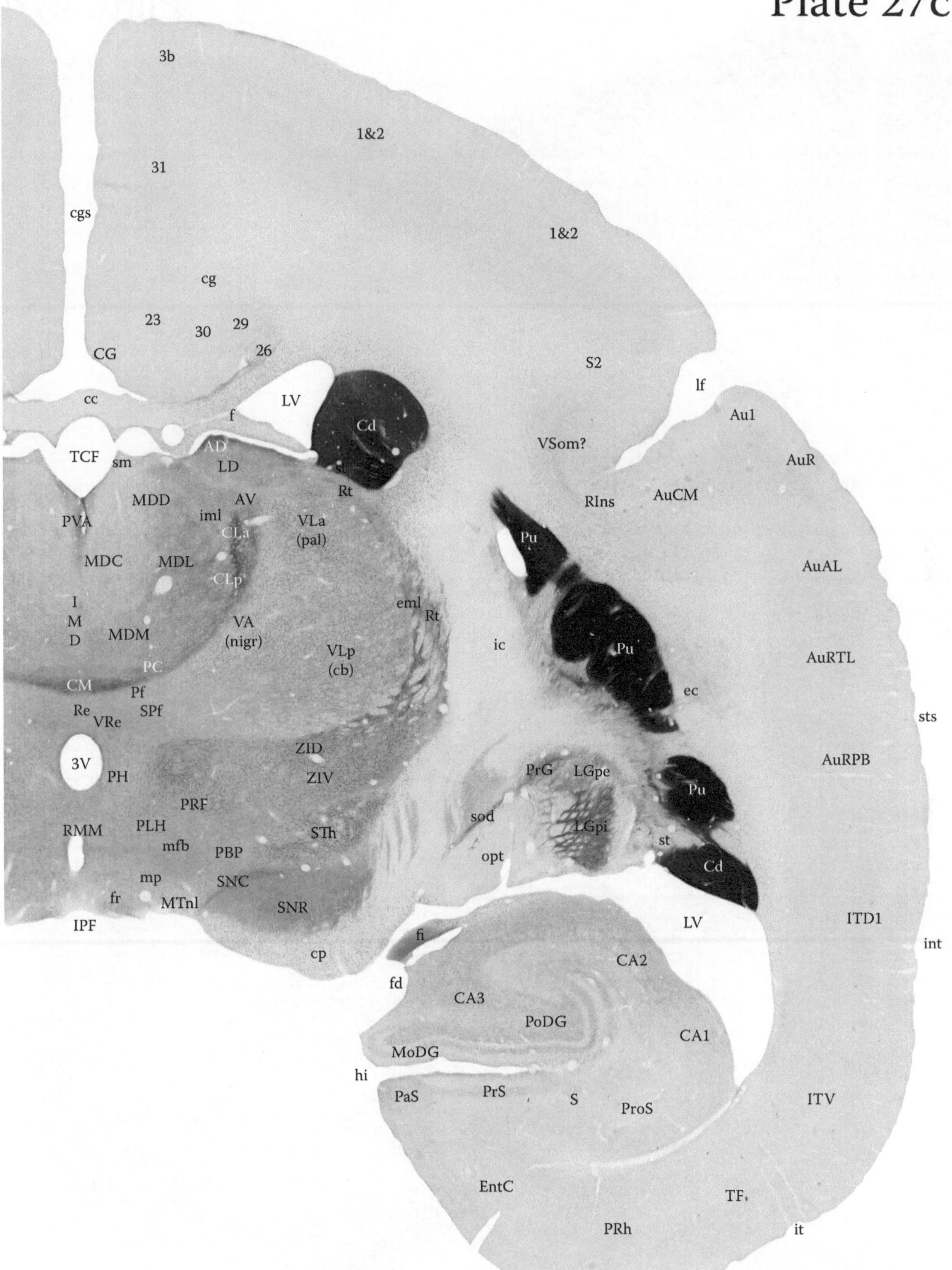


**Acetylcholinesterase**
Interaural +05.70 mm

# Plate 27d

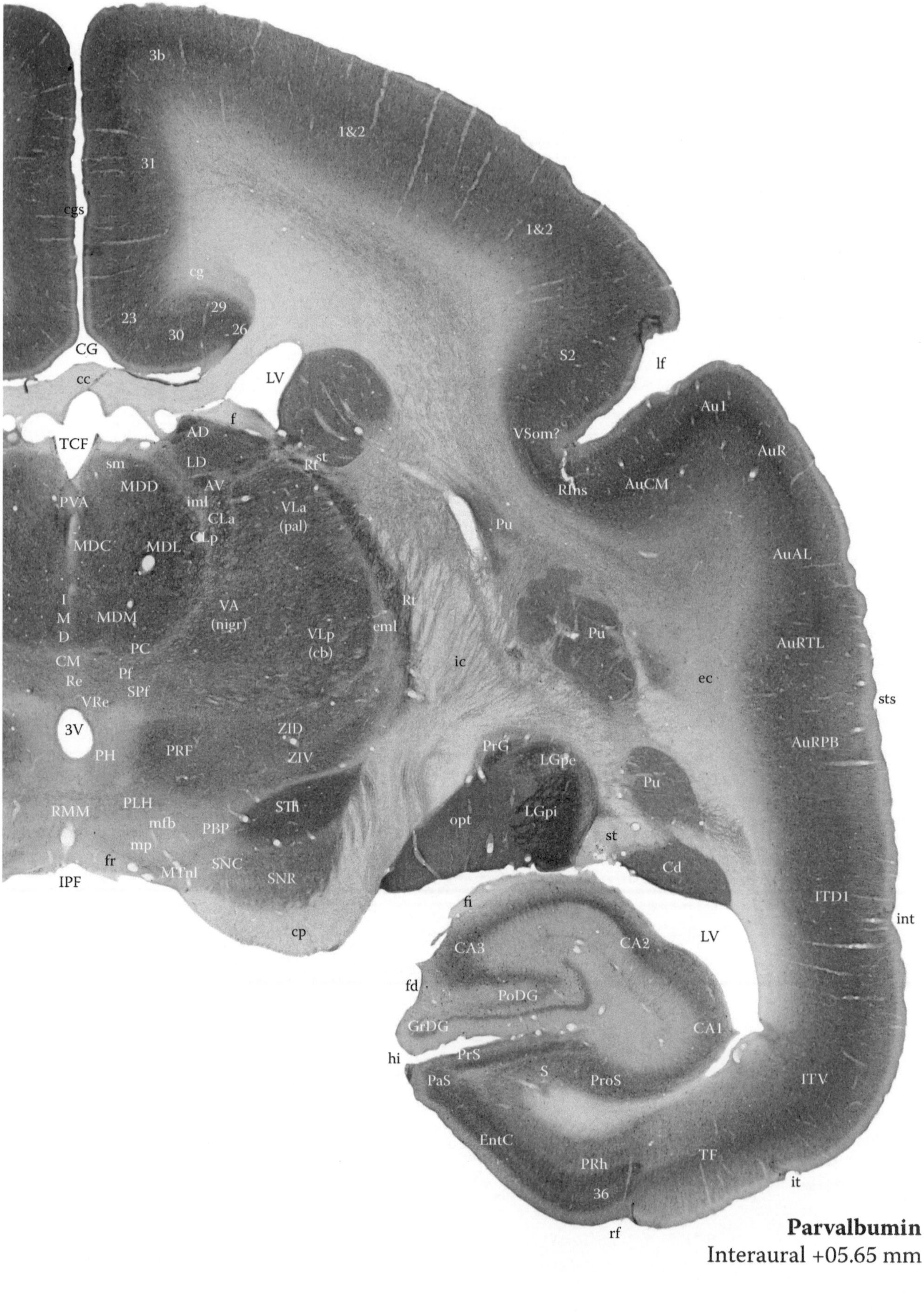


**Parvalbumin**
Interaural +05.65 mm

# Plate 27e

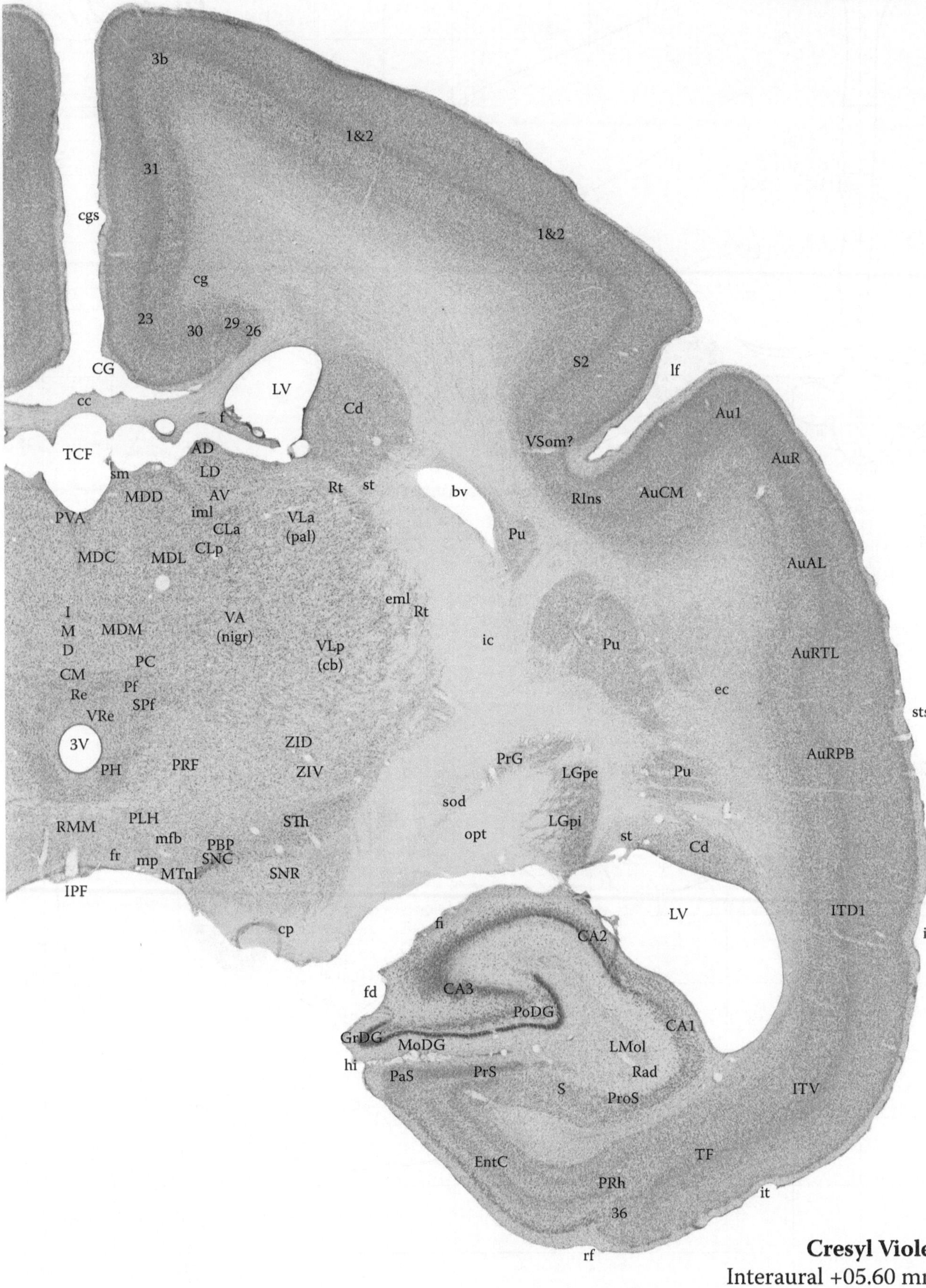


**Cresyl Violet**
Interaural +05.60 mm

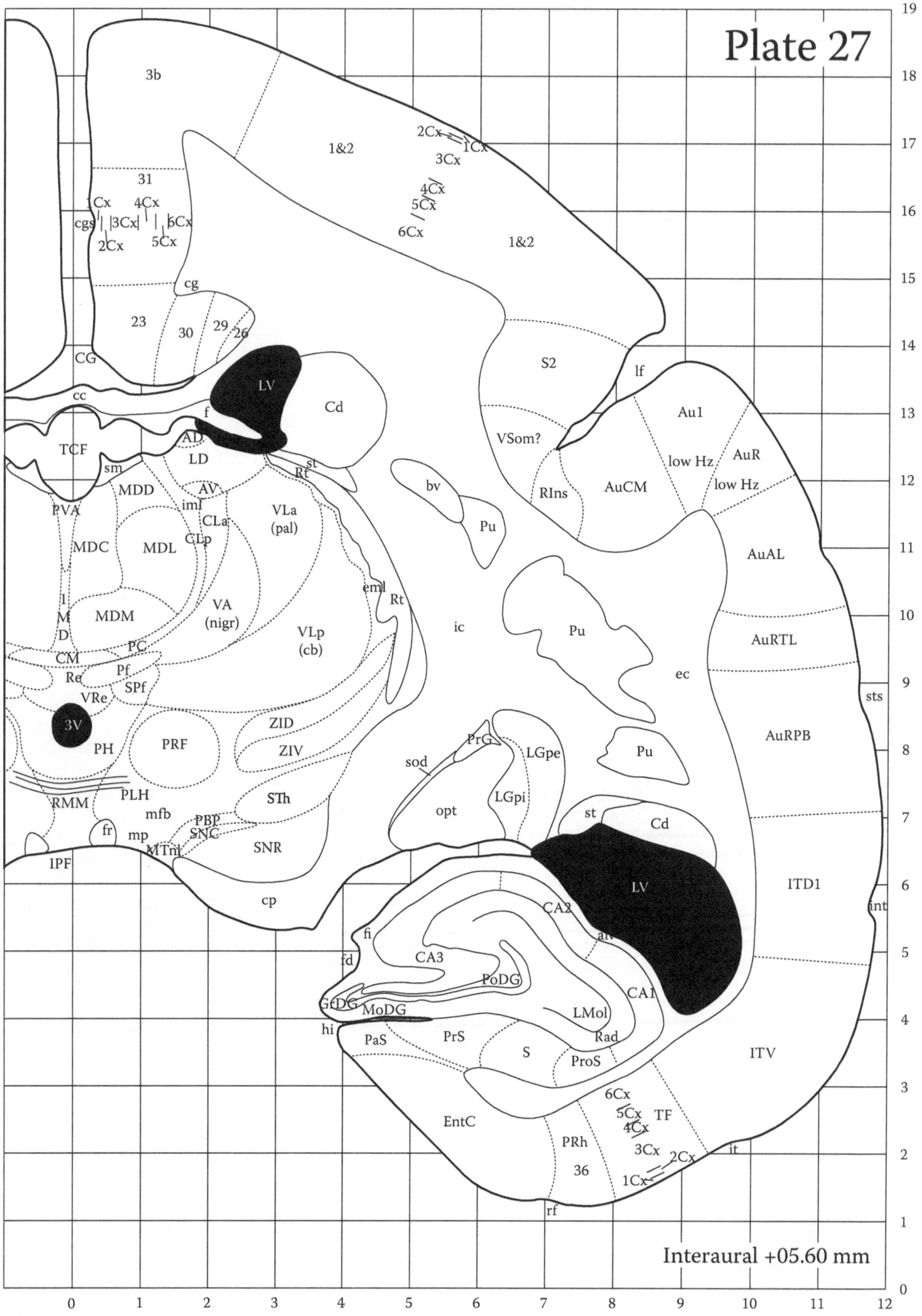

Plate 27
3b
1&2
2Cx
1Cx
3Cx
31
1Cx
4Cx
4Cx
5Cx
cgs
3Cx
6Cx
6Cx
2Cx
5Cx
1&2
cg
23
30
29
26
S2
CG
LV
lf
cc
Cd
Au1
f
VSom?
AuR
AD
TCF
LD
st
low Hz
Rt
sm
low Hz
AV
MDD
bv
RIns
AuCM
PVA
iml
VLa
(pal)
CLa
Pu
MDC
MDL
CLp
AuAL
eml
I
M
D
MDM
VA
(nigr)
Rt
ic
Pu
VLp
(cb)
AuRTL
CM
PC
Pf
Re
ec
SPf
VRe
sts
ZID
3V
PH
PRF
ZIV
PrG
LGpe
Pu
AuRPB
sod
STh
PLH
opt
LGpi
RMM
mfb
st
Cd
fr
PBP
SNC
mp
MTnt
SNR
IPF
LV
ITD1
cp
CA2
int
fi
alv
fd
CA3
PoDG
CA1
GrDG
MoDG
LMol
hi
PaS
PrS
Rad
S
ProS
ITV
6Cx
5Cx
TF
4Cx
EntC
PRh
3Cx
2Cx
it
36
1Cx
rf
Interaural +05.60 mm

| | |
|---|---|
| 1&2 | areas 1/2 Cx |
| 3b | area 3b Cx |
| 3V | third ventricle |
| 23 | area 23 Cx |
| 26 | area 26 Cx |
| 29 | area 29 Cx |
| 30 | area 30 Cx |
| 31 | area 31 Cx |
| 36 | area 36 Cx |
| AD | anterodors. thal. nu. |
| alv | alveus |
| Au1 | aud. core Cx, primary |
| AuAL | aud. belt Cx, anterolat. |
| AuCM | aud. belt Cx, caudomed. |
| AuR | aud. core Cx, rost. |
| AuRPB | aud. parabelt Cx, rost. |
| AuRTL | aud. belt Cx, rostrotemplat. |
| AV | anterovent. thal. nu. |
| bv | blood vessel |
| CA1 | cornu ammonis 1 |
| CA2 | cornu ammonis 2 |
| CA3 | cornu ammonis 3 |
| cc | corp. callosum |
| Cd | caudate nu. |
| cg | cingulum |
| CG | cingulate gyrus Cx |
| cgs | cingulate sulc. |
| CLa | centrolat. thal. nu., ant. |
| CLp | centrolat. thal. nu., post. |
| CM | cent. medial thal. nu. |
| cp | cerebral peduncle |
| ec | external capsule |
| eml | extern. medull. lamina |
| EntC | entorhin. Cx, caud. |
| f | fornix |
| fd | fimbriodent. sulcus |
| fi | fimbria |
| fr | fasciculus retroflexus |
| GrDG | gran. cell layer DG |
| hi | hippocamp. fissure |
| ic | intern. caps. |
| IMD | intermediodors. thal. nu. |
| iml | intern. medull. lamina |
| int | intermed. temp. sulc. |
| IPF | interpedunc. fossa |
| it | inf. temporal sulcus |
| ITD1 | inf. temp. dors. Cx 1 |
| ITV | inf. temp. vent. Cx |
| LD | laterodors. thal. nu. |
| lf | lateral fissure |
| LGpe | LG, parvicell. external |
| LGpi | LG, parvicell. internal |
| LMol | strat. lacun. mol. CA |
| LV | lateral ventricle |
| MDC | mediodors. thal. nu., cent. |
| MDD | mediodors. thal. nu., dors. |
| MDL | mediodors. thal. nu., lat. |
| MDM | mediodors. thal. nu., med. |
| mfb | med. forebrain bundle |
| MoDG | molecular layer DG |
| mp | mamill. peduncle |
| MTnl | med. terminal nu. |
| opt | optic tract |
| PaS | parasubiculum |
| PBP | parabrach. pigm. nu. |
| PC | paracent. thal. nu. |
| Pf | parafascicular nu. |
| PH | post. hypothal. area |
| PLH | pedunc. lat. hypothal. |
| PoDG | polymorphic layer DG |
| PRF | prerubral field |
| PrG | pregeniculate nu. |
| PRh | perirhinal Cx |
| ProS | prosubiculum |
| PrS | presubiculum |
| Pu | putamen |
| PVA | paravent. thal. nu., ant. |
| Rad | strat. radiat. CA |
| Re | reuniens nu. |
| rf | rhinal fiss. |
| RIns | retroinsular Cx |
| RMM | retromamill. nu., med. |
| Rt | reticular nu. |
| S | subiculum |
| S2 | second. somat. area Cx |
| sm | stria medullaris |
| SNC | subst. nigra, comp. |
| SNR | subst. nigra, retic. |
| sod | suproptic decn |
| SPf | subparafasc. thal. nu. |
| st | stria terminalis |
| STh | subthalamic nu. |
| sts | sup. temp. sulcus |
| TCF | transverse cerebral fissure |
| TF | temp. field F Cx |
| VA | vent. ant. thal. nu. |
| VLa | ventrolat. thal. nu., ant. |
| VLp | ventrolat. thal. nu., post. |
| VRe | vent. reuniens nu. |
| VSom? | vent. somatosens. Cx |
| ZID | zona incerta, dors. |
| ZIV | zona incerta, vent. |

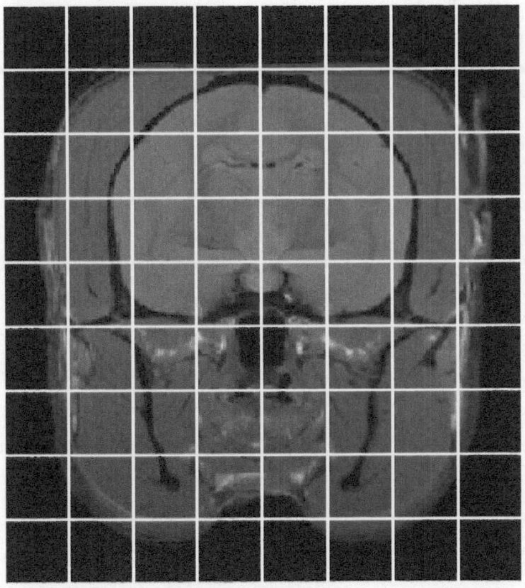

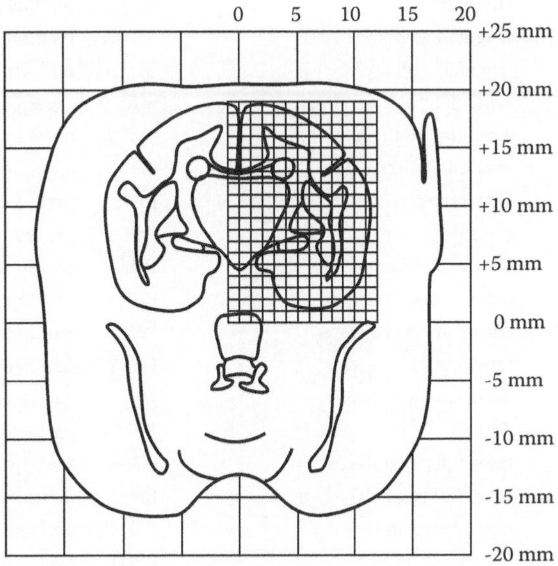


Interaural +05.85 mm

# Plate 28

Interaural +05.10 mm

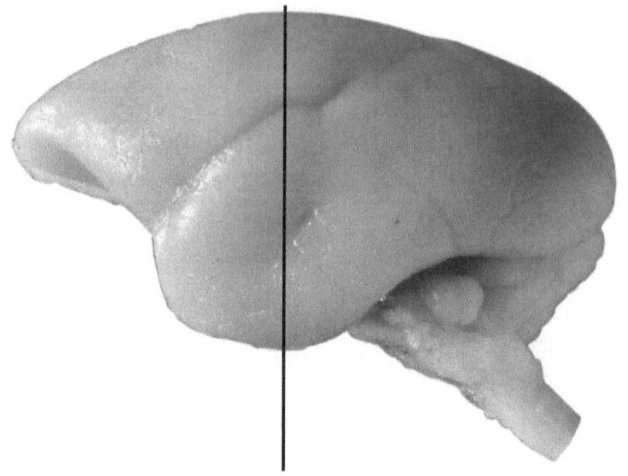

| | | | | | |
|---|---|---|---|---|---|
| 1&2 | areas 1/2 Cx | int | intermed. temp. sulc. | PPV | post. par. Cx, vent. |
| 3n | oculomotor nerve | ip | intraparietal sulc. | PrG | pregeniculate nu. |
| 3V | third ventricle | IPF | interpedunc. fossa | PRh | perirhinal Cx |
| 23 | area 23 Cx | it | inf. temporal sulcus | ProS | prosubiculum |
| 26 | area 26 Cx | ITD1 | inf. temp. dors. Cx 1 | PrS | presubiculum |
| 29 | area 29 Cx | ITV | inf. temp. vent. Cx | Pu | putamen |
| 30 | area 30 Cx | LD | laterodors. thal. nu. | PVA | paravent. thal. nu., ant. |
| 31 | area 31 Cx | lf | lateral fissure | Rad | strat. radiat. CA |
| Au1 | aud. core Cx, primary | LGpe | LG, parvicell. external | rf | rhinal fiss. |
| AuCM | aud. belt Cx, caudomed. | LGpi | LG, parvicell. internal | RI | rost. interstit. nu. |
| AuCPB | aud. parabelt Cx, caud. | LGei | LG, external interlaminar | RIns | retroinsular Cx |
| AuML | aud. belt Cx, mid. lat. | LGme | LG, magnocell. external | RLi | rost. linear nu. |
| CA1 | cornu ammonis 1 | LGmi | LG, magnocell. internal | RPC | red nu., parvicellular |
| CA2 | cornu ammonis 2 | LMol | strat. lacun. mol. CA | Rt | reticular nu. |
| CA3 | cornu ammonis 3 | LV | lateral ventricle | S | subiculum |
| cc | corp. callosum | MDC | mediodors. thal. nu., cent. | sm | stria medullaris |
| Cd | caudate nu. | MDD | mediodors. thal. nu., dors. | SNC | subst. nigra, comp. |
| cg | cingulum | MDL | mediodors. thal. nu., lat. | SNR | subst. nigra, retic. |
| CG | cingulate gyrus Cx | MDM | mediodors. thal. nu., med. | sod | suproptic decn |
| cgs | cingulate sulc. | MoDG | molecular layer DG | st | stria terminalis |
| CLa | centrolat. thal. nu., ant. | mRt | mesen. retic. formn | STh | subthalamic nu. |
| CLp | centrolat. thal. nu., post. | opt | optic tract | sts | sup. temp. sulcus |
| CM | cent. medial thal. nu. | Or | strat. oriens CA | TF | temp. field F Cx |
| CMn | centromedian thal. nu. | p1Rt | pros. 1 retic. formn | vhc | vent. hipp. comm. |
| cp | cerebral peduncle | PaP | parapeduncular nu. | VLa | ventrolat. thal. nu., ant. |
| DR | dors. raphe nu. | PaS | parasubiculum | VLp | ventrolat. thal. nu., post. |
| eml | extern. medull. lamina | PBP | parabrach. pigm. nu. | VPI | vent. post. thal. nu., inf. |
| EntC | entorhin. Cx, caud. | PC | paracent. thal. nu. | VPL | vent. post. thal. nu., lat. |
| f | fornix | PE | par. field E Cx | VPM | vent. post. thal. nu., med. |
| fd | fimbriodent. sulcus | Pf | parafascicular nu. | VPMPC | vent. post. thal. nu., med. parv. |
| fi | fimbria | PF | par. field F Cx | VPS | vent. post. thal. nu., sup. |
| fr | fasciculus retroflexus | PH | post. hypothal. area | VSom? | vent. somatosens. Cx |
| GrDG | gran. cell layer DG | Pn | pontine nu. | VTA | vent. tegment. area |
| hi | hippocamp. fiss. | PoDG | polymorphic layer DG | ZID | zona incerta, dors. |
| IF | interfascic. nu. | PPD1 | post. par. Cx, dors. 1 | ZIV | zona incerta, vent. |
| IMD | intermediodors. thal. nu. | | | | |

# Plate 28a

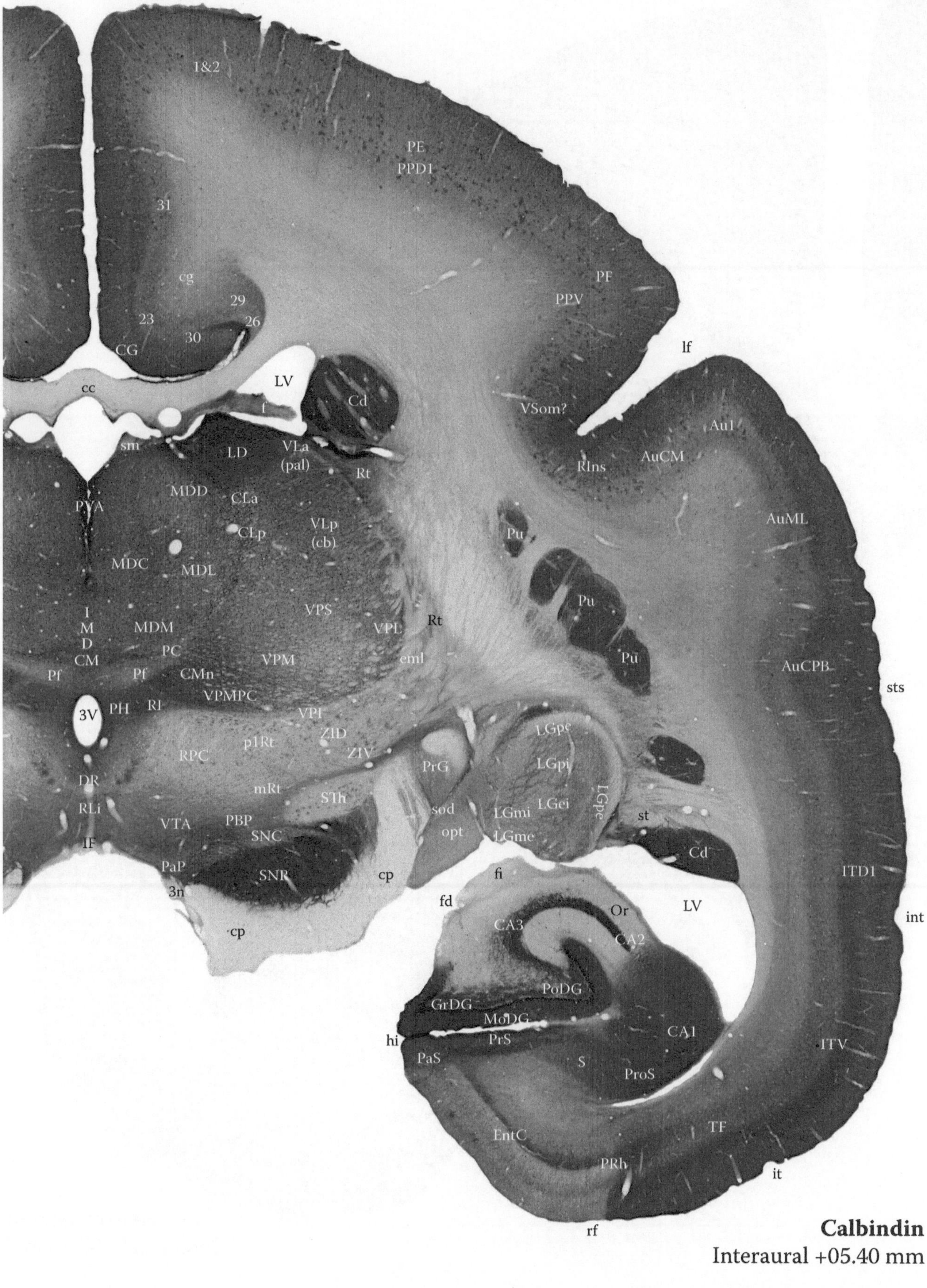


**Calbindin**
Interaural +05.40 mm

# Plate 28b

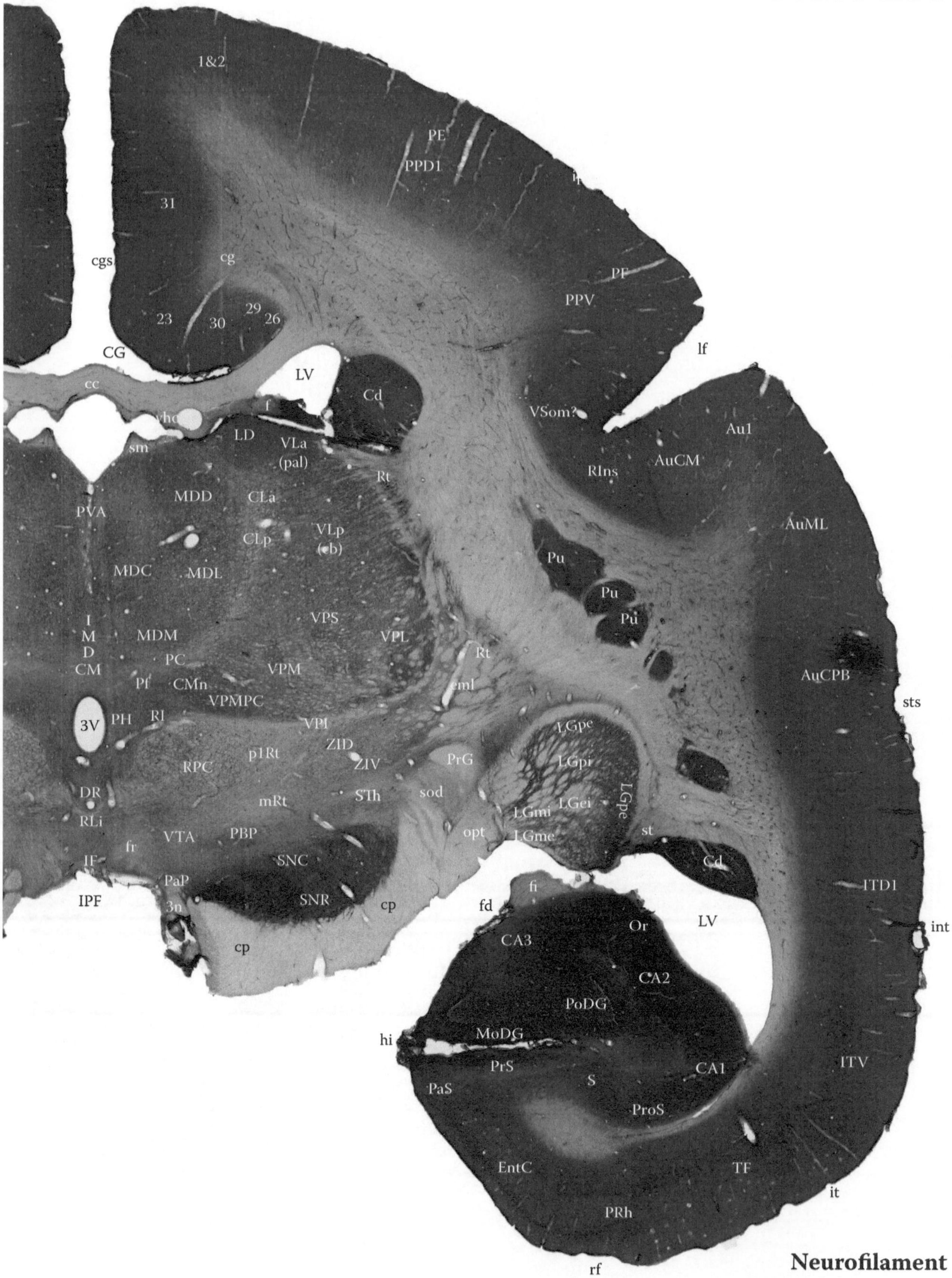


**Neurofilament**
Interaural +05.25 mm

# Plate 28c

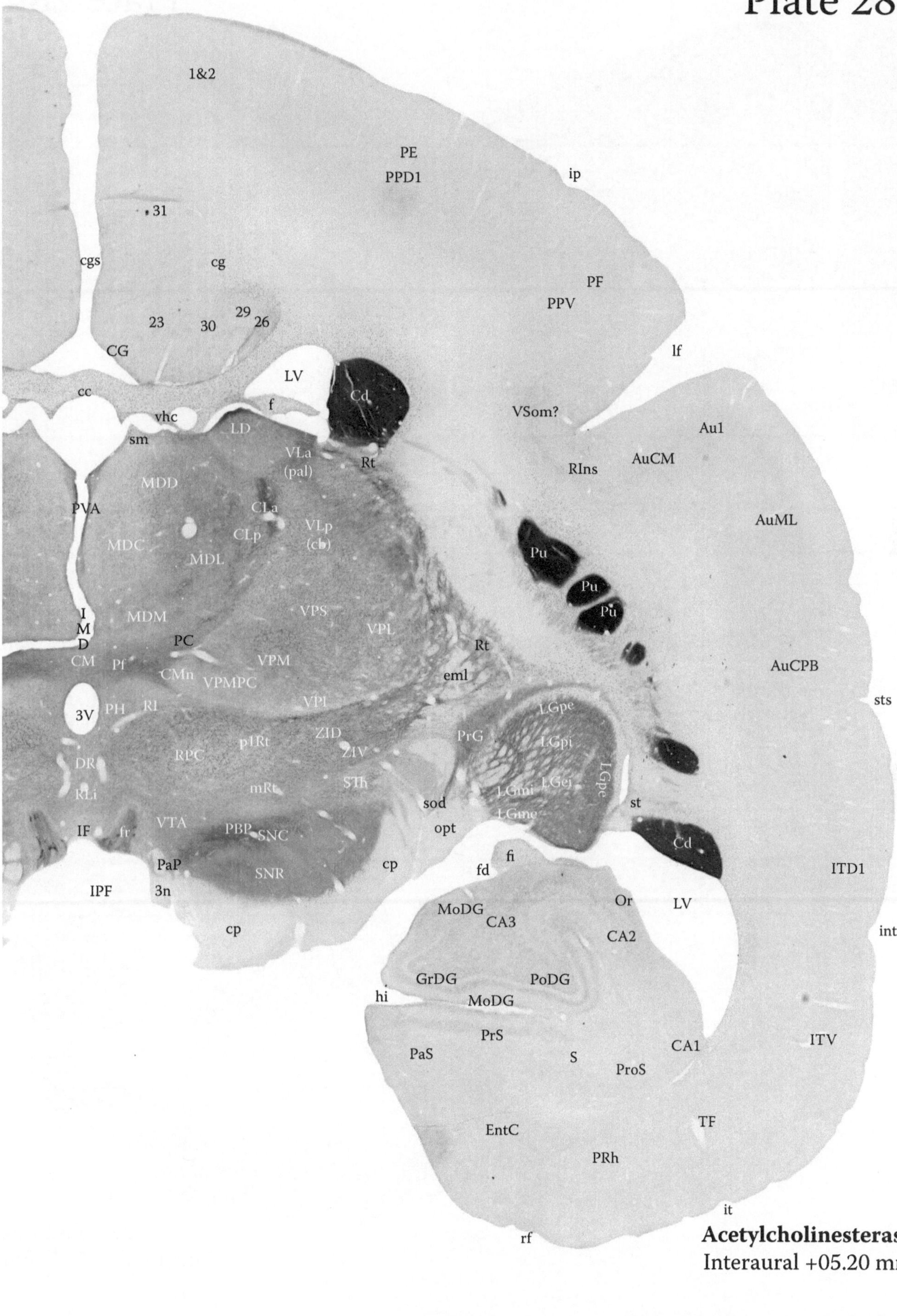


**Acetylcholinesterase**
Interaural +05.20 mm

# Plate 28d

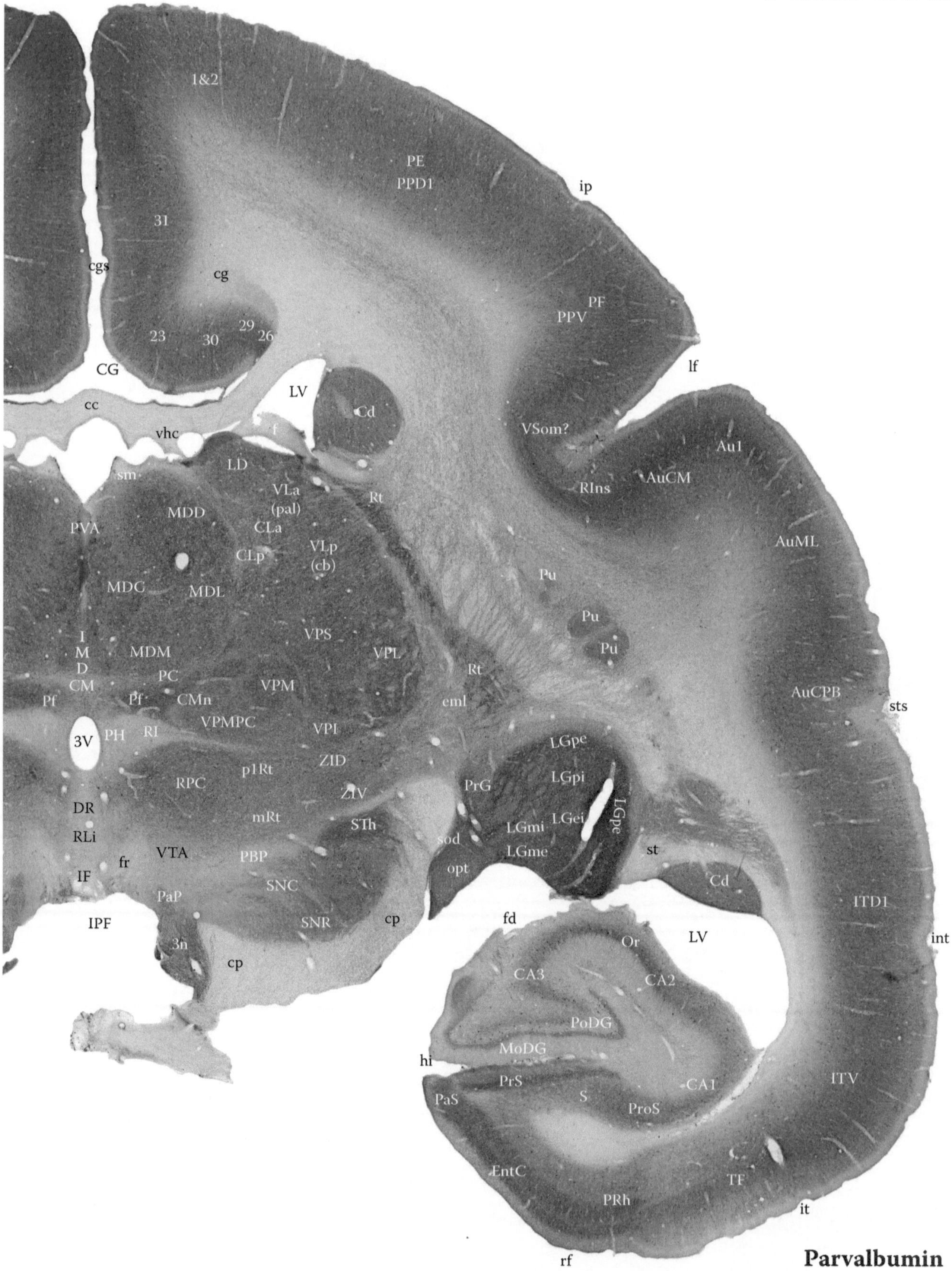


**Parvalbumin**
Interaural +05.15 mm

# Plate 28e

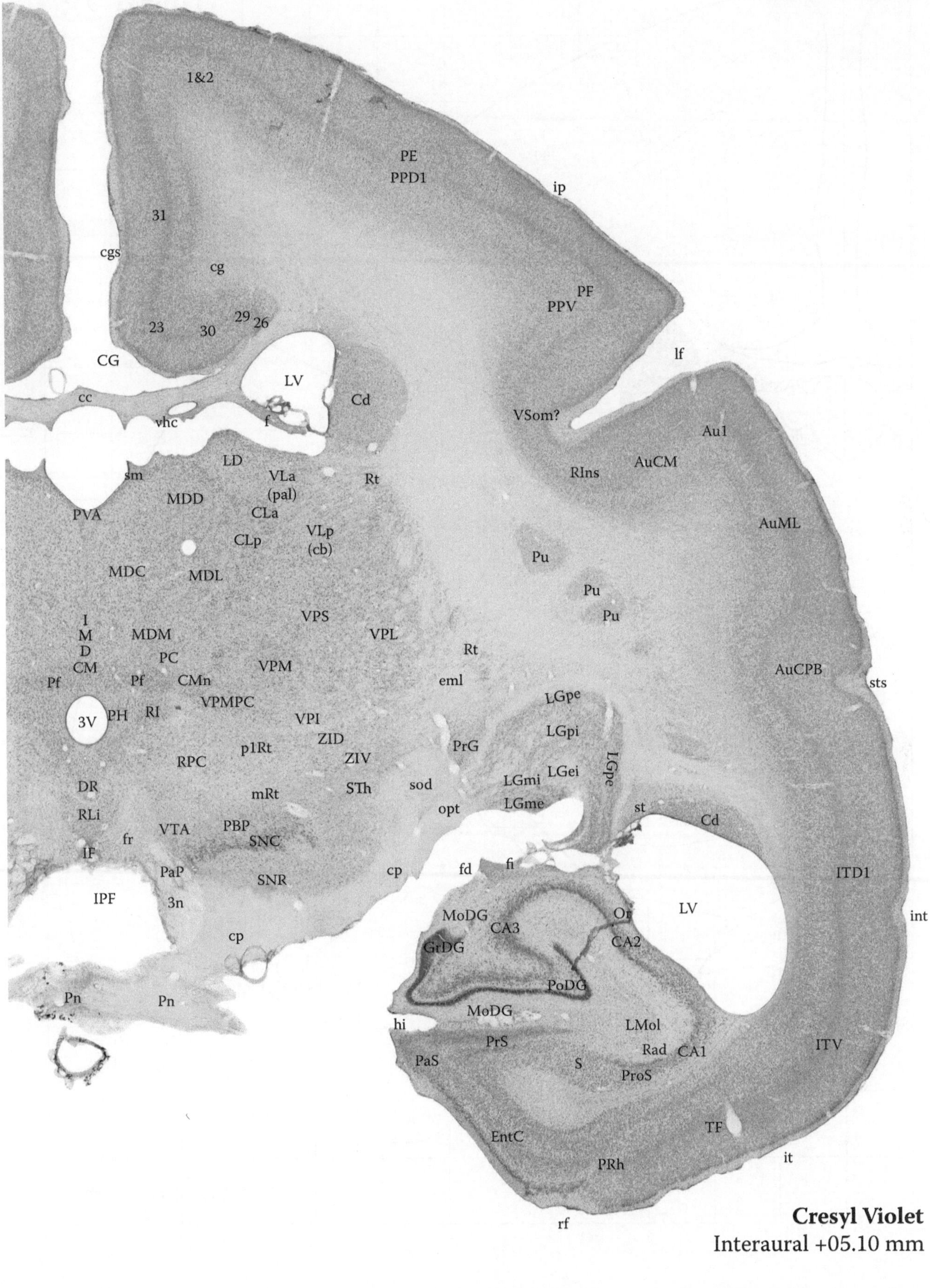


**Cresyl Violet**
Interaural +05.10 mm

# Plate 28

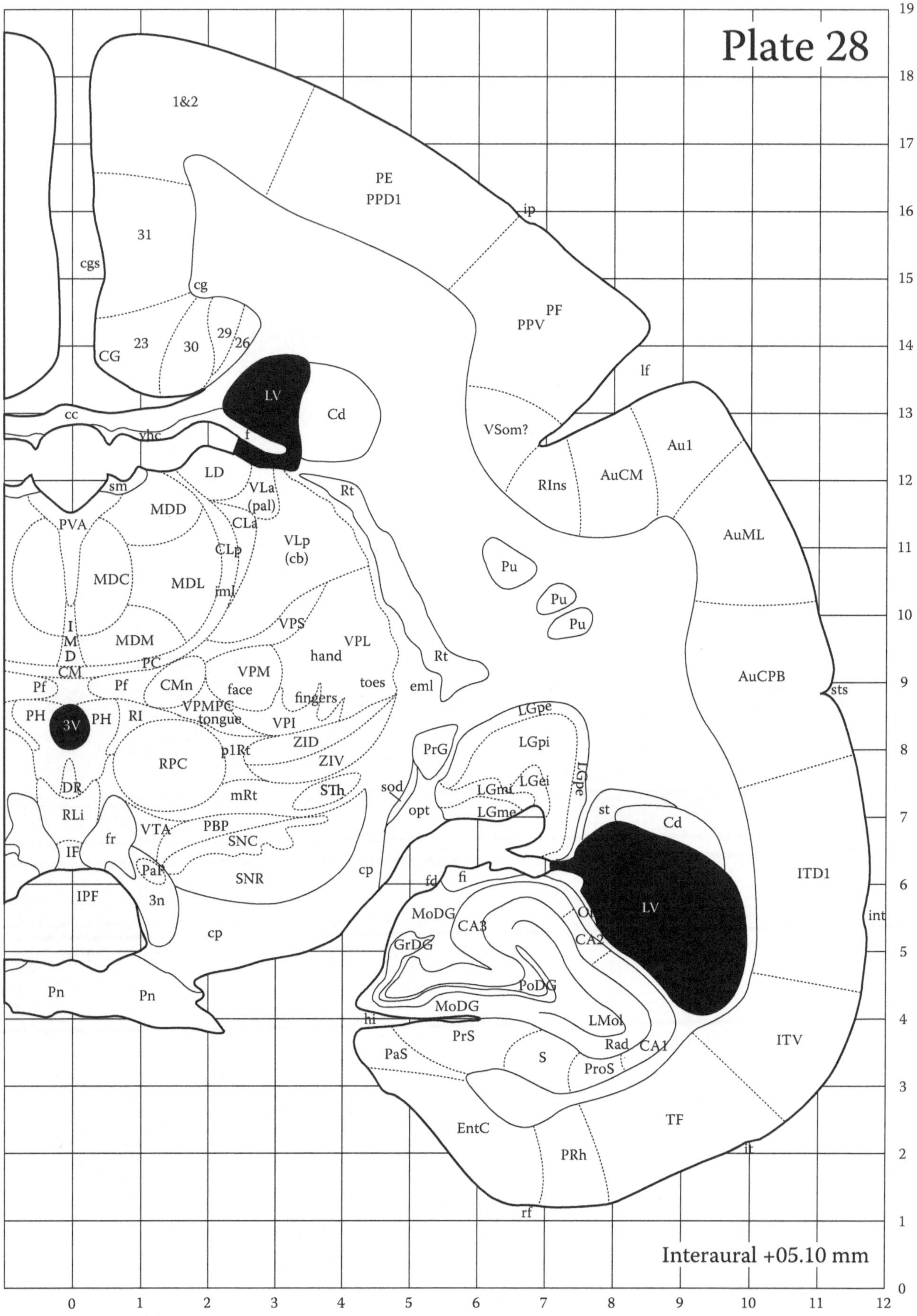
1&2
PE
PPD1
ip
31
cgs
cg
PF
PPV
23
30
29
26
CG
LV
Cd
lf
cc
vhc
f
VSom?
Au1
LD
Rt
RIns
AuCM
sm
VLa
(pal)
MDD
CLa
PVA
AuML
VLp
(cb)
CLp
Pu
MDC
MDL
iml
Pu
Pu
I
M
D
MDM
VPS
VPL
hand
Rt
PC
CM
VPM
AuCPB
Pf
Pf
CMn
face
toes
eml
sts
fingers
PH
3V
PH
RI
VPMPC
tongue
VPI
LGpe
ZID
PrG
LGpi
RPC
p1Rt
ZIV
DR
sod
LGmi
LGei
LGpe
STh
mRt
RLi
opt
LGme
st
Cd
fr
VTA
PBP
SNC
IF
PaP
cp
ITD1
SNR
fi
fd
IPF
3n
LV
MoDG
CA3
Or
int
cp
GrDG
CA2
PoDG
Pn
Pn
MoDG
hi
LMol
PrS
ITV
Rad
CA1
PaS
S
ProS
TF
EntC
it
PRh
rf
Interaural +05.10 mm

| | |
|---|---|
| 1&2 | areas 1/2 Cx |
| 3n | oculomotor nerve |
| 3V | third ventricle |
| 23 | area 23 Cx |
| 26 | area 26 Cx |
| 29 | area 29 Cx |
| 30 | area 30 Cx |
| 31 | area 31 Cx |
| Au1 | aud. core Cx, primary |
| AuCM | aud. belt Cx, caudomed. |
| AuCPB | aud. parabelt Cx, caud. |
| AuML | aud. belt Cx, mid. lat. |
| CA1 | cornu ammonis 1 |
| CA2 | cornu ammonis 2 |
| CA3 | cornu ammonis 3 |
| cc | corp. callosum |
| Cd | caudate nu. |
| cg | cingulum |
| CG | cingulate gyrus Cx |
| cgs | cingulate sulc. |
| CLa | centrolat. thal. nu., ant. |
| CLp | centrolat. thal. nu., post. |
| CM | cent. medial thal. nu. |
| CMn | centromedian thal. nu. |
| cp | cerebral peduncle |
| DR | dors. raphe nu. |
| eml | extern. medull. lamina |
| EntC | entorhin. Cx, caud. |
| f | fornix |
| fd | fimbriodent. sulcus |
| fi | fimbria |
| fr | fasciculus retroflexus |
| GrDG | gran. cell layer DG |
| hi | hippocamp. fiss. |
| IF | interfascic. nu. |
| IMD | intermediodors. thal. nu. |
| int | intermed. temp. sulc. |
| ip | intraparietal sulc. |
| IPF | interpedunc. fossa |
| it | inf. temporal sulcus |
| ITD1 | inf. temp. dors. Cx 1 |
| ITV | inf. temp. vent. Cx |
| LD | laterodors. thal. nu. |
| lf | lateral fissure |
| LGpe | LG, parvicell. external |
| LGpi | LG, parvicell. internal |
| LGei | LG, external interlaminar |
| LGme | LG, magnocell. external |
| LGmi | LG, magnocell. internal |
| LMol | strat. lacun. mol. CA |
| LV | lateral ventricle |
| MDC | mediodors. thal. nu., cent. |
| MDD | mediodors. thal. nu., dors. |
| MDL | mediodors. thal. nu., lat. |
| MDM | mediodors. thal. nu., med. |
| MoDG | molecular layer DG |
| mRt | mesen. retic. formn |
| opt | optic tract |
| Or | strat. oriens CA |
| p1Rt | pros. 1 retic. formn |
| PaP | parapeduncular nu. |
| PaS | parasubiculum |
| PBP | parabrach. pigm. nu. |
| PC | paracent. thal. nu. |
| PE | par. field E Cx |
| Pf | parafascicular nu. |
| PF | par. field F Cx |
| PH | post. hypothal. area |
| Pn | pontine nu. |
| PoDG | polymorphic layer DG |
| PPD1 | post. par. Cx, dors. 1 |
| PPV | post. par. Cx, vent. |
| PrG | pregeniculate nu. |
| PRh | perirhinal Cx |
| ProS | prosubiculum |
| PrS | presubiculum |
| Pu | putamen |
| PVA | paravent. thal. nu., ant. |
| Rad | strat. radiat. CA |
| rf | rhinal fiss. |
| RI | rost. interstit. nu. |
| RIns | retroinsular Cx |
| RLi | rost. linear nu. |
| RPC | red nu., parvicellular |
| Rt | reticular nu. |
| S | subiculum |
| sm | stria medullaris |
| SNC | subst. nigra, comp. |
| SNR | subst. nigra, retic. |
| sod | suproptic decn |
| st | stria terminalis |
| STh | subthalamic nu. |
| sts | sup. temp. sulcus |
| TF | temp. field F Cx |
| vhc | vent. hipp. comm. |
| VLa | ventrolat. thal. nu., ant. |
| VLp | ventrolat. thal. nu., post. |
| VPI | vent. post. thal. nu., inf. |
| VPL | vent. post. thal. nu., lat. |
| VPM | vent. post. thal. nu., med. |
| VPMPC | vent. post. thal. nu., med. parv. |
| VPS | vent. post. thal. nu., sup. |
| VSom? | vent. somatosens. Cx |
| VTA | vent. tegment. area |
| ZID | zona incerta, dors. |
| ZIV | zona incerta, vent. |

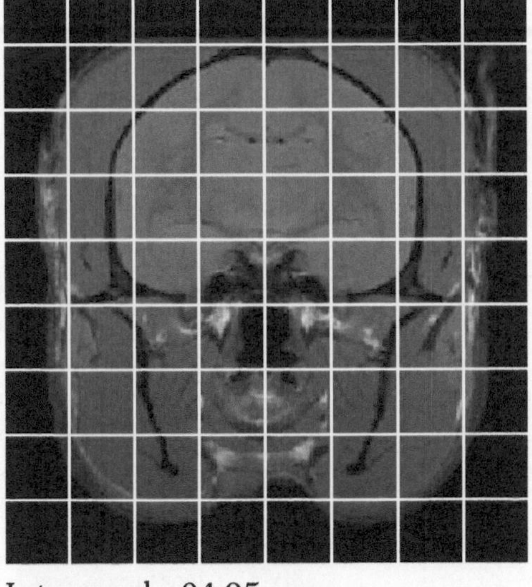

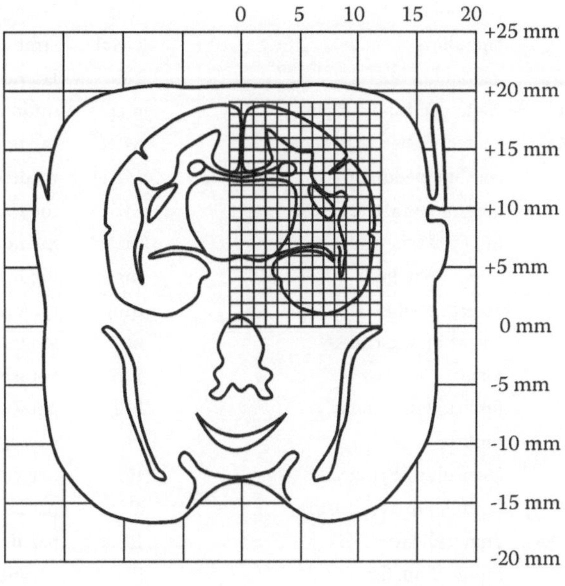


Interaural +04.95 mm

# Plate 29

Interaural +04.60 mm

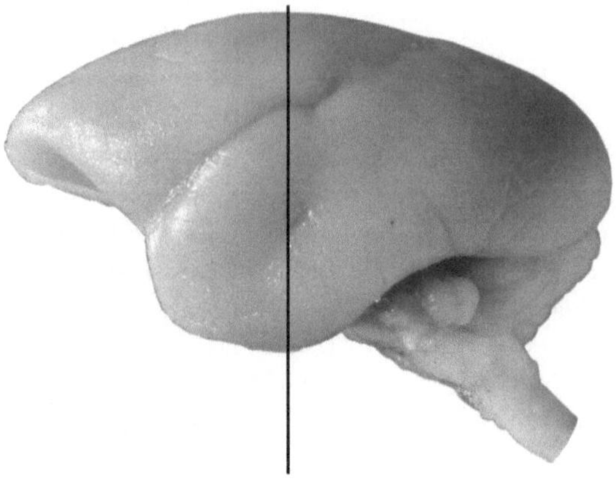

| | |
|---|---|
| 1&2 | areas 1/2 Cx |
| 3n | oculomotor nerve |
| 3V | third ventricle |
| 23 | area 23 Cx |
| 26 | area 26 Cx |
| 29 | area 29 Cx |
| 30 | area 30 Cx |
| 31 | area 31 Cx |
| alv | alveus of hippocampus |
| Au1L | aud. core Cx, prim., lat. |
| Au1M | aud. core Cx, prim., med. |
| AuCL | aud. belt Cx, caudolat. |
| AuCM | aud. belt Cx, caudomed. |
| AuCPB | aud. parabelt Cx, caud. |
| AuML | aud. belt Cx, mid. lat. |
| CA1 | cornu ammonis 1 |
| CA2 | cornu ammonis 2 |
| CA3 | cornu ammonis 3 |
| Cd | caudate nu. |
| cg | cingulum |
| CLi | caudal lin. nu. |
| CLp | centrolat. thal. nu., post. |
| CMn | centromedian thal. nu. |
| cp | cerebral peduncle |
| ctg | central tegm. tract |
| Dk | nu. Darkschewitsch |
| DR | dors. raphe nu. |
| eml | extern. medull. lamina |
| EntC | entorhin. Cx, caud. |
| f | fornix |
| fd | fimbriodent. sulcus |
| fi | fimbria |
| fr | fasciculus retroflexus |
| FST | fundus sup. temp. Cx |
| GrDG | gran. cell layer DG |
| hi | hippocamp. fiss. |
| IMD | intermediodors. thal. nu. |
| iml | intern. medull. lamina |
| ip | intraparietal sulc. |
| IPC | interped. nu., caud. |
| IPD | interped. nu., dorsal |
| IPI | interped. nu., interfasc. |
| IPL | interped. nu., lat. |
| IPR | interped. nu., rost. |
| it | inf. temporal sulcus |
| ITD1 | inf. temp. dors. Cx 1 |
| ITV | inf. temp. vent. Cx |
| LD | laterodors. thal. nu. |
| lf | lateral fissure |
| LGei | LG, external interlaminar |
| LGii | LG, internal interlaminar |
| LGme | LG, magnocell. external |
| LGmi | LG, magnocell. internal |
| LGpe | LG, parvicell. external |
| LGpi | LG, parvicell. internal |
| LGs | lat. habenular nu., dorsal |
| LMol | strat. lacun. mol. CA |
| LV | lateral ventricle |
| mcp | middle cerebell. ped. |
| MDC | mediodors. thal. nu., cent. |
| MDD | mediodors. thal. nu., dors. |
| MDL | mediodors. thal. nu., lat. |
| MDM | mediodors. thal. nu., med. |
| ml | medial lemniscus |
| MoDG | molecular layer DG |
| Or | strat. oriens CA |
| PaS | parasubiculum |
| PBP | parabrach. pigm. nu. |
| PC | paracent. thal. nu. |
| PE | par. field E Cx |
| Pf | parafascicular nu. |
| PF | par. field F Cx |
| PN | paranigral nu. |
| PoDG | polymorphic layer DG |
| PPD1 | post. par. Cx, dors. 1 |
| PPV | post. par. Cx, vent. |
| PRh | perirhinal Cx |
| ProS | prosubiculum |
| PrS | presubiculum |
| PulA | pulvinar nu., ant. |
| PVA | paravent. thal. nu., ant. |
| PVG | periventricular gray |
| PVP | paravent. thal. nu., post. |
| Rad | strat. radiat. CA |
| rf | rhinal fiss. |
| RIns | retroinsular Cx |
| RLi | rost. linear nu. |
| RMC | red nu., magnocell. |
| RPC | red nu., parvicellular |
| Rt | reticular nu. |
| S | subiculum |
| sm | stria medullaris |
| SNCD | subst. nigra, comp., dors. |
| SNCV | subst. nigra, comp., vent. |
| SNR | subst. nigra, retic. |
| st | stria terminalis |
| sts | sup. temp. sulcus |
| TCF | transverse cerebral fiss. |
| TF | temp. field F Cx |
| tfp | transv. fibers pons |
| vhc | vent. hipp. comm. |
| VLp | ventrolat. thal. nu., post. |
| VPI | vent. post. thal. nu., inf. |
| VPL | vent. post. thal. nu., lat. |
| VPM | vent. post. thal. nu., med. |
| VPMPC | vent. post. thal. nu., med. parv. |
| VPS | vent. post. thal. nu., sup. |
| VTA | vent. tegment. area |
| ZID | zona incerta, dors. |
| ZIV | zona incerta, vent. |

# Plate 29a

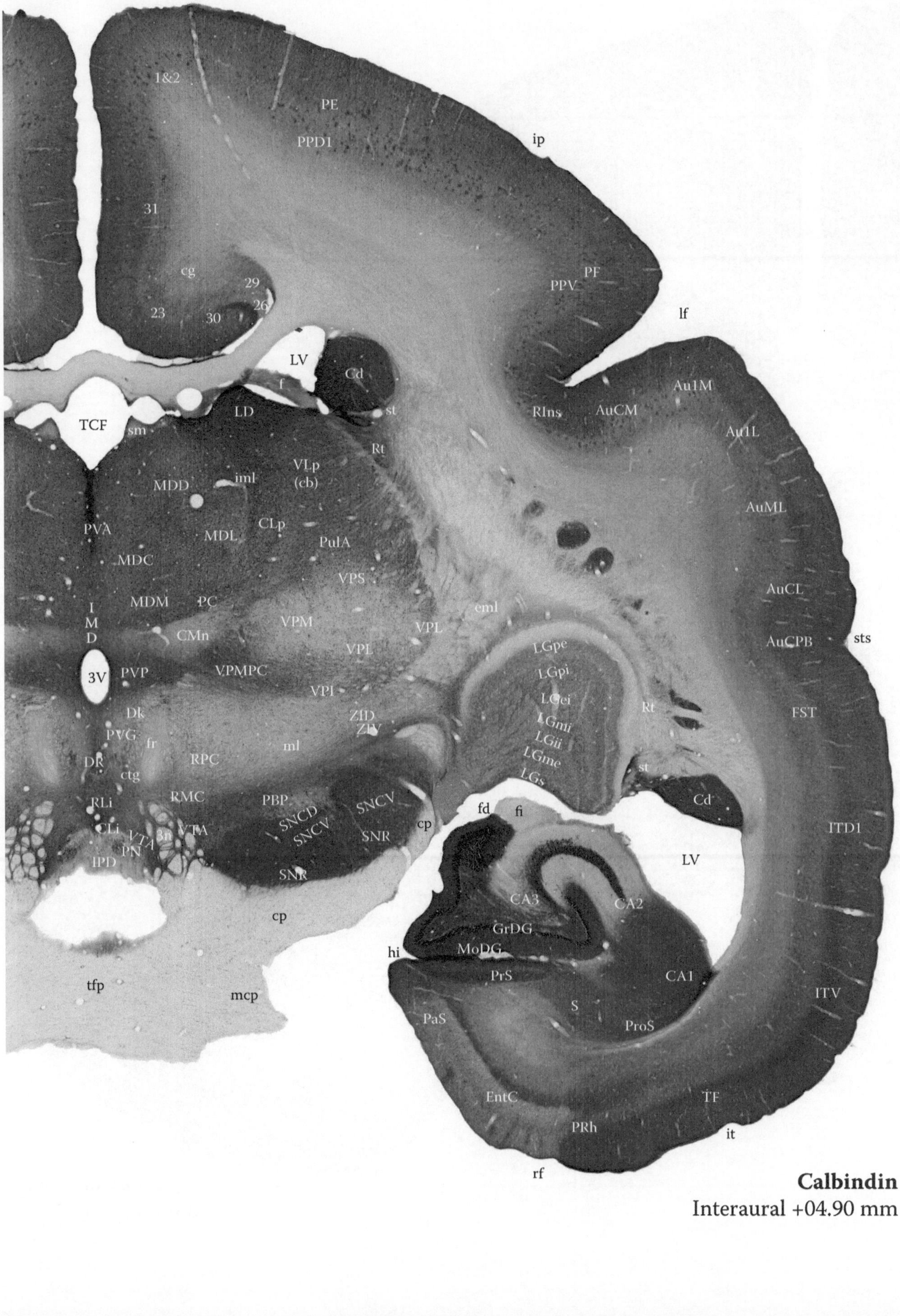


**Calbindin**
Interaural +04.90 mm

# Plate 29b

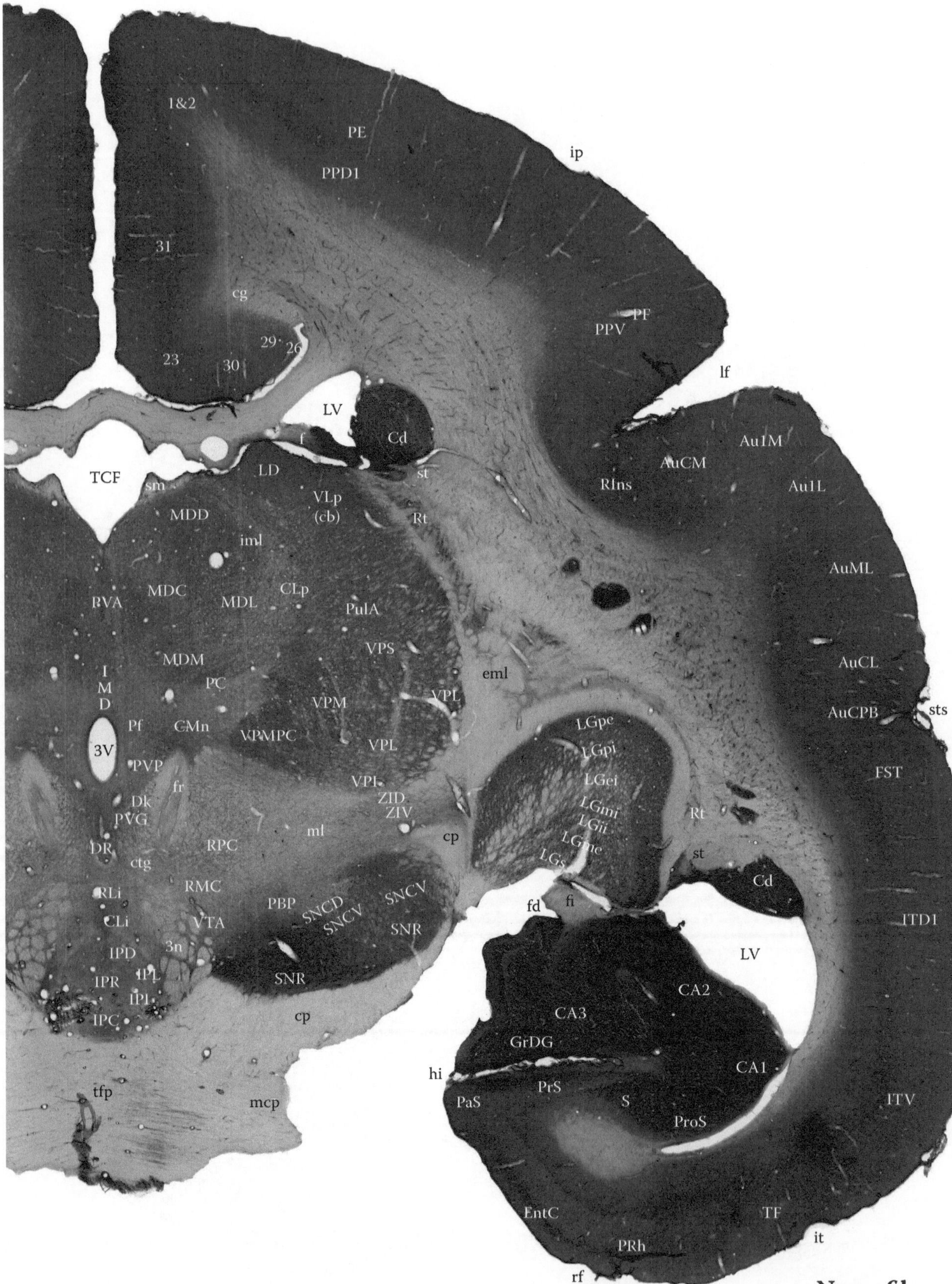


**Neurofilament**
Interaural +04.75 mm

# Plate 29c

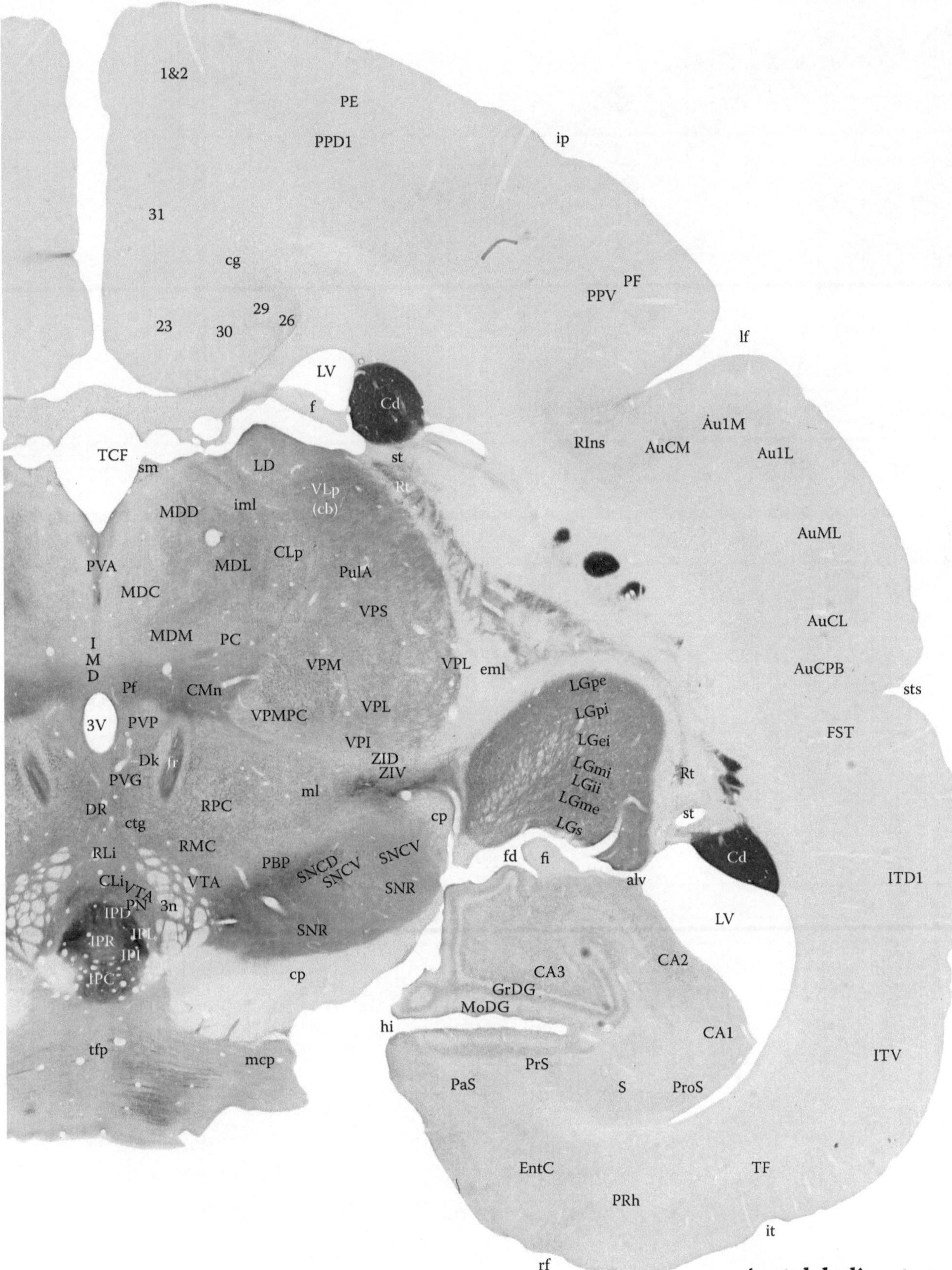


**Acetylcholinesterase**
Interaural +04.70 mm

# Plate 29d

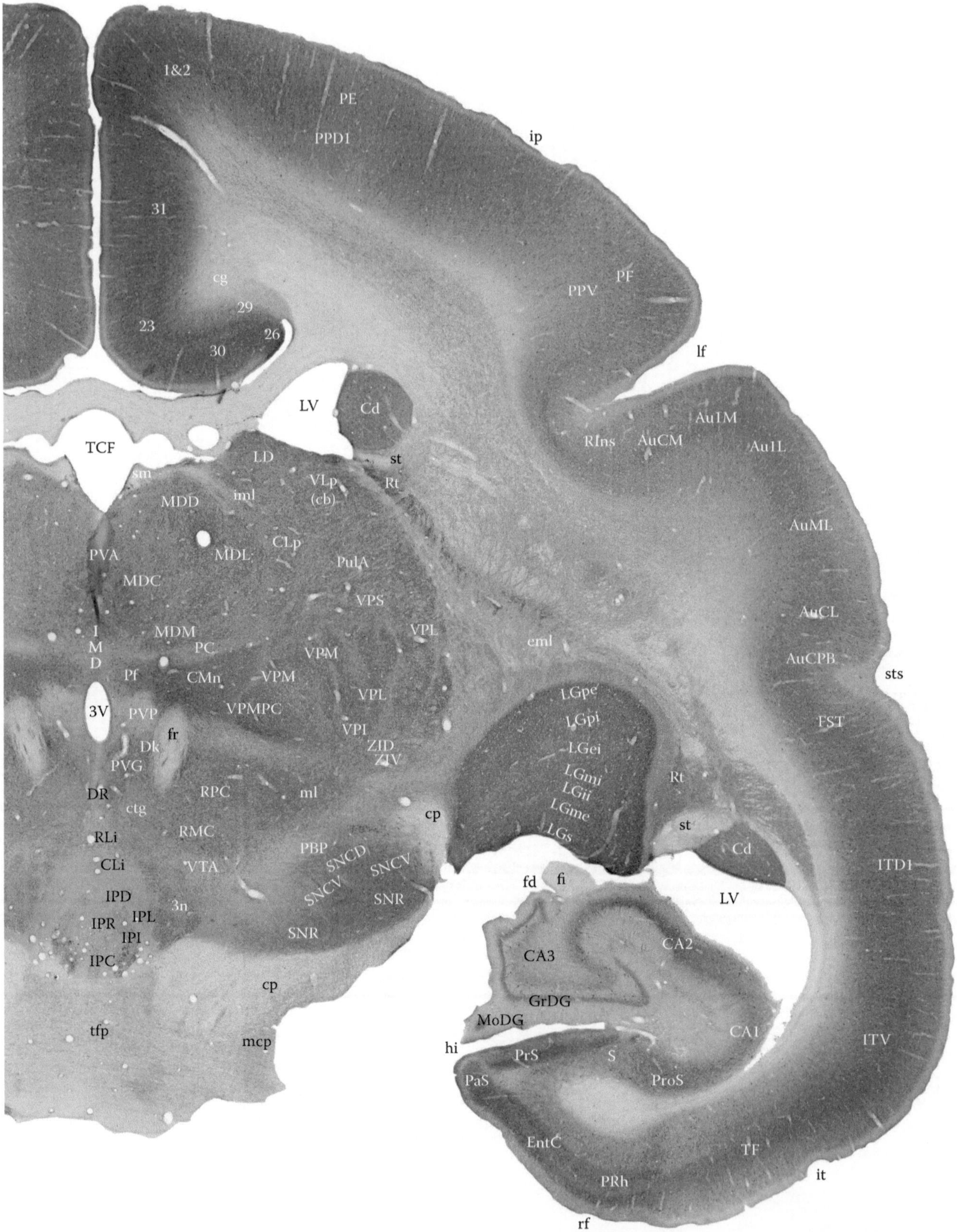


**Parvalbumin**
Interaural +04.65 mm

# Plate 29e

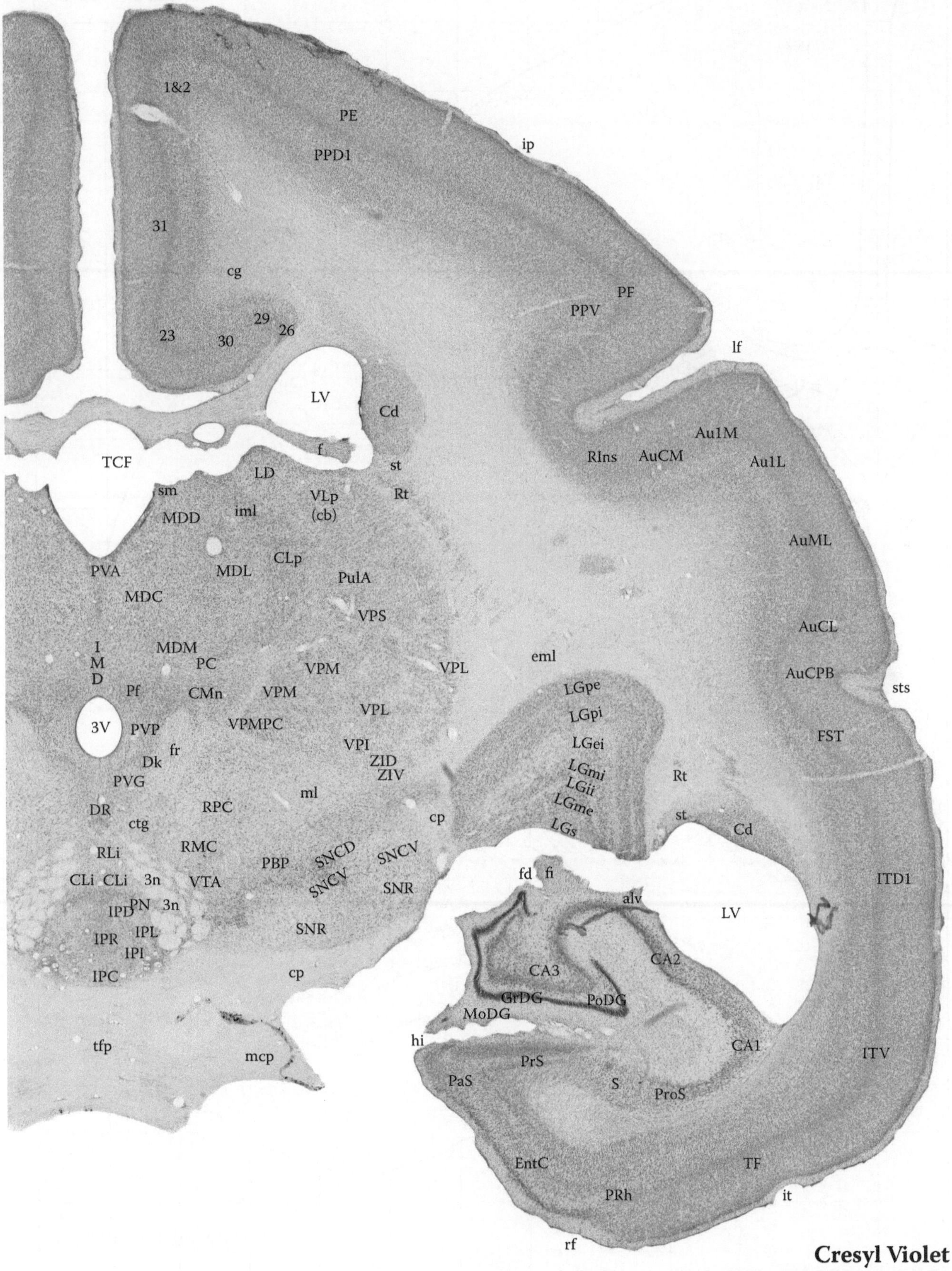


**Cresyl Violet**
Interaural +04.60 mm

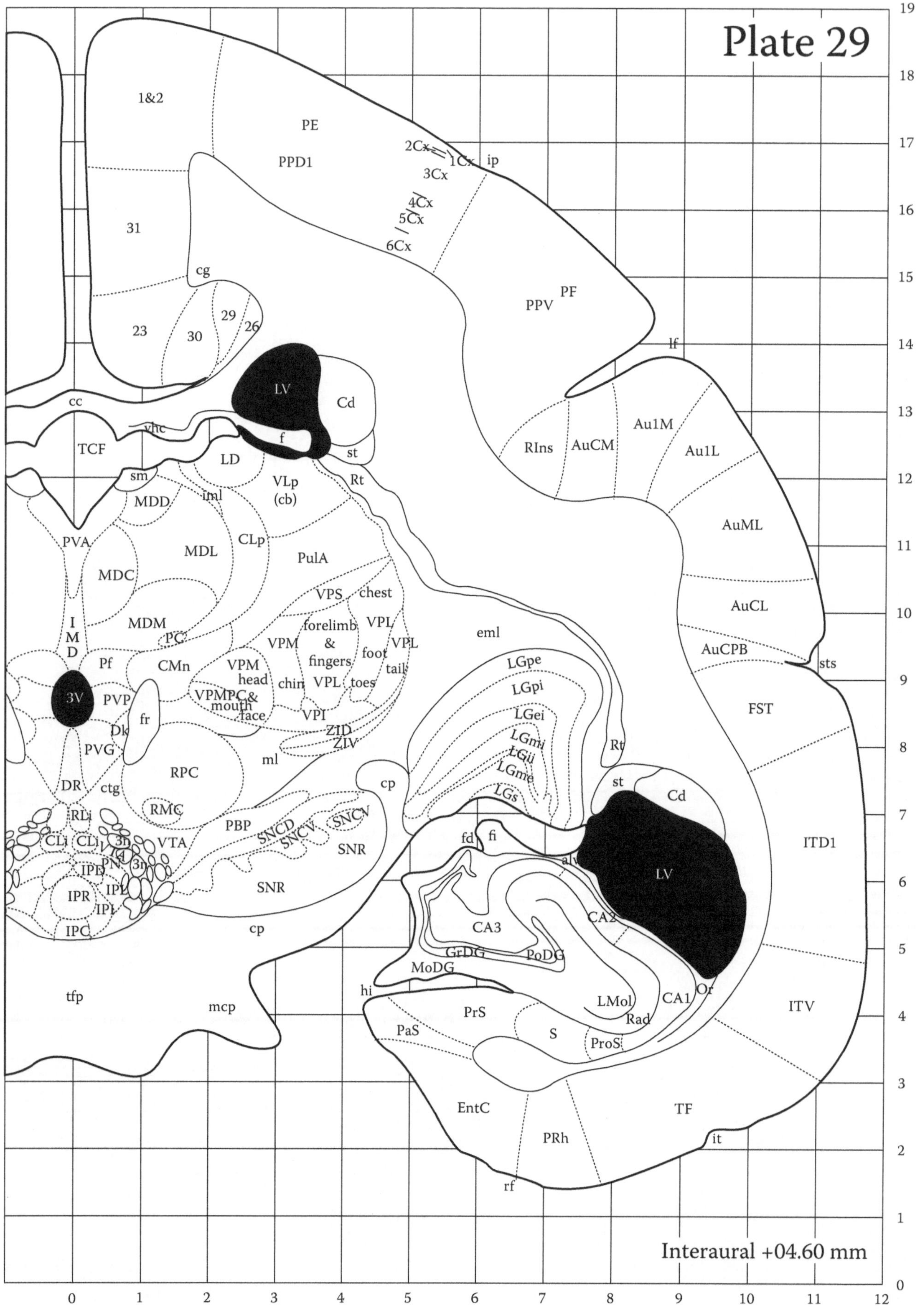
Plate 29
1&2
PE
PPD1
2Cx
1Cx
ip
3Cx
4Cx
5Cx
6Cx
31
cg
23
30
29
26
PPV
PF
lf
LV
Cd
cc
vhc
TCF
f
st
LD
sm
MDD
iml
VLp
(cb)
Rt
RIns
AuCM
Au1M
Au1L
AuML
PVA
MDL
CLp
PulA
MDC
AuCL
VPS
chest
I
M
D
MDM
PC
forelimb
&
fingers
VPM
VPL
VPL
foot
tail
eml
AuCPB
sts
Pf
CMn
VPM
head
chin
VPL
toes
LGpe
LGpi
3V
PVP
VPMPC&
mouth
face
VPI
fr
LGei
FST
Dk
ZID
ZIV
LGmi
LGii
Rt
PVG
ml
LGme
RPC
LGs
DR
ctg
cp
st
Cd
RMC
RLi
PBP
SNCD
SNCV
SNCV
fd
fi
ITD1
CLi
CLi
3n
VTA
SNR
alv
PN
3n
IPD
LV
IPR
IPL
IPL
SNR
CA3
CA2
IPC
cp
GrDG
PoDG
MoDG
tfp
hi
mcp
PrS
LMol
CA1
Or
ITV
Rad
PaS
S
ProS
EntC
TF
PRh
it
rf
Interaural +04.60 mm
0
1
2
3
4
5
6
7
8
9
10
11
12
19
18
17
16
15
14
13

| | | | | | |
|---|---|---|---|---|---|
| 1&2 | areas 1/2 Cx | iml | intern. medull. lamina | PPD1 | post. par. Cx, dors. 1 |
| 3n | oculomotor nerve | ip | intraparietal sulc. | PPV | post. par. Cx, vent. |
| 3V | third ventricle | IPC | interped. nu., caud. | PRh | perirhinal Cx |
| 23 | area 23 Cx | IPD | interped. nu., dorsal | ProS | prosubiculum |
| 26 | area 26 Cx | IPI | interped. nu., interfasc. | PrS | presubiculum |
| 29 | area 29 Cx | IPL | interped. nu., lat. | PulA | pulvinar nu., ant. |
| 30 | area 30 Cx | IPR | interped. nu., rost. | PVA | paravent. thal. nu., ant. |
| 31 | area 31 Cx | it | inf. temporal sulcus | PVG | periventricular gray |
| alv | alveus of hippocampus | ITD1 | inf. temp. dors. Cx 1 | PVP | paravent. thal. nu., post. |
| Au1L | aud. core Cx, prim., lat. | ITV | inf. temp. vent. Cx | Rad | strat. radiat. CA |
| Au1M | aud. core Cx, prim., med. | LD | laterodors. thal. nu. | rf | rhinal fiss. |
| AuCL | aud. belt Cx, caudolat. | lf | lateral fissure | RIns | retroinsular Cx |
| AuCM | aud. belt Cx, caudomed. | LGei | LG, external interlaminar | RLi | rost. linear nu. |
| AuCPB | aud. parabelt Cx, caud. | LGii | LG, internal interlaminar | RMC | red nu., magnocell. |
| AuML | aud. belt Cx, mid. lat. | LGme | LG, magnocell. external | RPC | red nu., parvicellular |
| CA1 | cornu ammonis 1 | LGmi | LG, magnocell. internal | Rt | reticular nu. |
| CA2 | cornu ammonis 2 | LGpe | LG, parvicell. external | S | subiculum |
| CA3 | cornu ammonis 3 | LGpi | LG, parvicell. internal | sm | stria medullaris |
| Cd | caudate nu. | LGs | lat. habenular nu., dorsal | SNCD | subst. nigra, comp., dors. |
| cg | cingulum | LMol | strat. lacun. mol. CA | SNCV | subst. nigra, comp., vent. |
| CLi | caudal lin. nu. | LV | lateral ventricle | SNR | subst. nigra, retic. |
| CLp | centrolat. thal. nu., post. | mcp | middle cerebell. ped. | st | stria terminalis |
| CMn | centromedian thal. nu. | MDC | mediodors. thal. nu., cent. | sts | sup. temp. sulcus |
| cp | cerebral peduncle | MDD | mediodors. thal. nu., dors. | TCF | transverse cerebral fiss. |
| ctg | central tegm. tract | MDL | mediodors. thal. nu., lat. | TF | temp. field F Cx |
| Dk | nu. Darkschewitsch | MDM | mediodors. thal. nu., med. | tfp | transv. fibers pons |
| DR | dors. raphe nu. | ml | medial lemniscus | vhc | vent. hipp. comm. |
| eml | extern. medull. lamina | MoDG | molecular layer DG | VLp | ventrolat. thal. nu., post. |
| EntC | entorhin. Cx, caud. | Or | strat. oriens CA | VPI | vent. post. thal. nu., inf. |
| f | fornix | PaS | parasubiculum | VPL | vent. post. thal. nu., lat. |
| fd | fimbriodent. sulcus | PBP | parabrach. pigm. nu. | VPM | vent. post. thal. nu., med. |
| fi | fimbria | PC | paracent. thal. nu. | VPMPC | vent. post. thal. nu., med. parv. |
| fr | fasciculus retroflexus | PE | par. field E Cx | VPS | vent. post. thal. nu., sup. |
| FST | fundus sup. temp. Cx | Pf | parafascicular nu. | VTA | vent. tegment. area |
| GrDG | gran. cell layer DG | PF | par. field F Cx | ZID | zona incerta, dors. |
| hi | hippocamp. fiss. | PN | paranigral nu. | ZIV | zona incerta, vent. |
| IMD | intermediodors. thal. nu. | PoDG | polymorphic layer DG | | |

# Plate 30

Interaural +04.05 mm

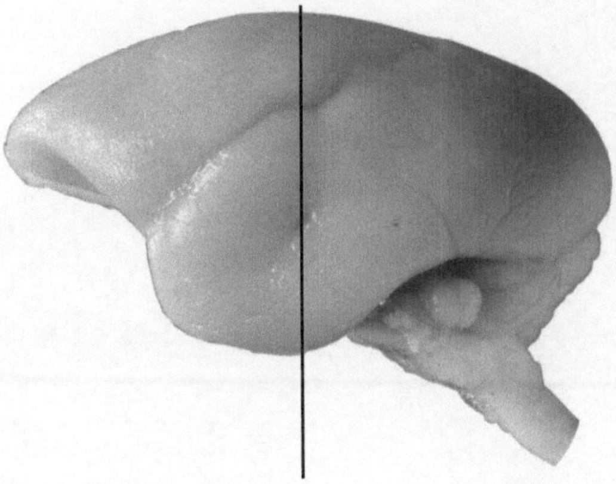

| | |
|---|---|
| 3n | oculomotor nerve |
| 3N | oculomotor nu. |
| 3V | third ventricle |
| 23 | area 23 Cx |
| 26 | area 26 Cx |
| 29a-c | area 29 Cx |
| 29d | area 29d Cx |
| 30 | area 30 Cx |
| alv | alveus of hippocampus |
| Au1 | aud. core Cx, primary |
| AuCM | aud. belt Cx, caudomed. |
| AuCPB | aud. parabelt Cx, caud. |
| AuML | aud. belt Cx, mid. lat. |
| CA1 | cornu ammonis 1 |
| CA2 | cornu ammonis 2 |
| CA3 | cornu ammonis 3 |
| cc | corp. callosum |
| Cd | caudate nu. |
| cg | cingulum |
| CLi | caudal lin. nu. |
| CLp | centrolat. thal. nu., post. |
| CMn | centromedian thal. nu. |
| cp | cerebral peduncle |
| ctg | central tegm. tract |
| Dk | nu. Darkschewitsch |
| DR | dors. raphe nu. |
| eml | extern. medull. lamina |
| EntC | entorhin. Cx, caud. |
| EW | Edinger Westphal nu. |
| f | fornix |
| fd | fimbriodent. sulcus |
| fi | fimbria |
| fr | fasciculus retroflexus |
| FST | fundus sup. temp. Cx |
| GrDG | gran. cell layer DG |
| hi | hippocamp. fiss. |
| IMD | intermediodors. thal. nu. |
| iml | intern. medull. lamina |
| InC | interst. nu. Cajal |
| InCSh | interst. nu. Cajal, shell |
| int | intermed. temp. sulc. |
| IPC | interped. nu., caud. |
| IPD | interped. nu., dorsal |
| IPI | interped. nu., interfasc. |
| IPL | interped. nu., lat. |
| IPR | interped. nu., rost. |
| it | inf. temporal sulcus |
| ITD1 | inf. temp. dors. Cx 1 |
| ITV | inf. temp. vent. Cx |
| LD | laterodors. thal. nu. |
| lf | lateral fissure |
| LGei | LG, external interlaminar |
| LGme | LG, magnocell. external |
| LGmi | LG, magnocell. internal |
| LGpe | LG, parvicell. external |
| LGpi | LG, parvicell. internal |
| LGs | lat. habenular nu., dorsal |
| LHb | lat. habenular nu., ventral |
| LMol | strat. lacun. mol. CA |
| LTnl | lat. terminal nu. |
| LV | lateral ventricle |
| mcp | middle cerebell. ped. |
| MDC | mediodors. thal. nu., cent. |
| MDD | mediodors. thal. nu., dors. |
| MDL | mediodors. thal. nu., lat. |
| MDM | mediodors. thal. nu., med. |
| MGAD | med. gen. nu., anterodors. |
| MGM | med. gen. nu., med. |
| MGV | med. gen. nu., vent. |
| MHb | med. habenular nu. |
| ml | medial lemniscus |
| MoDG | molecular layer DG |
| mRt | mesen. retic. formn |
| MST | med. sup. temp. area Cx |
| MT | mid. temp. area Cx |
| MTc | mid. temp. cresc. Cx |
| opt | optic tract |
| Or | strat. oriens CA |
| p1Rt | pros. 1 retic. formn |
| PaS | parasubiculum |
| Pf | parabrach. pigm. nu. |
| PBP | par. field E Cx |
| PE | parafascicular nu. |
| PF | par. field F Cx |
| Pn | pontine nu. |
| PN | paranigral nu. |
| PoDG | polymorphic layer DG |
| PPD1 | post. par. Cx, dors. 1 |
| PPD2 | post. par. Cx, dors. 2 |
| PPV | post. par. Cx, vent. |
| PRh | perirhinal Cx |
| ProS | prosubiculum |
| PrS | presubiculum |
| PulA | pulvinar nu., ant. |
| PVG | periventricular gray |
| PVP | paravent. thal. nu., post. |
| Rad | strat. radiat. CA |
| rf | rhinal fiss. |
| RMC | red nu., magnocell. |
| RPC | red nu., parvicellular |
| RRF | retrorubral field |
| Rt | reticular nu. |
| S | subiculum |
| SG | supragenic. thal. nu. |
| sm | stria medullaris |
| SNCD | subst. nigra, comp., dors. |
| SNCV | subst. nigra, comp., vent. |
| SNR | subst. nigra, retic. |
| st | stria terminalis |
| sts | sup. temp. sulcus |
| TCF | transverse cerebral fiss. |
| TF | temp. field F Cx |
| TPt | temp. paritemp. area Cx |
| vhc | vent. hipp. comm. |
| VLp | ventrolat. thal. nu., post. |
| VPI | vent. post. thal. nu., inf. |
| VPL | vent. post. thal. nu., lat. |
| VPM | vent. post. thal. nu., med. |
| VPS | vent. post. thal. nu., sup. |
| VTA | vent. tegment. area |

# Plate 30a

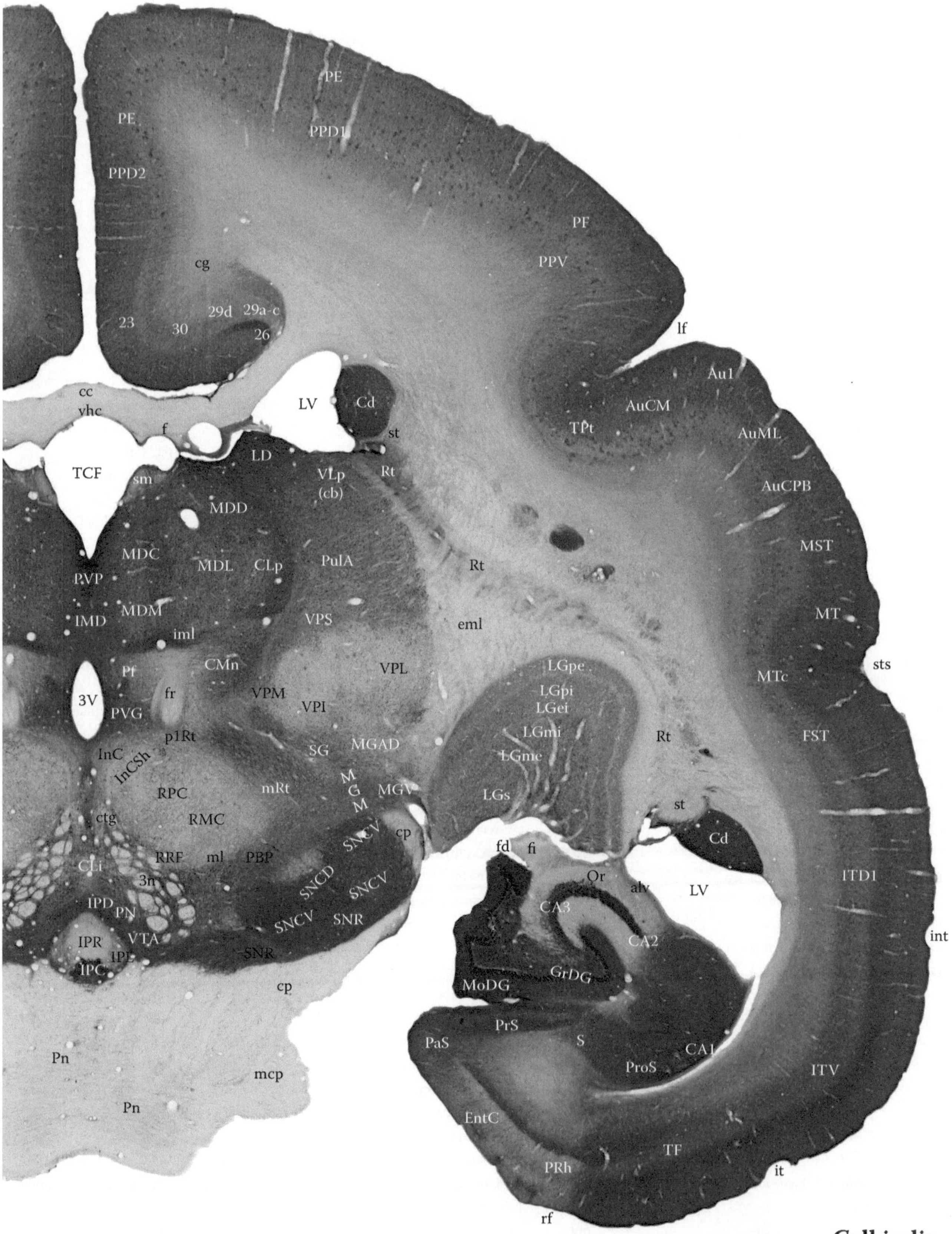


**Calbindin**
Interaural +04.35 mm

# Plate 30b

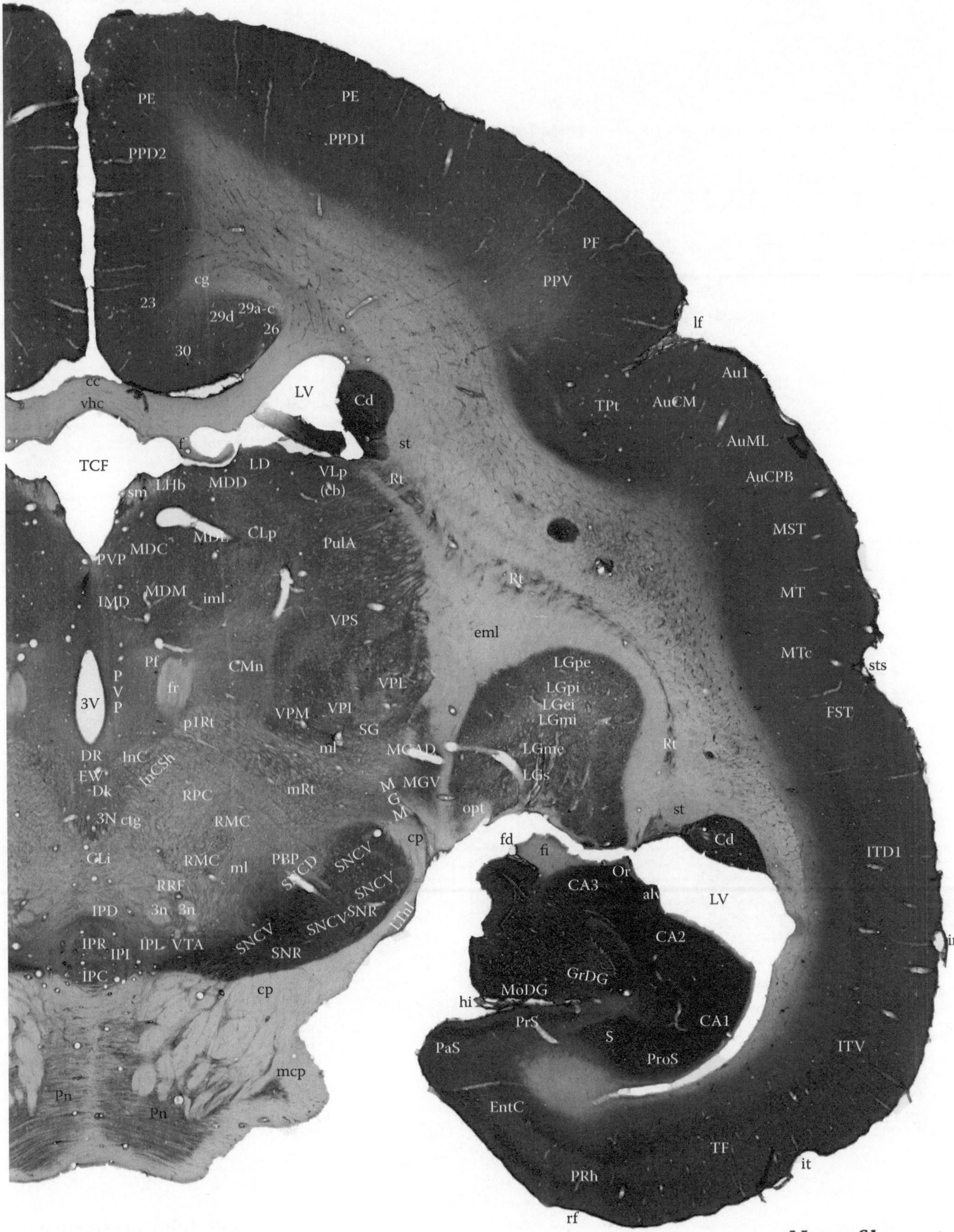


**Neurofilament**
Interaural +04.20 mm

# Plate 30c

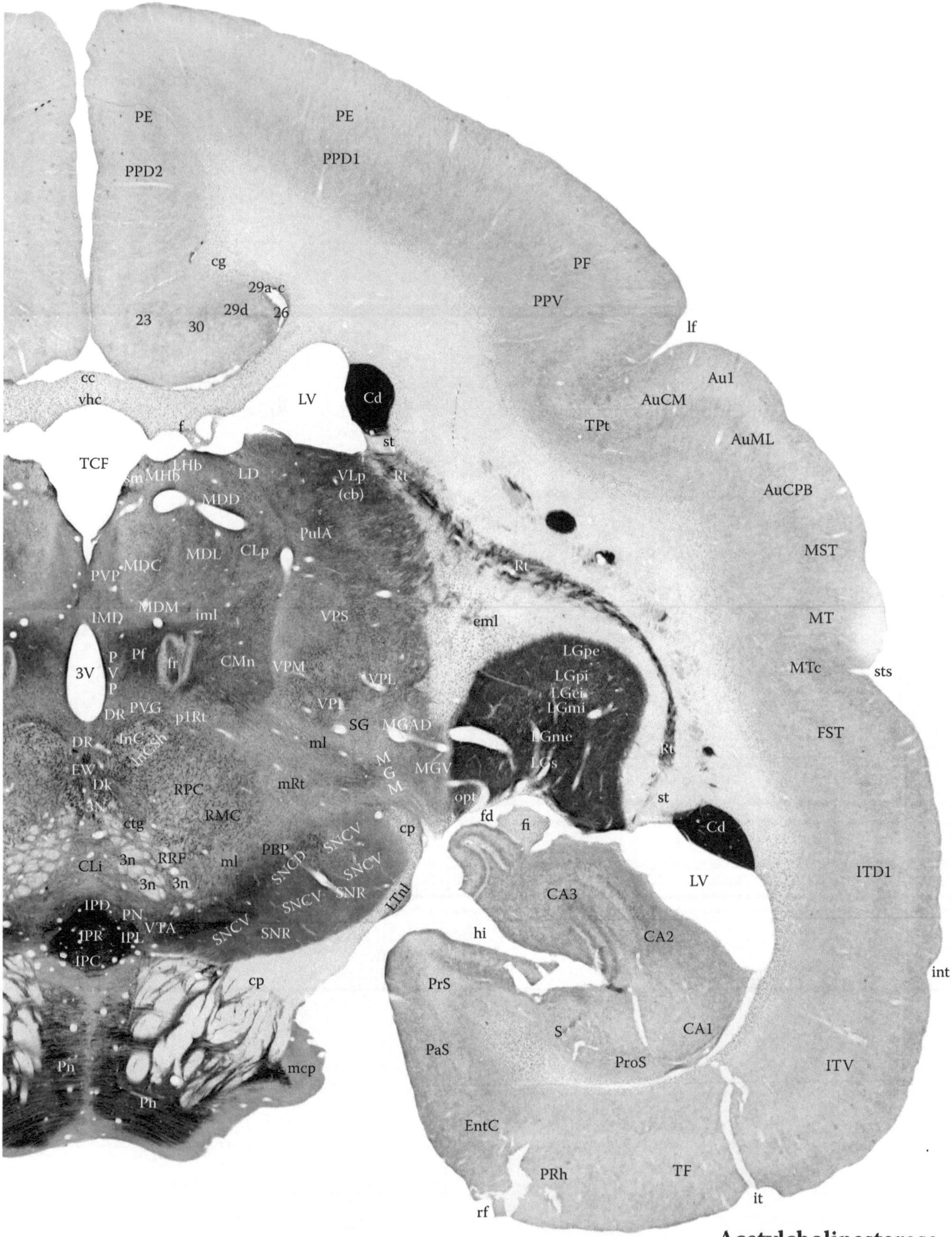


**Acetylcholinesterase**
Interaural +04.15 mm

# Plate 30d

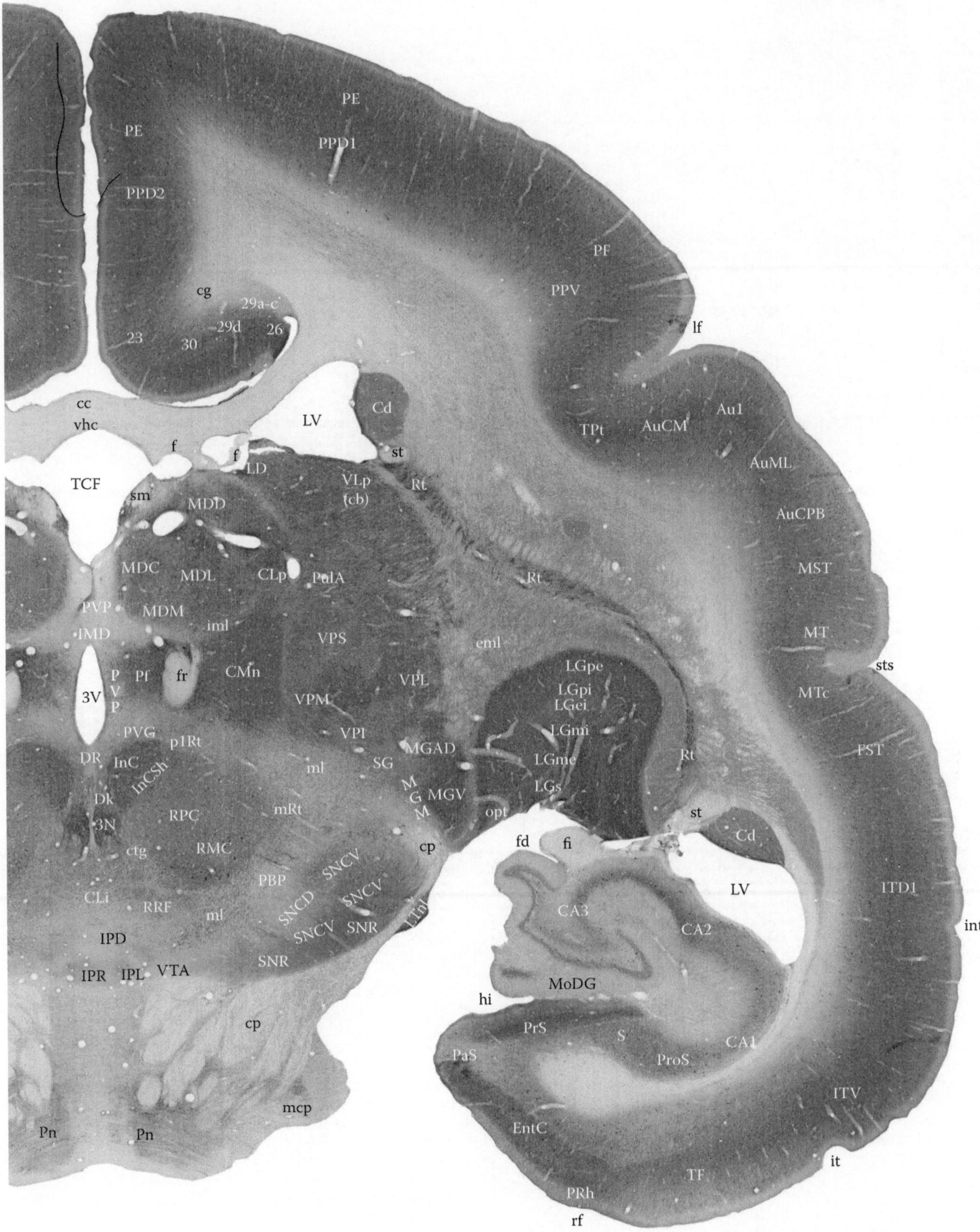


**Parvalbumin**
Interaural +04.10 mm

# Plate 30e

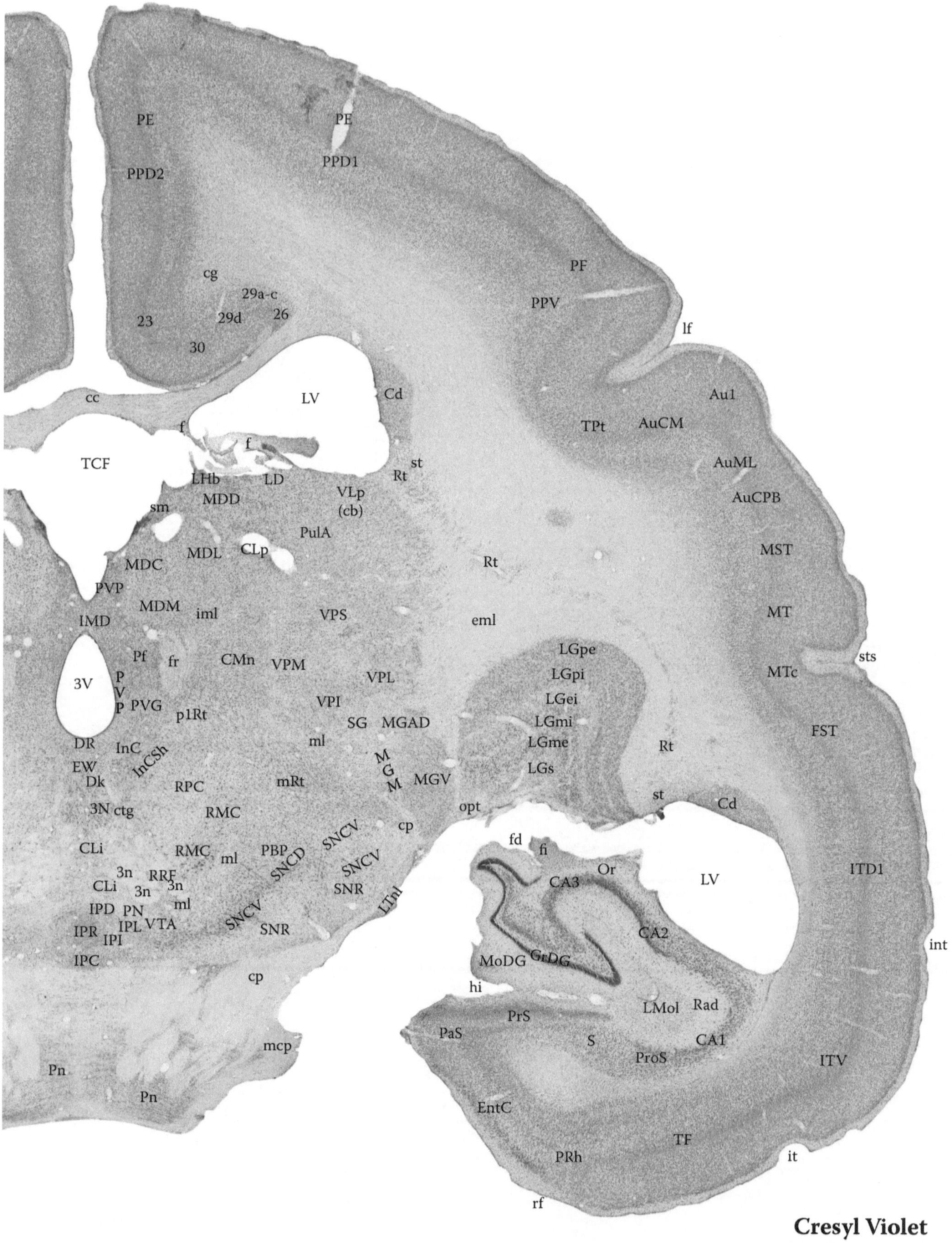


**Cresyl Violet**
Interaural +04.05 mm

# Plate 30

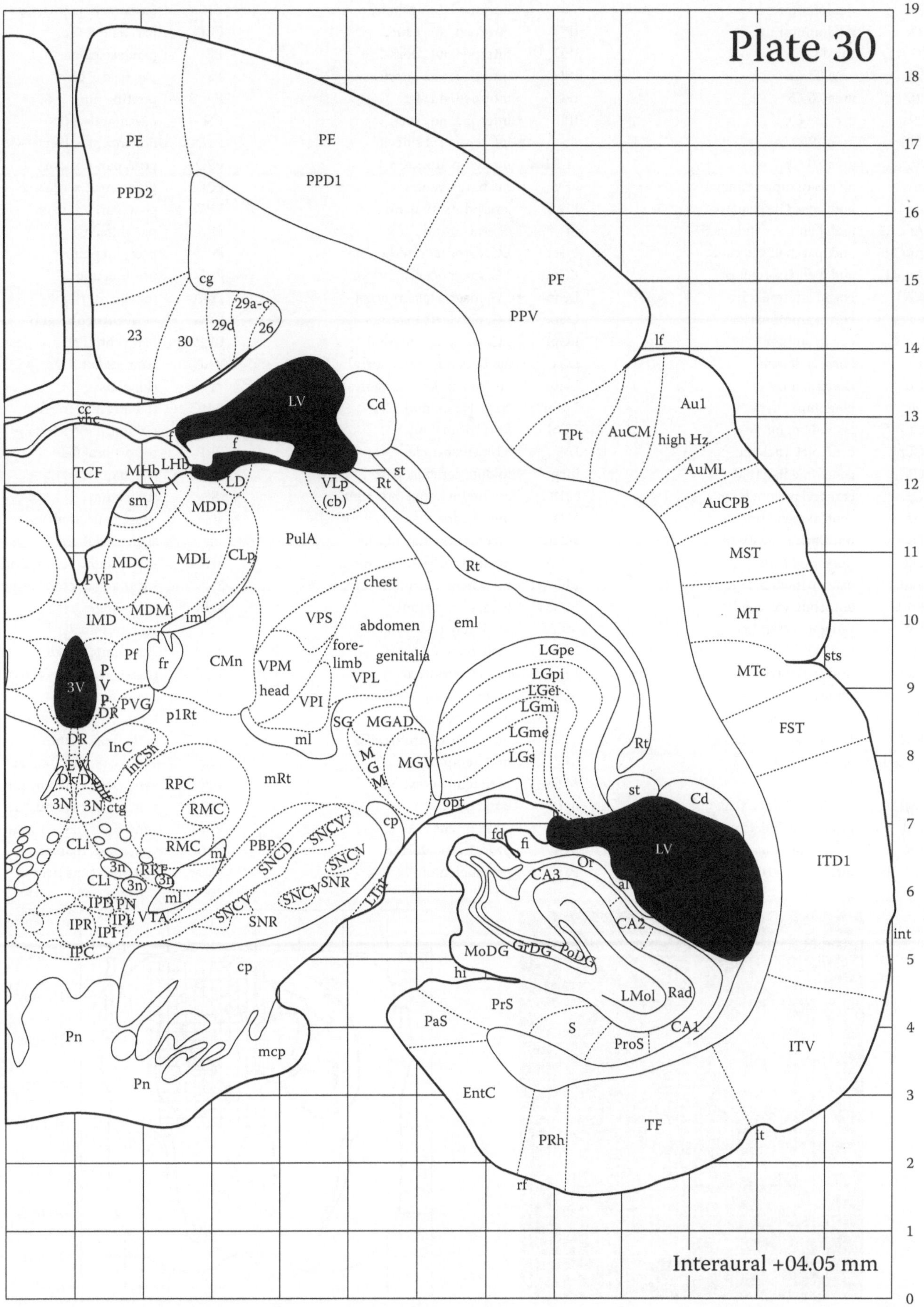


Interaural +04.05 mm

| | |
|---|---|
| 3n | oculomotor nerve |
| 3N | oculomotor nu. |
| 3V | third ventricle |
| 23 | area 23 Cx |
| 26 | area 26 Cx |
| 29a-c | area 29 Cx |
| 29d | area 29d Cx |
| 30 | area 30 Cx |
| alv | alveus of hippocampus |
| Au1 | aud. core Cx, primary |
| AuCM | aud. belt Cx, caudomed. |
| AuCPB | aud. parabelt Cx, caud. |
| AuML | aud. belt Cx, mid. lat. |
| CA1 | cornu ammonis 1 |
| CA2 | cornu ammonis 2 |
| CA3 | cornu ammonis 3 |
| cc | corp. callosum |
| Cd | caudate nu. |
| cg | cingulum |
| CLi | caudal lin. nu. |
| CLp | centrolat. thal. nu., post. |
| CMn | centromedian thal. nu. |
| cp | cerebral peduncle |
| ctg | central tegm. tract |
| Dk | nu. Darkschewitsch |
| DR | dors. raphe nu. |
| eml | extern. medull. lamina |
| EntC | entorhin. Cx, caud. |
| EW | Edinger Westphal nu. |
| f | fornix |
| fd | fimbriodent. sulcus |
| fi | fimbria |
| fr | fasciculus retroflexus |
| FST | fundus sup. temp. Cx |
| GrDG | gran. cell layer DG |
| hi | hippocamp. fiss. |
| IMD | intermediodors. thal. nu. |
| iml | intern. medull. lamina |
| InC | interst. nu. Cajal |
| InCSh | interst. nu. Cajal, shell |
| int | intermed. temp. sulc. |
| IPC | interped. nu., caud. |
| IPD | interped. nu., dorsal |
| IPI | interped. nu., interfasc. |
| IPL | interped. nu., lat. |
| IPR | interped. nu., rost. |
| it | inf. temporal sulcus |
| ITD1 | inf. temp. dors. Cx 1 |
| ITV | inf. temp. vent. Cx |
| LD | laterodors. thal. nu. |
| lf | lateral fissure |
| LGei | LG, external interlaminar |
| LGme | LG, magnocell. external |
| LGmi | LG, magnocell. internal |
| LGpe | LG, parvicell. external |
| LGpi | LG, parvicell. internal |
| LGs | lat. habenular nu., dorsal |
| LHb | lat. habenular nu., ventral |
| LMol | strat. lacun. mol. CA |
| LTnl | lat. terminal nu. |
| LV | lateral ventricle |
| mcp | middle cerebell. ped. |
| MDC | mediodors. thal. nu., cent. |
| MDD | mediodors. thal. nu., dors. |
| MDL | mediodors. thal. nu., lat. |
| MDM | mediodors. thal. nu., med. |
| MGAD | med. gen. nu., anterodors. |
| MGM | med. gen. nu., med. |
| MGV | med. gen. nu., vent. |
| MHb | med. habenular nu. |
| ml | medial lemniscus |
| MoDG | molecular layer DG |
| mRt | mesen. retic. formn |
| MST | med. sup. temp. area Cx |
| MT | mid. temp. area Cx |
| MTc | mid. temp. cresc. Cx |
| opt | optic tract |
| Or | strat. oriens CA |
| p1Rt | pros. 1 retic. formn |
| PaS | parasubiculum |
| Pf | parabrach. pigm. nu. |
| PBP | par. field E Cx |
| PE | parafascicular nu. |
| PF | par. field F Cx |
| Pn | pontine nu. |
| PN | paranigral nu. |
| PoDG | polymorphic layer DG |
| PPD1 | post. par. Cx, dors. 1 |
| PPD2 | post. par. Cx, dors. 2 |
| PPV | post. par. Cx, vent. |
| PRh | perirhinal Cx |
| ProS | prosubiculum |
| PrS | presubiculum |
| PulA | pulvinar nu., ant. |
| PVG | periventricular gray |
| PVP | paravent. thal. nu., post. |
| Rad | strat. radiat. CA |
| rf | rhinal fiss. |
| RMC | red nu., magnocell. |
| RPC | red nu., parvicellular |
| RRF | retrorubral field |
| Rt | reticular nu. |
| S | subiculum |
| SG | suprageniс. thal. nu. |
| sm | stria medullaris |
| SNCD | subst. nigra, comp., dors. |
| SNCV | subst. nigra, comp., vent. |
| SNR | subst. nigra, retic. |
| st | stria terminalis |
| sts | sup. temp. sulcus |
| TCF | transverse cerebral fiss. |
| TF | temp. field F Cx |
| TPt | temp. paritemp. area Cx |
| vhc | vent. hipp. comm. |
| VLp | ventrolat. thal. nu., post. |
| VPI | vent. post. thal. nu., inf. |
| VPL | vent. post. thal. nu., lat. |
| VPM | vent. post. thal. nu., med. |
| VPS | vent. post. thal. nu., sup. |
| VTA | vent. tegment. area |

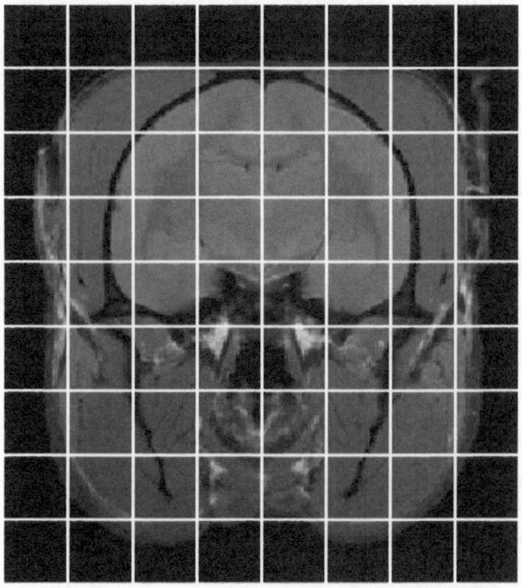

Interaural +04.00 mm

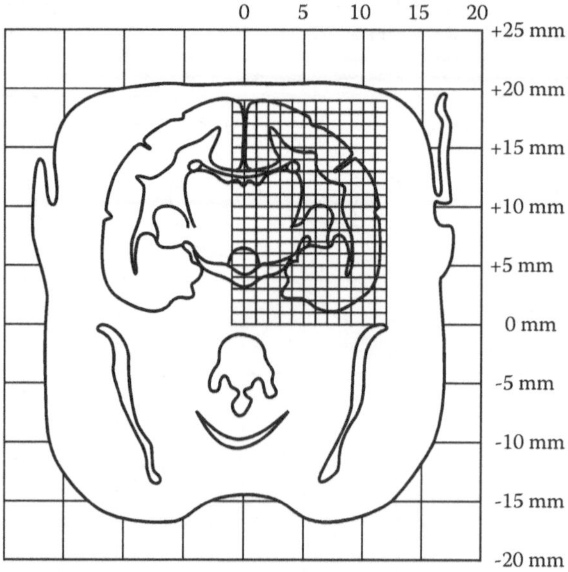

# Plate 31

Interaural +03.55 mm

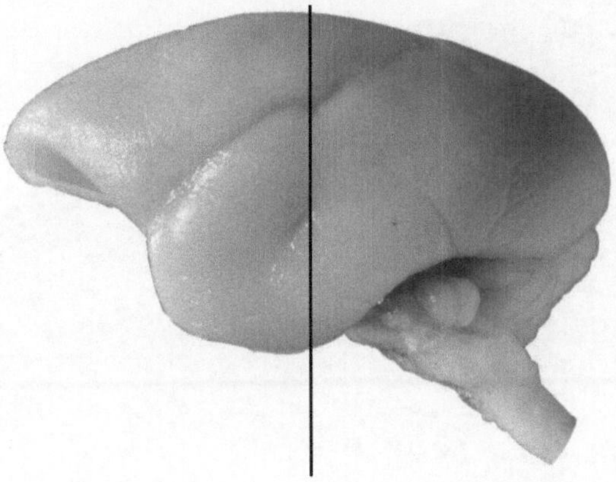

| | |
|---|---|
| 3n | oculomotor nerve |
| 3N | oculomotor nu. |
| 3V | third ventricle |
| 23 | area 23 Cx |
| 26 | area 26 Cx |
| 29a-c | area 29 Cx |
| 29d | area 29d Cx |
| 30 | area 30 Cx |
| alv | alveus of hippocampus |
| APT | ant. pretect. nu. |
| CA1 | cornu ammonis 1 |
| CA2 | cornu ammonis 2 |
| CA3 | cornu ammonis 3 |
| cc | corp. callosum |
| Cd | caudate nu. |
| CL | centrolat. thal. nu. |
| CLi | caudal lin. nu. |
| CMn | centromedian thal. nu. |
| cp | cerebral peduncle |
| ctg | central tegm. tract |
| Dk | nu. Darkschewitsch |
| DR | dors. raphe nu. |
| eml | extern. medull. lamina |
| EntC | entorhin. Cx, caud. |
| EW | Edinger Westphal nu. |
| f | fornix |
| fd | fimbriodent. sulcus |
| fi | fimbria |
| fr | fasciculus retroflexus |
| GrDG | gran. cell layer DG |
| hi | hippocamp. fiss. |
| IG | induseum gris. |
| iml | intern. medull. lamina |
| InC | interst. nu. Cajal |
| InCSh | interst. nu. Cajal, shell |
| int | intermed. temp. sulc. |
| ip | intraparietal sulc. |
| IPC | interped. nu., caud. |
| it | inf. temporal sulcus |
| ITD1 | inf. temp. dors. Cx 1 |
| ITV | inf. temp. vent. Cx |
| lf | lateral fissure |
| lfp | longit. fibers pons |
| LGei | LG, external interlaminar |
| LGme | LG, magnocell. external |
| LGmi | LG, magnocell. internal |
| LGpe | LG, parvicell. external |
| LGpi | LG, parvicell. internal |
| LHbD | lat. habenular nu., dorsal |
| LHbV | lat. habenular nu., ventral |
| LMol | strat. lacun. mol. CA |
| LTnl | lat. terminal nu. |
| LV | lateral ventricle |
| MA3 | med. access. 3N |
| mcp | middle cerebell. ped. |
| MDC | mediodors. thal. nu., cent. |
| MDL | mediodors. thal. nu., lat. |
| MGAD | med. gen. nu., anterodors. |
| MGM | med. gen. nu., med. |
| MGPD | med. gen. nu., posterodors. |
| MGV | med. gen. nu., vent. |
| MHb | med. habenular nu. |
| ml | medial lemniscus |
| mlf | med. longit. fasc. |
| MnR | median raphe nu. |
| MoDG | molecular layer DG |
| mRt | mesen. retic. formn |
| MTc | mid. temp. cresc. Cx |
| MT | mid. temp. area Cx |
| Or | strat. oriens CA |
| p1Rt | pros. 1 retic. formn |
| PaS | parasubiculum |
| PBP | parabrach. pigm. nu. |
| PCom | paracomm. nu. |
| PE | par. field E Cx |
| Pf | parafascicular nu. |
| PG | par. field G Cx |
| PMnR | paramedian raphe nu. |
| Pn | pontine nu. |
| Po | post. thal. nu. |
| PP | peripedunc. nu. |
| PPD1 | post. par. Cx, dors. 1 |
| PPD2 | post. par. Cx, dors. 2 |
| PPV | post. par. Cx, vent. |
| ProS | prosubiculum |
| PrS | presubiculum |
| PulA | pulvinar nu., ant. |
| PulI | pulvinar nu., inf. |
| PVG | periventricular gray |
| Rad | strat. radiat. CA |
| RMC | red nu., magnocell. |
| RPC | red nu., parvicellular |
| RRF | retrorubral field |
| Rt | reticular nu. |
| S | subiculum |
| SG | supragenic. thal. nu. |
| SNCD | subst. nigra, comp., dors. |
| SNCV | subst. nigra, comp., vent. |
| SNR | subst. nigra, retic. |
| st | stria terminalis |
| sts | sup. temp. sulcus |
| TCF | transverse cerebral fiss. |
| TF | temp. field F Cx |
| vhc | vent. hipp. comm. |
| VPL | vent. post. thal. nu., lat. |
| VPM | vent. post. thal. nu., med. |
| VTA | vent. tegment. area |
| vtgd | vent. tegm. decn |

# Plate 31a

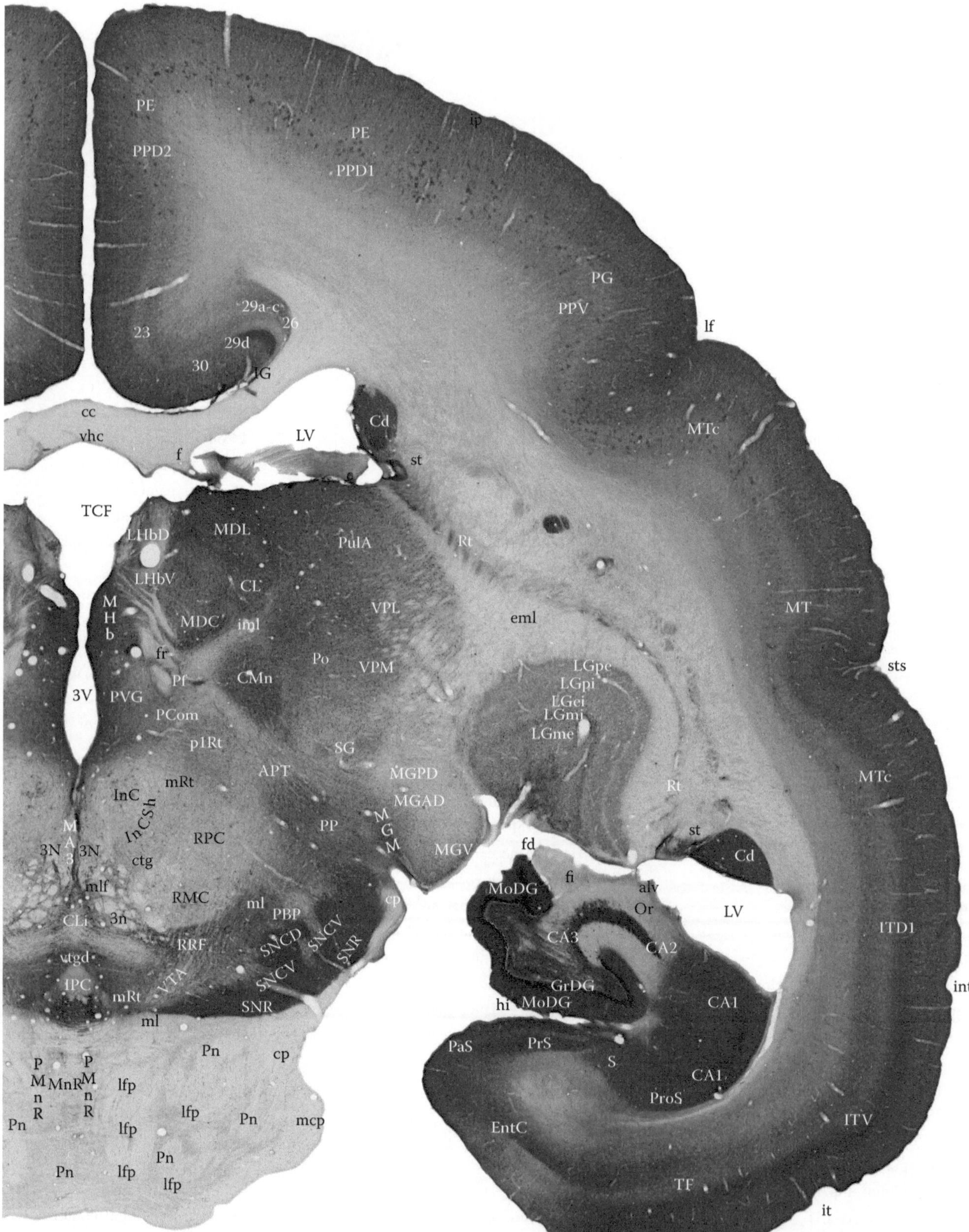


**Calbindin**
Interaural +03.85 mm

# Plate 31b

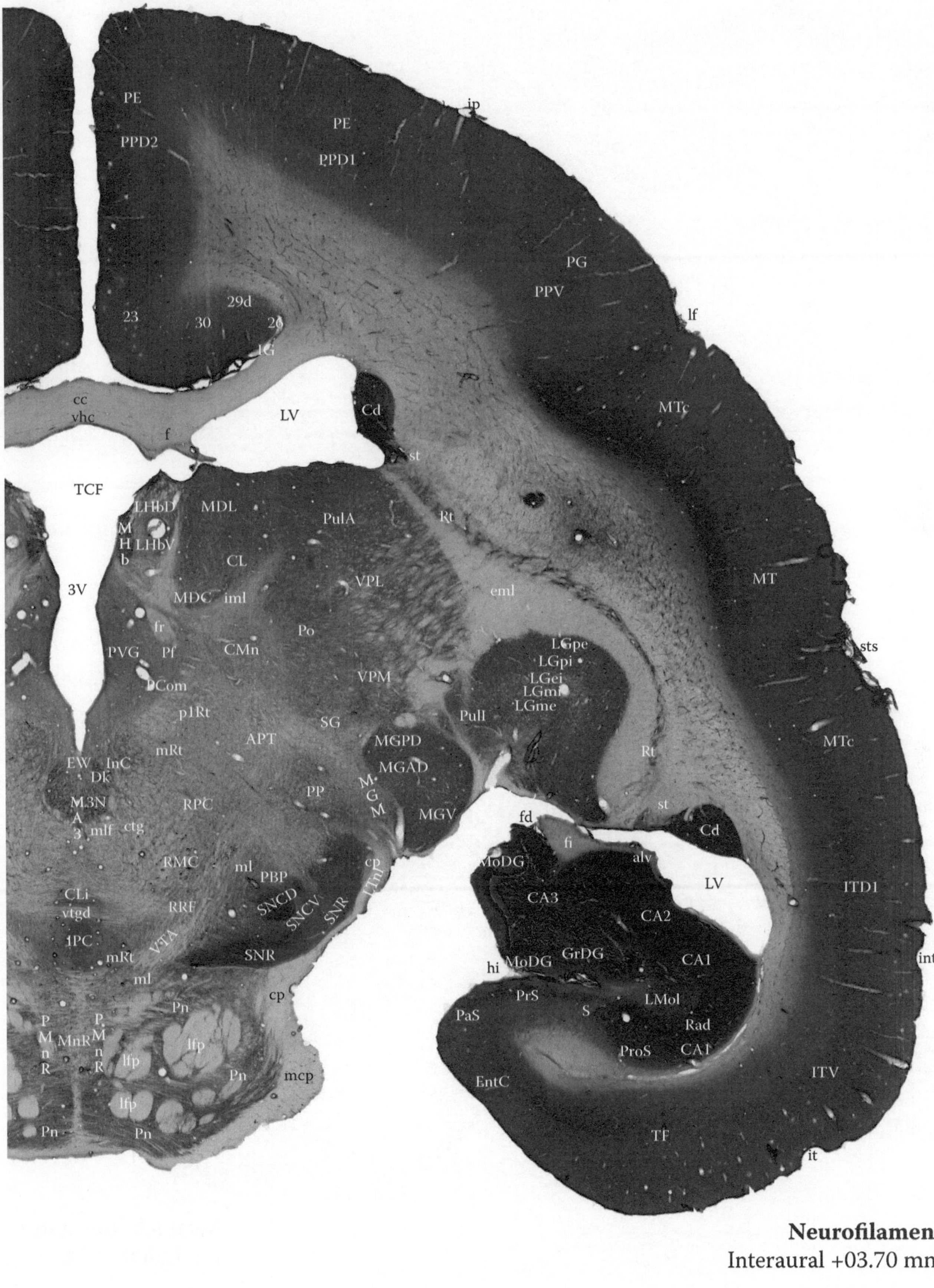


**Neurofilament**
Interaural +03.70 mm

# Plate 31c

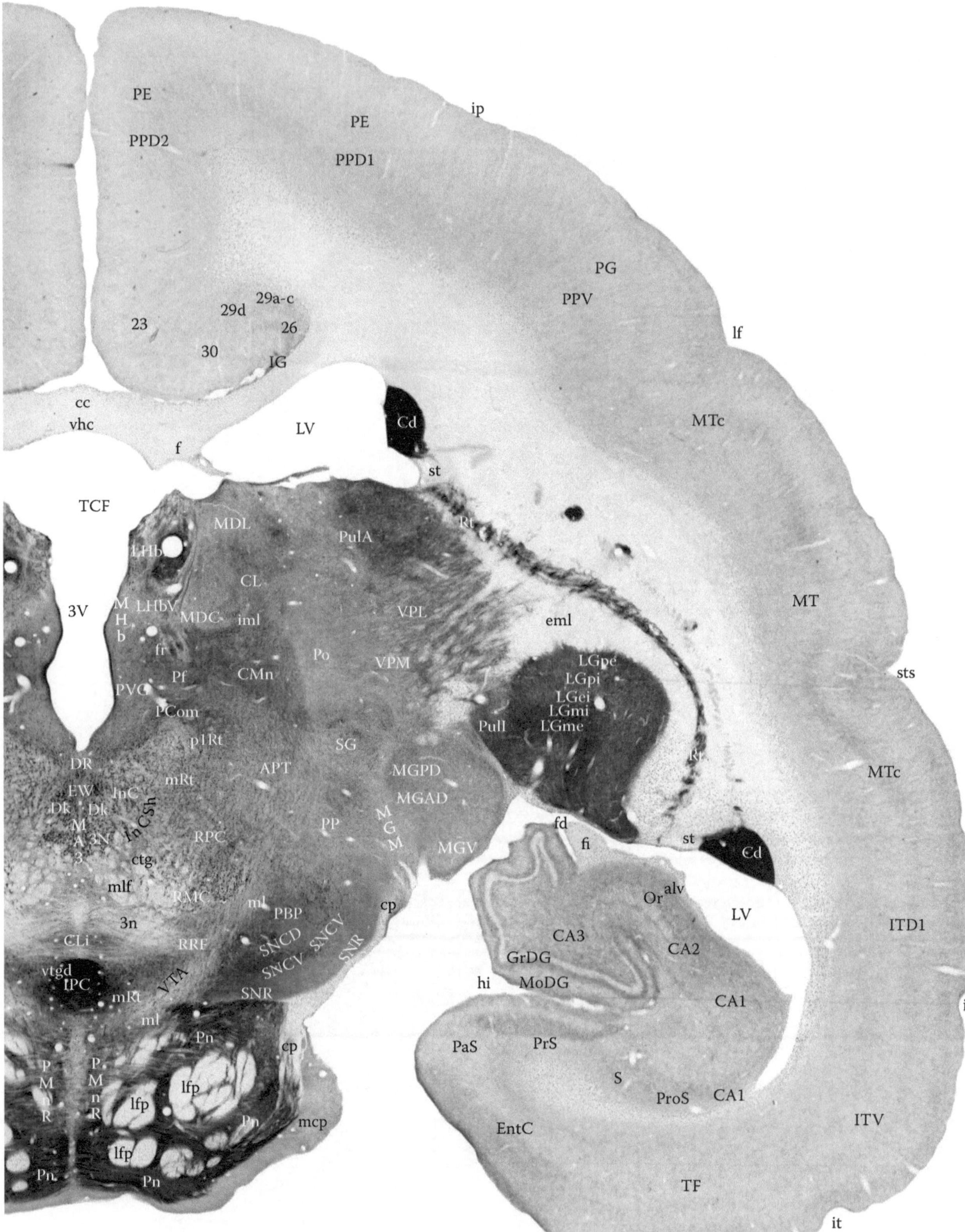


**Acetylcholinesterase**
Interaural +03.65 mm

# Plate 31d

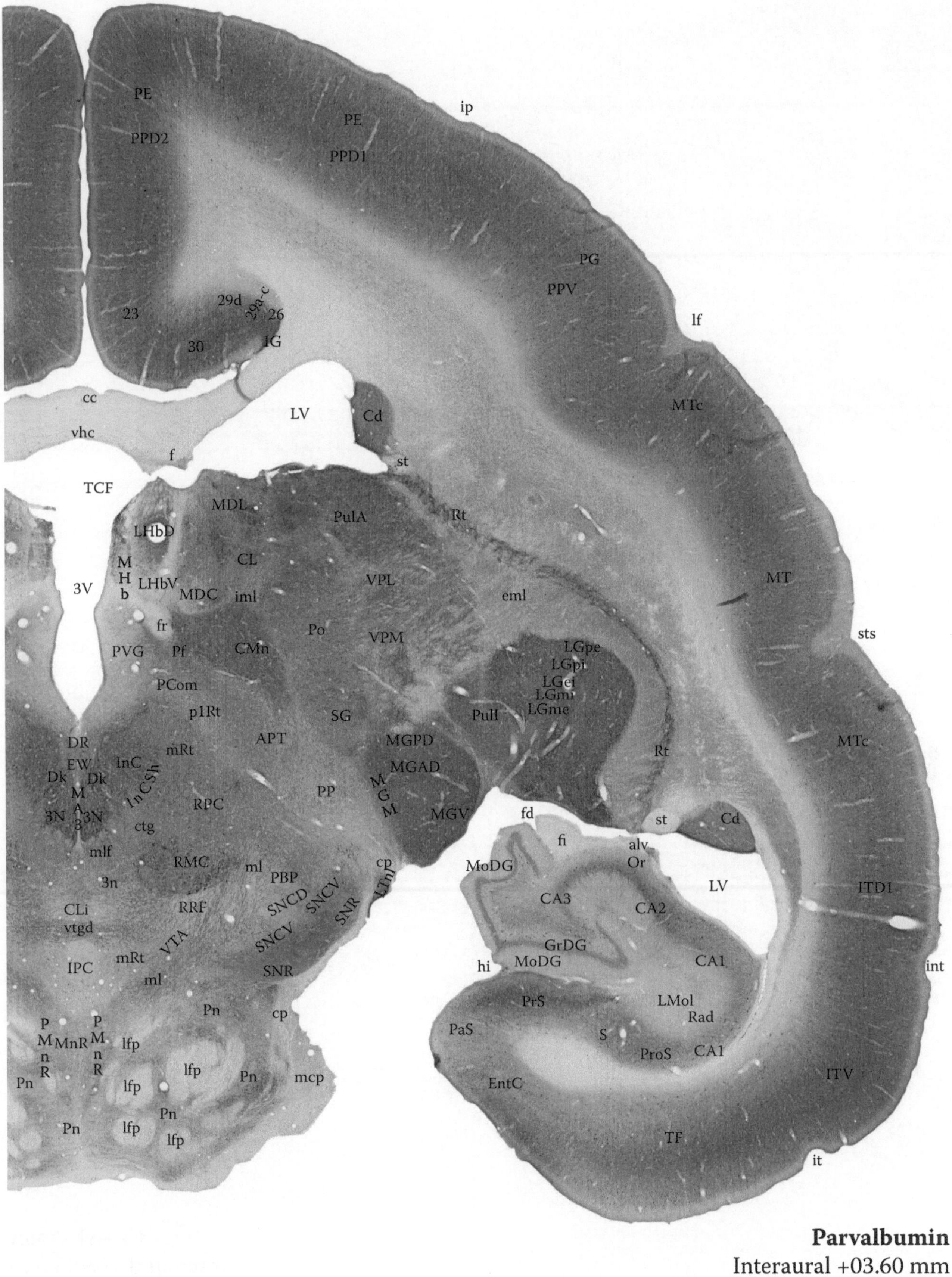


**Parvalbumin**
Interaural +03.60 mm

# Plate 31e

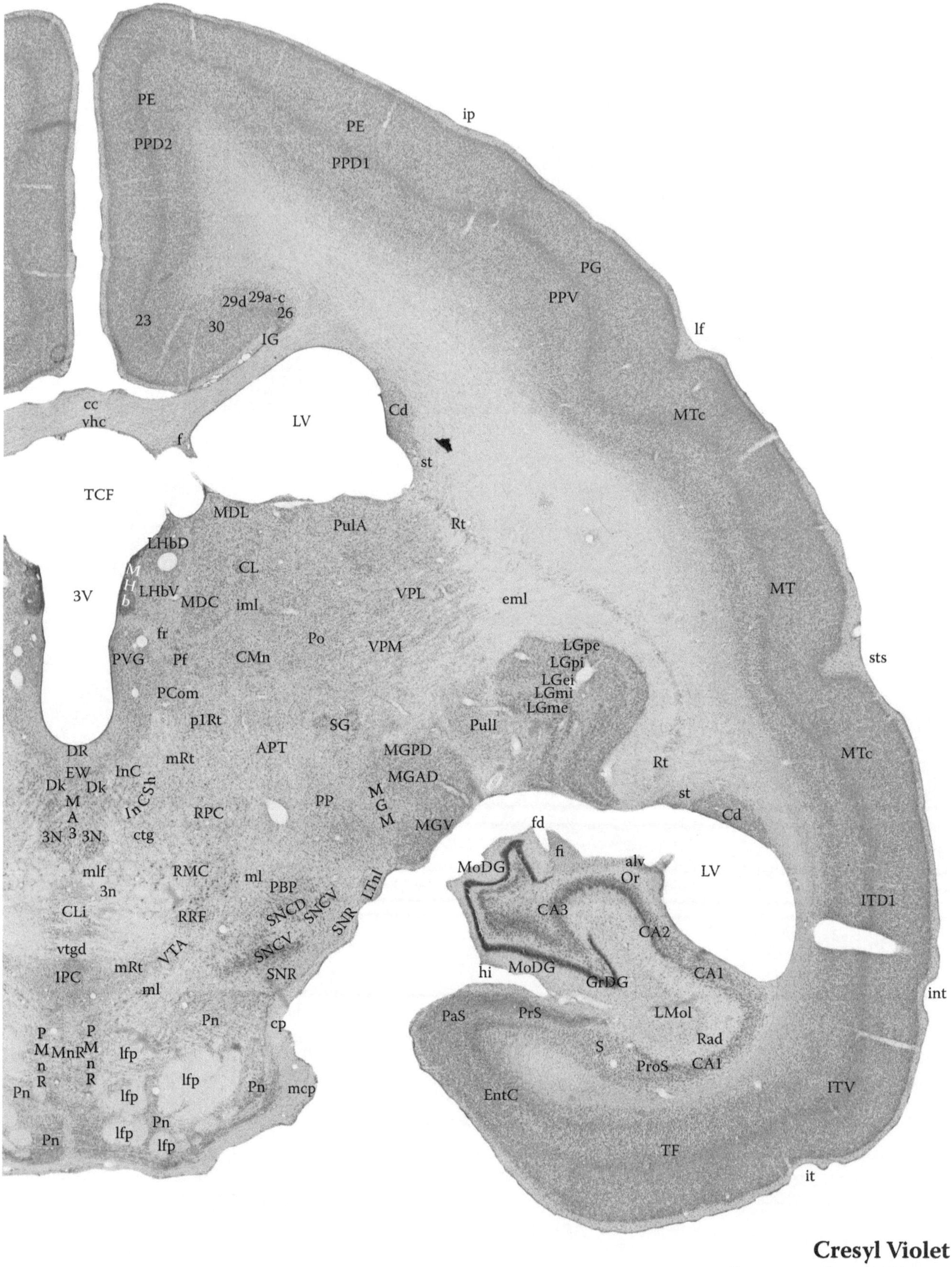


**Cresyl Violet**
Interaural +03.55 mm

# Plate 31

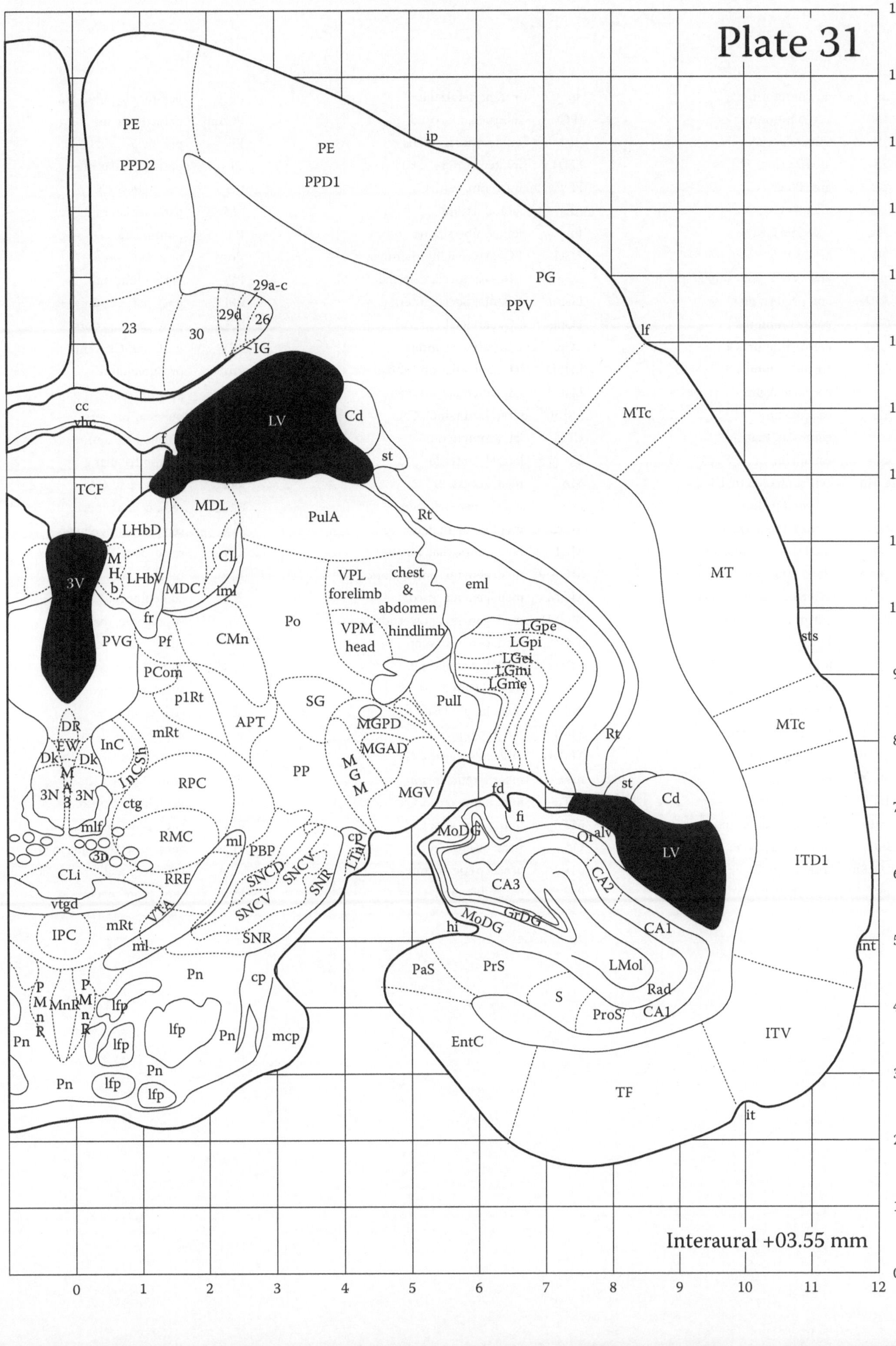
PE
PPD2
PE
ip
PPD1
PG
PPV
29a-c
29d
26
23
30
IG
lf
cc
vhc
LV
Cd
f
st
MTc
TCF
MDL
PulA
Rt
LHbD
CL
M
H
b
3V
LHbV
MDC
iml
VPL
forelimb
chest
&
abdomen
eml
MT
fr
Pf
CMn
Po
VPM
head
hindlimb
LGpe
LGpi
LGei
LGmi
LGme
PVG
PCom
sts
p1Rt
SG
PulI
DR
InC
mRt
APT
MGPD
MGAD
MTc
EW
Dk
Dk
M
A
3
InCSh
RPC
PP
M
G
M
Rt
3N
3N
ctg
MGV
fd
st
Cd
fi
mlf
RMC
ml
cp
MoDG
Or
alv
3n
LV
ITD1
PBP
CLi
RRF
SNCD
SNCV
SNR
LTh
CA3
CA2
vtgd
VTA
SNCV
GrDG
MoDG
hi
CA1
IPC
mRt
ml
SNR
int
PaS
PrS
LMol
Pn
cp
Rad
S
P
M
n
R
MnR
P
M
n
R
lfp
ProS
CA1
lfp
Pn
mcp
Pn
lfp
EntC
ITV
Pn
Pn
lfp
lfp
TF
it
Interaural +03.55 mm
0
1
2
3
4
5
6
7
8
9
10
11
12
19
18
17
16
15
14
13
12
11
10
9
8
7
6
5
4
3
2
1
0

| | |
|---|---|
| 3n | oculomotor nerve |
| 3N | oculomotor nu. |
| 3V | third ventricle |
| 23 | area 23 Cx |
| 26 | area 26 Cx |
| 29a-c | area 29 Cx |
| 29d | area 29d Cx |
| 30 | area 30 Cx |
| alv | alveus of hippocampus |
| APT | ant. pretect. nu. |
| CA1 | cornu ammonis 1 |
| CA2 | cornu ammonis 2 |
| CA3 | cornu ammonis 3 |
| cc | corp. callosum |
| Cd | caudate nu. |
| CL | centrolat. thal. nu. |
| CLi | caudal lin. nu. |
| CMn | centromedian thal. nu. |
| cp | cerebral peduncle |
| ctg | central tegm. tract |
| Dk | nu. Darkschewitsch |
| DR | dors. raphe nu. |
| eml | extern. medull. lamina |
| EntC | entorhin. Cx, caud. |
| EW | Edinger Westphal nu. |
| f | fornix |
| fd | fimbriodent. sulcus |
| fi | fimbria |
| fr | fasciculus retroflexus |
| GrDG | gran. cell layer DG |
| hi | hippocamp. fiss. |
| IG | induseum gris. |
| iml | intern. medull. lamina |
| InC | interst. nu. Cajal |
| InCSh | interst. nu. Cajal, shell |
| int | intermed. temp. sulc. |
| ip | intraparietal sulc. |
| IPC | interped. nu., caud. |
| it | inf. temporal sulcus |
| ITD1 | inf. temp. dors. Cx 1 |
| ITV | inf. temp. vent. Cx |
| lf | lateral fissure |
| lfp | longit. fibers pons |
| LGei | LG, external interlaminar |
| LGme | LG, magnocell. external |
| LGmi | LG, magnocell. internal |
| LGpe | LG, parvicell. external |
| LGpi | LG, parvicell. internal |
| LHbD | lat. habenular nu., dorsal |
| LHbV | lat. habenular nu., ventral |
| LMol | strat. lacun. mol. CA |
| LTnl | lat. terminal nu. |
| LV | lateral ventricle |
| MA3 | med. access. 3N |
| mcp | middle cerebell. ped. |
| MDC | mediodors. thal. nu., cent. |
| MDL | mediodors. thal. nu., lat. |
| MGAD | med. gen. nu., anterodors. |
| MGM | med. gen. nu., med. |
| MGPD | med. gen. nu., posterodors. |
| MGV | med. gen. nu., vent. |
| MHb | med. habenular nu. |
| ml | medial lemniscus |
| mlf | med. longit. fasc. |
| MnR | median raphe nu. |
| MoDG | molecular layer DG |
| mRt | mesen. retic. formn |
| MTc | mid. temp. cresc. Cx |
| MT | mid. temp. area Cx |
| Or | strat. oriens CA |
| p1Rt | pros. 1 retic. formn |
| PaS | parasubiculum |
| PBP | parabrach. pigm. nu. |
| PCom | paracomm. nu. |
| PE | par. field E Cx |
| Pf | parafascicular nu. |
| PG | par. field G Cx |
| PMnR | paramedian raphe nu. |
| Pn | pontine nu. |
| Po | post. thal. nu. |
| PP | peripedunc. nu. |
| PPD1 | post. par. Cx, dors. 1 |
| PPD2 | post. par. Cx, dors. 2 |
| PPV | post. par. Cx, vent. |
| ProS | prosubiculum |
| PrS | presubiculum |
| PulA | pulvinar nu., ant. |
| PulI | pulvinar nu., inf. |
| PVG | periventricular gray |
| Rad | strat. radiat. CA |
| RMC | red nu., magnocell. |
| RPC | red nu., parvicellular |
| RRF | retrorubral field |
| Rt | reticular nu. |
| S | subiculum |
| SG | supragenic. thal. nu. |
| SNCD | subst. nigra, comp., dors. |
| SNCV | subst. nigra, comp., vent. |
| SNR | subst. nigra, retic. |
| st | stria terminalis |
| sts | sup. temp. sulcus |
| TCF | transverse cerebral fiss. |
| TF | temp. field F Cx |
| vhc | vent. hipp. comm. |
| VPL | vent. post. thal. nu., lat. |
| VPM | vent. post. thal. nu., med. |
| VTA | vent. tegment. area |
| vtgd | vent. tegm. decn |

# Plate 32

Interaural +03.05 mm

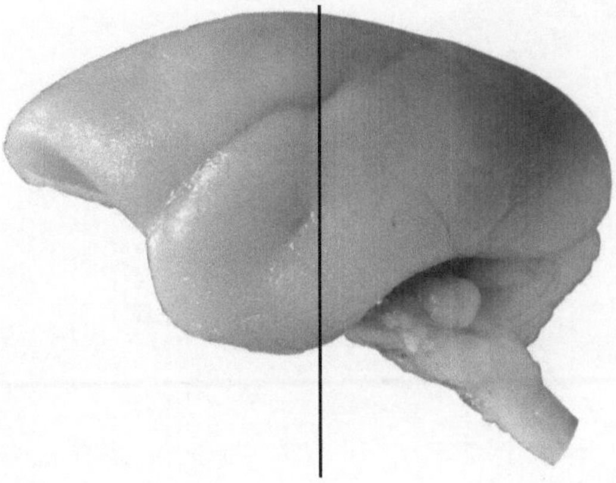

| | |
|---|---|
| 3N | oculomotor nu. |
| 3V | third ventricle |
| 23 | area 23 Cx |
| 26 | area 26 Cx |
| 29a-c | area 29 Cx |
| 29d | area 29d Cx |
| 30 | area 30 Cx |
| APT | ant. pretect. nu. |
| bic | brach. inf. coll. |
| bsc | brach. sup. coll. |
| CA1 | cornu ammonis 1 |
| CA2 | cornu ammonis 2 |
| CA3 | cornu ammonis 3 |
| cc | corp. callosum |
| Cd | caudate nu. |
| cg | cingulum |
| CG | cingulate gyrus Cx |
| CLi | caudal lin. nu. |
| cm | callosomarginal sulc. |
| cp | cerebral peduncle |
| ctg | central tegm. tract |
| Dk | nu. Darkschewitsch |
| DR | dors. raphe nu. |
| dtgd | dors. tegm. decn |
| eml | extern. medull. lamina |
| EW | Edinger Westphal nu. |
| f | fornix |
| fd | fimbriodent. sulcus |
| fi | fimbria |
| GrDG | gran. cell layer DG |
| hi | hippocamp. fiss. |
| IG | induseum gris. |
| InC | interst. nu. Cajal |
| InCSh | interst. nu. Cajal, shell |
| ip | intraparietal sulc. |
| IPC | interped. nu., caud. |
| it | inf. temporal sulcus |
| ITD1 | inf. temp. dors. Cx 1 |
| ITD2 | inf. temp. dors. Cx 2 |
| lf | lateral fissure |
| lfp | longit. fibers pons |
| LGei | LG, external interlaminar |
| LGmi | LG, magnocell. internal |
| LGpe | LG, parvicell. external |
| LGpi | LG, parvicell. internal |
| LHb | lat. habenular nu. |
| Lim | limitans nu. |
| LMol | strat. lacun. mol. CA |
| LPAG | lat. periaqued. gr. |
| LTnl | lat. terminal nu. |
| mcp | middle cerebell. ped. |
| MD | mediodors. thal. nu. |
| MGAD | med. gen. nu., anterodors. |
| MGM | med. gen. nu., med. |
| MGPD | med. gen. nu., posterodors. |
| MGV | med. gen. nu., vent. |
| MHb | med. habenular nu. |
| ml | medial lemniscus |
| mlf | med. longit. fasc. |
| MnR | median raphe nu. |
| MoDG | molecular layer DG |
| mRt | mesen. retic. formn |
| MT | mid. temp. area Cx |
| MTc | mid. temp. cresc. Cx |
| mte | med. temp. sulc. |
| OPT | olivary pretect. nu. |
| Or | strat. oriens CA |
| p1Rt | pros. 1 retic. formn |
| PaR | pararubral area |
| PaS | parasubiculum |
| PBP | parabrach. pigm. nu. |
| pc | post. commissure |
| PCom | paracomm. nu. |
| PE | par. field E Cx |
| PG | par. field G Cx |
| PMnR | paramedian raphe nu. |
| Pn | pontine nu. |
| Po | post. thal. nu. |
| PP | peripedunc. nu. |
| PPD1 | post. par. Cx, dors. 1 |
| PPD2 | post. par. Cx, dors. 2 |
| PPM | post. par. Cx, medial |
| PPV | post. par. Cx, vent. |
| PrC | precomm. nu. |
| ProS | prosubiculum |
| PrS | presubiculum |
| PulI | pulvinar nu., inf. |
| PulL | pulvinar nu., lat. |
| PulM | pulvinar nu., med. |
| Rad | strat. radiat. CA |
| RMC | red nu., magnocell. |
| RRF | retrorubral field |
| Rt | reticular nu. |
| RtTg | reticulotegm. nu. |
| S | subiculum |
| SCO | subcommiss. organ |
| scpd | sup. cb. ped. decn |
| SG | supragenic. thal. nu. |
| SNCV | subst. nigra, comp., vent. |
| SNR | subst. nigra, retic. |
| st | stria terminalis |
| sts | sup. temp. sulcus |
| TCF | transverse cerebral fiss. |
| TF | temp. field F Cx |
| tfp | transv. fibers pons |
| TH | temp. field H Cx |
| vhc | vent. hipp. comm. |

# Plate 32a



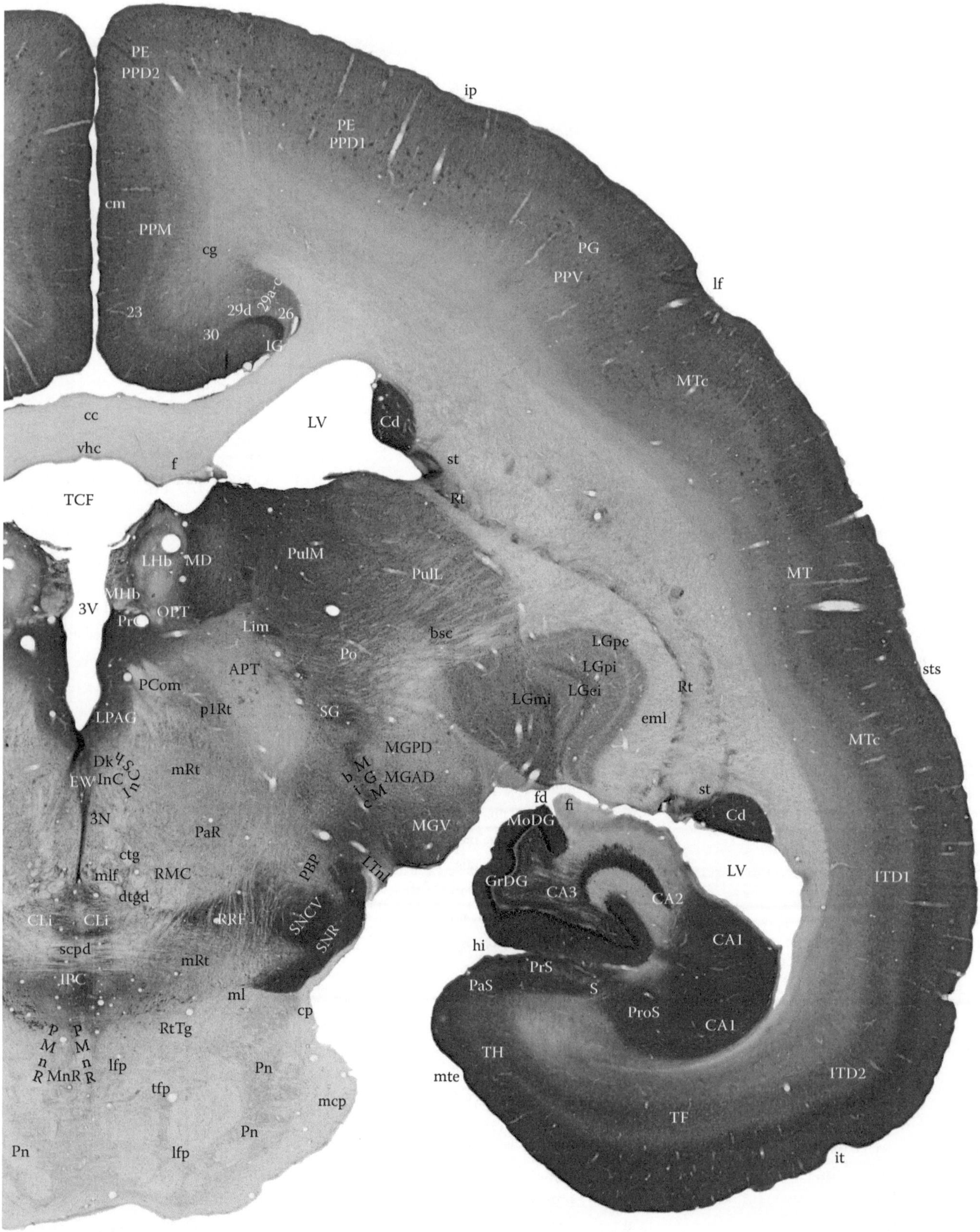


**Calbindin**
Interaural +03.35 mm

# Plate 32b

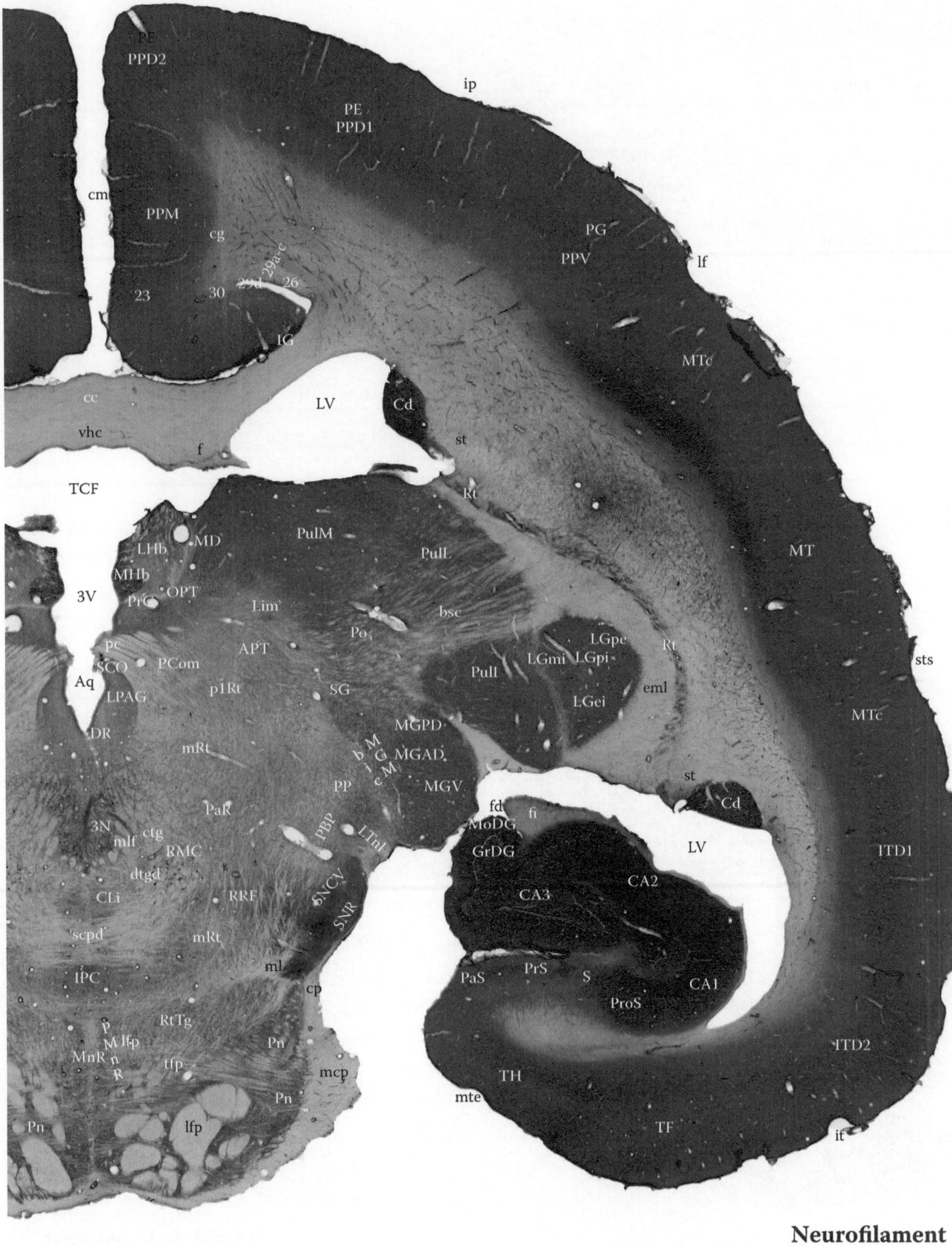


**Neurofilament**
Interaural +03.20 mm

# Plate 32c

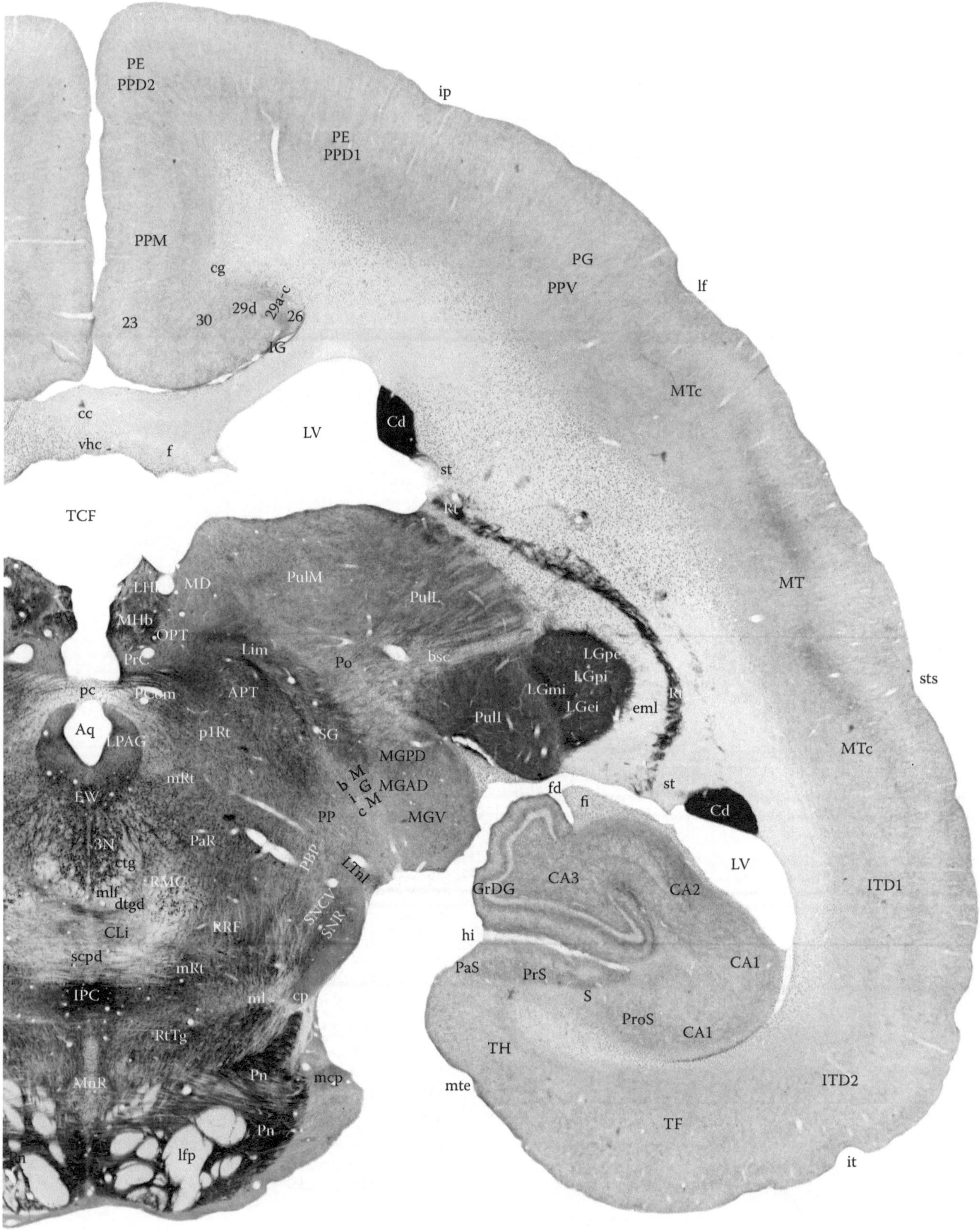


**Acetylcholinesterase**
Interaural +03.15 mm

# Plate 32d

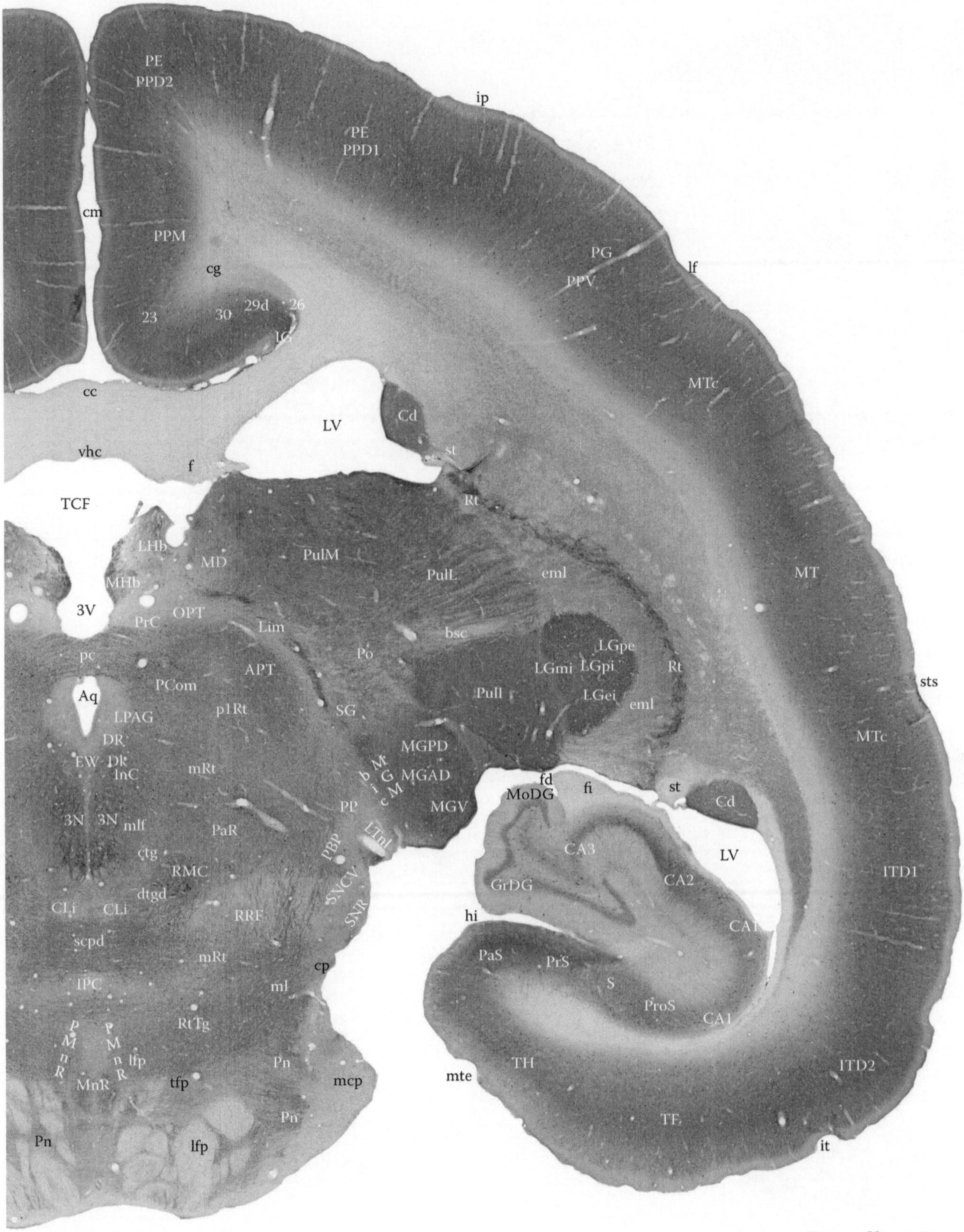


**Parvalbumin**

Interaural +03.10 mm

# Plate 32e

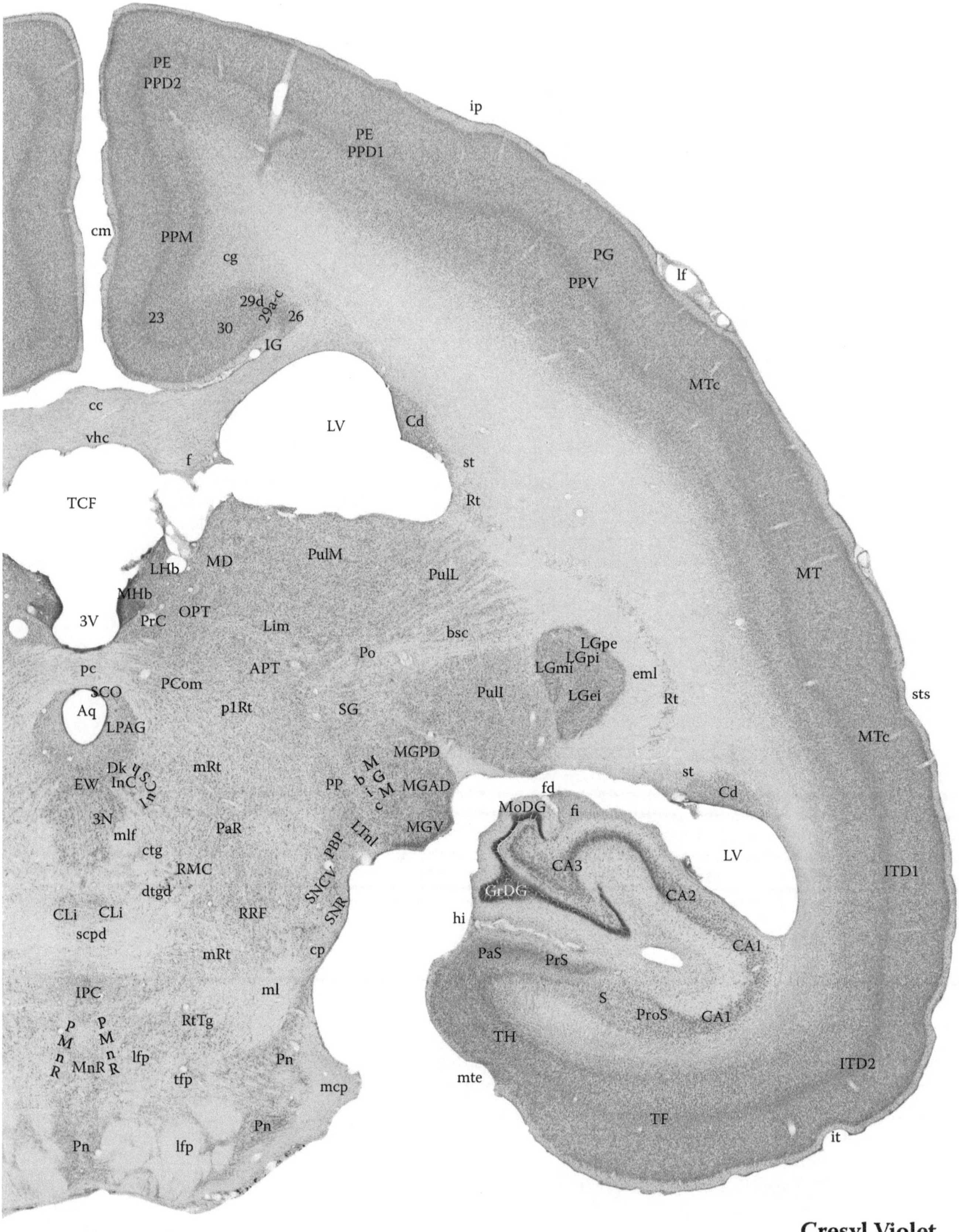


**Cresyl Violet**
Interaural +03.05 mm

# Plate 32

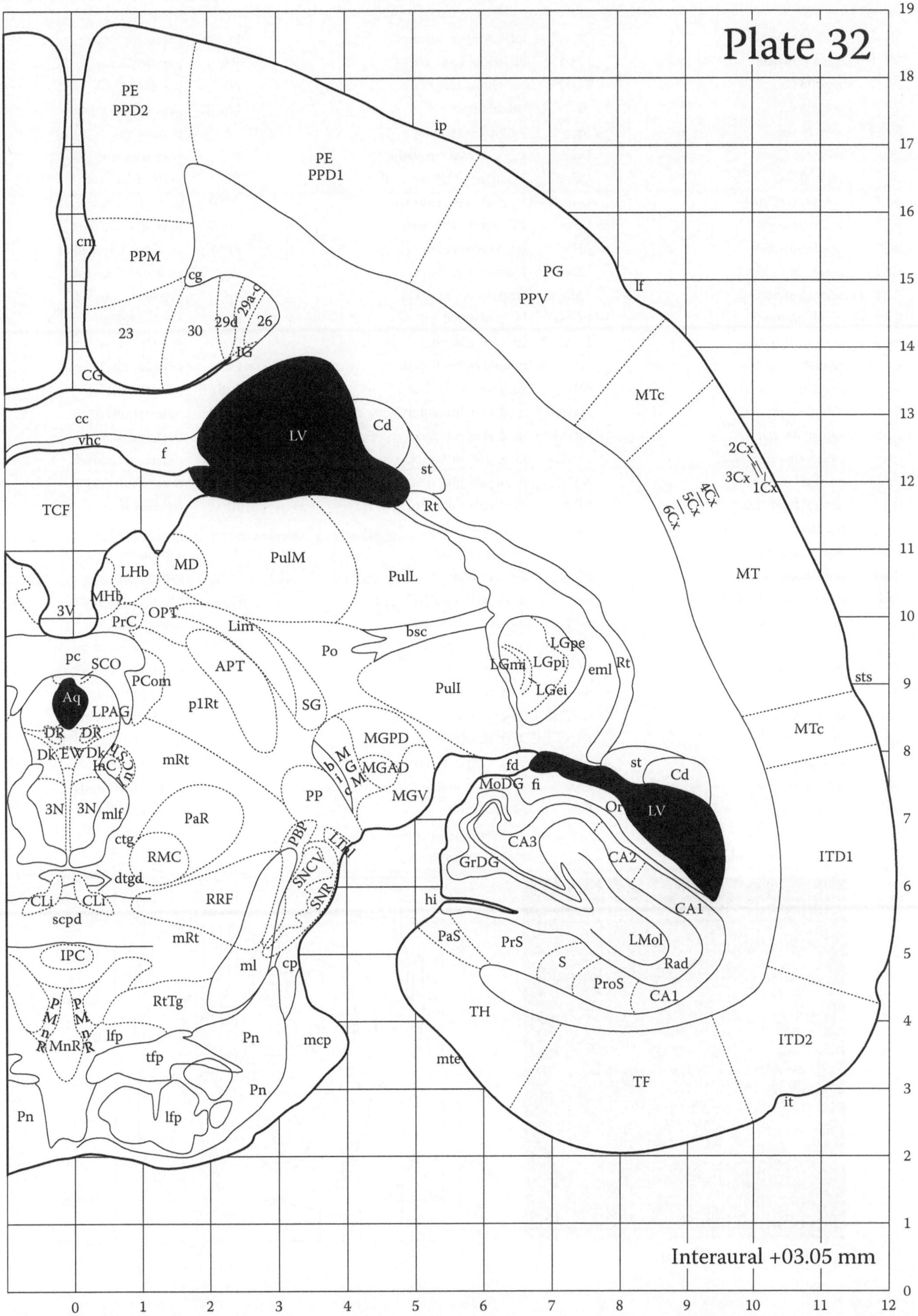
PE
PPD2
PE
PPD1
ip
cm
PPM
cg
29a-c
29d
26
23
30
IG
CG
PG
PPV
lf
MTc
cc
vhc
f
LV
Cd
st
Rt
TCF
2Cx
3Cx
1Cx
4Cx
5Cx
6Cx
LHb
MD
PulM
PulL
MT
MHb
3V
PrC
OPT
Lim
bsc
APT
Po
LGpe
LGmi
LGpi
LGei
eml
Rt
sts
pc
SCO
PCom
Aq
LPAG
p1Rt
SG
PulI
MTc
DR
DR
Dk
EW
Dk
InC
ISC
InCS
mRt
MGPD
b
i
c
M
G
M
MGAD
fd
st
Cd
fi
ModG
PP
MGV
3N
3N
mlf
PaR
Or
LV
ctg
RMC
PBP
LTRI
SNCV
CA3
CA2
ITD1
GrDG
dtgd
CLi
CLi
RRF
SNR
hi
CA1
scpd
PaS
PrS
LMol
mRt
IPC
ml
cp
S
Rad
ProS
CA1
PMnR
PMnR
MnR
RtTg
lfp
Pn
TH
mcp
ITD2
tfp
mte
TF
Pn
it
Pn
lfp
Interaural +03.05 mm
0
1
2
3
4
5
6
7
8
9
10
11
12
19
18
17
16
15
14
13
12
11
10
9
8
7
6
5
4
3
2
1
0

| | |
|---|---|
| 3N | oculomotor nu. |
| 3V | third ventricle |
| 23 | area 23 Cx |
| 26 | area 26 Cx |
| 29a-c | area 29 Cx |
| 29d | area 29d Cx |
| 30 | area 30 Cx |
| APT | ant. pretect. nu. |
| bic | brach. inf. coll. |
| bsc | brach. sup. coll. |
| CA1 | cornu ammonis 1 |
| CA2 | cornu ammonis 2 |
| CA3 | cornu ammonis 3 |
| cc | corp. callosum |
| Cd | caudate nu. |
| cg | cingulum |
| CG | cingulate gyrus Cx |
| CLi | caudal lin. nu. |
| cm | callosomarginal sulc. |
| cp | cerebral peduncle |
| ctg | central tegm. tract |
| Dk | nu. Darkschewitsch |
| DR | dors. raphe nu. |
| dtgd | dors. tegm. decn |
| eml | extern. medull. lamina |
| EW | Edinger Westphal nu. |
| f | fornix |
| fd | fimbriodent. sulcus |
| fi | fimbria |
| GrDG | gran. cell layer DG |
| hi | hippocamp. fiss. |
| IG | induseum gris. |
| InC | interst. nu. Cajal |
| InCSh | interst. nu. Cajal, shell |
| ip | intraparietal sulc. |
| IPC | interped. nu., caud. |
| it | inf. temporal sulcus |
| ITD1 | inf. temp. dors. Cx 1 |
| ITD2 | inf. temp. dors. Cx 2 |
| lf | lateral fissure |
| lfp | longit. fibers pons |
| LGei | LG, external interlaminar |
| LGmi | LG, magnocell. internal |
| LGpe | LG, parvicell. external |
| LGpi | LG, parvicell. internal |
| LHb | lat. habenular nu. |
| Lim | limitans nu. |
| LMol | strat. lacun. mol. CA |
| LPAG | lat. periaqued. gr. |
| LTnl | lat. terminal nu. |
| mcp | middle cerebell. ped. |
| MD | mediodors. thal. nu. |
| MGAD | med. gen. nu., anterodors. |
| MGM | med. gen. nu., med. |
| MGPD | med. gen. nu., posterodors. |
| MGV | med. gen. nu., vent. |
| MHb | med. habenular nu. |
| ml | medial lemniscus |
| mlf | med. longit. fasc. |
| MnR | median raphe nu. |
| MoDG | molecular layer DG |
| mRt | mesen. retic. formn |
| MT | mid. temp. area Cx |
| MTc | mid. temp. cresc. Cx |
| mte | med. temp. sulc. |
| OPT | olivary pretect. nu. |
| Or | strat. oriens CA |
| p1Rt | pros. 1 retic. formn |
| PaR | pararubral area |
| PaS | parasubiculum |
| PBP | parabrach. pigm. nu. |
| pc | post. commissure |
| PCom | paracomm. nu. |
| PE | par. field E Cx |
| PG | par. field G Cx |
| PMnR | paramedian raphe nu. |
| Pn | pontine nu. |
| Po | post. thal. nu. |
| PP | peripedunc. nu. |
| PPD1 | post. par. Cx, dors. 1 |
| PPD2 | post. par. Cx, dors. 2 |
| PPM | post. par. Cx, medial |
| PPV | post. par. Cx, vent. |
| PrC | precomm. nu. |
| ProS | prosubiculum |
| PrS | presubiculum |
| PulI | pulvinar nu., inf. |
| PulL | pulvinar nu., lat. |
| PulM | pulvinar nu., med. |
| Rad | strat. radiat. CA |
| RMC | red nu., magnocell. |
| RRF | retrorubral field |
| Rt | reticular nu. |
| RtTg | reticulotegm. nu. |
| S | subiculum |
| SCO | subcommiss. organ |
| scpd | sup. cb. ped. decn |
| SG | supragenic. thal. nu. |
| SNCV | subst. nigra, comp., vent. |
| SNR | subst. nigra, retic. |
| st | stria terminalis |
| sts | sup. temp. sulcus |
| TCF | transverse cerebral fiss. |
| TF | temp. field F Cx |
| tfp | transv. fibers pons |
| TH | temp. field H Cx |
| vhc | vent. hipp. comm. |

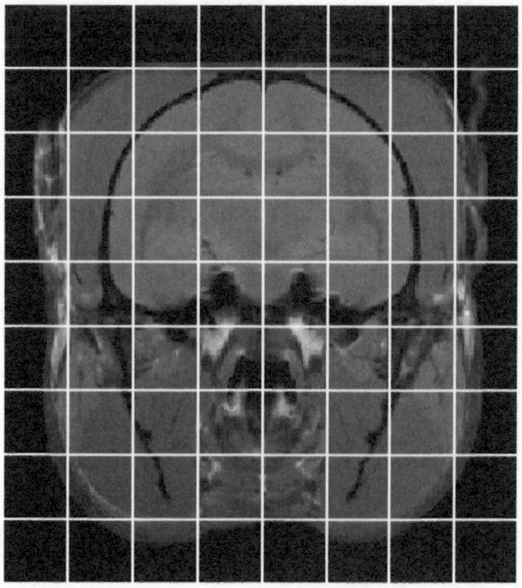

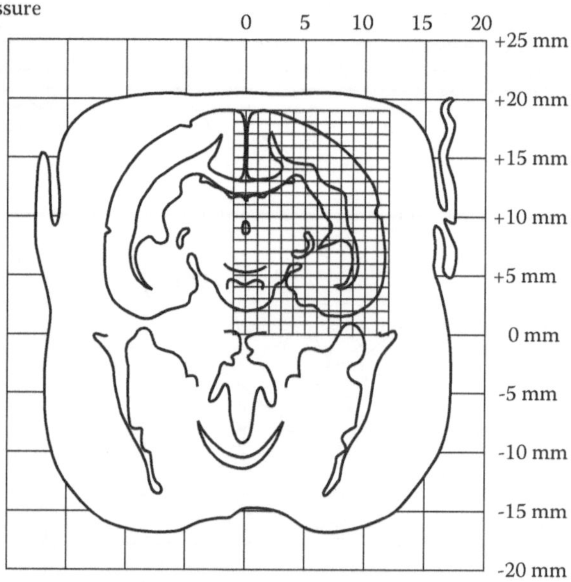


Interaural +03.05 mm

# Plate 33

Interaural +02.55 mm

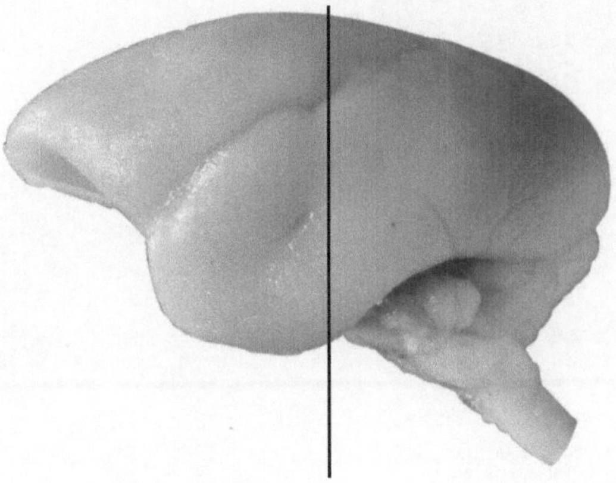

| | |
|---|---|
| 3N | oculomotor nu. |
| 3PC | oculomotor nu., parv. |
| 23a | area 23a Cx |
| 23b | area 23b Cx |
| 26 | area 26 Cx |
| 29a-c | area 29 Cx |
| 29d | area 29d Cx |
| 30 | area 30 Cx |
| alv | alveus of hippocampus |
| Aq | cerebral aqueduct |
| APT | ant. pretect. nu. |
| bic | brach. inf. coll. |
| BIC | brach. inf. coll. nu. |
| bsc | brach. sup. coll. |
| CA1 | cornu ammonis 1 |
| CA2 | cornu ammonis 2 |
| CA3 | cornu ammonis 3 |
| cc | corp. callosum |
| Cd | caudate nu. |
| cg | cingulum |
| CG | cingulate gyrus Cx |
| cm | callosomarginal sulc. |
| csc | comm. sup. coll. |
| ctg | central tegm. tract |
| DLPAG | dorsolat. periaq. gr. |
| DR | dors. raphe nu. |
| DRI | dors. raphe nu., interf. |
| eml | extern. medull. lamina |
| EW | Edinger Westphal nu. |
| f | fornix |
| fd | fimbriodent. sulcus |
| fi | fimbria |
| GrDG | gran. cell layer DG |
| hbc | habenular comm. |
| hi | hippocamp. fiss. |
| IG | induseum gris. |
| ip | intraparietal sulc. |
| it | inf. temporal sulcus |
| ITD1 | inf. temp. dors. Cx 1 |
| ITD2 | inf. temp. dors. Cx 2 |
| lf | lateral fissure |
| lfp | longit. fibers pons |
| Lim | limitans nu. |
| LMol | strat. lacun. mol. CA |
| LPAG | lat. periaqued. gr. |
| LV | lateral ventricle |
| mcp | middle cerebell. ped. |
| MCPC | magnocell. nu. pc |
| MGM | med. gen. nu., med. |
| MGPD | med. gen. nu., posterodors. |
| MGV | med. gen. nu., vent. |
| MHb | med. habenular nu. |
| ml | medial lemniscus |
| mlf | med. longit. fasc. |
| MnR | median raphe nu. |
| MoDG | molecular layer DG |
| MPT | med. pretect. nu. |
| mRt | mesen. retic. formn |
| MT | mid. temp. area Cx |
| MTc | mid. temp. cresc. Cx |
| mte | med. temp. sulc. |
| OPT | olivary pretect. nu. |
| Or | strat. oriens CA |
| p1Rt | pros. 1 retic. formn |
| PaR | pararubral area |
| PaS | parasubiculum |
| PBG | parabigeminal nu. |
| PBP | parabrach. pigm. nu. |
| PCom | paracomm. nu. |
| PEc | par. field E Cx, caud. |
| PG | par. field G Cx |
| Pn | pontine nu. |
| PnO | pontine oral nu. |
| Po | post. thal. nu. |
| PPD1 | post. par. Cx, dors. 1 |
| PPD2 | post. par. Cx, dors. 2 |
| PPM | post. par. Cx, medial |
| PPT | post. pretectum |
| PPTg | peduncpont. tegm. nu. |
| PPV | post. par. Cx, vent. |
| ProS | prosubiculum |
| PrS | presubiculum |
| PulI | pulvinar nu., inf. |
| PulL | pulvinar nu., lat. |
| PulM | pulvinar nu., med. |
| Rad | strat. radiat. CA |
| Rbd | rhabdoid nu. |
| RRF | retrorubral field |
| Rt | reticular nu. |
| RtTg | reticulotegm. nu. |
| S | subiculum |
| s5 | sensory root 5n |
| scp | sup. cerebell. ped. |
| scpd | sup. cb. ped. decn |
| SG | supragenic. thal. nu. |
| st | stria terminalis |
| sts | sup. temp. sulcus |
| Su3 | supraoculomotor PAG |
| Su3C | supraoculomotor cap |
| TCF | transverse cerebral fiss. |
| TF | temp. field F Cx |
| tfp | transv. fibers pons |
| TH | temp. field H Cx |
| vhc | vent. hipp. comm. |

# Plate 33a

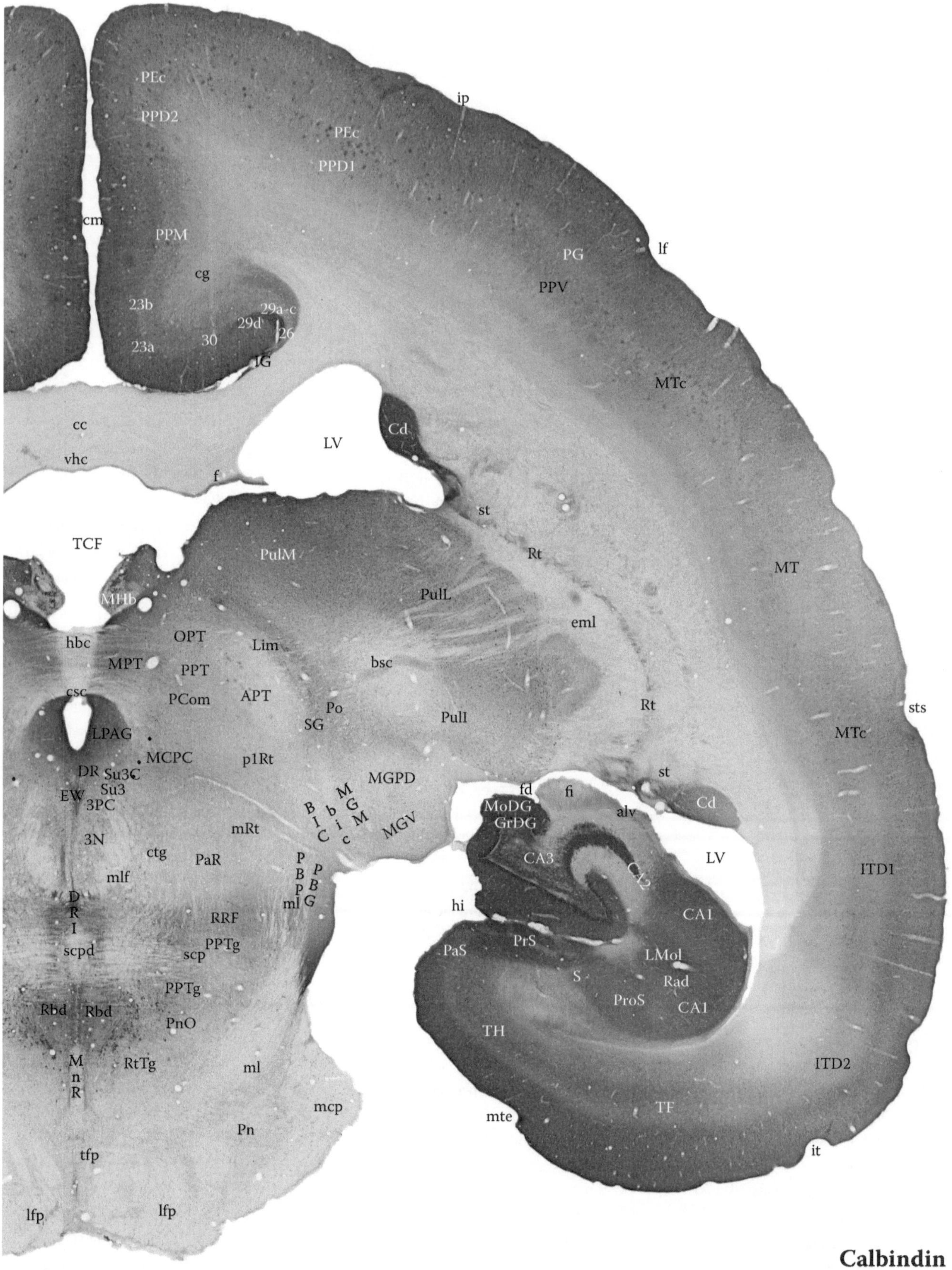


**Calbindin**
Interaural +02.85 mm

# Plate 33b

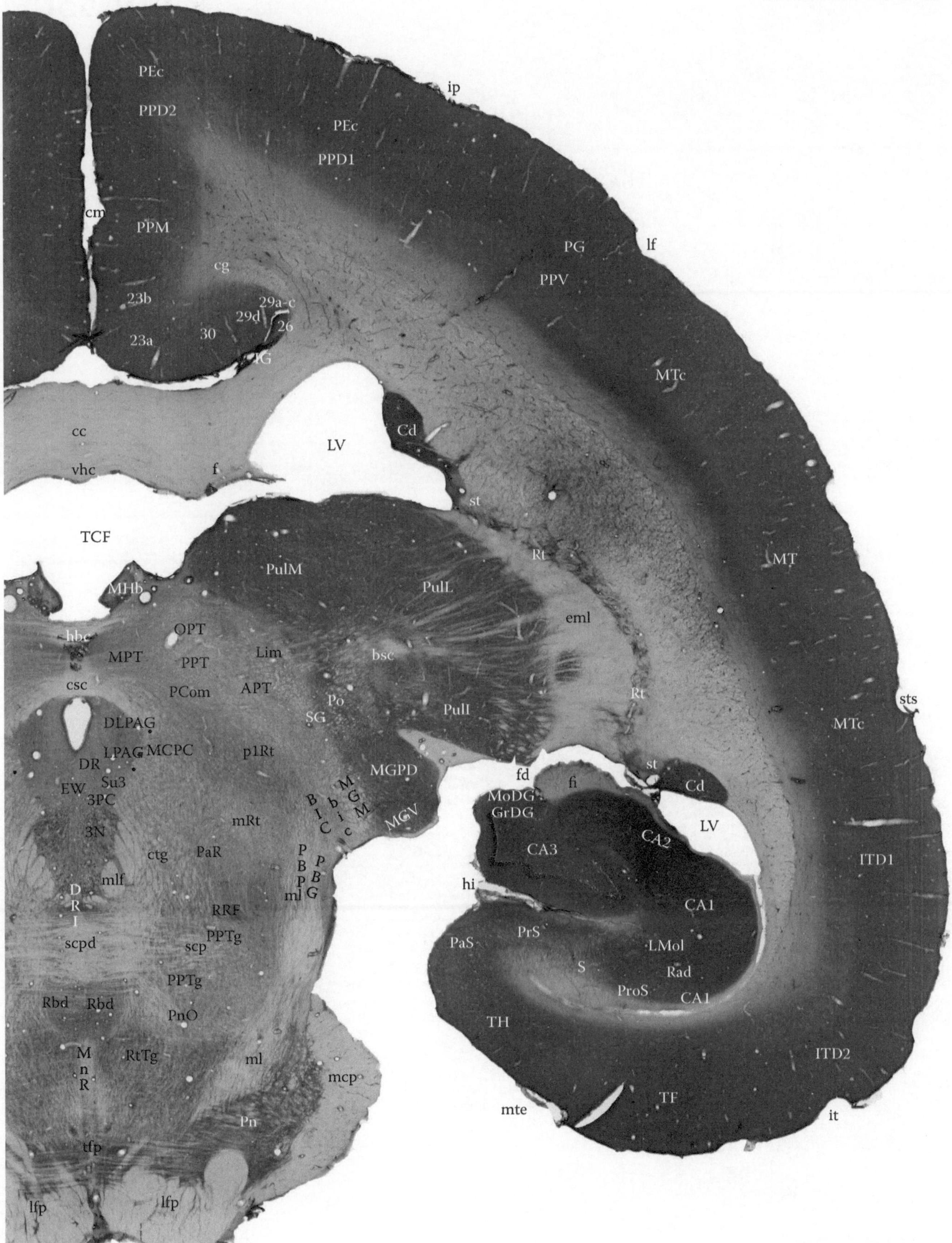


**Neurofilament**
Interaural +02.70 mm

# Plate 33c

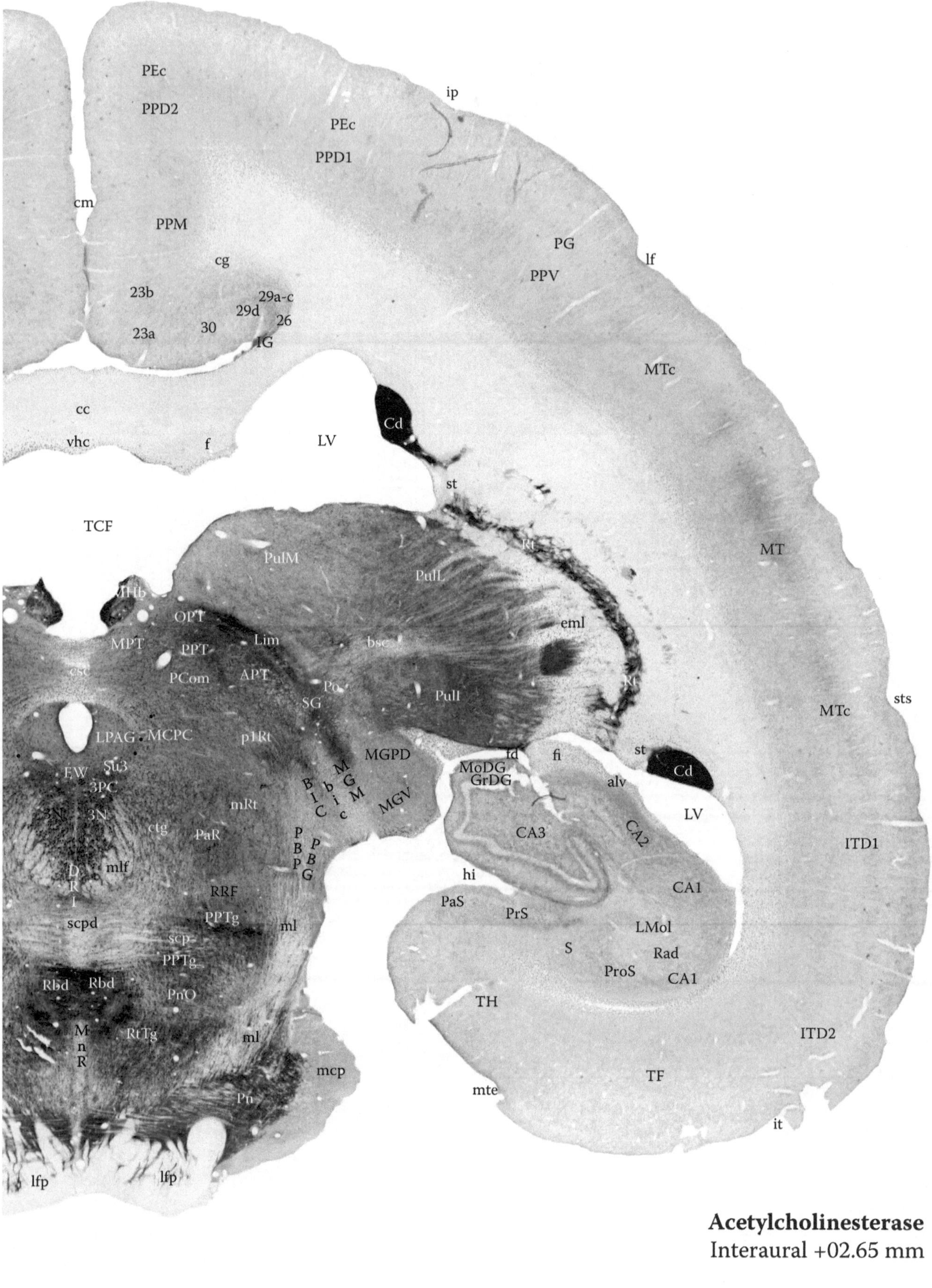


**Acetylcholinesterase**
Interaural +02.65 mm

# Plate 33d

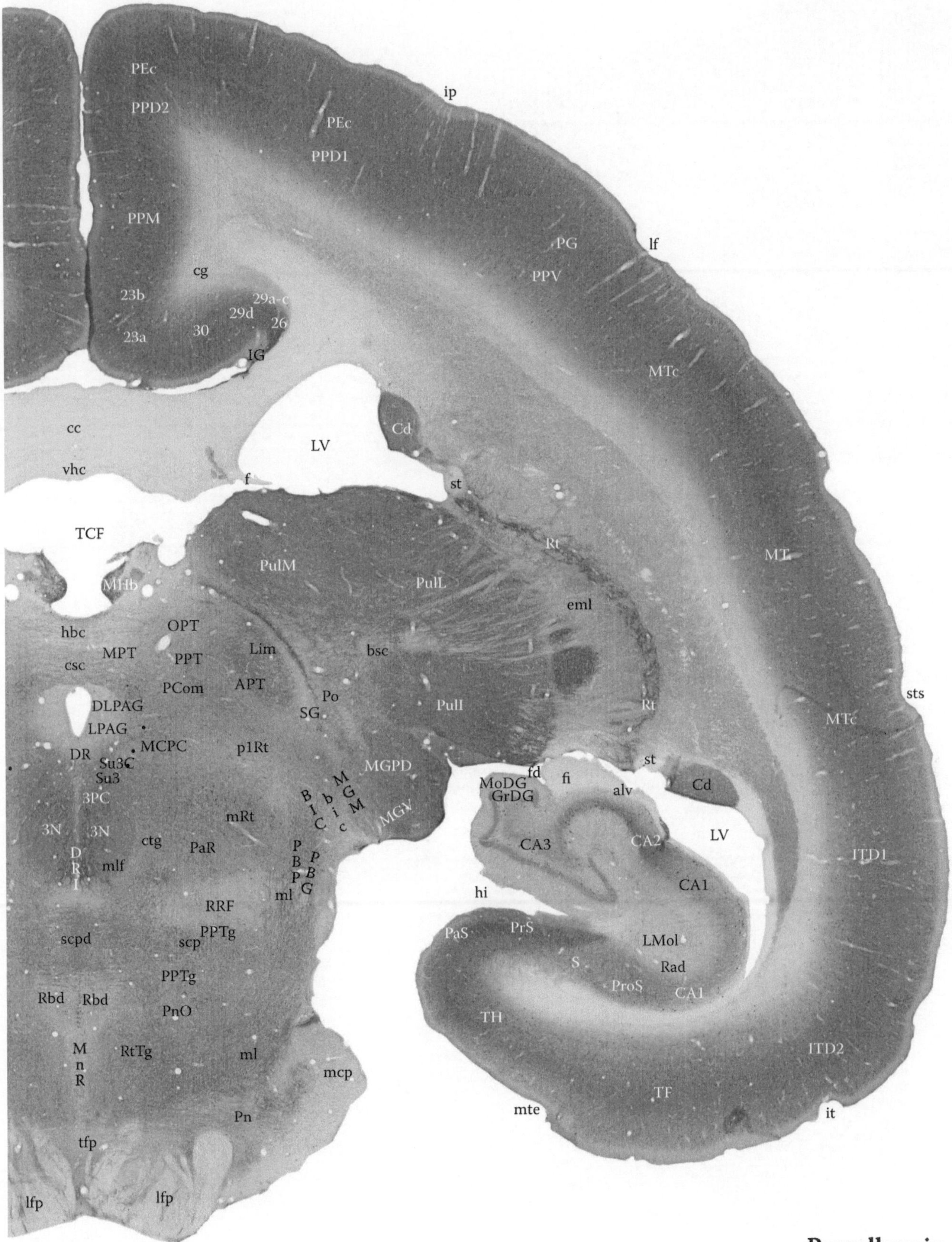


**Parvalbumin**
Interaural +02.60 mm

# Plate 33e

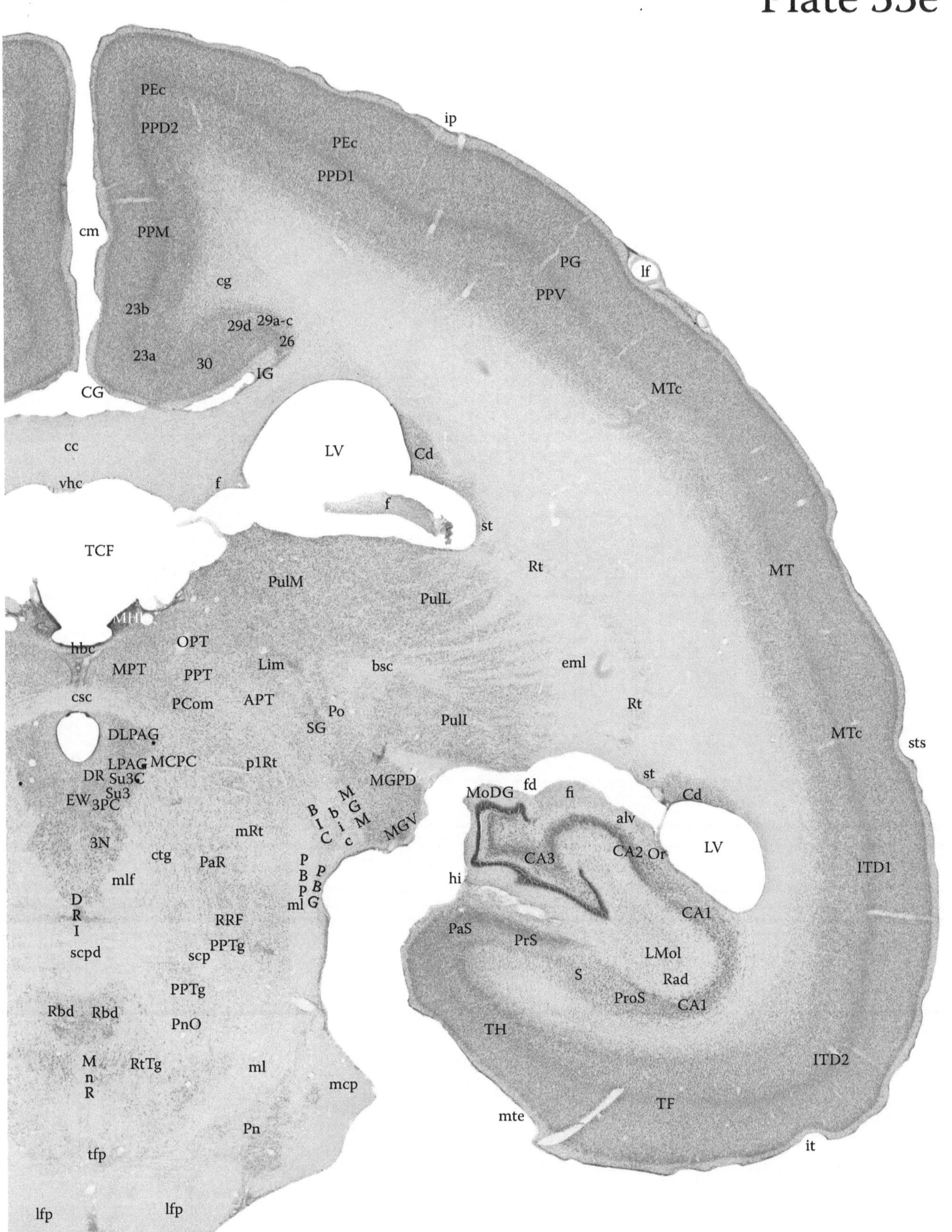


**Cresyl Violet**
Interaural +02.55 mm

# Plate 33

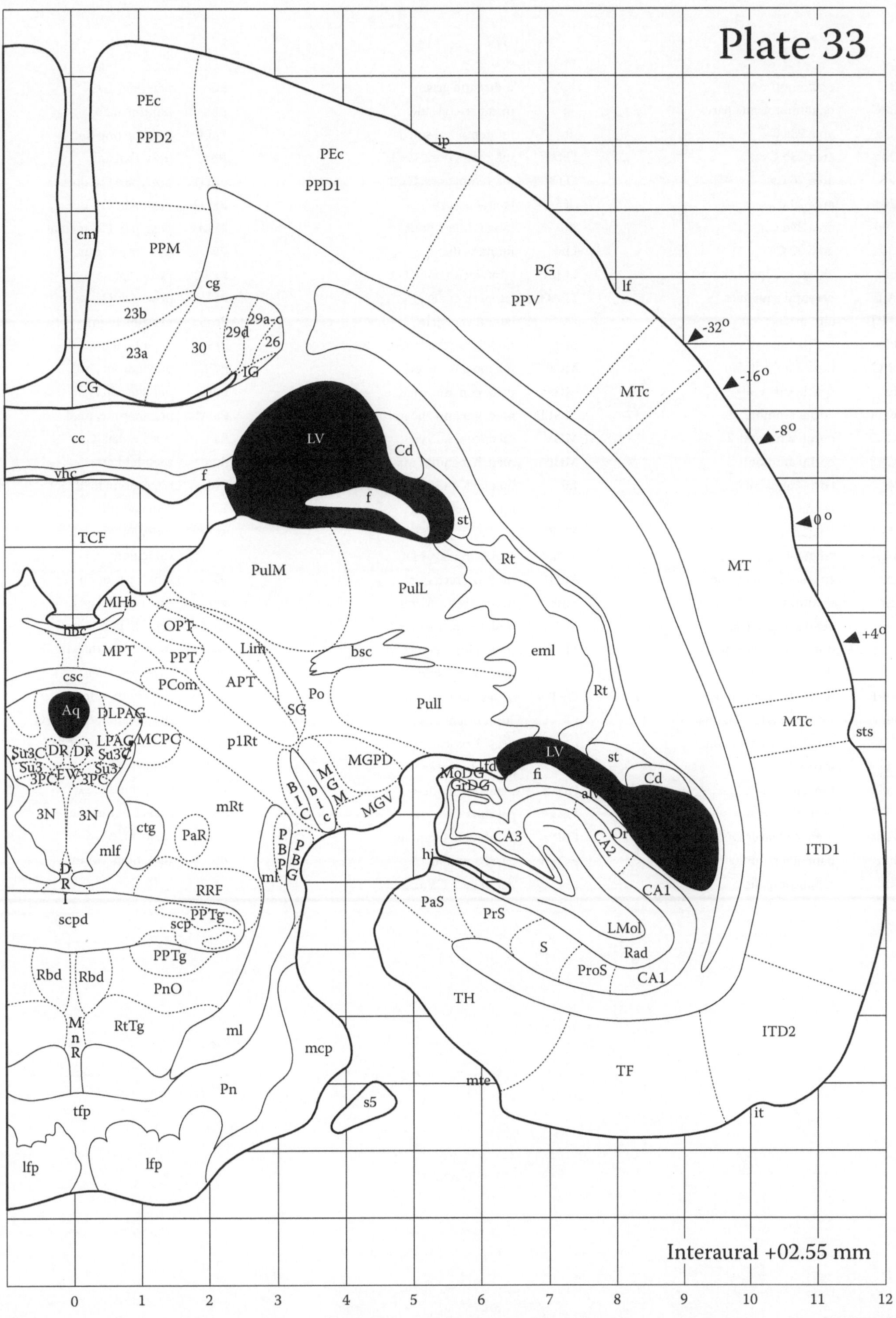


Interaural +02.55 mm

| | |
|---|---|
| 3N | oculomotor nu. |
| 3PC | oculomotor nu., parv. |
| 23a | area 23a Cx |
| 23b | area 23b Cx |
| 26 | area 26 Cx |
| 29a-c | area 29 Cx |
| 29d | area 29d Cx |
| 30 | area 30 Cx |
| alv | alveus of hippocampus |
| Aq | cerebral aqueduct |
| APT | ant. pretect. nu. |
| bic | brach. inf. coll. |
| BIC | brach. inf. coll. nu. |
| bsc | brach. sup. coll. |
| CA1 | cornu ammonis 1 |
| CA2 | cornu ammonis 2 |
| CA3 | cornu ammonis 3 |
| cc | corp. callosum |
| Cd | caudate nu. |
| cg | cingulum |
| CG | cingulate gyrus Cx |
| cm | callosomarginal sulc. |
| csc | comm. sup. coll. |
| ctg | central tegm. tract |
| DLPAG | dorsolat. periaq. gr. |
| DR | dors. raphe nu. |
| DRI | dors. raphe nu., interf. |
| eml | extern. medull. lamina |
| EW | Edinger Westphal nu. |
| f | fornix |
| fd | fimbriodent. sulcus |
| fi | fimbria |
| GrDG | gran. cell layer DG |
| hbc | habenular comm. |
| hi | hippocamp. fiss. |
| IG | induseum gris. |
| ip | intraparietal sulc. |
| it | inf. temporal sulcus |
| ITD1 | inf. temp. dors. Cx 1 |
| ITD2 | inf. temp. dors. Cx 2 |
| lf | lateral fissure |
| lfp | longit. fibers pons |
| Lim | limitans nu. |
| LMol | strat. lacun. mol. CA |
| LPAG | lat. periaqued. gr. |
| LV | lateral ventricle |
| mcp | middle cerebell. ped. |
| MCPC | magnocell. nu. pc |
| MGM | med. gen. nu., med. |
| MGPD | med. gen. nu., posterodors. |
| MGV | med. gen. nu., vent. |
| MHb | med. habenular nu. |
| ml | medial lemniscus |
| mlf | med. longit. fasc. |
| MnR | median raphe nu. |
| MoDG | molecular layer DG |
| MPT | med. pretect. nu. |
| mRt | mesen. retic. formn |
| MT | mid. temp. area Cx |
| MTc | mid. temp. cresc. Cx |
| mte | med. temp. sulc. |
| OPT | olivary pretect. nu. |
| Or | strat. oriens CA |
| p1Rt | pros. 1 retic. formn |
| PaR | pararubral area |
| PaS | parasubiculum |
| PBG | parabigeminal nu. |
| PBP | parabrach. pigm. nu. |
| PCom | paracomm. nu. |
| PEc | par. field E Cx, caud. |
| PG | par. field G Cx |
| Pn | pontine nu. |
| PnO | pontine oral nu. |
| Po | post. thal. nu. |
| PPD1 | post. par. Cx, dors. 1 |
| PPD2 | post. par. Cx, dors. 2 |
| PPM | post. par. Cx, medial |
| PPT | post. pretectum |
| PPTg | peduncpont. tegm. nu. |
| PPV | post. par. Cx, vent. |
| ProS | prosubiculum |
| PrS | presubiculum |
| PulI | pulvinar nu., inf. |
| PulL | pulvinar nu., lat. |
| PulM | pulvinar nu., med. |
| Rad | strat. radiat. CA |
| Rbd | rhabdoid nu. |
| RRF | retrorubral field |
| Rt | reticular nu. |
| RtTg | reticulotegm. nu. |
| S | subiculum |
| s5 | sensory root 5n |
| scp | sup. cerebell. ped. |
| scpd | sup. cb. ped. decn |
| SG | supragenic. thal. nu. |
| st | stria terminalis |
| sts | sup. temp. sulcus |
| Su3 | supraoculomotor PAG |
| Su3C | supraoculomotor cap |
| TCF | transverse cerebral fiss. |
| TF | temp. field F Cx |
| tfp | transv. fibers pons |
| TH | temp. field H Cx |
| vhc | vent. hipp. comm. |

# Plate 34

Interaural +02.00 mm

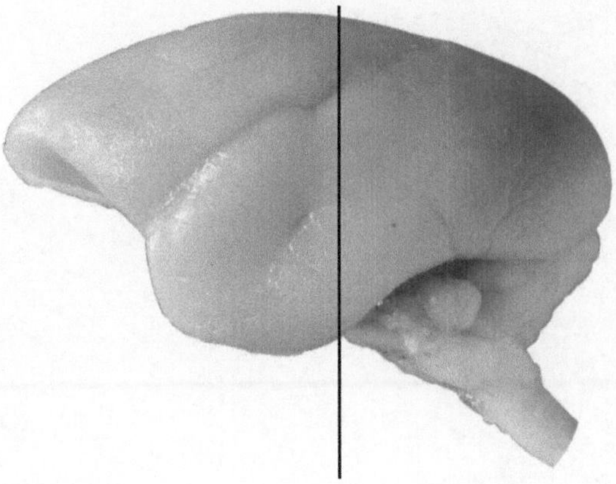

| | |
|---|---|
| 3N | oculomotor nu. |
| 3PC | oculomotor nu., parv. |
| 23a | area 23a Cx |
| 23b | area 23b Cx |
| 26 | area 26 Cx |
| 29a-c | area 29a-c Cx |
| 29d | area 29d Cx |
| 30 | area 30 Cx |
| Aq | cerebral aqueduct |
| bic | brach. inf. coll. |
| BIC | brach. inf. coll. nu. |
| bsc | brach. sup. coll. |
| CA1 | cornu ammonis 1 |
| CA2 | cornu ammonis 2 |
| CA3 | cornu ammonis 3 |
| cc | corp. callosum |
| Cd | caudate nu. |
| cg | cingulum |
| CG | cingulate gyrus Cx |
| cm | callosomarginal sulc. |
| csc | comm. sup. coll. |
| ctg | central tegm. tract |
| DLL | dorsal nu. lat. lemn. |
| DLPAG | dorsolat. periaq. gr. |
| DMPAG | dorsomed. periaq. gr |
| DpG | deep gray sup. coll. |
| DpWh | deep white sup. coll. |
| DRD | dors. raphe nu., dors. |
| DRI | dors. raphe nu., interf. |
| DTnl | dors. term. nu. |
| eml | extern. medull. lamina |
| EW | Edinger Westphal nu. |
| f | fornix |
| fi | fimbria |
| GrDG | gran. cell layer DG |
| IG | induseum gris. |
| ip | intraparietal sulc. |
| it | inf. temporal sulcus |
| ITD1 | inf. temp. dors. Cx 1 |
| ITD2 | inf. temp. dors. Cx 2 |
| lf | lateral fissure |
| Lim | limitans nu. |
| ll | lateral lemniscus |
| LPAG | lat. periaqued. gr. |
| LV | lateral ventricle |
| mcp | middle cerebell. ped. |
| Me5 | mesen. trigem. nu. |
| MiTg | microcell. tegm. nu. |
| ml | medial lemniscus |
| mlf | med. longit. fasc. |
| MnR | median raphe nu. |
| MoDG | molecular layer DG |
| MPT | med. pretect. nu. |
| mRt | mesen. retic. formn |
| MT | mid. temp. area Cx |
| MTc | mid. temp. cresc. Cx |
| mte | med. temp. sulc. |
| OT | nu. of optic tract |
| PaS | parasubiculum |
| PBG | parabigeminal nu. |
| PEc | par. field E Cx, caud. |
| PG | par. field G Cx |
| Pi | pineal gland |
| Pn | pontine nu. |
| PnO | pontine oral nu. |
| PPD1 | post. par. Cx, dors. 1 |
| PPD2 | post. par. Cx, dors. 2 |
| PPM | post. par. Cx, medial |
| PPT | post. pretectum |
| PPTgC | peduncpont. tegm. nu., compact |
| PPTgD | peduncpont. tegm. nu., diff. |
| PPV | post. par. Cx, vent. |
| ProS | prosubiculum |
| PrS | presubiculum |
| PulI | pulvinar nu., inf. |
| PulL | pulvinar nu., lat. |
| PulM | pulvinar nu., med. |
| py | pyramidal tract |
| RRF | retrorubral field |
| Rt | reticular nu. |
| RtTg | reticulotegm. nu. |
| S | subiculum |
| s5 | sensory root 5n |
| scpd | sup. cb. ped. decn |
| SG | supragenic. thal. nu. |
| st | stria terminalis |
| sts | sup. temp. sulcus |
| Su3 | supraoculomotor PAG |
| Su3C | supraoculomotor cap |
| TF | temp. field F Cx |
| tfp | transv. fibers pons |
| TH | temp. field H Cx |
| tth | trigeminothal. tr. |
| vhc | vent. hipp. comm. |
| VLL | vent. nu. lat. lemn. |
| VLP (V3v) | ventlatpost. (V3v) Cx |
| VLPAG | ventrolat. periaq. gr. |

# Plate 34a

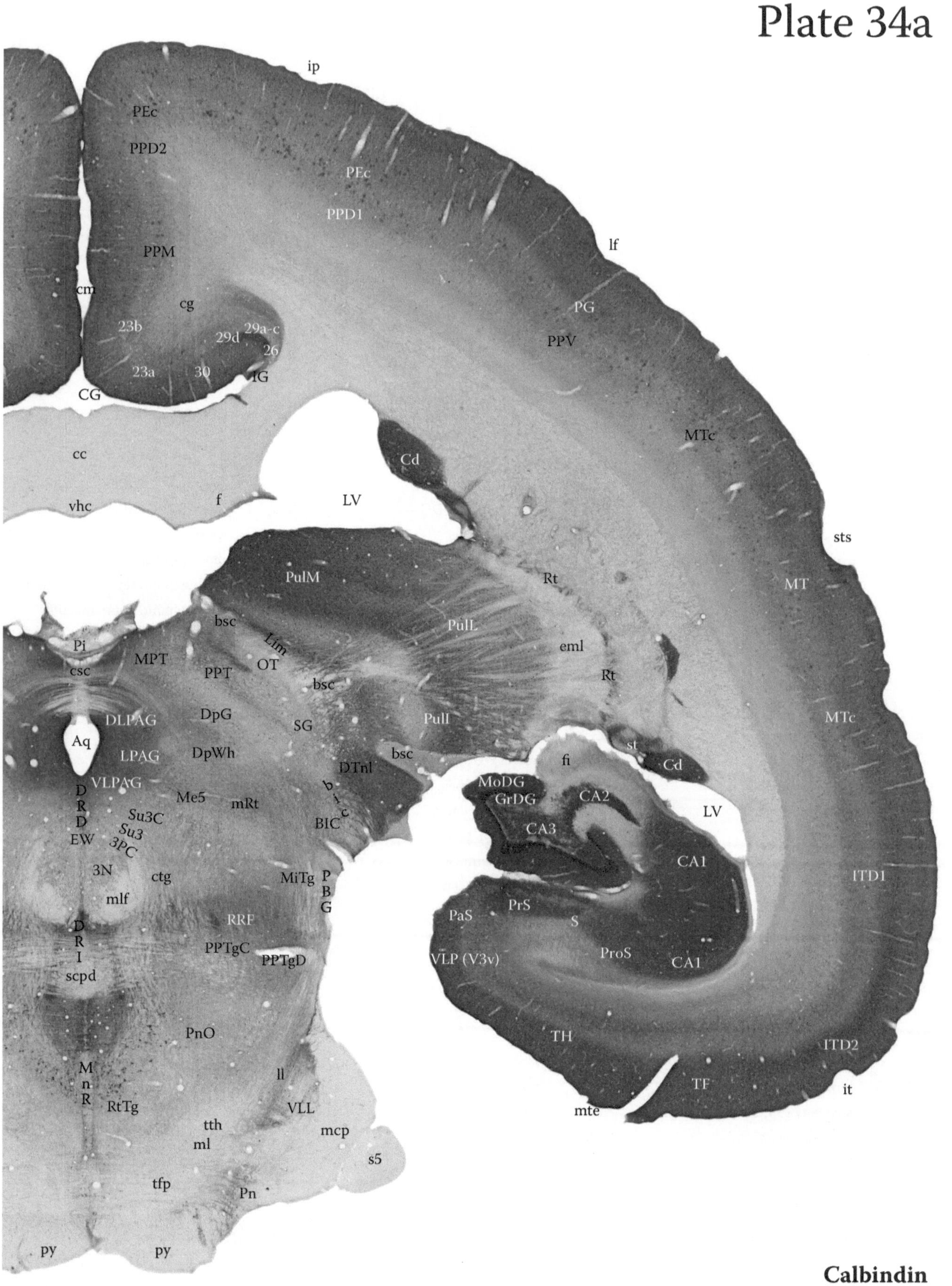


**Calbindin**
Interaural +02.30 mm

# Plate 34b

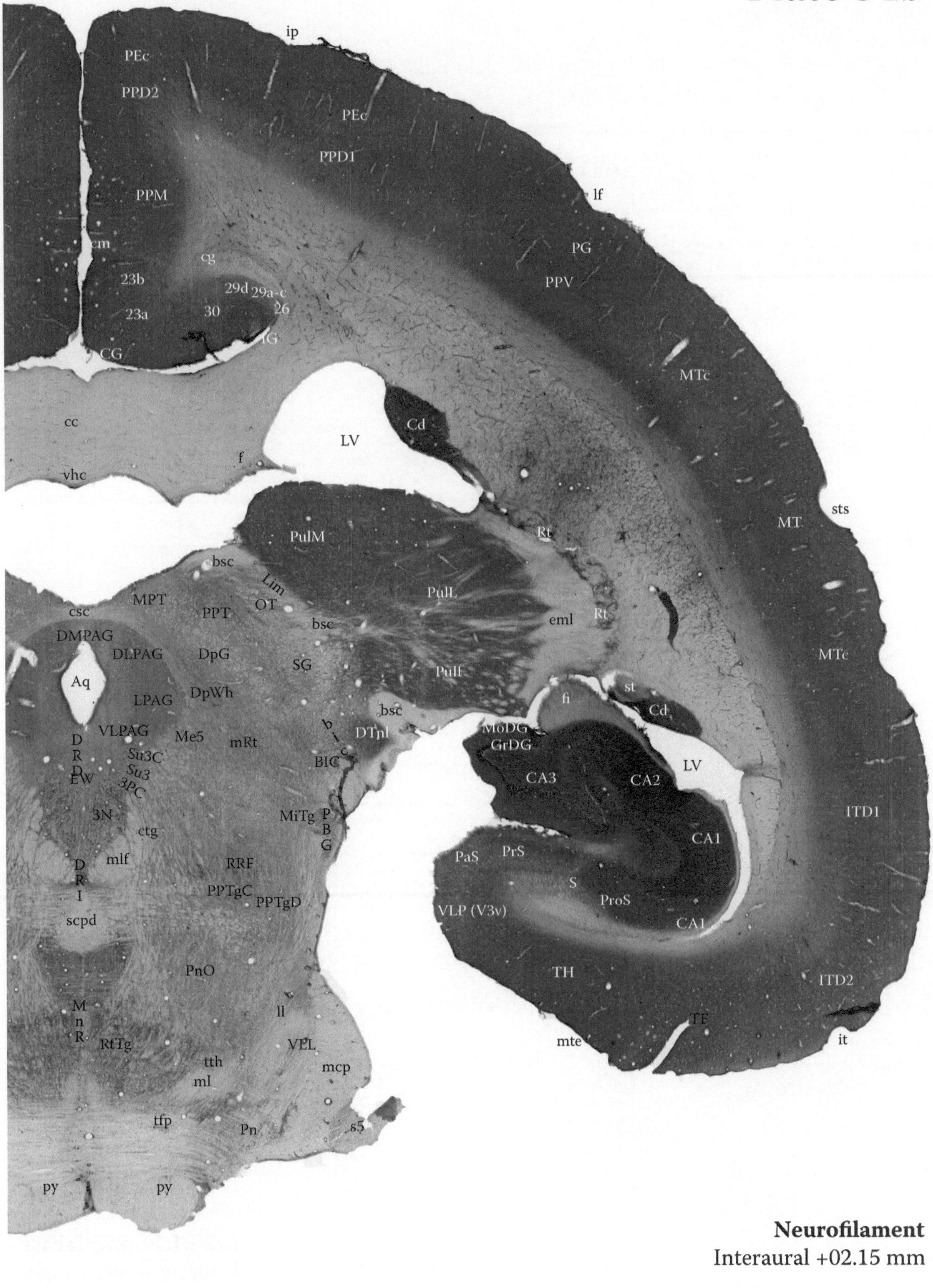


**Neurofilament**
Interaural +02.15 mm

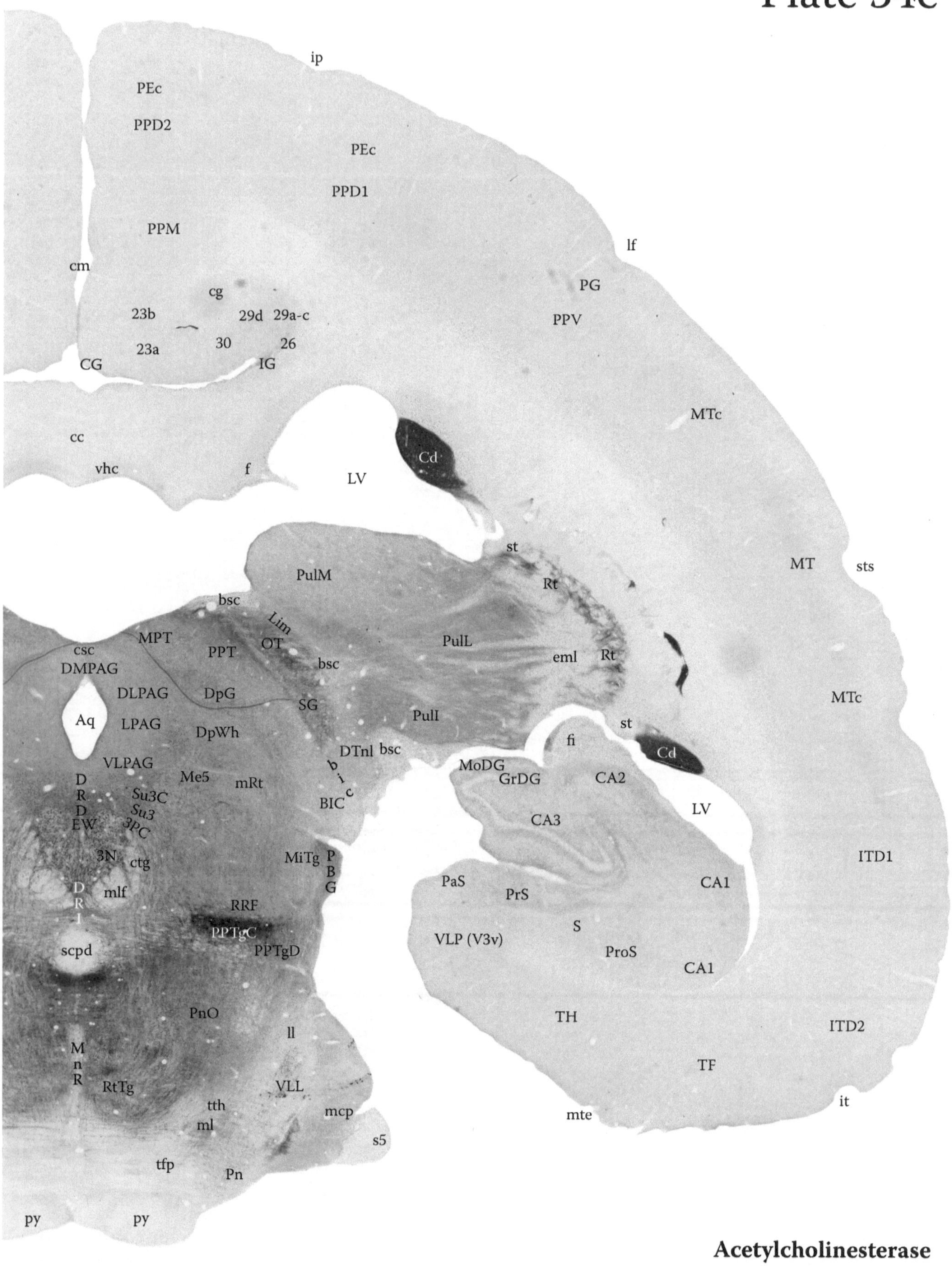


**Acetylcholinesterase**
Interaural +02.10 mm

# Plate 34d

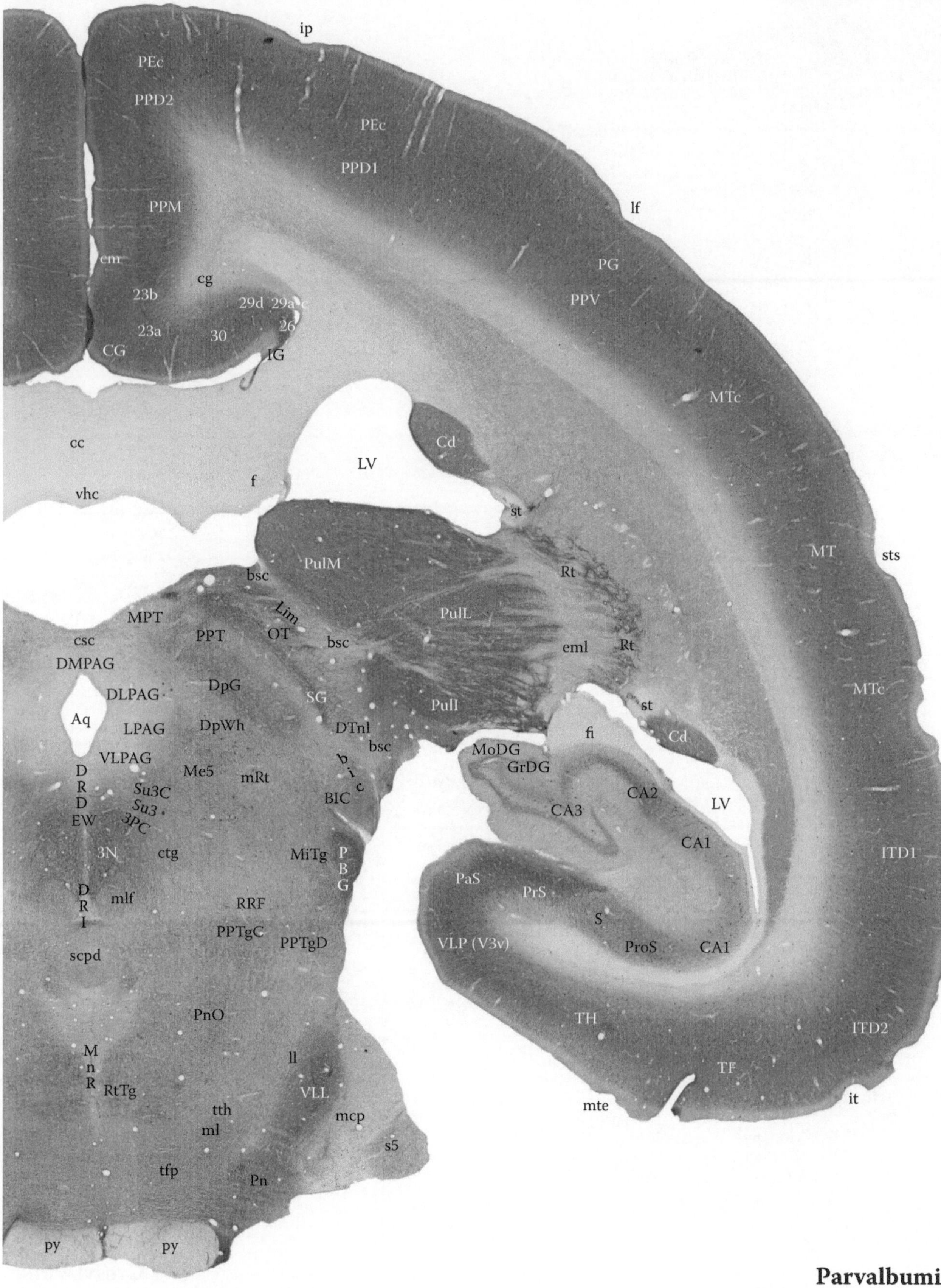


**Parvalbumin**
Interaural +02.05 mm

# Plate 34e

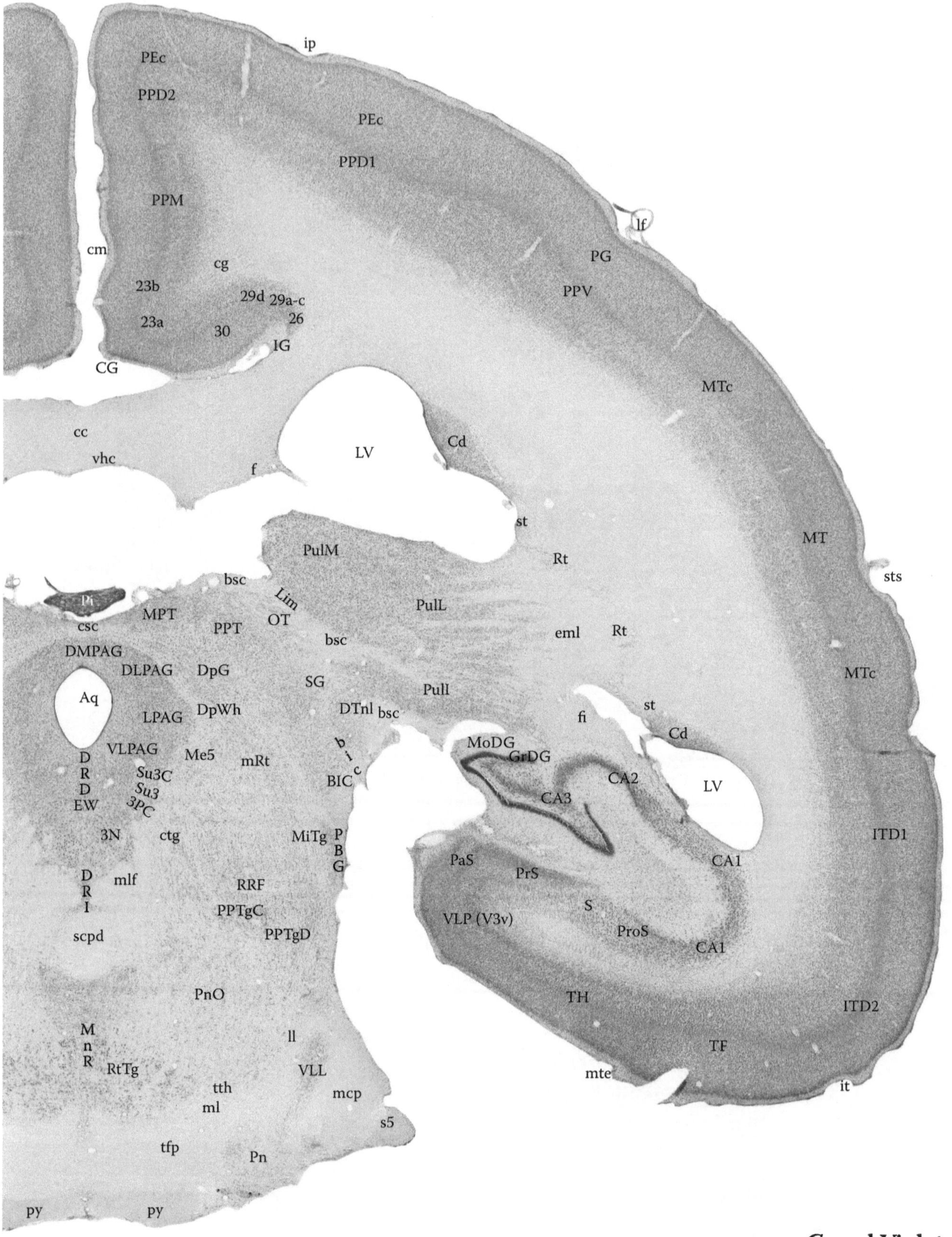


**Cresyl Violet**
Interaural +02.00 mm

# Plate 34

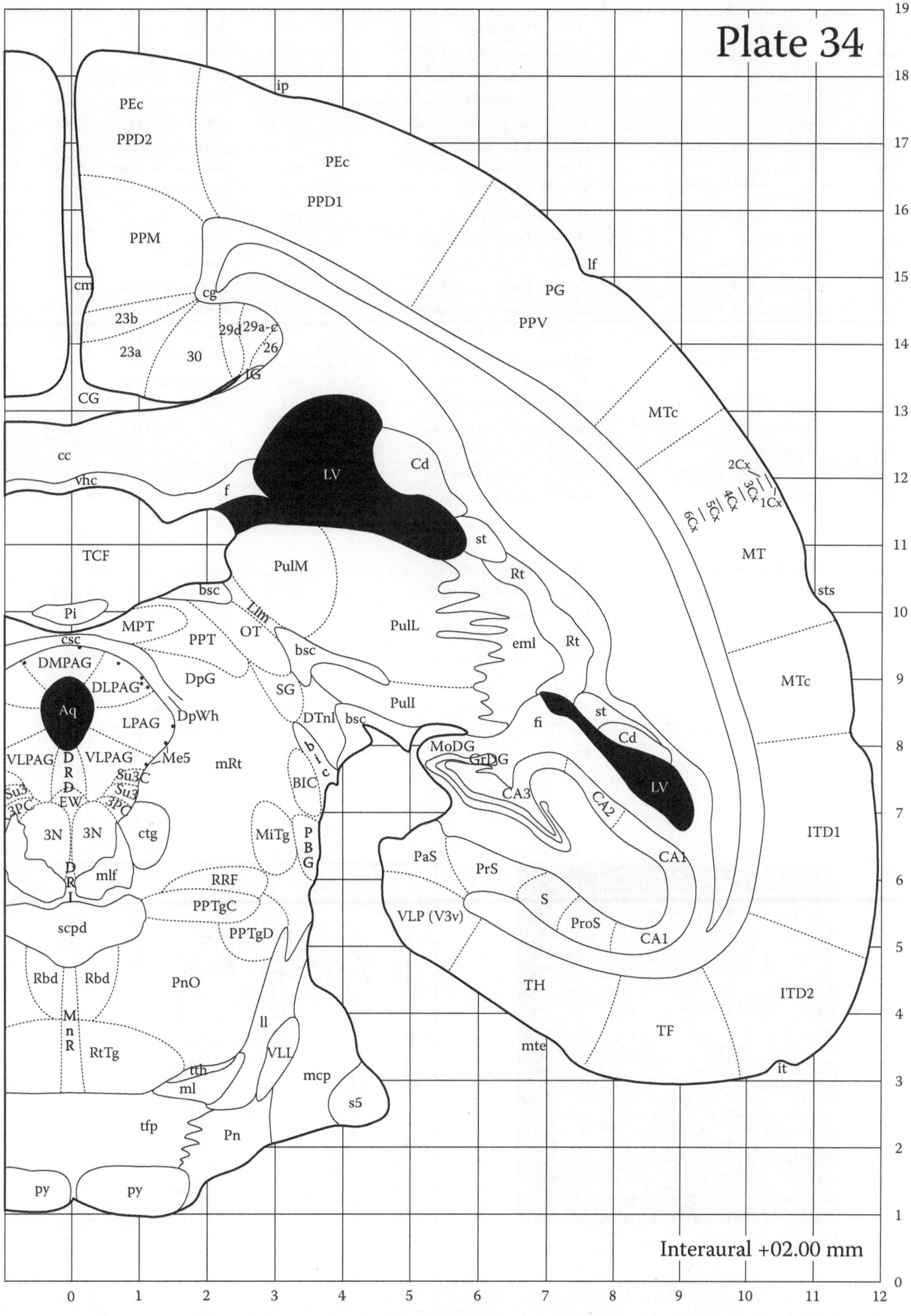


Interaural +02.00 mm

| | |
|---|---|
| 3N | oculomotor nu. |
| 3PC | oculomotor nu., parv. |
| 23a | area 23a Cx |
| 23b | area 23b Cx |
| 26 | area 26 Cx |
| 29a-c | area 29a-c Cx |
| 29d | area 29d Cx |
| 30 | area 30 Cx |
| Aq | cerebral aqueduct |
| bic | brach. inf. coll. |
| BIC | brach. inf. coll. nu. |
| bsc | brach. sup. coll. |
| CA1 | cornu ammonis 1 |
| CA2 | cornu ammonis 2 |
| CA3 | cornu ammonis 3 |
| cc | corp. callosum |
| Cd | caudate nu. |
| cg | cingulum |
| CG | cingulate gyrus Cx |
| cm | callosomarginal sulc. |
| csc | comm. sup. coll. |
| ctg | central tegm. tract |
| DLL | dorsal nu. lat. lemn. |
| DLPAG | dorsolat. periaq. gr. |
| DMPAG | dorsomed. periaq. gr |
| DpG | deep gray sup. coll. |
| DpWh | deep white sup. coll. |
| DRD | dors. raphe nu., dors. |
| DRI | dors. raphe nu., interf. |
| DTnl | dors. term. nu. |
| eml | extern. medull. lamina |
| EW | Edinger Westphal nu. |
| f | fornix |
| fi | fimbria |
| GrDG | gran. cell layer DG |
| IG | induseum gris. |
| ip | intraparietal sulc. |
| it | inf. temporal sulcus |
| ITD1 | inf. temp. dors. Cx 1 |
| ITD2 | inf. temp. dors. Cx 2 |
| lf | lateral fissure |
| Lim | limitans nu. |
| ll | lateral lemniscus |
| LPAG | lat. periaqued. gr. |
| LV | lateral ventricle |
| mcp | middle cerebell. ped. |
| Me5 | mesen. trigem. nu. |
| MiTg | microcell. tegm. nu. |
| ml | medial lemniscus |
| mlf | med. longit. fasc. |
| MnR | median raphe nu. |
| MoDG | molecular layer DG |
| MPT | med. pretect. nu. |
| mRt | mesen. retic. formn |
| MT | mid. temp. area Cx |
| MTc | mid. temp. cresc. Cx |
| mte | med. temp. sulc. |
| OT | nu. of optic tract |
| PaS | parasubiculum |
| PBG | parabigeminal nu. |
| PEc | par. field E Cx, caud. |
| PG | par. field G Cx |
| Pi | pineal gland |
| Pn | pontine nu. |
| PnO | pontine oral nu. |
| PPD1 | post. par. Cx, dors. 1 |
| PPD2 | post. par. Cx, dors. 2 |
| PPM | post. par. Cx, medial |
| PPT | post. pretectum |
| PPTgC | peduncpont. tegm. nu., compact |
| PPTgD | peduncpont. tegm. nu., diff. |
| PPV | post. par. Cx, vent. |
| ProS | prosubiculum |
| PrS | presubiculum |
| PulI | pulvinar nu., inf. |
| PulL | pulvinar nu., lat. |
| PulM | pulvinar nu., med. |
| py | pyramidal tract |
| RRF | retrorubral field |
| Rt | reticular nu. |
| RtTg | reticulotegm. nu. |
| S | subiculum |
| s5 | sensory root 5n |
| scpd | sup. cb. ped. decn |
| SG | supragenic. thal. nu. |
| st | stria terminalis |
| sts | sup. temp. sulcus |
| Su3 | supraoculomotor PAG |
| Su3C | supraoculomotor cap |
| TF | temp. field F Cx |
| tfp | transv. fibers pons |
| TH | temp. field H Cx |
| tth | trigeminothal. tr. |
| vhc | vent. hipp. comm. |
| VLL | vent. nu. lat. lemn. |
| VLP (V3v) | ventlatpost. (V3v) Cx |
| VLPAG | ventrolat. periaq. gr. |

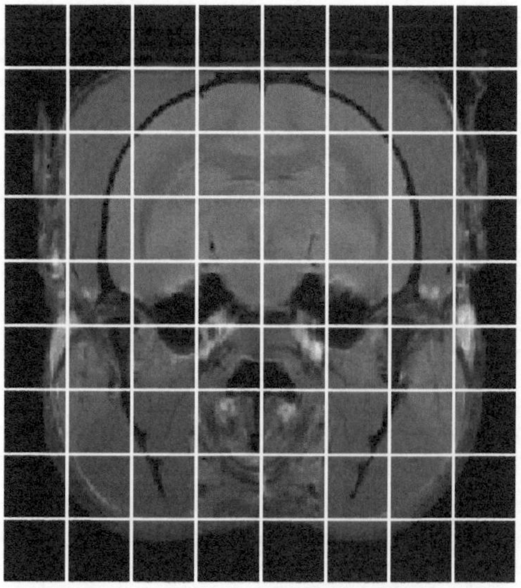

Interaural +02.10 mm

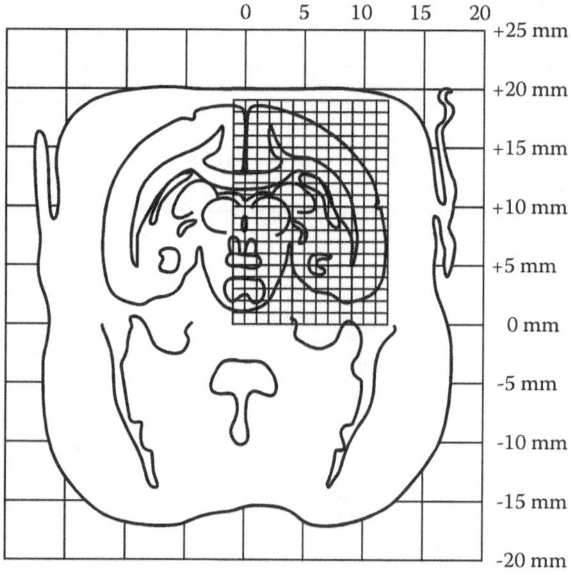

# Plate 35

Interaural +01.50 mm

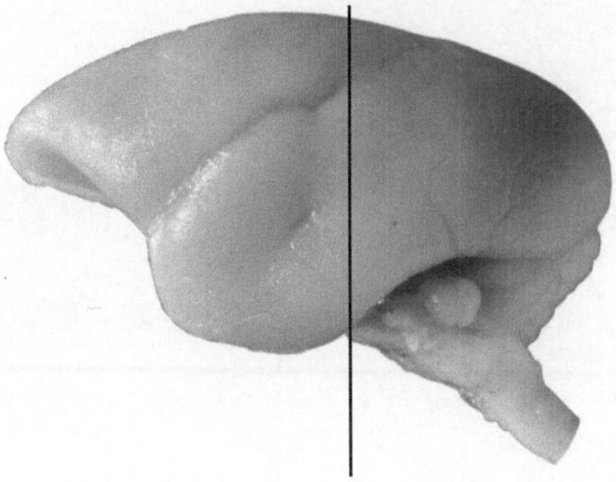

| | | | | | |
|---|---|---|---|---|---|
| 4n | trochlear nerve | GrDG | gran. cell layer DG | PnO | pontine oral nu. |
| 4N | trochlear nu. | IG | induseum gris. | PPD1 | post. par. Cx, dors. 1 |
| 23a | area 23a Cx | ILL | intermed. nu. lat. lemn. | PPD2 | post. par. Cx, dors. 2 |
| 23b | area 23b Cx | InCo | intercoll. nu. | PPM | post. par. Cx, medial |
| 26 | area 26 Cx | InG | intermed. gr. sup. coll. | PPTgC | peduncpont. tegm. nu., compact |
| 29 | area 29 Cx | InWh | intermed. wh. sup coll. | PPTgD | peduncpont. tegm. nu., diff. |
| 30 | area 30 Cx | ip | intraparietal sulc. | PPV | post. par. Cx, vent. |
| Aq | cerebral aqueduct | it | inf. temporal sulcus | ProS | prosubiculum |
| ATg | ant. tegmental nu. | ITD1 | inf. temp. dors. Cx 1 | PrS | presubiculum |
| bic | brach. inf. coll. | ITD2 | inf. temp. dors. Cx 2 | pt | post. temp. sulc. |
| BIC | brach. inf. coll. nu. | lf | lateral fissure | PulL | pulvinar nu., lat. |
| bsc | brach. sup. coll. | ll | lateral lemniscus | PulM | pulvinar nu., med. |
| CA1 | cornu ammonis 1 | LPAG | lat. periaqued. gr. | py | pyramidal tract |
| CA2 | cornu ammonis 2 | LV | lateral ventricle | Rbd | rhabdoid nu. |
| CA3 | cornu ammonis 3 | m5 | motor root 5n | RtTg | reticulotegm. nu. |
| cc | corp. callosum | mcp | middle cerebell. ped. | S | subiculum |
| Cd | caudate nu. | Me5 | mesen. trigem. nu. | s5 | sensory root 5n |
| cg | cingulum | MiTg | microcell. tegm. nu. | scp | sup. cerebell. ped. |
| CG | cingulate gyrus Cx | ml | medial lemniscus | scpd | sup. cb. ped. decn |
| cm | callosomarginal sulc. | mlf | med. longit. fasc. | SPTg | subpedunc. tegm. nu. |
| CnF | cuneiform nu. | MnR | median raphe nu. | st | stria terminalis |
| csc | comm. sup. coll. | MoDG | molecular layer DG | Su3 | supraoculomotor PAG |
| ctg | central tegm. tract | MPT | med. pretect. nu. | Su3C | supraoculomotor cap |
| DLPAG | dorsolat. periaq. gr. | MTc | mid. temp. cresc. Cx | SuG | sup. gray sup. coll. |
| DMPAG | dorsomed. periaq. gr | mte | med. temp. sulc. | TCF | transverse cerebral fiss. |
| DpG | deep gray sup. coll. | Op | strat. opt. sup. coll. | TF | temp. field F Cx |
| DpWh | deep white sup. coll. | OT | nu. of optic tract | tfp | transv. fibers pons |
| DRD | dors. raphe nu., dors. | PaS | parasubiculum | TH | temp. field H Cx |
| DRI | dors. raphe nu., interf. | PBG | parabigeminal nu. | V2v | V2 (vent.) Cx |
| DRV | dors. raphe nu., vent. | PEc | par. field E Cx, caud. | vhc | vent. hipp. comm. |
| DRVL | dors. raphe nu., ventrolat. | PG | par. field G Cx | VLA (V4) | ventlatant. (V4) Cx |
| DTnl | dors. term. nu. | Pi | pineal gland | VLL | vent. nu. lat. lemn. |
| EW | Edinger Westphal nu. | Pn | pontine nu. | VLP (V3v) | ventlatpost. (V3v) Cx |
| f | fornix | PnC | pontine caudal nu. | VLPAG | ventrolat. periaq. gr. |
| fi | fimbria | | | | |

# Plate 35a

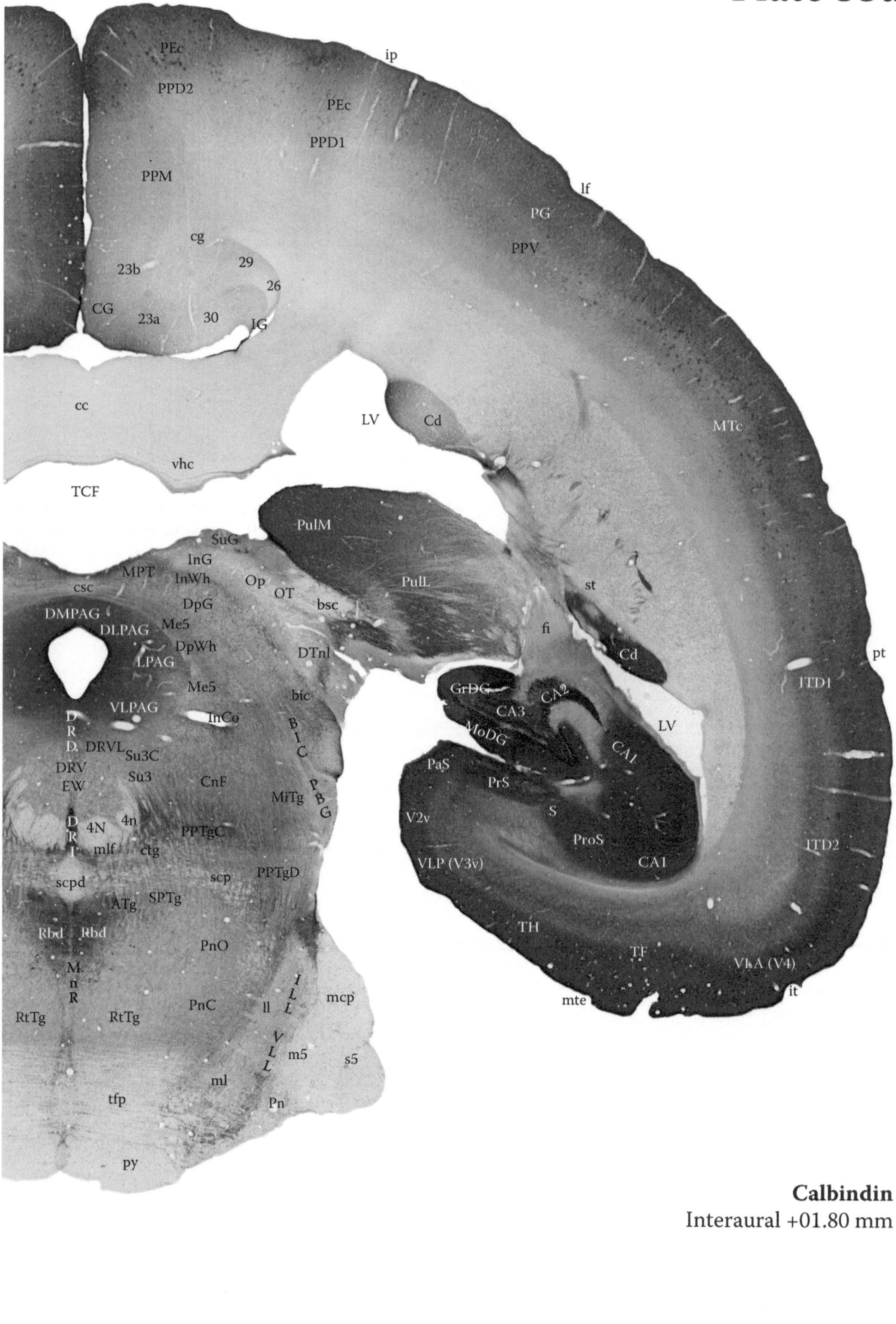


**Calbindin**
Interaural +01.80 mm

# Plate 35b

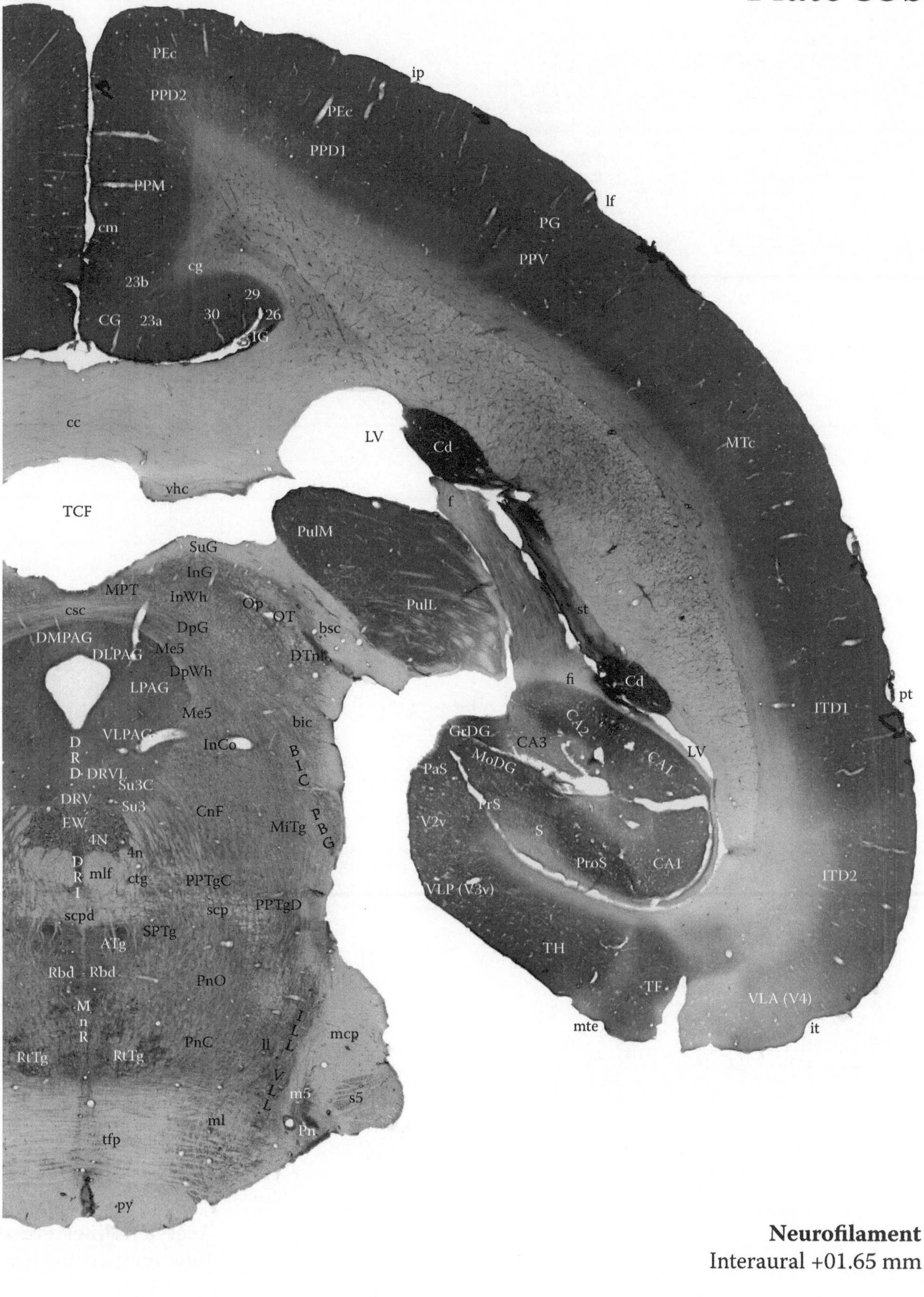


**Neurofilament**
Interaural +01.65 mm

# Plate 35c

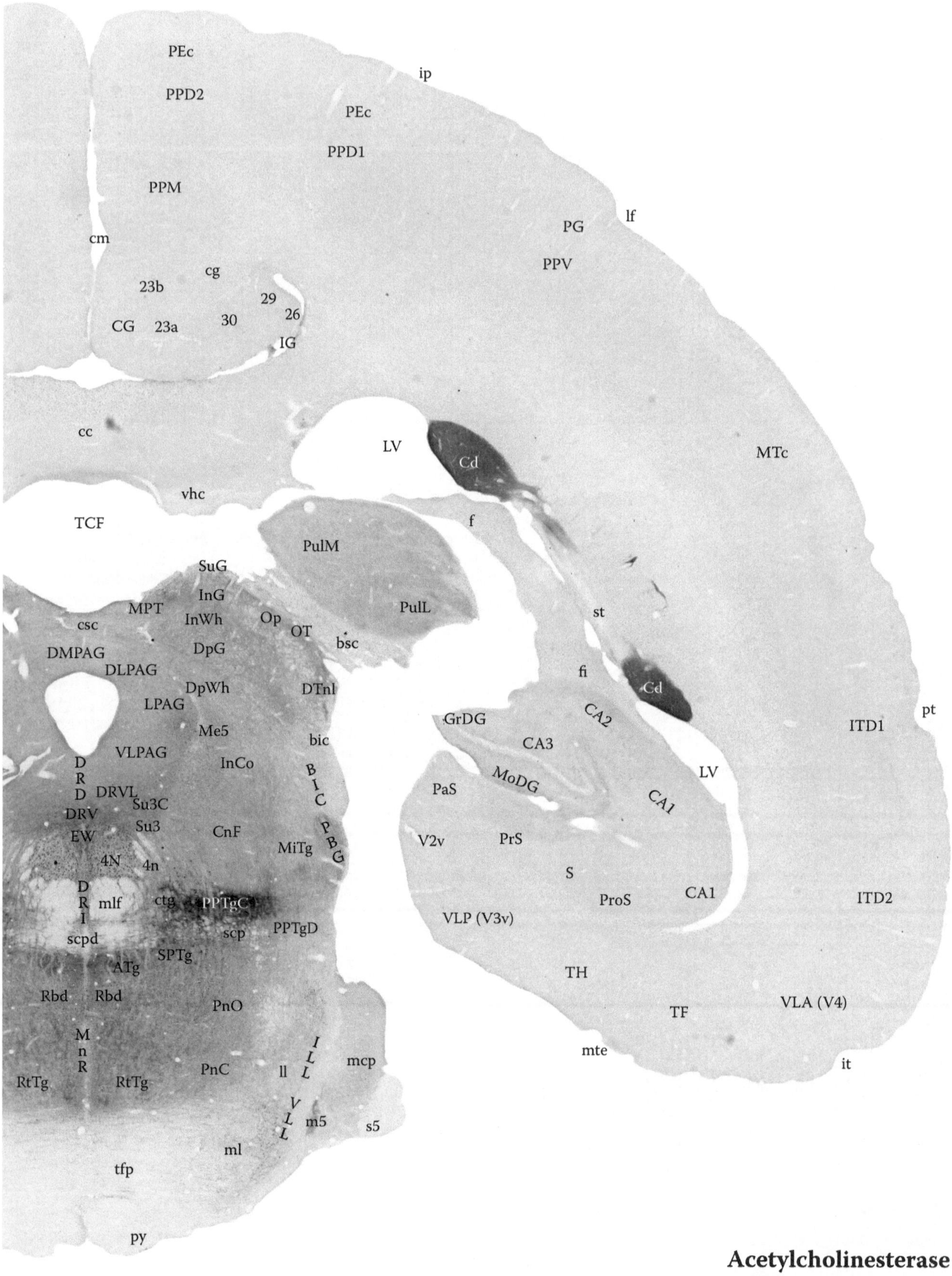


**Acetylcholinesterase**
Interaural +01.60 mm

# Plate 35d

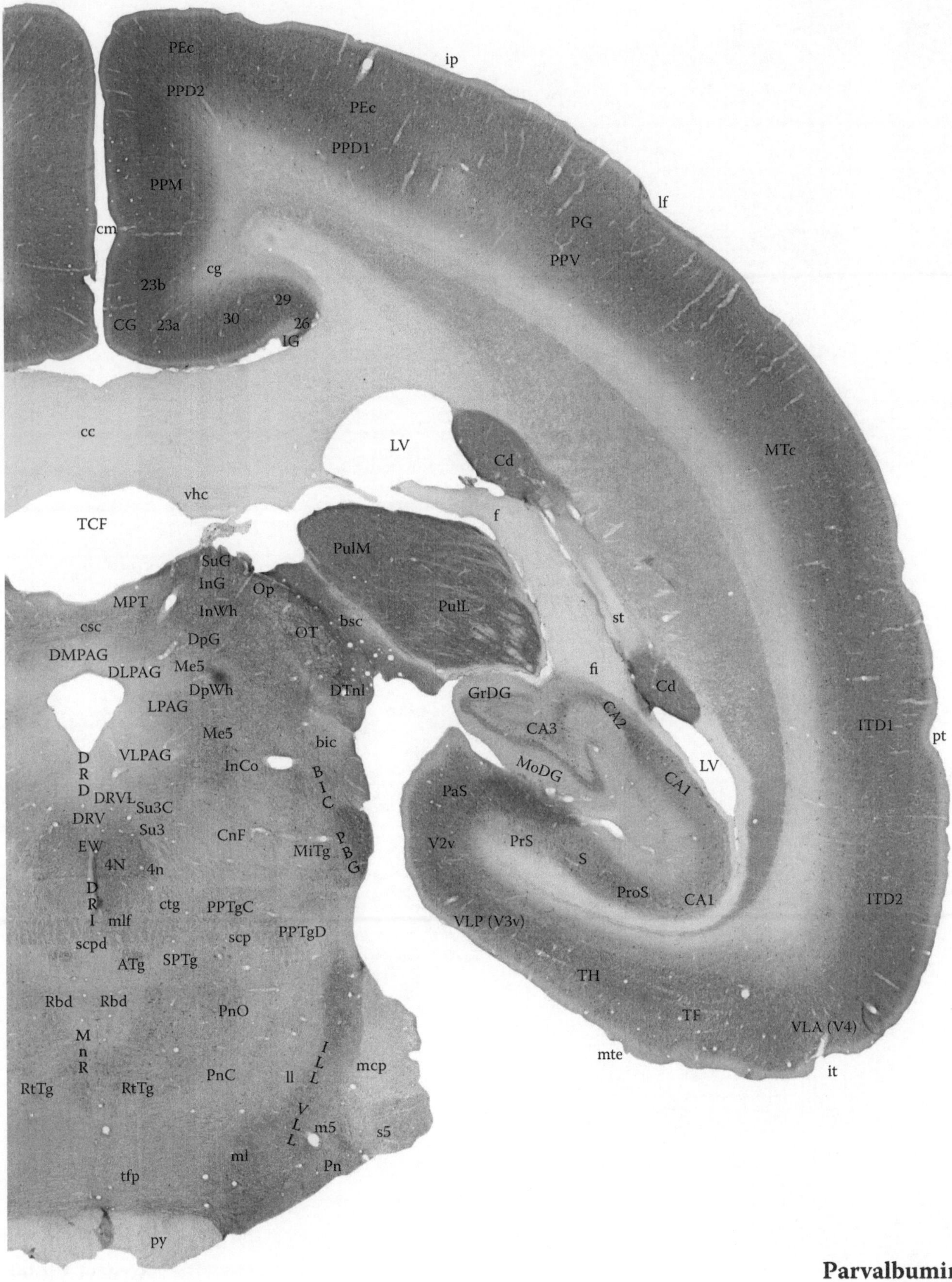


**Parvalbumin**
Interaural +01.55 mm

# Plate 35e

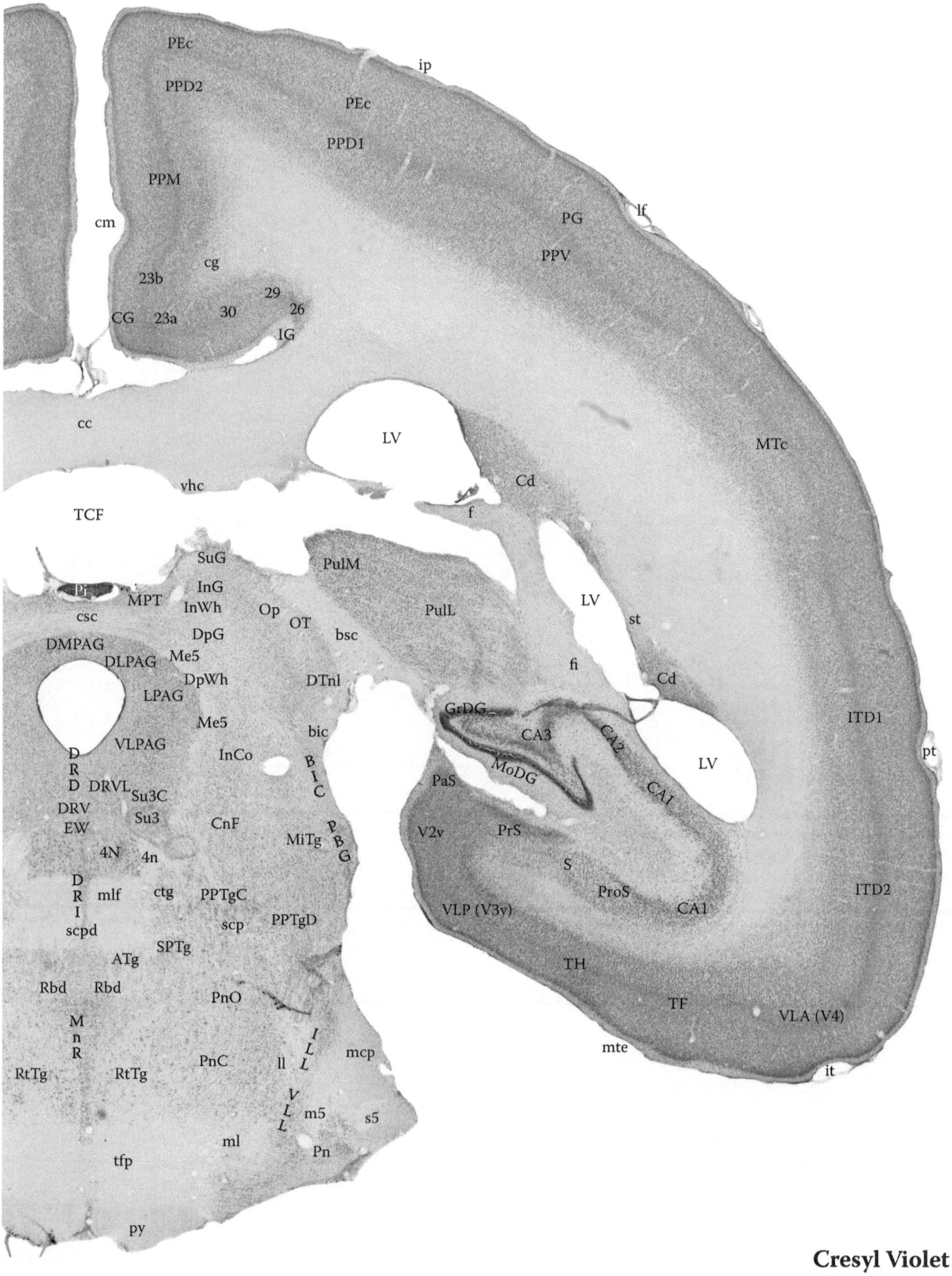


**Cresyl Violet**
Interaural +01.50 mm

# Plate 35

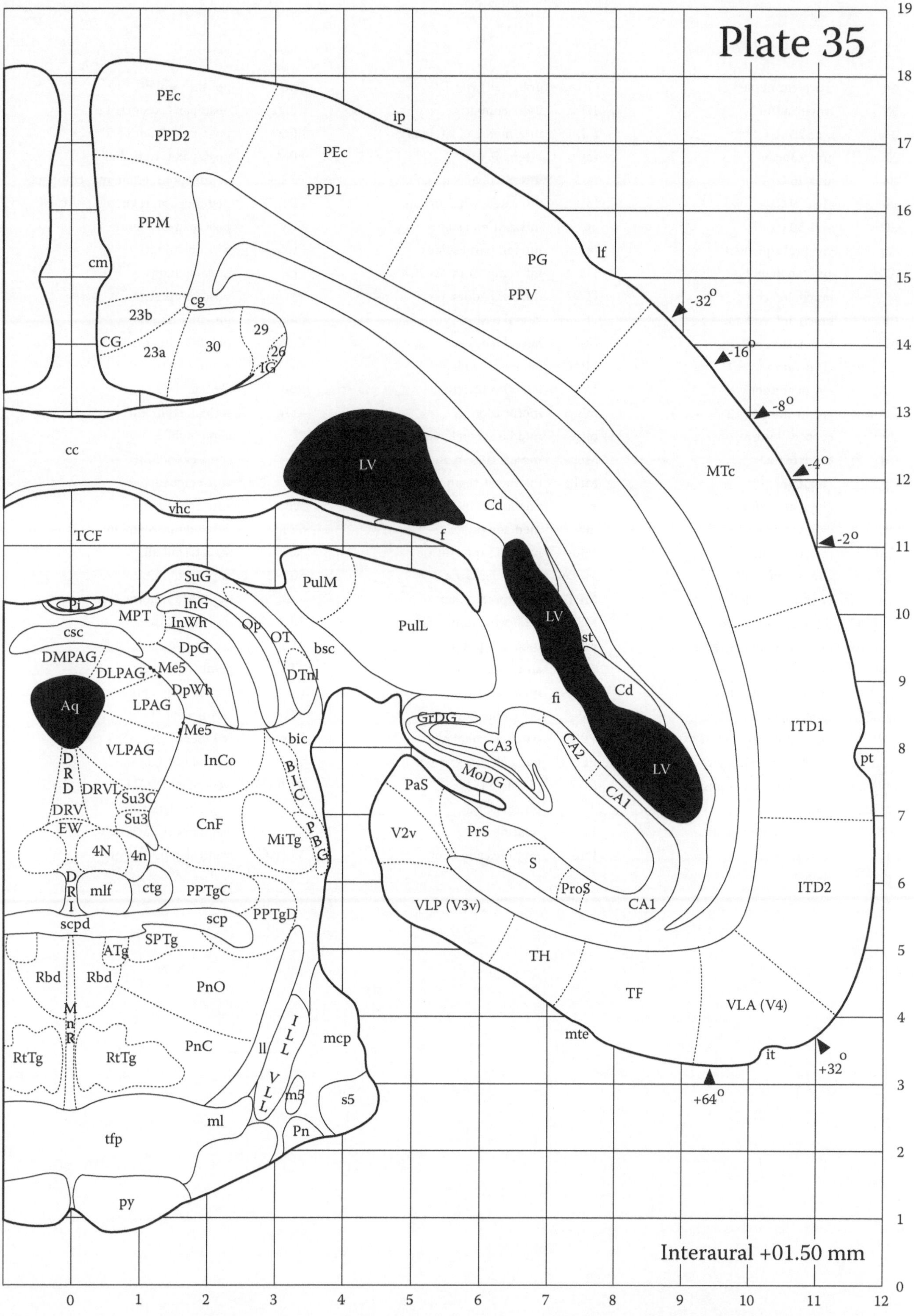


Interaural +01.50 mm

| | |
|---|---|
| 4n | trochlear nerve |
| 4N | trochlear nu. |
| 23a | area 23a Cx |
| 23b | area 23b Cx |
| 26 | area 26 Cx |
| 29 | area 29 Cx |
| 30 | area 30 Cx |
| Aq | cerebral aqueduct |
| ATg | ant. tegmental nu. |
| bic | brach. inf. coll. |
| BIC | brach. inf. coll. nu. |
| bsc | brach. sup. coll. |
| CA1 | cornu ammonis 1 |
| CA2 | cornu ammonis 2 |
| CA3 | cornu ammonis 3 |
| cc | corp. callosum |
| Cd | caudate nu. |
| cg | cingulum |
| CG | cingulate gyrus Cx |
| cm | callosomarginal sulc. |
| CnF | cuneiform nu. |
| csc | comm. sup. coll. |
| ctg | central tegm. tract |
| DLPAG | dorsolat. periaq. gr. |
| DMPAG | dorsomed. periaq. gr |
| DpG | deep gray sup. coll. |
| DpWh | deep white sup. coll. |
| DRD | dors. raphe nu., dors. |
| DRI | dors. raphe nu., interf. |
| DRV | dors. raphe nu., vent. |
| DRVL | dors. raphe nu., ventrolat. |
| DTnl | dors. term. nu. |
| EW | Edinger Westphal nu. |
| f | fornix |
| fi | fimbria |
| GrDG | gran. cell layer DG |
| IG | induseum gris. |
| ILL | intermed. nu. lat. lemn. |
| InCo | intercoll. nu. |
| InG | intermed. gr. sup. coll. |
| InWh | intermed. wh. sup coll. |
| ip | intraparietal sulc. |
| it | inf. temporal sulcus |
| ITD1 | inf. temp. dors. Cx 1 |
| ITD2 | inf. temp. dors. Cx 2 |
| lf | lateral fissure |
| ll | lateral lemniscus |
| LPAG | lat. periaqued. gr. |
| LV | lateral ventricle |
| m5 | motor root 5n |
| mcp | middle cerebell. ped. |
| Me5 | mesen. trigem. nu. |
| MiTg | microcell. tegm. nu. |
| ml | medial lemniscus |
| mlf | med. longit. fasc. |
| MnR | median raphe nu. |
| MoDG | molecular layer DG |
| MPT | med. pretect. nu. |
| MTc | mid. temp. cresc. Cx |
| mte | med. temp. sulc. |
| Op | strat. opt. sup. coll. |
| OT | nu. of optic tract |
| PaS | parasubiculum |
| PBG | parabigeminal nu. |
| PEc | par. field E Cx, caud. |
| PG | par. field G Cx |
| Pi | pineal gland |
| Pn | pontine nu. |
| PnC | pontine caudal nu. |
| PnO | pontine oral nu. |
| PPD1 | post. par. Cx, dors. 1 |
| PPD2 | post. par. Cx, dors. 2 |
| PPM | post. par. Cx, medial |
| PPTgC | peduncpont. tegm. nu., compact |
| PPTgD | peduncpont. tegm. nu., diff. |
| PPV | post. par. Cx, vent. |
| ProS | prosubiculum |
| PrS | presubiculum |
| pt | post. temp. sulc. |
| PulL | pulvinar nu., lat. |
| PulM | pulvinar nu., med. |
| py | pyramidal tract |
| Rbd | rhabdoid nu. |
| RtTg | reticulotegm. nu. |
| S | subiculum |
| s5 | sensory root 5n |
| scp | sup. cerebell. ped. |
| scpd | sup. cb. ped. decn |
| SPTg | subpedunc. tegm. nu. |
| st | stria terminalis |
| Su3 | supraoculomotor PAG |
| Su3C | supraoculomotor cap |
| SuG | sup. gray sup. coll. |
| TCF | transverse cerebral fiss. |
| TF | temp. field F Cx |
| tfp | transv. fibers pons |
| TH | temp. field H Cx |
| V2v | V2 (vent.) Cx |
| vhc | vent. hipp. comm. |
| VLA (V4) | ventlatant. (V4) Cx |
| VLL | vent. nu. lat. lemn. |
| VLP (V3v) | ventlatpost. (V3v) Cx |
| VLPAG | ventrolat. periaq. gr. |

# Plate 36

Interaural +01.00 mm

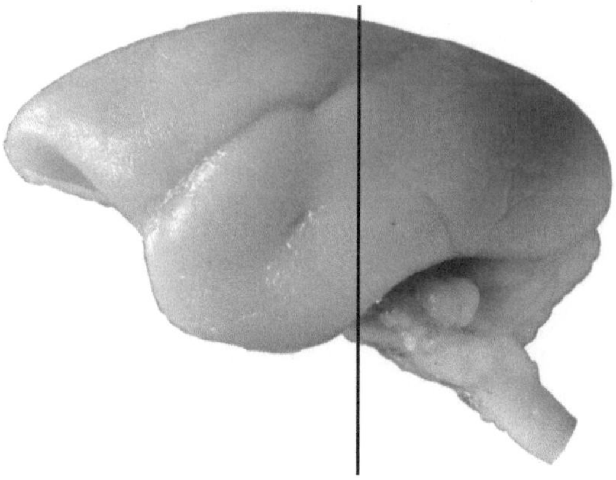

| | |
|---|---|
| 4n | trochlear nerve |
| 5n | trigeminal nerve |
| 19m | area 19m Cx |
| 23a | area 23a Cx |
| 23b | area 23b Cx |
| 26 | area 26 Cx |
| 29 | area 29 Cx |
| 30 | area 30 Cx |
| alv | alveus |
| Aq | cerebral aqueduct |
| ATg | ant. tegmental nu. |
| bic | brach. inf. coll. |
| BIC | brach. inf. coll. nu. |
| CA1 | cornu ammonis 1 |
| CA2 | cornu ammonis 2 |
| CA3 | cornu ammonis 3 |
| cc | corp. callosum |
| Cd | caudate nu. |
| cg | cingulum |
| cm | callosomarginal sulc. |
| CnF | cuneiform nu. |
| ctg | central tegm. tract |
| DA | dorsoant. visual Cx |
| DLL | dorsal nu. lat. lemn. |
| DLPAG | dorsolat. periaq. gr. |
| DM | dorsomed. visual Cx |
| DMPAG | dorsomed. periaq. gr |
| DpG | deep gray sup. coll. |
| DpWh | deep white sup. coll. |
| DRD | dors. raphe nu., dors. |
| DRI | dors. raphe nu., interf. |
| DRV | dors. raphe nu., vent. |
| DRVL | dors. raphe nu., ventrolat. |
| DTnl | dors. term. nu. |
| fd | fimbriodent. sulcus |
| fi | fimbria |
| gc | geniculocalcarine tr. |
| GrDG | gran. cell layer DG |
| IG | induseum gris. |
| ILL | intermed. nu. lat. lemn. |
| InCo | intercoll. nu. |
| InG | intermed. gr. sup. coll. |
| InWh | intermed. wh. sup coll. |
| ip | intraparietal sulc. |
| IsRt | isthmic retic. formn |
| it | inf. temporal sulcus |
| ITD | inf. temp. dors. Cx |
| LaPO | lat. periolivary nu. |
| LDTgD | laterod. tegm nu., dors. |
| lf | lateral fissure |
| ll | lateral lemniscus |
| LMol | strat. lacun. mol. CA |
| LPAG | lat. periaqued. gr. |
| LSO | lat. sup. olivary nu. |
| LV | lateral ventricle |
| LVPO | laterovent. perioliv. nu. |
| m5 | motor root 5n |
| mcp | middle cerebell. ped. |
| me5 | mesen. trigem. tr. |
| Me5 | mesen. trigem. nu. |
| MePO | med. periolivary nu. |
| MiTg | microcell. tegm. nu. |
| ml | medial lemniscus |
| mlf | med. longit. fasc. |
| MnR | median raphe nu. |
| MolDG | molecular layer DG |
| MPT | med. pretect. nu. |
| MSO | med. sup. olivary nu. |
| MTc | mid. temp. cresc. Cx |
| MVPO | mediovent. perioliv. nu. |
| Op | strat. opt. sup. coll. |
| Or | strat. oriens CA |
| PaS | parasubiculum |
| PBG | parabigeminal nu. |
| PFl | paraflocculus |
| PMnR | paramedian raphe nu. |
| PnC | pontine caudal nu. |
| PnO | pontine oral nu. |
| PoDG | polymorphic layer DG |
| PPM | post. par. Cx, medial |
| PPTgC | peduncpont. tegm. nu., compact |
| PPTgD | peduncpont. tegm. nu., diff. |
| PPV | post. par. Cx, vent. |
| ProS | prosubiculum |
| PrS | presubiculum |
| pt | post. temp. sulc. |
| py | pyramidal tract |
| Py | strat. pyramid. CA |
| Rad | strat. radiat. CA |
| Rbd | rhabdoid nu. |
| RPO | rostral perioliv. nu. |
| rs | rubrospinal tract |
| RtTg | reticulotegm. nu. |
| S | subiculum |
| s5 | sensory root 5n |
| Sag | sagulum nu. |
| scp | sup. cerebell. ped. |
| SPTg | subpedunc. tegm. nu. |
| SuG | sup. gray sup. coll. |
| TCF | transverse cerebral fiss. |
| TF | temp. field F Cx |
| ts | tectospinal tract |
| tz | trapezoid body |
| Tz | nu. of trapezoid body |
| V2v | V2 (vent.) Cx |
| vhc | vent. hipp. comm. |
| VLA (V4) | ventlatant. (V4) Cx |
| VLP (V3v) | ventlatpost. (V3v) Cx |
| VLPAG | ventrolat. periaq. gr. |

# Plate 36a

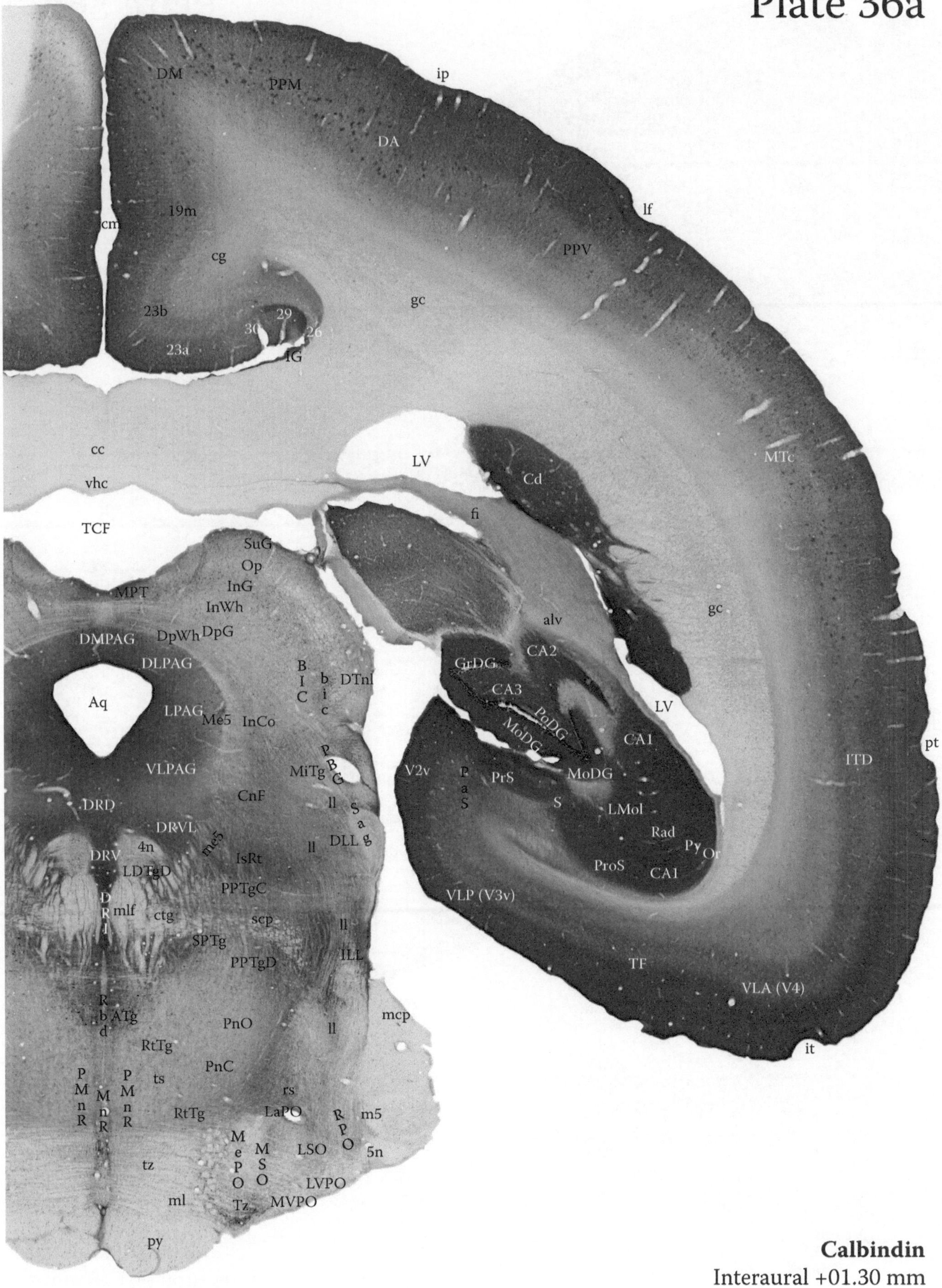


**Calbindin**
Interaural +01.30 mm

# Plate 36b

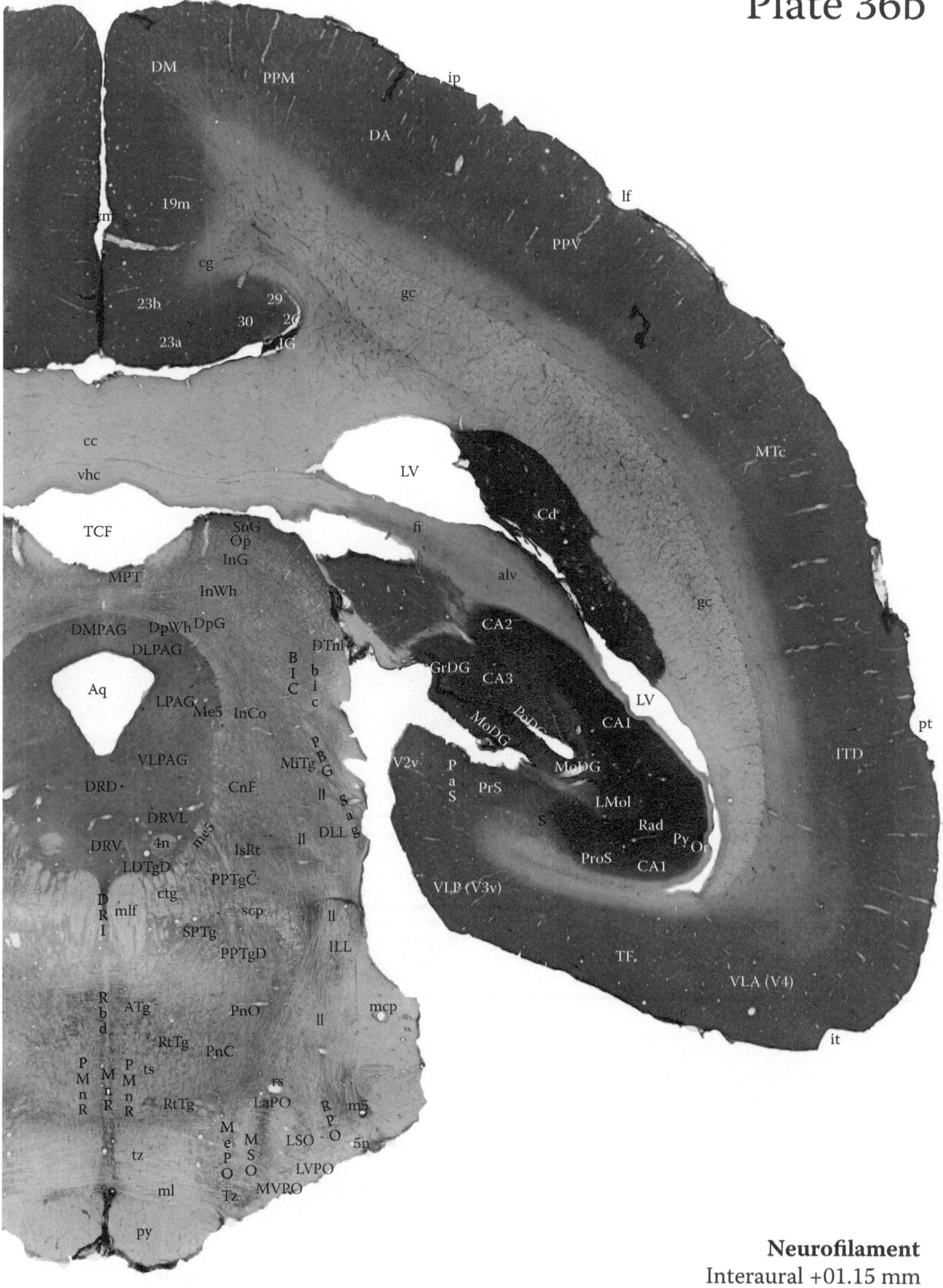


**Neurofilament**
Interaural +01.15 mm

# Plate 36c



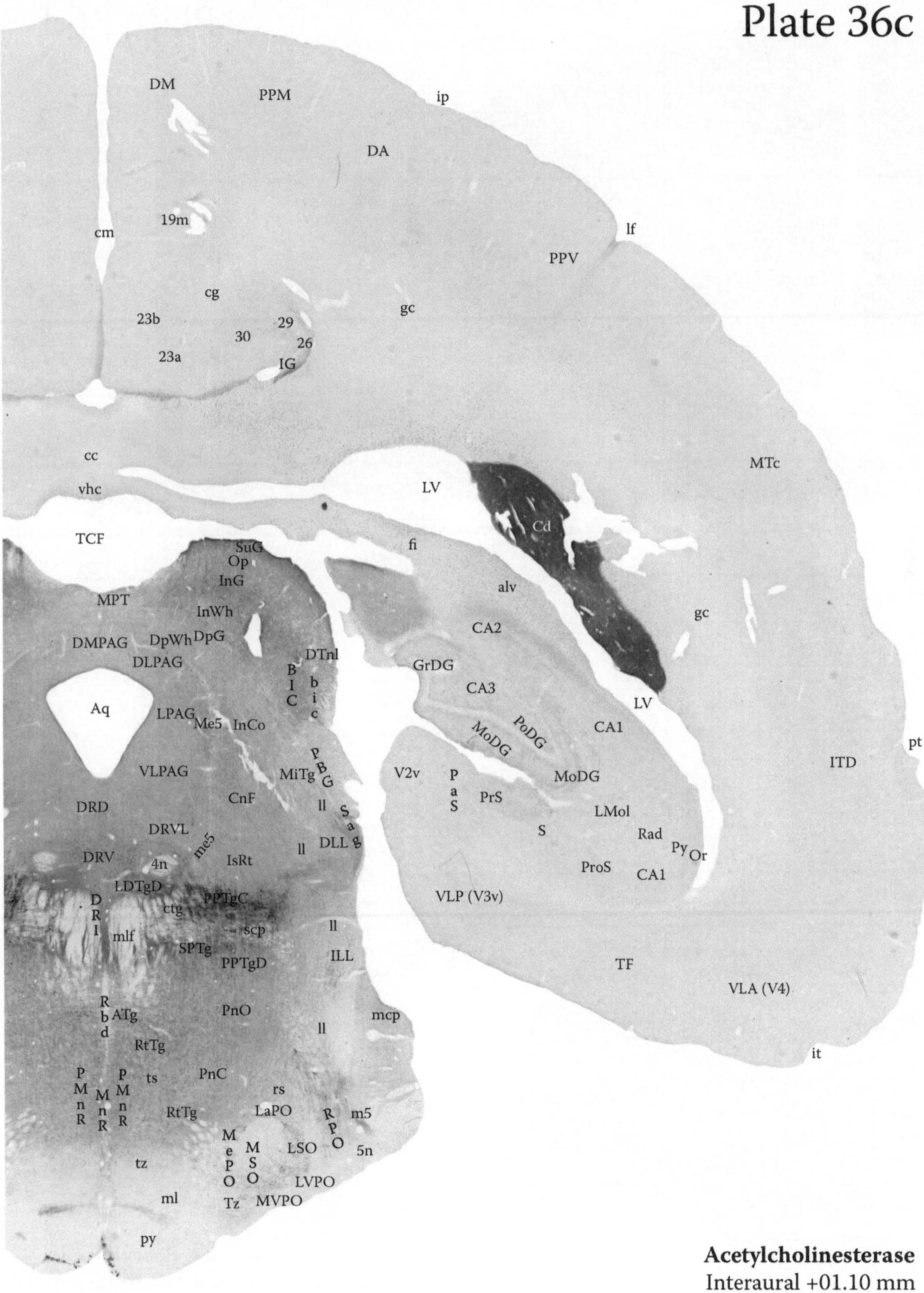


**Acetylcholinesterase**
Interaural +01.10 mm

# Plate 36d

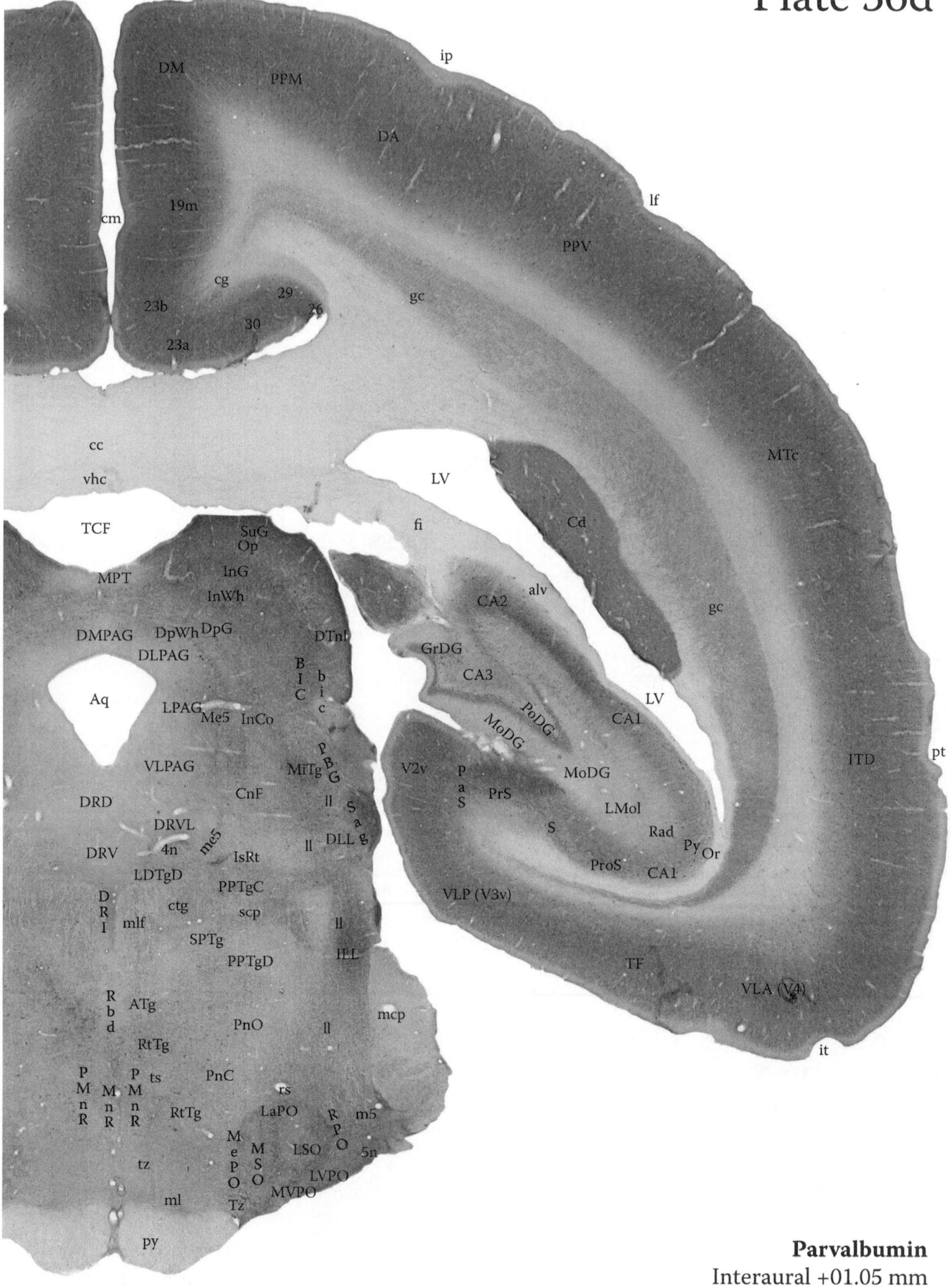


**Parvalbumin**
Interaural +01.05 mm

# Plate 36e

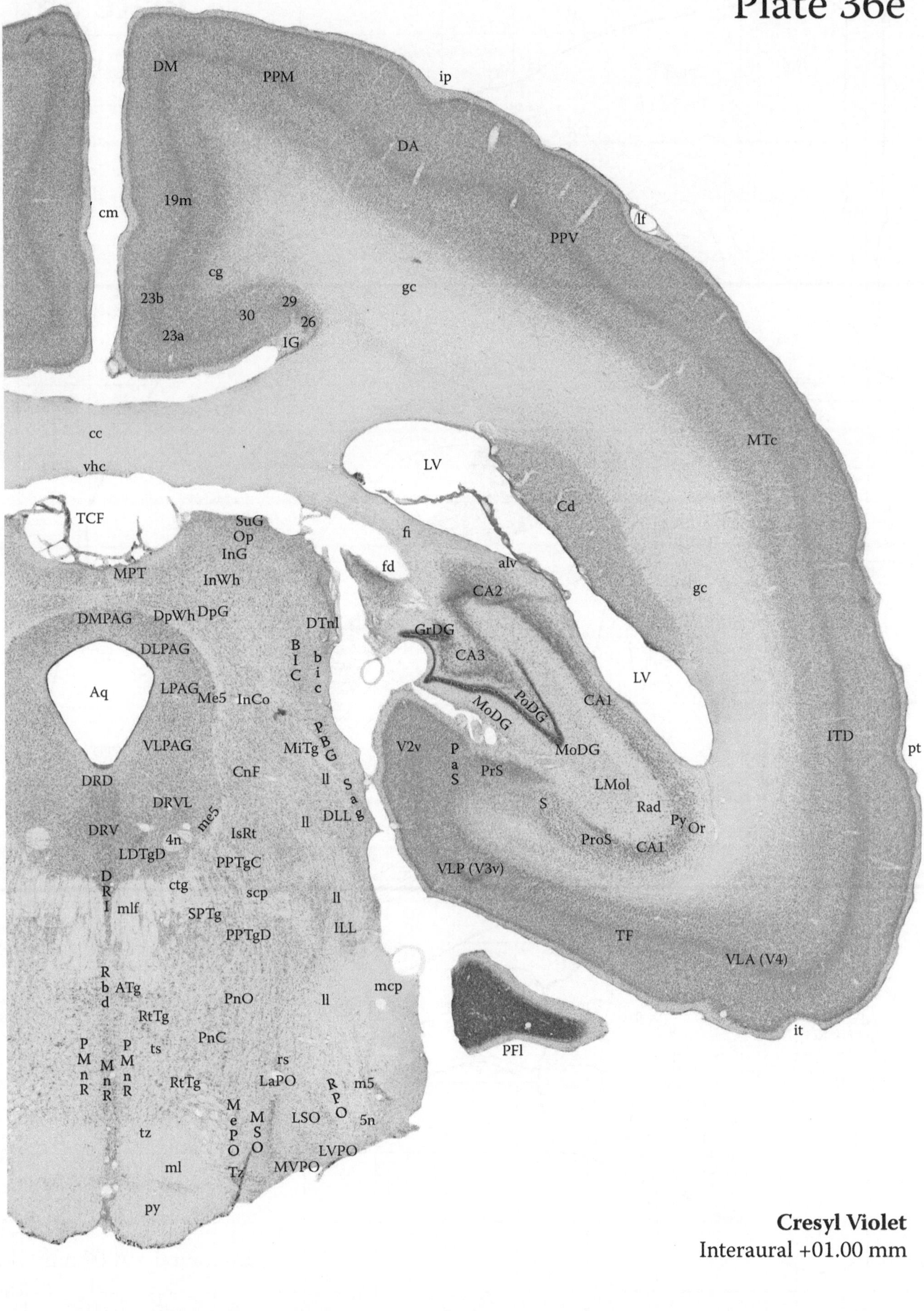


**Cresyl Violet**
Interaural +01.00 mm

# Plate 36

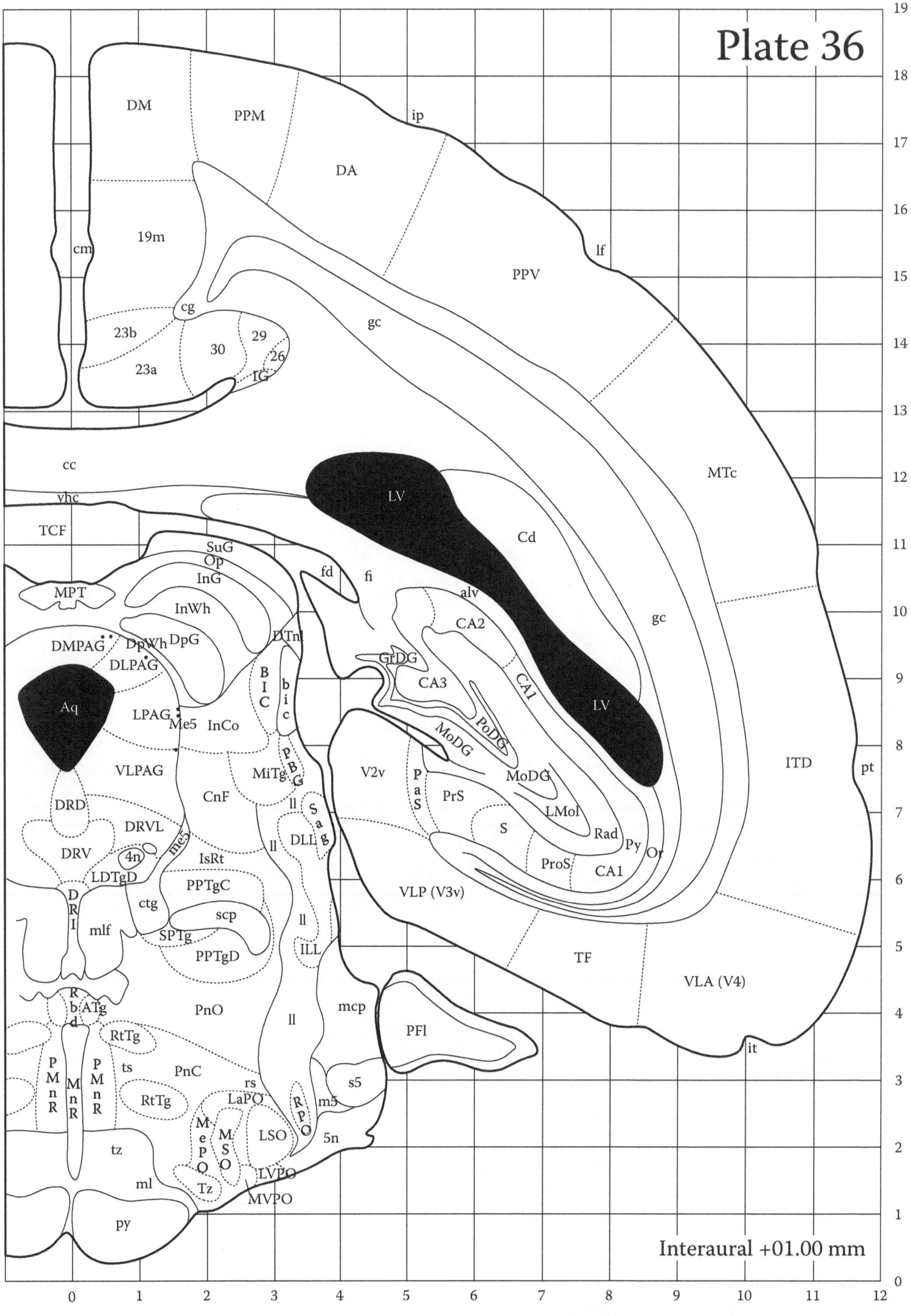


Interaural +01.00 mm

| | |
|---|---|
| 4n | trochlear nerve |
| 5n | trigeminal nerve |
| 19m | area 19m Cx |
| 23a | area 23a Cx |
| 23b | area 23b Cx |
| 26 | area 26 Cx |
| 29 | area 29 Cx |
| 30 | area 30 Cx |
| alv | alveus |
| Aq | cerebral aqueduct |
| ATg | ant. tegmental nu. |
| bic | brach. inf. coll. |
| BIC | brach. inf. coll. nu. |
| CA1 | cornu ammonis 1 |
| CA2 | cornu ammonis 2 |
| CA3 | cornu ammonis 3 |
| cc | corp. callosum |
| Cd | caudate nu. |
| cg | cingulum |
| cm | callosomarginal sulc. |
| CnF | cuneiform nu. |
| ctg | central tegm. tract |
| DA | dorsoant. visual Cx |
| DLL | dorsal nu. lat. lemn. |
| DLPAG | dorsolat. periaq. gr. |
| DM | dorsomed. visual Cx |
| DMPAG | dorsomed. periaq. gr |
| DpG | deep gray sup. coll. |
| DpWh | deep white sup. coll. |
| DRD | dors. raphe nu., dors. |
| DRI | dors. raphe nu., interf. |
| DRV | dors. raphe nu., vent. |
| DRVL | dors. raphe nu., ventrolat. |
| DTnl | dors. term. nu. |
| fd | fimbriodent. sulcus |
| fi | fimbria |
| gc | geniculocalcarine tr. |
| GrDG | gran. cell layer DG |
| IG | induseum gris. |
| ILL | intermed. nu. lat. lemn. |
| InCo | intercoll. nu. |
| InG | intermed. gr. sup. coll. |
| InWh | intermed. wh. sup coll. |
| ip | intraparietal sulc. |
| IsRt | isthmic retic. formn |
| it | inf. temporal sulcus |
| ITD | inf. temp. dors. Cx |
| LaPO | lat. periolivary nu. |
| LDTgD | laterod. tegm nu., dors. |
| lf | lateral fissure |
| ll | lateral lemniscus |
| LMol | strat. lacun. mol. CA |
| LPAG | lat. periaqued. gr. |
| LSO | lat. sup. olivary nu. |
| LV | lateral ventricle |
| LVPO | laterovent. perioliv. nu. |
| m5 | motor root 5n |
| mcp | middle cerebell. ped. |
| me5 | mesen. trigem. tr. |
| Me5 | mesen. trigem. nu. |
| MePO | med. periolivary nu. |
| MiTg | microcell. tegm. nu. |
| ml | medial lemniscus |
| mlf | med. longit. fasc. |
| MnR | median raphe nu. |
| MolDG | molecular layer DG |
| MPT | med. pretect. nu. |
| MSO | med. sup. olivary nu. |
| MTc | mid. temp. cresc. Cx |
| MVPO | mediovent. perioliv. nu. |
| Op | strat. opt. sup. coll. |
| Or | strat. oriens CA |
| PaS | parasubiculum |
| PBG | parabigeminal nu. |
| PFl | paraflocculus |
| PMnR | paramedian raphe nu. |
| PnC | pontine caudal nu. |
| PnO | pontine oral nu. |
| PoDG | polymorphic layer DG |
| PPM | post. par. Cx, medial |
| PPTgC | peduncpont. tegm. nu., compact |
| PPTgD | peduncpont. tegm. nu., diff. |
| PPV | post. par. Cx, vent. |
| ProS | prosubiculum |
| PrS | presubiculum |
| pt | post. temp. sulc. |
| py | pyramidal tract |
| Py | strat. pyramid. CA |
| Rad | strat. radiat. CA |
| Rbd | rhabdoid nu. |
| RPO | rostral perioliv. nu. |
| rs | rubrospinal tract |
| RtTg | reticulotegm. nu. |
| S | subiculum |
| s5 | sensory root 5n |
| Sag | sagulum nu. |
| scp | sup. cerebell. ped. |
| SPTg | subpedunc. tegm. nu. |
| SuG | sup. gray sup. coll. |
| TCF | transverse cerebral fiss. |
| TF | temp. field F Cx |
| ts | tectospinal tract |
| tz | trapezoid body |
| Tz | nu. of trapezoid body |
| V2v | V2 (vent.) Cx |
| vhc | vent. hipp. comm. |
| VLA (V4) | ventlatant. (V4) Cx |
| VLP (V3v) | ventlatpost. (V3v) Cx |
| VLPAG | ventrolat. periaq. gr. |

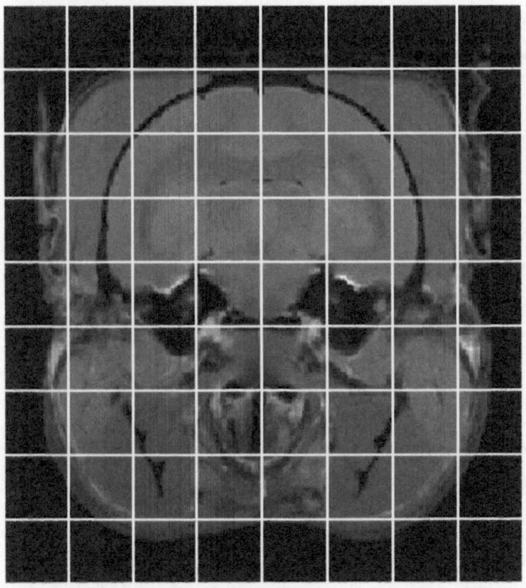

Interaural +01.15 mm

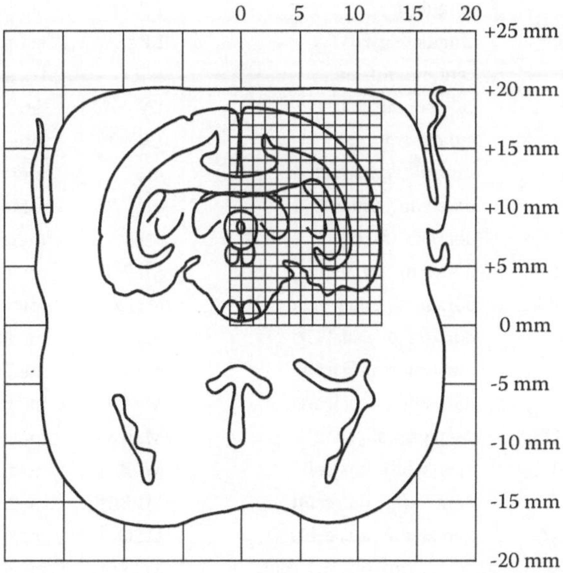

# Plate 37

Interaural +00.45 mm

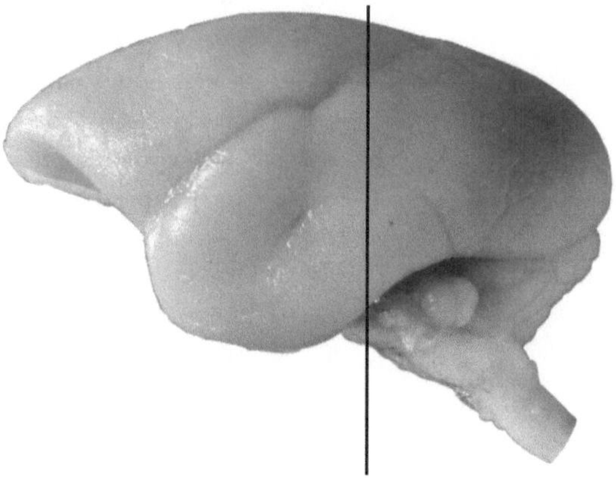

| | |
|---|---|
| 4n | trochlear nerve |
| 5Ma | trigem. nu., masset. |
| 5n | trigeminal nerve |
| 5Pt | trigem. nu., pteryg. |
| 5Te | trigem. nu., temp. |
| 19m | area 19m Cx |
| 23 | area 23 Cx |
| 29 | area 29 Cx |
| 30 | area 30 Cx |
| A5 | A5 noradr. cells |
| alv | alveus |
| Aq | cerebral aqueduct |
| bic | brach. inf. coll. |
| BIC | brach. inf. coll. nu. |
| CA1 | cornu ammonis 1 |
| CA2 | cornu ammonis 2 |
| CA3 | cornu ammonis 3 |
| cc | corp. callosum |
| cg | cingulum |
| CG | cingulate gyrus Cx |
| CIC | cent. nu. inf. coll. |
| cm | callosomarginal sulc. |
| CnF | cuneiform nu. |
| ctg | central tegm. tract |
| DA | dorsoant. visual Cx |
| DCIC | dors. Cx inf. coll. |
| DLL | dorsal nu. lat. lemn. |
| DLPAG | dorsolat. periaq. gr. |
| DM | dorsomed. visual Cx |
| DMPAG | dorsomed. periaq. gr |
| DMTg | dorsomed. tegm. area |
| DpG | deep gray sup. coll. |
| DpWh | deep white sup. coll. |
| DRC | dors. raphe nu., caud. |
| DTgC | dors. tegm. nu., cent. |
| DTgP | dors. tegm. nu., pericent. |
| ECIC | extern Cx inf. coll. |
| fd | fimbriodent. sulcus |
| fi | fimbria |
| gc | geniculocalcarine tr. |
| GrDG | gran. cell layer DG |
| I5 | intertrigem. nu. |
| IG | induseum gris. |
| InG | intermed. gr. sup. coll. |
| InWh | intermed. wh. sup coll. |
| ip | intraparietal sulc. |
| it | inf. temporal sulcus |
| KF | Kölliker-Fuse nu. |
| LaPO | lat. periolivary nu. |
| LDTgD | laterod. tegm nu., dors. |
| LDTgV | laterod. tegm nu., vent. |
| ll | lateral lemniscus |
| LMol | strat. lacun. mol. CA |
| LPAG | lat. periaqued. gr. |
| LPBC | lat. parabr. nu., cent. |
| LPBD | lat. parabr. nu., dors. |
| LPBE | lat. parabr. nu., ext. |
| LSO | lat. sup. olivary nu. |
| LV | lateral ventricle |
| LVPO | laterovent. perioliv. nu. |
| m5 | motor root 5n |
| mcp | middle cerebell. ped. |
| Me5 | mesen. trigem. nu. |
| MePO | med. periolivary nu. |
| MiTg | microcell. tegm. nu. |
| ml | medial lemniscus |
| mlf | med. longit. fasc. |
| MnR | median raphe nu. |
| MolDG | molecular layer DG |
| MPB | med. parabrach. nu. |
| MPBE | med. parabr. nu., ext. |
| MSO | med. sup. olivary nu. |
| MVPO | mediovent. perioliv. nu. |
| Op | strat. opt. sup. coll. |
| Or | strat. oriens CA |
| P5 | peritrigeminal nu. |
| PaS | parasubiculum |
| PC5 | parv. trigem. nu. |
| PFlV | paraflocculus vent. |
| PMnR | paramedian raphe nu. |
| PnC | pontine caudal nu. |
| PPM | post. par. Cx, medial |
| PPTgC | peduncpont. tegm. nu., compact |
| PPV | post. par. Cx, vent. |
| Pr5 | princip. trigem. nu. |
| ProS | prosubiculum |
| PrS | presubiculum |
| pt | post. temp. sulc. |
| py | pyramidal tract |
| Rad | strat. radiat. CA |
| rs | rubrospinal tract |
| S | subiculum |
| s5 | sensory root 5n |
| Sag | sagulum nu. |
| scp | sup. cerebell. ped. |
| Su5 | supratrigeminal nu. |
| SubCD | subcoer. nu., dors. |
| SubCV | subcoer. nu., vent. |
| SuG | sup. gray sup. coll. |
| TCF | transverse cerebral fiss. |
| ts | tectospinal tract |
| tz | trapezoid body |
| Tz | nu. of trapezoid body |
| V2v | V2 (vent.) Cx |
| VLA (V4) | ventlatant. (V4) Cx |
| VLP (V3v) | ventlatpost. (V3v) Cx |
| VLPAG | ventrolat. periaq. gr. |
| vsc | vent. spinocereb. tr. |
| Zo | strat. zonale SC |

# Plate 37a

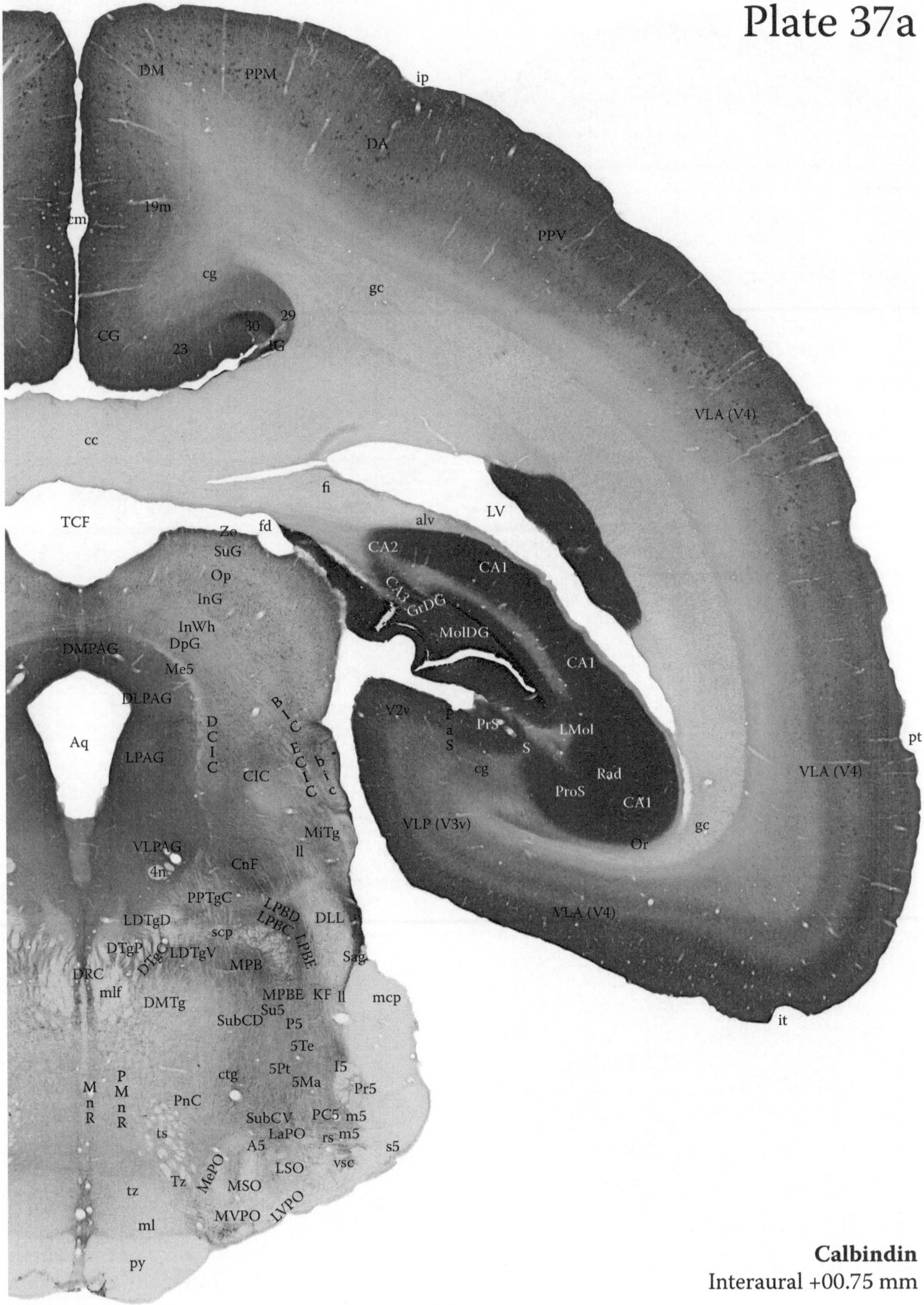


**Calbindin**
Interaural +00.75 mm

# Plate 37b

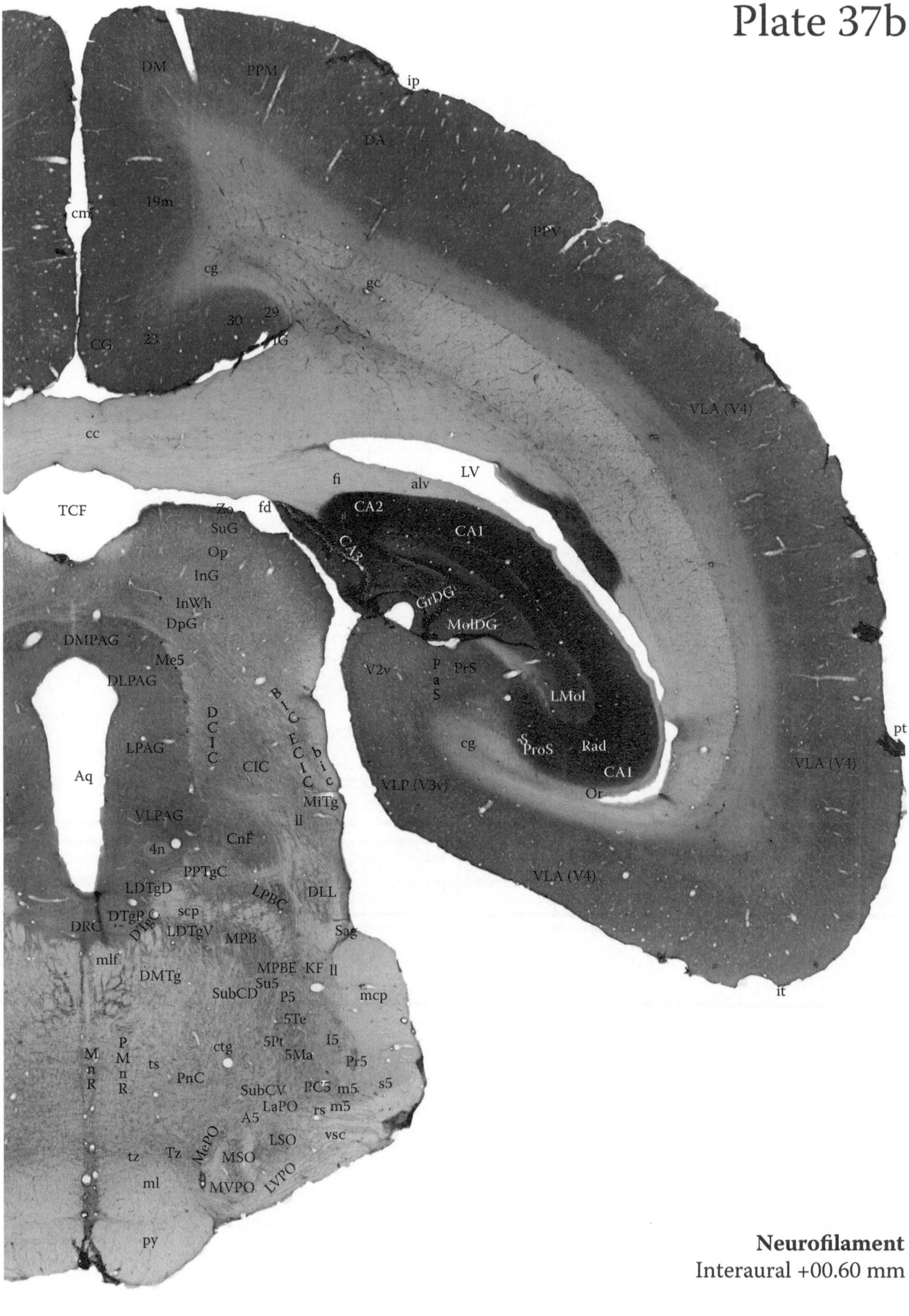


**Neurofilament**
Interaural +00.60 mm

# Plate 37c

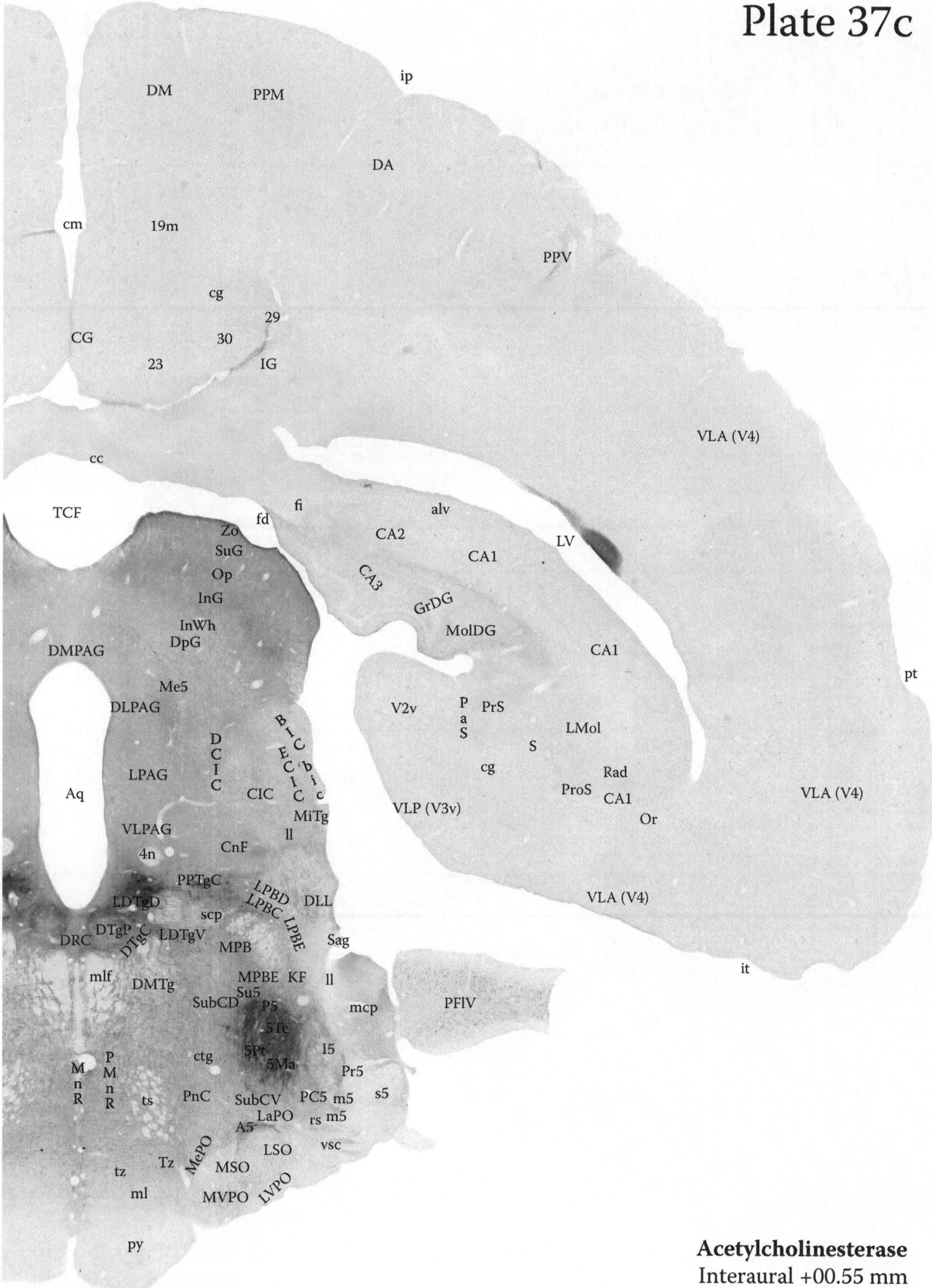


**Acetylcholinesterase**
Interaural +00.55 mm

# Plate 37d

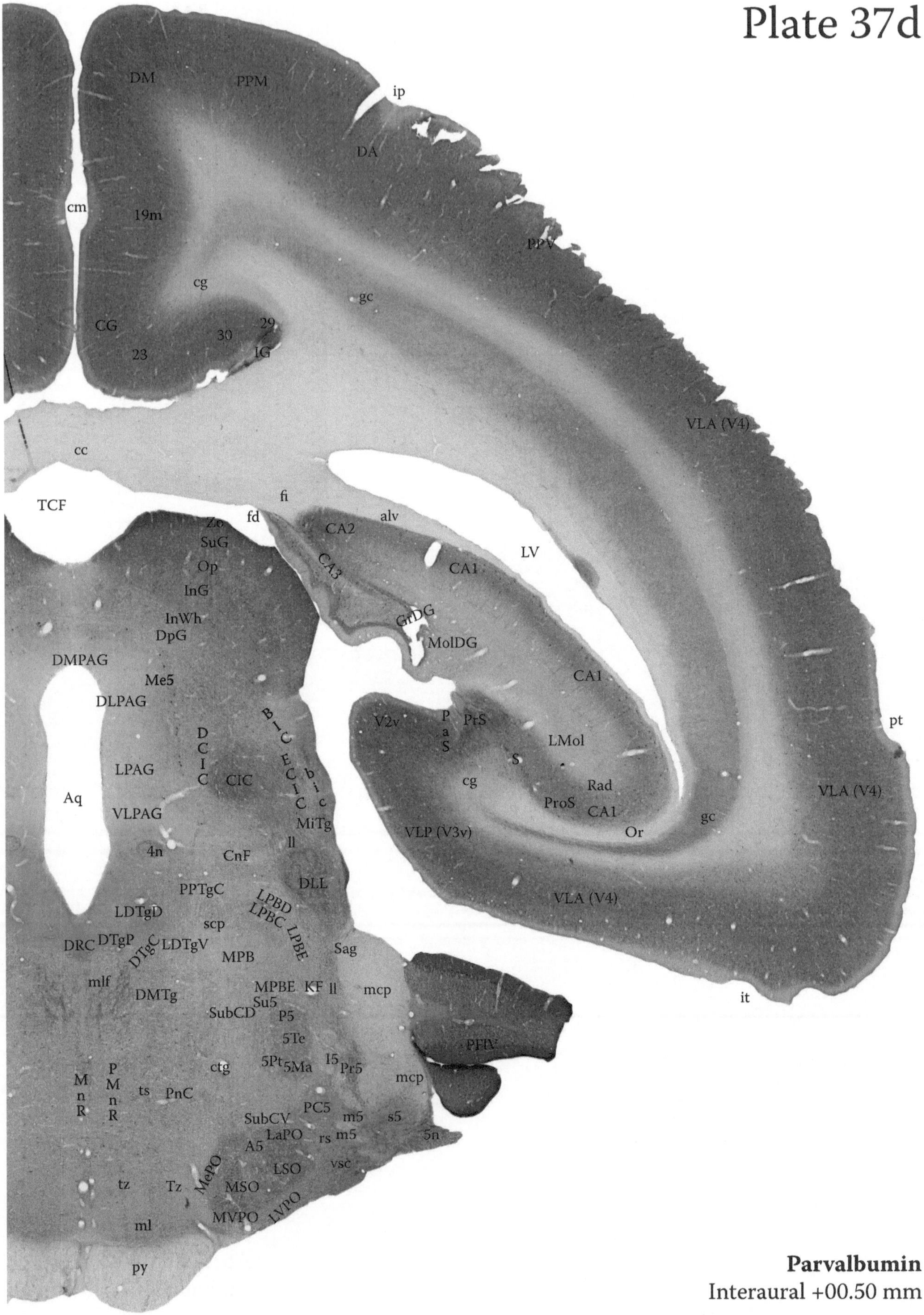


**Parvalbumin**
Interaural +00.50 mm

# Plate 37e

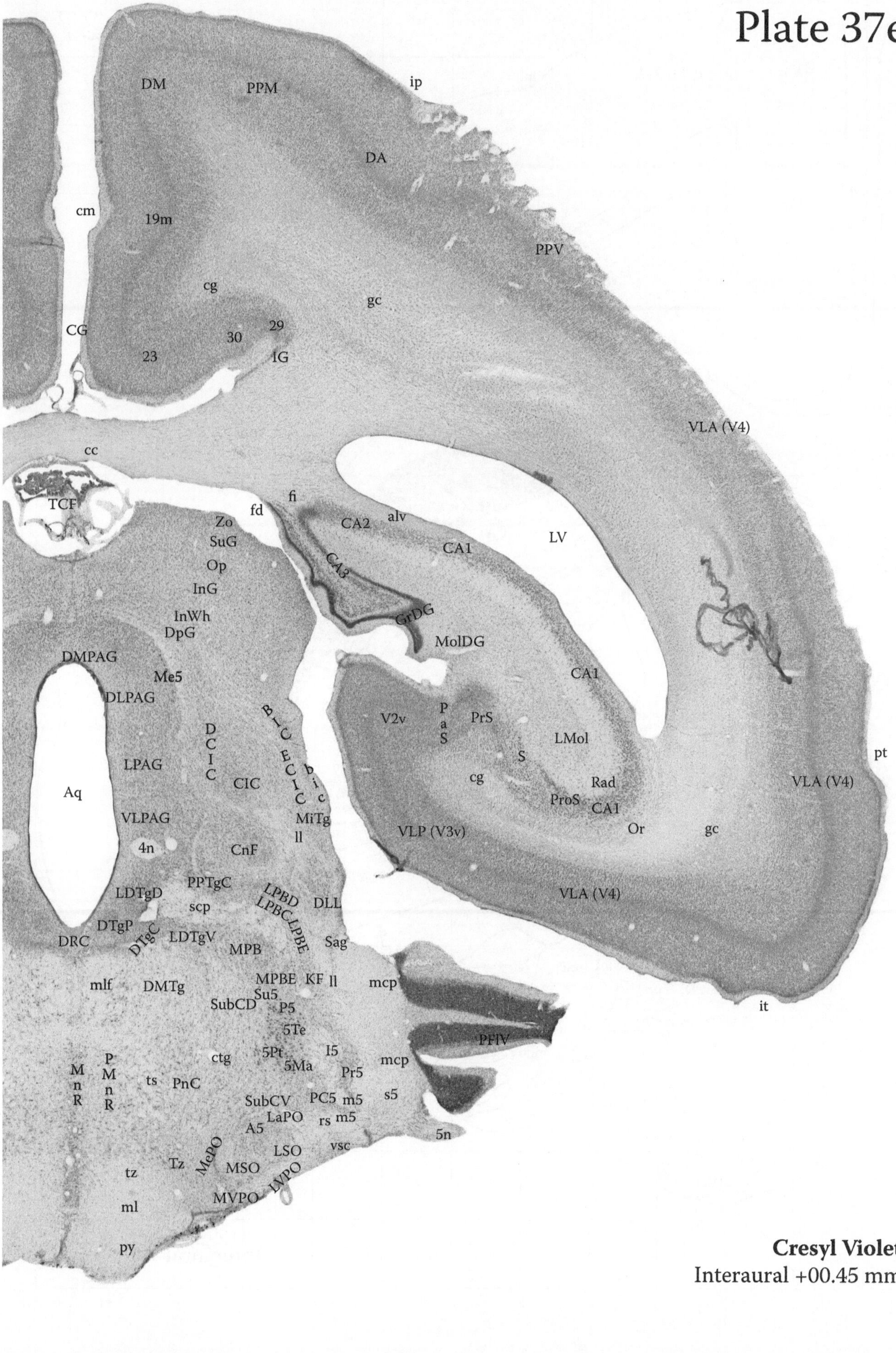


**Cresyl Violet**
Interaural +00.45 mm

# Plate 37

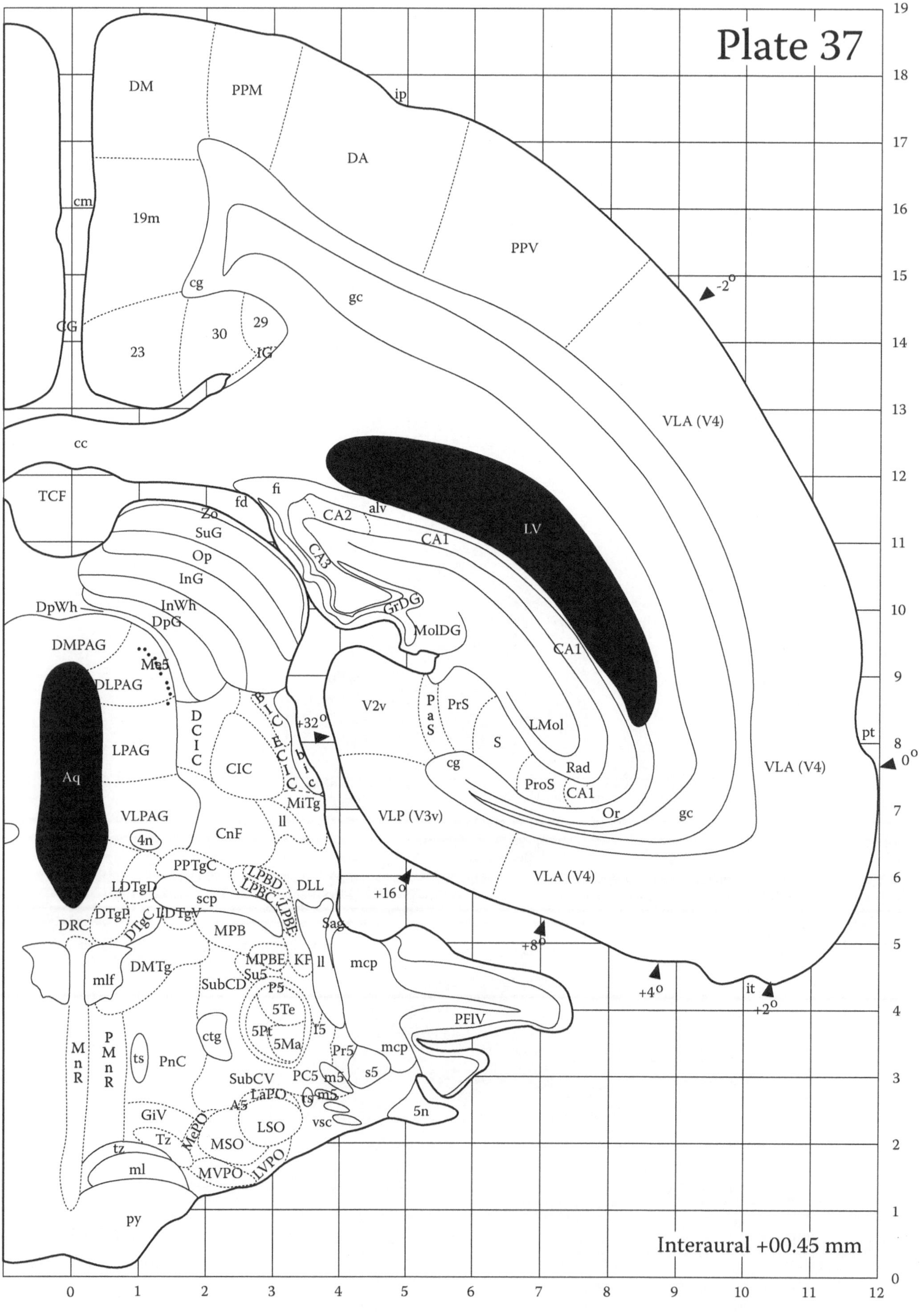


Interaural +00.45 mm

| | | | | | |
|---|---|---|---|---|---|
| 4n | trochlear nerve | fd | fimbriodent. sulcus | Op | strat. opt. sup. coll. |
| 5Ma | trigem. nu., masset. | fi | fimbria | Or | strat. oriens CA |
| 5n | trigeminal nerve | gc | geniculocalcarine tr. | P5 | peritrigeminal nu. |
| 5Pt | trigem. nu., pteryg. | GrDG | gran. cell layer DG | PaS | parasubiculum |
| 5Te | trigem. nu., temp. | I5 | intertrigem. nu. | PC5 | parv. trigem. nu. |
| 19m | area 19m Cx | IG | induseum gris. | PFlV | paraflocculus vent. |
| 23 | area 23 Cx | InG | intermed. gr. sup. coll. | PMnR | paramedian raphe nu. |
| 29 | area 29 Cx | InWh | intermed. wh. sup coll. | PnC | pontine caudal nu. |
| 30 | area 30 Cx | ip | intraparietal sulc. | PPM | post. par. Cx, medial |
| A5 | A5 noradr. cells | it | inf. temporal sulcus | PPTgC | peduncpont. tegm. nu., compact |
| alv | alveus | KF | Kölliker-Fuse nu. | PPV | post. par. Cx, vent. |
| Aq | cerebral aqueduct | LaPO | lat. periolivary nu. | Pr5 | princip. trigem. nu. |
| bic | brach. inf. coll. | LDTgD | laterod. tegm nu., dors. | ProS | prosubiculum |
| BIC | brach. inf. coll. nu. | LDTgV | laterod. tegm nu., vent. | PrS | presubiculum |
| CA1 | cornu ammonis 1 | ll | lateral lemniscus | pt | post. temp. sulc. |
| CA2 | cornu ammonis 2 | LMol | strat. lacun. mol. CA | py | pyramidal tract |
| CA3 | cornu ammonis 3 | LPAG | lat. periaqued. gr. | Rad | strat. radiat. CA |
| cc | corp. callosum | LPBC | lat. parabr. nu., cent. | rs | rubrospinal tract |
| cg | cingulum | LPBD | lat. parabr. nu., dors. | S | subiculum |
| CG | cingulate gyrus Cx | LPBE | lat. parabr. nu., ext. | s5 | sensory root 5n |
| CIC | cent. nu. inf. coll. | LSO | lat. sup. olivary nu. | Sag | sagulum nu. |
| cm | callosomarginal sulc. | LV | lateral ventricle | scp | sup. cerebell. ped. |
| CnF | cuneiform nu. | LVPO | laterovent. perioliv. nu. | Su5 | supratrigeminal nu. |
| ctg | central tegm. tract | m5 | motor root 5n | SubCD | subcoer. nu., dors. |
| DA | dorsoant. visual Cx | mcp | middle cerebell. ped. | SubCV | subcoer. nu., vent. |
| DCIC | dors. Cx inf. coll. | Me5 | mesen. trigem. nu. | SuG | sup. gray sup. coll. |
| DLL | dorsal nu. lat. lemn. | MePO | med. periolivary nu. | TCF | transverse cerebral fiss. |
| DLPAG | dorsolat. periaq. gr. | MiTg | microcell. tegm. nu. | ts | tectospinal tract |
| DM | dorsomed. visual Cx | ml | medial lemniscus | tz | trapezoid body |
| DMPAG | dorsomed. periaq. gr | mlf | med. longit. fasc. | Tz | nu. of trapezoid body |
| DMTg | dorsomed. tegm. area | MnR | median raphe nu. | V2v | V2 (vent.) Cx |
| DpG | deep gray sup. coll. | MolDG | molecular layer DG | VLA (V4) | ventlatant. (V4) Cx |
| DpWh | deep white sup. coll. | MPB | med. parabrach. nu. | VLP (V3v) | ventlatpost. (V3v) Cx |
| DRC | dors. raphe nu., caud. | MPBE | med. parabr. nu., ext. | VLPAG | ventrolat. periaq. gr. |
| DTgC | dors. tegm. nu., cent. | MSO | med. sup. olivary nu. | vsc | vent. spinocereb. tr. |
| DTgP | dors. tegm. nu., pericent. | MVPO | mediovent. perioliv. nu. | Zo | strat. zonale SC |
| ECIC | extern Cx inf. coll. | | | | |

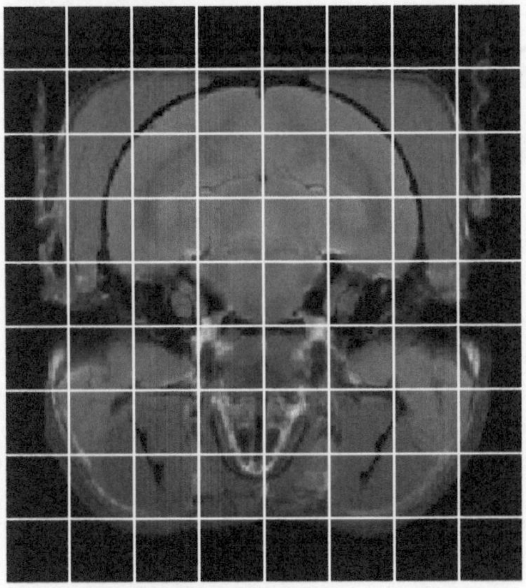

Interaural +00.20 mm

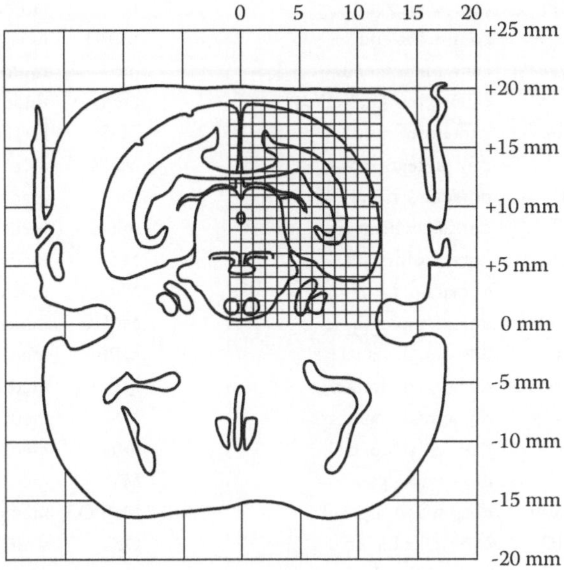

# Plate 38

Interaural -00.05 mm

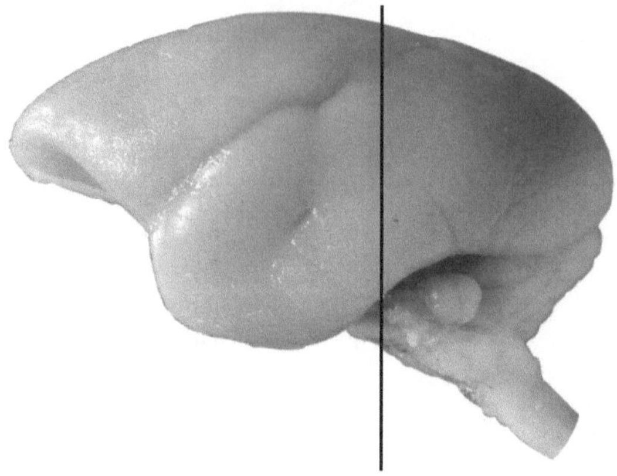

| | |
|---|---|
| 2Cb | cerebell. lobule 2b |
| 4n | trochlear nerve |
| 4nd | trochlear nerve decn |
| 4V | fourth ventricle |
| 5n | trigeminal nerve |
| 6N | abducens nu. |
| 7n | facial nerve |
| 7N | facial (motor) nu. |
| 8n | vestibulocochlear nerve |
| 19m | area 19m Cx |
| 23 | area 23 Cx |
| 23v | area 23v Cx |
| 29 | area 29 Cx |
| 30 | area 30 Cx |
| A5 | A5 noradr. cells |
| Bar | Barrington's nu. |
| bic | brach. inf. coll. |
| BIC | brach. inf. coll. nu. |
| C1 | C1 adrenaline cells |
| ca | calcarine sulcus |
| CA1 | cornu ammonis 1 |
| cc | corp. callosum |
| cg | cingulum |
| CG | cingulate gyrus Cx |
| CGPn | central gray pons |
| cic | comm. inf. coll. |
| CIC | cent. nu. inf. coll. |
| cm | callosomarginal sulc. |
| CnF | cuneiform nu. |
| ctg | central tegm. tract |
| DA | dorsoant. visual Cx |
| DC | dorsal cochlear nu. |
| DCIC | dors. Cx inf. coll. |
| DLL | dorsal nu. lat. lemn. |
| DLPAG | dorsolat. periaq. gr. |
| DM | dorsomed. visual Cx |
| DMPAG | dorsomed. periaq. gr |
| DMTg | dorsomed. tegm. area |
| DpG | deep gray sup. coll. |
| DPGi | dors. paragigant. nu. |
| DpWh | deep white sup. coll. |
| ECIC | extern Cx inf. coll. |
| g7 | genu of facial n. |
| gc | geniculocalcarine tr. |
| Gi | gigantocell. retic. nu. |
| GiV | gigantocell. retic. nu., vent. |
| I5 | intertrigem. nu. |
| IG | induseum gris. |
| InCo | intercoll. nu. |
| InG | intermed. gr. sup. coll. |
| InWh | intermed. wh. sup coll. |
| IOD | inf. oliv. nu., dors. |
| IODM | inf. oliv. nu., dorsomed. |
| IOPr | inf. oliv. nu., princ. |
| ip | intraparietal sulc. |
| it | inf. temporal sulcus |
| KF | Kölliker-Fuse nu. |
| LaPO | lat. periolivary nu. |
| LC | locus coeruleus |
| LMol | strat. lacun. mol. CA |
| LPAG | lat. periaqued. gr. |
| LPBC | lat. parabr. nu., cent. |
| LPBD | lat. parabr. nu., dors. |
| LPBE | lat. parabr. nu., ext. |
| LPBV | lat. parabr. nu., vent. |
| LPGi | lat. paragigant. nu. |
| LSO | lat. sup. olivary nu. |
| LV | lateral ventricle |
| LVPO | laterovent. perioliv. nu. |
| m5 | motor root 5n |
| mcp | middle cerebell. ped. |
| Me5 | mesen. trigem. nu. |
| MePO | med. periolivary nu. |
| ml | medial lemniscus |
| mlf | med. longit. fasc. |
| MnR | median raphe nu. |
| Mo5 | motor trigeminal nu. |
| MoDG | molecular layer DG |
| MPB | med. parabrach. nu. |
| MPBE | med. parabr. nu., ext. |
| MSO | med. sup. olivary nu. |
| mte | med. temp. sulc. |
| MVe | med. vestib. nu. |
| MVPO | mediovent. perioliv. nu. |
| Op | strat. opt. sup. coll. |
| Or | strat. oriens CA |
| P5 | peritrigeminal nu. |
| P7 | perifacial nu. |
| Pa6 | parabducens nu. |
| PBG | parabigeminal nu. |
| PCRtα | parv. retic. nu., alpha |
| PCTg | paracollicular tegm. |
| PFl | perifornical nu., lat. |
| PMnR | paramedian raphe nu. |
| PnC | pontine caudal nu. |
| PnR | pontine raphe nu. |
| ProS | prosubiculum |
| ProSt | prostriate cortex |
| PrS | presubiculum |
| pt | post. temp. sulc. |
| py | pyramidal tract |
| Rad | strat. radiat. CA |
| ReIC | recess inf. coll. |
| RMg | raphe magnus nu. |
| RPa | raphe pallidus nu. |
| rs | rubrospinal tract |
| RVL | rost. ventrolat. area |
| S | subiculum |
| s5 | sensory root 5n |
| scp | sup. cerebell. ped. |
| smv | sup. medull. velum |
| sp5 | spinal trigeminal tr. |
| Sp5I | spin. trigem. nu., interpol. |
| Sp5O | spin. trigem. nu., oral. |
| Su5 | supratrigeminal nu. |
| SubCD | subcoer. nu., dors. |
| SubCV | subcoer. nu., vent. |
| SuG | sup. gray sup. coll. |
| SuVe | sup. vestib. nu. |
| ts | tectospinal tract |
| tz | trapezoid body |
| Tz | nu. of trapezoid body |
| V2v | V2 (vent.) Cx |
| VCA | vent. coch. nu., ant. |
| VLA (V4) | ventlatant. (V4) Cx |
| VLP (V3v) | ventlatpost. (V3v) Cx |
| VLPAG | ventrolat. periaq. gr. |
| vsc | vent. spinocereb. tr. |
| Zo | strat. zonale SC |

# Plate 38a

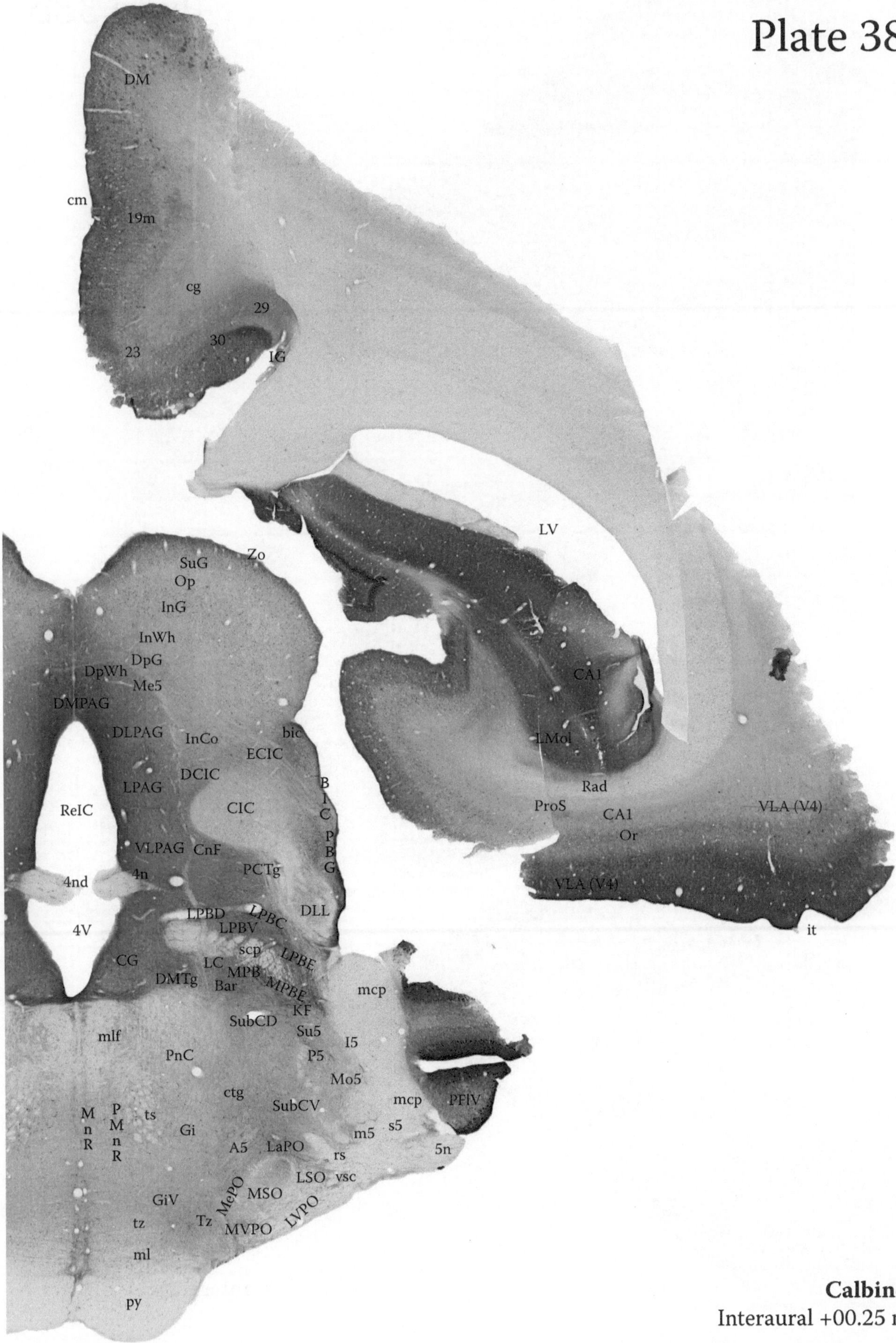


**Calbindin**
Interaural +00.25 mm

# Plate 38b

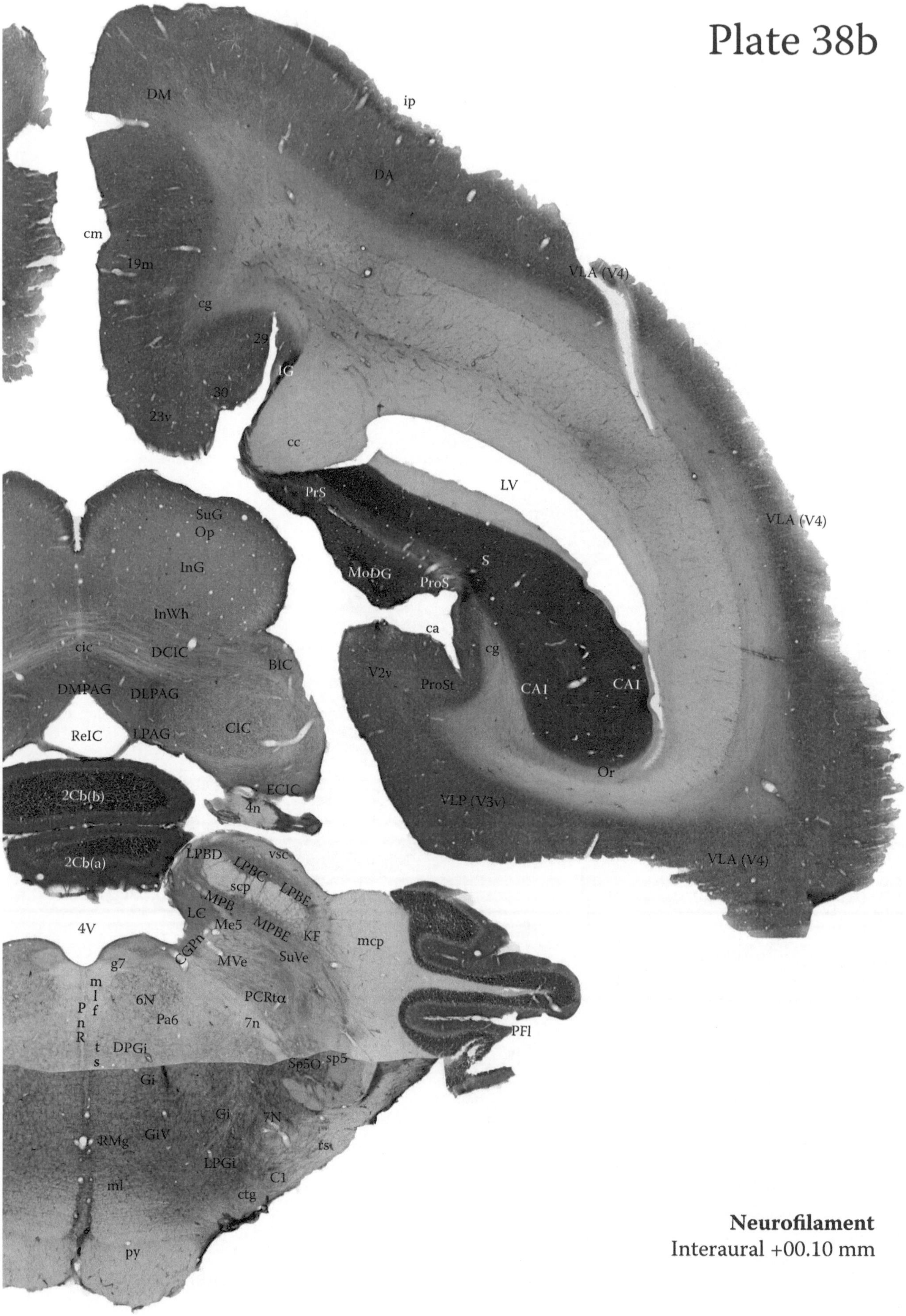


**Neurofilament**
Interaural +00.10 mm

# Plate 38c

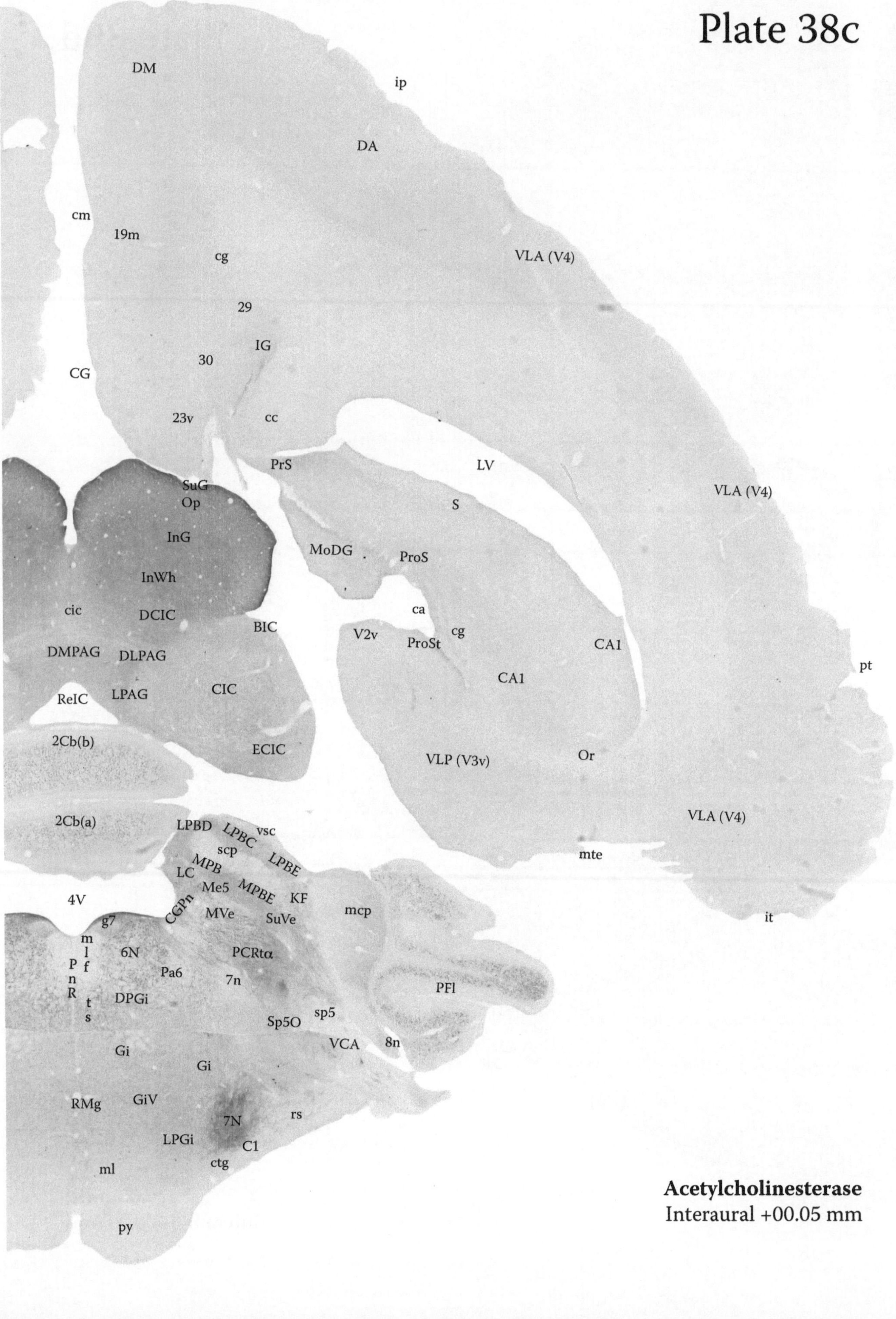


**Acetylcholinesterase**
Interaural +00.05 mm

# Plate 38d

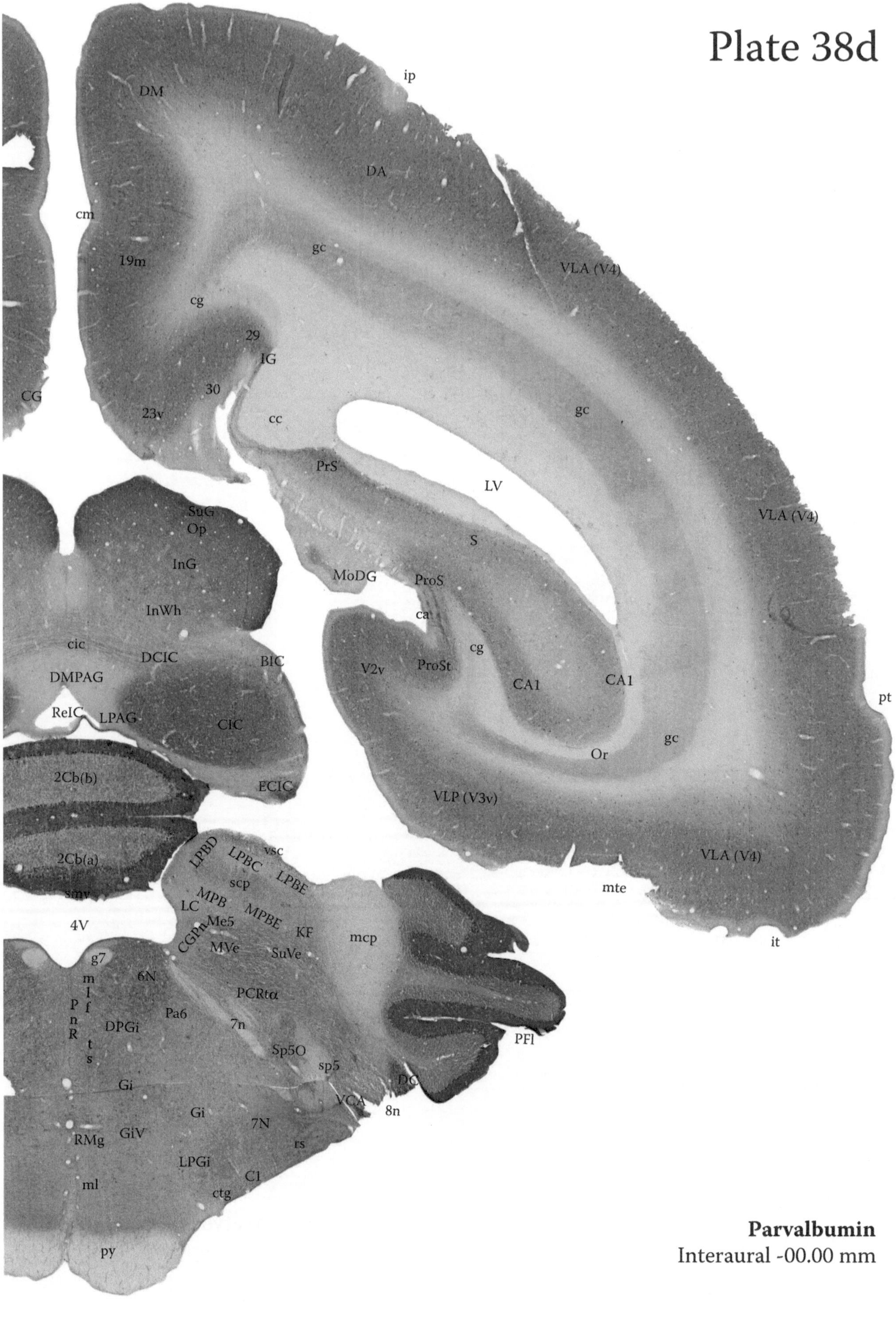


**Parvalbumin**
Interaural -00.00 mm

# Plate 38e

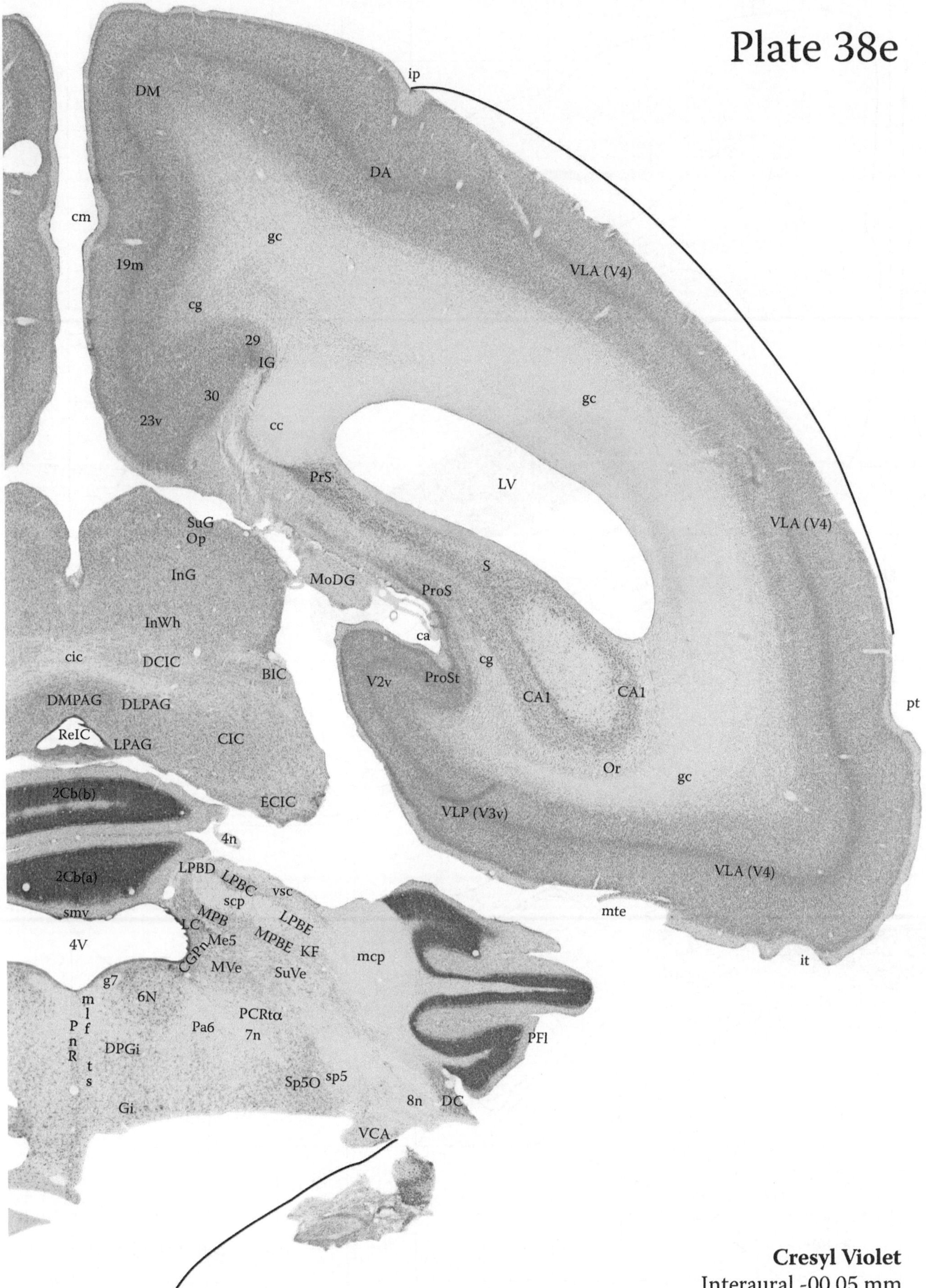


**Cresyl Violet**
Interaural -00.05 mm

# Plate 38

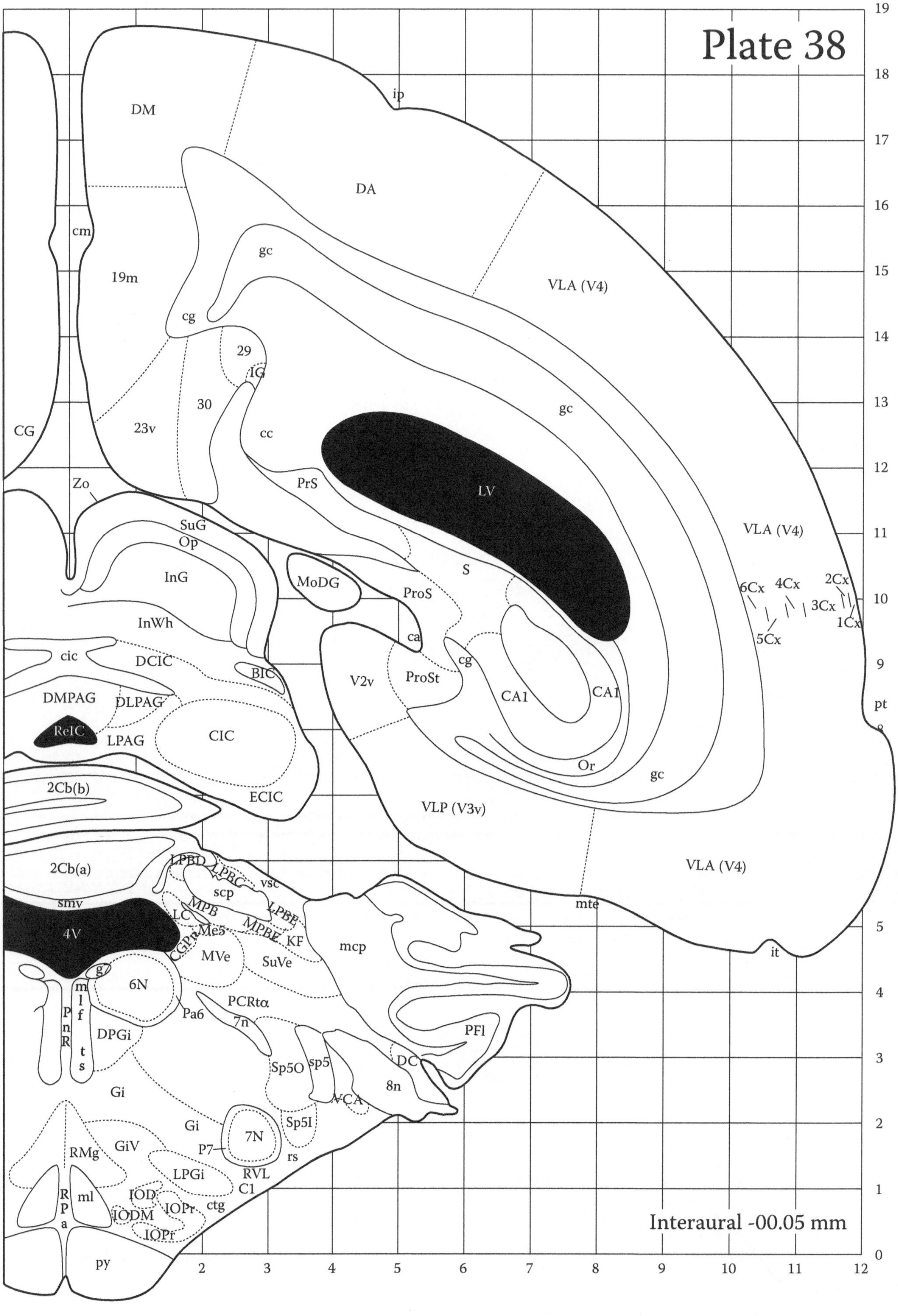
DM
ip
DA
cm
19m
gc
VLA (V4)
cg
29
IG
30
23v
CG
cc
gc
LV
PrS
Zo
SuG
Op
InG
MoDG
ProS
S
InWh
ca
cic
DCIC
BIC
V2v
ProSt
cg
CA1
CA1
DMPAG
DLPAG
ReIC
LPAG
CIC
Or
gc
2Cb(b)
ECIC
VLP (V3v)
VLA (V4)
6Cx
4Cx
2Cx
3Cx
1Cx
5Cx
pt
2Cb(a)
LPBD
LPBC
vsc
scp
smv
LC
MPB
LPBE
mte
4V
Me5
MPBE
KF
CGPn
MVe
SuVe
mcp
it
g7
6N
m
l
f
PnR
ts
Pa6
PCRtα
7n
DPGi
PFl
Sp5O
sp5
DC
8n
Gi
VCA
Gi
7N
Sp5I
RMg
GiV
P7
rs
LPGi
RVL
C1
RPa
ml
IOD
IODM
IOPr
ctg
IOPf
py
Interaural -00.05 mm
2
3
4
5
6
7
8
9
10
11
12
0
1
2
3
4
5
9
10
11
12
13
14
15
16
17
18
19

| | |
|---|---|
| 2Cb | cerebell. lobule 2b |
| 4n | trochlear nerve |
| 4nd | trochlear nerve decn |
| 4V | fourth ventricle |
| 5n | trigeminal nerve |
| 6N | abducens nu. |
| 7n | facial nerve |
| 7N | facial (motor) nu. |
| 8n | vestibulocochlear nerve |
| 19m | area 19m Cx |
| 23 | area 23 Cx |
| 23v | area 23v Cx |
| 29 | area 29 Cx |
| 30 | area 30 Cx |
| A5 | A5 noradr. cells |
| Bar | Barrington's nu. |
| bic | brach. inf. coll. |
| BIC | brach. inf. coll. nu. |
| C1 | C1 adrenaline cells |
| ca | calcarine sulcus |
| CA1 | cornu ammonis 1 |
| cc | corp. callosum |
| cg | cingulum |
| CG | cingulate gyrus Cx |
| CGPn | central gray pons |
| cic | comm. inf. coll. |
| CIC | cent. nu. inf. coll. |
| cm | callosomarginal sulc. |
| CnF | cuneiform nu. |
| ctg | central tegm. tract |
| DA | dorsoant. visual Cx |
| DC | dorsal cochlear nu. |
| DCIC | dors. Cx inf. coll. |
| DLL | dorsal nu. lat. lemn. |
| DLPAG | dorsolat. periaq. gr. |
| DM | dorsomed. visual Cx |
| DMPAG | dorsomed. periaq. gr |
| DMTg | dorsomed. tegm. area |
| DpG | deep gray sup. coll. |
| DPGi | dors. paragigant. nu. |
| DpWh | deep white sup. coll. |
| ECIC | extern Cx inf. coll. |
| g7 | genu of facial n. |
| gc | geniculocalcarine tr. |
| Gi | gigantocell. retic. nu. |
| GiV | gigantocell. retic. nu., vent. |
| I5 | intertrigem. nu. |
| IG | induseum gris. |
| InCo | intercoll. nu. |
| InG | intermed. gr. sup. coll. |
| InWh | intermed. wh. sup coll. |
| IOD | inf. oliv. nu., dors. |
| IODM | inf. oliv. nu., dorsomed. |
| IOPr | inf. oliv. nu., princ. |
| ip | intraparietal sulc. |
| it | inf. temporal sulcus |
| KF | Kölliker-Fuse nu. |
| LaPO | lat. periolivary nu. |
| LC | locus coeruleus |
| LMol | strat. lacun. mol. CA |
| LPAG | lat. periaqued. gr. |
| LPBC | lat. parabr. nu., cent. |
| LPBD | lat. parabr. nu., dors. |
| LPBE | lat. parabr. nu., ext. |
| LPBV | lat. parabr. nu., vent. |
| LPGi | lat. paragigant. nu. |
| LSO | lat. sup. olivary nu. |
| LV | lateral ventricle |
| LVPO | laterovent. perioliv. nu. |
| m5 | motor root 5n |
| mcp | middle cerebell. ped. |
| Me5 | mesen. trigem. nu. |
| MePO | med. periolivary nu. |
| ml | medial lemniscus |
| mlf | med. longit. fasc. |
| MnR | median raphe nu. |
| Mo5 | motor trigeminal nu. |
| MoDG | molecular layer DG |
| MPB | med. parabrach. nu. |
| MPBE | med. parabr. nu., ext. |
| MSO | med. sup. olivary nu. |
| mte | med. temp. sulc. |
| MVe | med. vestib. nu. |
| MVPO | mediovent. perioliv. nu. |
| Op | strat. opt. sup. coll. |
| Or | strat. oriens CA |
| P5 | peritrigeminal nu. |
| P7 | perifacial nu. |
| Pa6 | parabducens nu. |
| PBG | parabigeminal nu. |
| PCRtα | parv. retic. nu., alpha |
| PCTg | paracollicular tegm. |
| PFl | perifornical nu., lat. |
| PMnR | paramedian raphe nu. |
| PnC | pontine caudal nu. |
| PnR | pontine raphe nu. |
| ProS | prosubiculum |
| ProSt | prostriate cortex |
| PrS | presubiculum |
| pt | post. temp. sulc. |
| py | pyramidal tract |
| Rad | strat. radiat. CA |
| ReIC | recess inf. coll. |
| RMg | raphe magnus nu. |
| RPa | raphe pallidus nu. |
| rs | rubrospinal tract |
| RVL | rost. ventrolat. area |
| S | subiculum |
| s5 | sensory root 5n |
| scp | sup. cerebell. ped. |
| smv | sup. medull. velum |
| sp5 | spinal trigeminal tr. |
| Sp5I | spin. trigem. nu., interpol. |
| Sp5O | spin. trigem. nu., oral. |
| Su5 | supratrigeminal nu. |
| SubCD | subcoer. nu., dors. |
| SubCV | subcoer. nu., vent. |
| SuG | sup. gray sup. coll. |
| SuVe | sup. vestib. nu. |
| ts | tectospinal tract |
| tz | trapezoid body |
| Tz | nu. of trapezoid body |
| V2v | V2 (vent.) Cx |
| VCA | vent. coch. nu., ant. |
| VLA (V4) | ventlatant. (V4) Cx |
| VLP (V3v) | ventlatpost. (V3v) Cx |
| VLPAG | ventrolat. periaq. gr. |
| vsc | vent. spinocereb. tr. |
| Zo | strat. zonale SC |

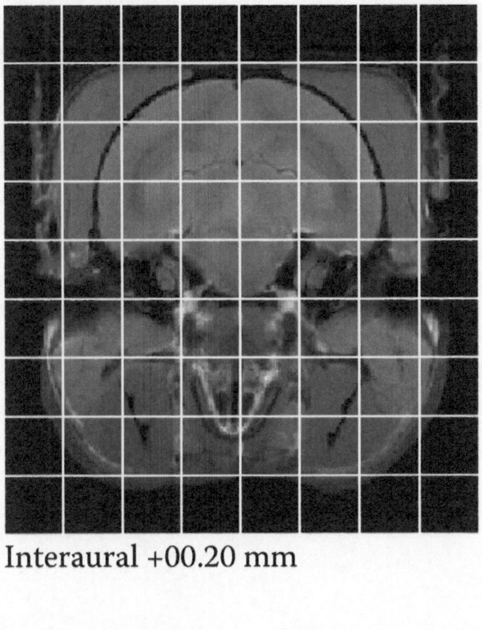

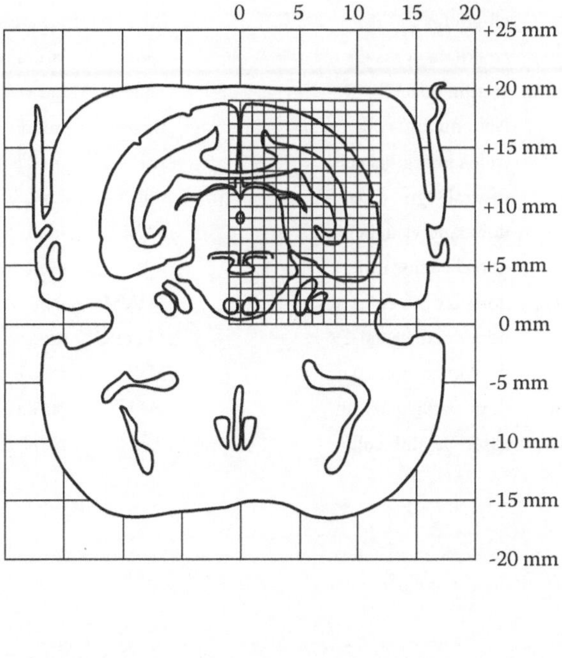


Interaural +00.20 mm

# Plate 39

Interaural -00.55 mm

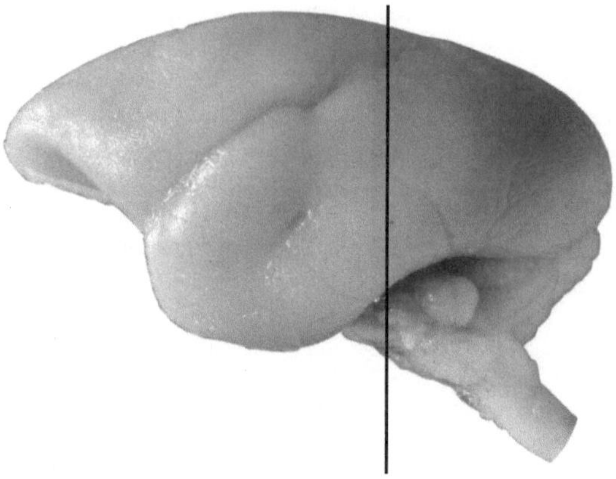

| | | | | | |
|---|---|---|---|---|---|
| 2Cb(a) | cerebell. lobule 2a | g7 | genu of facial n. | PCRt | parv. retic. nu. |
| 2Cb(b) | cerebell. lobule 2b | gc | geniculocalcarine tr. | PFl | paraflocculus |
| 3Cb(a) | cerebell. lobule 3a | Gi | gigantocell. retic. nu. | PnR | pontine raphe nu. |
| 4V | fourth ventricle | GiV | gigantocell. retic. nu., vent. | prcf | precentral fissure Cb |
| 6N | abducens nu. | I8Ve | interst. nu. 8vn | ProSt | prostriate cortex |
| 7N | facial (motor) nu. | IOD | inf. oliv. nu., dors. | pt | post. temp. sulc. |
| 8vn | vestibular divn 8n | IODM | inf. oliv. nu., dorsomed. | py | pyramidal tract |
| 19m | area 19m Cx | IOPr | inf. oliv. nu., princ. | RMg | raphe magnus nu. |
| 23v | area 23v Cx | ip | intraparietal sulc. | ROb | raphe obscurus nu. |
| asc7 | ascending 7n | IRt | intermed. retic. formn | RPa | raphe pallidus nu. |
| C1 | C1 adrenaline cells | it | inf. temporal sulcus | rs | rubrospinal tract |
| ca | calcarine sulcus | LC | locus coeruleus | RVL | rost. ventrolat. area |
| CA1 | cornu ammonis 1 | LPGi | lat. paragigant. nu. | SC | superior colliculus |
| cg | cingulum | lu | lunate sulcus | scp | sup. cerebell. ped. |
| CG | cingulate gyrus Cx | LV | lateral ventricle | Sol | nu. of solitary tr. |
| cic | comm. inf. coll. | LVe | lat. vestib. nu. | sp5 | spinal trigeminal tr. |
| CIC | cent. nu. inf. coll. | mcp | middle cerebell. ped. | Sp5I | spin. trigem. nu., interpol. |
| cm | callosomarginal sulc. | ml | medial lemniscus | SuVe | sup. vestib. nu. |
| ctg | central tegm. tract | mlf | med. longit. fasc. | ts | tectospinal tract |
| DA | dorsoant. visual Cx | MPB | med. parabrach. nu. | V2m | V2 (med.) Cx |
| DC | dorsal cochlear nu. | mte | med. temp. sulc. | V2v | V2 (vent.) Cx |
| DCIC | dors. Cx inf. coll. | MVeMC | med. vestib. nu., magn. | VCA | vent. coch. nu., ant. |
| DM | dorsomed. visual Cx | MVePC | med. vestib. nu., parv. | VLA (V4) | ventlatant. (V4) Cx |
| DMSp5 | dorsomed. spin. 5 nu. | Mx | matrix sensory area | VLP (V3v) | ventlatpost. (V3v) Cx |
| DPGi | dors. paragigant. nu. | P7 | perifacial nu. | vsc | vent. spinocereb. tr. |
| ECIC | extern Cx inf. coll. | Pa6 | parabducens nu. | Zo | strat. zonale SC |

# Plate 39a

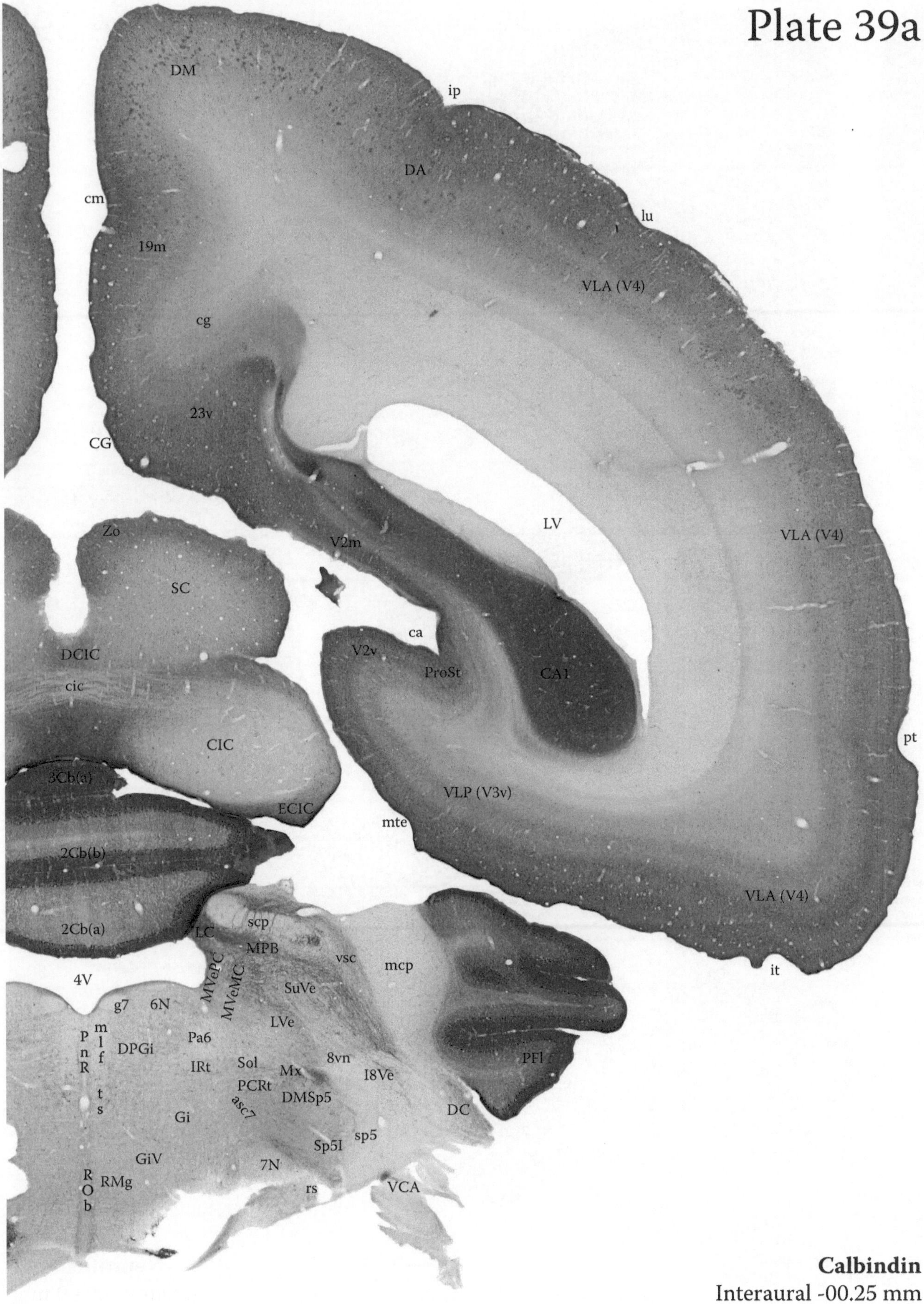


**Calbindin**
Interaural -00.25 mm

# Plate 39b

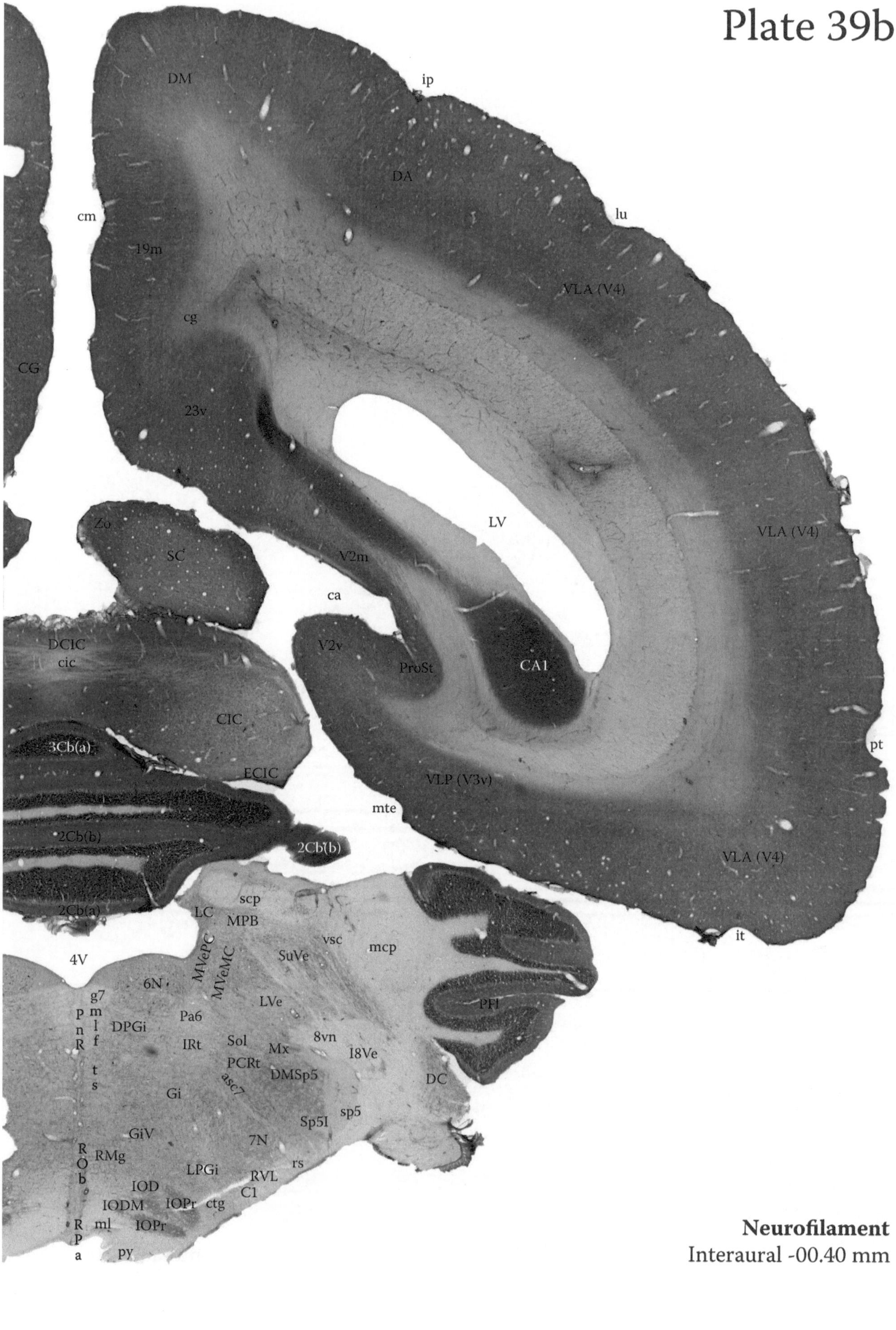


**Neurofilament**
Interaural -00.40 mm

# Plate 39c

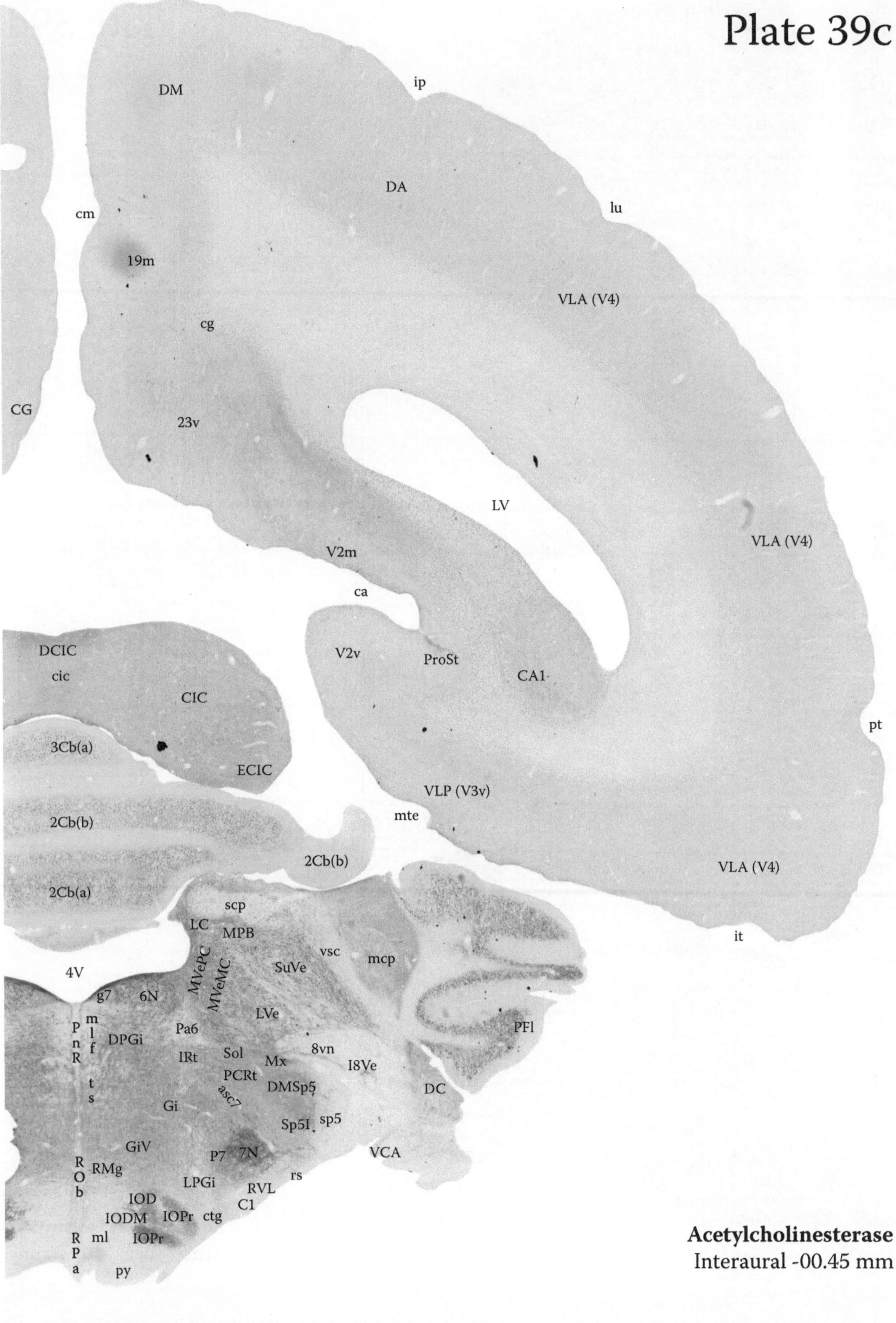


**Acetylcholinesterase**
Interaural -00.45 mm

# Plate 39d

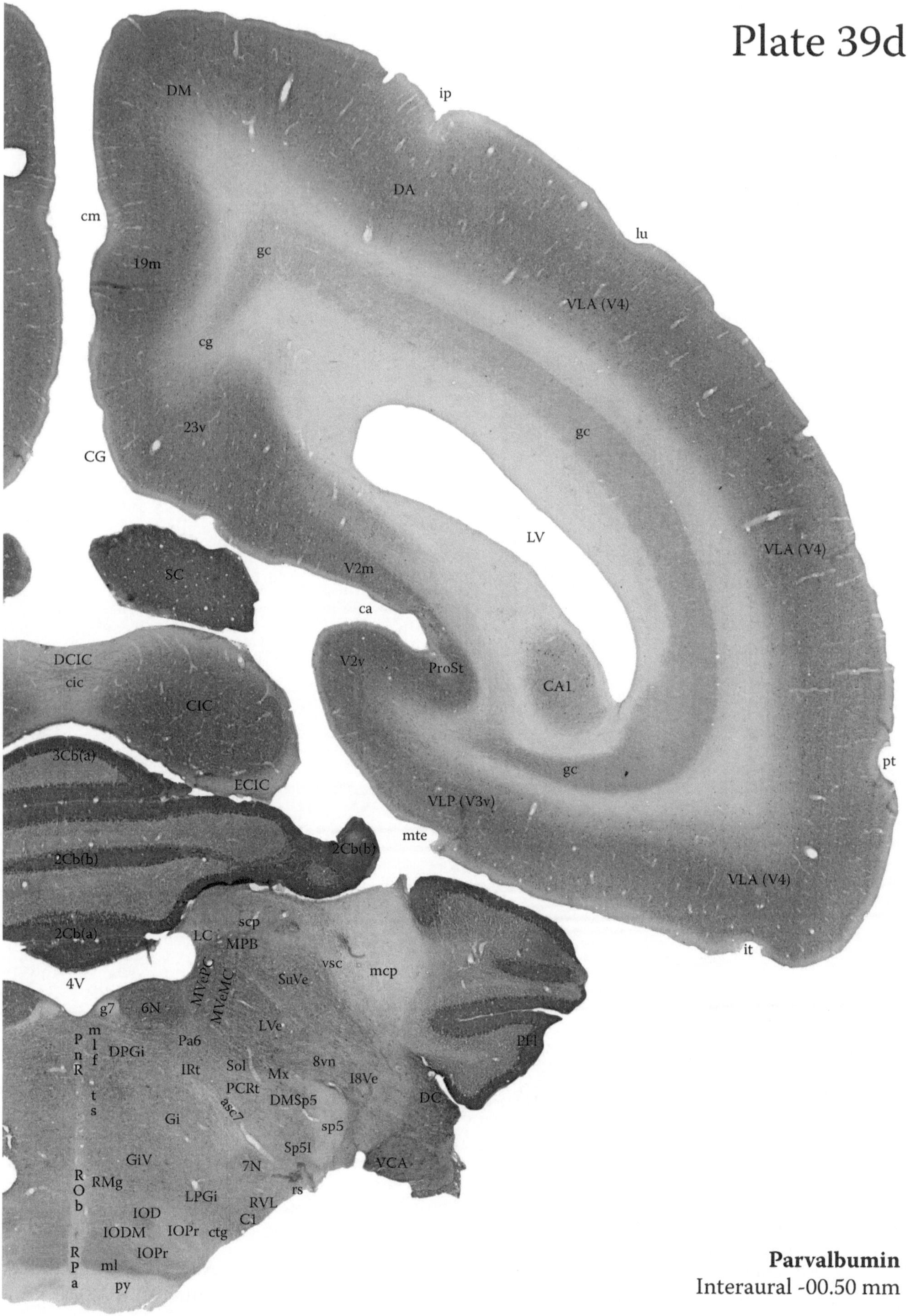


**Parvalbumin**
Interaural -00.50 mm

# Plate 39e

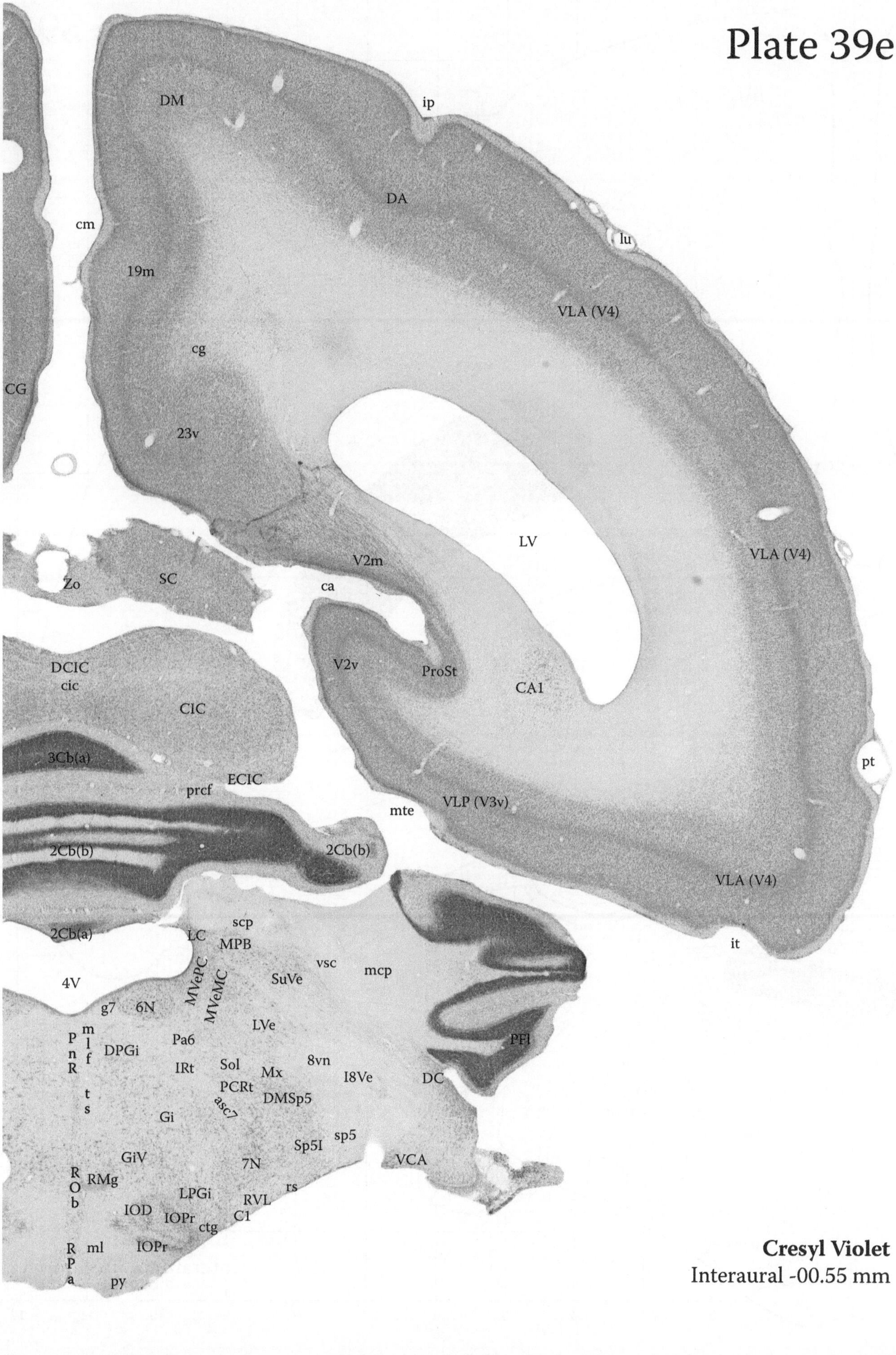


**Cresyl Violet**
Interaural -00.55 mm

# Plate 39

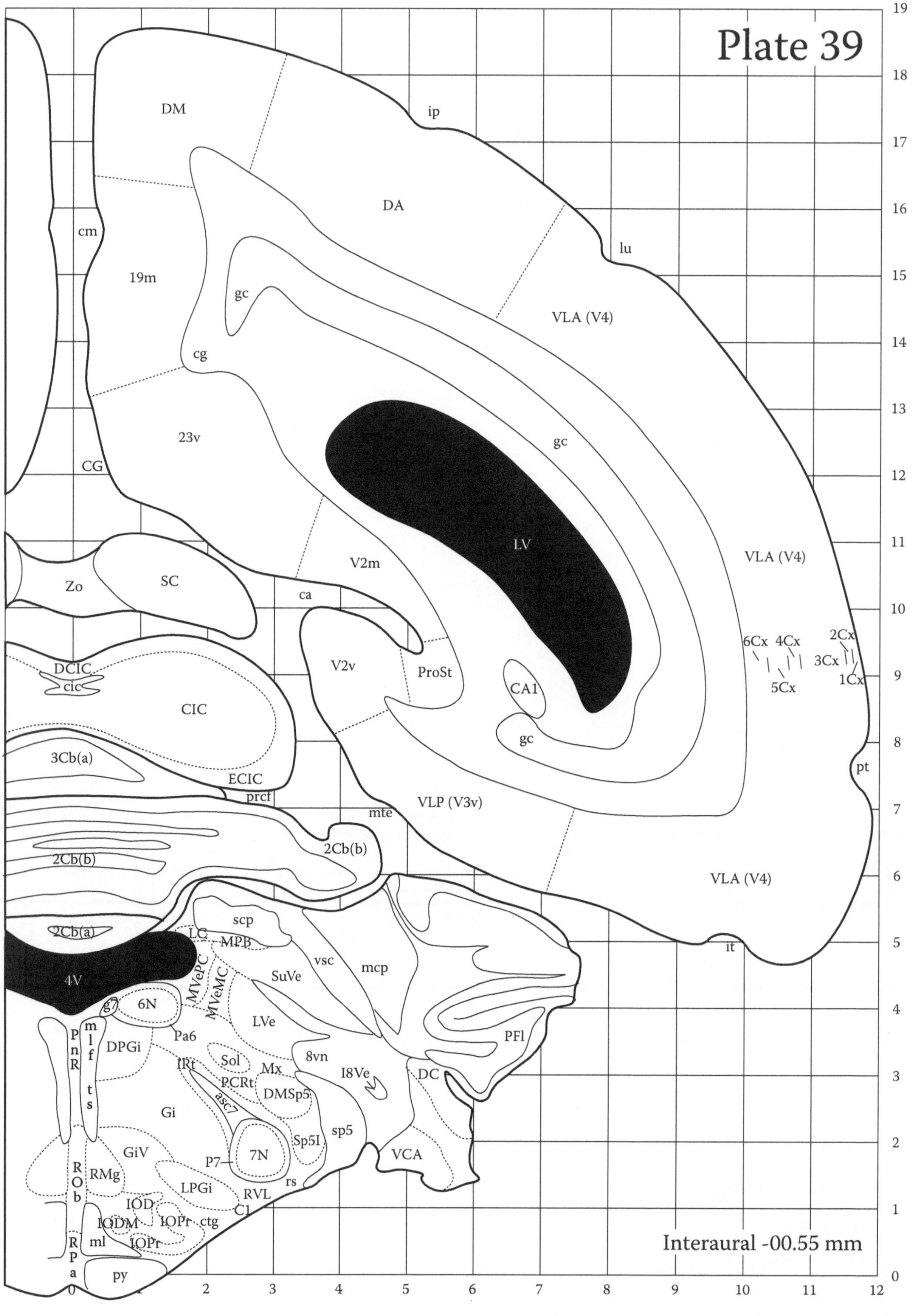


Interaural -00.55 mm

| | |
|---|---|
| 2Cb(a) | cerebell. lobule 2a |
| 2Cb(b) | cerebell. lobule 2b |
| 3Cb(a) | cerebell. lobule 3a |
| 4V | fourth ventricle |
| 6N | abducens nu. |
| 7N | facial (motor) nu. |
| 8vn | vestibular divn 8n |
| 19m | area 19m Cx |
| 23v | area 23v Cx |
| asc7 | ascending 7n |
| C1 | C1 adrenaline cells |
| ca | calcarine sulcus |
| CA1 | cornu ammonis 1 |
| cg | cingulum |
| CG | cingulate gyrus Cx |
| cic | comm. inf. coll. |
| CIC | cent. nu. inf. coll. |
| cm | callosomarginal sulc. |
| ctg | central tegm. tract |
| DA | dorsoant. visual Cx |
| DC | dorsal cochlear nu. |
| DCIC | dors. Cx inf. coll. |
| DM | dorsomed. visual Cx |
| DMSp5 | dorsomed. spin. 5 nu. |
| DPGi | dors. paragigant. nu. |
| ECIC | extern Cx inf. coll. |
| g7 | genu of facial n. |
| gc | geniculocalcarine tr. |
| Gi | gigantocell. retic. nu. |
| GiV | gigantocell. retic. nu., vent. |
| I8Ve | interst. nu. 8vn |
| IOD | inf. oliv. nu., dors. |
| IODM | inf. oliv. nu., dorsomed. |
| IOPr | inf. oliv. nu., princ. |
| ip | intraparietal sulc. |
| IRt | intermed. retic. formn |
| it | inf. temporal sulcus |
| LC | locus coeruleus |
| LPGi | lat. paragigant. nu. |
| lu | lunate sulcus |
| LV | lateral ventricle |
| LVe | lat. vestib. nu. |
| mcp | middle cerebell. ped. |
| ml | medial lemniscus |
| mlf | med. longit. fasc. |
| MPB | med. parabrach. nu. |
| mte | med. temp. sulc. |
| MVeMC | med. vestib. nu., magn. |
| MVePC | med. vestib. nu., parv. |
| Mx | matrix sensory area |
| P7 | perifacial nu. |
| Pa6 | parabducens nu. |
| PCRt | parv. retic. nu. |
| PFl | paraflocculus |
| PnR | pontine raphe nu. |
| prcf | precentral fissure Cb |
| ProSt | prostriate cortex |
| pt | post. temp. sulc. |
| py | pyramidal tract |
| RMg | raphe magnus nu. |
| ROb | raphe obscurus nu. |
| RPa | raphe pallidus nu. |
| rs | rubrospinal tract |
| RVL | rost. ventrolat. area |
| SC | superior colliculus |
| scp | sup. cerebell. ped. |
| Sol | nu. of solitary tr. |
| sp5 | spinal trigeminal tr. |
| Sp5I | spin. trigem. nu., interpol. |
| SuVe | sup. vestib. nu. |
| ts | tectospinal tract |
| V2m | V2 (med.) Cx |
| V2v | V2 (vent.) Cx |
| VCA | vent. coch. nu., ant. |
| VLA (V4) | ventlatant. (V4) Cx |
| VLP (V3v) | ventlatpost. (V3v) Cx |
| vsc | vent. spinocereb. tr. |
| Zo | strat. zonale SC |

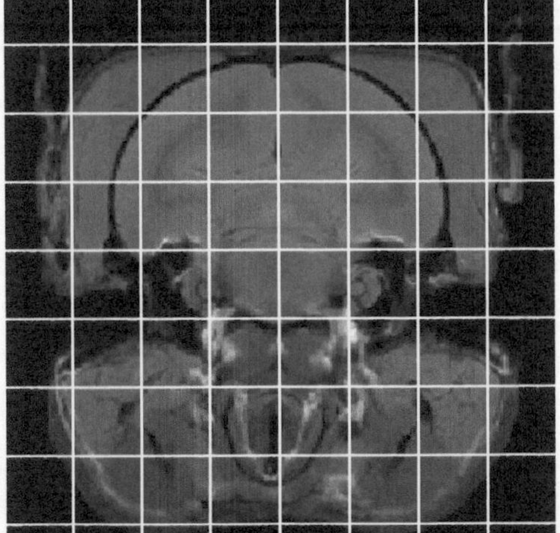

Interaural +00.75 mm

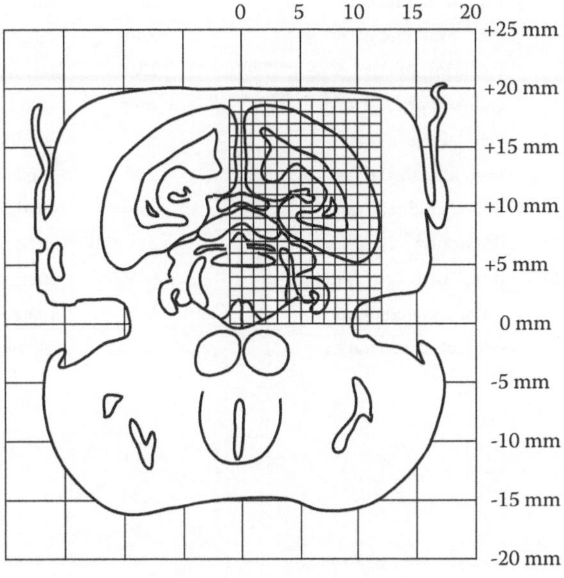

# Plate 40

Interaural -01.10 mm

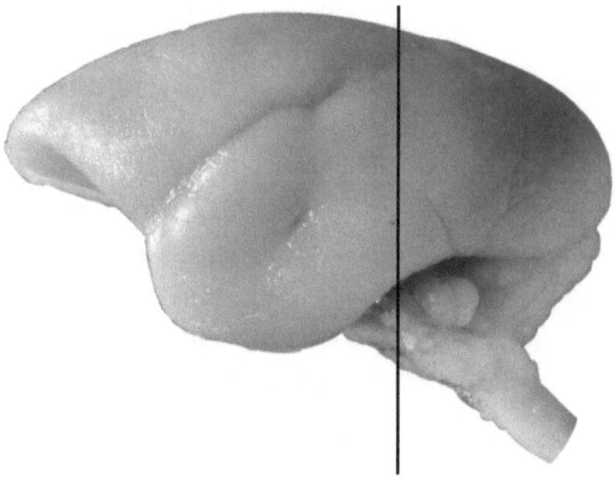

| | | | | | |
|---|---|---|---|---|---|
| 2Cb(a) | cerebell. lobule 2a | Gi | gigantocell. retic. nu. | PCRt | parv. retic. nu. |
| 2Cb(b) | cerebell. lobule 2b | GiV | gigantocell. retic. nu., vent. | PFl | paraflocculus |
| 3Cb(a) | cerebell. lobule 3a | IC | inferior colliculus | Pr | prepositus hypogl. nu. |
| 3Cb(b) | cerebell. lobule 3b | icp | inf. cerebell. ped. | ProSt | prostriate cortex |
| 4Cb(a) | cerebell. lobule 4a | IOD | inf. oliv. nu., dors. | pt | post. temp. sulc. |
| 4V | fourth ventricle | IODM | inf. oliv. nu., dorsomed. | py | pyramidal tract |
| 5Cb | cerebell. lobule 5 | IOPr | inf. oliv. nu., princ. | RMg | raphe magnus nu. |
| 5Sol | trigemsol. transn zone | ip | intraparietal sulc. | ROb | raphe obscurus nu. |
| 8n | vestibulocochlear nerve | IPo | interpositus nu. | RPa | raphe pallidus nu. |
| 19m | area 19m Cx | IRt | intermed. retic. formn | rs | rubrospinal tract |
| 23v | area 23v Cx | IS | inf. salivatory nu. | scp | sup. cerebell. ped. |
| AmbC | ambig. nu., compact | it | inf. temporal sulcus | smv | sup. medull. velum |
| C1 | C1 adrenaline cells | LPGi | lat. paragigant. nu. | Sol | nu. of solitary tr. |
| C2 | C2 adrenaline cells | lu | lunate sulcus | sp5 | spinal trigeminal tr. |
| C3 | C3 adrenaline cells | LV | lateral ventricle | Sp5I | spin. trigem. nu., interpol. |
| ca | calcarine sulcus | LVe | lat. vestib. nu. | SuVe | sup. vestib. nu. |
| cm | callosomarginal sulc. | mcp | middle cerebell. ped. | ts | tectospinal tract |
| ctg | central tegm. tract | ml | medial lemniscus | V2m | V2 (med.) Cx |
| DA | dorsoant. visual Cx | mlf | med. longit. fasc. | V2v | V2 (vent.) Cx |
| DC | dorsal cochlear nu. | mte | med. temp. sulc. | VCP | vent. coch. nu., post. |
| DM | dorsomed. visual Cx | MVeMC | med. vestib. nu., magn. | VeCb | vestibulocerebell. nu. |
| DMSp5 | dorsomed. spin. 5 nu. | MVePC | med. vestib. nu., parv. | VLA (V4) | ventlatant. (V4) Cx |
| DPGi | dors. paragigant. nu. | Mx | matrix sensory area | VLP (V3v) | ventlatpost. (V3v) Cx |
| gc | geniculocalcarine tr. | oc | olivocerebellar tr. | vsc | vent. spinocereb. tr. |

# Plate 40a

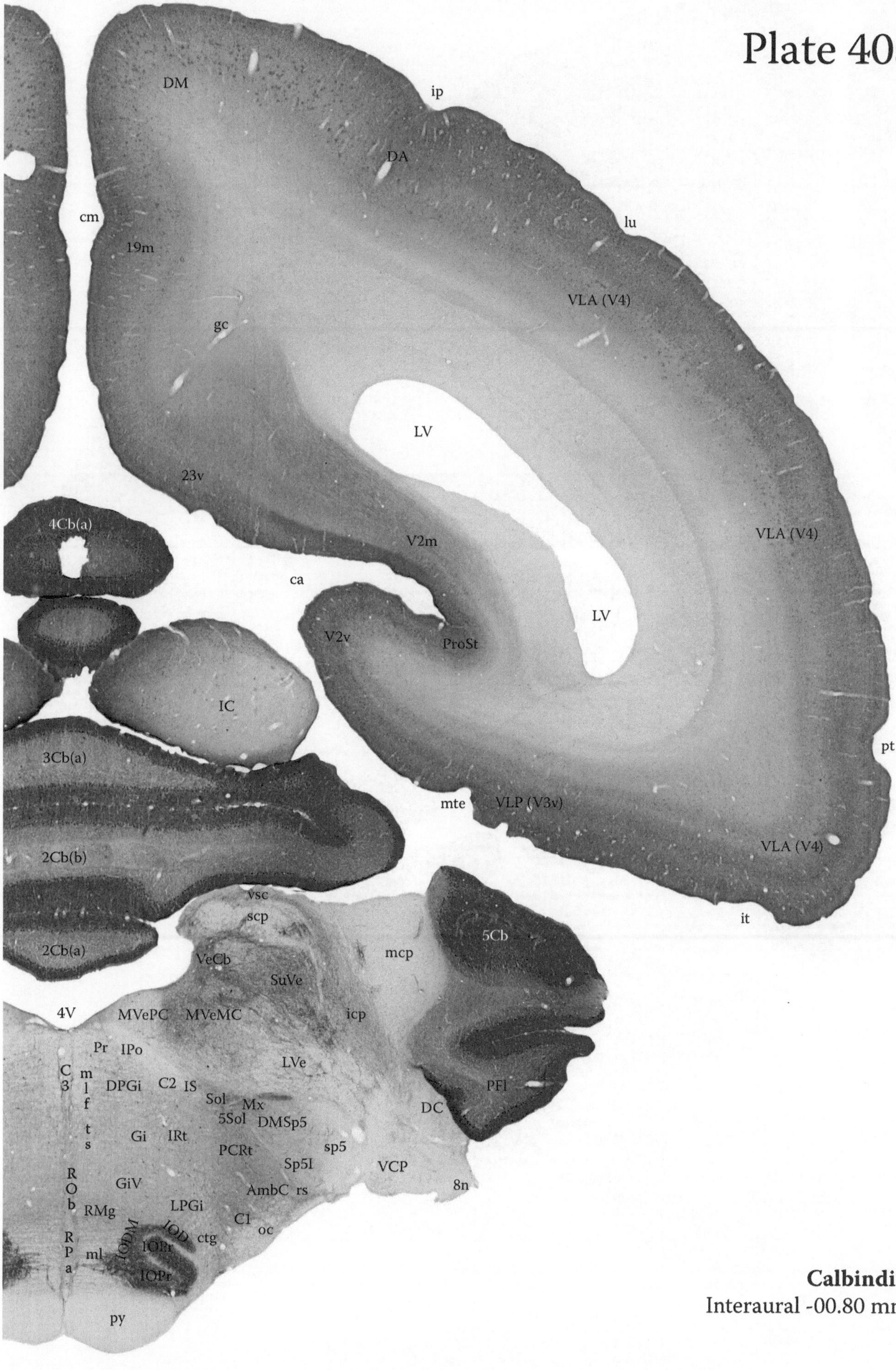


**Calbindin**
Interaural -00.80 mm

# Plate 40b

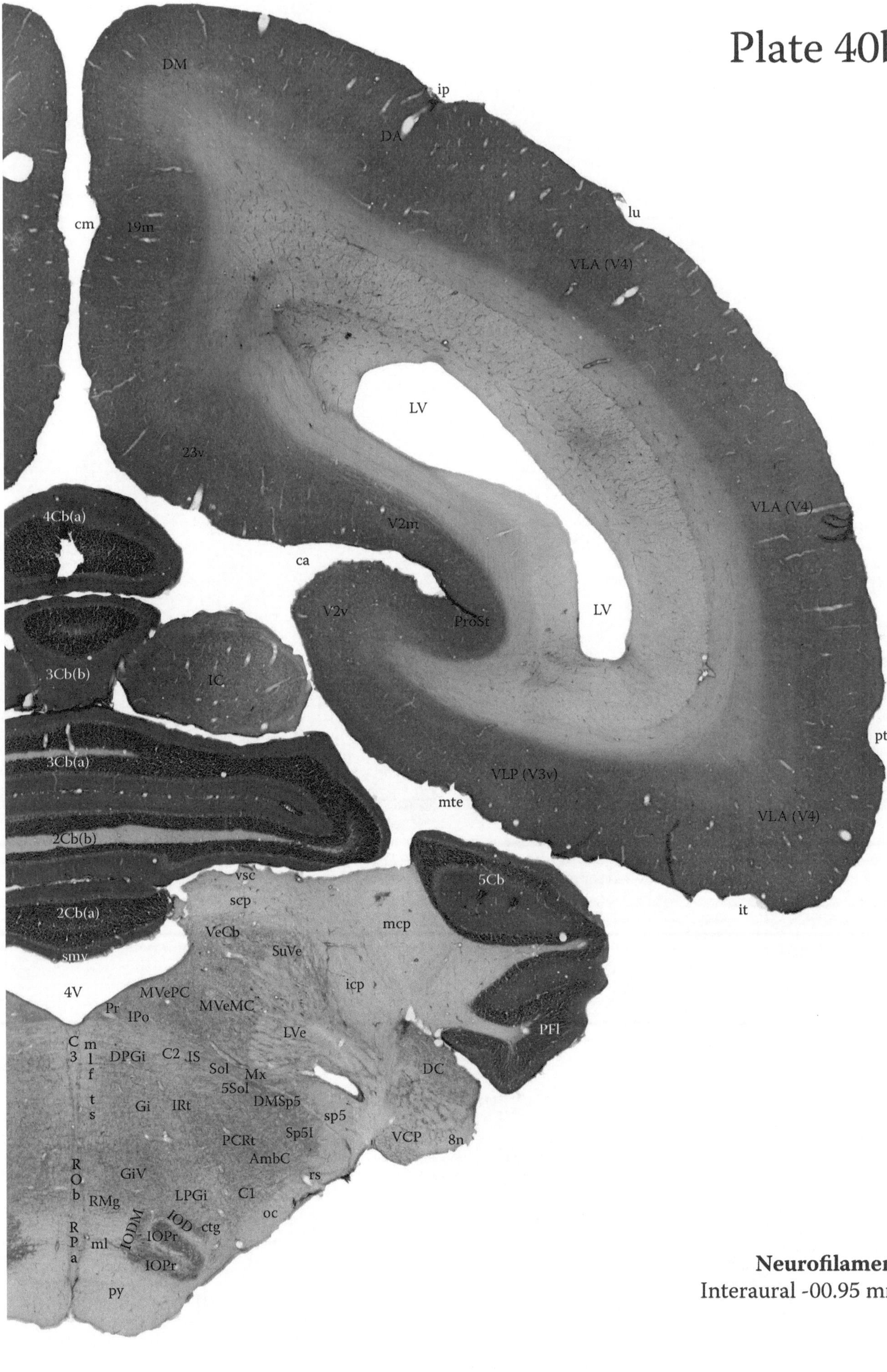


**Neurofilament**
Interaural -00.95 mm

# Plate 40c

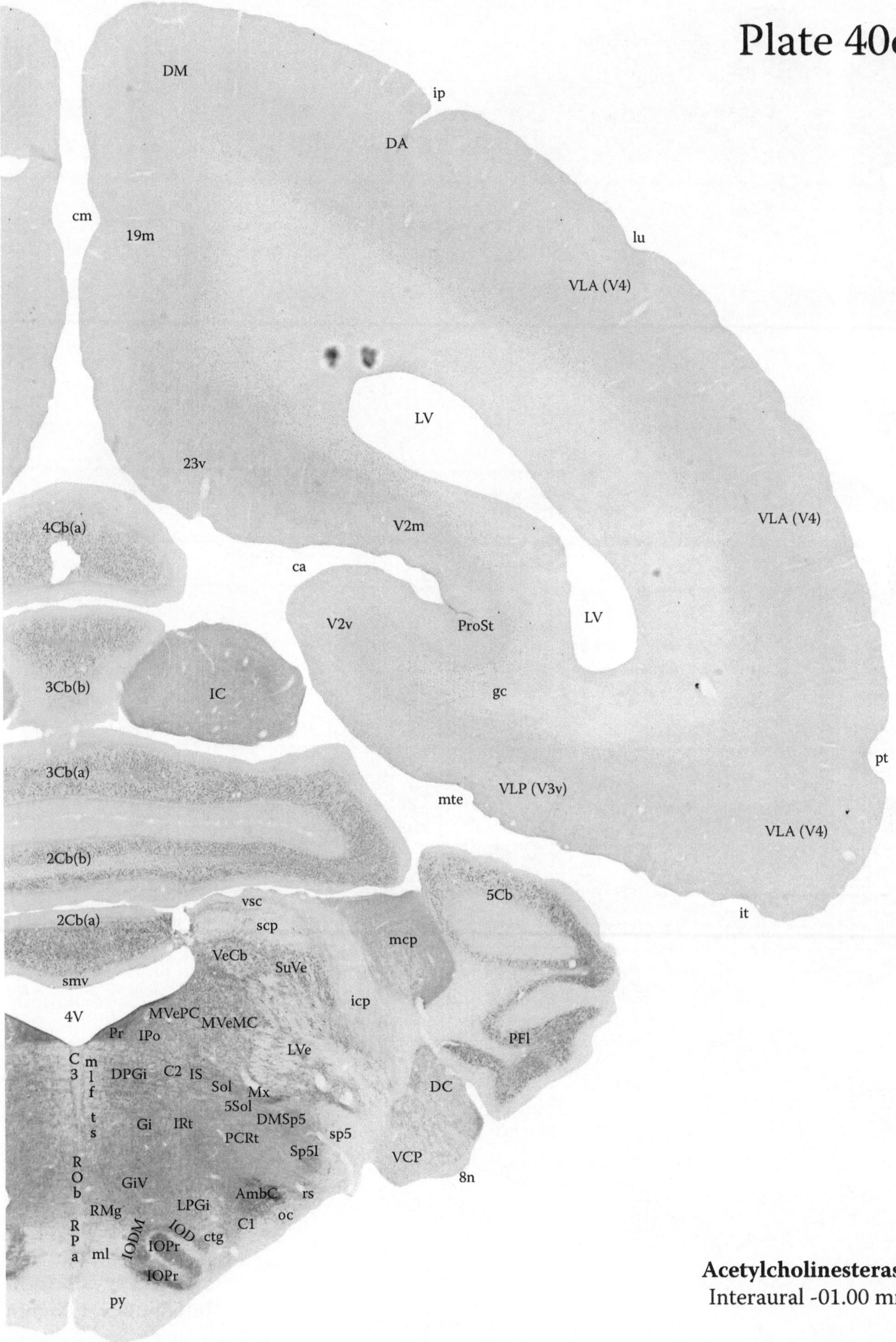


**Acetylcholinesterase**
Interaural -01.00 mm

# Plate 40d

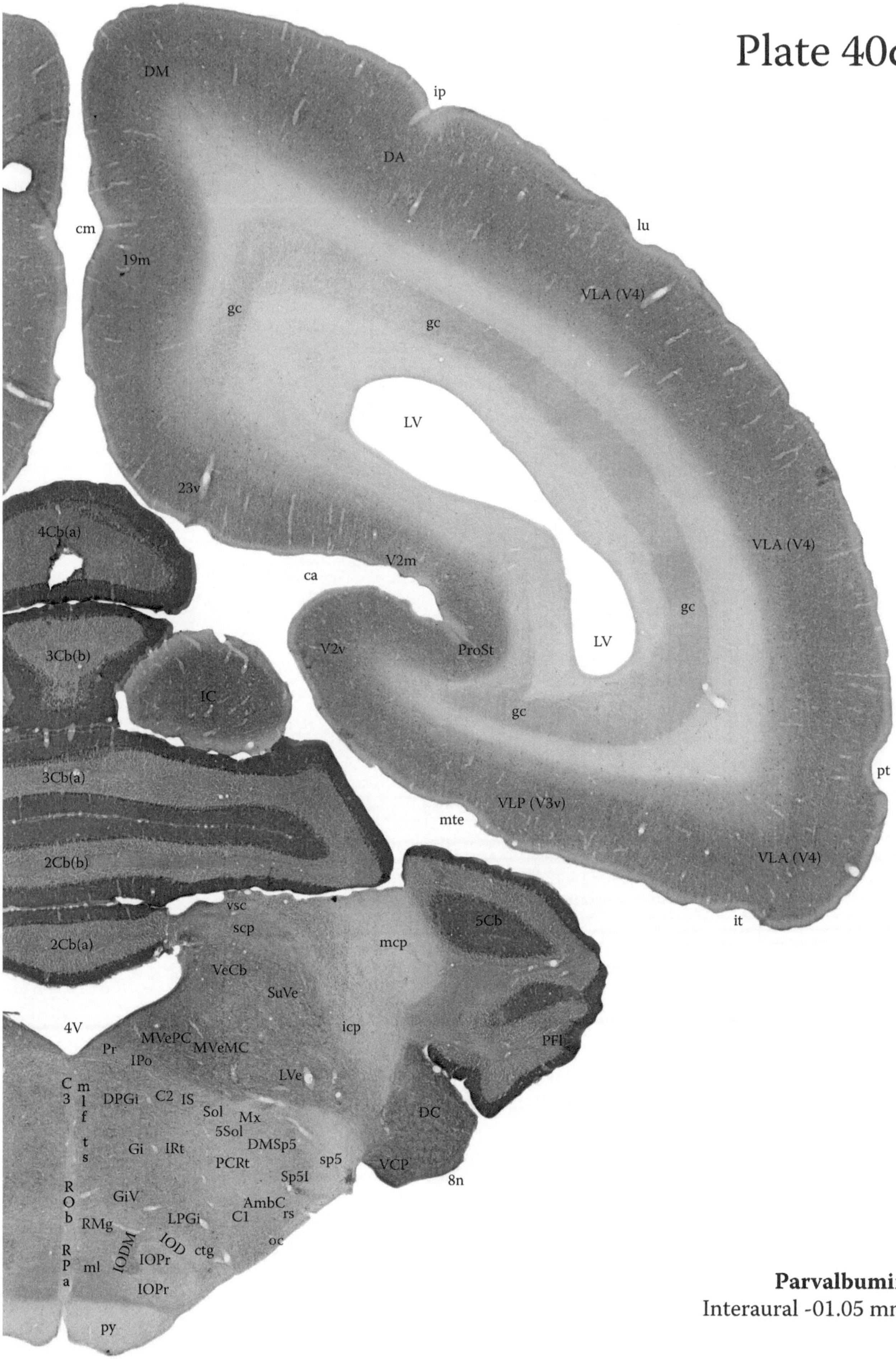


**Parvalbumin**
Interaural -01.05 mm

# Plate 40e

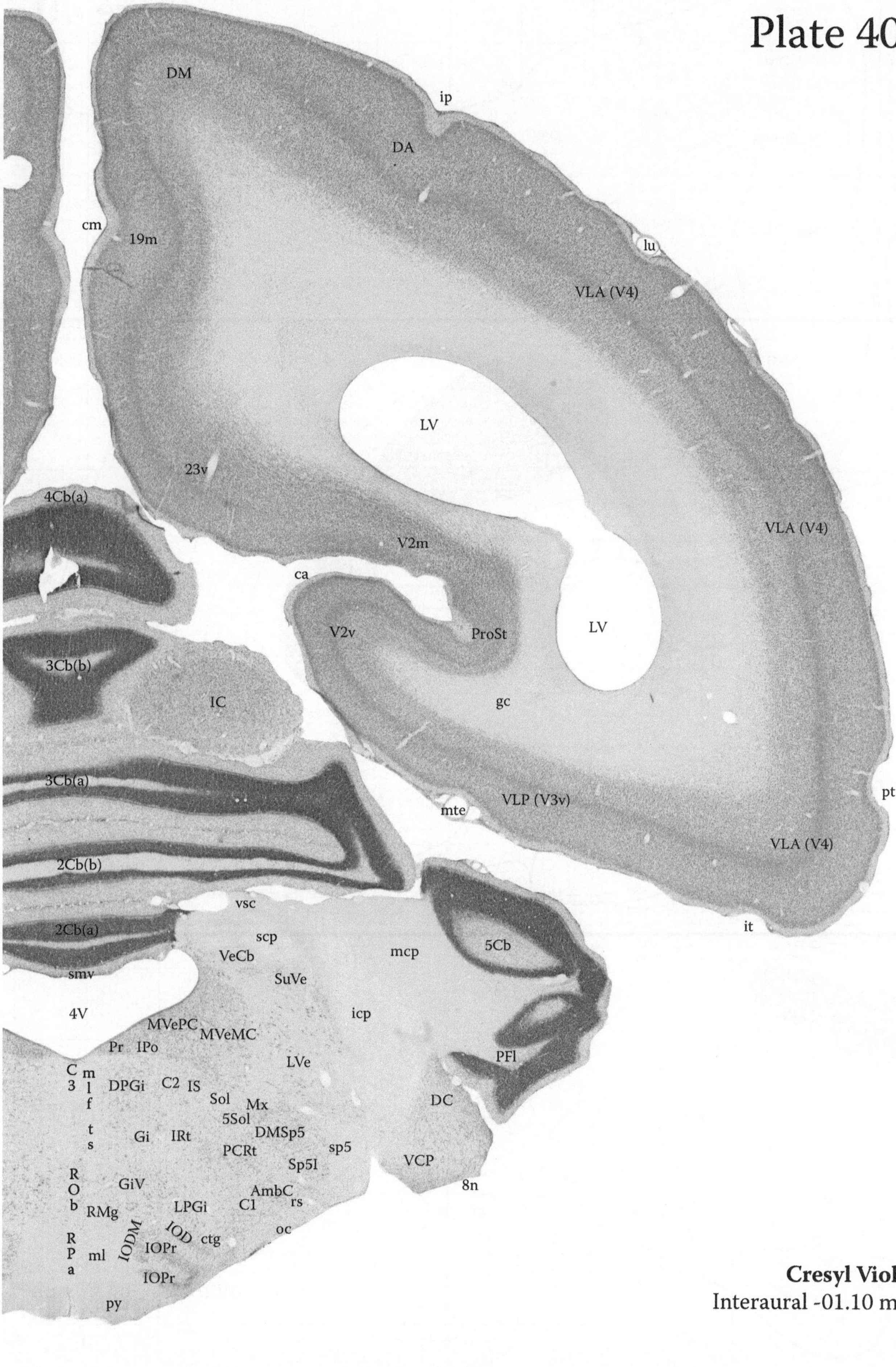


**Cresyl Violet**
Interaural -01.10 mm

# Plate 40

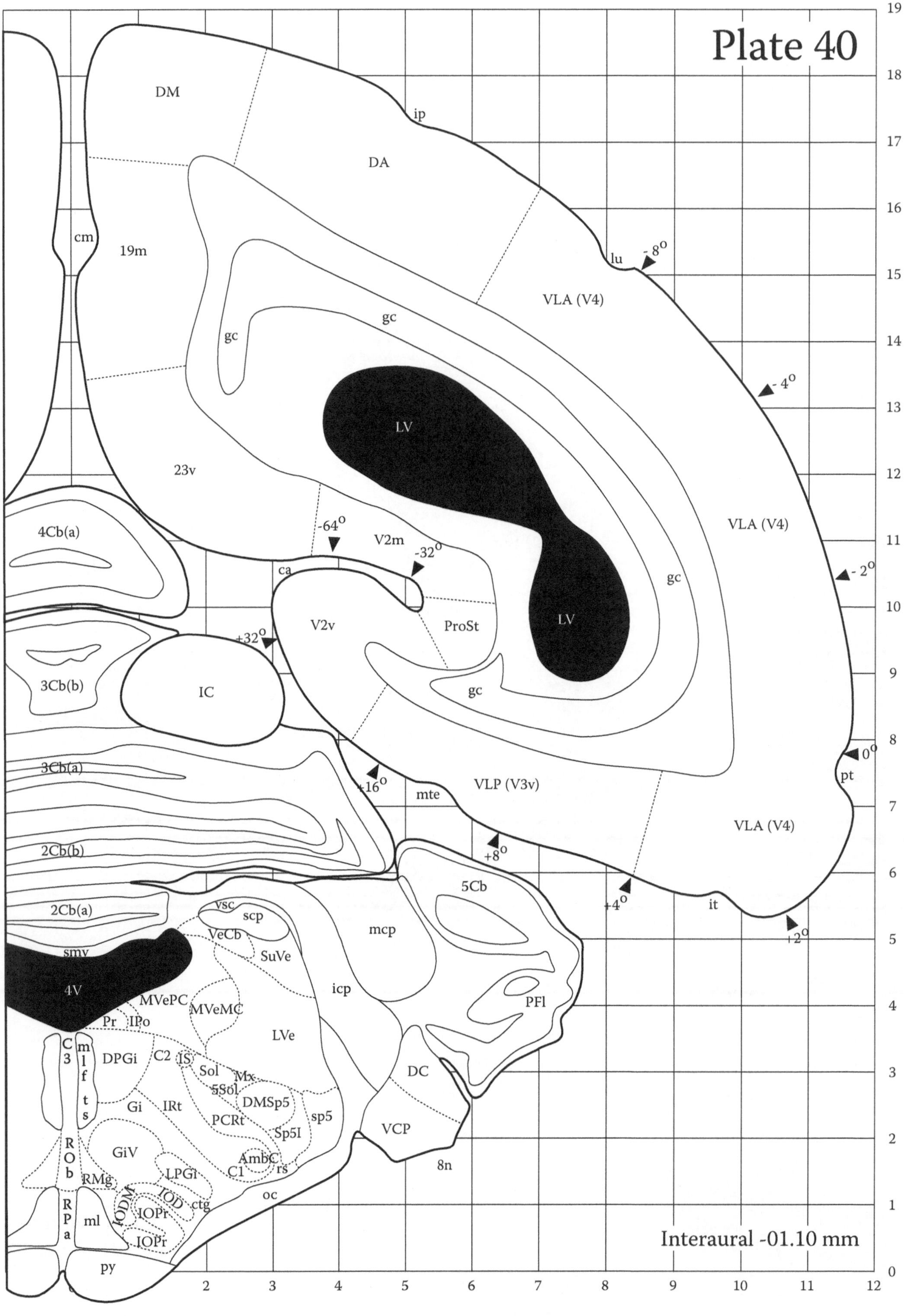


Interaural -01.10 mm

| | |
|---|---|
| 2Cb(a) | cerebell. lobule 2a |
| 2Cb(b) | cerebell. lobule 2b |
| 3Cb(a) | cerebell. lobule 3a |
| 3Cb(b) | cerebell. lobule 3b |
| 4Cb(a) | cerebell. lobule 4a |
| 4V | fourth ventricle |
| 5Cb | cerebell. lobule 5 |
| 5Sol | trigemsol. transn zone |
| 8n | vestibulocochlear nerve |
| 19m | area 19m Cx |
| 23v | area 23v Cx |
| AmbC | ambig. nu., compact |
| C1 | C1 adrenaline cells |
| C2 | C2 adrenaline cells |
| C3 | C3 adrenaline cells |
| ca | calcarine sulcus |
| cm | callosomarginal sulc. |
| ctg | central tegm. tract |
| DA | dorsoant. visual Cx |
| DC | dorsal cochlear nu. |
| DM | dorsomed. visual Cx |
| DMSp5 | dorsomed. spin. 5 nu. |
| DPGi | dors. paragigant. nu. |
| gc | geniculocalcarine tr. |
| Gi | gigantocell. retic. nu. |
| GiV | gigantocell. retic. nu., vent. |
| IC | inferior colliculus |
| icp | inf. cerebell. ped. |
| IOD | inf. oliv. nu., dors. |
| IODM | inf. oliv. nu., dorsomed. |
| IOPr | inf. oliv. nu., princ. |
| ip | intraparietal sulc. |
| IPo | interpositus nu. |
| IRt | intermed. retic. formn |
| IS | inf. salivatory nu. |
| it | inf. temporal sulcus |
| LPGi | lat. paragigant. nu. |
| lu | lunate sulcus |
| LV | lateral ventricle |
| LVe | lat. vestib. nu. |
| mcp | middle cerebell. ped. |
| ml | medial lemniscus |
| mlf | med. longit. fasc. |
| mte | med. temp. sulc. |
| MVeMC | med. vestib. nu., magn. |
| MVePC | med. vestib. nu., parv. |
| Mx | matrix sensory area |
| oc | olivocerebellar tr. |
| PCRt | parv. retic. nu. |
| PFl | paraflocculus |
| Pr | prepositus hypogl. nu. |
| ProSt | prostriate cortex |
| pt | post. temp. sulc. |
| py | pyramidal tract |
| RMg | raphe magnus nu. |
| ROb | raphe obscurus nu. |
| RPa | raphe pallidus nu. |
| rs | rubrospinal tract |
| scp | sup. cerebell. ped. |
| smv | sup. medull. velum |
| Sol | nu. of solitary tr. |
| sp5 | spinal trigeminal tr. |
| Sp5I | spin. trigem. nu., interpol. |
| SuVe | sup. vestib. nu. |
| ts | tectospinal tract |
| V2m | V2 (med.) Cx |
| V2v | V2 (vent.) Cx |
| VCP | vent. coch. nu., post. |
| VeCb | vestibulocerebell. nu. |
| VLA (V4) | ventlatant. (V4) Cx |
| VLP (V3v) | ventlatpost. (V3v) Cx |
| vsc | vent. spinocereb. tr. |

# Plate 41

Interaural -01.60 mm

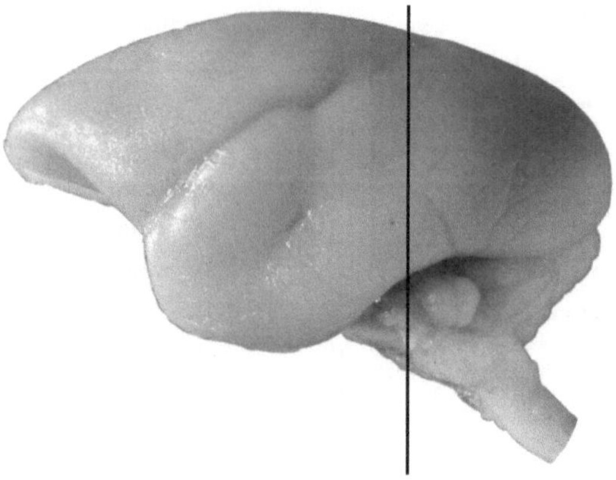

| | |
|---|---|
| 1Cb | cerebell. lobule 1 |
| 2Cb | cerebell. lobule 2 |
| 3Cb(a) | cerebell. lobule 3a |
| 4Cb(a) | cerebell. lobule 4a |
| 4Cb(b) | cerebell. lobule 4b |
| 4V | fourth ventricle |
| 5Sol | trigemsol. transn zone |
| 19m | area 19m Cx |
| 23v | area 23v Cx |
| AmbC | ambig. nu., compact |
| bv | blood vessel |
| C1 | C1 adrenaline cells |
| C3 | C3 adrenaline cells |
| ca | calcarine sulcus |
| cm | callosomarginal sulc. |
| ctg | central tegm. tract |
| DI | dorsoint. visual Cx |
| DM | dorsomed. visual Cx |
| DMSp5 | dorsomed. spin. 5 nu. |
| DPGi | dors. paragigant. nu. |
| Fl | flocculus |
| gc | geniculocalcarine tr. |
| Gi | gigantocell. retic. nu. |
| GiV | gigantocell. retic. nu., vent. |
| icp | inf. cerebell. ped. |
| IOA | inf. oliv. nu., subn. A |
| IOβ | inf. oliv. nu., subn. β |
| IOB | inf. oliv. nu., subn. B |
| IOC | inf. oliv. nu., subn. C |
| IOD | inf. oliv. nu., dors. |
| IOPr | inf. oliv. nu., princ. |
| ip | intraparietal sulc. |
| IPo | interpositus nu. |
| IRt | intermed. retic. formn |
| it | inf. temporal sulcus |
| LPGi | lat. paragigant. nu. |
| lu | lunate sulcus |
| LV | lateral ventricle |
| LVe | lat. vestib. nu. |
| mcp | middle cerebell. ped. |
| Med | med. deep Cb nu. |
| ml | medial lemniscus |
| mlf | med. longit. fasc. |
| mte | med. temp. sulc. |
| MVeMC | med. vestib. nu., magn. |
| MVePC | med. vestib. nu., parv. |
| Mx | matrix sensory area |
| oc | olivocerebellar tr. |
| pcf | preculminate fiss. Cb. |
| PCRt | parv. retic. nu. |
| PFl | paraflocculus |
| plf | posterolat. fissure |
| Pr | prepositus hypogl. nu. |
| pt | post. temp. sulc. |
| py | pyramidal tract |
| ROb | raphe obscurus nu. |
| RPa | raphe pallidus nu. |
| rs | rubrospinal tract |
| scp | sup. cerebell. ped. |
| Sim | simplex lobule Cb |
| Sol | nu. of solitary tr. |
| SolI | nu. of solitary tr., intrafasc. |
| sp5 | spinal trigeminal tr. |
| Sp5I | spin. trigem. nu., interpol. |
| SpVe | spinal vestib. nu. |
| SuVe | sup. vestib. nu. |
| ts | tectospinal tract |
| V1 | V1 Cx |
| V2m | V2 (med.) Cx |
| V2v | V2 (vent.) Cx |
| V3d | V3 (dors.) Cx |
| VCP | vent. coch. nu., post. |
| VeCb | vestibulocerebell. nu. |
| VLA (V4) | ventlatant. (V4) Cx |
| VLP (V3v) | ventlatpost. (V3v) Cx |
| vsc | vent. spinocereb. tr. |

# Plate 41a

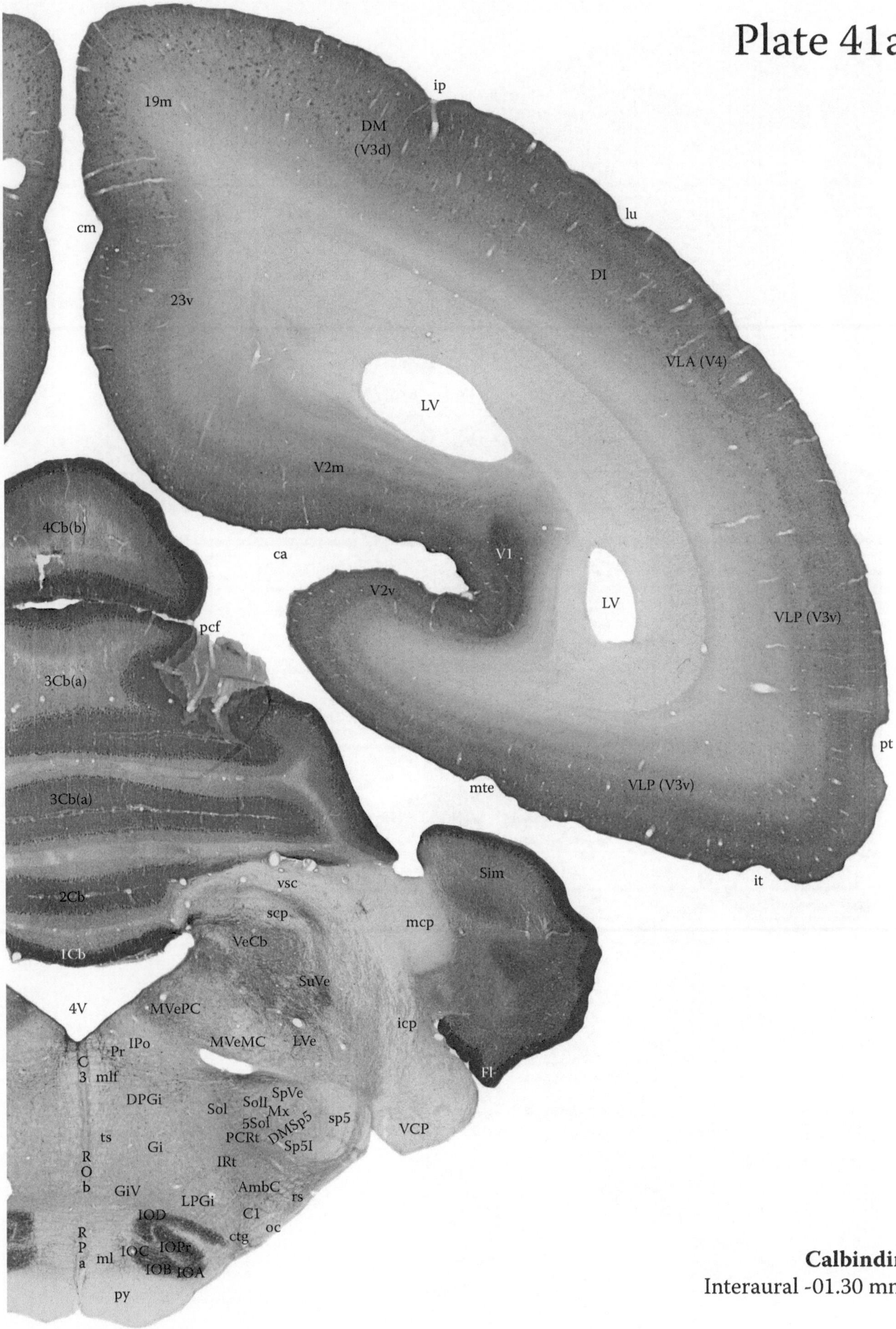


**Calbindin**
Interaural -01.30 mm

# Plate 41b

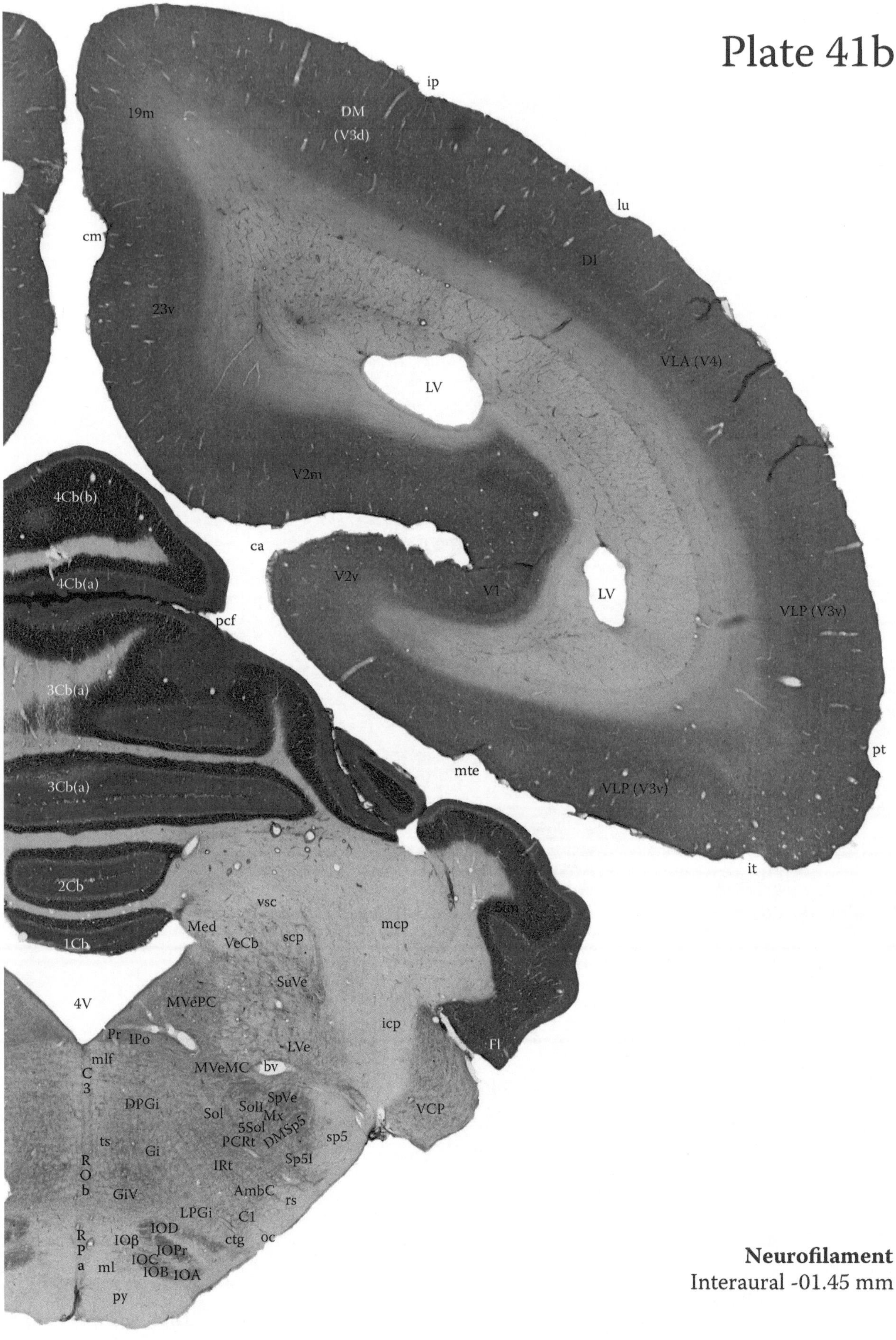


**Neurofilament**
Interaural -01.45 mm

# Plate 41c

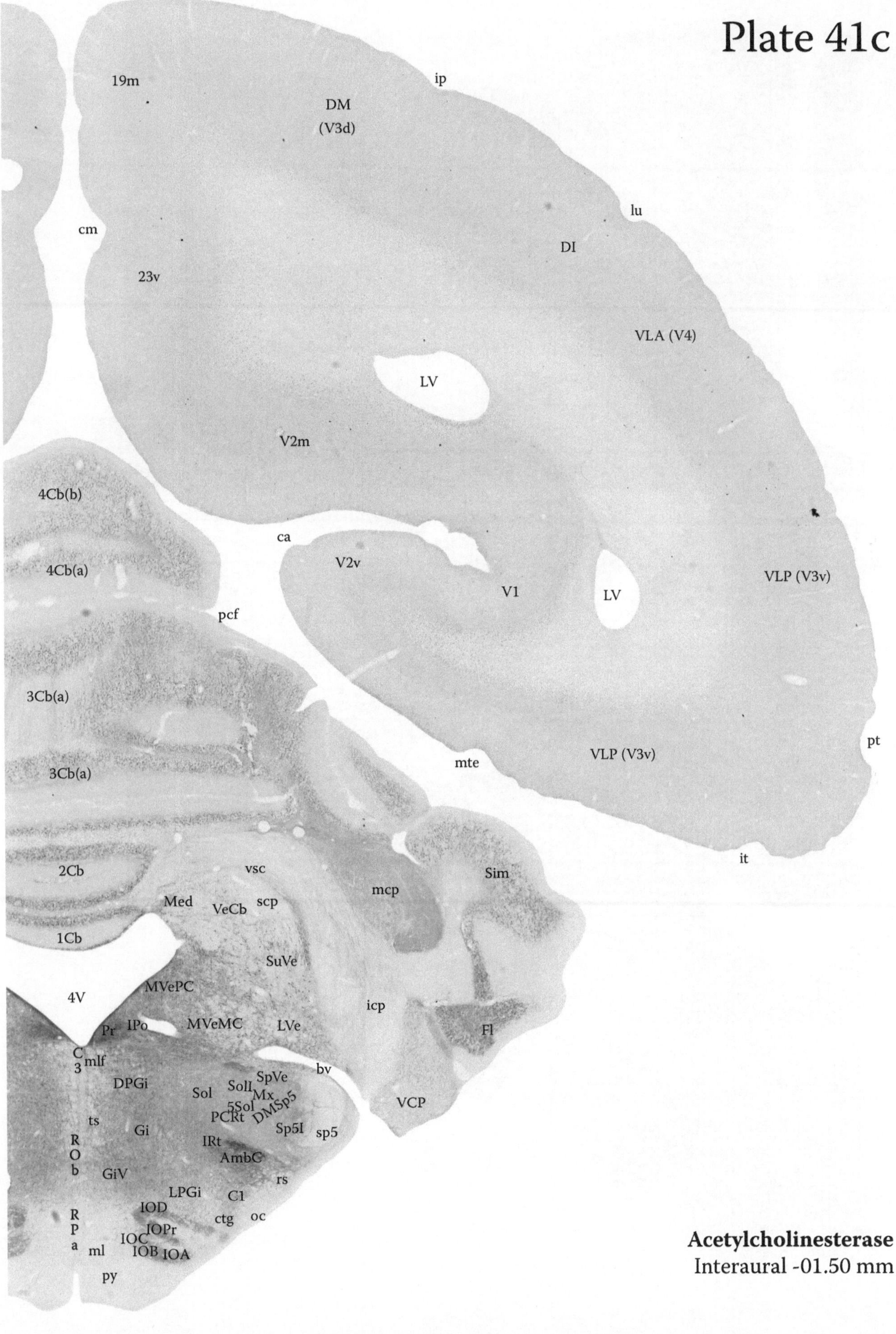


**Acetylcholinesterase**
Interaural -01.50 mm

# Plate 41d

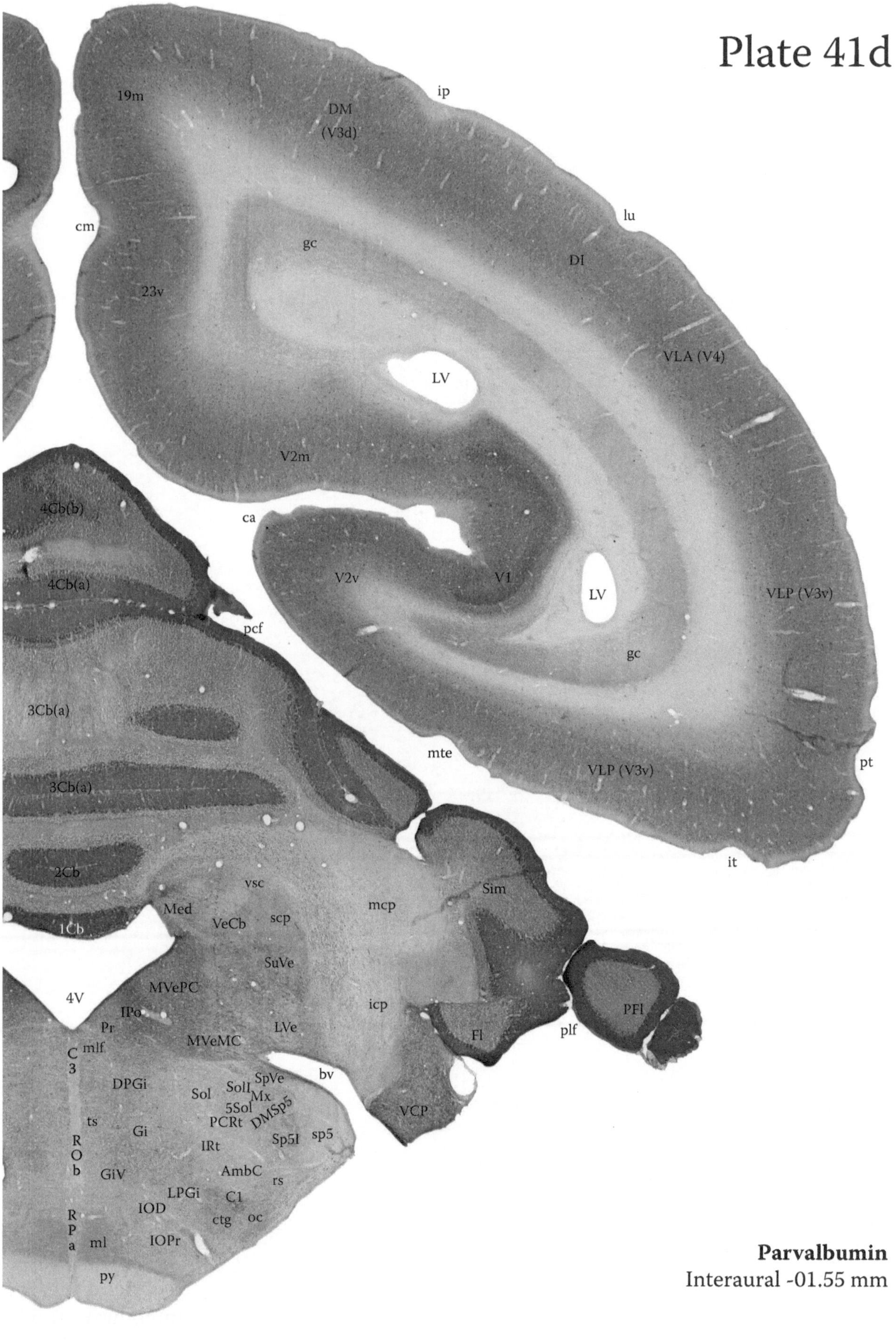


**Parvalbumin**
Interaural -01.55 mm

# Plate 41e

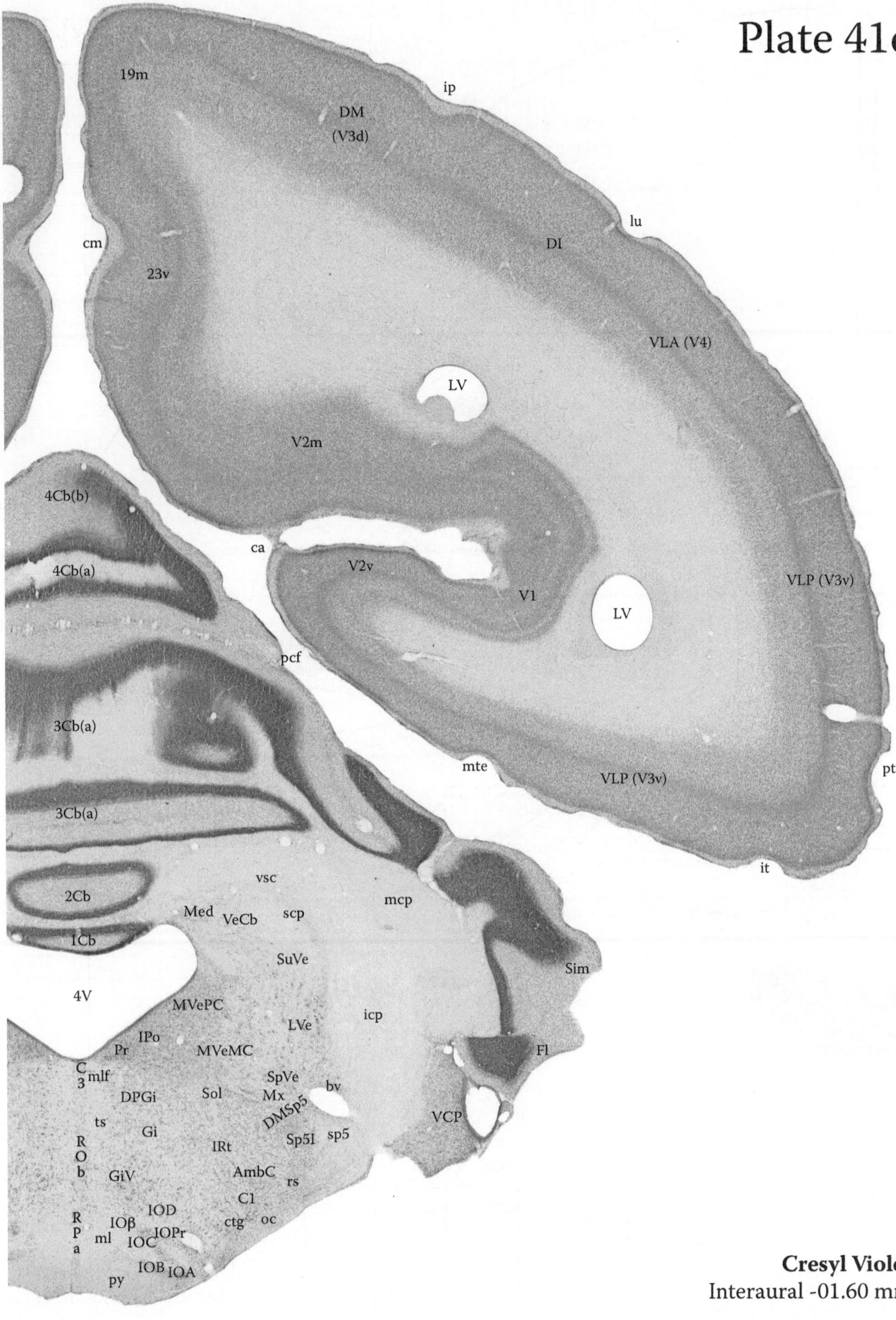


**Cresyl Violet**
Interaural -01.60 mm

# Plate 41

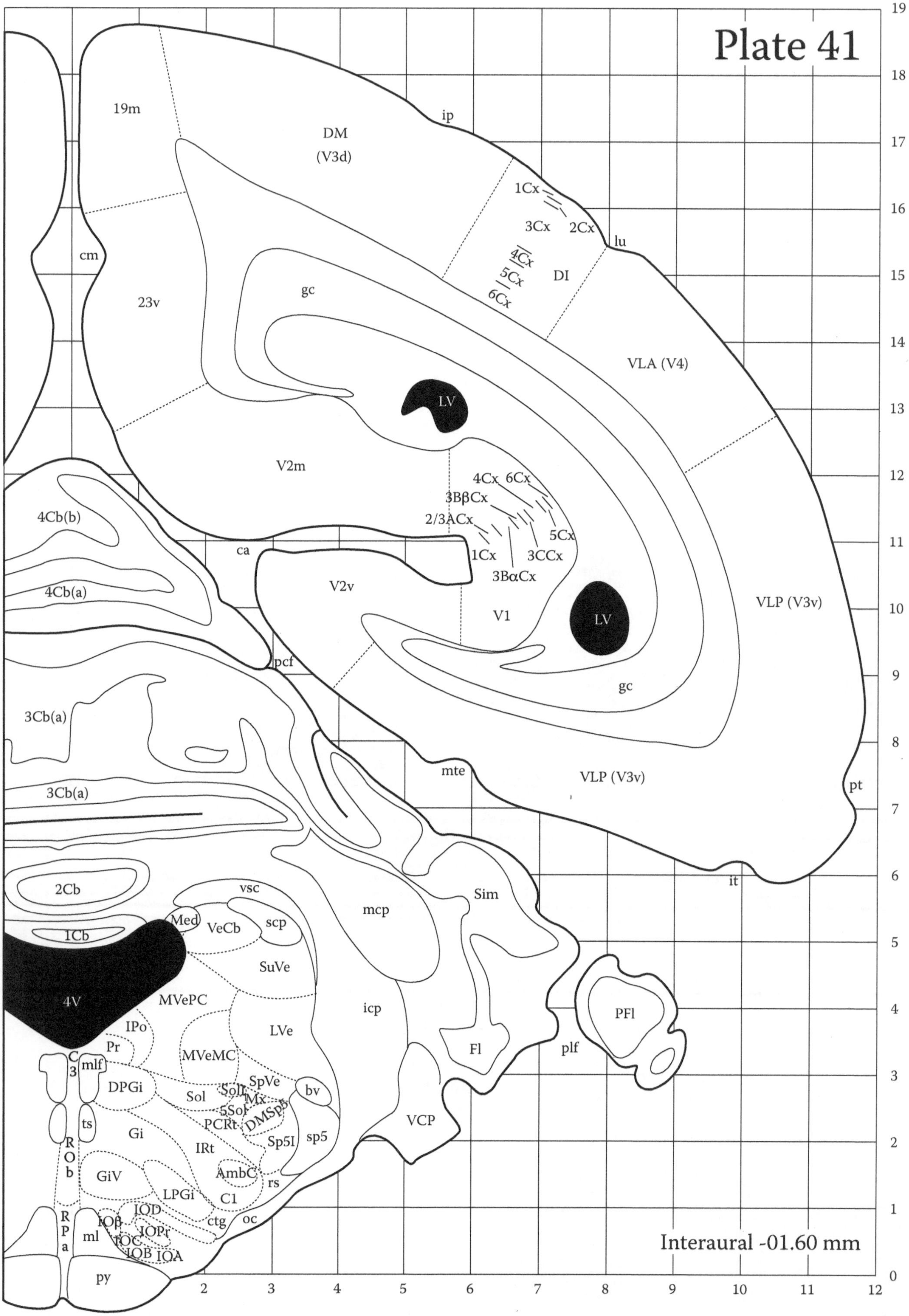
19m
DM
(V3d)
ip
1Cx
3Cx
2Cx
lu
4Cx
5Cx
6Cx
DI
cm
23v
gc
VLA (V4)
LV
V2m
4Cx
6Cx
3BβCx
2/3ACx
5Cx
1Cx
3CCx
3BαCx
4Cb(b)
4Cb(a)
ca
V2v
V1
LV
VLP (V3v)
pcf
gc
3Cb(a)
3Cb(a)
mte
VLP (V3v)
pt
it
2Cb
vsc
Med
VeCb
scp
mcp
Sim
1Cb
SuVe
4V
MVePC
icp
IPo
LVe
PFl
Pr
MVeMC
Fl
plf
C3
mlf
DPGi
Sol
SolI
SpVe
Mx
bv
5Sol
DMSp5
PCRt
ts
Gi
Sp5I
sp5
VCP
ROb
IRt
GiV
AmbC
rs
LPGi
C1
RPa
IOD
IOβ
ctg
oc
ml
IOC
IOPr
IOB
IOA
py
Interaural -01.60 mm

| | |
|---|---|
| 1Cb | cerebell. lobule 1 |
| 2Cb | cerebell. lobule 2 |
| 3Cb(a) | cerebell. lobule 3a |
| 4Cb(a) | cerebell. lobule 4a |
| 4Cb(b) | cerebell. lobule 4b |
| 4V | fourth ventricle |
| 5Sol | trigemsol. transn zone |
| 19m | area 19m Cx |
| 23v | area 23v Cx |
| AmbC | ambig. nu., compact |
| bv | blood vessel |
| C1 | C1 adrenaline cells |
| C3 | C3 adrenaline cells |
| ca | calcarine sulcus |
| cm | callosomarginal sulc. |
| ctg | central tegm. tract |
| DI | dorsoint. visual Cx |
| DM | dorsomed. visual Cx |
| DMSp5 | dorsomed. spin. 5 nu. |
| DPGi | dors. paragigant. nu. |
| Fl | flocculus |
| gc | geniculocalcarine tr. |
| Gi | gigantocell. retic. nu. |
| GiV | gigantocell. retic. nu., vent. |
| icp | inf. cerebell. ped. |
| IOA | inf. oliv. nu., subn. A |
| IOβ | inf. oliv. nu., subn. β |
| IOB | inf. oliv. nu., subn. B |
| IOC | inf. oliv. nu., subn. C |
| IOD | inf. oliv. nu., dors. |
| IOPr | inf. oliv. nu., princ. |
| ip | intraparietal sulc. |
| IPo | interpositus nu. |
| IRt | intermed. retic. formn |
| it | inf. temporal sulcus |
| LPGi | lat. paragigant. nu. |
| lu | lunate sulcus |
| LV | lateral ventricle |
| LVe | lat. vestib. nu. |
| mcp | middle cerebell. ped. |
| Med | med. deep Cb nu. |
| ml | medial lemniscus |
| mlf | med. longit. fasc. |
| mte | med. temp. sulc. |
| MVeMC | med. vestib. nu., magn. |
| MVePC | med. vestib. nu., parv. |
| Mx | matrix sensory area |
| oc | olivocerebellar tr. |
| pcf | preculminate fiss. Cb. |
| PCRt | parv. retic. nu. |
| PFl | paraflocculus |
| plf | posterolat. fissure |
| Pr | prepositus hypogl. nu. |
| pt | post. temp. sulc. |
| py | pyramidal tract |
| ROb | raphe obscurus nu. |
| RPa | raphe pallidus nu. |
| rs | rubrospinal tract |
| scp | sup. cerebell. ped. |
| Sim | simplex lobule Cb |
| Sol | nu. of solitary tr. |
| SolI | nu. of solitary tr., intrafasc. |
| sp5 | spinal trigeminal tr. |
| Sp5I | spin. trigem. nu., interpol. |
| SpVe | spinal vestib. nu. |
| SuVe | sup. vestib. nu. |
| ts | tectospinal tract |
| V1 | V1 Cx |
| V2m | V2 (med.) Cx |
| V2v | V2 (vent.) Cx |
| V3d | V3 (dors.) Cx |
| VCP | vent. coch. nu., post. |
| VeCb | vestibulocerebell. nu. |
| VLA (V4) | ventlatant. (V4) Cx |
| VLP (V3v) | ventlatpost. (V3v) Cx |
| vsc | vent. spinocereb. tr. |

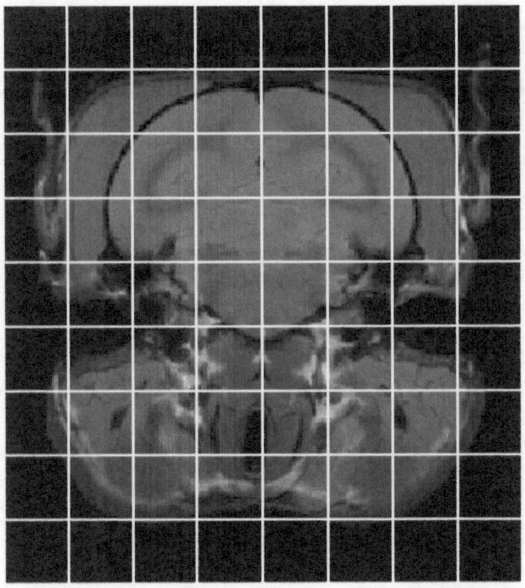

Interaural -01.70 mm

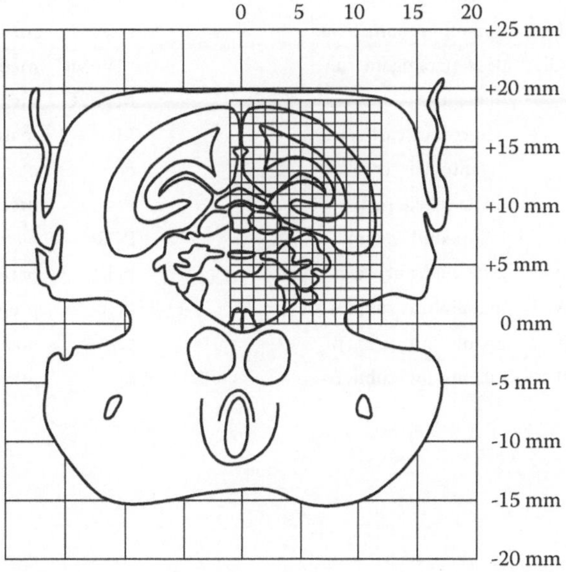

# Plate 42

Interaural -02.10 mm

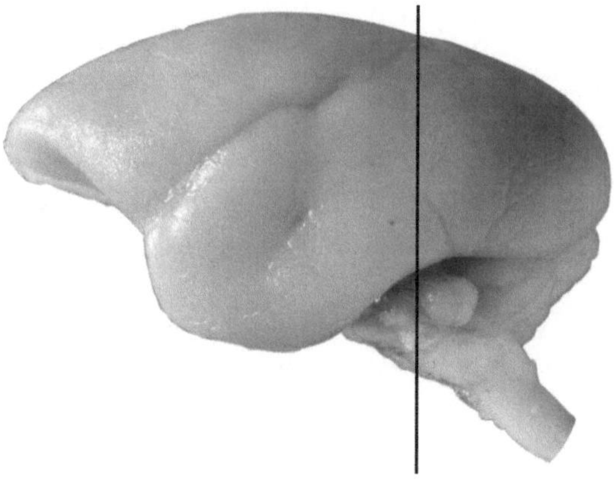

| | | | | | |
|---|---|---|---|---|---|
| 3Cb(a) | cerebell. lobule 3a | IOC | inf. oliv. nu., subn. C | ROb | raphe obscurus nu. |
| 4Cb(a) | cerebell. lobule 4a | IOD | inf. oliv. nu., dors. | RPa | raphe pallidus nu. |
| 4Cb(b) | cerebell. lobule 4b | IPo | interpositus nu. | rs | rubrospinal tract |
| 4V | fourth ventricle | IRt | intermed. retic. formn | scp | sup. cerebell. ped. |
| 5Cb | cerebell. lobule 5 | it | inf. temporal sulcus | Sim | simplex lobule Cb |
| 5Sol | trigemsol. transn zone | Lat | lat. deep Cb nu. | SolCe | nu. of solitary tr., cent. |
| AmbC | ambig. nu., compact | LRt | lat. reticular nu. | SolDM | nu. of solitary tr., dorsomed. |
| C1 | C1 adrenaline cells | LRtS5 | lat. retic. nu., subtrigem. | SolI | nu. of solitary tr., intrafasc. |
| C3 | C3 adrenaline cells | LVe | lat. vestib. nu. | SolIM | nu. of solitary tr., int. |
| ca | calcarine sulcus | mcp | middle cerebell. ped. | SolL | nu. of solitary tr., lat. |
| cbc | cerebellar commiss. | Med | med. deep Cb nu. | SolV | nu. of solitary tr., vent. |
| cm | callosomarginal sulc. | ml | medial lemniscus | SolVL | nu. of solitary tr., ventlat. |
| DM | dorsomed. visual Cx | mlf | med. longit. fasc. | sp5 | spinal trigeminal tr. |
| DMSp5 | dorsomed. spin. 5 nu. | mte | med. temp. sulc. | Sp5I | spin. trigem. nu., interpol. |
| DPGi | dors. paragigant. nu. | MVeMC | med. vestib. nu., magn. | SpVe | spinal vestib. nu. |
| Fl | flocculus | MVePC | med. vestib. nu., parv. | SuVe | sup. vestib. nu. |
| gc | geniculocalcarine tr. | Mx | matrix sensory area | ts | tectospinal tract |
| Gi | gigantocell. retic. nu. | oc | olivocerebellar tr. | V1 | V1 Cx |
| icp | inf. cerebell. ped. | pcf | preculminate fiss. Cb. | V2m | V2 (med.) Cx |
| imv | inf. medull. velum | PCRt | parv. retic. nu. | V2v | V2 (vent.) Cx |
| IntA | intermed. ant. deep Cb nu. | PFlP | paraflocculus post. | VCP | vent. coch. nu., post. |
| IOA | inf. oliv. nu., subn. A | Pr | prepositus hypogl. nu. | VeCb | vestibulocerebell. nu. |
| IOβ | inf. oliv. nu., subn. β | pt | post. temp. sulc. | VeR | ventrorost. 10N |
| IOB | inf. oliv. nu., subn. B | py | pyramidal tract | VLP (V3v) | ventlatpost. (V3v) Cx |

# Plate 42a

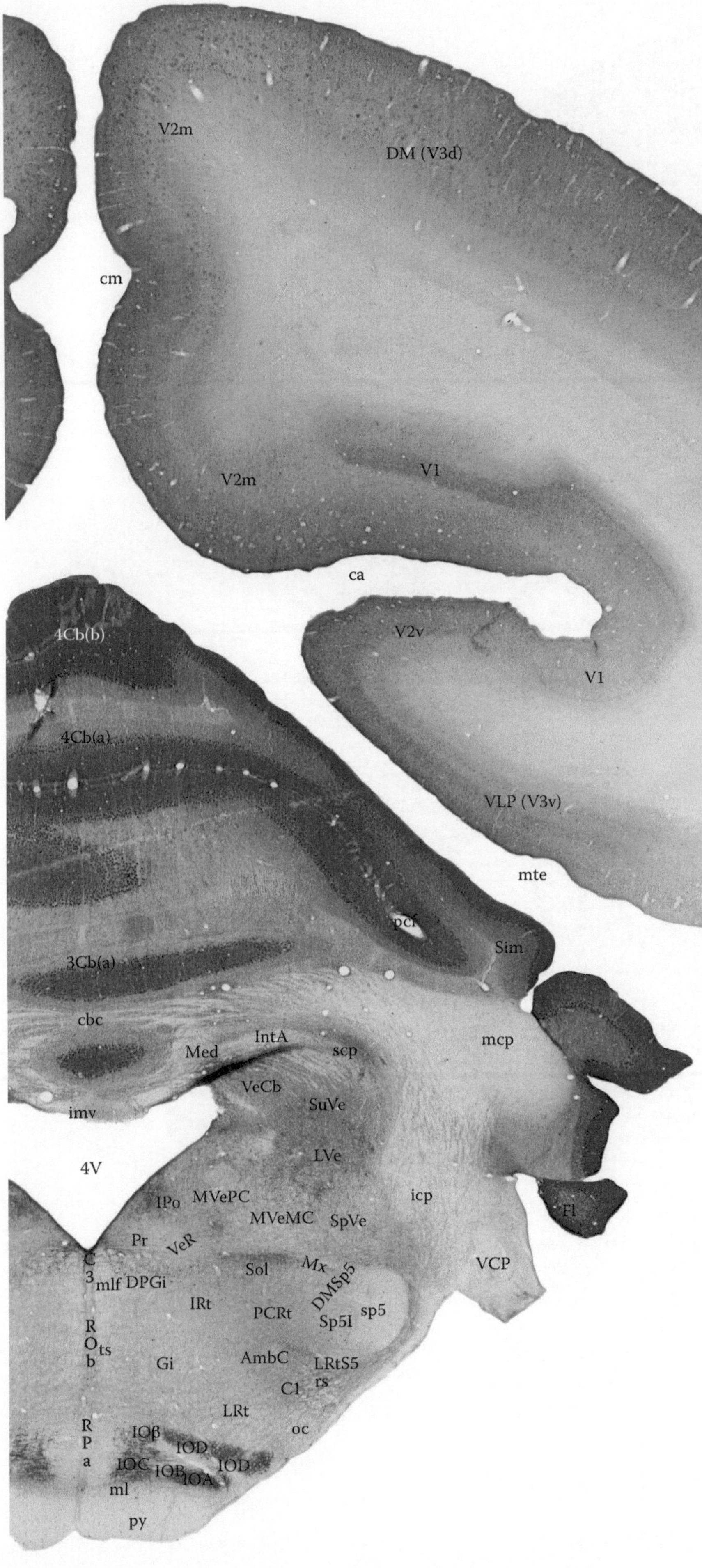


**Calbindin**
Interaural -01.80 mm

# Plate 42b

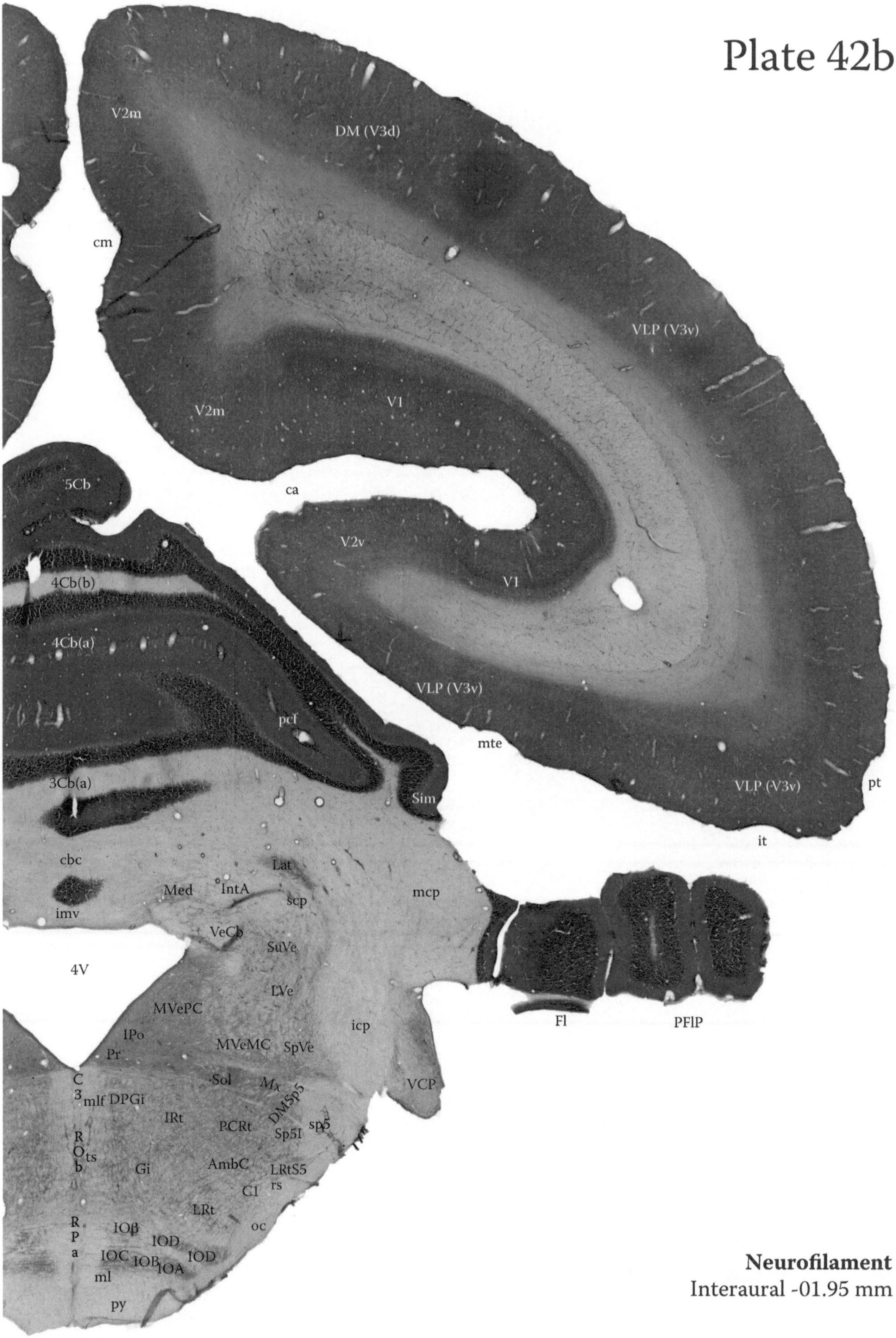


**Neurofilament**
Interaural -01.95 mm

# Plate 42c

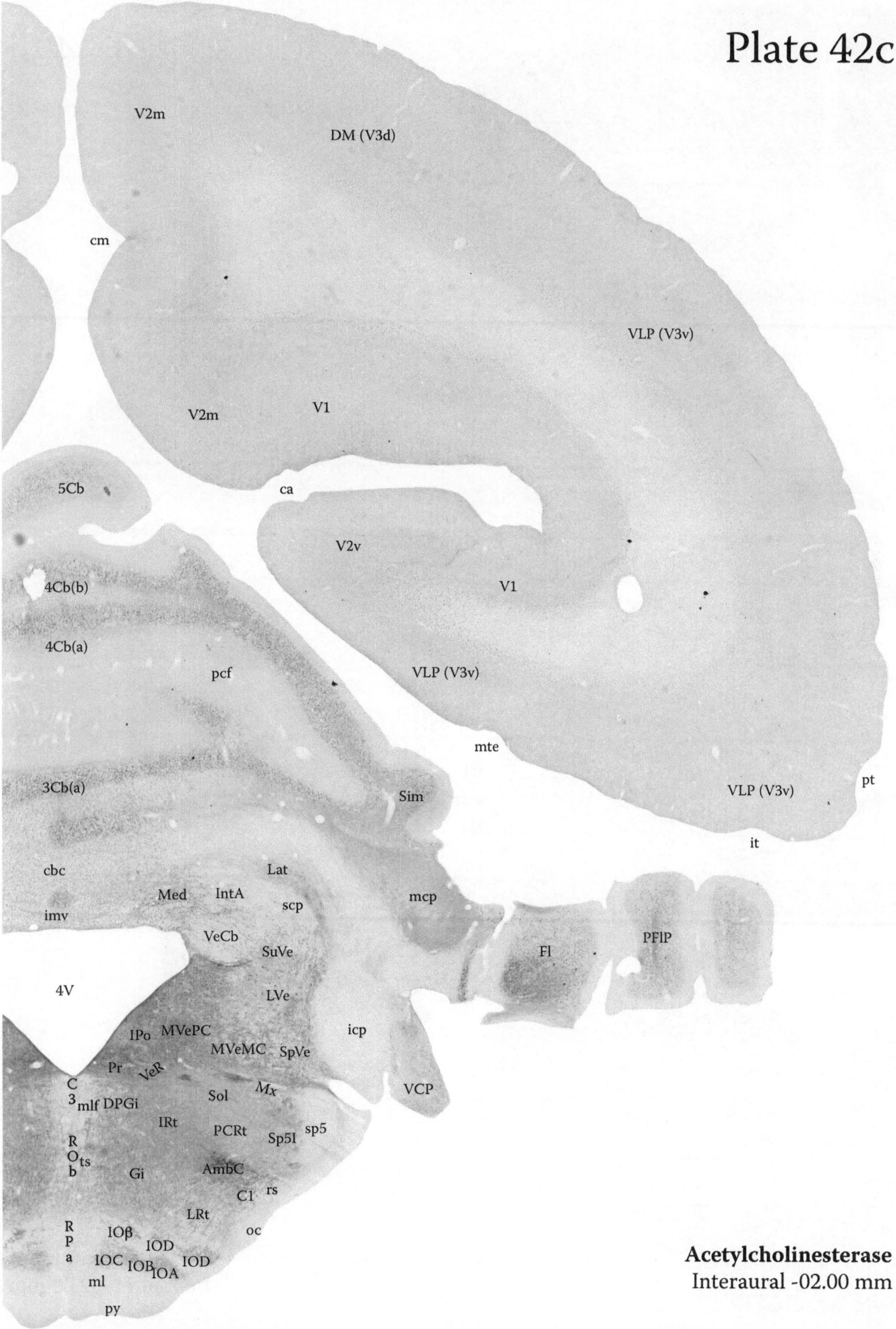


**Acetylcholinesterase**
Interaural -02.00 mm

# Plate 42d

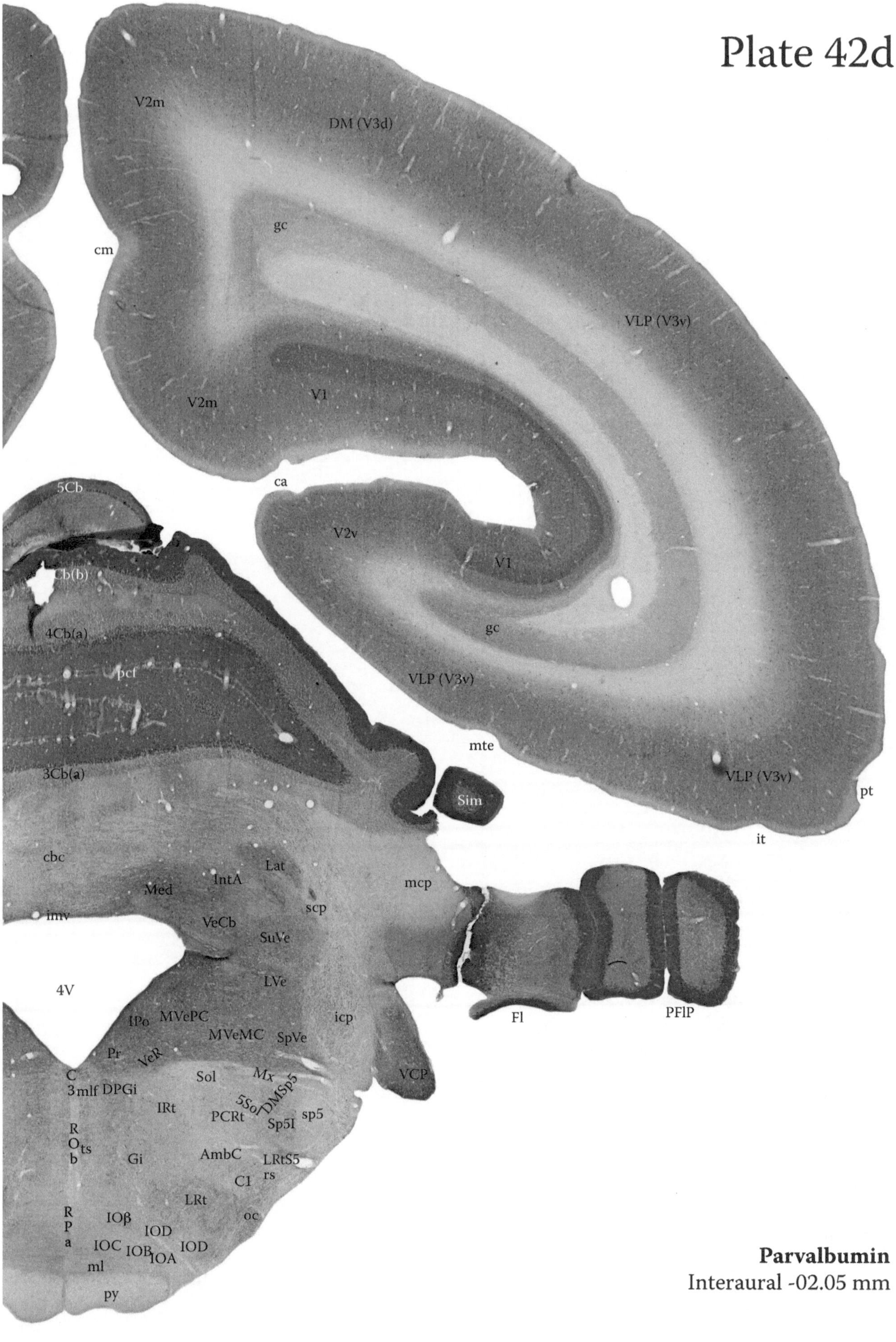


**Parvalbumin**
Interaural -02.05 mm

# Plate 42e

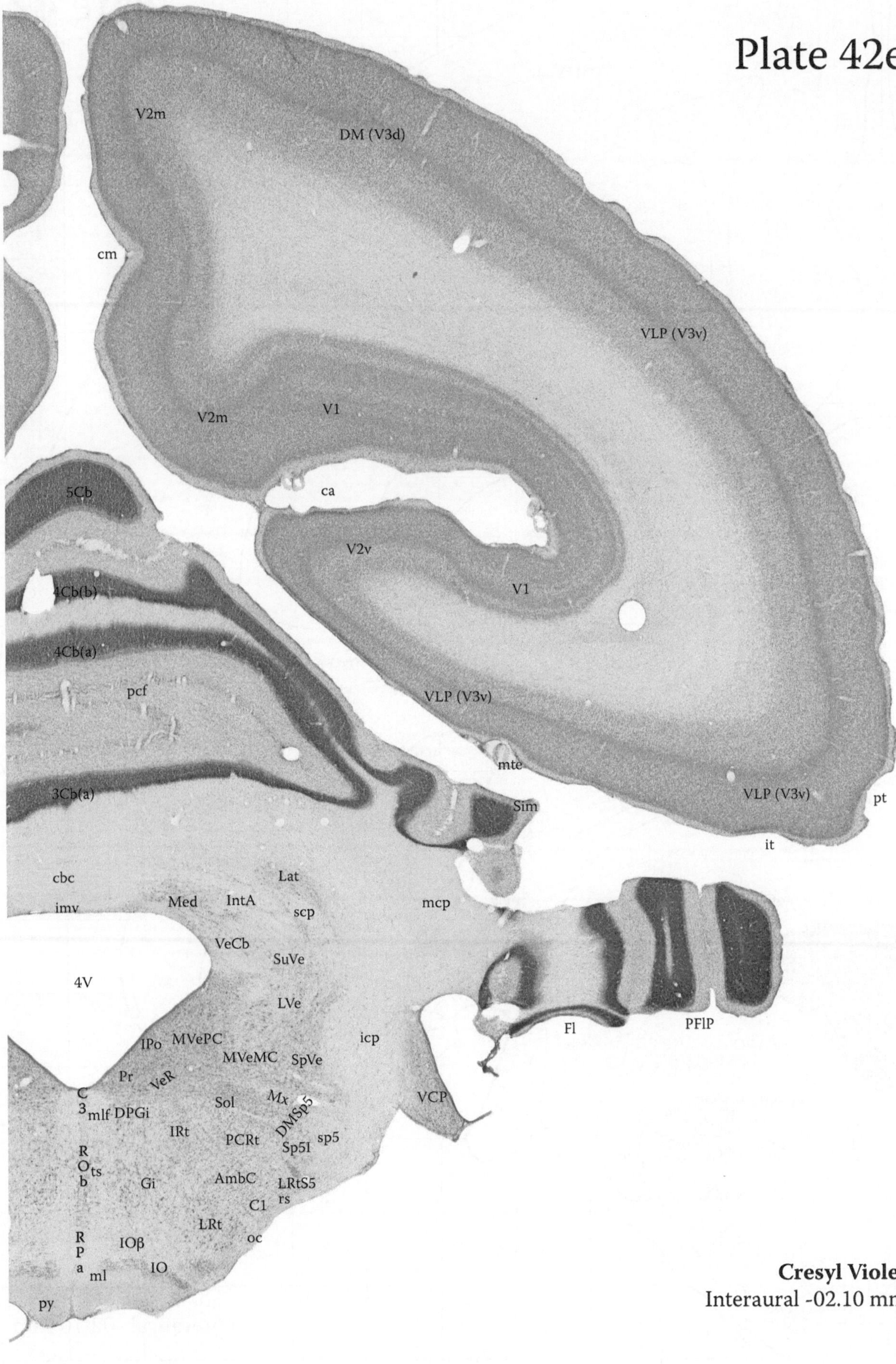


**Cresyl Violet**
Interaural -02.10 mm

# Plate 42

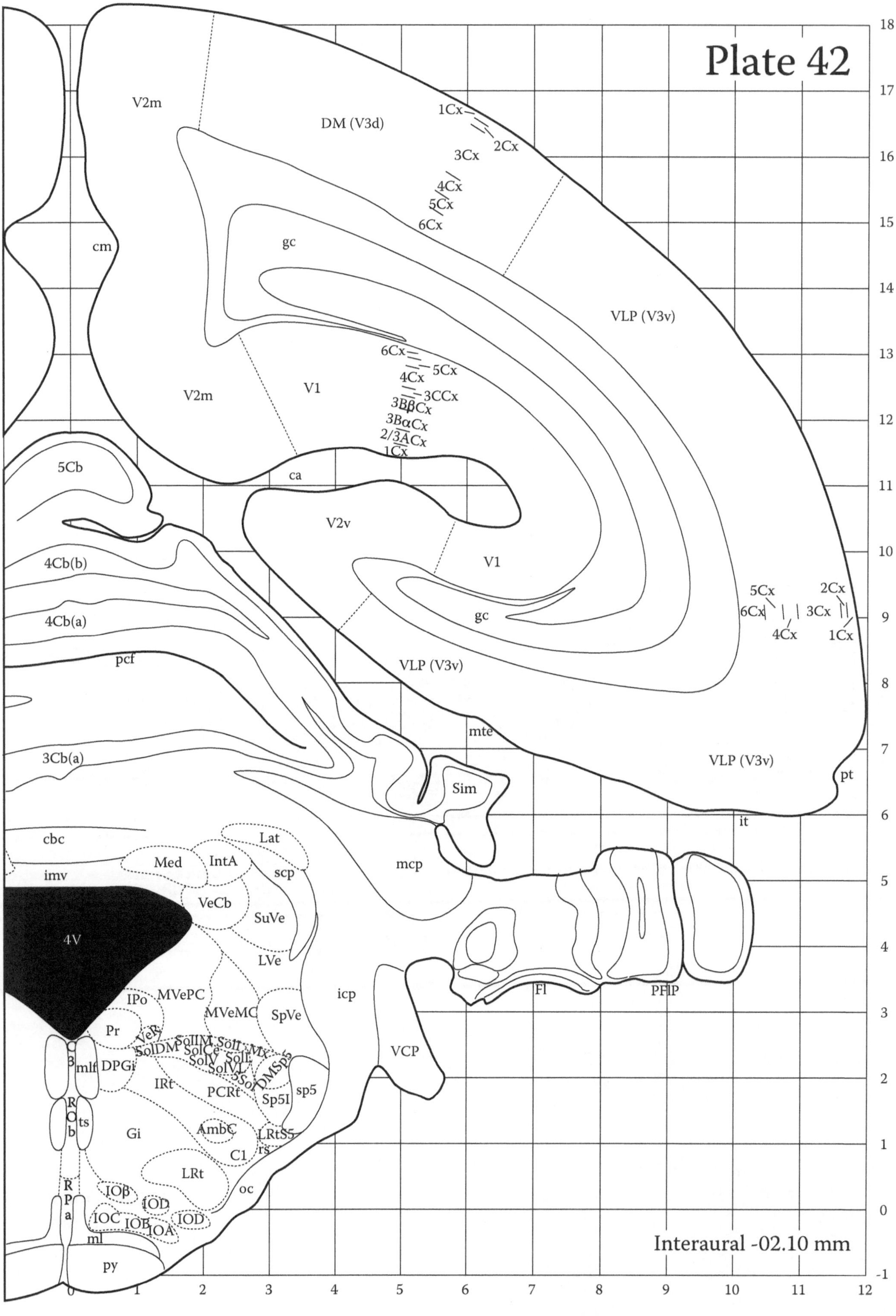

V2m
DM (V3d)
1Cx
2Cx
3Cx
4Cx
5Cx
6Cx
cm
gc
VLP (V3v)
6Cx
5Cx
4Cx
3CCx
3BβCx
3BαCx
2/3ACx
1Cx
V2m
V1
5Cb
ca
V2v
V1
4Cb(b)
4Cb(a)
5Cx
2Cx
6Cx
3Cx
4Cx
1Cx
gc
pcf
VLP (V3v)
mte
3Cb(a)
VLP (V3v)
pt
Sim
it
cbc
Lat
Med
IntA
imv
scp
mcp
VeCb
SuVe
4V
LVe
IPo
MVePC
MVeMC
SpVe
icp
Pr
VeR
SolIM
SolDM
SolCe
SolI
SolL
Mx
SolV
SolVL
DMSp5
5Sol
mlf
DPGi
IRt
PCRt
Sp5I
sp5
VCP
Fl
PFlP
ROb
ts
Gi
AmbC
LRtS5
rs
C1
LRt
oc
RPa
IOβ
IOD
IOC
IOB
IOA
IOD
ml
py
Interaural -02.10 mm
18
17
16
15
14
13
12
11
10
9
8
7
6
5
4
3
2
1
0
-1
0
1
2
3
4
5
6
7
8
9
10
11
12

| | |
|---|---|
| 3Cb(a) | cerebell. lobule 3a |
| 4Cb(a) | cerebell. lobule 4a |
| 4Cb(b) | cerebell. lobule 4b |
| 4V | fourth ventricle |
| 5Cb | cerebell. lobule 5 |
| 5Sol | trigemsol. transn zone |
| AmbC | ambig. nu., compact |
| C1 | C1 adrenaline cells |
| C3 | C3 adrenaline cells |
| ca | calcarine sulcus |
| cbc | cerebellar commiss. |
| cm | callosomarginal sulc. |
| DM | dorsomed. visual Cx |
| DMSp5 | dorsomed. spin. 5 nu. |
| DPGi | dors. paragigant. nu. |
| Fl | flocculus |
| gc | geniculocalcarine tr. |
| Gi | gigantocell. retic. nu. |
| icp | inf. cerebell. ped. |
| imv | inf. medull. velum |
| IntA | intermed. ant. deep Cb nu. |
| IOA | inf. oliv. nu., subn. A |
| IOβ | inf. oliv. nu., subn. β |
| IOB | inf. oliv. nu., subn. B |
| IOC | inf. oliv. nu., subn. C |
| IOD | inf. oliv. nu., dors. |
| IPo | interpositus nu. |
| IRt | intermed. retic. formn |
| it | inf. temporal sulcus |
| Lat | lat. deep Cb nu. |
| LRt | lat. reticular nu. |
| LRtS5 | lat. retic. nu., subtrigem. |
| LVe | lat. vestib. nu. |
| mcp | middle cerebell. ped. |
| Med | med. deep Cb nu. |
| ml | medial lemniscus |
| mlf | med. longit. fasc. |
| mte | med. temp. sulc. |
| MVeMC | med. vestib. nu., magn. |
| MVePC | med. vestib. nu., parv. |
| Mx | matrix sensory area |
| oc | olivocerebellar tr. |
| pcf | preculminate fiss. Cb. |
| PCRt | parv. retic. nu. |
| PFlP | paraflocculus post. |
| Pr | prepositus hypogl. nu. |
| pt | post. temp. sulc. |
| py | pyramidal tract |
| ROb | raphe obscurus nu. |
| RPa | raphe pallidus nu. |
| rs | rubrospinal tract |
| scp | sup. cerebell. ped. |
| Sim | simplex lobule Cb |
| SolCe | nu. of solitary tr., cent. |
| SolDM | nu. of solitary tr., dorsomed. |
| SolI | nu. of solitary tr., intrafasc. |
| SolIM | nu. of solitary tr., int. |
| SolL | nu. of solitary tr., lat. |
| SolV | nu. of solitary tr., vent. |
| SolVL | nu. of solitary tr., ventlat. |
| sp5 | spinal trigeminal tr. |
| Sp5I | spin. trigem. nu., interpol. |
| SpVe | spinal vestib. nu. |
| SuVe | sup. vestib. nu. |
| ts | tectospinal tract |
| V1 | V1 Cx |
| V2m | V2 (med.) Cx |
| V2v | V2 (vent.) Cx |
| VCP | vent. coch. nu., post. |
| VeCb | vestibulocerebell. nu. |
| VeR | ventrorost. 10N |
| VLP (V3v) | ventlatpost. (V3v) Cx |

# Plate 43

Interaural -02.65 mm

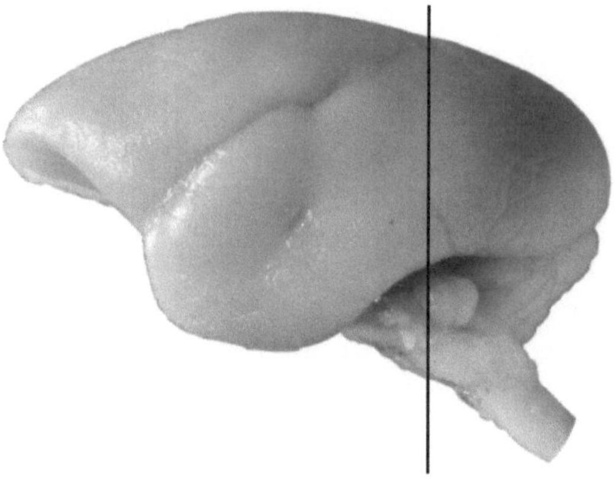

| | | | | | |
|---|---|---|---|---|---|
| 3Cb | cerebell. lobule 3 | IntP | intermed. post. deep Cb nu. | Ro | nu. of Roller |
| 4Cb | cerebell. lobule 4 | IOA | inf. oliv. nu., subn. A | ROb | raphe obscurus nu. |
| 4V | fourth ventricle | IOβ | inf. oliv. nu., subn. β | RPa | raphe pallidus nu. |
| 5Cb | cerebell. lobule 5 | IOB | inf. oliv. nu., subn. B | rs | rubrospinal tract |
| 5Sol | trigemsol. transn zone | IOC | inf. oliv. nu., subn. C | scp | sup. cerebell. ped. |
| 8vn | vestibular divn 8n | IOD | inf. oliv. nu., dors. | Sima | simplex lobule Cb, ant. |
| 10CeI | dors. nu. 10, cent. int. | IOK | inf. oliv. nu., Kooy. | Simp | simplex lobule Cb, post. |
| 10DI | dors. nu. 10, dors. int. | IRt | intermed. retic. formn | sol | solitary tract |
| 12n | hypoglossal nerve | Lat | lat. deep Cb nu. | SolCe | nu. of solitary tr., cent. |
| 12N | hypoglossal nu. | LRt | lat. reticular nu. | SolDM | nu. of solitary tr., dorsomed. |
| AmbC | ambig. nu., compact | LRtPC | lat. retic. nu., parv. | SolIM | nu. of solitary tr., int. |
| AP | area postrema | LRtS5 | lat. retic. nu., subtrigem. | SolL | nu. of solitary tr., lat. |
| BI | basal interstitial nu. | mcp | middle cerebell. ped. | SolV | nu. of solitary tr., vent. |
| C1 | C1 adrenaline cells | Med | med. deep Cb nu. | SolVL | nu. of solitary tr., ventlat. |
| ca | calcarine sulcus | ml | medial lemniscus | sp5 | spinal trigeminal tr. |
| cbc | cerebellar commiss. | mlf | med. longit. fasc. | Sp5I | spin. trigem. nu., interpol. |
| DM (V3d) | dorsomed. visual Cx | MVeMC | med. vestib. nu., magn. | SpVe | spinal vestib. nu. |
| DMSp5 | dorsomed. spin. 5 nu. | MVePC | med. vestib. nu., parv. | ts | tectospinal tract |
| DPGi | dors. paragigant. nu. | Mx | matrix sensory area | V1 | V1 Cx |
| dsc | dors. spinocerebell. tr | oc | olivocerebellar tr. | V2m | V2 (med.) Cx |
| Fl | flocculus | pcf | preculminate fiss. Cb. | V2v | V2 (vent.) Cx |
| gc | geniculocalcarine tr. | PCRt | parv. retic. nu. | VCP | vent. coch. nu., post. |
| Gi | gigantocell. retic. nu. | PFlP | paraflocculus post. | VLP (V3v) | ventlatpost. (V3v) Cx |
| icp | inf. cerebell. ped. | plf | posterolat. fissure | X | nucleus X |
| In | intercal. nu. medulla | prf | prim. cerebell. fiss. | Y | nucleus Y |
| IntA | intermed. ant. deep Cb nu. | py | pyramidal tract | | |

# Plate 43a

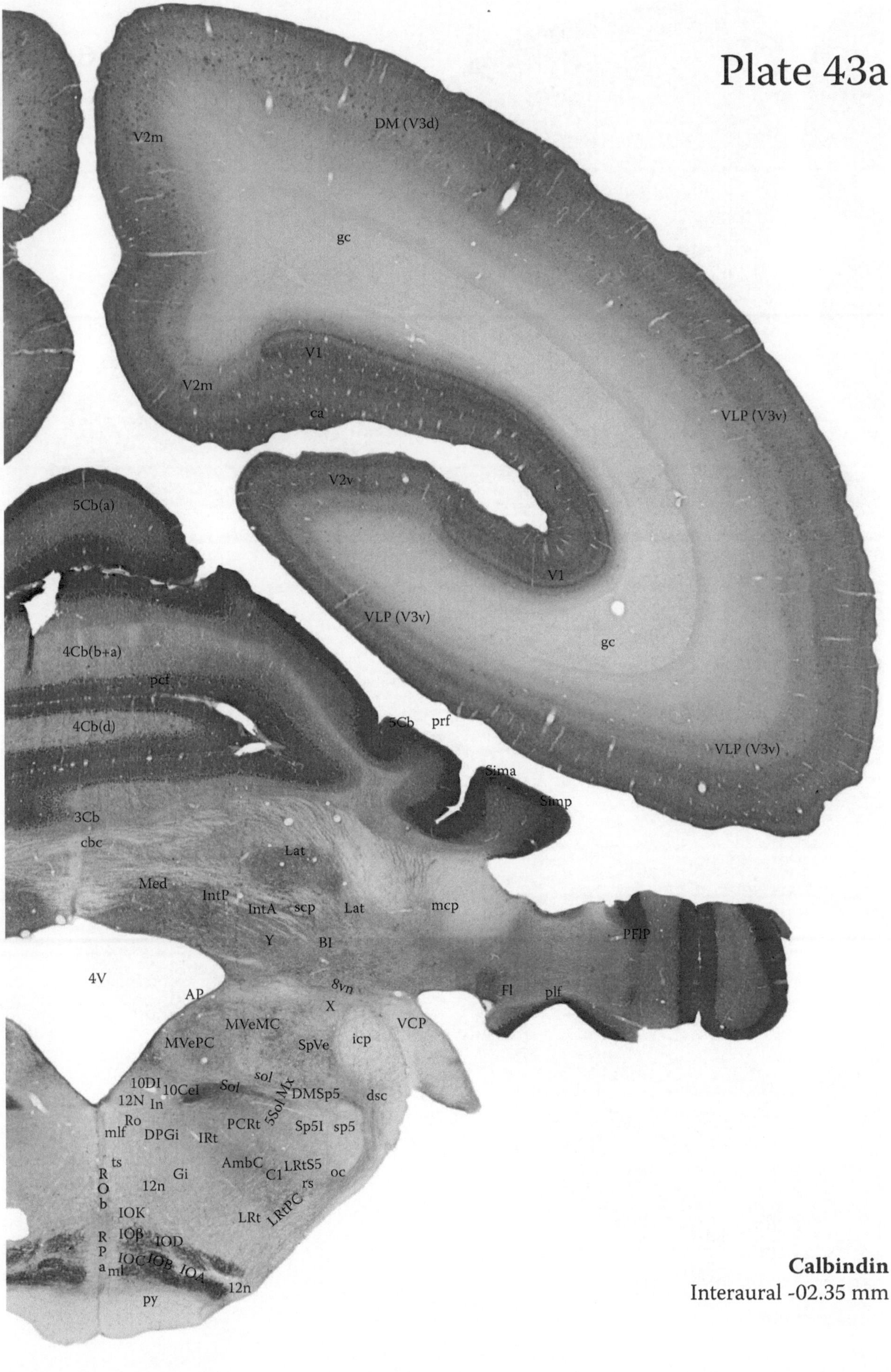


**Calbindin**
Interaural -02.35 mm

# Plate 43b

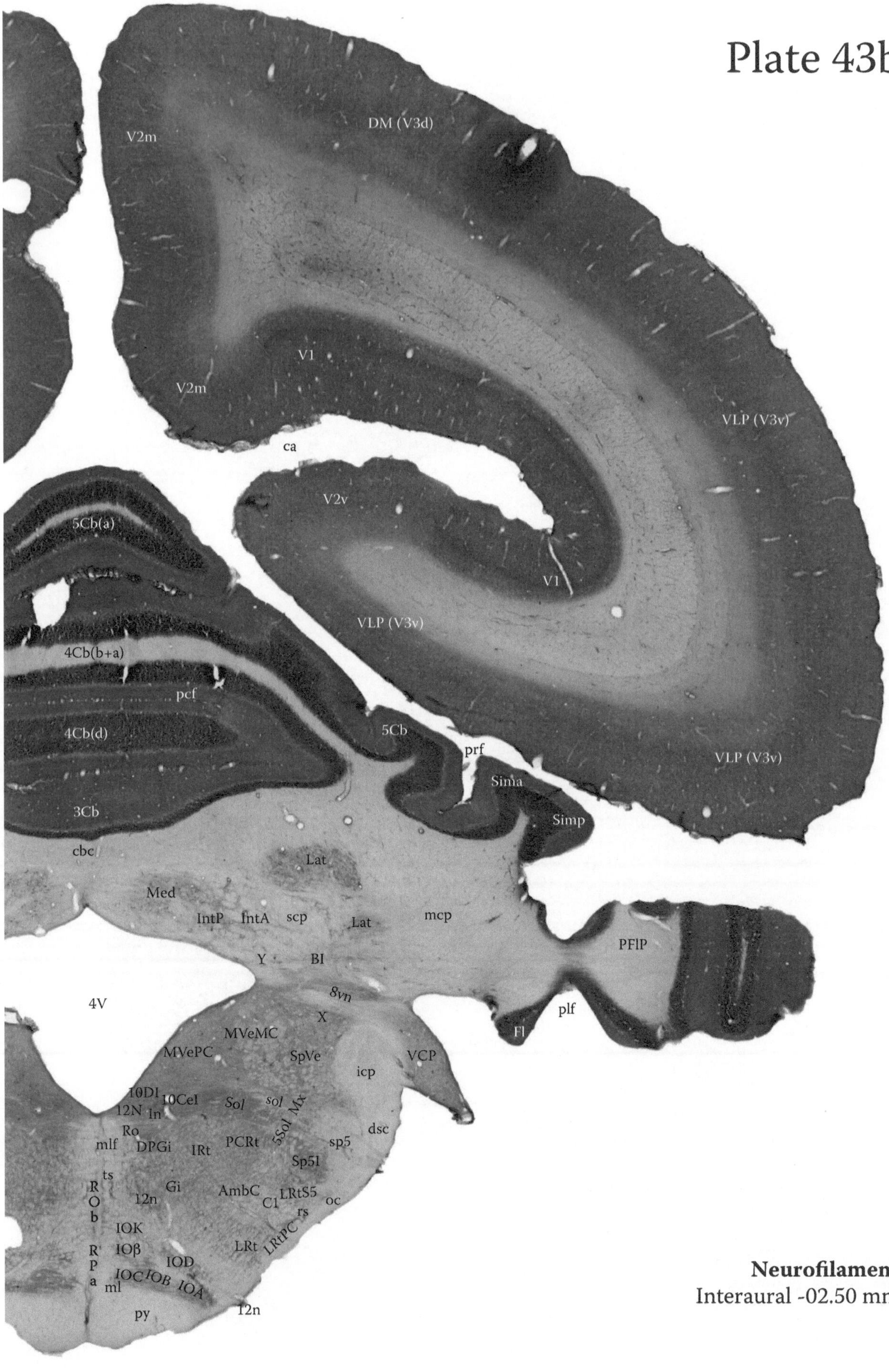


**Neurofilament**
Interaural -02.50 mm

# Plate 43c

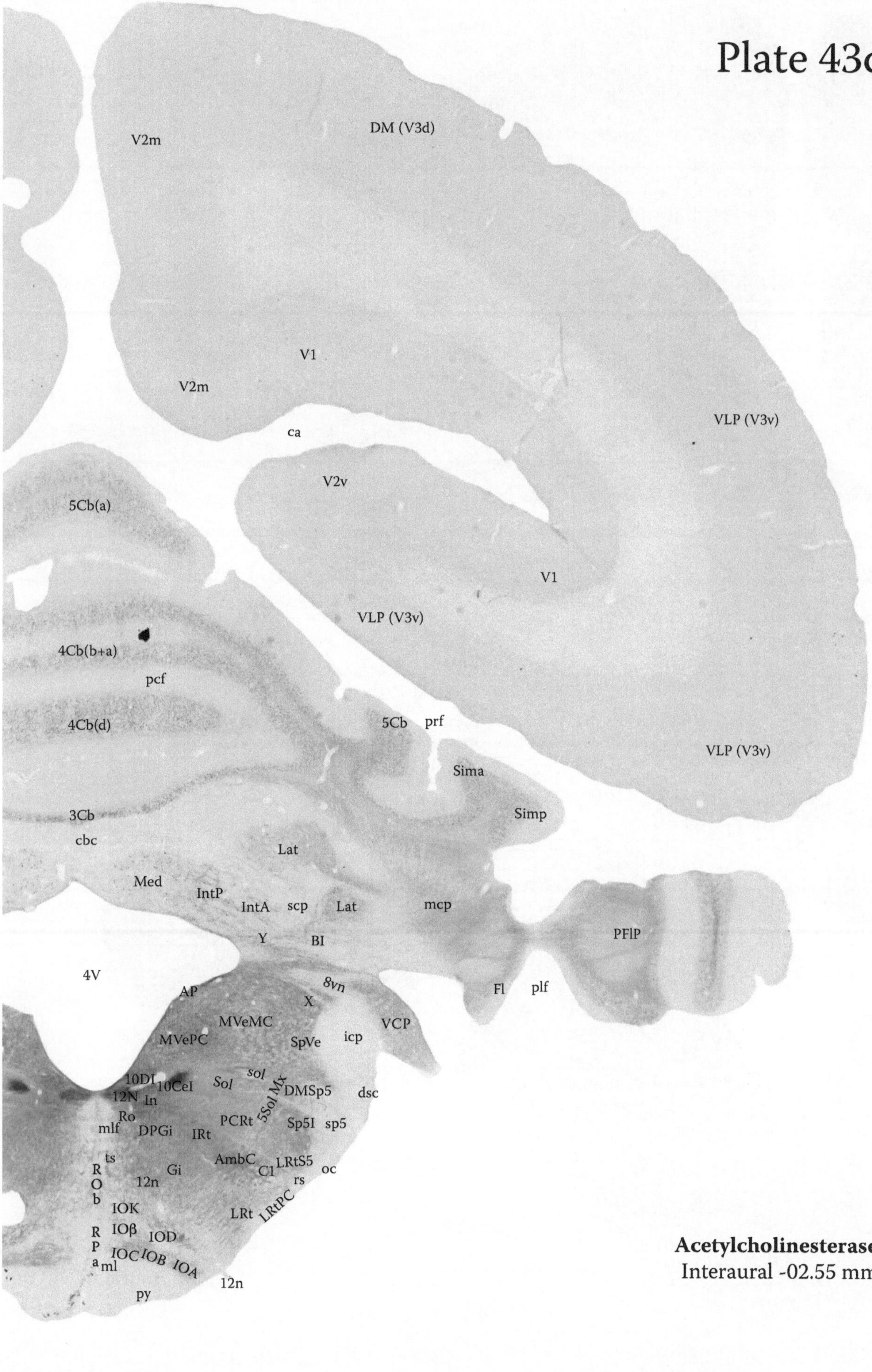


**Acetylcholinesterase**
Interaural -02.55 mm

# Plate 43d

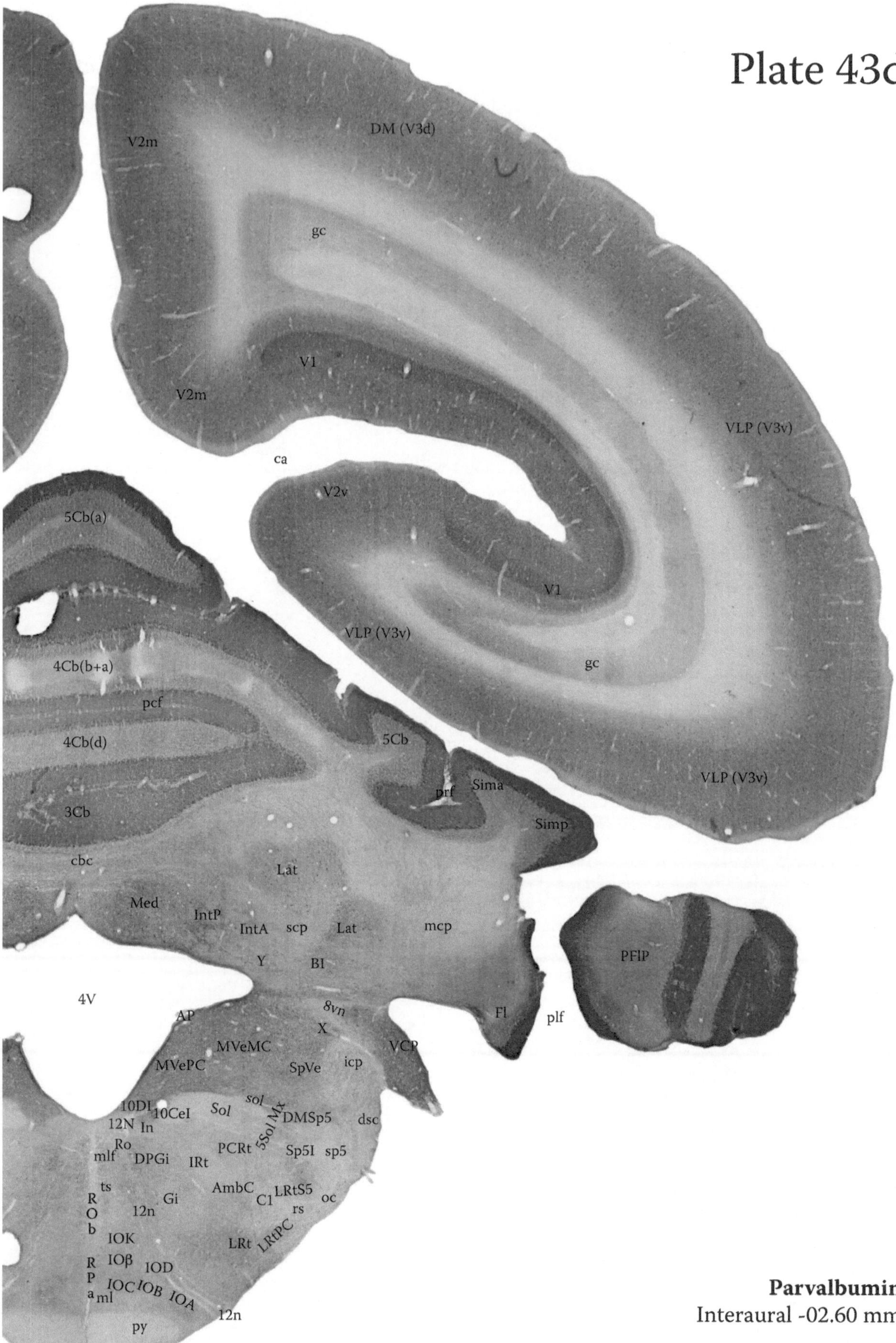


**Parvalbumin**
Interaural -02.60 mm

# Plate 43e

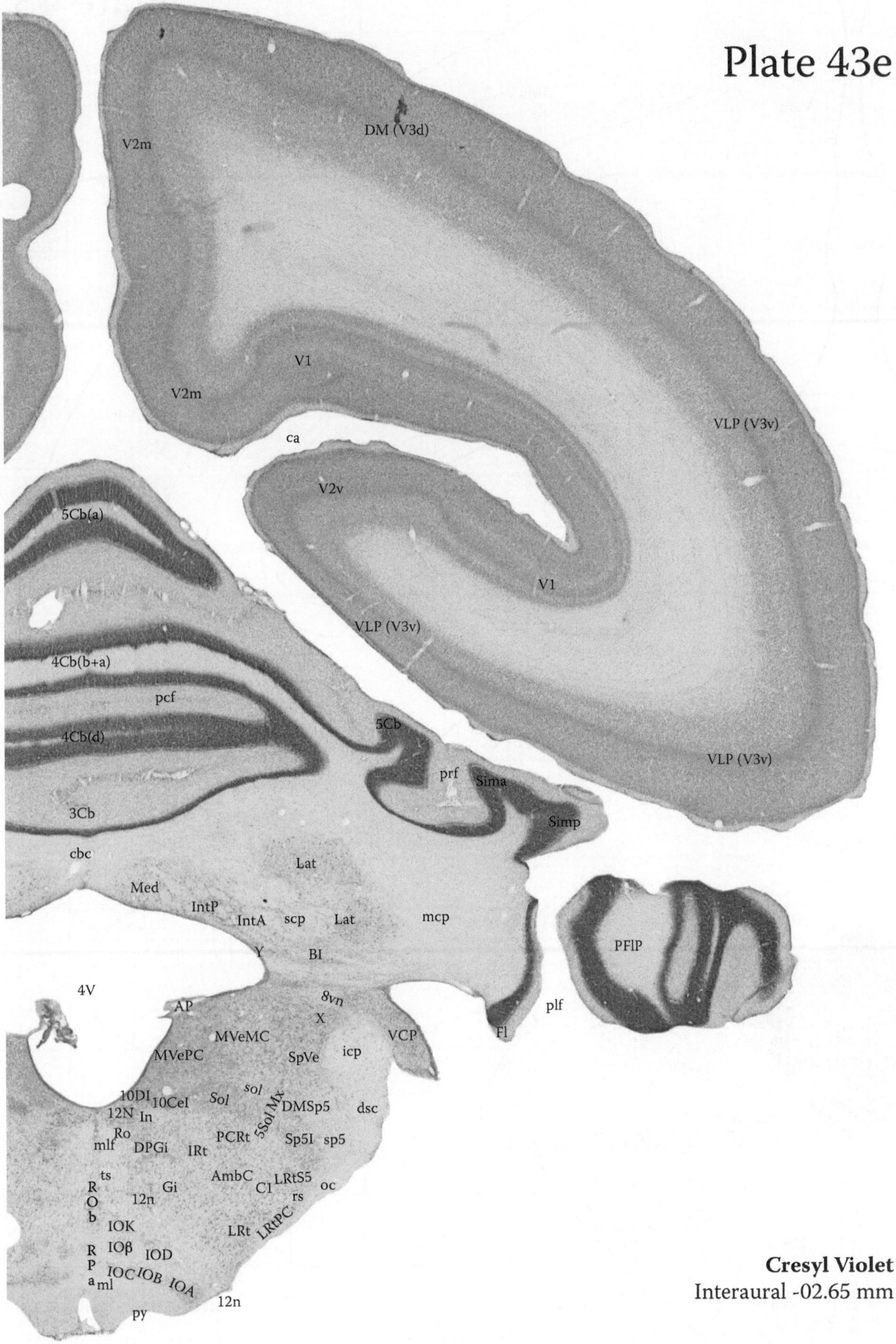


**Cresyl Violet**
Interaural -02.65 mm

# Plate 43

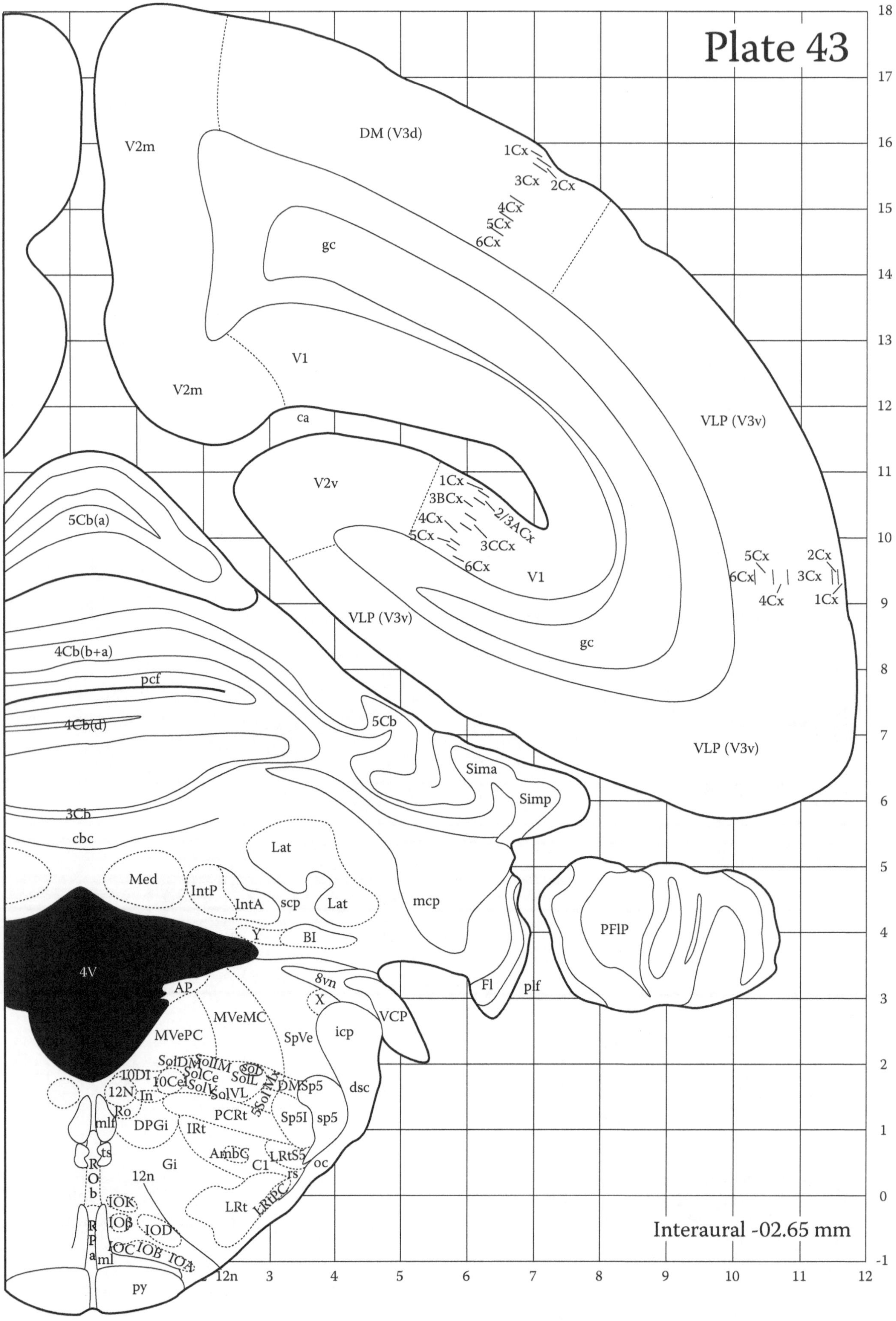
V2m
DM (V3d)
1Cx
3Cx
2Cx
4Cx
5Cx
6Cx
gc
V1
V2m
ca
VLP (V3v)
V2v
1Cx
3BCx
2/3ACx
4Cx
5Cx
3CCx
6Cx
V1
5Cx
2Cx
6Cx
3Cx
4Cx
1Cx
5Cb(a)
VLP (V3v)
gc
4Cb(b+a)
pcf
4Cb(d)
5Cb
VLP (V3v)
Sima
Simp
3Cb
cbc
Lat
Med
IntP
IntA
scp
Lat
mcp
Y
BI
PFlP
4V
AP
8vn
X
Fl
plf
MVeMC
VCP
MVePC
SpVe
icp
SolDM
SolIM
SolD
SolL
SolIMA
10DI
10Ce
SolCe
SolV
SolVL
DMSp5
dsc
12N
In
Ro
mlf
PCRt
Sp5I
sp5
DPGi
IRt
ts
AmbC
LRtS5
C1
oc
ROb
Gi
12n
rs
IOK
LRt
LRtPC
IOβ
IOD
RPa
IOC
IOB
IOA
ml
py
12n
Interaural -02.65 mm
18
17
16
15
14
13
12
11
10
9
8
7
6
5
4
3
2
1
0
-1
3
4
5
6
7
8
9
10
11
12

| | |
|---|---|
| 3Cb | cerebell. lobule 3 |
| 4Cb | cerebell. lobule 4 |
| 4V | fourth ventricle |
| 5Cb | cerebell. lobule 5 |
| 5Sol | trigemsol. transn zone |
| 8vn | vestibular divn 8n |
| 10CeI | dors. nu. 10, cent. int. |
| 10DI | dors. nu. 10, dors. int. |
| 12n | hypoglossal nerve |
| 12N | hypoglossal nu. |
| AmbC | ambig. nu., compact |
| AP | area postrema |
| BI | basal interstitial nu. |
| C1 | C1 adrenaline cells |
| ca | calcarine sulcus |
| cbc | cerebellar commiss. |
| DM (V3d) | dorsomed. visual Cx |
| DMSp5 | dorsomed. spin. 5 nu. |
| DPGi | dors. paragigant. nu. |
| dsc | dors. spinocerebell. tr |
| Fl | flocculus |
| gc | geniculocalcarine tr. |
| Gi | gigantocell. retic. nu. |
| icp | inf. cerebell. ped. |
| In | intercal. nu. medulla |
| IntA | intermed. ant. deep Cb nu. |
| IntP | intermed. post. deep Cb nu. |
| IOA | inf. oliv. nu., subn. A |
| IOβ | inf. oliv. nu., subn. β |
| IOB | inf. oliv. nu., subn. B |
| IOC | inf. oliv. nu., subn. C |
| IOD | inf. oliv. nu., dors. |
| IOK | inf. oliv. nu., Kooy. |
| IRt | intermed. retic. formn |
| Lat | lat. deep Cb nu. |
| LRt | lat. reticular nu. |
| LRtPC | lat. retic. nu., parv. |
| LRtS5 | lat. retic. nu., subtrigem. |
| mcp | middle cerebell. ped. |
| Med | med. deep Cb nu. |
| ml | medial lemniscus |
| mlf | med. longit. fasc. |
| MVeMC | med. vestib. nu., magn. |
| MVePC | med. vestib. nu., parv. |
| Mx | matrix sensory area |
| oc | olivocerebellar tr. |
| pcf | preculminate fiss. Cb. |
| PCRt | parv. retic. nu. |
| PFlP | paraflocculus post. |
| plf | posterolat. fissure |
| prf | prim. cerebell. fiss. |
| py | pyramidal tract |
| Ro | nu. of Roller |
| ROb | raphe obscurus nu. |
| RPa | raphe pallidus nu. |
| rs | rubrospinal tract |
| scp | sup. cerebell. ped. |
| Sima | simplex lobule Cb, ant. |
| Simp | simplex lobule Cb, post. |
| sol | solitary tract |
| SolCe | nu. of solitary tr., cent. |
| SolDM | nu. of solitary tr., dorsomed. |
| SolIM | nu. of solitary tr., int. |
| SolL | nu. of solitary tr., lat. |
| SolV | nu. of solitary tr., vent. |
| SolVL | nu. of solitary tr., ventlat. |
| sp5 | spinal trigeminal tr. |
| Sp5I | spin. trigem. nu., interpol. |
| SpVe | spinal vestib. nu. |
| ts | tectospinal tract |
| V1 | V1 Cx |
| V2m | V2 (med.) Cx |
| V2v | V2 (vent.) Cx |
| VCP | vent. coch. nu., post. |
| VLP (V3v) | ventlatpost. (V3v) Cx |
| X | nucleus X |
| Y | nucleus Y |

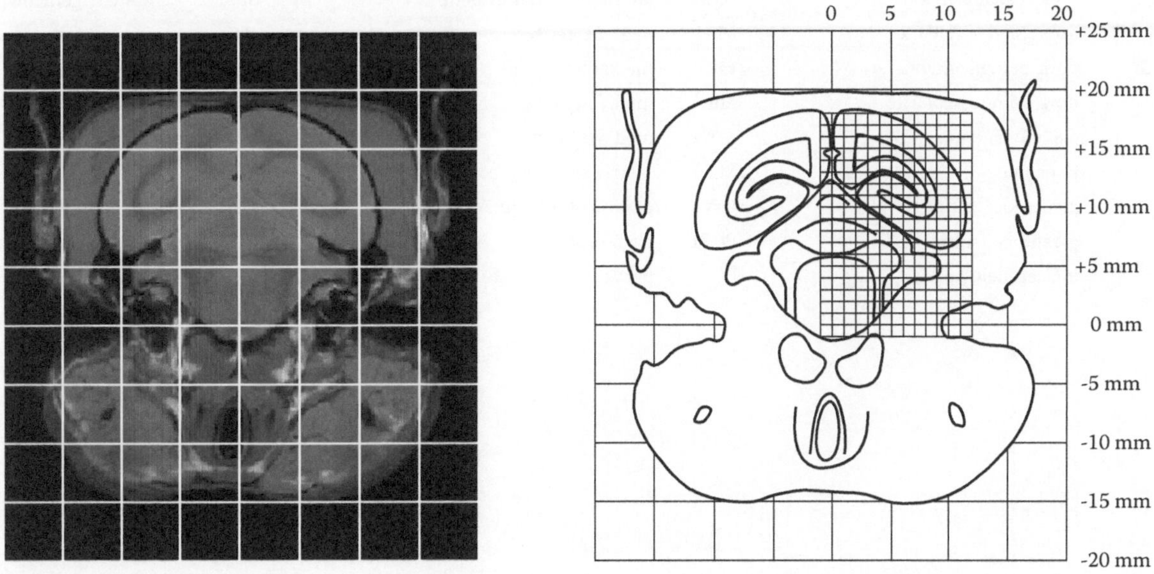


Interaural -02.65 mm

# Plate 44

Interaural -03.15 mm

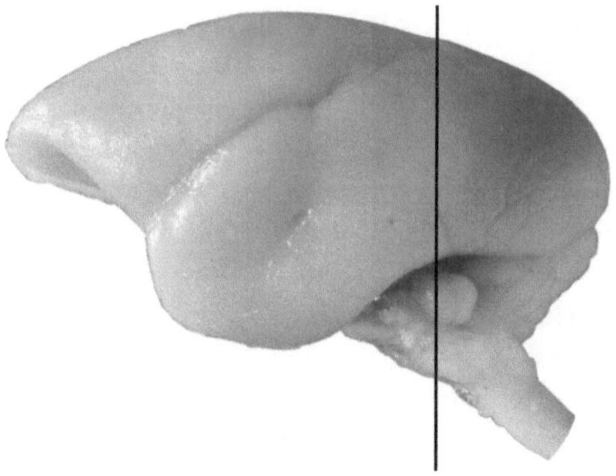

| | | | | | |
|---|---|---|---|---|---|
| 4Cb | cerebell. lobule 4 | In | intercal. nu. medulla | prf | prim. cerebell. fiss. |
| 4V | fourth ventricle | IntA | intermed. ant. deep Cb nu. | PSol | parasolitary nu. |
| 5Cb | cerebell. lobule 5 | IntP | intermed. post. deep Cb nu. | py | pyramidal tract |
| 5Sol | trigemsol. transn zone | IOA | inf. oliv. nu., subn. A | ROb | raphe obscurus nu. |
| 10CaI | dors. nu. 10, caud. int. | IOβ | inf. oliv. nu., subn. β | rs | rubrospinal tract |
| 10CeI | dors. nu. 10, cent. int. | IOB | inf. oliv. nu., subn. B | Sim | simplex lobule Cb |
| 10DI | dors. nu. 10, dors. int. | IOC | inf. oliv. nu., subn. C | sol | solitary tract |
| 12n | hypoglossal nerve | IOD | inf. oliv. nu., dors. | SolCe | nu. of solitary tr., cent. |
| 12N | hypoglossal nu. | IRt | intermed. retic. formn | SolDL | nu. of solitary tr., dorsolat. |
| Amb | ambiguus nu. | Lat | lat. deep Cb nu. | SolG | nu. of solitary tr., gelat. |
| AP | area postrema | lf | lateral fissure | SolM | nu. of solitary tr., int. |
| BI | basal interstitial nu. | LR4V | lat. recess 4V | SolV | nu. of solitary tr., vent. |
| C1 | C1 adrenaline cells | LRt | lat. reticular nu. | SolVL | nu. of solitary tr., ventlat. |
| ca | calcarine sulcus | LRtPC | lat. retic. nu., parv. | sp5 | spinal trigeminal tr. |
| Crus1 | crus 1, ansiform lob. | LRtS5 | lat. retic. nu., subtrigem. | Sp5I | spin. trigem. nu., interpol. |
| DM (V3d) | dorsomed. visual Cx | Med | med. deep Cb nu. | spth | spinothalamic tr. |
| DPGi | dors. paragigant. nu. | ml | medial lemniscus | SpVe | spinal vestib. nu. |
| dsc | dors. spinocerebell. tr | mlf | med. longit. fasc. | ts | tectospinal tract |
| ECu | ext. cuneate nu. | MVe | med. vestib. nu. | V1 | V1 Cx |
| Fl | flocculus | Mx | matrix sensory area | V2m | V2 (med.) Cx |
| gc | geniculocalcarine tr. | PaVe | paravestibular nu. | V2v | V2 (vent.) Cx |
| Gi | gigantocell. retic. nu. | PCRt | parv. retic. nu. | VLP (V3v) | ventlatpost. (V3v) Cx |
| icp | inf. cerebell. ped. | PFlP | paraflocculus post. | vsc | vent. spinocereb. tr. |

# Plate 44a

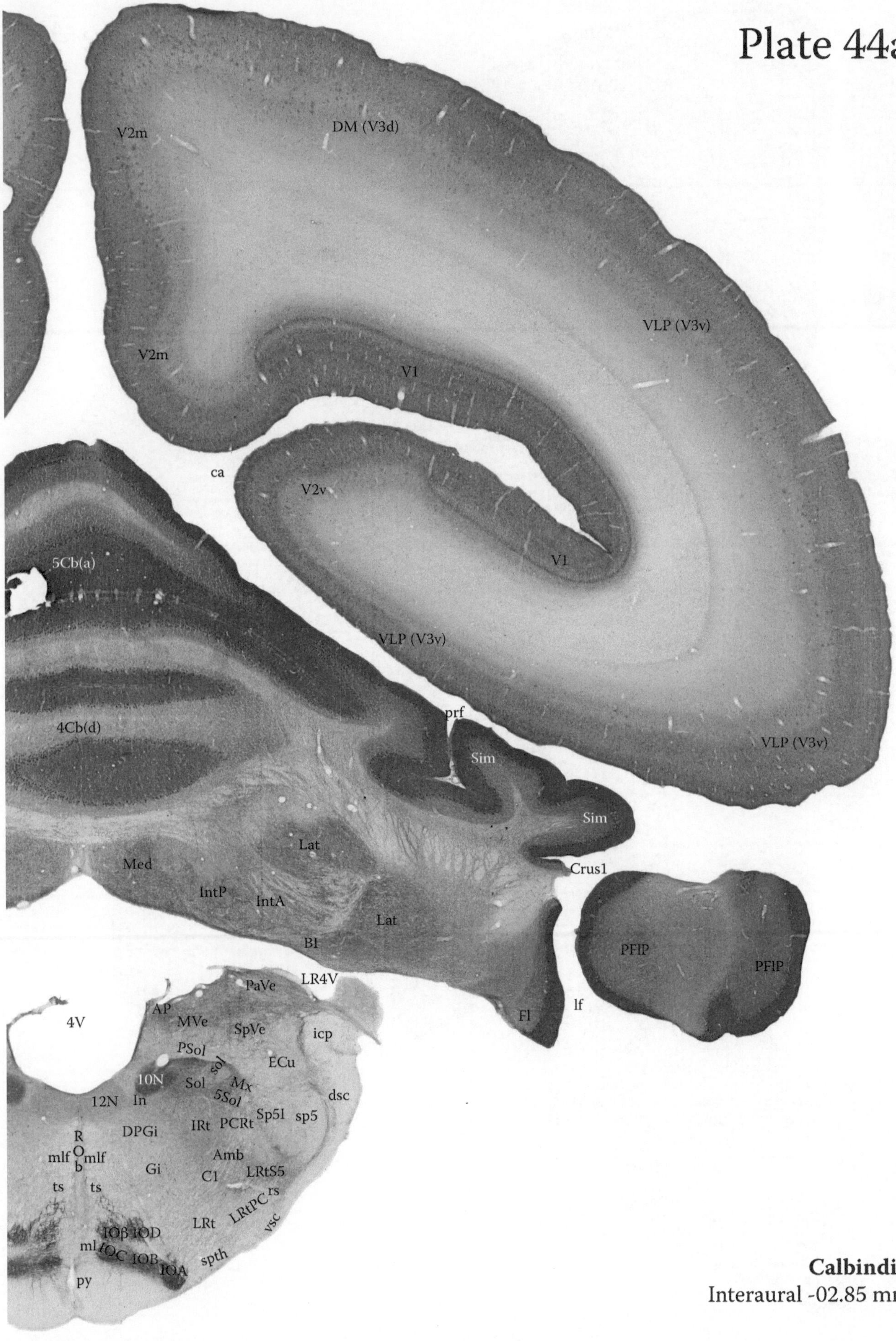


**Calbindin**
Interaural -02.85 mm

# Plate 44b

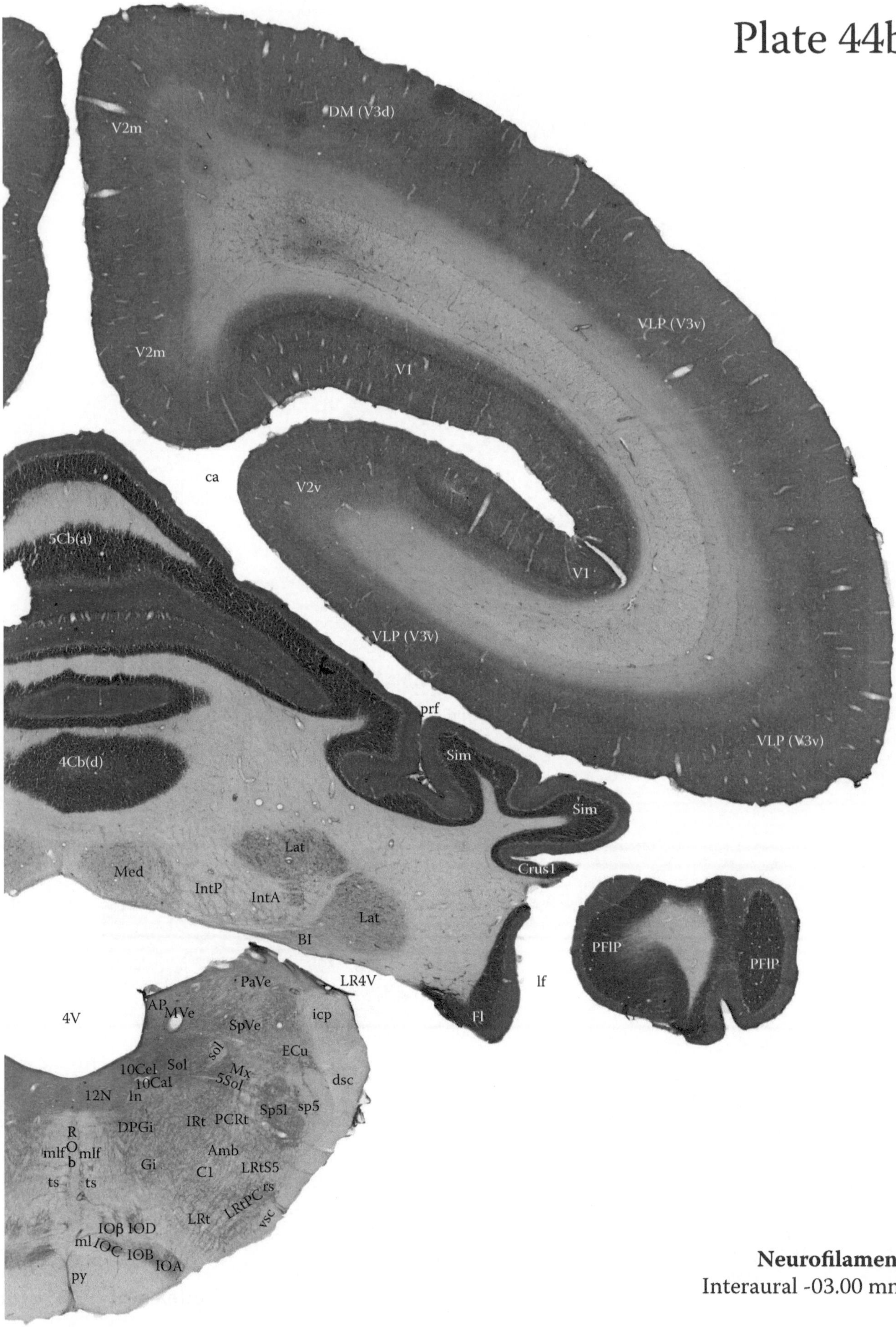


**Neurofilament**
Interaural -03.00 mm

# Plate 44c

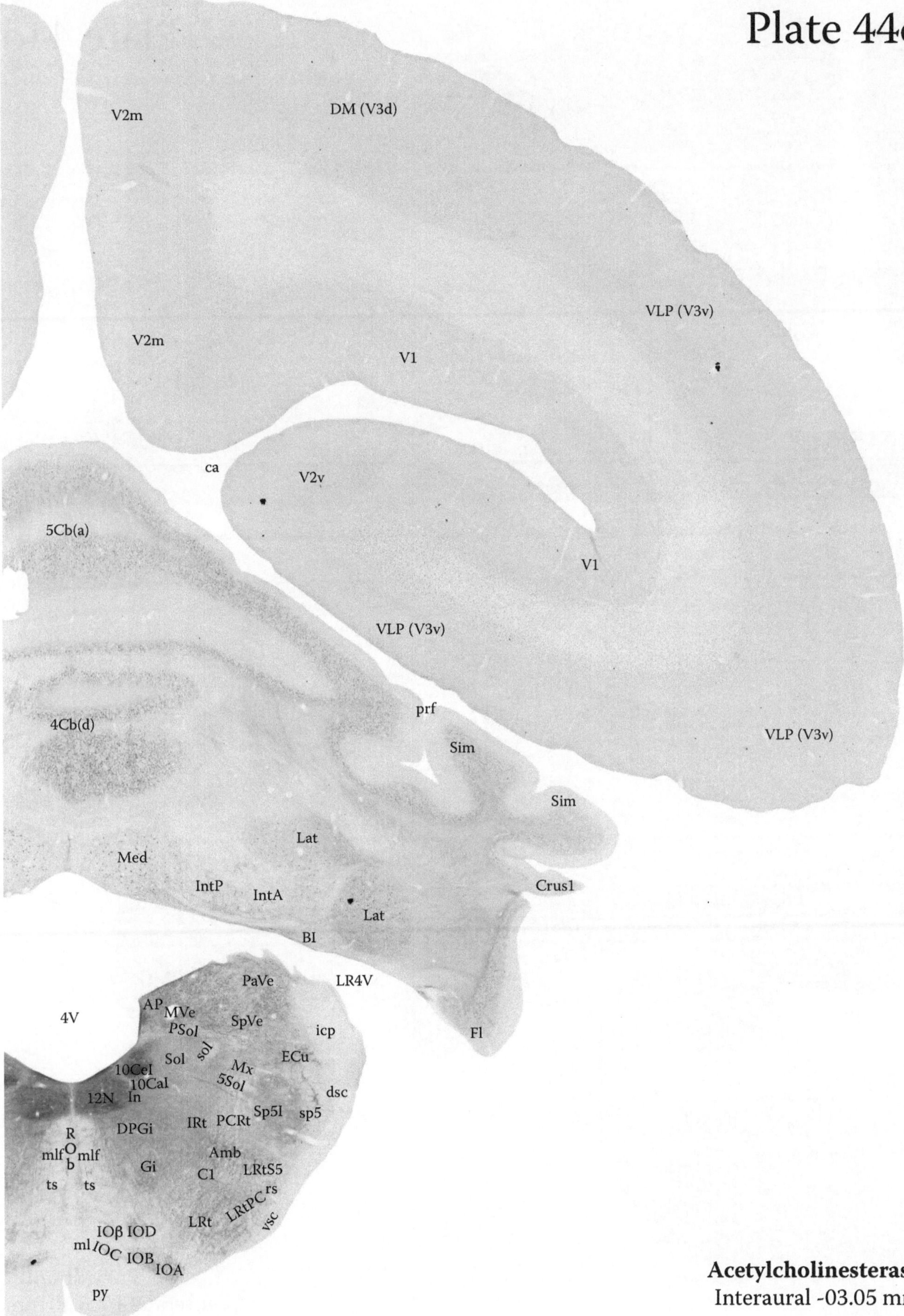


**Acetylcholinesterase**
Interaural -03.05 mm

# Plate 44d

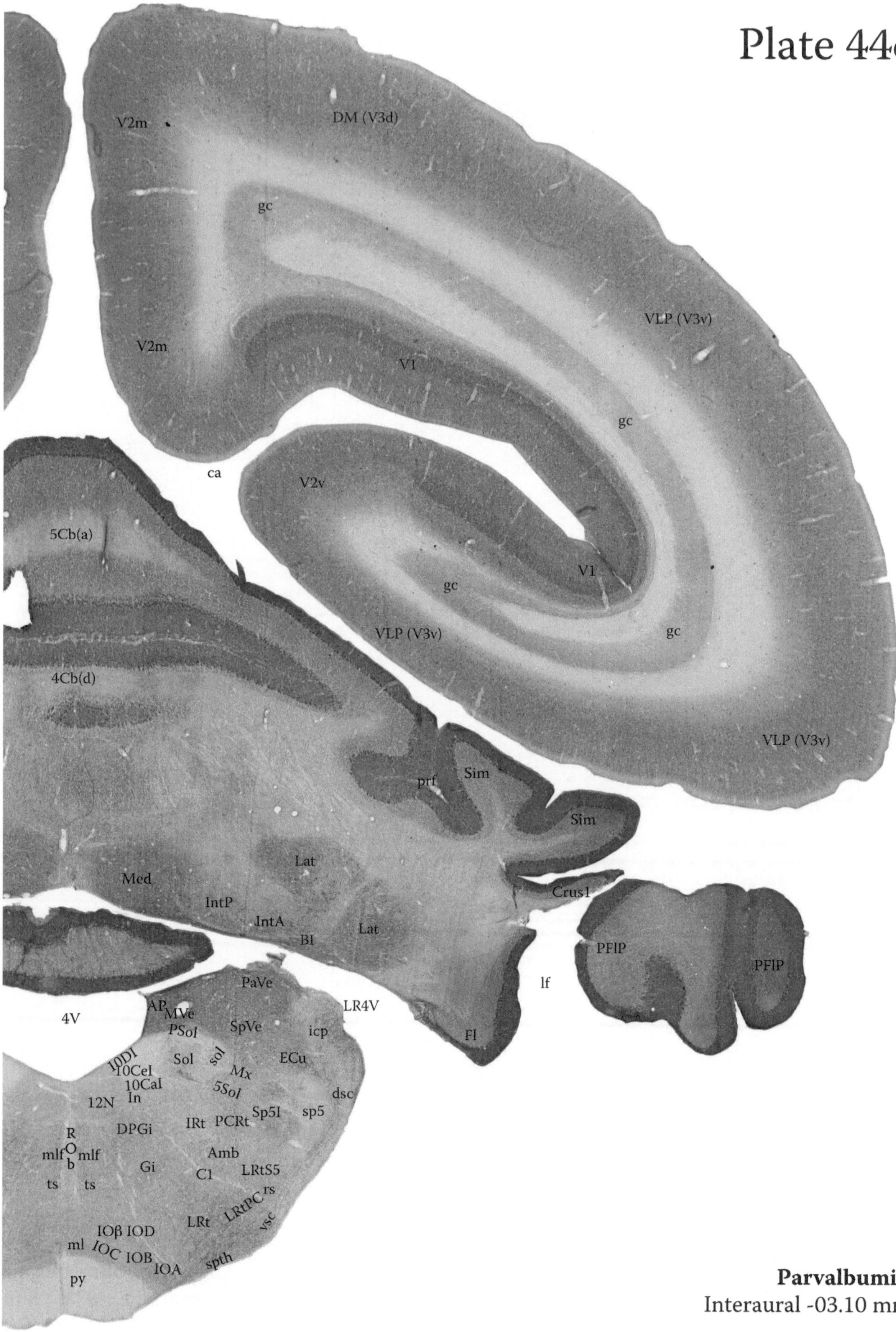


**Parvalbumin**
Interaural -03.10 mm

# Plate 44e

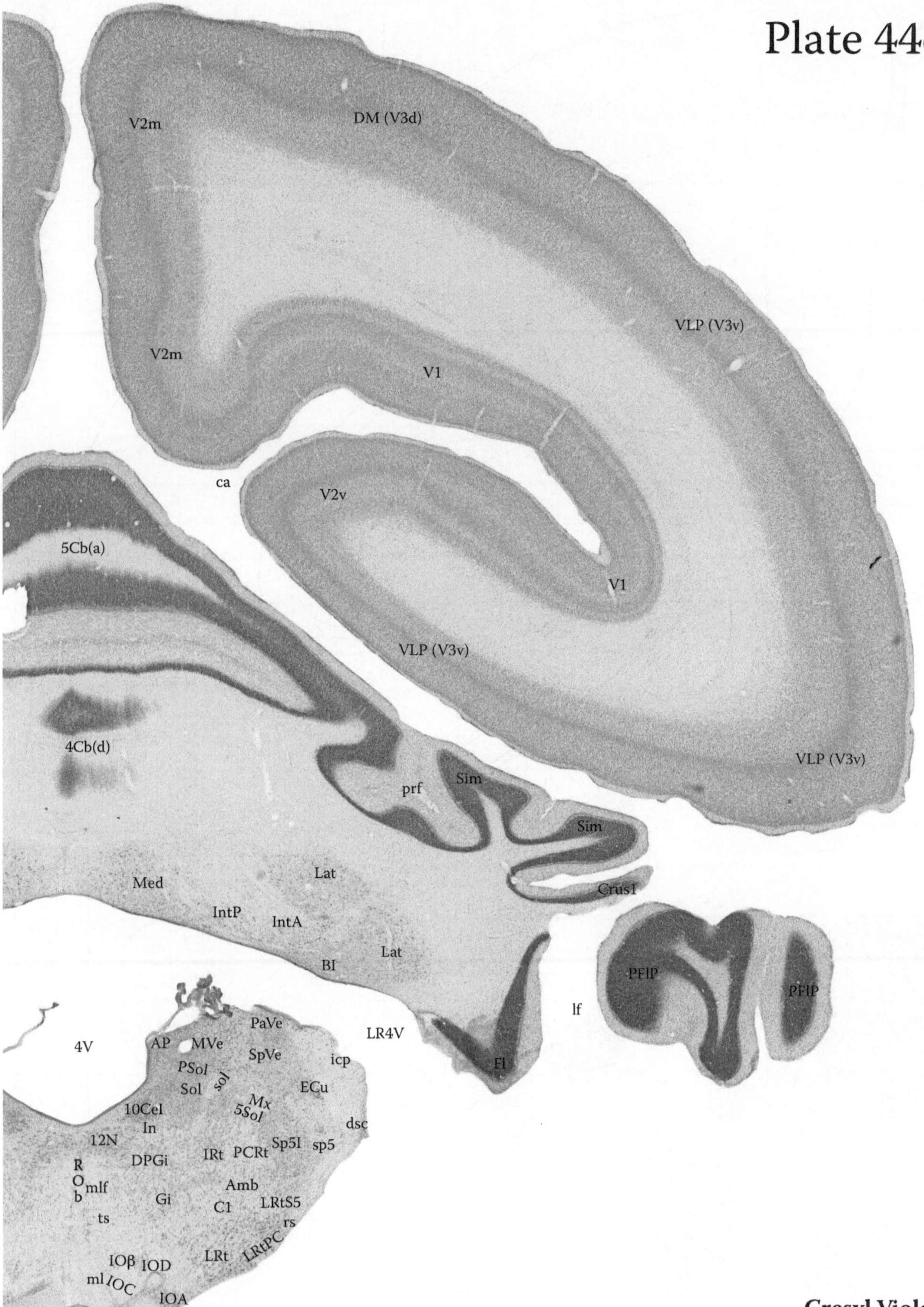


**Cresyl Violet**
Interaural -03.15 mm

# Plate 44

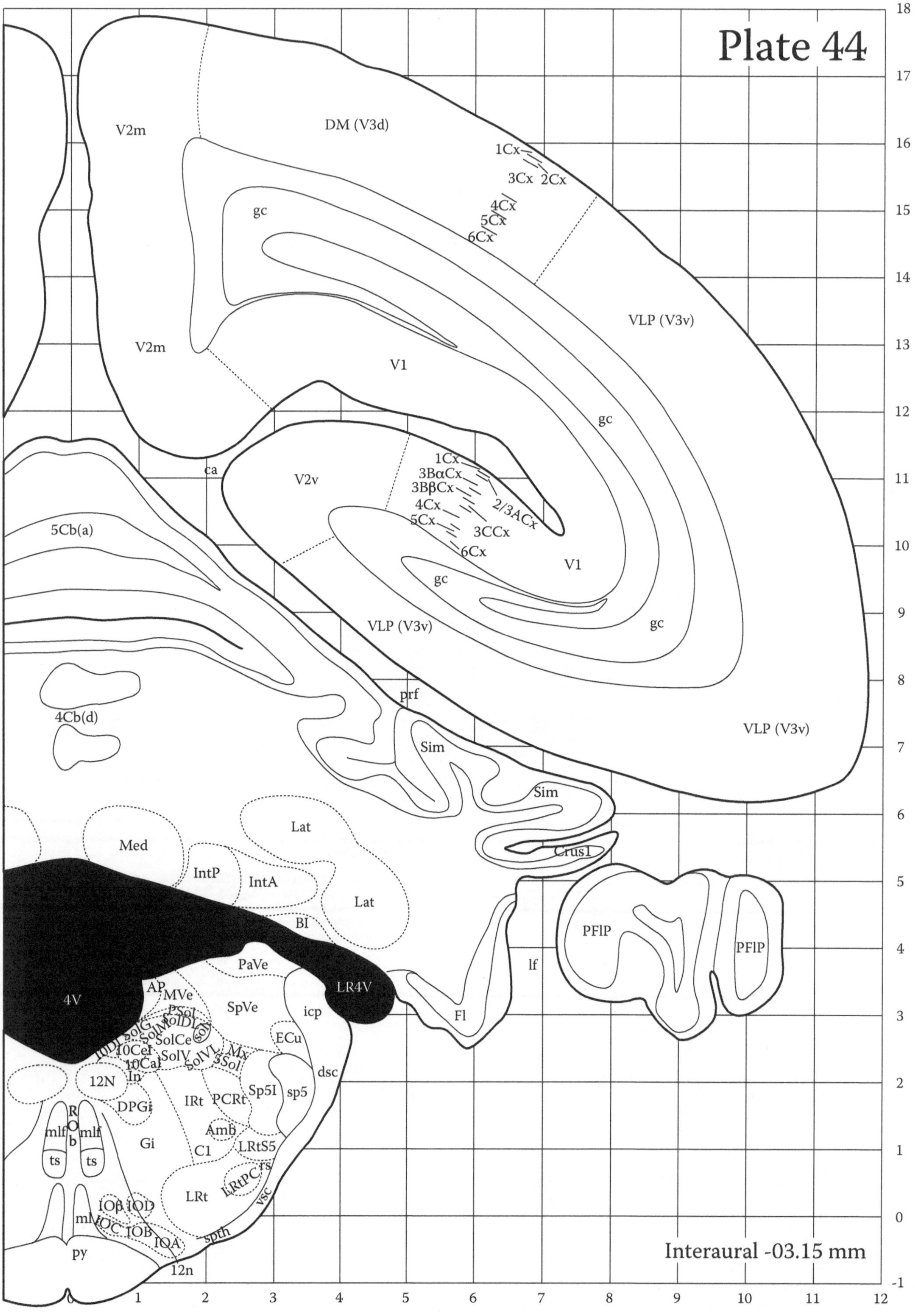


Interaural -03.15 mm

| | | | | | |
|---|---|---|---|---|---|
| 4Cb | cerebell. lobule 4 | In | intercal. nu. medulla | prf | prim. cerebell. fiss. |
| 4V | fourth ventricle | IntA | intermed. ant. deep Cb nu. | PSol | parasolitary nu. |
| 5Cb | cerebell. lobule 5 | IntP | intermed. post. deep Cb nu. | py | pyramidal tract |
| 5Sol | trigemsol. transn zone | IOA | inf. oliv. nu., subn. A | ROb | raphe obscurus nu. |
| 10CaI | dors. nu. 10, caud. int. | IOβ | inf. oliv. nu., subn. β | rs | rubrospinal tract |
| 10CeI | dors. nu. 10, cent. int. | IOB | inf. oliv. nu., subn. B | Sim | simplex lobule Cb |
| 10DI | dors. nu. 10, dors. int. | IOC | inf. oliv. nu., subn. C | sol | solitary tract |
| 12n | hypoglossal nerve | IOD | inf. oliv. nu., dors. | SolCe | nu. of solitary tr., cent. |
| 12N | hypoglossal nu. | IRt | intermed. retic. formn | SolDL | nu. of solitary tr., dorsolat. |
| Amb | ambiguus nu. | Lat | lat. deep Cb nu. | SolG | nu. of solitary tr., gelat. |
| AP | area postrema | lf | lateral fissure | SolM | nu. of solitary tr., int. |
| BI | basal interstitial nu. | LR4V | lat. recess 4V | SolV | nu. of solitary tr., vent. |
| C1 | C1 adrenaline cells | LRt | lat. reticular nu. | SolVL | nu. of solitary tr., ventlat. |
| ca | calcarine sulcus | LRtPC | lat. retic. nu., parv. | sp5 | spinal trigeminal tr. |
| Crus1 | crus 1, ansiform lob. | LRtS5 | lat. retic. nu., subtrigem. | Sp5I | spin. trigem. nu., interpol. |
| DM (V3d) | dorsomed. visual Cx | Med | med. deep Cb nu. | spth | spinothalamic tr. |
| DPGi | dors. paragigant. nu. | ml | medial lemniscus | SpVe | spinal vestib. nu. |
| dsc | dors. spinocerebell. tr | mlf | med. longit. fasc. | ts | tectospinal tract |
| ECu | ext. cuneate nu. | MVe | med. vestib. nu. | V1 | V1 Cx |
| Fl | flocculus | Mx | matrix sensory area | V2m | V2 (med.) Cx |
| gc | geniculocalcarine tr. | PaVe | paravestibular nu. | V2v | V2 (vent.) Cx |
| Gi | gigantocell. retic. nu. | PCRt | parv. retic. nu. | VLP (V3v) | ventlatpost. (V3v) Cx |
| icp | inf. cerebell. ped. | PFlP | paraflocculus post. | vsc | vent. spinocereb. tr. |

# Plate 45

Interaural -03.65 mm

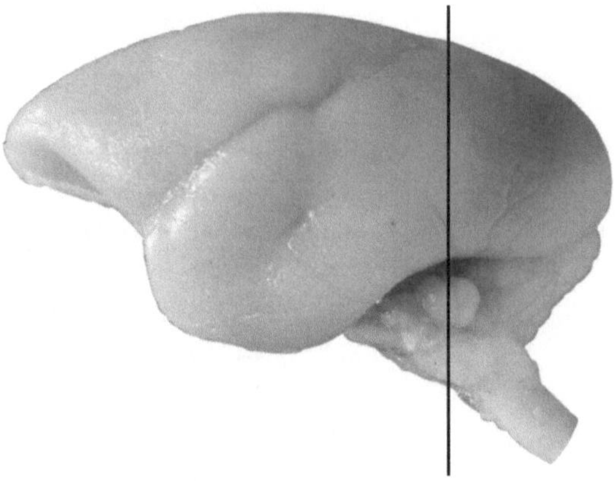

| | |
|---|---|
| 4V | fourth ventricle |
| 5Cb | cerebell. lobule 5 |
| 10CaI | dors. nu. 10, caud. int. |
| 10Cb | cerebell. lobule 10 |
| 10CeI | dors. nu. 10, cent. int. |
| 10DI | dors. nu. 10, dors. int. |
| 12N | hypoglossal nu. |
| 12VL | hypoglossal nu., ventlat. |
| A1 | A1 noradr. cells |
| Amb | ambiguus nu. |
| BI | basal interstitial nu. |
| ca | calcarine sulcus |
| Crus1 | crus 1, ansiform lob. |
| Cu | cuneate nu. |
| DM (V3d) | dorsomed. visual Cx |
| dsc | dors. spinocerebell. tr |
| ECu | ext. cuneate nu. |
| Fl | flocculus |
| gc | geniculocalcarine tr. |
| icp | inf. cerebell. ped. |
| In | intercal. nu. medulla |
| IntA | intermed. ant. deep Cb nu. |
| IntP | intermed. post. deep Cb nu. |
| IRt | intermed. retic. formn |
| Lat | lat. deep Cb nu. |
| LPCu | lat. pericuneate nu. |
| LR4V | lat. recess 4V |
| LRt | lat. reticular nu. |
| LRtS5 | lat. retic. nu., subtrigem. |
| MdD | medull. retic., nu. dors. |
| MdV | medull. retic., nu. vent. |
| Med | med. deep Cb nu. |
| ml | medial lemniscus |
| mlf | med. longit. fasc. |
| PaVe | paravestibular nu. |
| PCuMx | pericuneate matrix |
| PFlP | paraflocculus post. |
| prf | prim. cerebell. fiss. |
| PSol | parasolitary nu. |
| py | pyramidal tract |
| pyd | pyramidal decn |
| ROb | raphe obscurus nu. |
| rs | rubrospinal tract |
| Sim | simplex lobule Cb |
| sol | solitary tract |
| SolCe | nu. of solitary tr., cent. |
| SolDL | nu. of solitary tr., dorsolat. |
| SolG | nu. of solitary tr., gelat. |
| SolM | nu. of solitary tr., int. |
| SolV | nu. of solitary tr., vent. |
| SolVL | nu. of solitary tr., ventlat. |
| sp5 | spinal trigeminal tr. |
| Sp5C | spin. trigem. nu., caud. |
| Sp5I | spin. trigem. nu., interpol. |
| SpVe | spinal vestib. nu. |
| ts | tectospinal tract |
| V1 | V1 Cx |
| V2m | V2 (med.) Cx |
| V2v | V2 (vent.) Cx |
| VLP (V3v) | ventlatpost. (V3v) Cx |

# Plate 45a

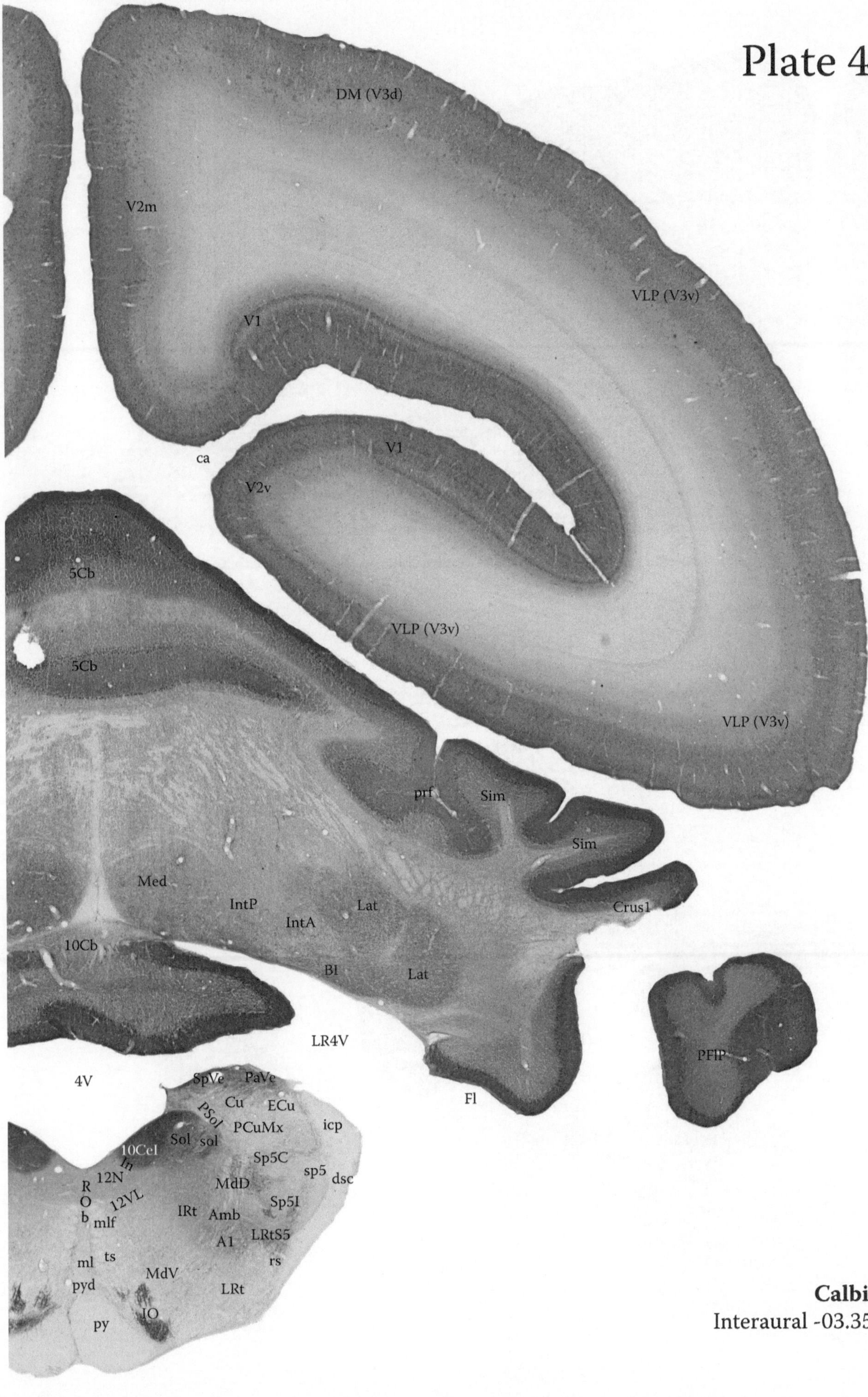


**Calbindin**
Interaural -03.35 mm

# Plate 45b

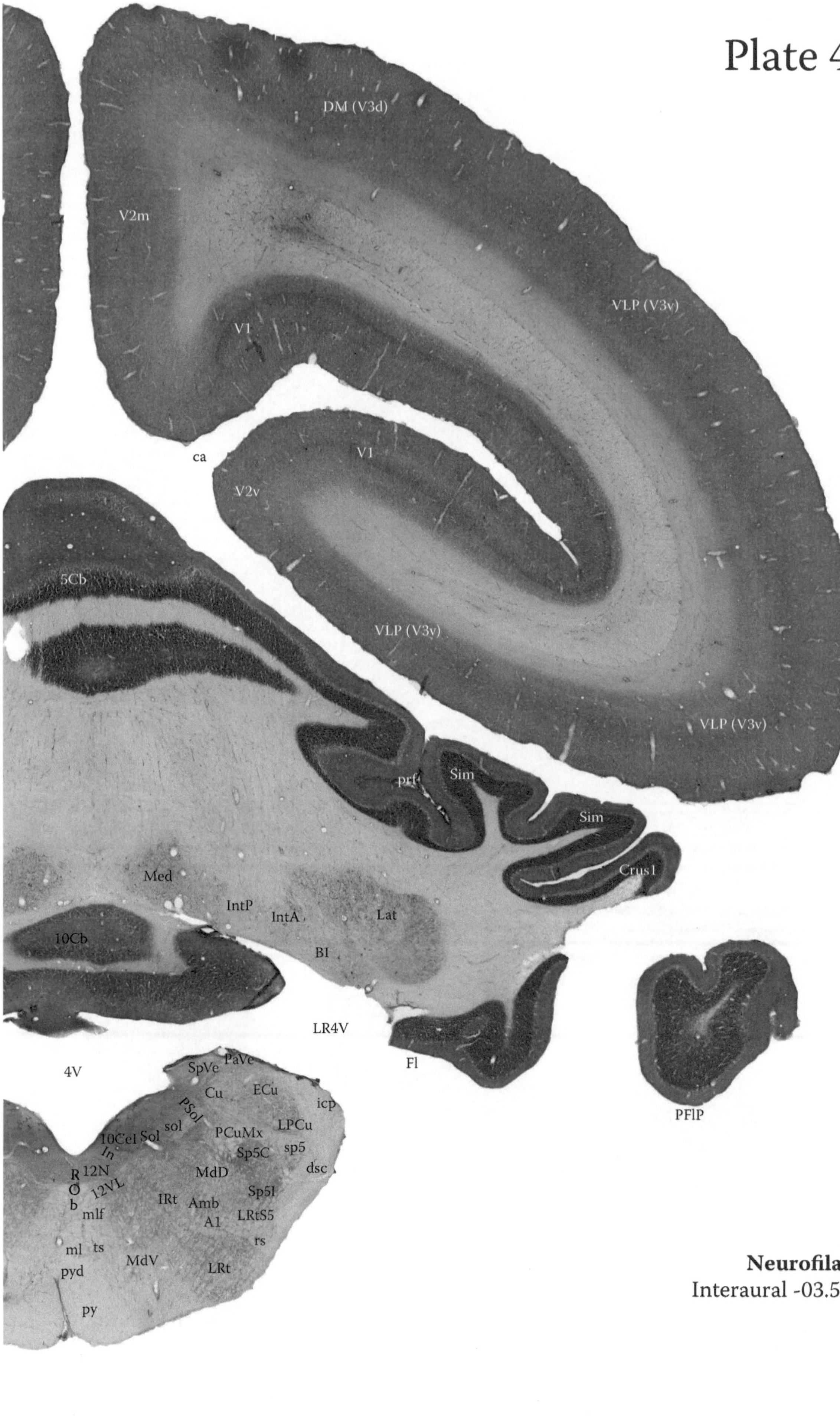


**Neurofilament**
Interaural -03.50 mm

# Plate 45c

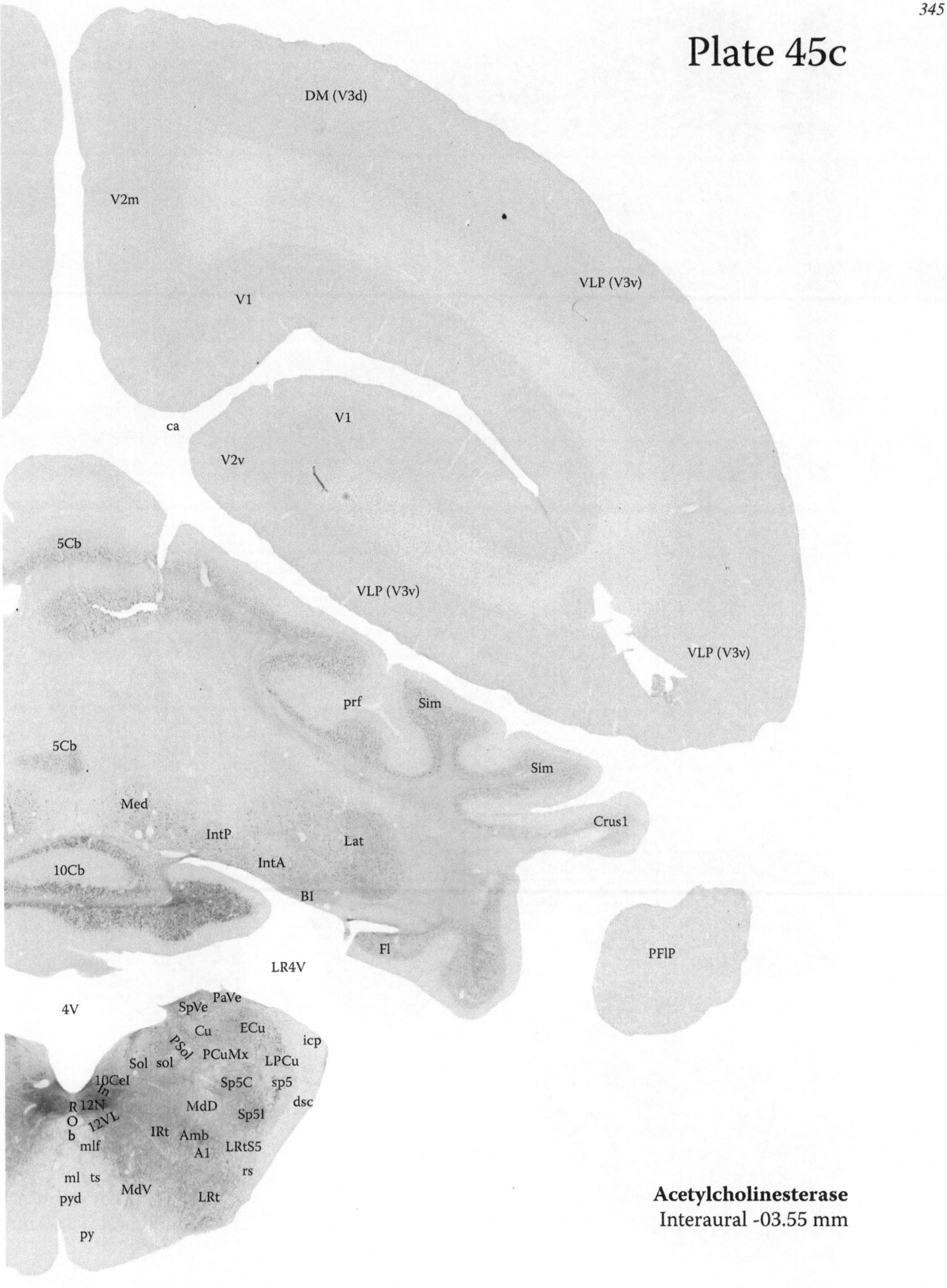


**Acetylcholinesterase**
Interaural -03.55 mm

# Plate 45d

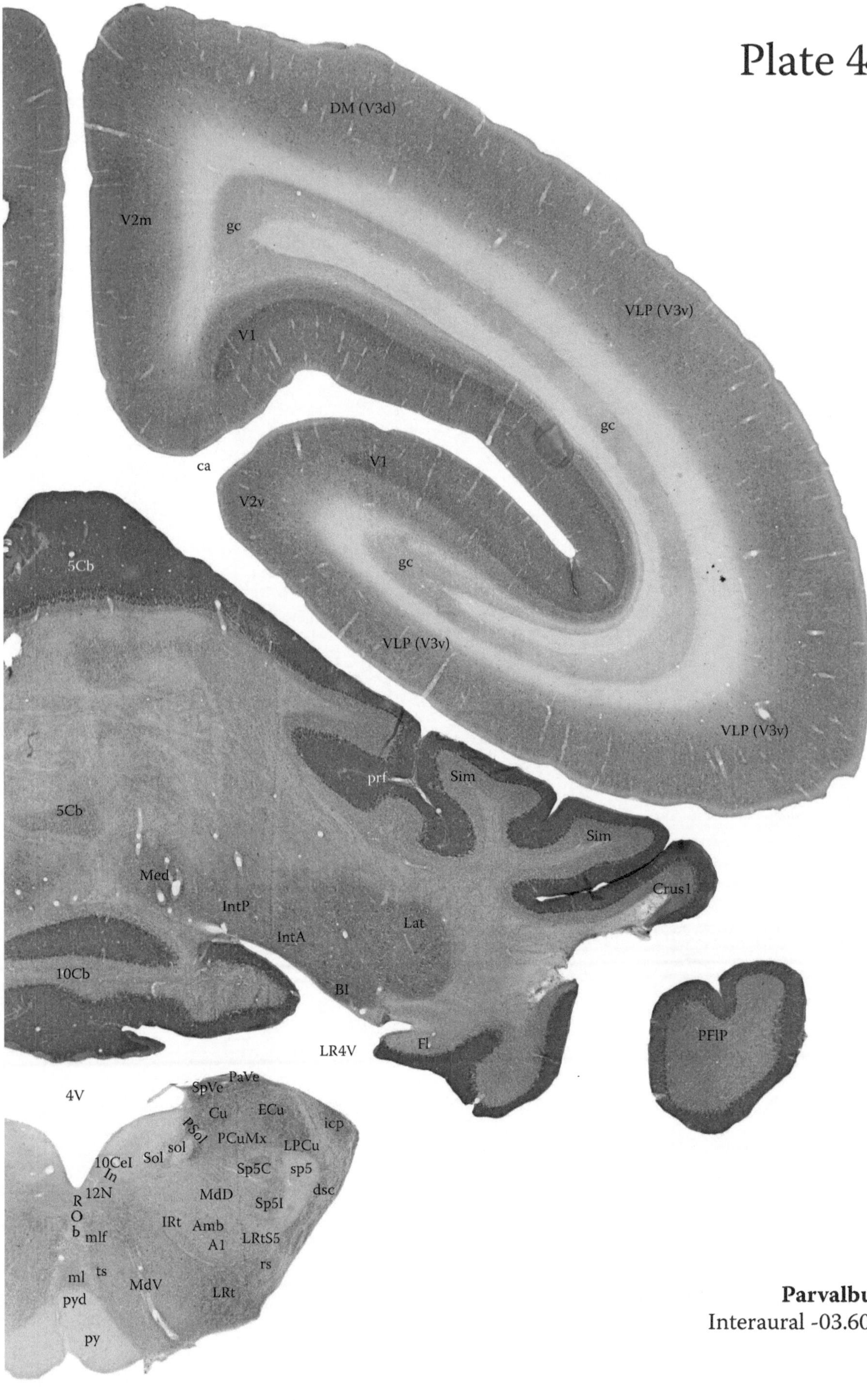


**Parvalbumin**
Interaural -03.60 mm

# Plate 45e

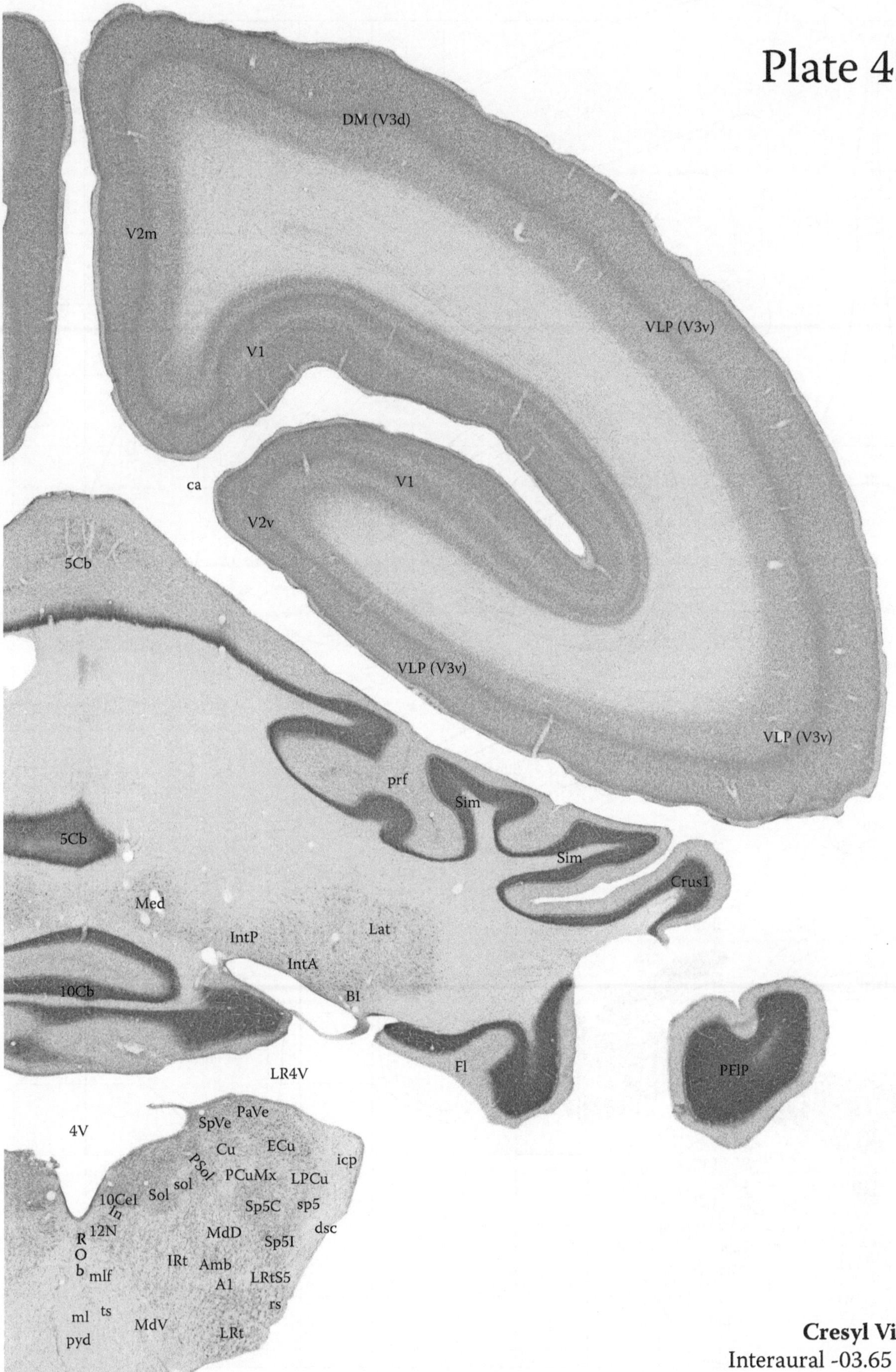


**Cresyl Violet**
Interaural -03.65 mm

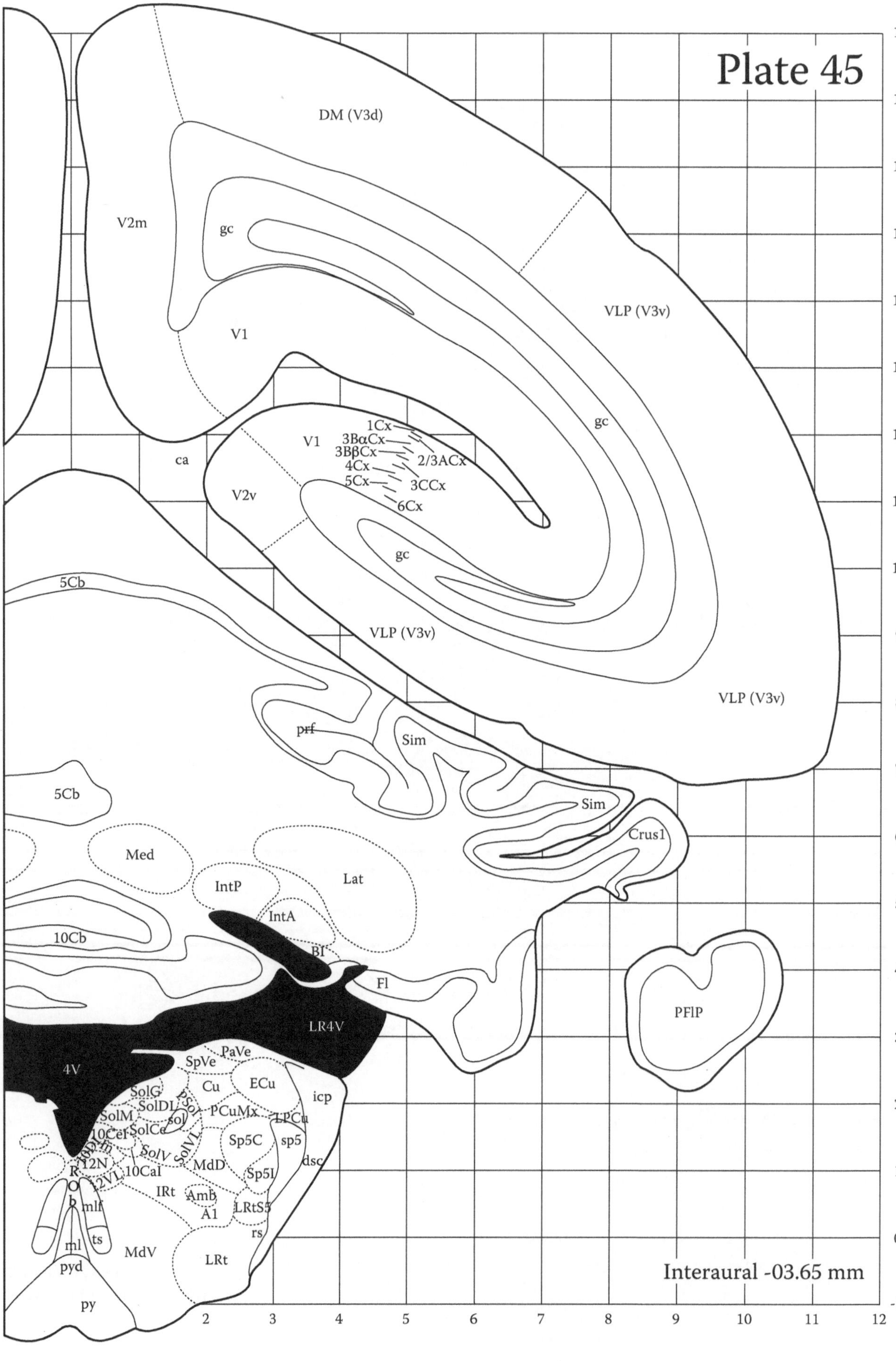
Plate 45
DM (V3d)
V2m
gc
V1
VLP (V3v)
gc
1Cx
3BαCx
3BβCx
4Cx
5Cx
2/3ACx
3CCx
6Cx
V1
ca
V2v
gc
5Cb
VLP (V3v)
VLP (V3v)
prf
Sim
Sim
Crus1
5Cb
Med
IntP
Lat
IntA
10Cb
BI
Fl
PFlP
LR4V
4V
PaVe
SpVe
ECu
Cu
icp
SolG
SolDL
PSol
PCuMx
LPCu
SolM
sol
10Cel
SolCe
SolVL
Sp5C
sp5
SolV
dsc
12N
10Cal
MdD
Sp5I
12VL
IRt
Amb
LRtS5
R
Ob
mlf
A1
ts
rs
ml
MdV
LRt
pyd
py
Interaural -03.65 mm
18
17
16
15
14
13
12
11
10
9
8
7
6
5
4
3
2
1
0
-1
2
3
4
5
6
7
8
9
10
11
12

| | | | | | |
|---|---|---|---|---|---|
| 4V | fourth ventricle | In | intercal. nu. medulla | pyd | pyramidal decn |
| 5Cb | cerebell. lobule 5 | IntA | intermed. ant. deep Cb nu. | ROb | raphe obscurus nu. |
| 10CaI | dors. nu. 10, caud. int. | IntP | intermed. post. deep Cb nu. | rs | rubrospinal tract |
| 10Cb | cerebell. lobule 10 | IRt | intermed. retic. formn | Sim | simplex lobule Cb |
| 10CeI | dors. nu. 10, cent. int. | Lat | lat. deep Cb nu. | sol | solitary tract |
| 10DI | dors. nu. 10, dors. int. | LPCu | lat. pericuneate nu. | SolCe | nu. of solitary tr., cent. |
| 12N | hypoglossal nu. | LR4V | lat. recess 4V | SolDL | nu. of solitary tr., dorsolat. |
| 12VL | hypoglossal nu., ventlat. | LRt | lat. reticular nu. | SolG | nu. of solitary tr., gelat. |
| A1 | A1 noradr. cells | LRtS5 | lat. retic. nu., subtrigem. | SolM | nu. of solitary tr., int. |
| Amb | ambiguus nu. | MdD | medull. retic., nu. dors. | SolV | nu. of solitary tr., vent. |
| BI | basal interstitial nu. | MdV | medull. retic., nu. vent. | SolVL | nu. of solitary tr., ventlat. |
| ca | calcarine sulcus | Med | med. deep Cb nu. | sp5 | spinal trigeminal tr. |
| Crus1 | crus 1, ansiform lob. | ml | medial lemniscus | Sp5C | spin. trigem. nu., caud. |
| Cu | cuneate nu. | mlf | med. longit. fasc. | Sp5I | spin. trigem. nu., interpol. |
| DM (V3d) | dorsomed. visual Cx | PaVe | paravestibular nu. | SpVe | spinal vestib. nu. |
| dsc | dors. spinocerebell. tr | PCuMx | pericuneate matrix | ts | tectospinal tract |
| ECu | ext. cuneate nu. | PFlP | paraflocculus post. | V1 | V1 Cx |
| Fl | flocculus | prf | prim. cerebell. fiss. | V2m | V2 (med.) Cx |
| gc | geniculocalcarine tr. | PSol | parasolitary nu. | V2v | V2 (vent.) Cx |
| icp | inf. cerebell. ped. | py | pyramidal tract | VLP (V3v) | ventlatpost. (V3v) Cx |

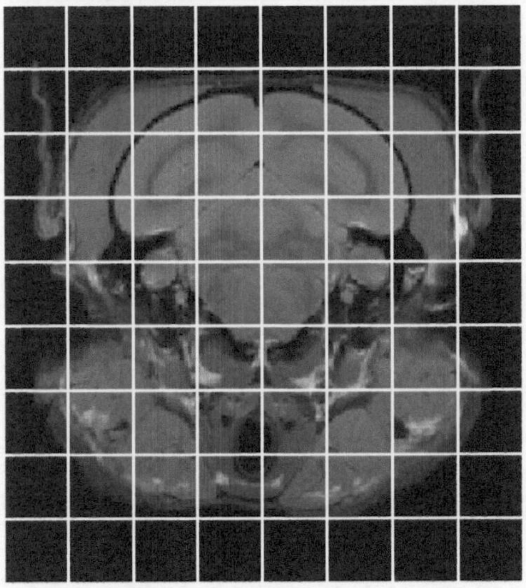

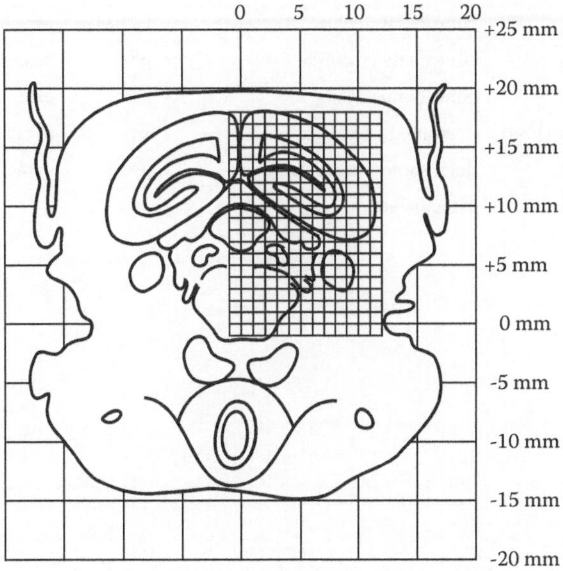


Interaural -03.60 mm

# Plate 46

Interaural -04.20 mm

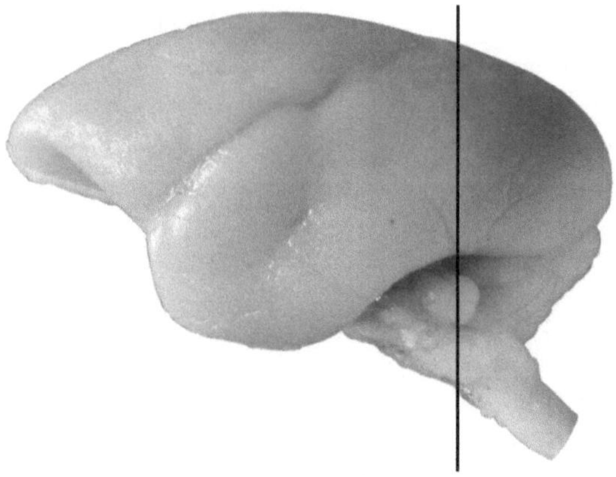

| | | | | | |
|---|---|---|---|---|---|
| 4V | fourth ventricle | gc | geniculocalcarine tr. | RAmb | retroambiguus nu. |
| 5Cb | cerebell. lobule 5 | Ge5 | gelat. layer Sp5C | rs | rubrospinal tract |
| 6Cb | cerebell. lobule 6 | Gr | gracile nu. | Sim | simplex lobule Cb |
| 9Cb | cerebell. lobule 9 | ia | intern. arcuate fibers | sol | solitary tract |
| 10Cb | cerebell. lobule 10 | icp | inf. cerebell. ped. | SolC | nu. of solitary tr., comm. |
| 10N | dorsal nu. 10n | IntP | intermed. post. deep Cb nu. | SolDL | nu. of solitary tr., dorsolat. |
| 11N | accessory motor nu. | IRt | intermed. retic. formn | SolG | nu. of solitary tr., gelat. |
| A1 | A1 noradr. cells | Lat | lat. deep Cb nu. | SolM | nu. of solitary tr., int. |
| BI | basal interstitial nu. | LR4V | lat. recess 4V | SolV | nu. of solitary tr., vent. |
| ca | calcarine sulcus | LRt | lat. reticular nu. | SolVL | nu. of solitary tr., ventlat. |
| CC | central canal | MdD | medull. retic., nu. dors. | sp5 | spinal trigeminal tr. |
| Cop | copula | MdV | medull. retic., nu. vent. | Sp5C | spin. trigem. nu., caud. |
| Crus1 | crus 1, ansiform lob. | Med | med. deep Cb nu. | spth | spinothalamic tr. |
| Crus2 | crus 2, ansiform lob. | mlf | med. longit. fasc. | ts | tectospinal tract |
| cu | cuneate fasciculus | PCuMx | pericuneate matrix | V1 | V1 Cx |
| CuR | cuneate nu., rotunda | plf | posterolat. fissure | V2m | V2 (med.) Cx |
| CuT | cuneate nu., triang. | prf | prim. cerebell. fiss. | V2v | V2 (vent.) Cx |
| DM (V3d) | dorsomed. visual Cx | py | pyramidal tract | VLP (V3v) | ventlatpost. (V3v) Cx |
| dsc | dors. spinocerebell. tr | pyd | pyramidal decn | vsc | vent. spinocereb. tr. |
| ECu | ext. cuneate nu. | | | | |

# Plate 46a

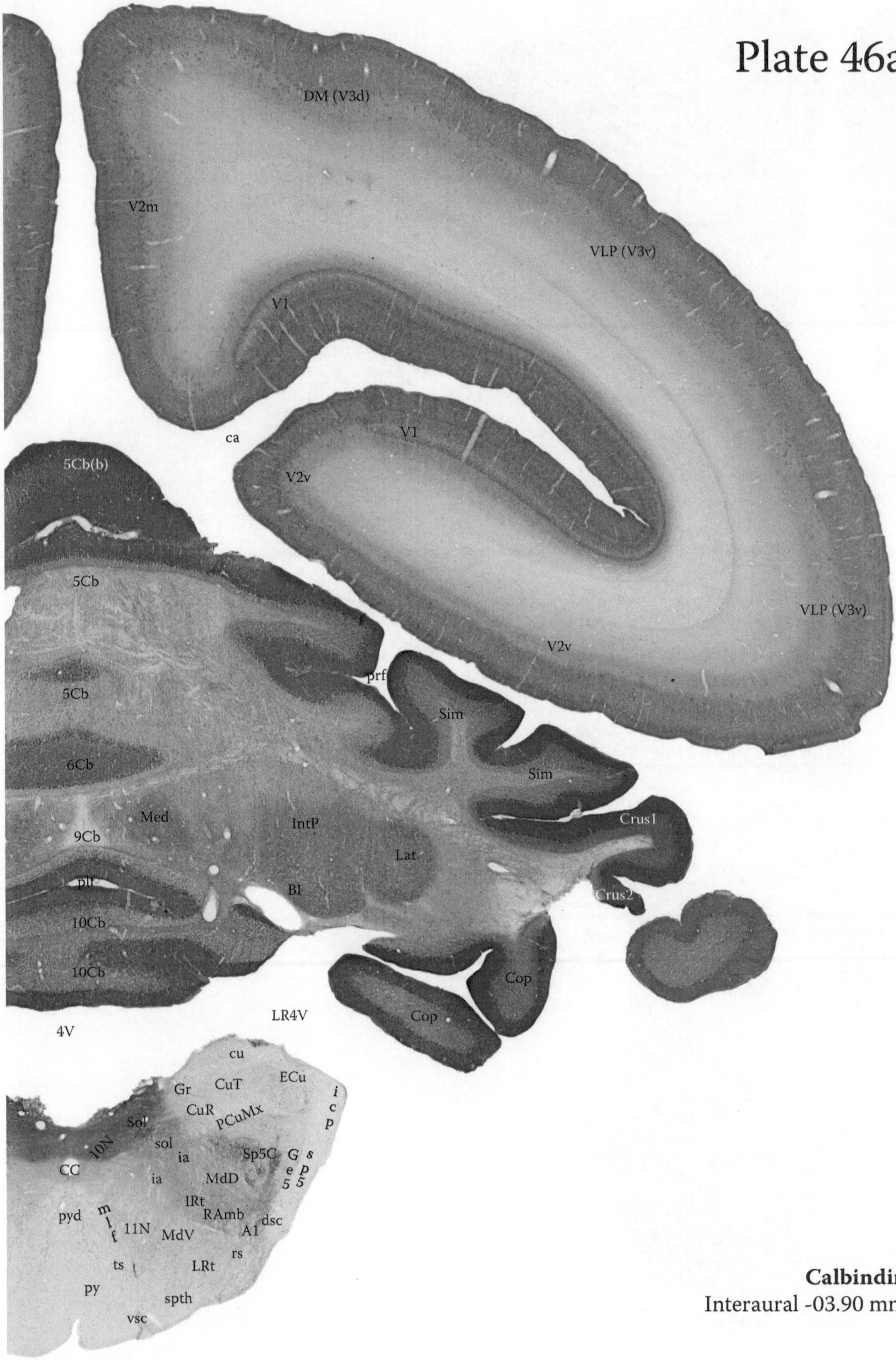


**Calbindin**
Interaural -03.90 mm

# Plate 46b

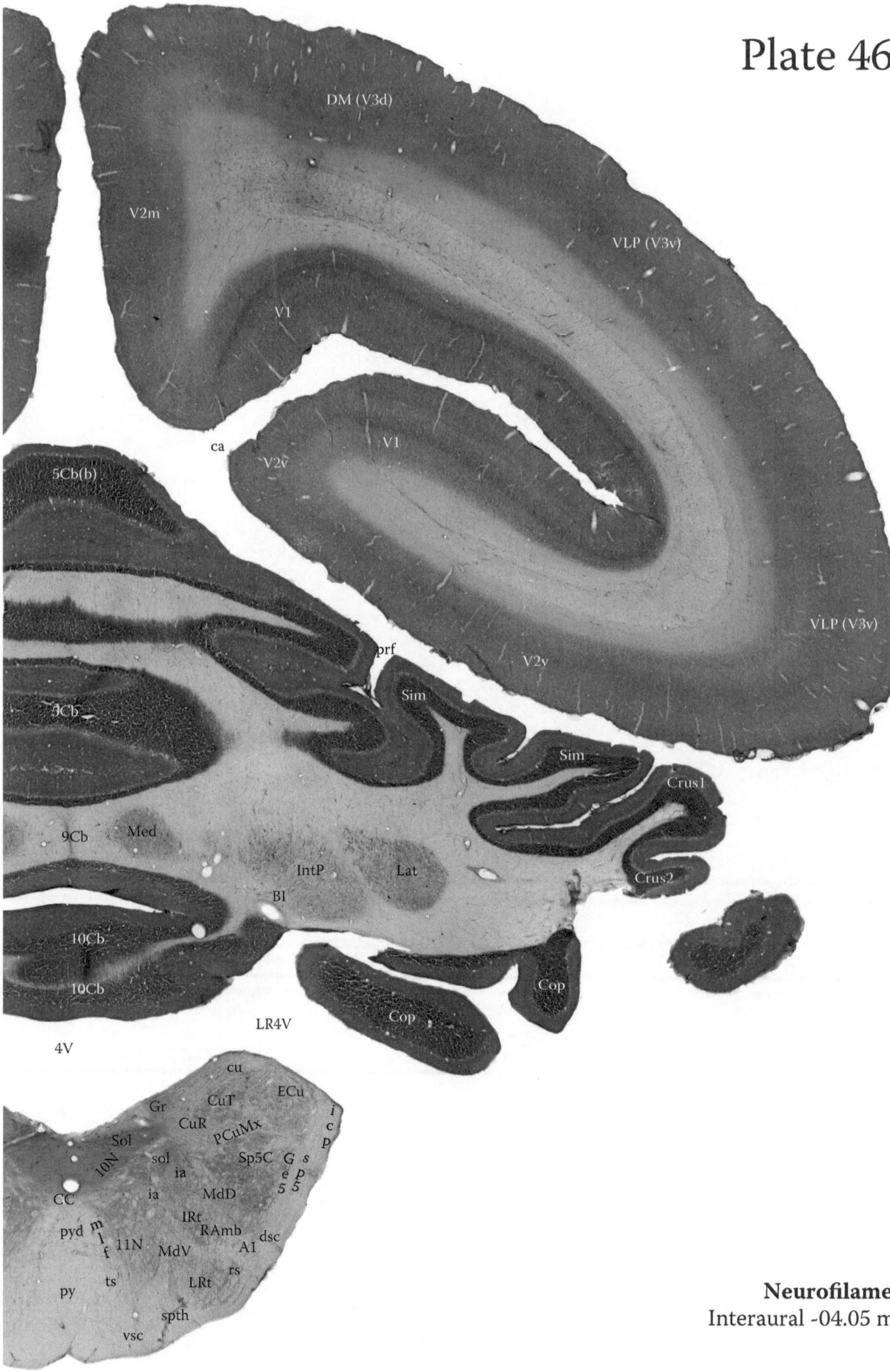


**Neurofilament**
Interaural -04.05 mm

# Plate 46c

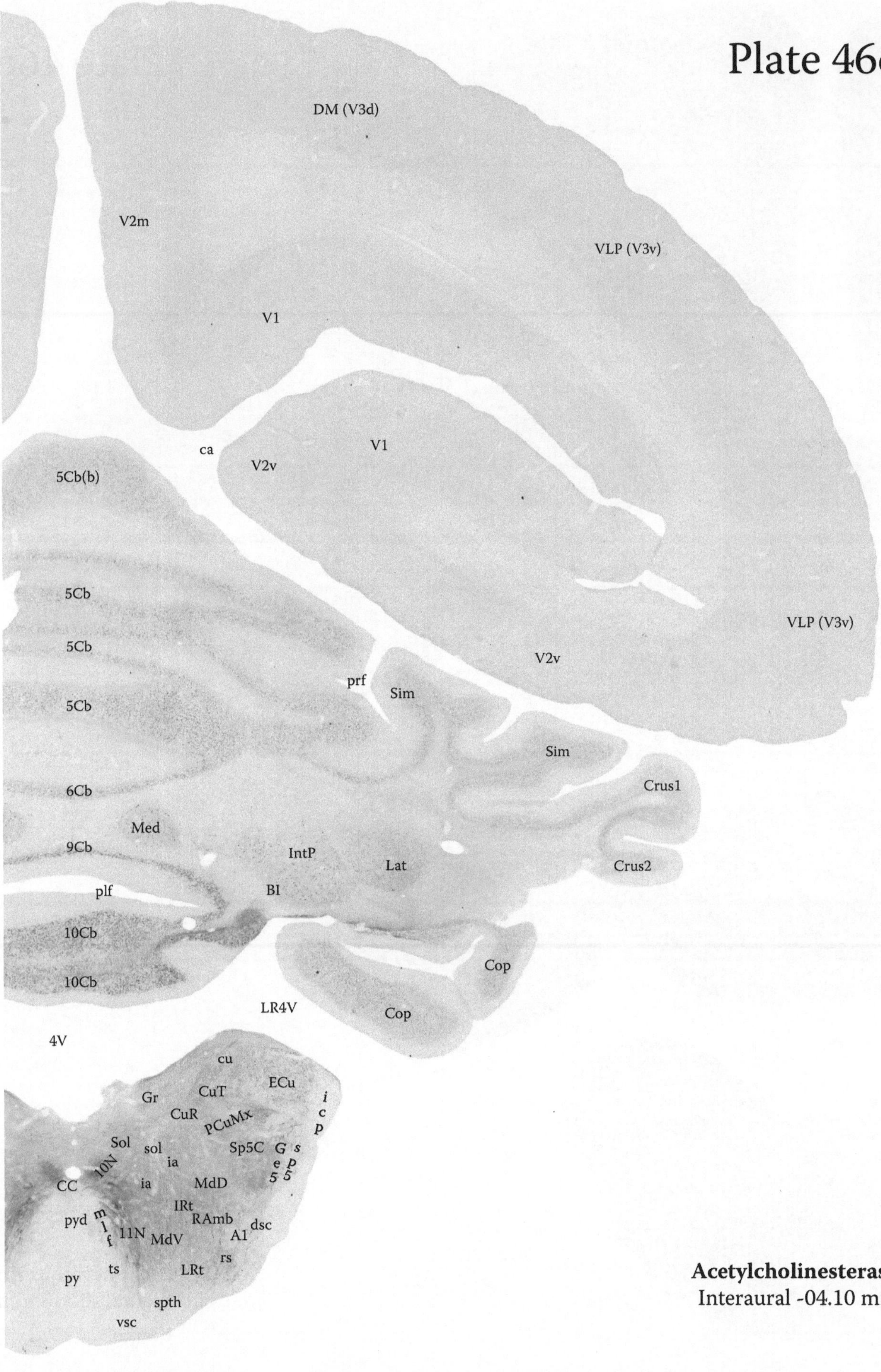


**Acetylcholinesterase**
Interaural -04.10 mm

# Plate 46d

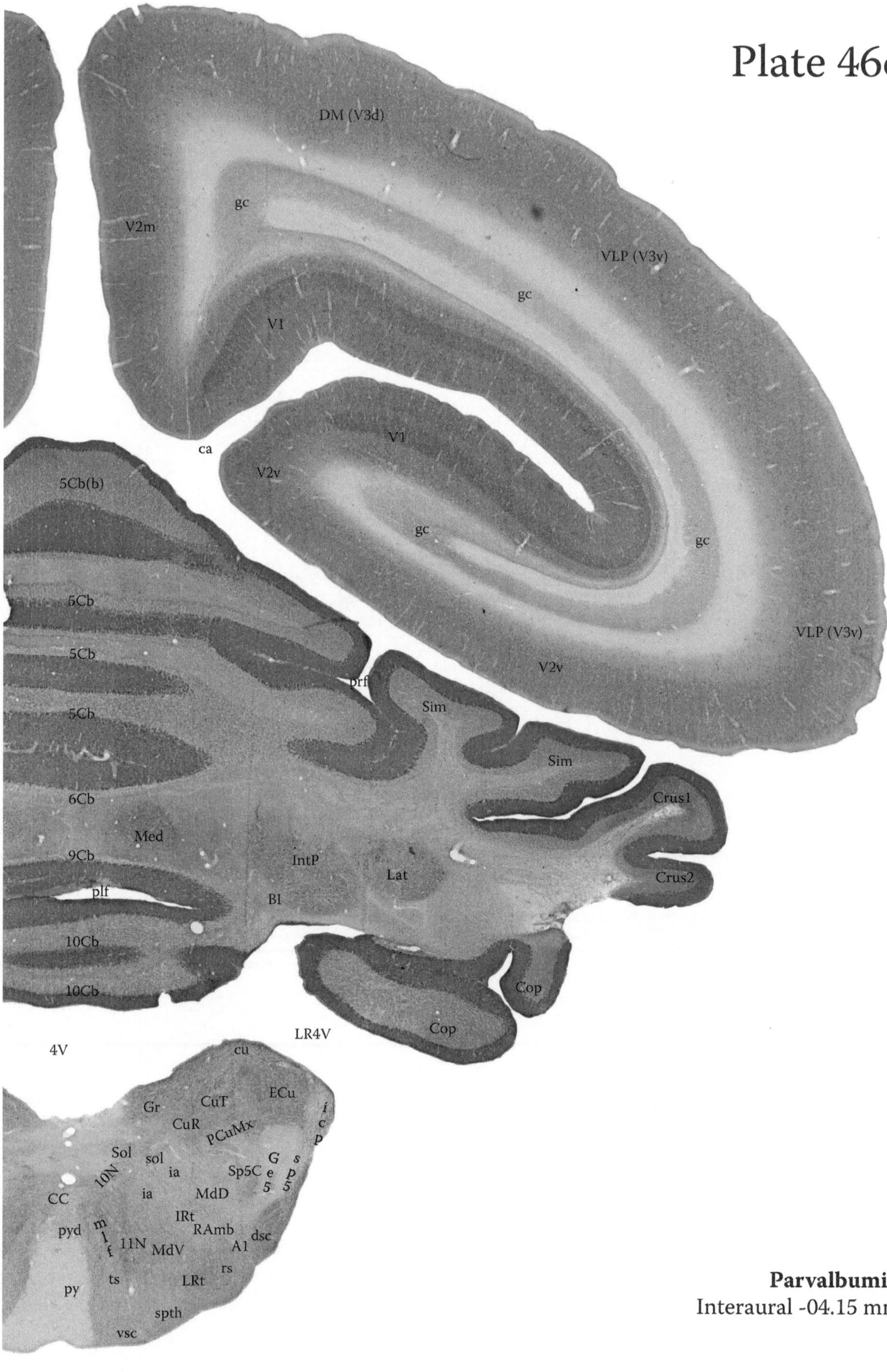


**Parvalbumin**
Interaural -04.15 mm

# Plate 46e

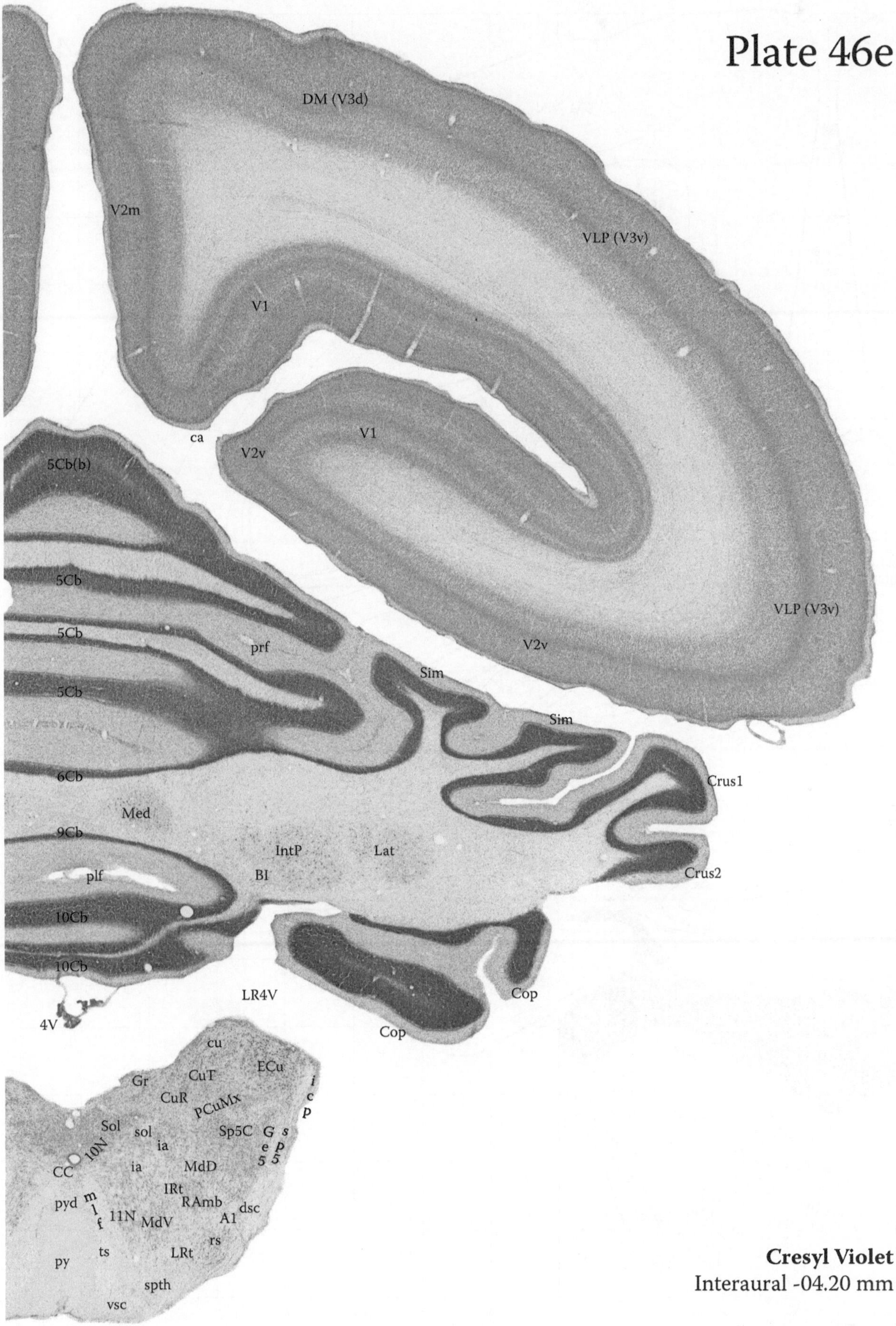


**Cresyl Violet**
Interaural -04.20 mm

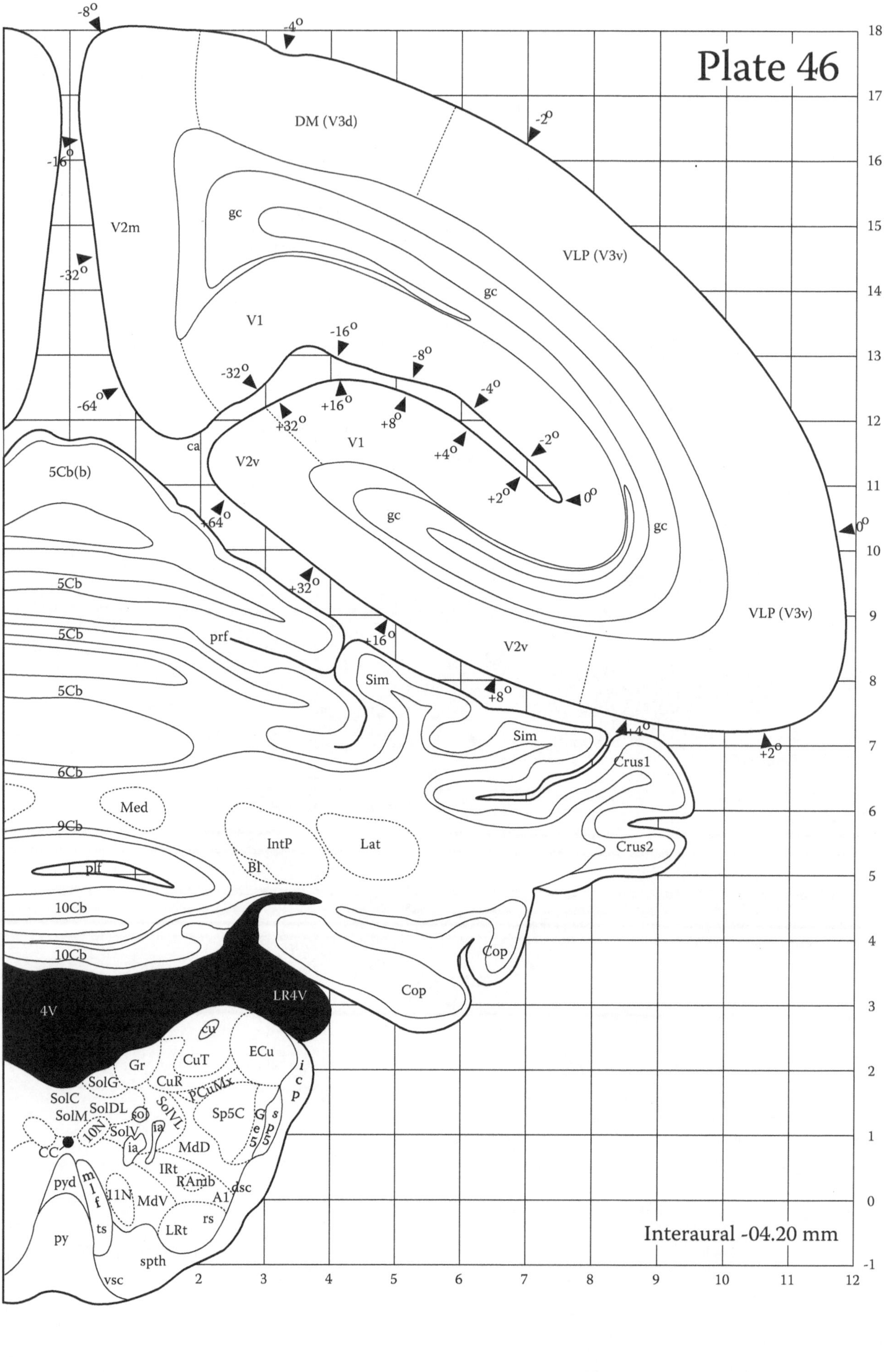
Plate 46
DM (V3d)
V2m
gc
VLP (V3v)
V1
V2v
ca
5Cb(b)
5Cb
6Cb
9Cb
10Cb
prf
plf
Sim
Crus1
Crus2
Med
IntP
Lat
Bl
Cop
4V
LR4V
cu
CuT
ECu
Gr
SolG
CuR
PCuMx
SolC
SolM
SolDL
sol
SolVL
Sp5C
SolV
10N
ia
MdD
CC
IRt
RAmb
dsc
pyd
11N
MdV
A1
py
ts
LRt
rs
spth
vsc
icp
Interaural -04.20 mm

| | |
|---|---|
| 4V | fourth ventricle |
| 5Cb | cerebell. lobule 5 |
| 6Cb | cerebell. lobule 6 |
| 9Cb | cerebell. lobule 9 |
| 10Cb | cerebell. lobule 10 |
| 10N | dorsal nu. 10n |
| 11N | accessory motor nu. |
| A1 | A1 noradr. cells |
| BI | basal interstitial nu. |
| ca | calcarine sulcus |
| CC | central canal |
| Cop | copula |
| Crus1 | crus 1, ansiform lob. |
| Crus2 | crus 2, ansiform lob. |
| cu | cuneate fasciculus |
| CuR | cuneate nu., rotunda |
| CuT | cuneate nu., triang. |
| DM (V3d) | dorsomed. visual Cx |
| dsc | dors. spinocerebell. tr |
| ECu | ext. cuneate nu. |
| gc | geniculocalcarine tr. |
| Ge5 | gelat. layer Sp5C |
| Gr | gracile nu. |
| ia | intern. arcuate fibers |
| icp | inf. cerebell. ped. |
| IntP | intermed. post. deep Cb nu. |
| IRt | intermed. retic. formn |
| Lat | lat. deep Cb nu. |
| LR4V | lat. recess 4V |
| LRt | lat. reticular nu. |
| MdD | medull. retic., nu. dors. |
| MdV | medull. retic., nu. vent. |
| Med | med. deep Cb nu. |
| mlf | med. longit. fasc. |
| PCuMx | pericuneate matrix |
| plf | posterolat. fissure |
| prf | prim. cerebell. fiss. |
| py | pyramidal tract |
| pyd | pyramidal decn |
| RAmb | retroambiguus nu. |
| rs | rubrospinal tract |
| Sim | simplex lobule Cb |
| sol | solitary tract |
| SolC | nu. of solitary tr., comm. |
| SolDL | nu. of solitary tr., dorsolat. |
| SolG | nu. of solitary tr., gelat. |
| SolM | nu. of solitary tr., int. |
| SolV | nu. of solitary tr., vent. |
| SolVL | nu. of solitary tr., ventlat. |
| sp5 | spinal trigeminal tr. |
| Sp5C | spin. trigem. nu., caud. |
| spth | spinothalamic tr. |
| ts | tectospinal tract |
| V1 | V1 Cx |
| V2m | V2 (med.) Cx |
| V2v | V2 (vent.) Cx |
| VLP (V3v) | ventlatpost. (V3v) Cx |
| vsc | vent. spinocereb. tr. |

# Plate 47

Interaural -04.70 mm

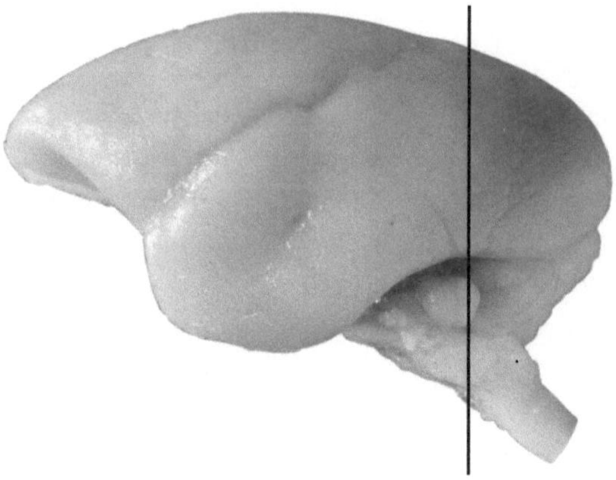

| | |
|---|---|
| 4V | fourth ventricle |
| 5Cb | cerebell. lobule 5 |
| 6Cb | cerebell. lobule 6 |
| 9Cb | cerebell. lobule 9 |
| 10Cb | cerebell. lobule 10 |
| 11N | accessory motor nu. |
| AP | area postrema |
| ca | calcarine sulcus |
| CC | central canal |
| CeCv | central cervical nu. |
| Cop | copula |
| Crus1 | crus 1, ansiform lob. |
| Crus2 | crus 2, ansiform lob. |
| cu | cuneate fasciculus |
| CuR | cuneate nu., rotunda |
| CuT | cuneate nu., triang. |
| DM (V3d) | dorsomed. visual Cx |
| dsc | dors. spinocerebell. tr |
| ECu | ext. cuneate nu. |
| gc | geniculocalcarine tr. |
| Ge5 | gelat. layer Sp5C |
| Gr | gracile nu. |
| ia | intern. arcuate fibers |
| IntP | intermed. post. deep Cb nu. |
| IRt | intermed. retic. formn |
| LR4V | lat. recess 4V |
| LSp | lat. spinal nu. |
| MdD | medull. retic., nu. dors. |
| MdV | medull. retic., nu. vent. |
| Med | med. deep Cb nu. |
| mlf | med. longit. fasc. |
| PCuMx | pericuneate matrix |
| plf | posterolat. fissure |
| prf | prim. cerebell. fiss. |
| py | pyramidal tract |
| pyd | pyramidal decn |
| rs | rubrospinal tract |
| Sim | simplex lobule Cb |
| sol | solitary tract |
| SolC | nu. of solitary tr., comm. |
| SolDL | nu. of solitary tr., dorsolat. |
| SolM | nu. of solitary tr., int. |
| SolV | nu. of solitary tr., vent. |
| SolVL | nu. of solitary tr., ventlat. |
| sp5 | spinal trigeminal tr. |
| Sp5C | spin. trigem. nu., caud. |
| spth | spinothalamic tr. |
| ts | tectospinal tract |
| V1 | V1 Cx |
| V2m | V2 (med.) Cx |
| V2v | V2 (vent.) Cx |
| VH | ventral horn |
| VLP (V3v) | ventlatpost. (V3v) Cx |

# Plate 47a

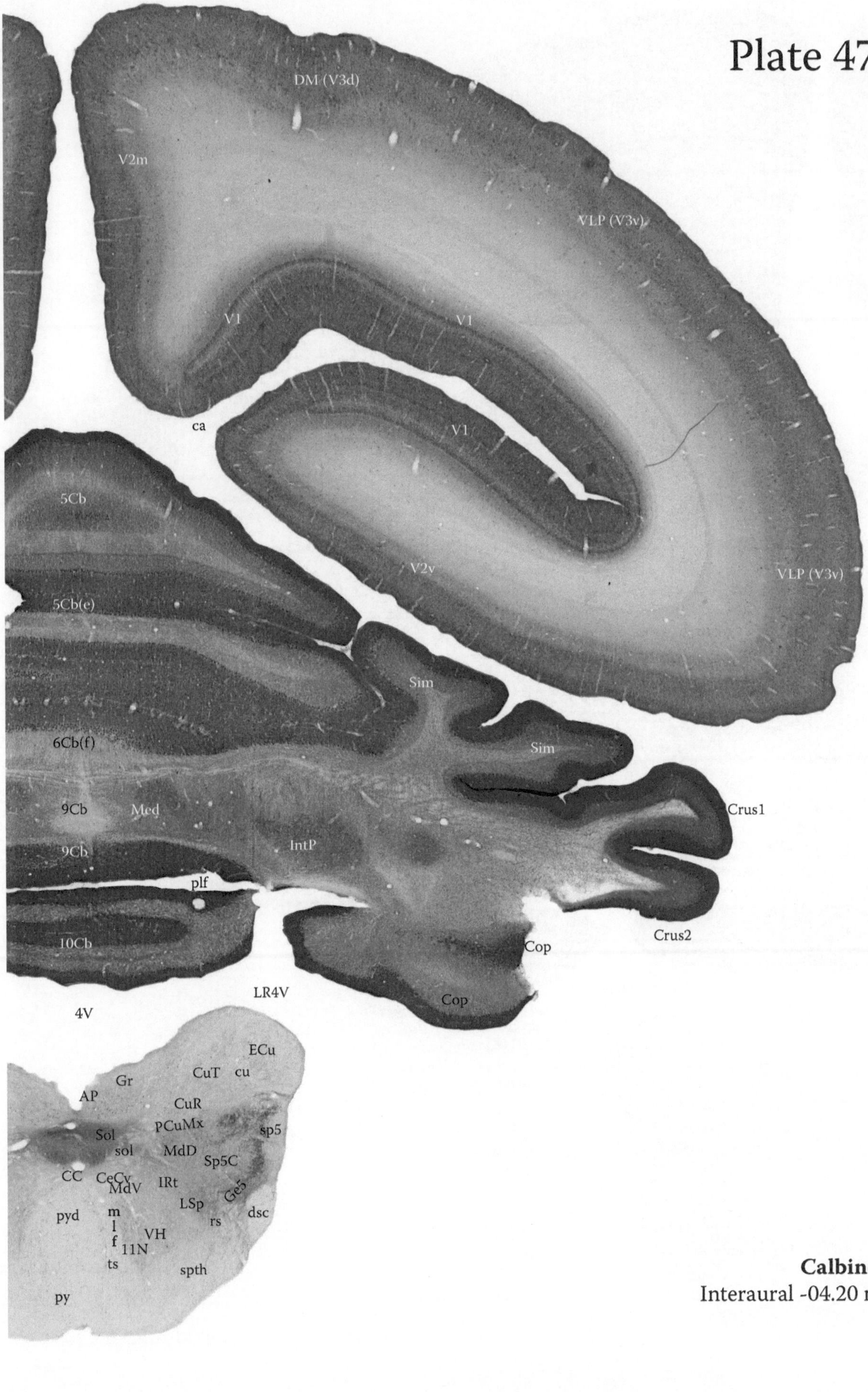


**Calbindin**
Interaural -04.20 mm

# Plate 47b

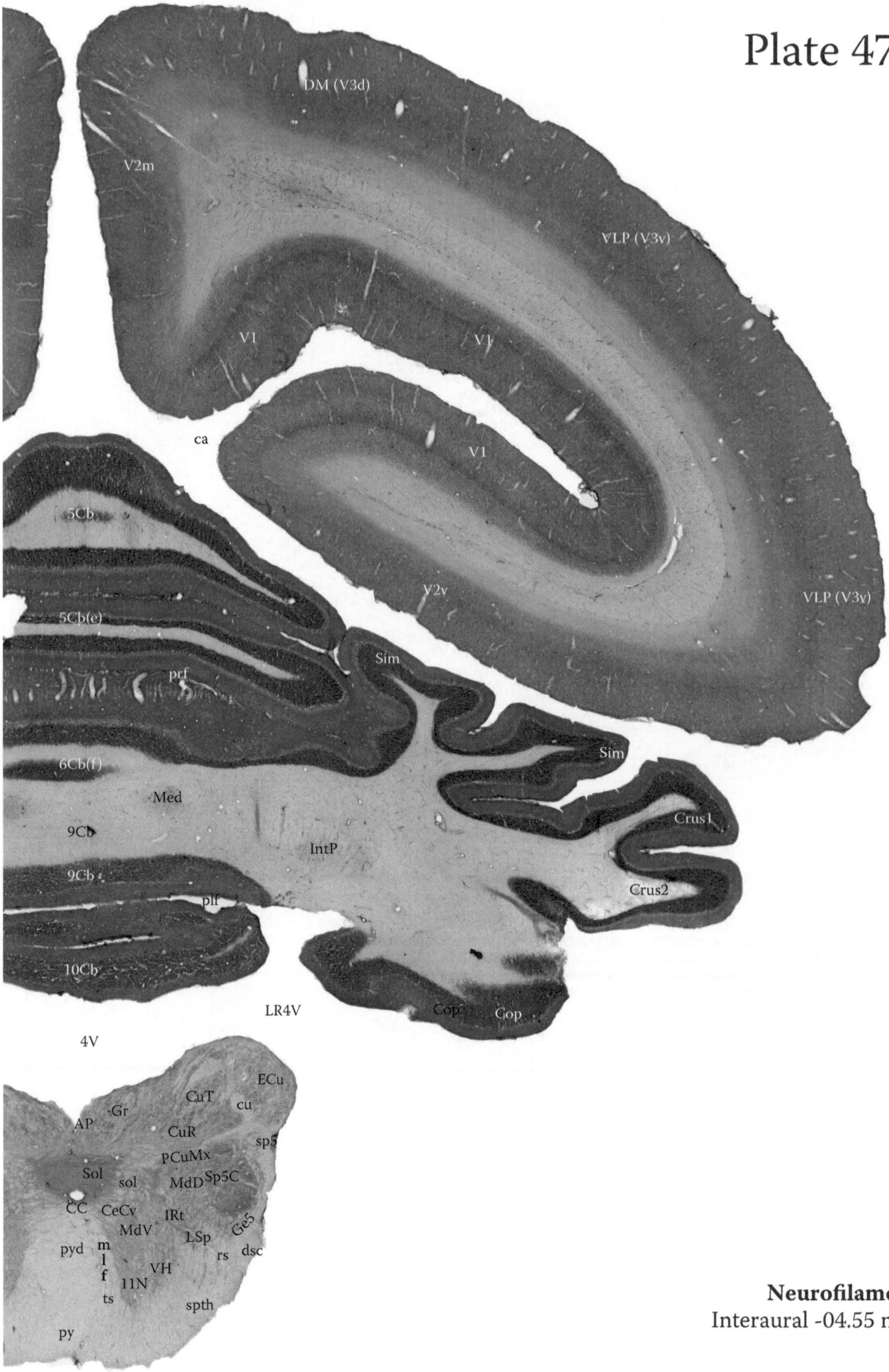


**Neurofilament**
Interaural -04.55 mm

# Plate 47c

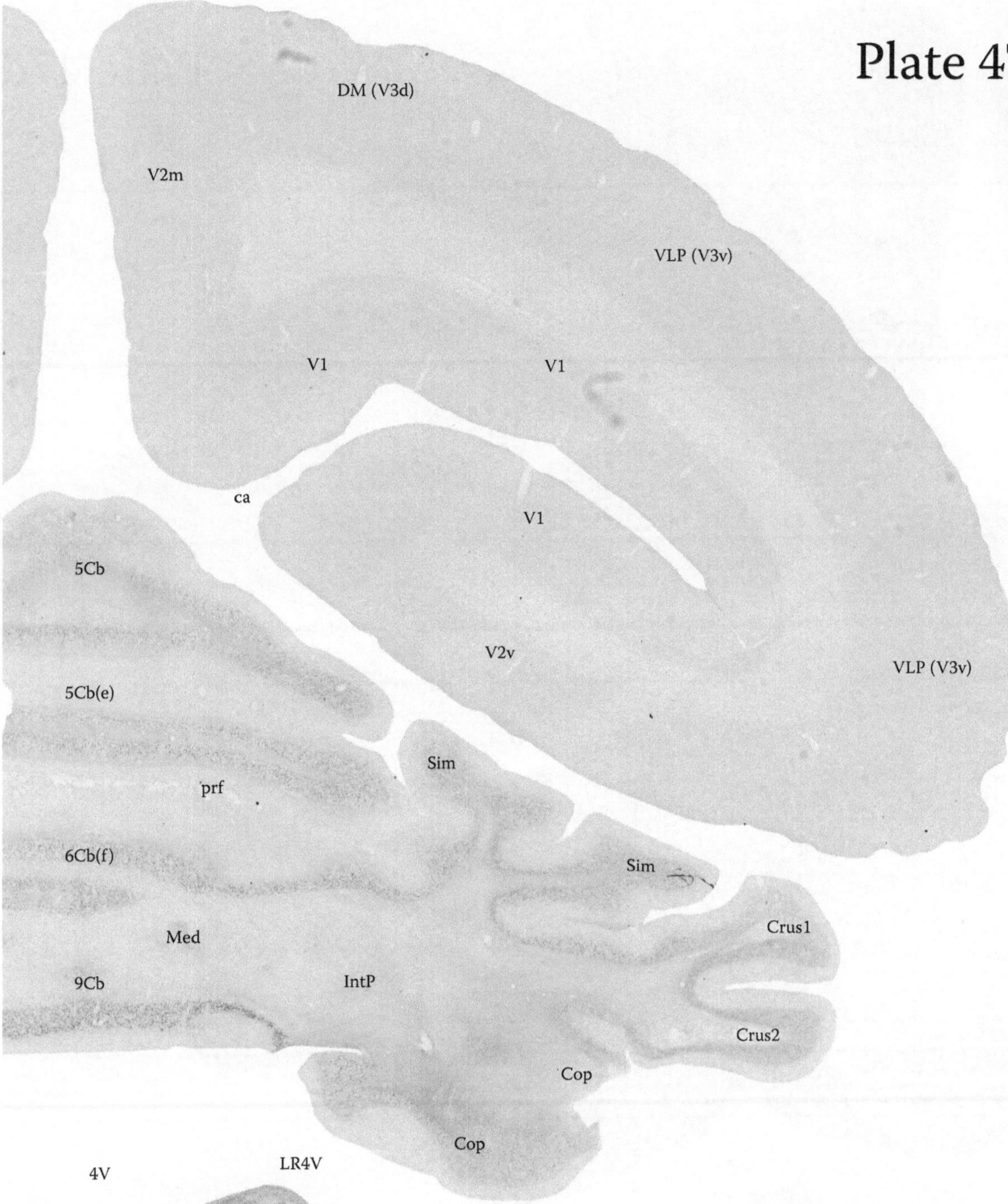


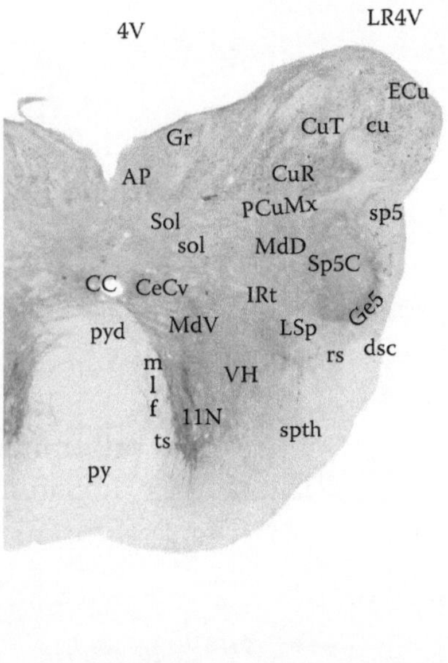


**Acetylcholinesterase**
Interaural -04.60 mm

# Plate 47d

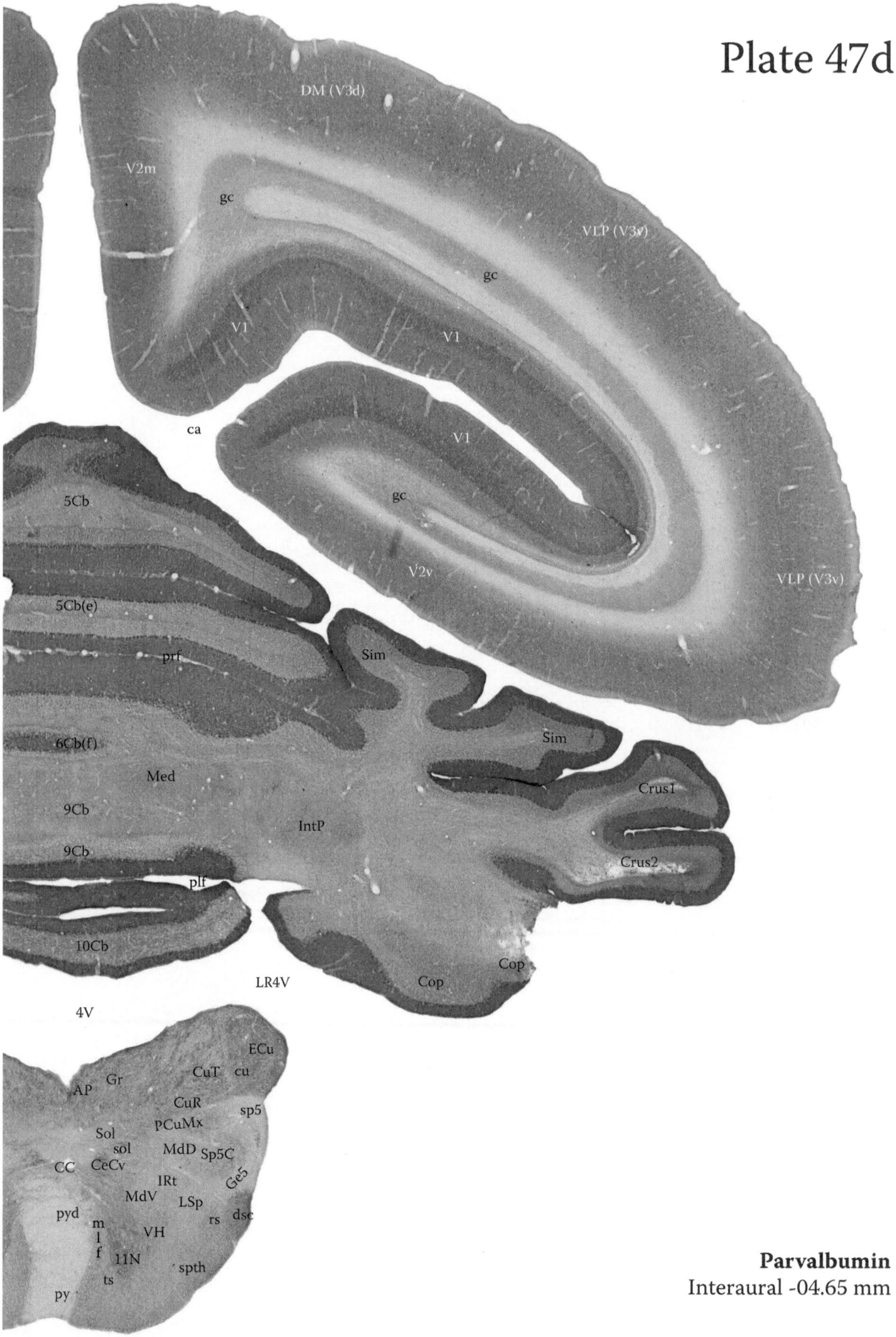


**Parvalbumin**
Interaural -04.65 mm

# Plate 47e

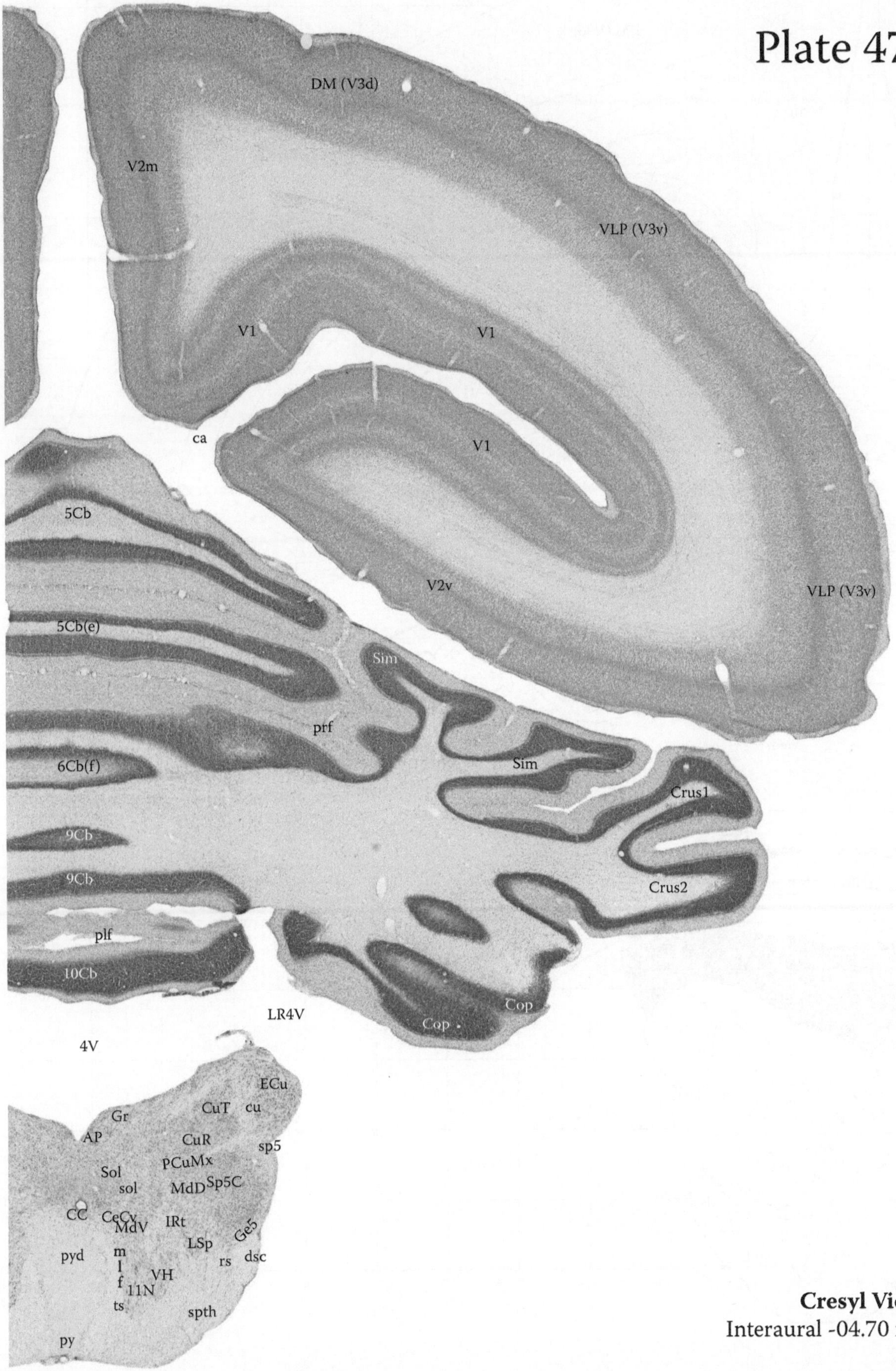


**Cresyl Violet**
Interaural -04.70 mm

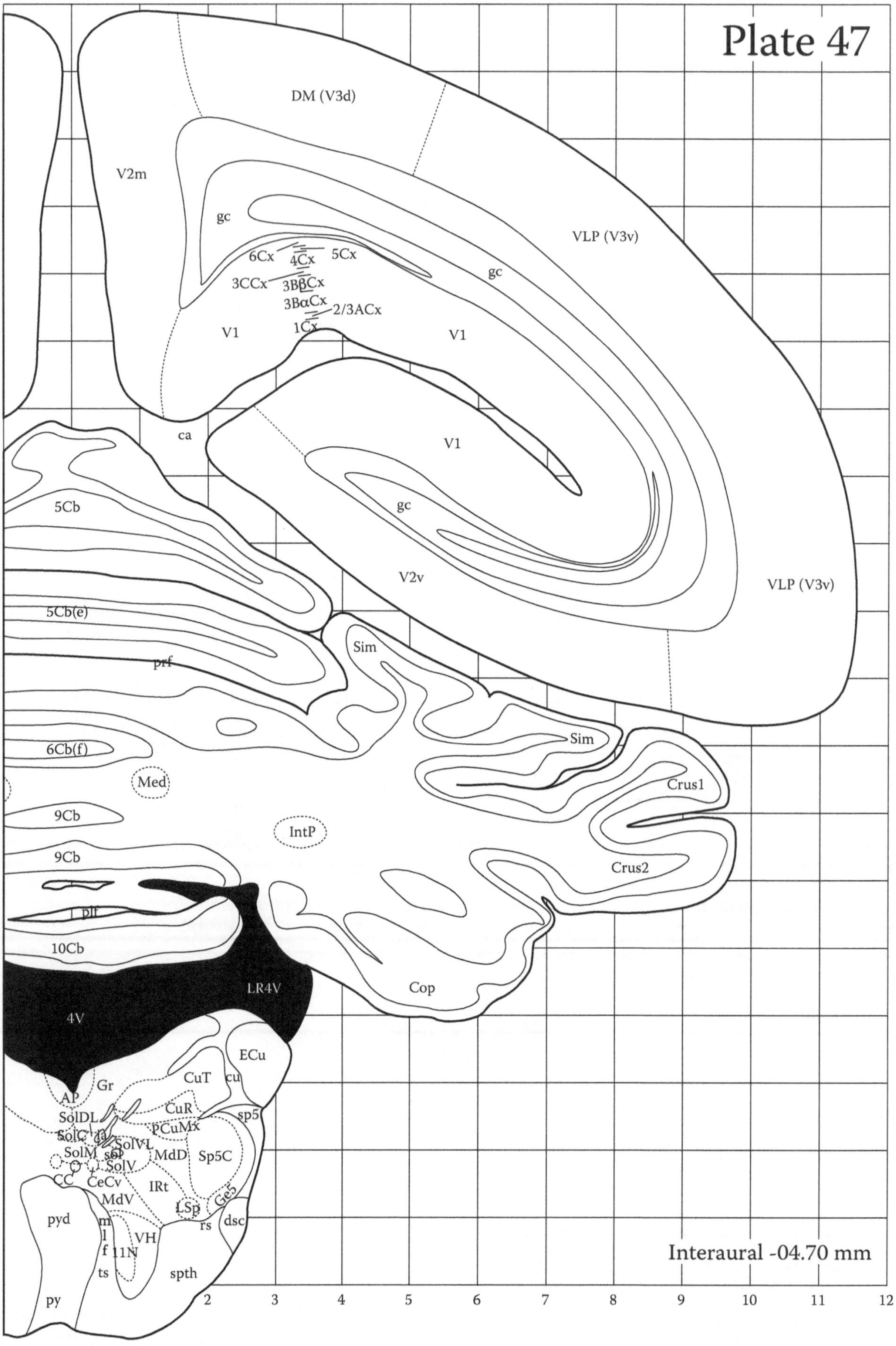
Plate 47
DM (V3d)
V2m
gc
VLP (V3v)
6Cx
4Cx
5Cx
3CCx
3BβCx
3BαCx
2/3ACx
1Cx
V1
ca
V2v
5Cb
5Cb(e)
prf
Sim
6Cb(f)
Med
IntP
Crus1
Crus2
9Cb
plf
10Cb
LR4V
4V
Cop
ECu
CuT
cu
Gr
AP
CuR
SolDL
PCuMx
sp5
SolC
da
SolVL
SolM
sol
MdD
Sp5C
SolV
CC
CeCv
IRt
MdV
LSp
Ge5
pyd
m
l
f
rs
dsc
VH
11N
ts
spth
py
Interaural -04.70 mm
18
17
16
15
14
13
12
11
10
9
8
7
6
5
4
3
2
1
0
-1
2
3
4
5
6
7
8
9
10
11
12

| | |
|---|---|
| 4V | fourth ventricle |
| 5Cb | cerebell. lobule 5 |
| 6Cb | cerebell. lobule 6 |
| 9Cb | cerebell. lobule 9 |
| 10Cb | cerebell. lobule 10 |
| 11N | accessory motor nu. |
| AP | area postrema |
| ca | calcarine sulcus |
| CC | central canal |
| CeCv | central cervical nu. |
| Cop | copula |
| Crus1 | crus 1, ansiform lob. |
| Crus2 | crus 2, ansiform lob. |
| cu | cuneate fasciculus |
| CuR | cuneate nu., rotunda |
| CuT | cuneate nu., triang. |
| DM (V3d) | dorsomed. visual Cx |
| dsc | dors. spinocerebell. tr |
| ECu | ext. cuneate nu. |
| gc | geniculocalcarine tr. |
| Ge5 | gelat. layer Sp5C |
| Gr | gracile nu. |
| ia | intern. arcuate fibers |
| IntP | intermed. post. deep Cb nu. |
| IRt | intermed. retic. formn |
| LR4V | lat. recess 4V |
| LSp | lat. spinal nu. |
| MdD | medull. retic., nu. dors. |
| MdV | medull. retic., nu. vent. |
| Med | med. deep Cb nu. |
| mlf | med. longit. fasc. |
| PCuMx | pericuneate matrix |
| plf | posterolat. fissure |
| prf | prim. cerebell. fiss. |
| py | pyramidal tract |
| pyd | pyramidal decn |
| rs | rubrospinal tract |
| Sim | simplex lobule Cb |
| sol | solitary tract |
| SolC | nu. of solitary tr., comm. |
| SolDL | nu. of solitary tr., dorsolat. |
| SolM | nu. of solitary tr., int. |
| SolV | nu. of solitary tr., vent. |
| SolVL | nu. of solitary tr., ventlat. |
| sp5 | spinal trigeminal tr. |
| Sp5C | spin. trigem. nu., caud. |
| spth | spinothalamic tr. |
| ts | tectospinal tract |
| V1 | V1 Cx |
| V2m | V2 (med.) Cx |
| V2v | V2 (vent.) Cx |
| VH | ventral horn |
| VLP (V3v) | ventlatpost. (V3v) Cx |

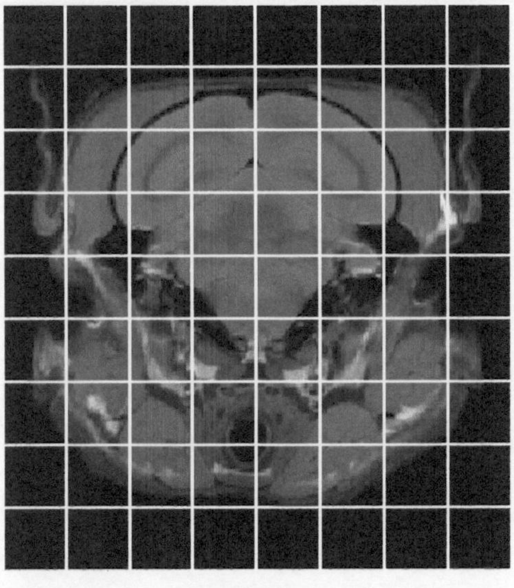

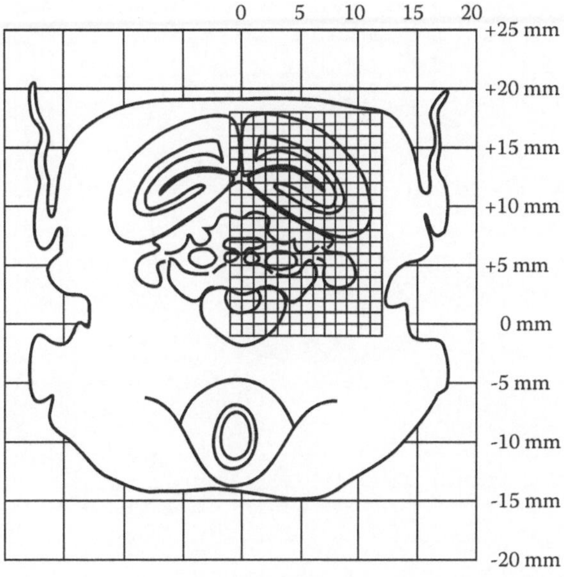


Interaural -04.55 mm

# Plate 48

Interaural -05.20 mm

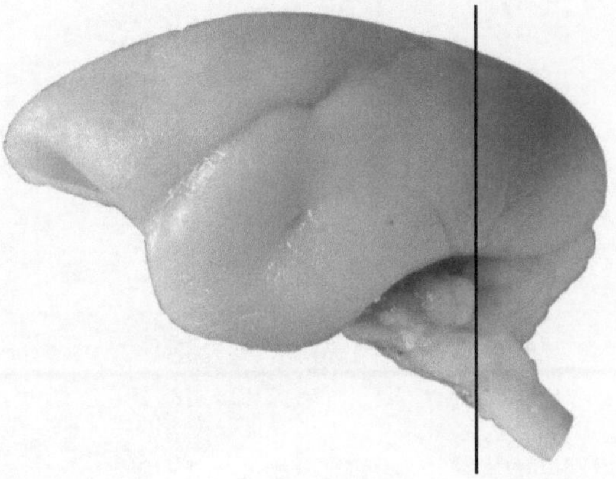

| | | | | | |
|---|---|---|---|---|---|
| 5Cb | cerebell. lobule 5 | ECu | ext. cuneate nu. | rs | rubrospinal tract |
| 6Cb | cerebell. lobule 6 | gc | geniculocalcarine tr. | Sim | simplex lobule Cb |
| 8Cb | cerebell. lobule 8 | Ge5 | gelat. layer Sp5C | SolC | nu. of solitary tr., comm. |
| 9Cb | cerebell. lobule 9 | gr | gracile fasciculus | SolM | nu. of solitary tr., int. |
| 10Cb | cerebell. lobule 10 | Gr | gracile nu. | SolVL | nu. of solitary tr., ventlat. |
| 11N | accessory motor nu. | IB | intern. basal nu. | sp5 | spinal trigeminal tr. |
| ca | calcarine sulcus | IRt | intermed. retic. formn | Sp5C | spin. trigem. nu., caud. |
| CC | central canal | LSp | lat. spinal nu. | spth | spinothalamic tr. |
| CeCv | central cervical nu. | lvs | lat. vestibulosp. tr. | ts | tectospinal tract |
| Cop | copula | MdD | medull. retic., nu. dors. | V1 | V1 Cx |
| Crus1 | crus 1, ansiform lob. | MdV | medull. retic., nu. vent. | V2m | V2 (med.) Cx |
| Crus2 | crus 2, ansiform lob. | mlf | med. longit. fasc. | V2v | V2 (vent.) Cx |
| cu | cuneate fasciculus | MnA | median access. nu. | VH | ventral horn |
| CuR | cuneate nu., rotunda | PCuMx | pericuneate matrix | VLP (V3v) | ventlatpost. (V3v) Cx |
| CuT | cuneate nu., triang. | PM | paramedian lobule Cb | vmf | vent. median fiss. |
| DM (V3d) | dorsomed. visual Cx | prf | prim. cerebell. fiss. | vsc | vent. spinocereb. tr. |
| dsc | dors. spinocerebell. tr | pyd | pyramidal decn | | |

# Plate 48a

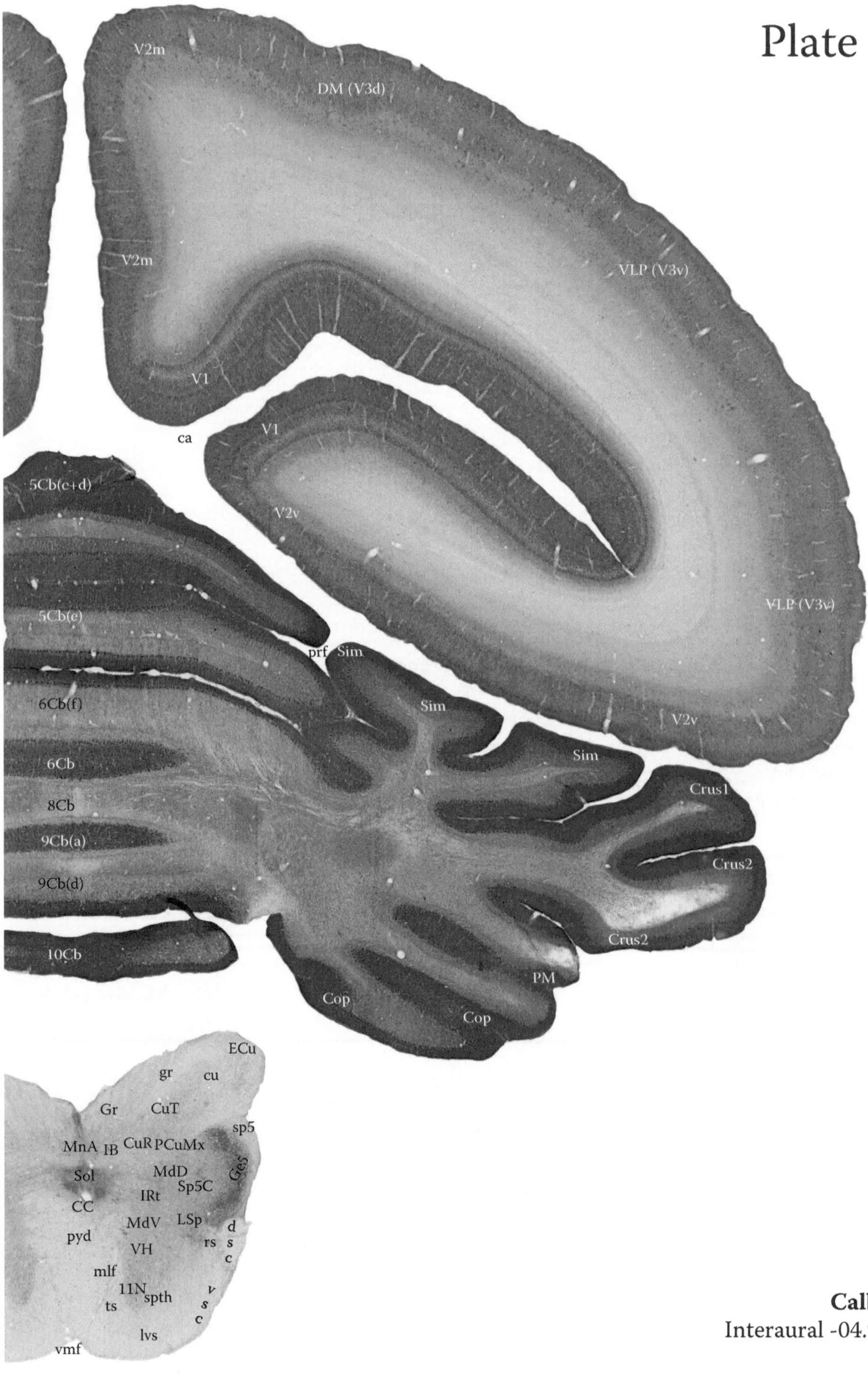


**Calbindin**
Interaural -04.90 mm

# Plate 48b

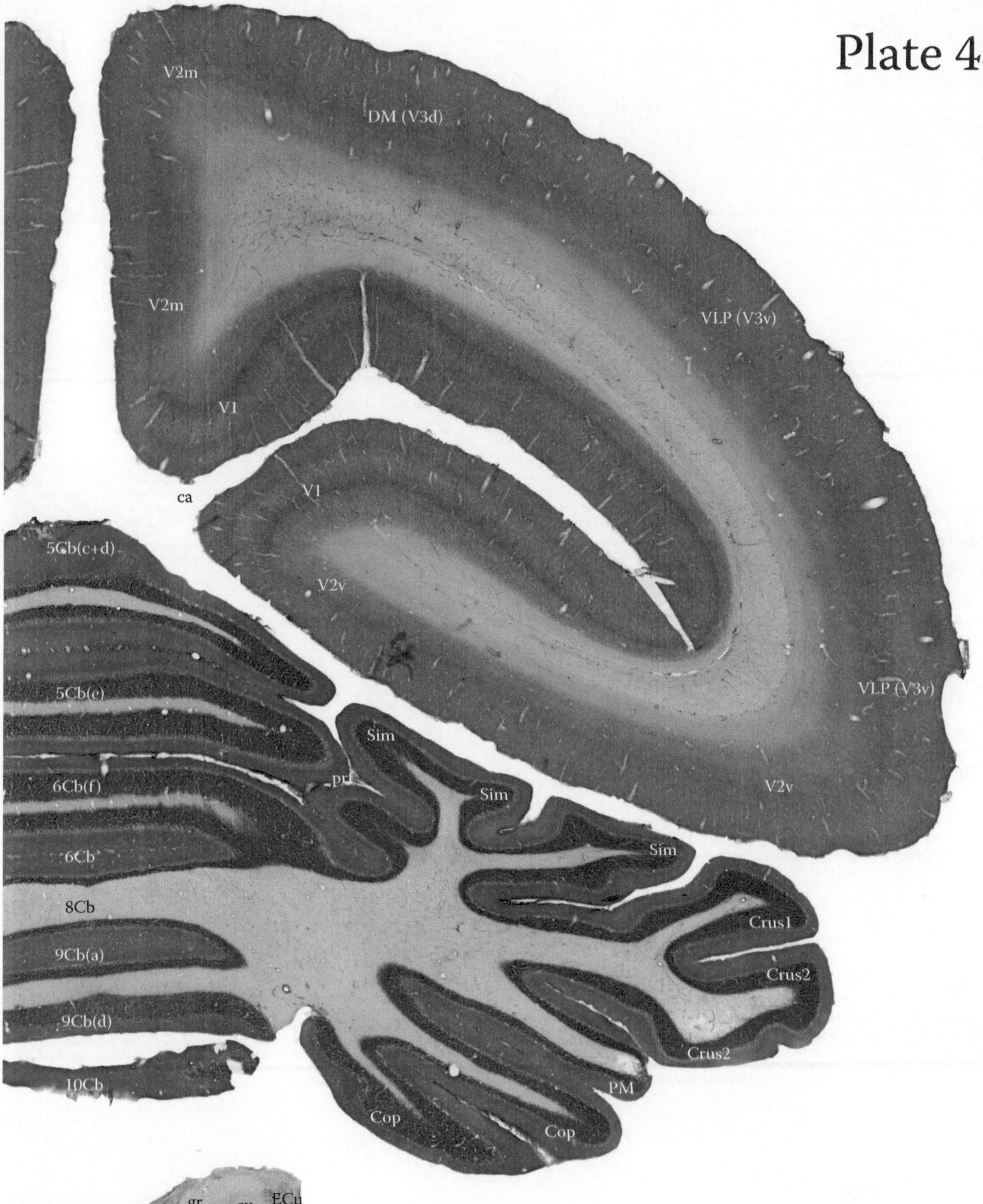


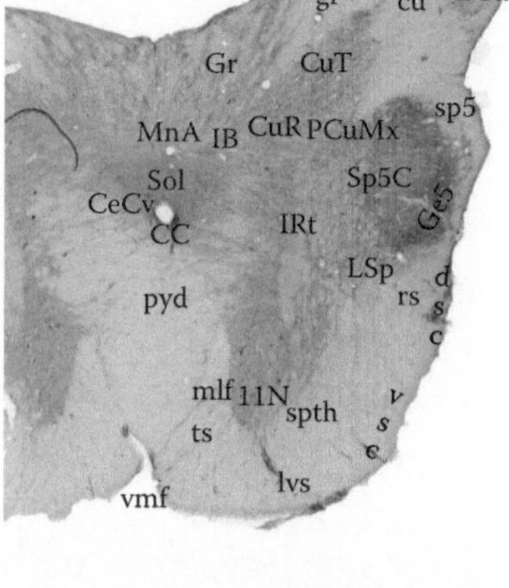


**Neurofilament**
Interaural -05.05 mm

# Plate 48c

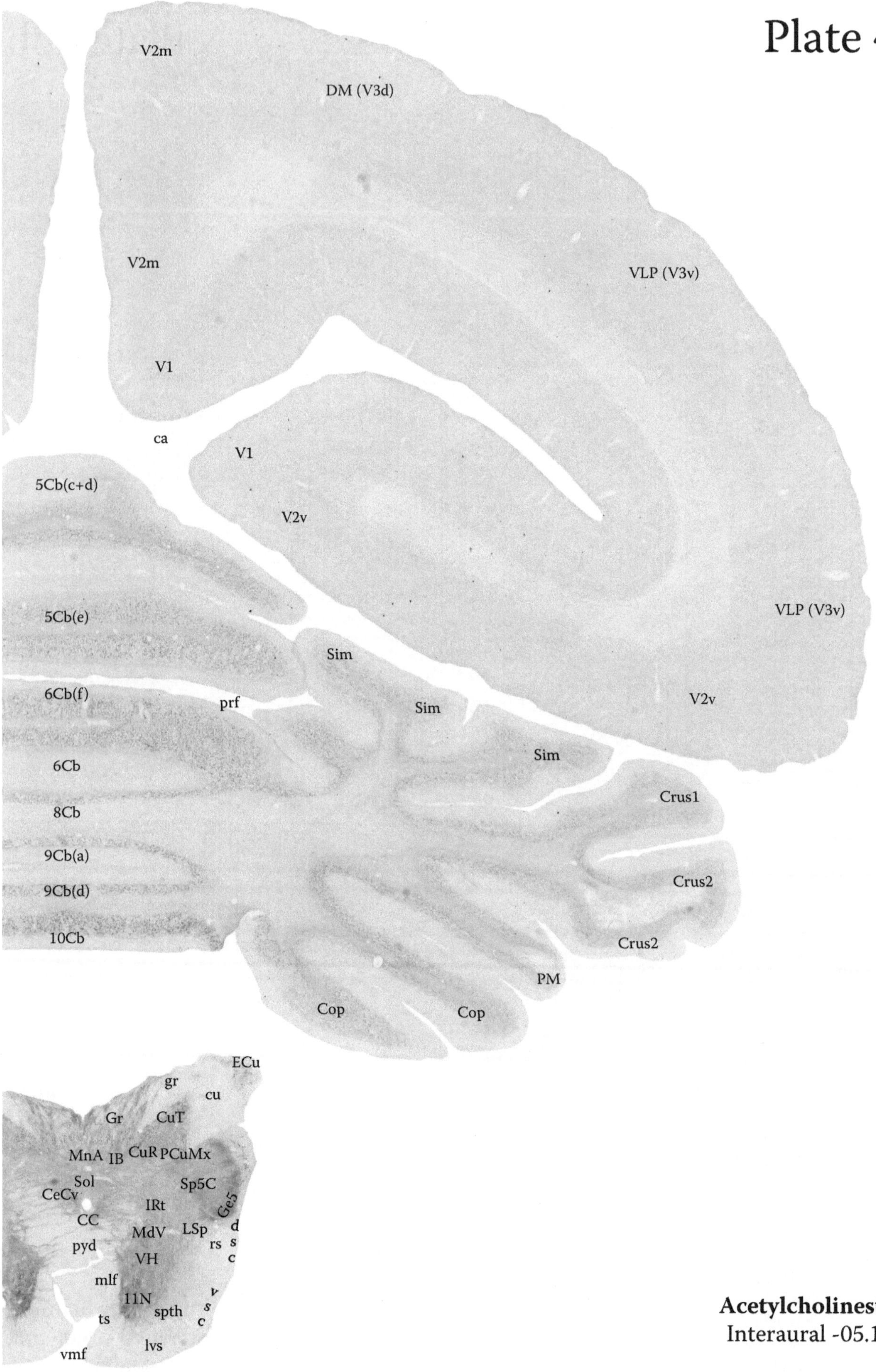


**Acetylcholinesterase**
Interaural -05.10 mm

# Plate 48d

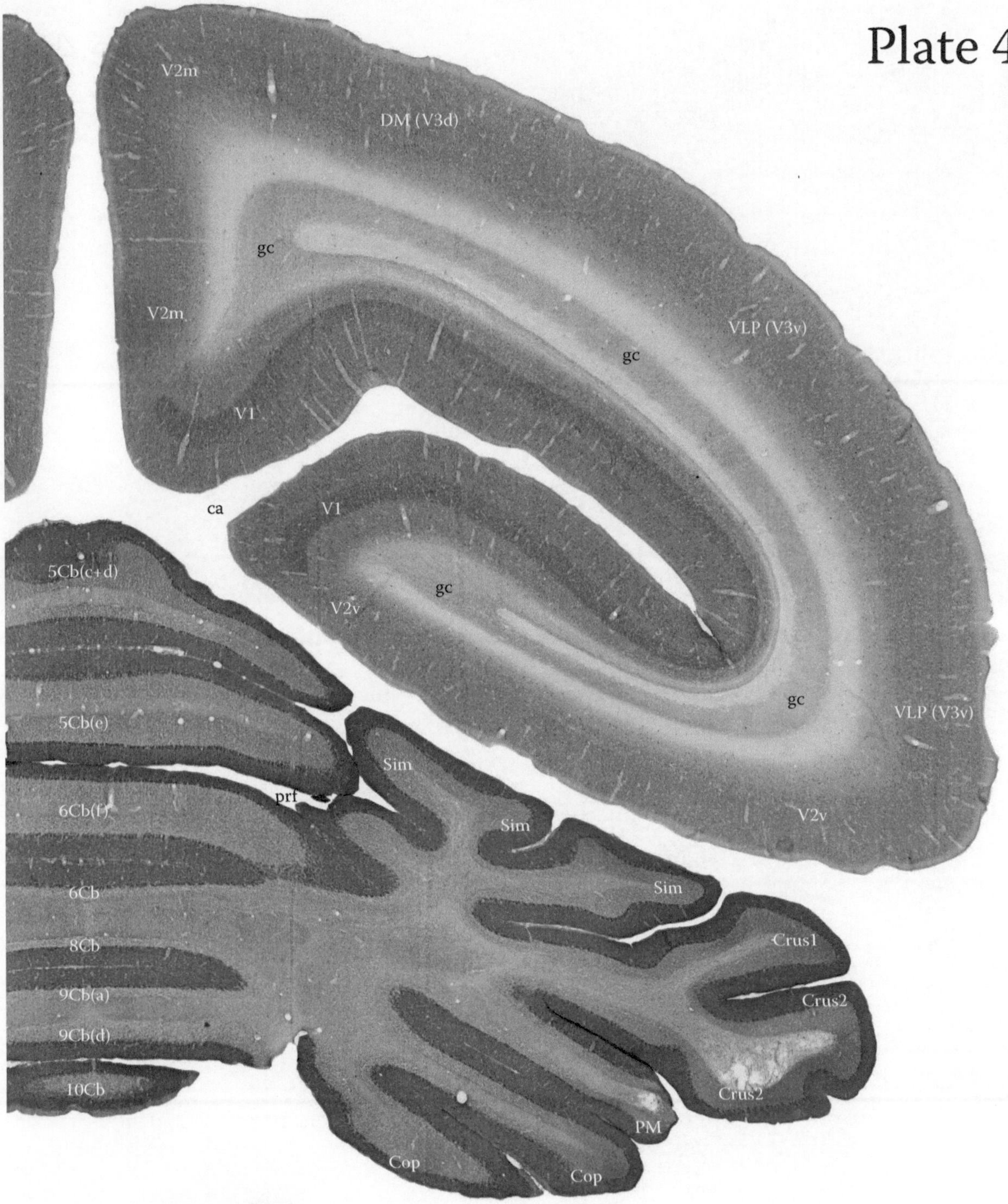


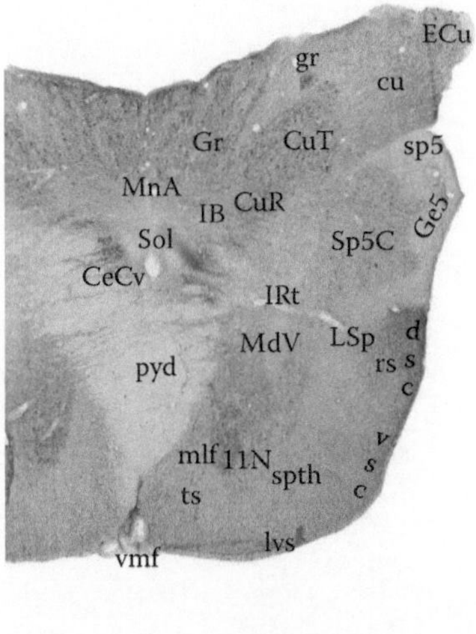


**Parvalbumin**
Interaural -05.15 mm

# Plate 48e

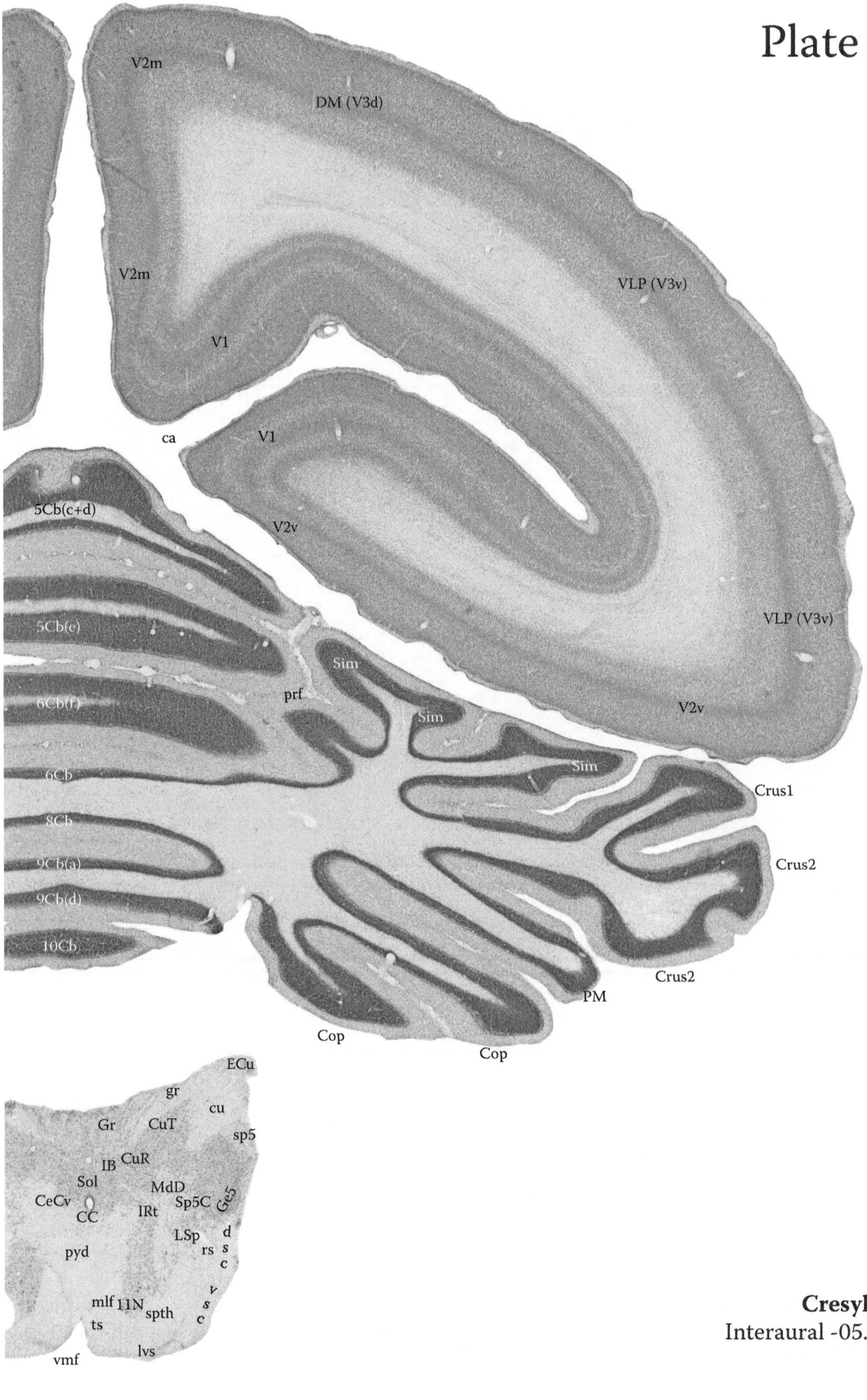


**Cresyl Violet**
Interaural -05.20 mm

# Plate 48

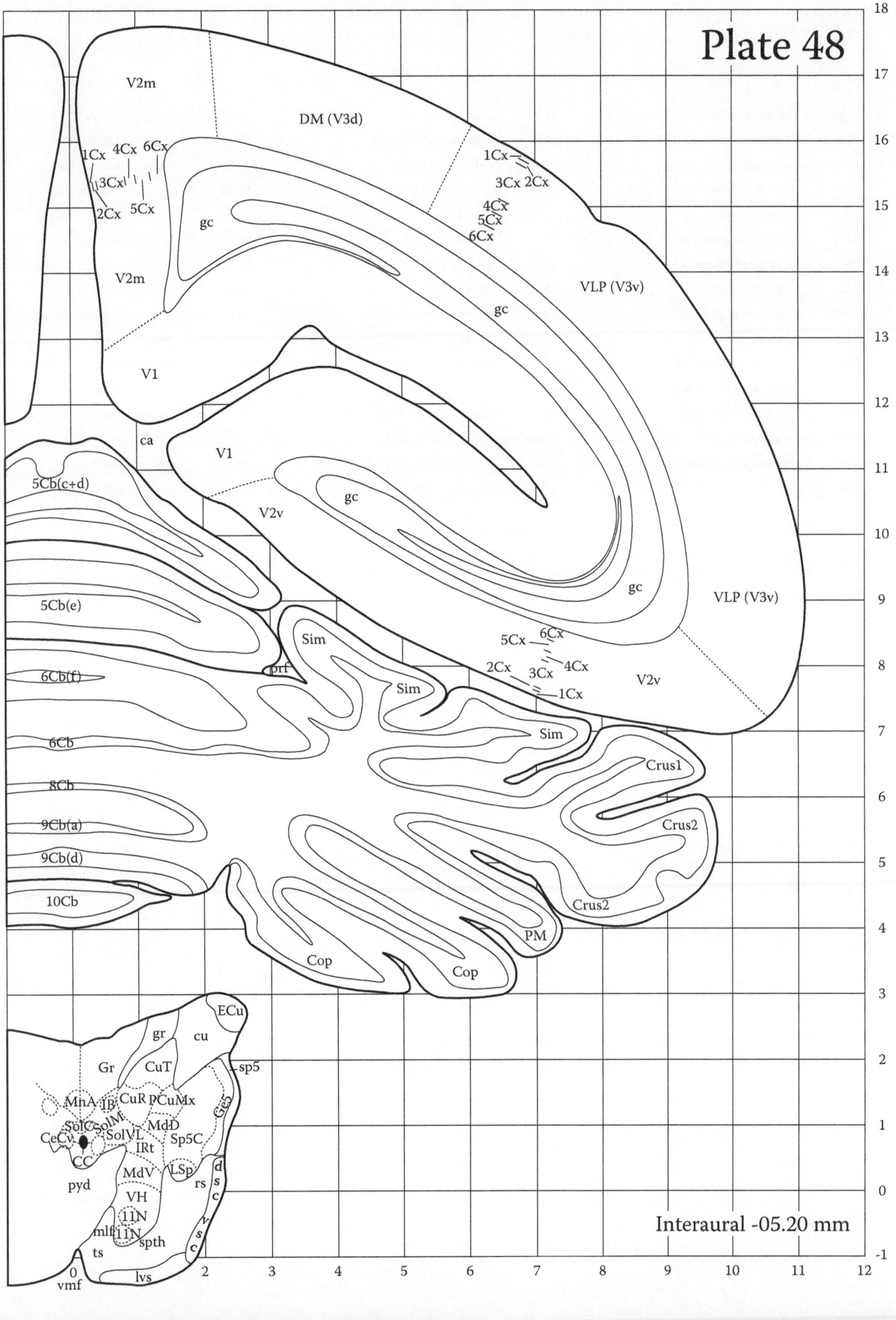

V2m
DM (V3d)
1Cx
4Cx
6Cx
3Cx
5Cx
2Cx
gc
V2m
V1
ca
V1
1Cx
3Cx
2Cx
4Cx
5Cx
6Cx
VLP (V3v)
gc
gc
V2v
gc
VLP (V3v)
5Cx
6Cx
4Cx
2Cx
3Cx
1Cx
V2v
5Cb(c+d)
5Cb(e)
Sim
prf
Sim
6Cb(f)
6Cb
Sim
Crus1
8Cb
9Cb(a)
Crus2
9Cb(d)
10Cb
Crus2
PM
Cop
Cop
ECu
gr
cu
Gr
CuT
sp5
MnA
IB
CuR
PCuMx
SolC
SolM
Ge5
CeCv
SolVL
MdD
Sp5C
IRt
CC
MdV
LSp
d
s
c
rs
pyd
VH
11N
mlf
11N
v
s
c
ts
spth
lvs
vmf
Interaural -05.20 mm
18
17
16
15
14
13
12
11
10
9
8
7
6
5
4
3
2
1
0
-1
0
2
3
4
5
6
7
8
9
10
11
12

| | |
|---|---|
| 5Cb | cerebell. lobule 5 |
| 6Cb | cerebell. lobule 6 |
| 8Cb | cerebell. lobule 8 |
| 9Cb | cerebell. lobule 9 |
| 10Cb | cerebell. lobule 10 |
| 11N | accessory motor nu. |
| ca | calcarine sulcus |
| CC | central canal |
| CeCv | central cervical nu. |
| Cop | copula |
| Crus1 | crus 1, ansiform lob. |
| Crus2 | crus 2, ansiform lob. |
| cu | cuneate fasciculus |
| CuR | cuneate nu., rotunda |
| CuT | cuneate nu., triang. |
| DM (V3d) | dorsomed. visual Cx |
| dsc | dors. spinocerebell. tr |
| ECu | ext. cuneate nu. |
| gc | geniculocalcarine tr. |
| Ge5 | gelat. layer Sp5C |
| gr | gracile fasciculus |
| Gr | gracile nu. |
| IB | intern. basal nu. |
| IRt | intermed. retic. formn |
| LSp | lat. spinal nu. |
| lvs | lat. vestibulosp. tr. |
| MdD | medull. retic., nu. dors. |
| MdV | medull. retic., nu. vent. |
| mlf | med. longit. fasc. |
| MnA | median access. nu. |
| PCuMx | pericuneate matrix |
| PM | paramedian lobule Cb |
| prf | prim. cerebell. fiss. |
| pyd | pyramidal decn |
| rs | rubrospinal tract |
| Sim | simplex lobule Cb |
| SolC | nu. of solitary tr., comm. |
| SolM | nu. of solitary tr., int. |
| SolVL | nu. of solitary tr., ventlat. |
| sp5 | spinal trigeminal tr. |
| Sp5C | spin. trigem. nu., caud. |
| spth | spinothalamic tr. |
| ts | tectospinal tract |
| V1 | V1 Cx |
| V2m | V2 (med.) Cx |
| V2v | V2 (vent.) Cx |
| VH | ventral horn |
| VLP (V3v) | ventlatpost. (V3v) Cx |
| vmf | vent. median fiss. |
| vsc | vent. spinocereb. tr. |

# Plate 49

Interaural -05.75 mm

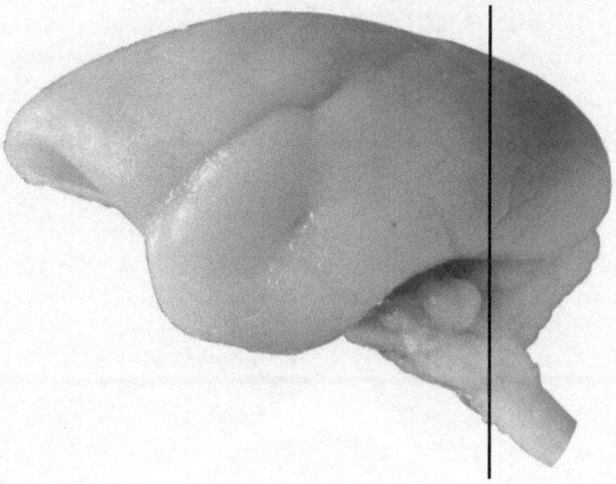

| | |
|---|---|
| 5Cb | cerebell. lobule 5 |
| 6Cb | cerebell. lobule 6 |
| 7Cb | cerebell. lobule 7 |
| 8Cb | cerebell. lobule 8 |
| 9Cb | cerebell. lobule 9 |
| ca | calcarine sulcus |
| CC | central canal |
| CeCv | central cervical nu. |
| Cop | copula |
| Crus1 | crus 1, ansiform lob. |
| Crus2 | crus 2, ansiform lob. |
| cu | cuneate fasciculus |
| Cu | cuneate nu. |
| DM (V3d) | dorsomed. visual Cx |
| dsc | dors. spinocerebell. tr |
| gc | geniculocalcarine tr. |
| Ge5 | gelat. layer Sp5C |
| gr | gracile fasciculus |
| Gr | gracile nu. |
| IB | intern. basal nu. |
| lcs | lat. corticospinal tr. |
| LSp | lat. spinal nu. |
| lvs | lat. vestibulosp. tr. |
| MdD | medull. retic., nu. dors. |
| MdV | medull. retic., nu. vent. |
| mlf | med. longit. fasc. |
| MnA | median access. nu. |
| PM | paramedian lobule Cb |
| ppf | prepyramidal fiss. |
| R8 | Rexed 8, spinal cord |
| R9 | Rexed 9, spinal cord |
| rs | rubrospinal tract |
| Sim | simplex lobule Cb |
| SolM | nu. of solitary tr., int. |
| sp5 | spinal trigeminal tr. |
| Sp5C | spin. trigem. nu., caud. |
| spth | spinothalamic tr. |
| ts | tectospinal tract |
| V1 | V1 Cx |
| V2m | V2 (med.) Cx |
| V2v | V2 (vent.) Cx |
| vcs | vent. corticospinal tr. |
| VH | ventral horn |
| VLP (V3v) | ventlatpost. (V3v) Cx |
| vmf | vent. median fiss. |
| vsc | vent. spinocereb. tr. |

# Plate 49a

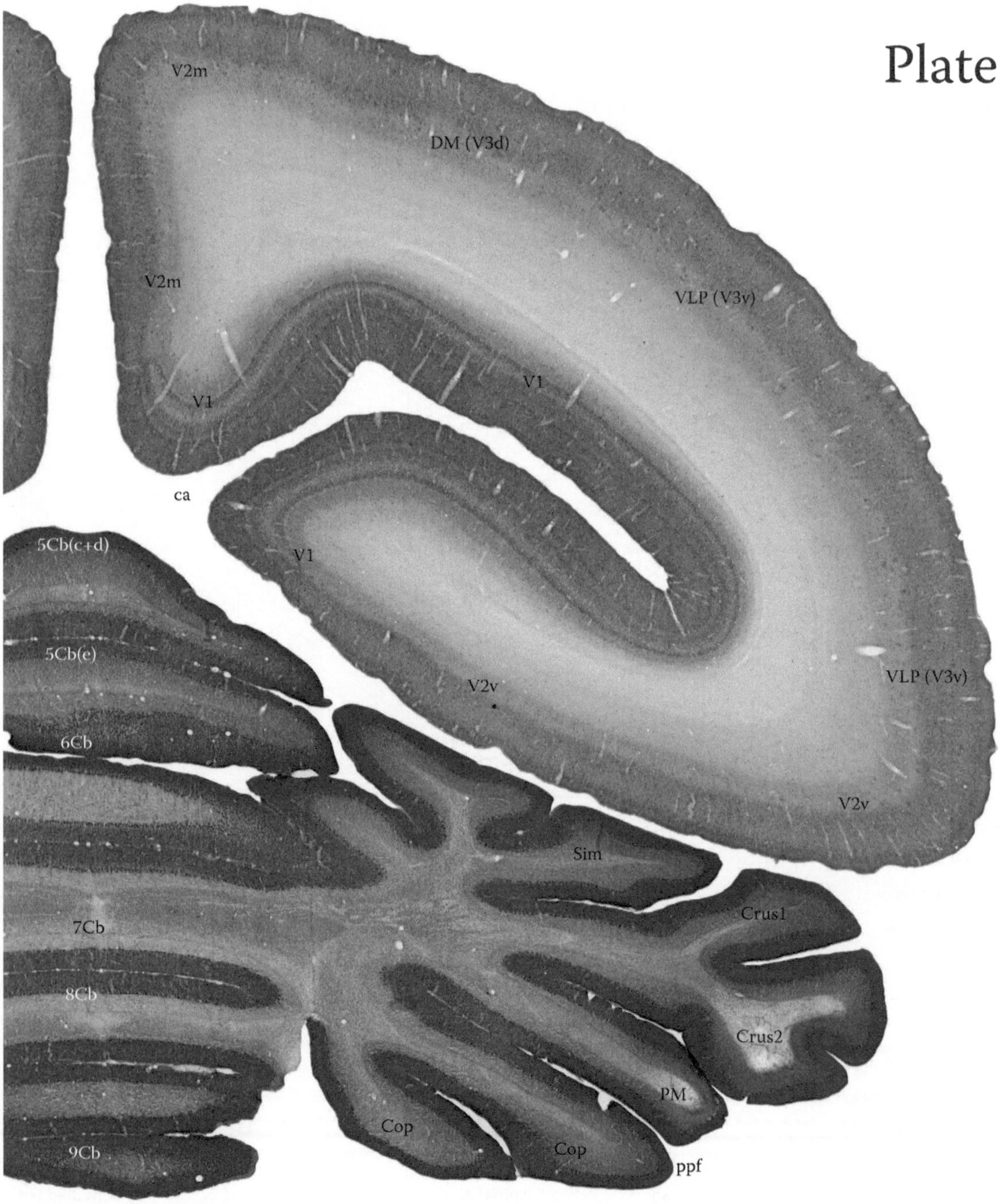


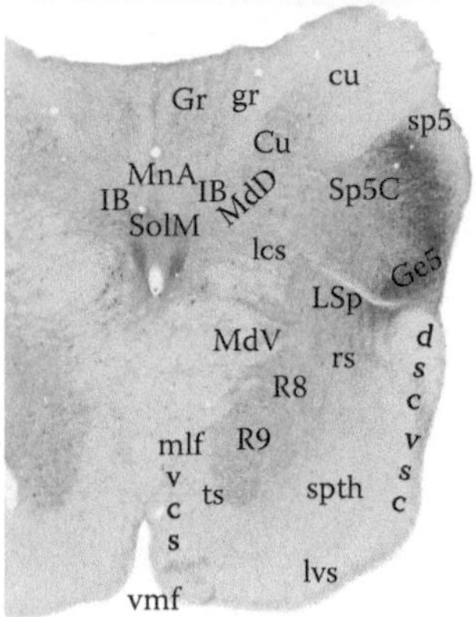


**Calbindin**
Interaural -05.45 mm

# Plate 49b

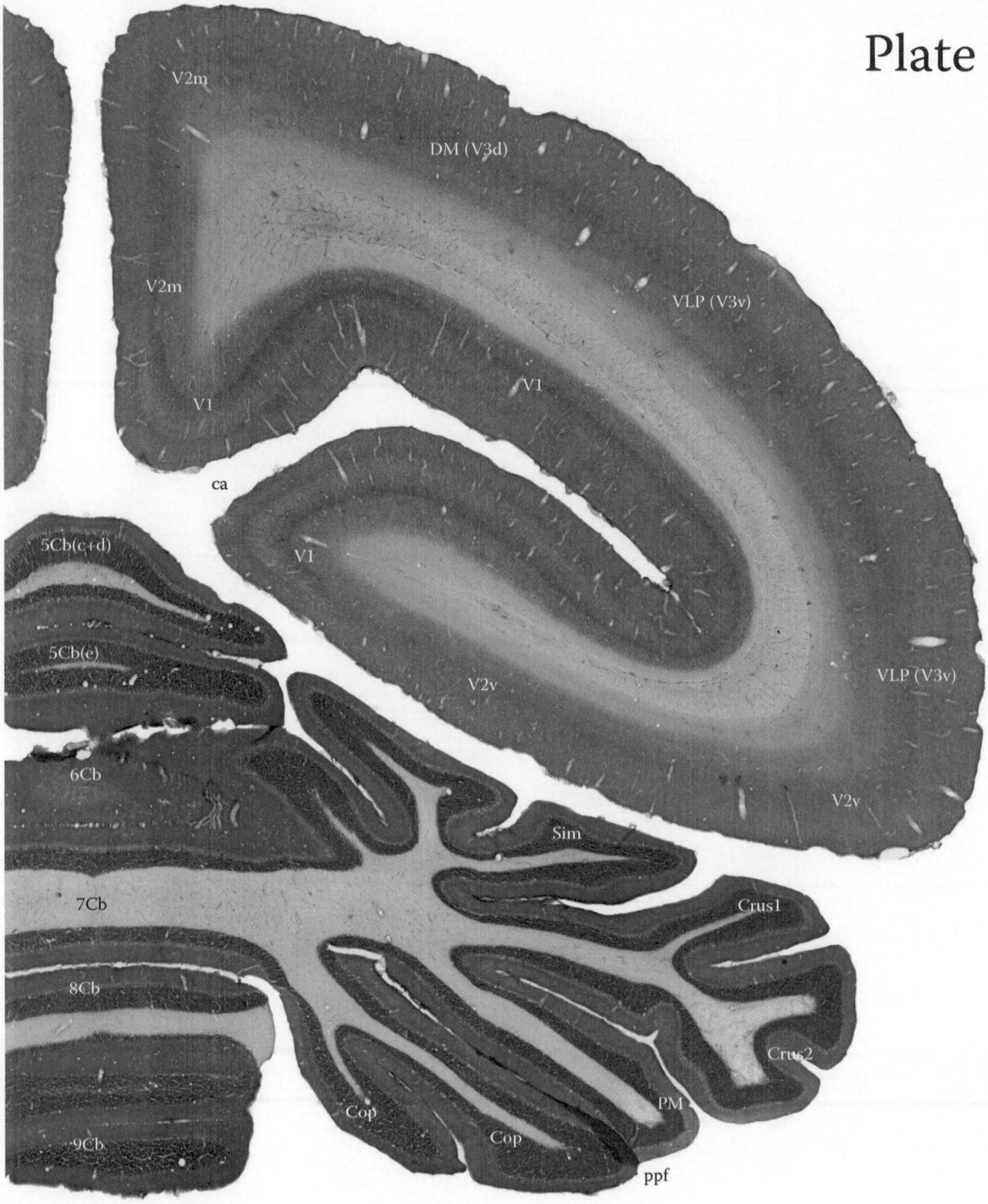


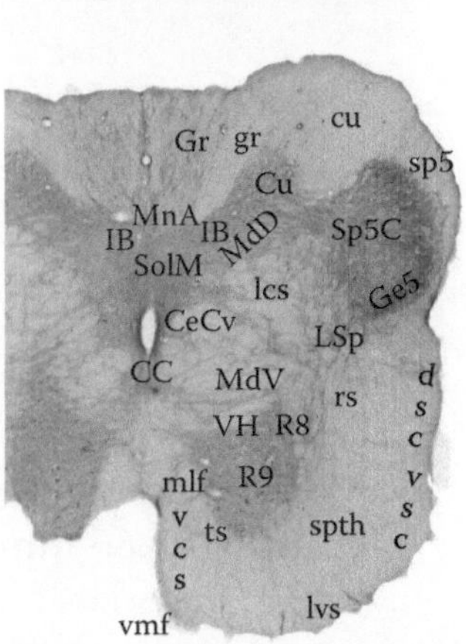


**Neurofilament**
Interaural -05.60 mm

# Plate 49c

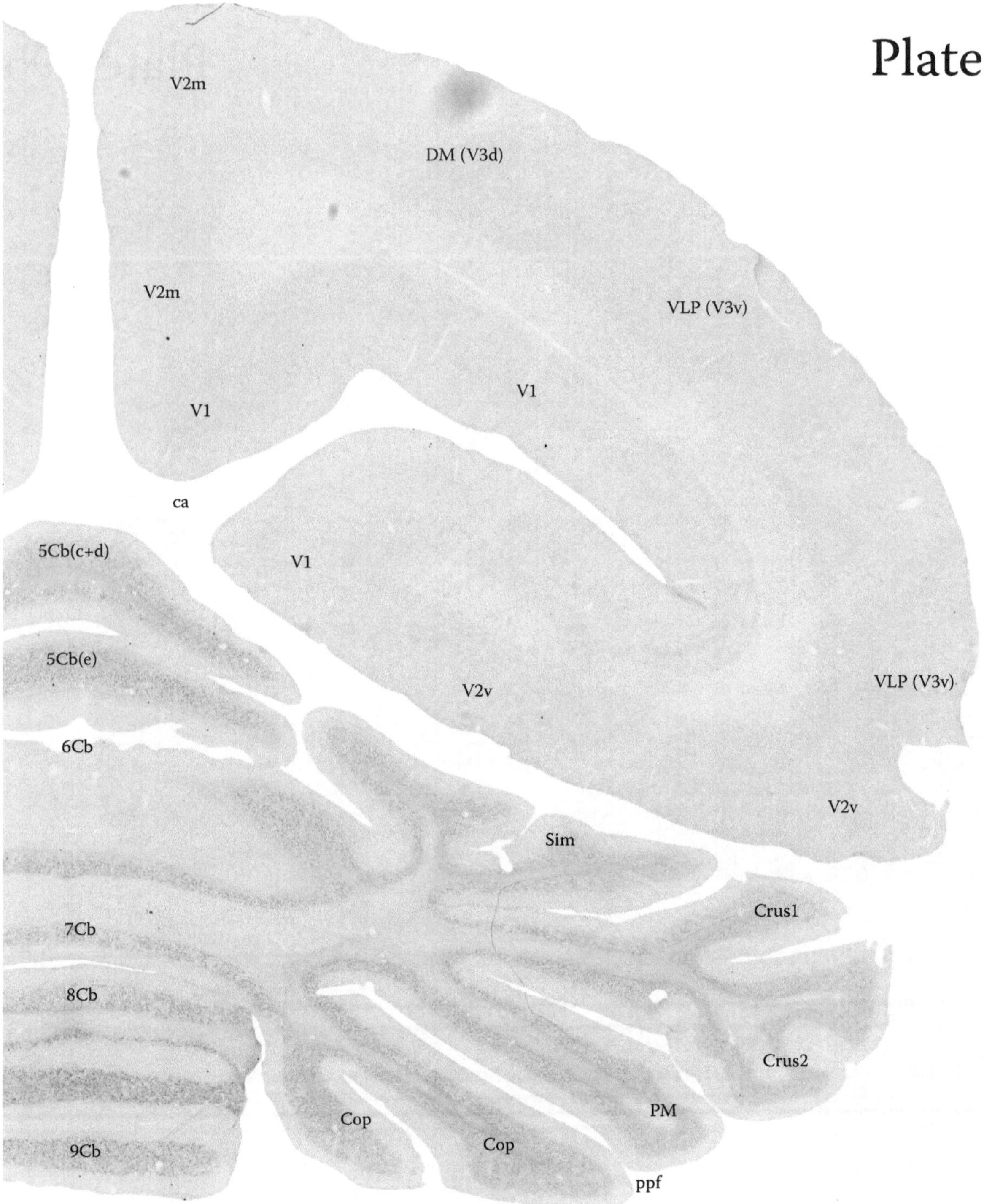


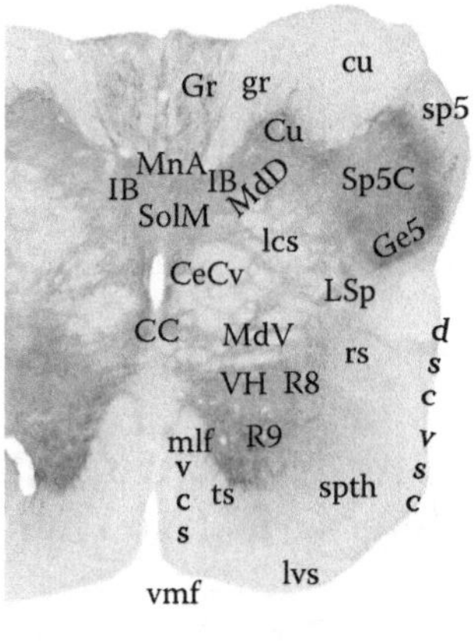


**Acetylcholinesterase**
Interaural -05.65 mm

# Plate 49d

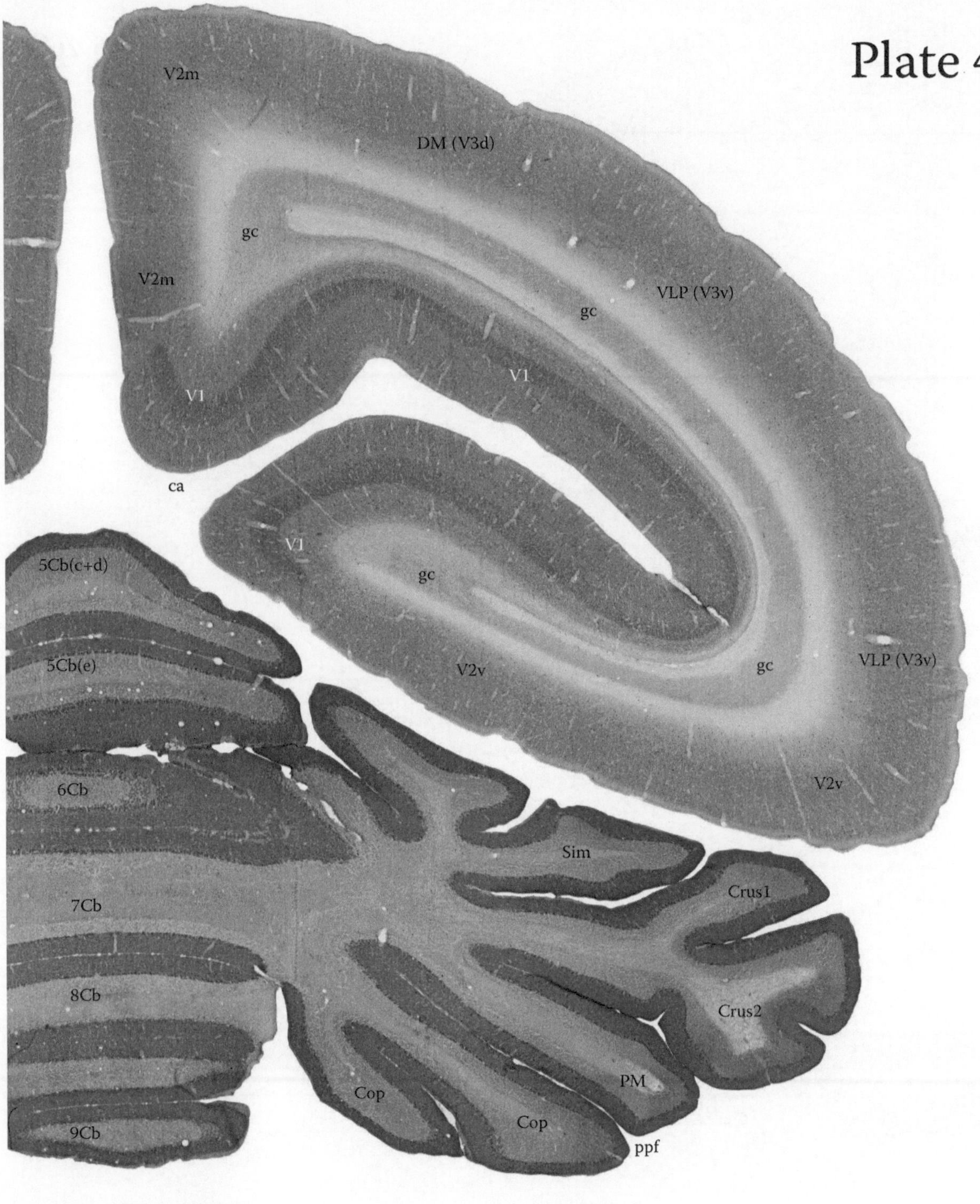


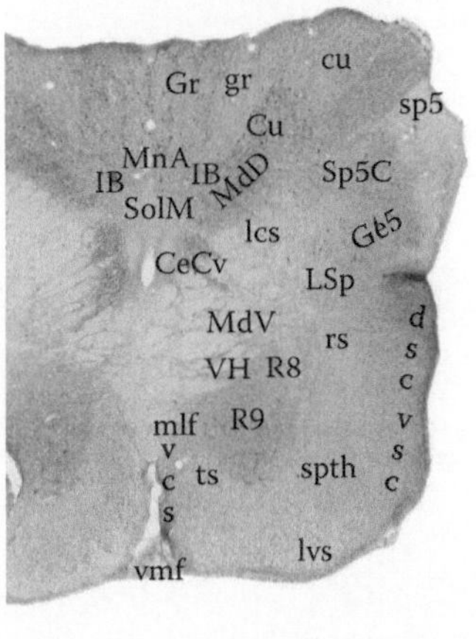


**Parvalbumin**
Interaural -05.70 mm

# Plate 49e

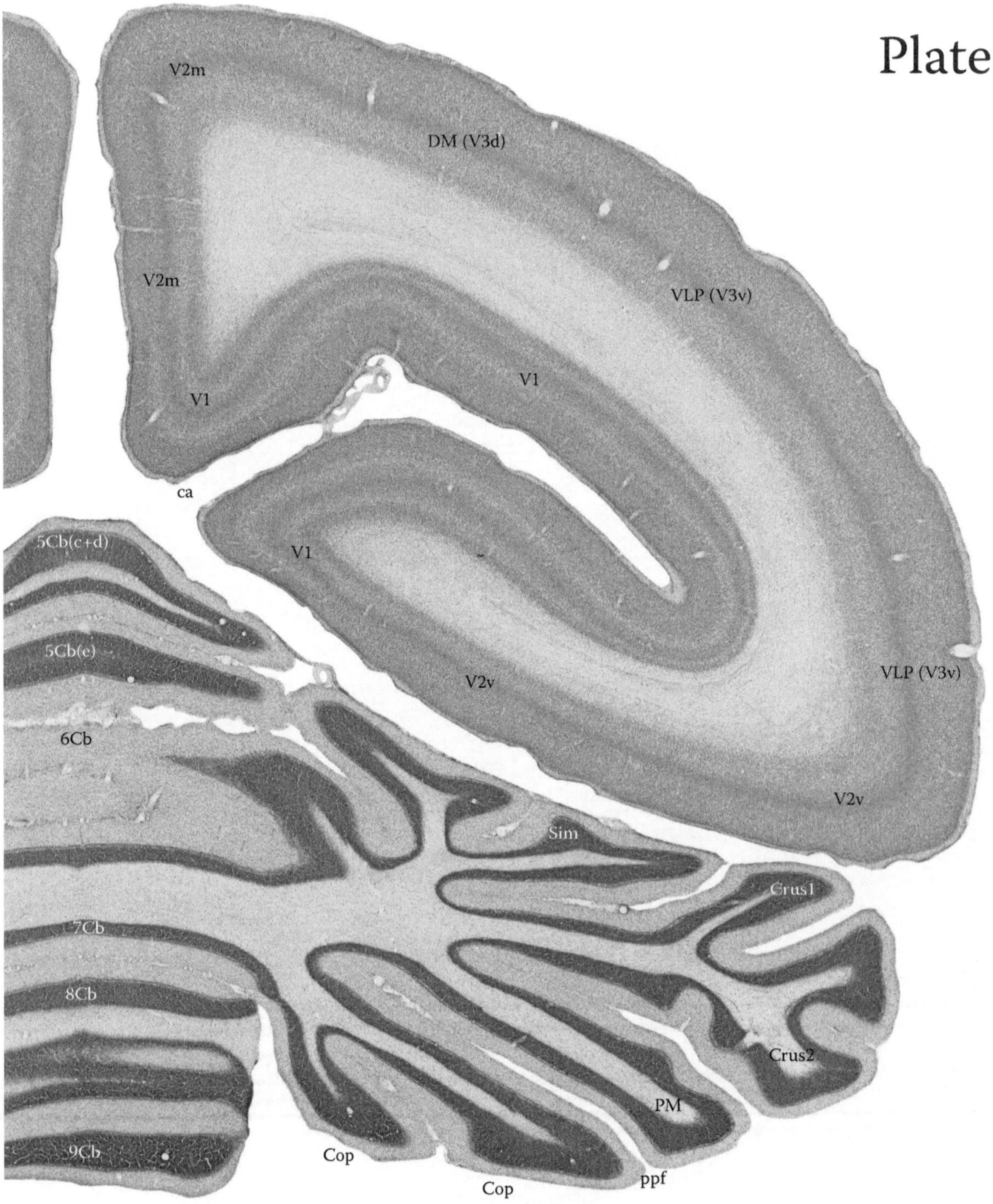


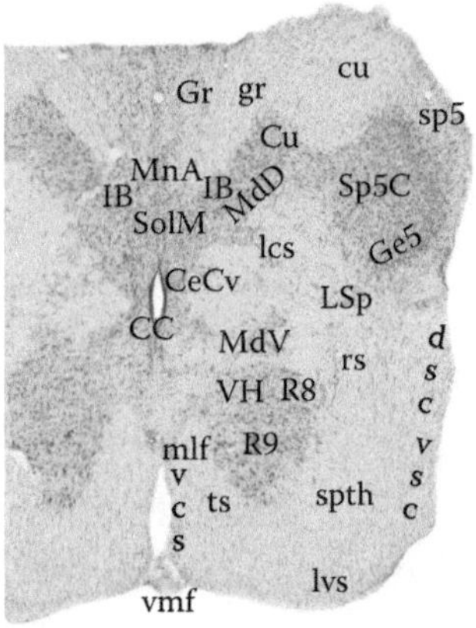


**Cresyl Violet**
Interaural -05.75 mm

# Plate 49

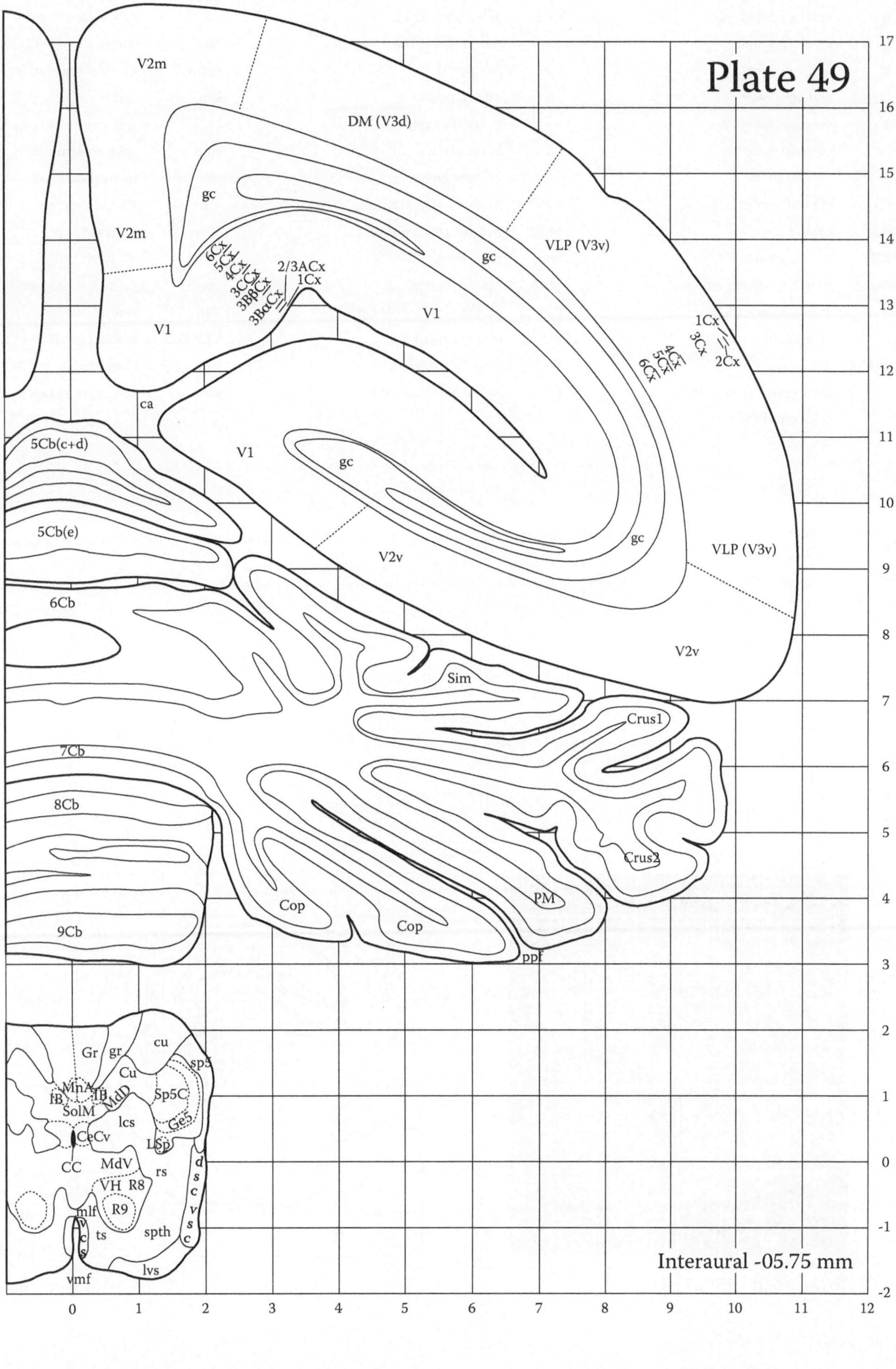


Interaural -05.75 mm

| | |
|---|---|
| 5Cb | cerebell. lobule 5 |
| 6Cb | cerebell. lobule 6 |
| 7Cb | cerebell. lobule 7 |
| 8Cb | cerebell. lobule 8 |
| 9Cb | cerebell. lobule 9 |
| ca | calcarine sulcus |
| CC | central canal |
| CeCv | central cervical nu. |
| Cop | copula |
| Crus1 | crus 1, ansiform lob. |
| Crus2 | crus 2, ansiform lob. |
| cu | cuneate fasciculus |
| Cu | cuneate nu. |
| DM (V3d) | dorsomed. visual Cx |
| dsc | dors. spinocerebell. tr |
| gc | geniculocalcarine tr. |
| Ge5 | gelat. layer Sp5C |
| gr | gracile fasciculus |
| Gr | gracile nu. |
| IB | intern. basal nu. |
| lcs | lat. corticospinal tr. |
| LSp | lat. spinal nu. |
| lvs | lat. vestibulosp. tr. |
| MdD | medull. retic., nu. dors. |
| MdV | medull. retic., nu. vent. |
| mlf | med. longit. fasc. |
| MnA | median access. nu. |
| PM | paramedian lobule Cb |
| ppf | prepyramidal fiss. |
| R8 | Rexed 8, spinal cord |
| R9 | Rexed 9, spinal cord |
| rs | rubrospinal tract |
| Sim | simplex lobule Cb |
| SolM | nu. of solitary tr., int. |
| sp5 | spinal trigeminal tr. |
| Sp5C | spin. trigem. nu., caud. |
| spth | spinothalamic tr. |
| ts | tectospinal tract |
| V1 | V1 Cx |
| V2m | V2 (med.) Cx |
| V2v | V2 (vent.) Cx |
| vcs | vent. corticospinal tr. |
| VH | ventral horn |
| VLP (V3v) | ventlatpost. (V3v) Cx |
| vmf | vent. median fiss. |
| vsc | vent. spinocereb. tr. |

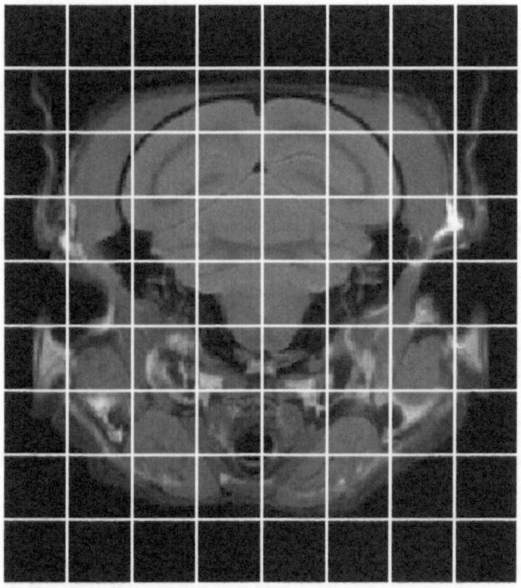

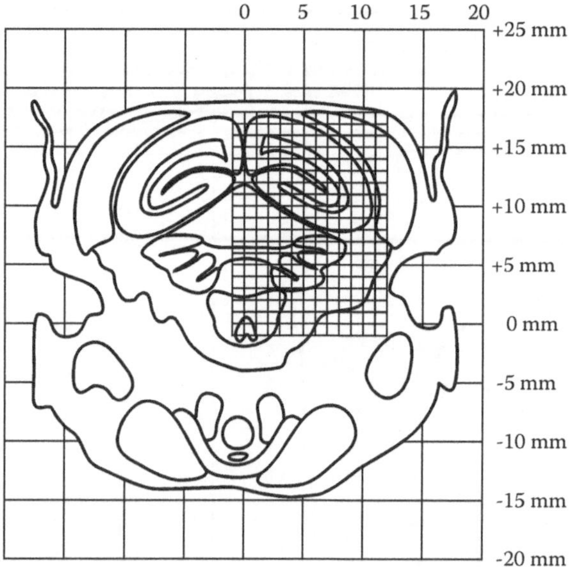


Interaural -05.50 mm

# Plate 50

Interaural -06.25 mm

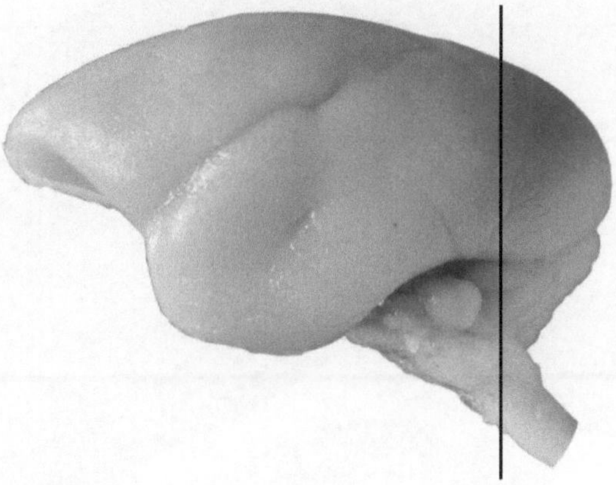

| | | | | | |
|---|---|---|---|---|---|
| 5Cb | cerebell. lobule 5 | gc | geniculocalcarine tr. | R8 | Rexed 8, spinal cord |
| 6Cb | cerebell. lobule 6 | gr | gracile fasciculus | R9 | Rexed 9, spinal cord |
| 7CbB | cerebell. lobule 7B | Gr | gracile nu. | rs | rubrospinal tract |
| 8CbA | cerebell. lobule 8A | IRt | intermed. retic. formn | sf | secondary fissure Cb |
| 8CbB | cerebell. lobule 8B | lcs | lat. corticospinal tr. | Sim | simplex lobule Cb |
| 9Cb | cerebell. lobule 9 | LSp | lat. spinal nu. | spth | spinothalamic tr. |
| ca | calcarine sulcus | lvs | lat. vestibulosp. tr. | ts | tectospinal tract |
| CC | central canal | mlf | med. longit. fasc. | V1 | V1 Cx |
| CeCv | central cervical nu. | PM | paramedian lobule Cb | V2l | V2 (lat.) Cx |
| Cop | copula | ppf | prepyramidal fiss. | V2m | V2 (med.) Cx |
| Crus1 | crus 1, ansiform lob. | prf | prim. cerebell. fiss. | V2v | V2 (vent.) Cx |
| Crus2 | crus 2, ansiform lob. | R1 | Rexed 1, spinal cord | vcs | vent. corticospinal tr. |
| cu | cuneate fasciculus | R2 | Rexed 2, spinal cord | VH | ventral horn |
| dlL | dorsolat. tr. Lissauer | R3 | Rexed 3, spinal cord | vsc | vent. spinocereb. tr. |
| dsc | dors. spinocerebell. tr | R4 | Rexed 4, spinal cord | | |

# Plate 50a

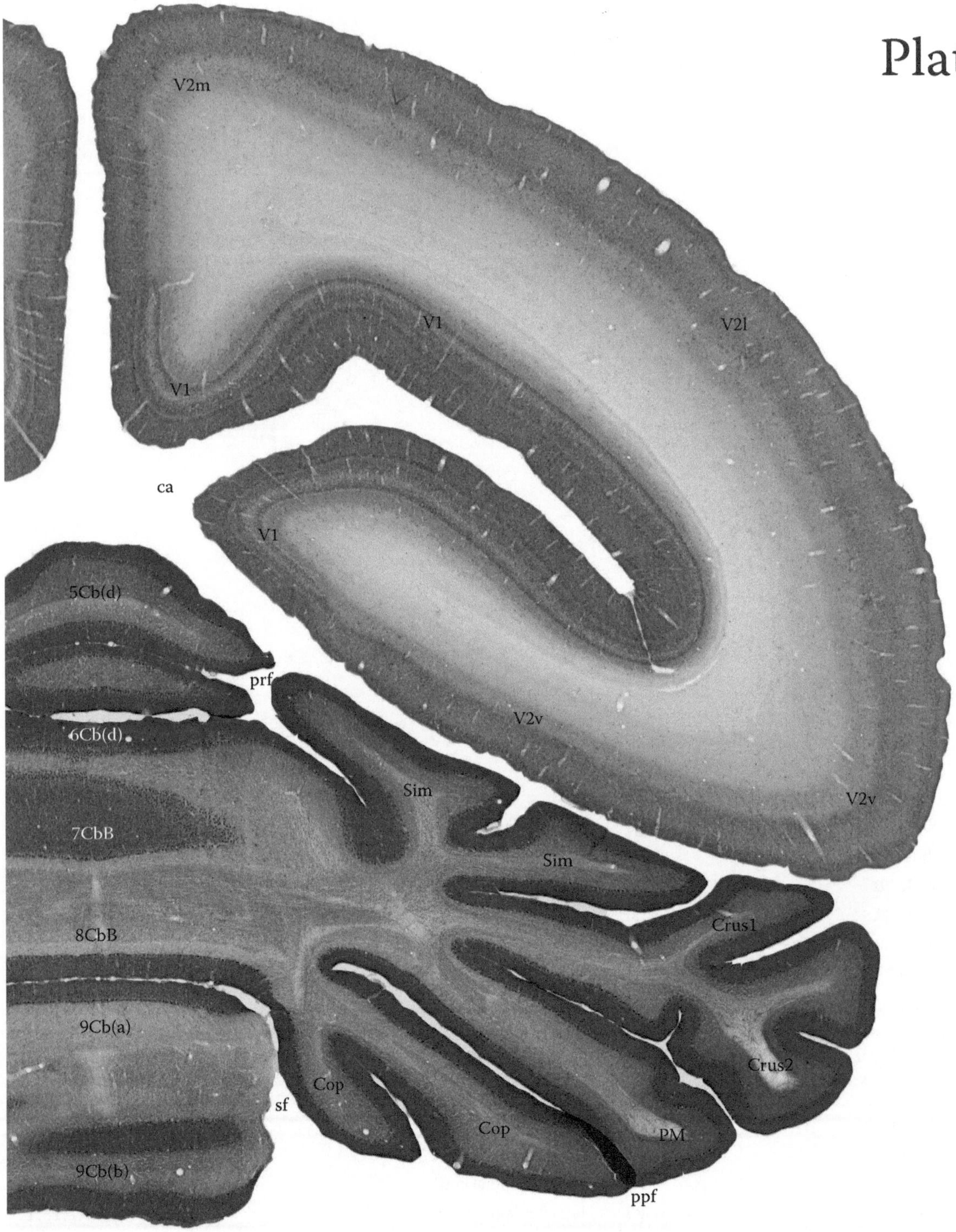


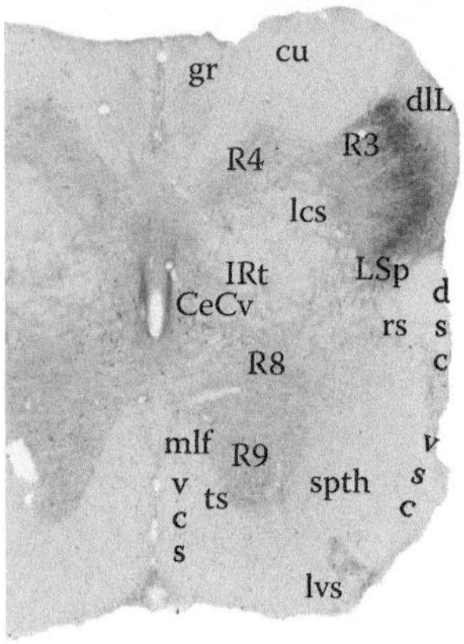


**Calbindin**
Interaural -05.95 mm

# Plate 50b

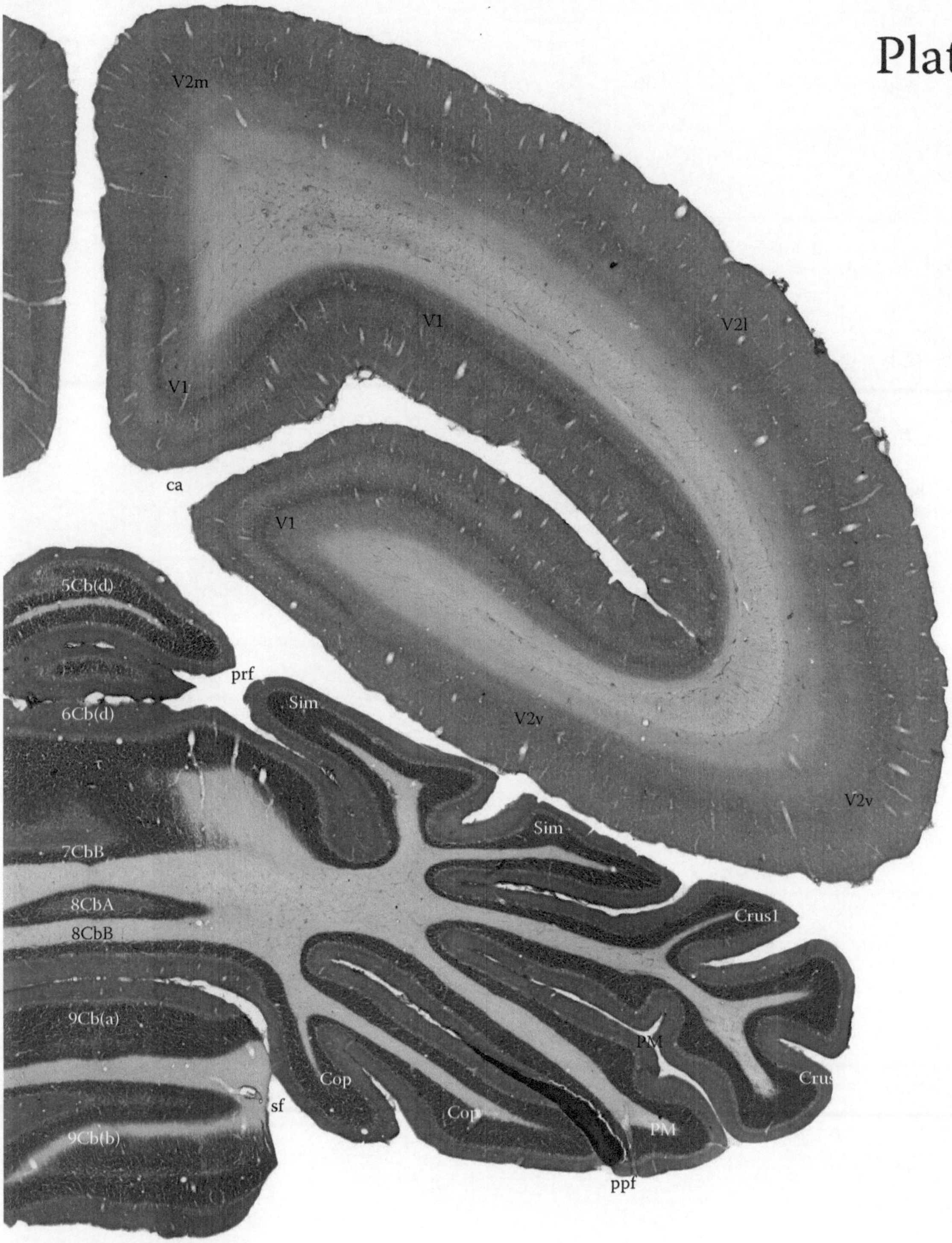


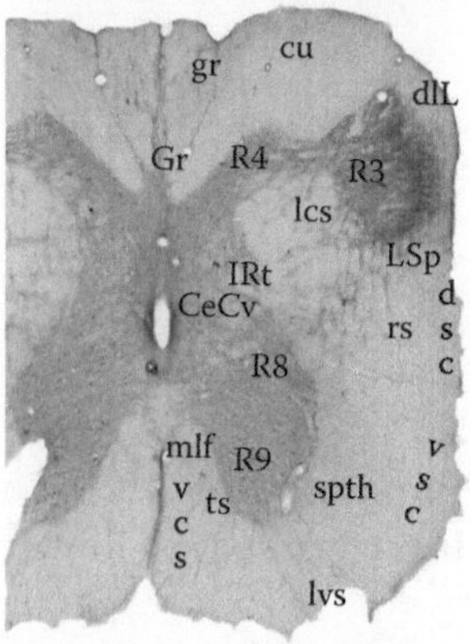


**Neurofilament**
Interaural -06.10 mm

# Plate 50c

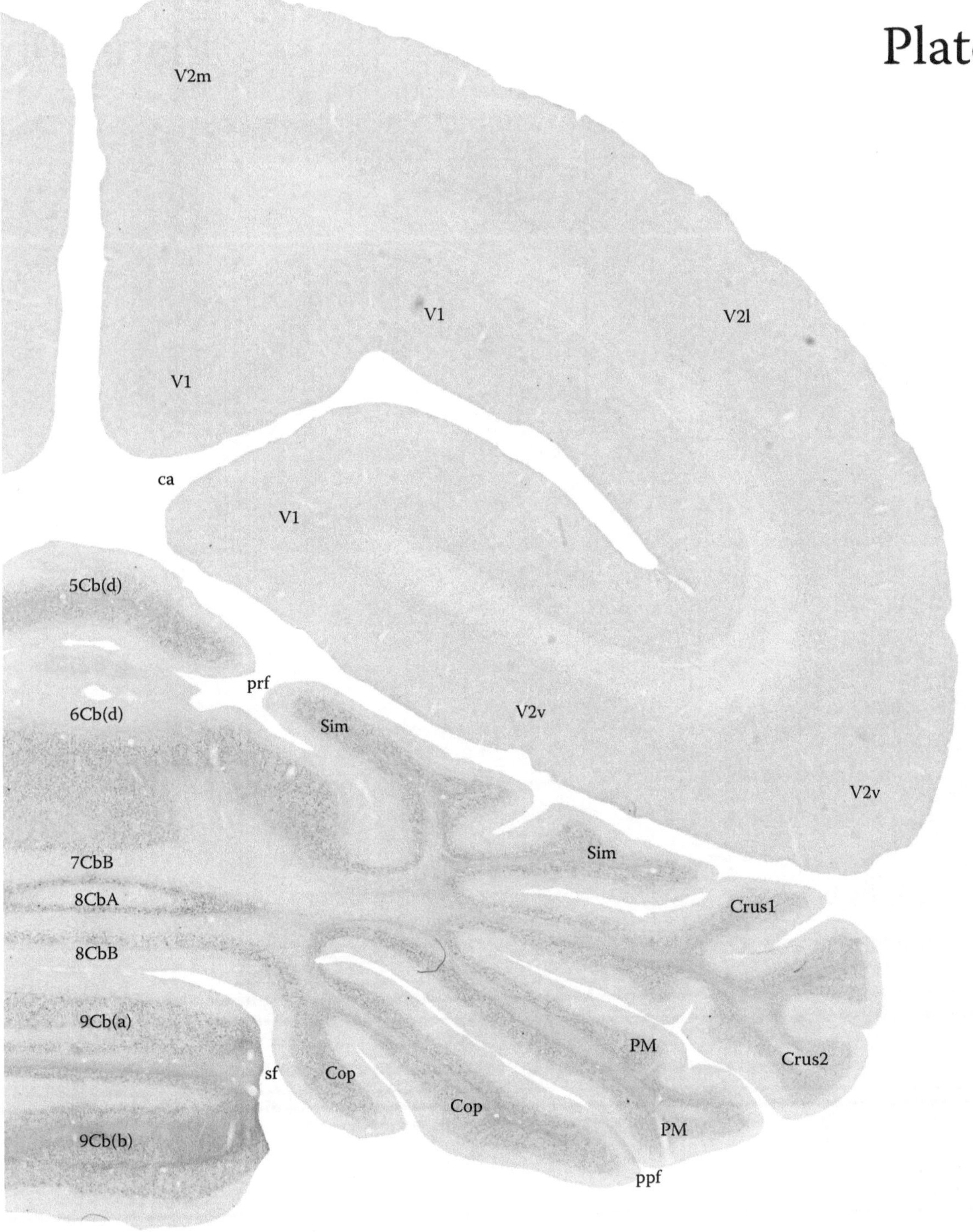


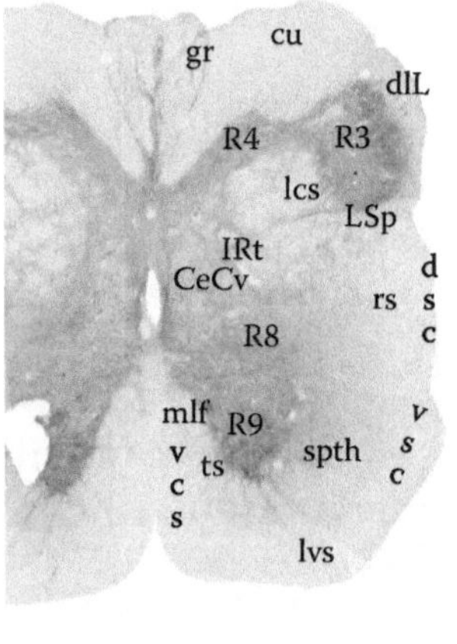


**Acetylcholinesterase**
Interaural -06.15 mm

# Plate 50d

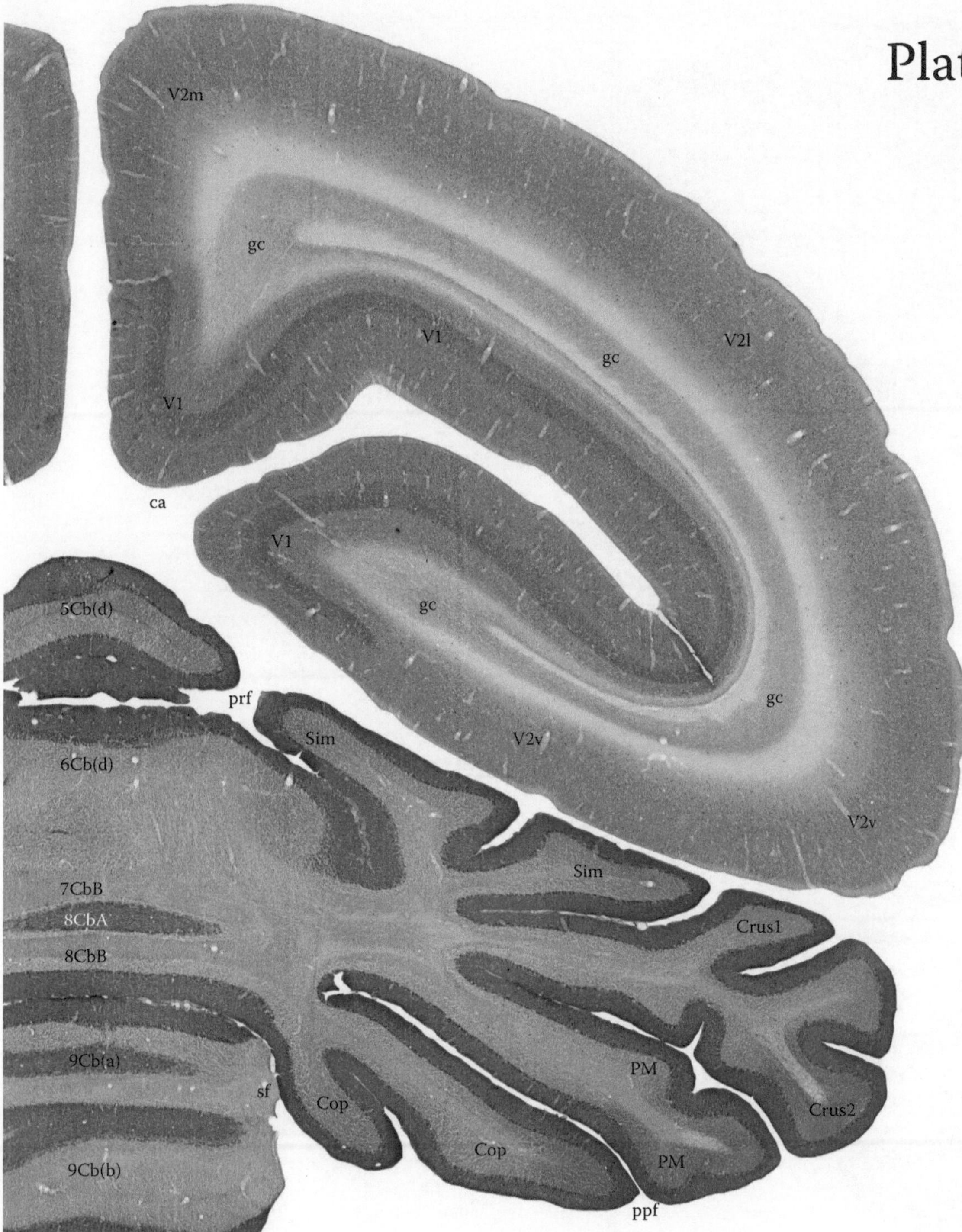


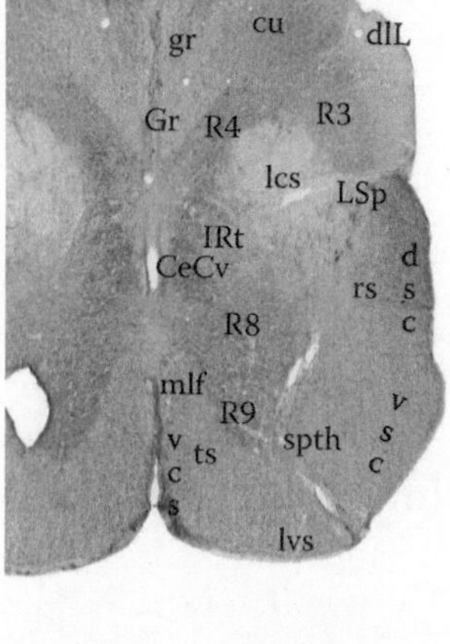


**Parvalbumin**
Interaural -06.20 mm

# Plate 50e

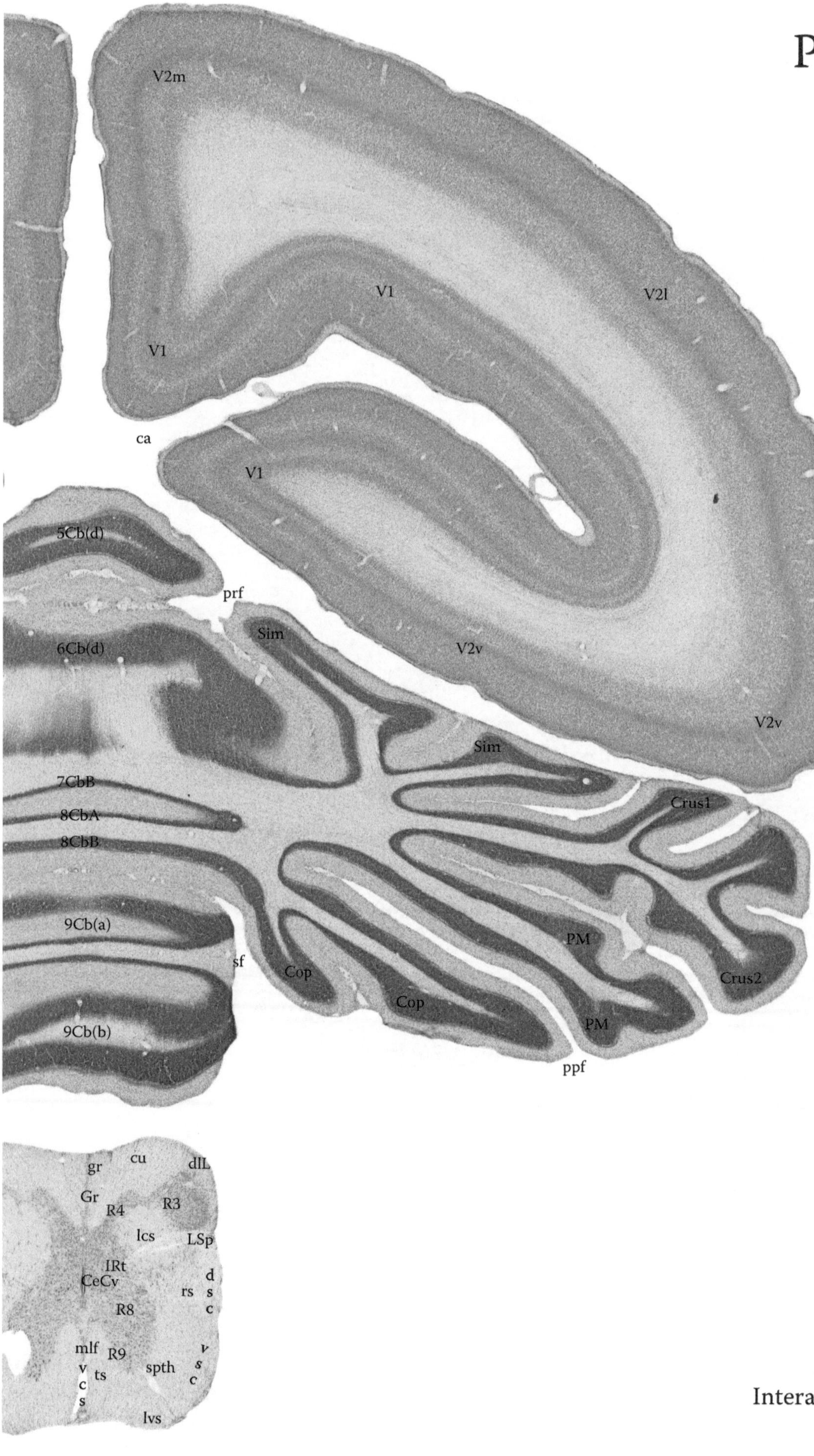


**Cresyl Violet**
Interaural -06.25 mm

# Plate 50

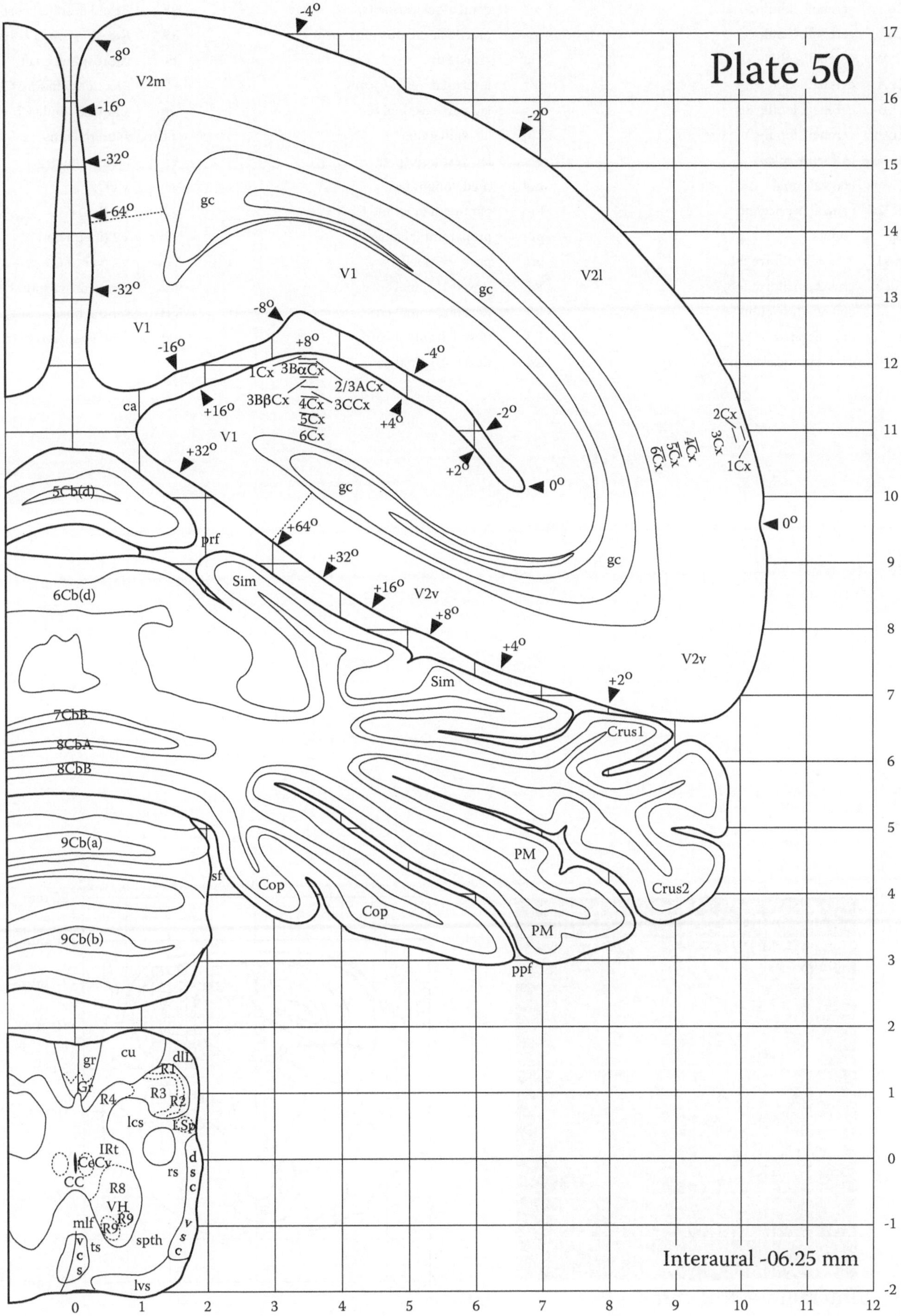
V2m
V1
gc
V2l
1Cx
3BαCx
2/3ACx
3BβCx
4Cx
3CCx
5Cx
6Cx
ca
prf
5Cb(d)
6Cb(d)
Sim
V2v
Crus1
Crus2
7CbB
8CbA
8CbB
9Cb(a)
9Cb(b)
sf
Cop
PM
ppf
2Cx
3Cx
gr
cu
dlL
R1
R2
R3
R4
Gr
lcs
ESp
IRt
CeCv
CC
rs
dsc
R8
VH
R9
mlf
ts
vcs
spth
vsc
lvs
Interaural -06.25 mm

| | |
|---|---|
| 5Cb | cerebell. lobule 5 |
| 6Cb | cerebell. lobule 6 |
| 7CbB | cerebell. lobule 7B |
| 8CbA | cerebell. lobule 8A |
| 8CbB | cerebell. lobule 8B |
| 9Cb | cerebell. lobule 9 |
| ca | calcarine sulcus |
| CC | central canal |
| CeCv | central cervical nu. |
| Cop | copula |
| Crus1 | crus 1, ansiform lob. |
| Crus2 | crus 2, ansiform lob. |
| cu | cuneate fasciculus |
| dlL | dorsolat. tr. Lissauer |
| dsc | dors. spinocerebell. tr |
| gc | geniculocalcarine tr. |
| gr | gracile fasciculus |
| Gr | gracile nu. |
| IRt | intermed. retic. formn |
| lcs | lat. corticospinal tr. |
| LSp | lat. spinal nu. |
| lvs | lat. vestibulosp. tr. |
| mlf | med. longit. fasc. |
| PM | paramedian lobule Cb |
| ppf | prepyramidal fiss. |
| prf | prim. cerebell. fiss. |
| R1 | Rexed 1, spinal cord |
| R2 | Rexed 2, spinal cord |
| R3 | Rexed 3, spinal cord |
| R4 | Rexed 4, spinal cord |
| R8 | Rexed 8, spinal cord |
| R9 | Rexed 9, spinal cord |
| rs | rubrospinal tract |
| sf | secondary fissure Cb |
| Sim | simplex lobule Cb |
| spth | spinothalamic tr. |
| ts | tectospinal tract |
| V1 | V1 Cx |
| V2l | V2 (lat.) Cx |
| V2m | V2 (med.) Cx |
| V2v | V2 (vent.) Cx |
| vcs | vent. corticospinal tr. |
| VH | ventral horn |
| vsc | vent. spinocereb. tr. |

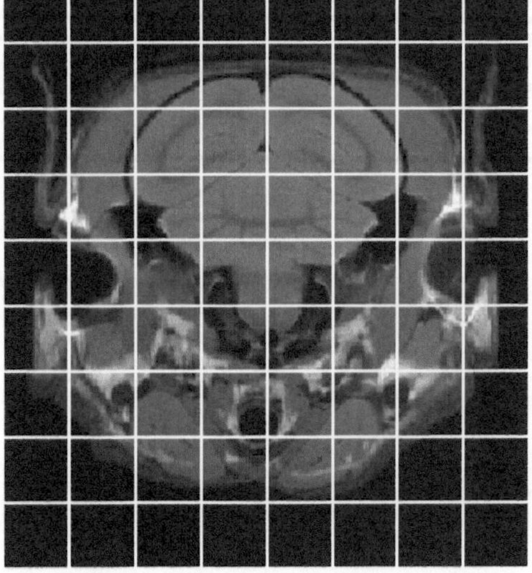

Interaural -06.45 mm

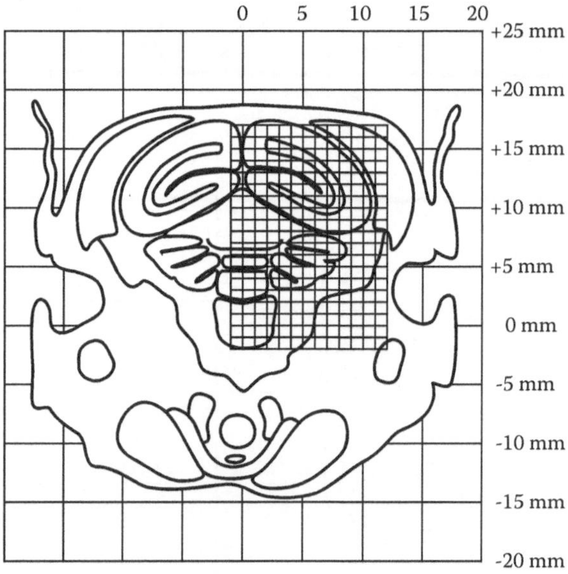

# Plate 51

Interaural -06.75 mm

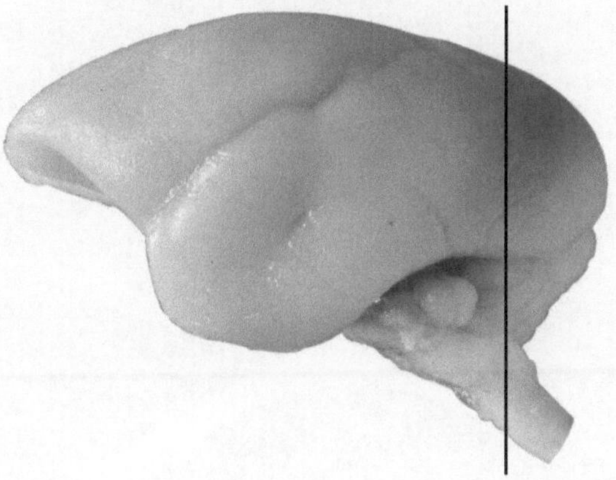

| | | | | | |
|---|---|---|---|---|---|
| 5Cb | cerebell. lobule 5 | gr | gracile fasciculus | R8 | Rexed 8, spinal cord |
| 6Cb | cerebell. lobule 6 | IB | intern. basal nu. | R9 | Rexed 9, spinal cord |
| 7Cb | cerebell. lobule 7 | lcs | lat. corticospinal tr. | rs | rubrospinal tract |
| 8CbA | cerebell. lobule 8A | lfu | lateral funiculus | sf | secondary fissure Cb |
| 8CbB | cerebell. lobule 8B | LSp | lat. spinal nu. | Sim | simplex lobule Cb |
| 9Cb | cerebell. lobule 9 | lvs | lat. vestibulosp. tr. | spth | spinothalamic tr. |
| ca | calcarine sulcus | mlf | med. longit. fasc. | ts | tectospinal tract |
| CC | central canal | PM | paramedian lobule Cb | V1 | V1 Cx |
| CeCv | central cervical nu. | ppf | prepyramidal fiss. | V2l | V2 (lat.) Cx |
| Cop | copula | prf | prim. cerebell. fiss. | V2m | V2 (med.) Cx |
| Crus1 | crus 1, ansiform lob. | R1 | Rexed 1, spinal cord | V2v | V2 (vent.) Cx |
| Crus2 | crus 2, ansiform lob. | R2 | Rexed 2, spinal cord | vcs | vent. corticospinal tr. |
| cu | cuneate fasciculus | R3 | Rexed 3, spinal cord | vfu | ventral funiculus |
| dlL | dorsolat. tr. Lissauer | R4 | Rexed 4, spinal cord | VH | ventral horn |
| dsc | dors. spinocerebell. tr | R5 | Rexed 5, spinal cord | vmf | vent. median fiss. |
| gc | geniculocalcarine tr. | R7 | Rexed 7, spinal cord | vsc | vent. spinocereb. tr. |

# Plate 51a

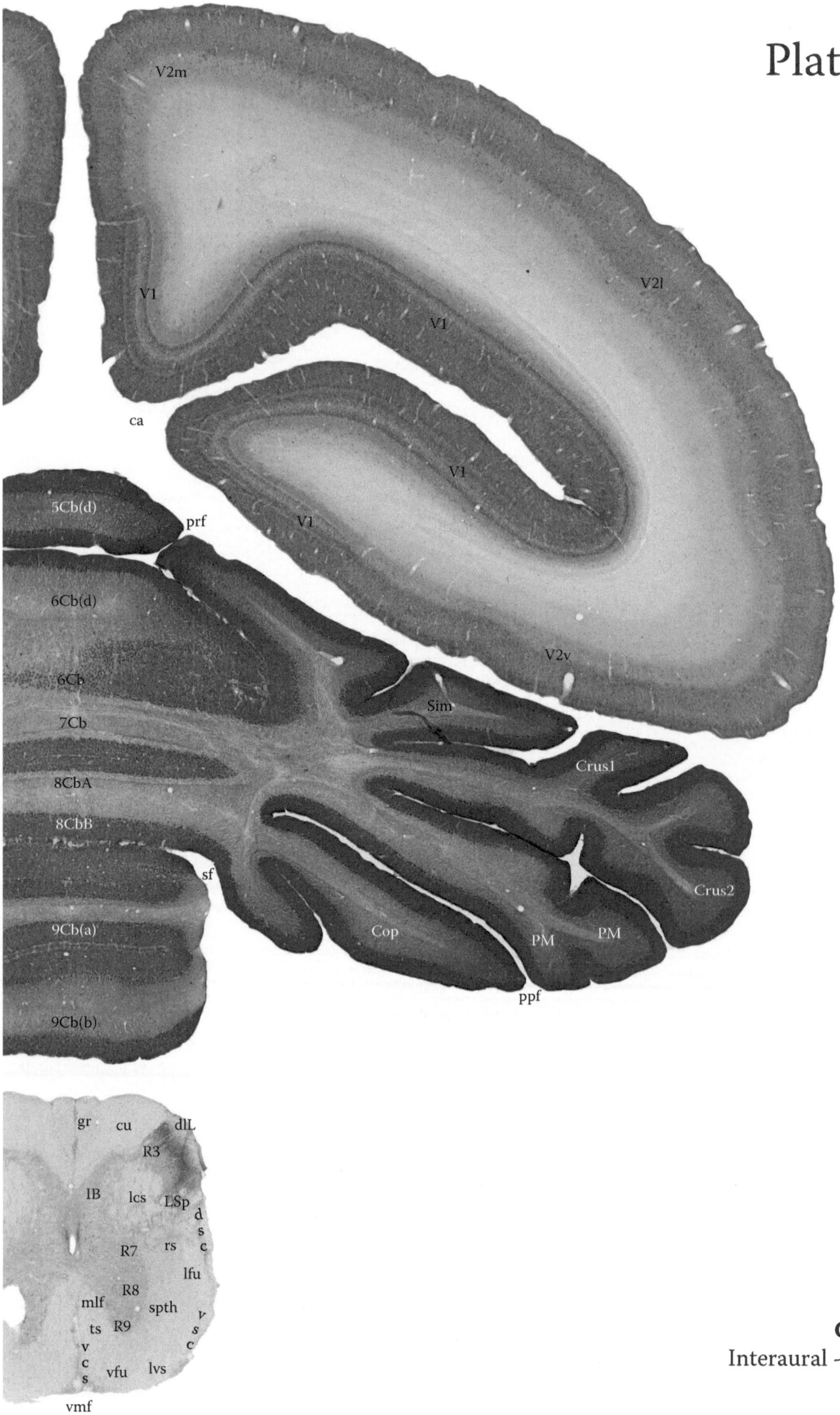


**Calbindin**
Interaural -06.45 mm

# Plate 51b

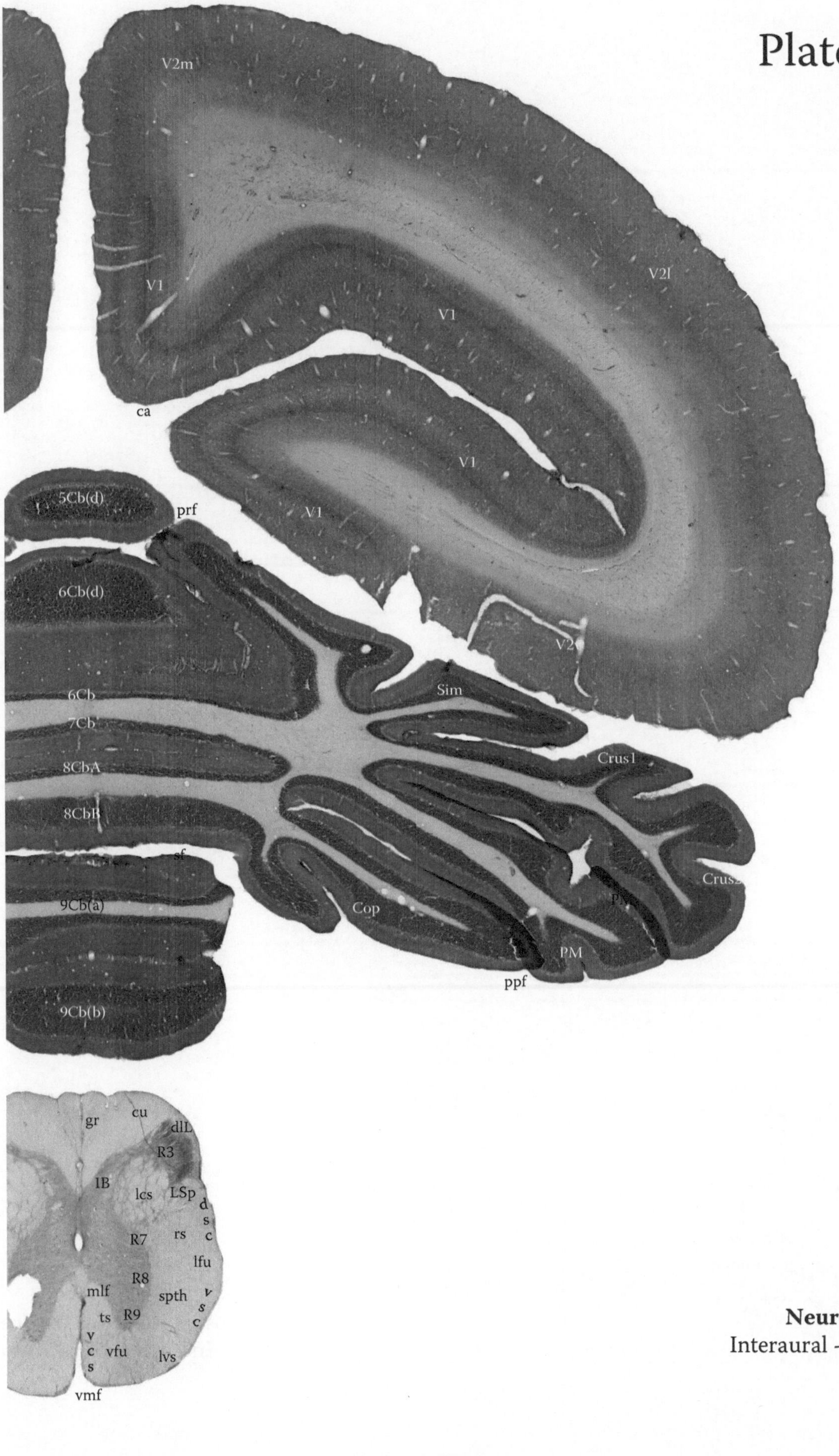


**Neurofilament**
Interaural -06.60 mm

# Plate 51c

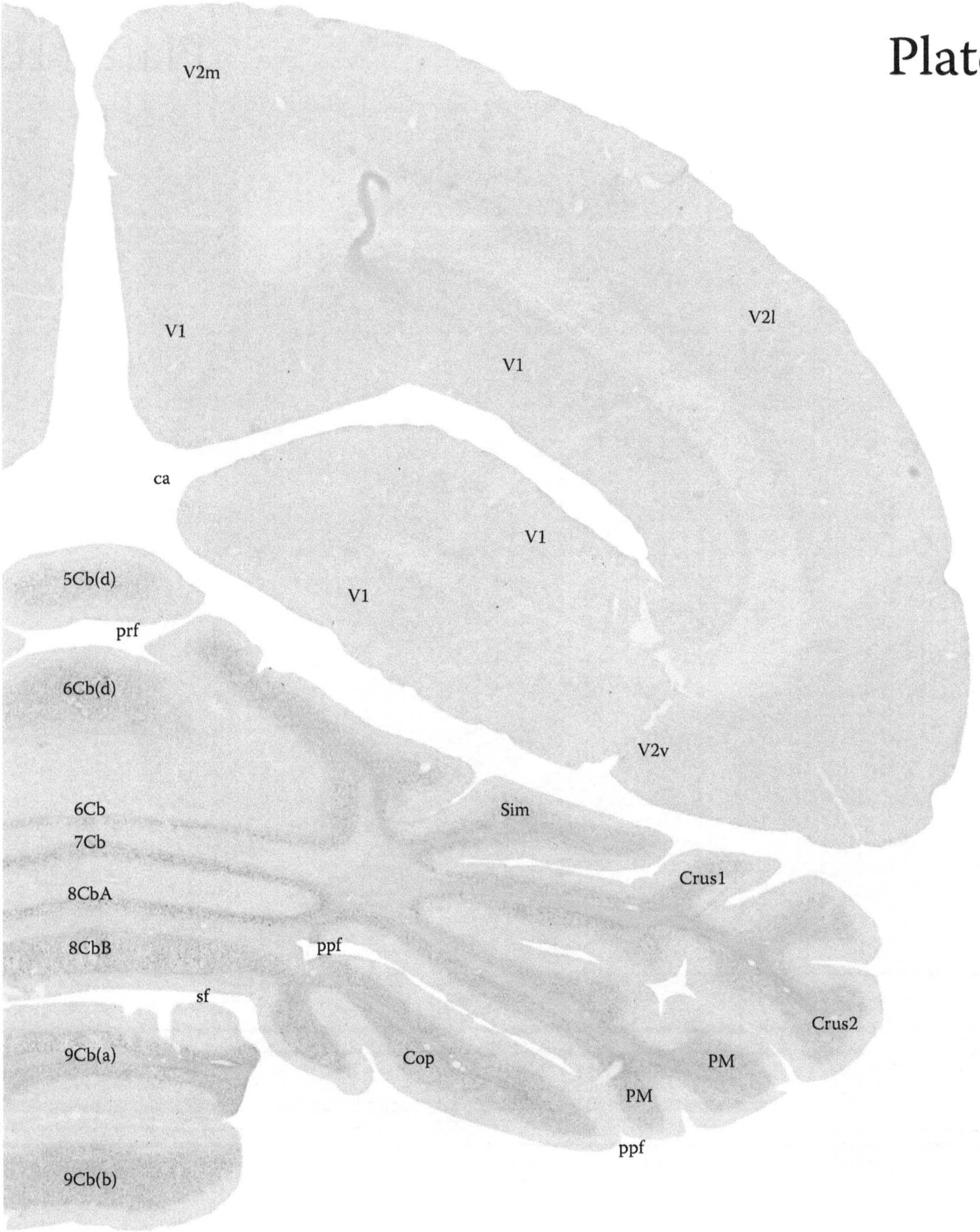


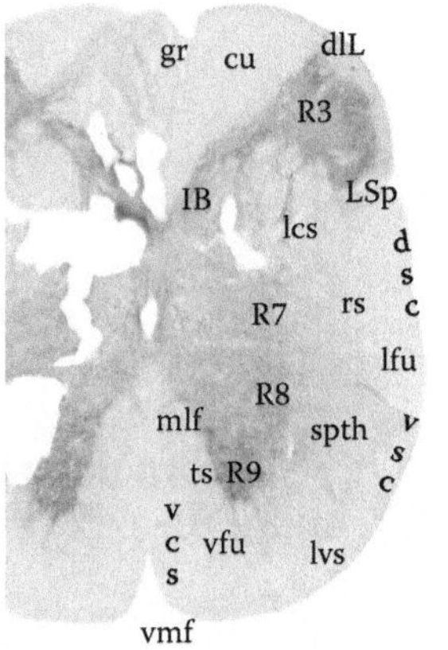


**Acetylcholinesterase**
Interaural -06.65 mm

# Plate 51d

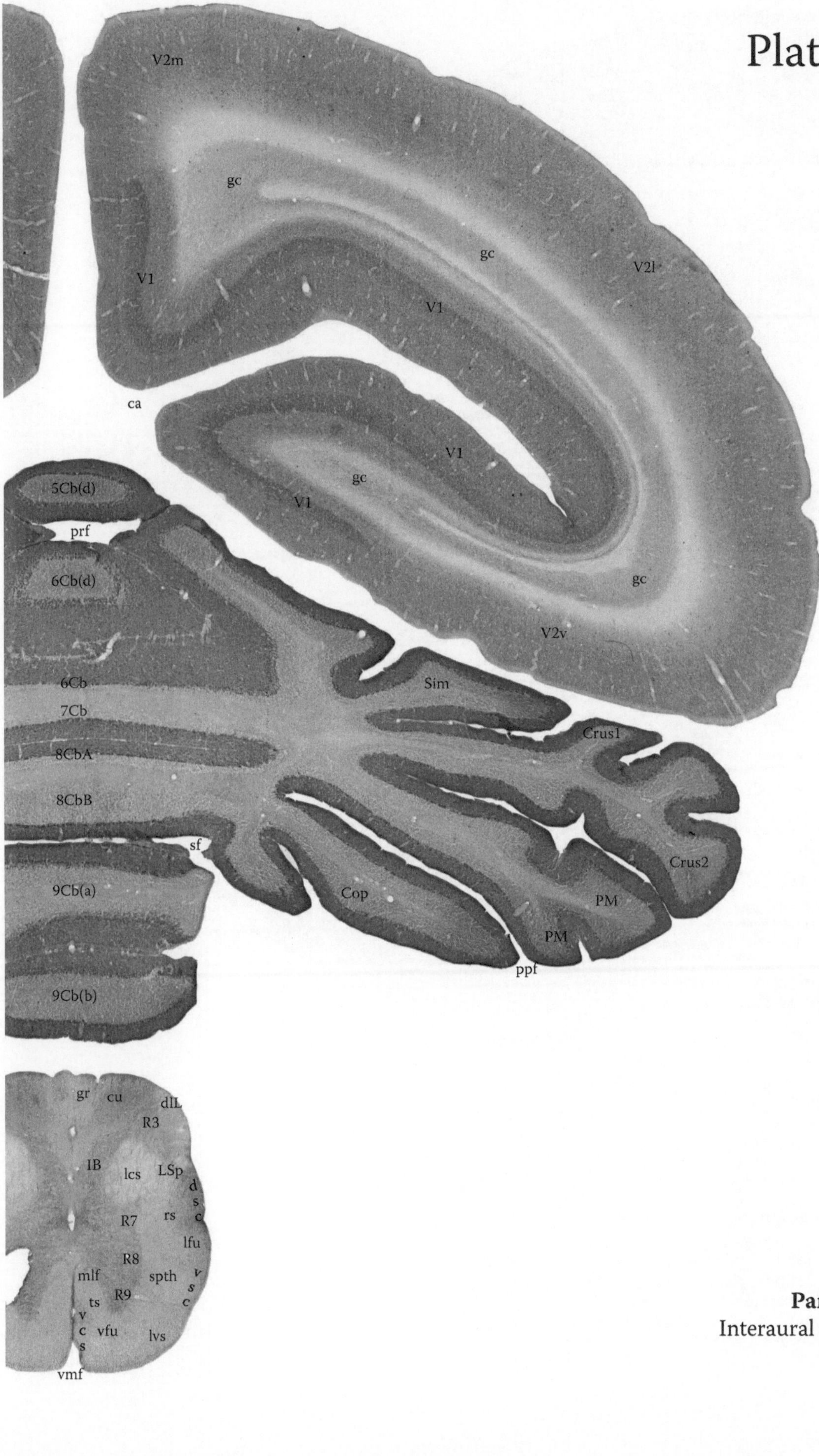


**Parvalbumin**
Interaural -06.70 mm

# Plate 51e

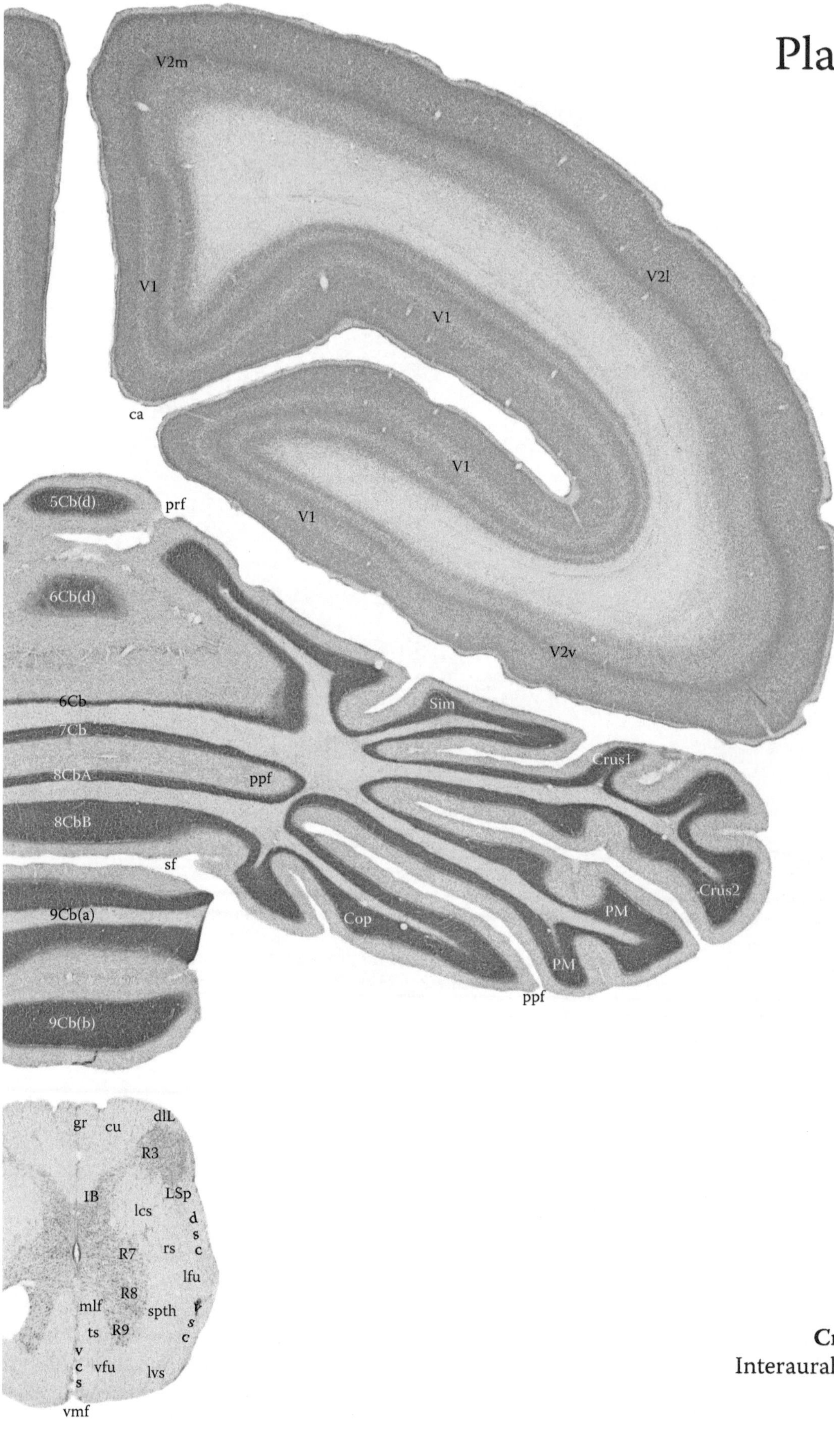


**Cresyl Violet**
Interaural -06.75 mm

# Plate 51

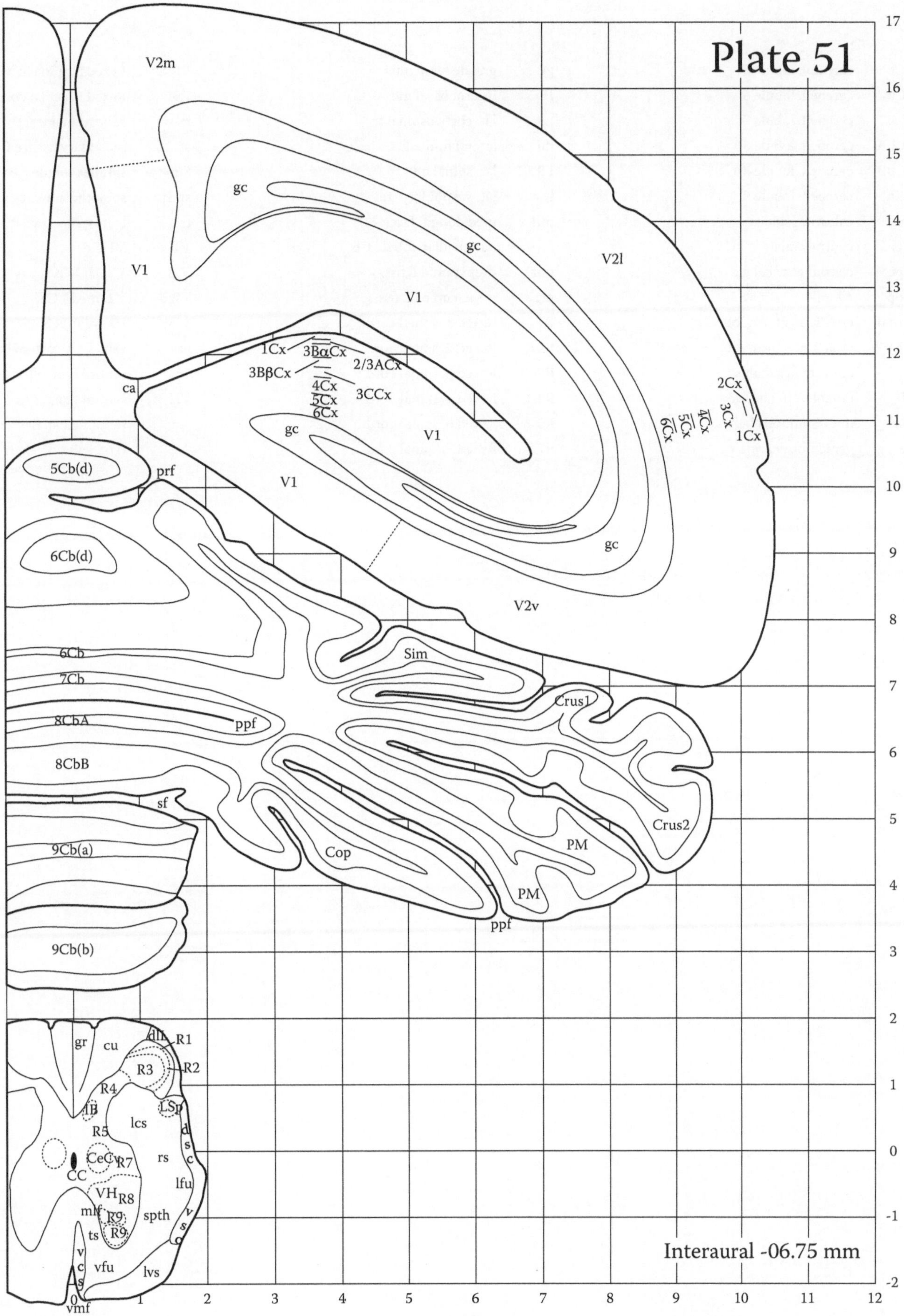
V2m
V1
gc
V2l
1Cx
3BαCx
2/3ACx
3BβCx
4Cx
5Cx
6Cx
3CCx
ca
2Cx
3Cx
4Cx
5Cx
6Cx
1Cx
5Cb(d)
prf
6Cb(d)
V2v
6Cb
7Cb
8CbA
ppf
8CbB
Sim
Crus1
Crus2
sf
9Cb(a)
9Cb(b)
Cop
PM
gr
cu
dlL
R1
R2
R3
R4
IB
LSp
lcs
R5
dsc
CeCv
R7
CC
rs
lfu
VH
R8
mlf
R9
spth
vsc
ts
vcs
vfu
lvs
vmf
Interaural -06.75 mm
17
16
15
14
13
12
11
10
9
8
7
6
5
4
3
2
1
0
-1
-2
0
1
2
3
4
5
6
7
8
9
10
11
12

| | |
|---|---|
| 5Cb | cerebell. lobule 5 |
| 6Cb | cerebell. lobule 6 |
| 7Cb | cerebell. lobule 7 |
| 8CbA | cerebell. lobule 8A |
| 8CbB | cerebell. lobule 8B |
| 9Cb | cerebell. lobule 9 |
| ca | calcarine sulcus |
| CC | central canal |
| CeCv | central cervical nu. |
| Cop | copula |
| Crus1 | crus 1, ansiform lob. |
| Crus2 | crus 2, ansiform lob. |
| cu | cuneate fasciculus |
| dlL | dorsolat. tr. Lissauer |
| dsc | dors. spinocerebell. tr |
| gc | geniculocalcarine tr. |
| gr | gracile fasciculus |
| IB | intern. basal nu. |
| lcs | lat. corticospinal tr. |
| lfu | lateral funiculus |
| LSp | lat. spinal nu. |
| lvs | lat. vestibulosp. tr. |
| mlf | med. longit. fasc. |
| PM | paramedian lobule Cb |
| ppf | prepyramidal fiss. |
| prf | prim. cerebell. fiss. |
| R1 | Rexed 1, spinal cord |
| R2 | Rexed 2, spinal cord |
| R3 | Rexed 3, spinal cord |
| R4 | Rexed 4, spinal cord |
| R5 | Rexed 5, spinal cord |
| R7 | Rexed 7, spinal cord |
| R8 | Rexed 8, spinal cord |
| R9 | Rexed 9, spinal cord |
| rs | rubrospinal tract |
| sf | secondary fissure Cb |
| Sim | simplex lobule Cb |
| spth | spinothalamic tr. |
| ts | tectospinal tract |
| V1 | V1 Cx |
| V2l | V2 (lat.) Cx |
| V2m | V2 (med.) Cx |
| V2v | V2 (vent.) Cx |
| vcs | vent. corticospinal tr. |
| vfu | ventral funiculus |
| VH | ventral horn |
| vmf | vent. median fiss. |
| vsc | vent. spinocereb. tr. |

# Plate 52

Interaural -07.25 mm

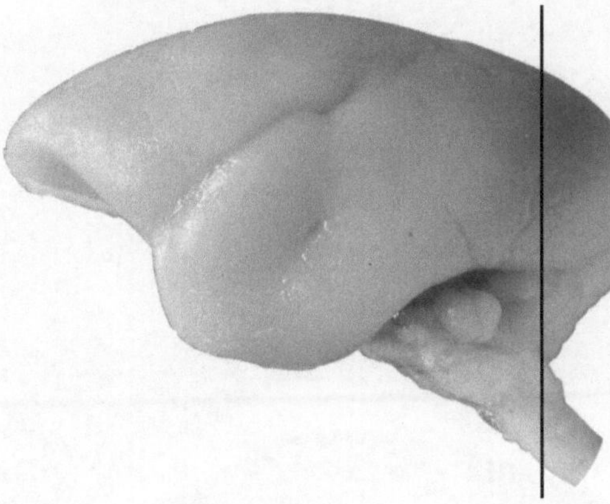

| | |
|---|---|
| 6Cb(a-c) | cerebell. lobule 6(a-c) |
| 6Cb(e) | cerebell. lobule 6(e) |
| 7CbB | cerebell. lobule 7B |
| 8Cb | cerebell. lobule 8 |
| 9Cb(a) | cerebell. lobule 9(a) |
| 9Cb(b) | cerebell. lobule 9(b) |
| ca | calcarine sulcus |
| CC | central canal |
| CeCv | central cervical nu. |
| Cop | copula |
| Crus1 | crus 1, ansiform lob. |
| Crus2 | crus 2, ansiform lob. |
| cu | cuneate fasciculus |
| dlL | dorsolat. tr. Lissauer |
| dsc | dors. spinocerebell. tr |
| gc | geniculocalcarine tr. |
| gr | gracile fasciculus |
| IB | intern. basal nu. |
| lcs | lat. corticospinal tr. |
| LSp | lat. spinal nu. |
| lvs | lat. vestibulosp. tr. |
| mlf | med. longit. fasc. |
| PM | paramedian lobule Cb |
| ppf | prepyramidal fiss. |
| R1 | Rexed 1, spinal cord |
| R2 | Rexed 2, spinal cord |
| R3 | Rexed 3, spinal cord |
| R4 | Rexed 4, spinal cord |
| R5 | Rexed 5, spinal cord |
| R7 | Rexed 7, spinal cord |
| R8 | Rexed 8, spinal cord |
| R9 | Rexed 9, spinal cord |
| rs | rubrospinal tract |
| sf | secondary fissure Cb |
| Sim | simplex lobule Cb |
| spth | spinothalamic tr. |
| ts | tectospinal tract |
| V1 | V1 Cx |
| V2l | V2 (lat.) Cx |
| V2m | V2 (med.) Cx |
| vcs | vent. corticospinal tr. |
| VH | ventral horn |
| vmf | vent. median fiss. |
| vsc | vent. spinocereb. tr. |

# Plate 52a

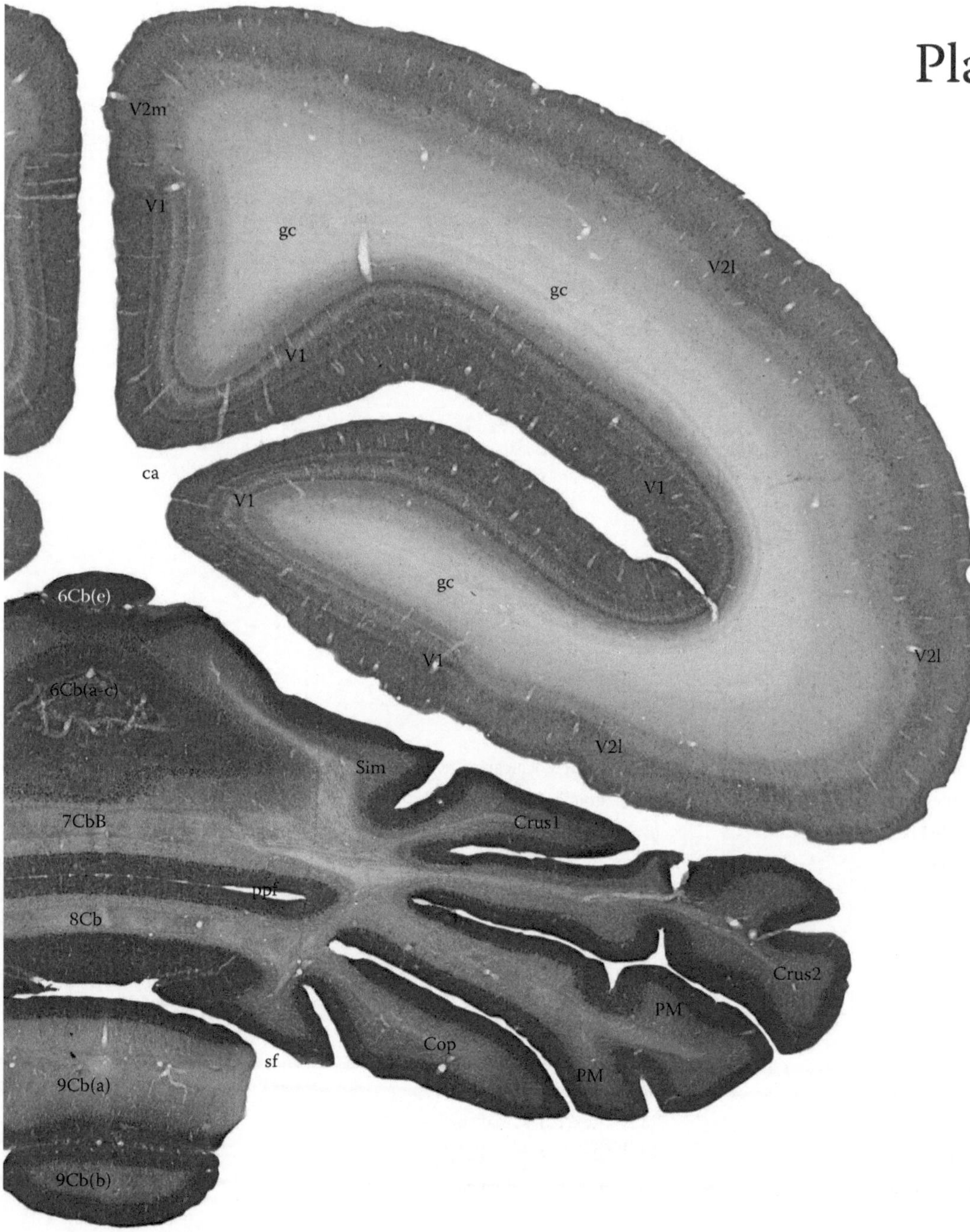


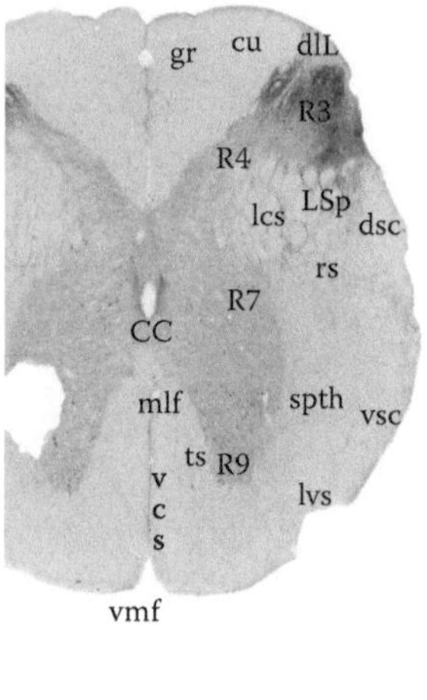


**Calbindin**
Interaural -06.95 mm

# Plate 52b

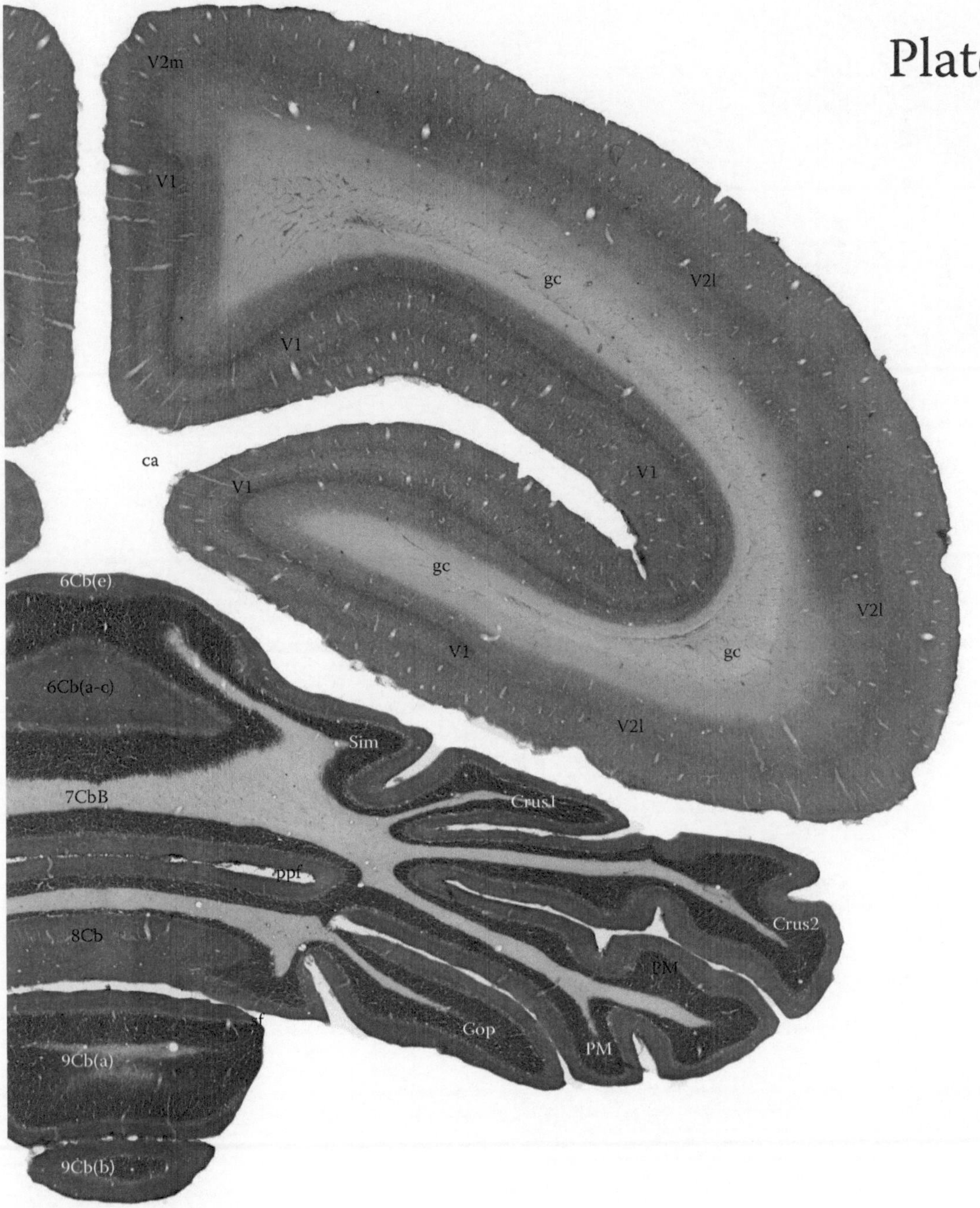


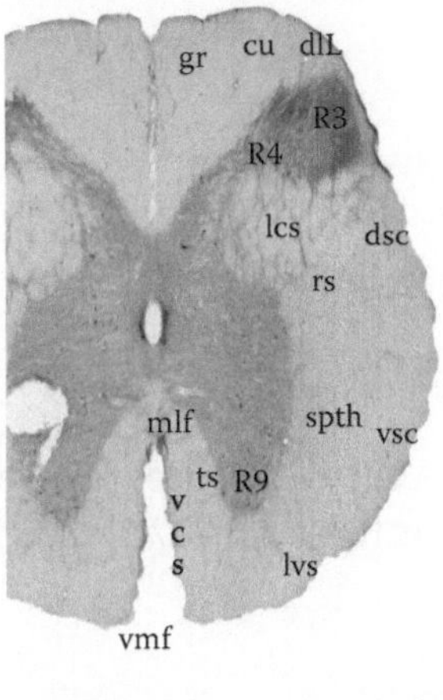


**Neurofilament**
Interaural -07.10 mm

# Plate 52c

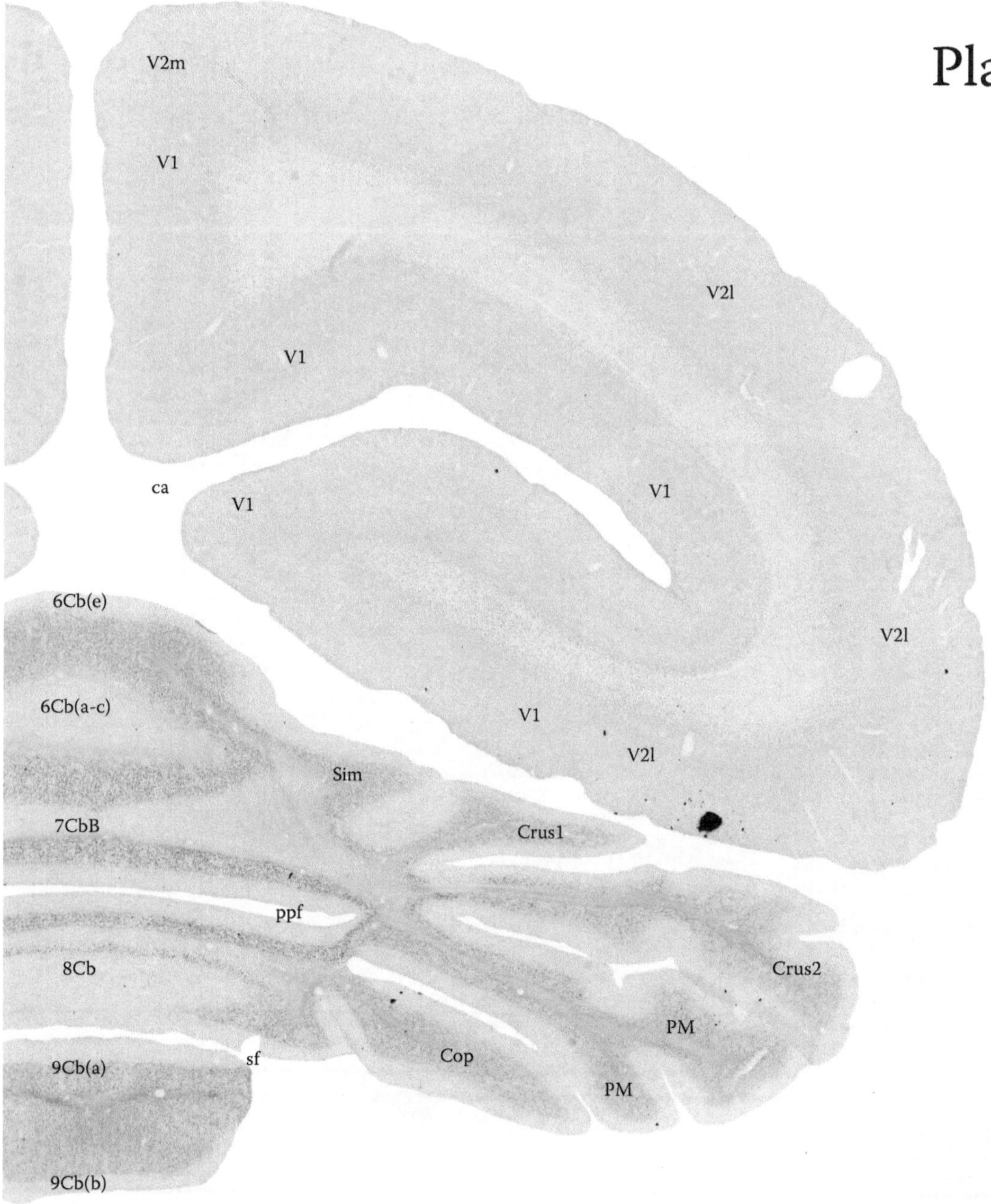


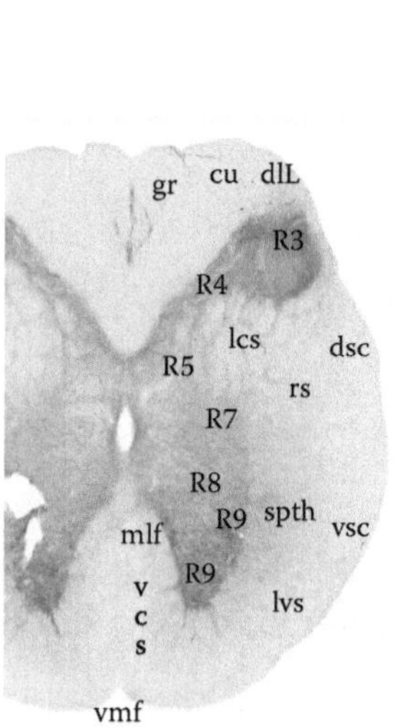


**Acetylcholinesterase**
Interaural -07.15 mm

# Plate 52d

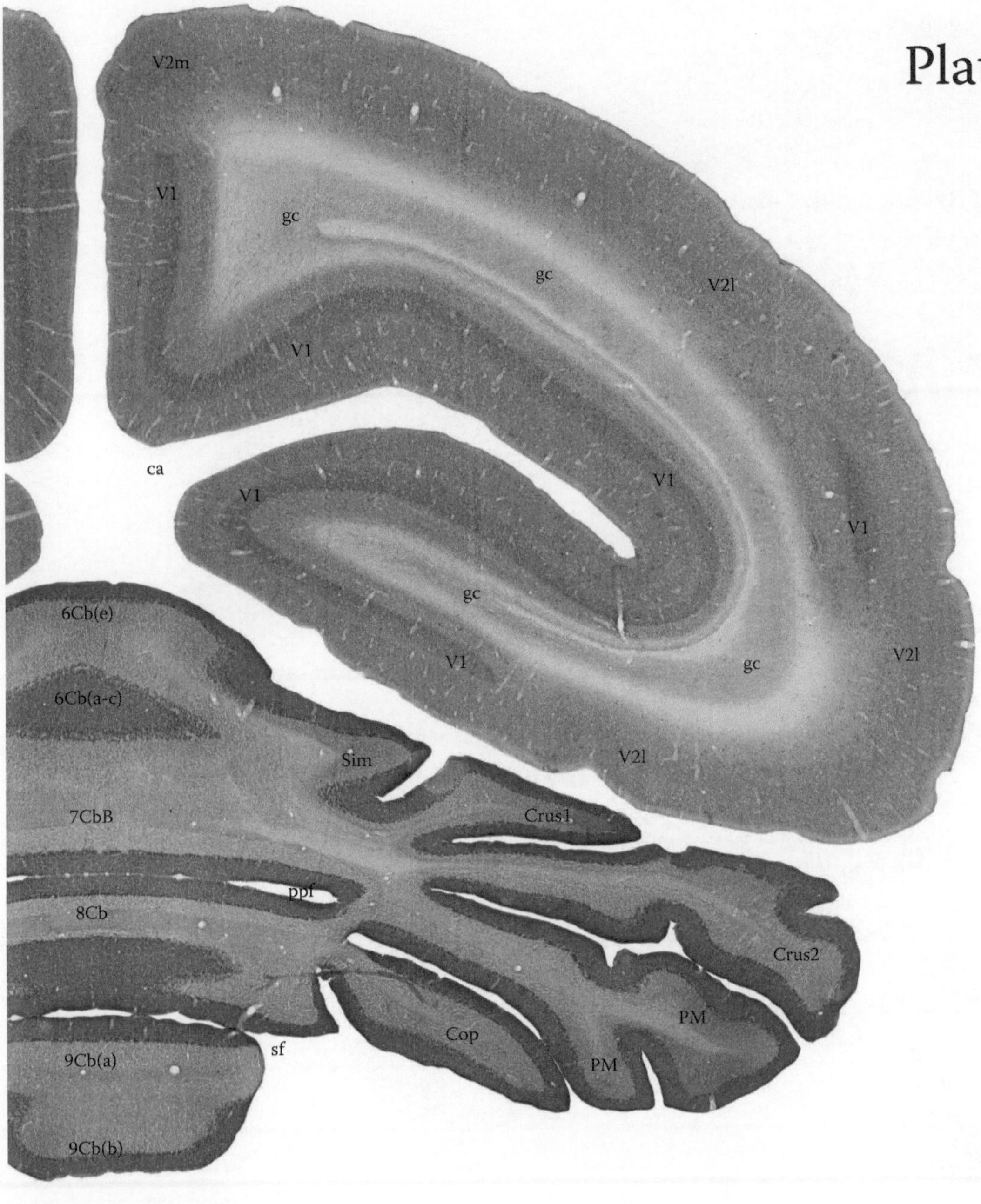


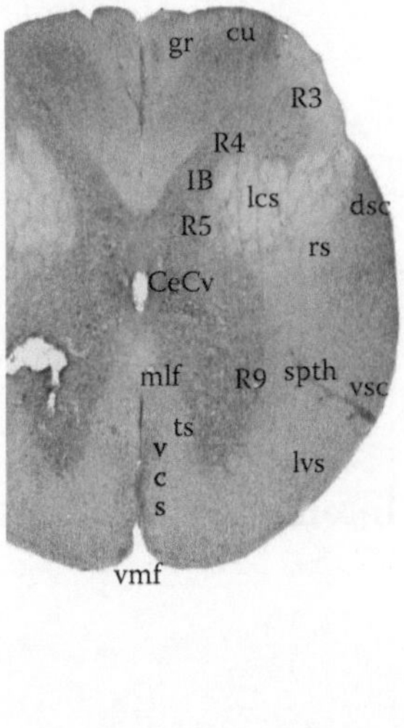


**Parvalbumin**
Interaural -07.20 mm

# Plate 52e

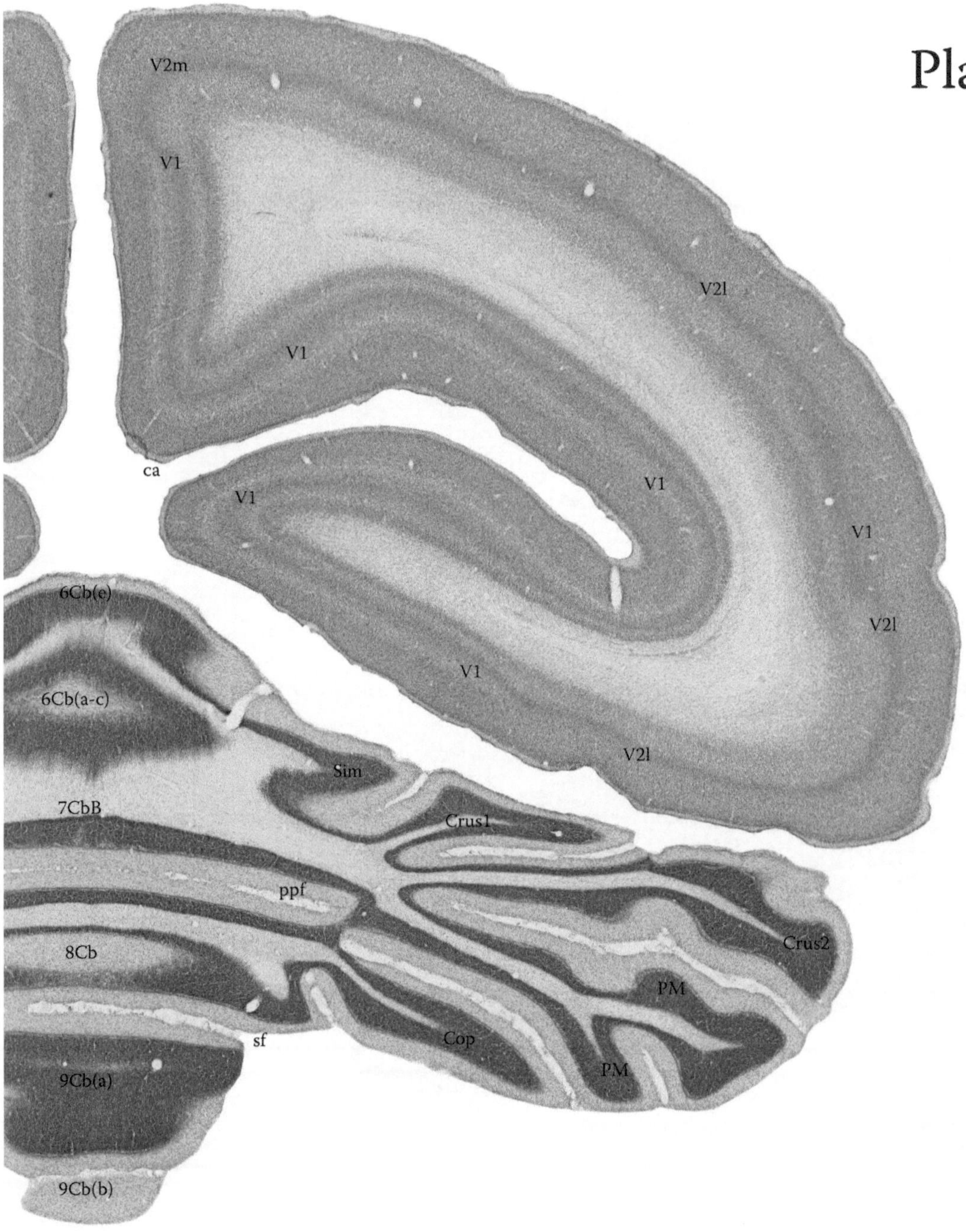


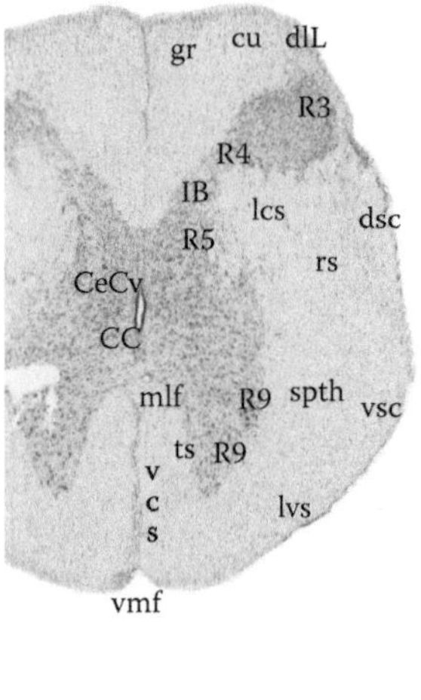


**Cresyl Violet**
Interaural -07.25 mm

# Plate 52

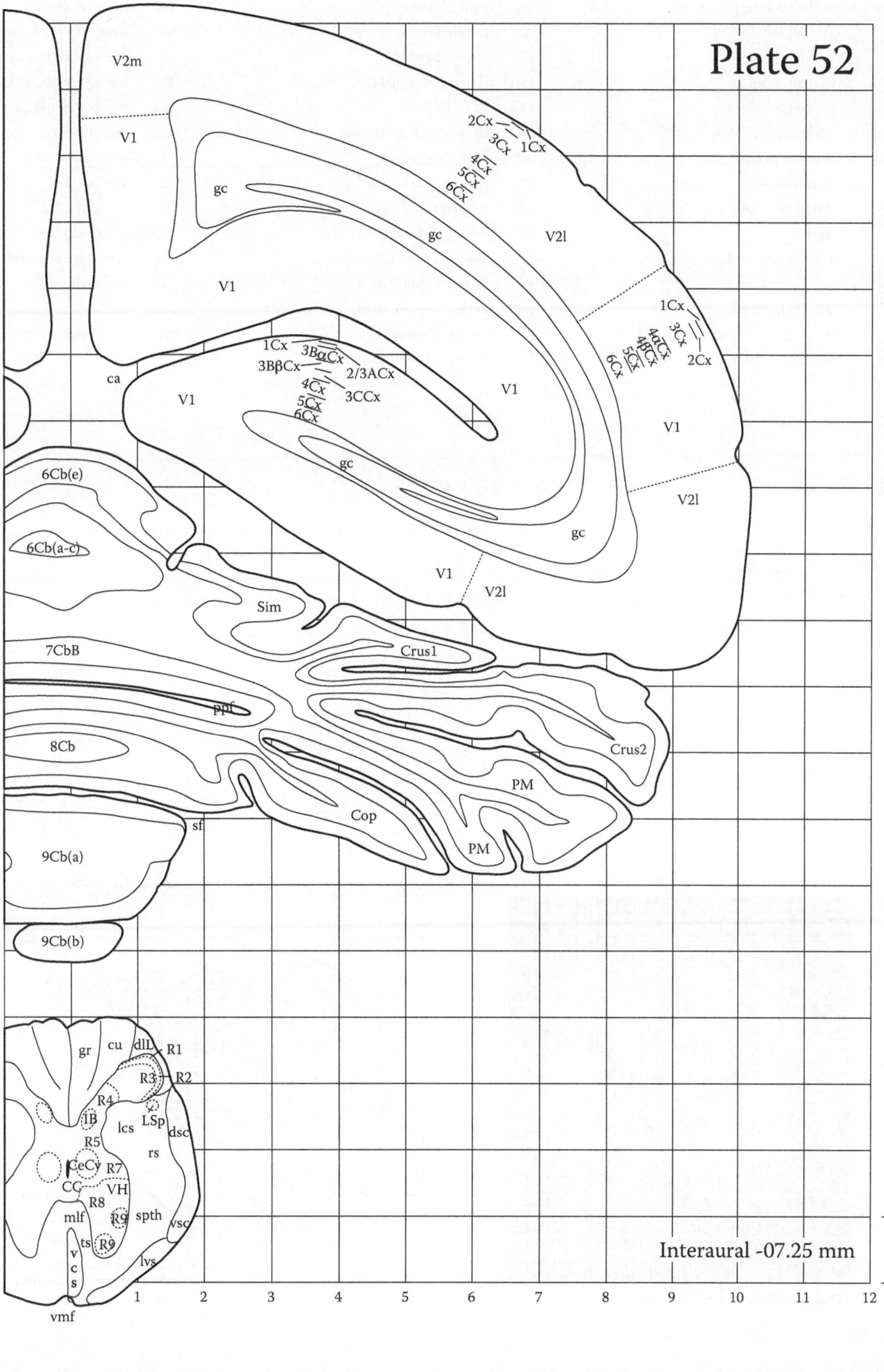


Interaural -07.25 mm

| | |
|---|---|
| 6Cb(a-c) | cerebell. lobule 6(a-c) |
| 6Cb(e) | cerebell. lobule 6(e) |
| 7CbB | cerebell. lobule 7B |
| 8Cb | cerebell. lobule 8 |
| 9Cb(a) | cerebell. lobule 9(a) |
| 9Cb(b) | cerebell. lobule 9(b) |
| ca | calcarine sulcus |
| CC | central canal |
| CeCv | central cervical nu. |
| Cop | copula |
| Crus1 | crus 1, ansiform lob. |
| Crus2 | crus 2, ansiform lob. |
| cu | cuneate fasciculus |
| dlL | dorsolat. tr. Lissauer |
| dsc | dors. spinocerebell. tr |
| gc | geniculocalcarine tr. |
| gr | gracile fasciculus |
| IB | intern. basal nu. |
| lcs | lat. corticospinal tr. |
| LSp | lat. spinal nu. |
| lvs | lat. vestibulosp. tr. |
| mlf | med. longit. fasc. |
| PM | paramedian lobule Cb |
| ppf | prepyramidal fiss. |
| R1 | Rexed 1, spinal cord |
| R2 | Rexed 2, spinal cord |
| R3 | Rexed 3, spinal cord |
| R4 | Rexed 4, spinal cord |
| R5 | Rexed 5, spinal cord |
| R7 | Rexed 7, spinal cord |
| R8 | Rexed 8, spinal cord |
| R9 | Rexed 9, spinal cord |
| rs | rubrospinal tract |
| sf | secondary fissure Cb |
| Sim | simplex lobule Cb |
| spth | spinothalamic tr. |
| ts | tectospinal tract |
| V1 | V1 Cx |
| V2l | V2 (lat.) Cx |
| V2m | V2 (med.) Cx |
| vcs | vent. corticospinal tr. |
| VH | ventral horn |
| vmf | vent. median fiss. |
| vsc | vent. spinocereb. tr. |

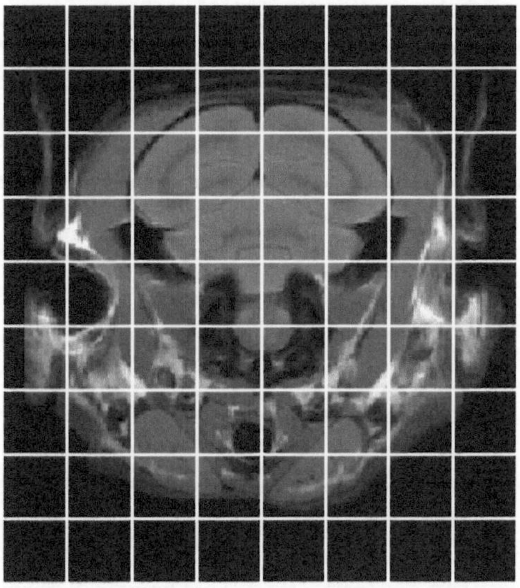

Interaural -07.40 mm

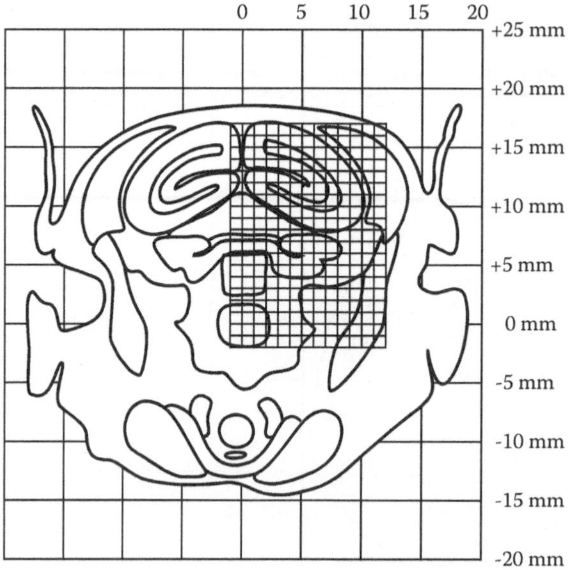

# Plate 53

Interaural -07.80 mm

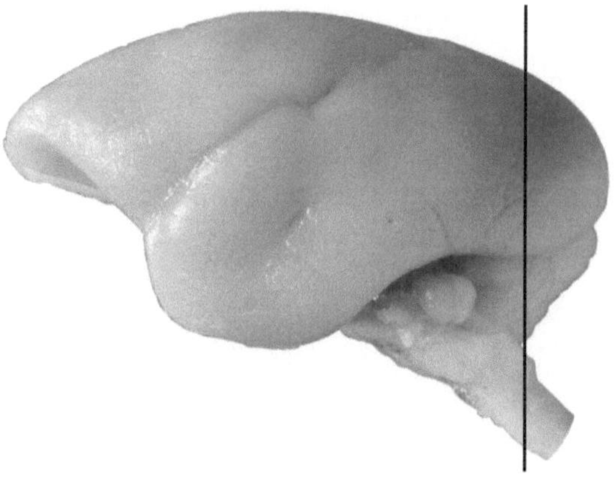

| | |
|---|---|
| 6Cb(a-c) | cerebell. lobule 6(a-c) |
| 7Cb | cerebell. lobule 7 |
| 8CbA | cerebell. lobule 8A |
| 8CbB | cerebell. lobule 8B |
| 9Cb(a) | cerebell. lobule 9(a) |
| ca | calcarine sulcus |
| CC | central canal |
| CeCv | central cervical nu. |
| Cop | copula |
| Crus1 | crus 1, ansiform lob. |
| cu | cuneate fasciculus |
| dlL | dorsolat. tr. Lissauer |
| dsc | dors. spinocerebell. tr |
| gc | geniculocalcarine tr. |
| gr | gracile fasciculus |
| IB | intern. basal nu. |
| lcs | lat. corticospinal tr. |
| LSp | lat. spinal nu. |
| lvs | lat. vestibulosp. tr. |
| mlf | med. longit. fasc. |
| PM | paramedian lobule Cb |
| ppf | prepyramidal fiss. |
| R2 | Rexed 2, spinal cord |
| R3 | Rexed 3, spinal cord |
| R4 | Rexed 4, spinal cord |
| R5 | Rexed 5, spinal cord |
| R7 | Rexed 7, spinal cord |
| R8 | Rexed 8, spinal cord |
| R9 | Rexed 9, spinal cord |
| rs | rubrospinal tract |
| sf | secondary fissure Cb |
| Sim | simplex lobule Cb |
| spth | spinothalamic tr. |
| ts | tectospinal tract |
| V1 | V1 Cx |
| V2 | V2 Cx |
| V2l | V2 (lat.) Cx |
| vcs | vent. corticospinal tr. |
| VH | ventral horn |
| vmf | vent. median fiss. |
| vsc | vent. spinocereb. tr. |

# Plate 53a

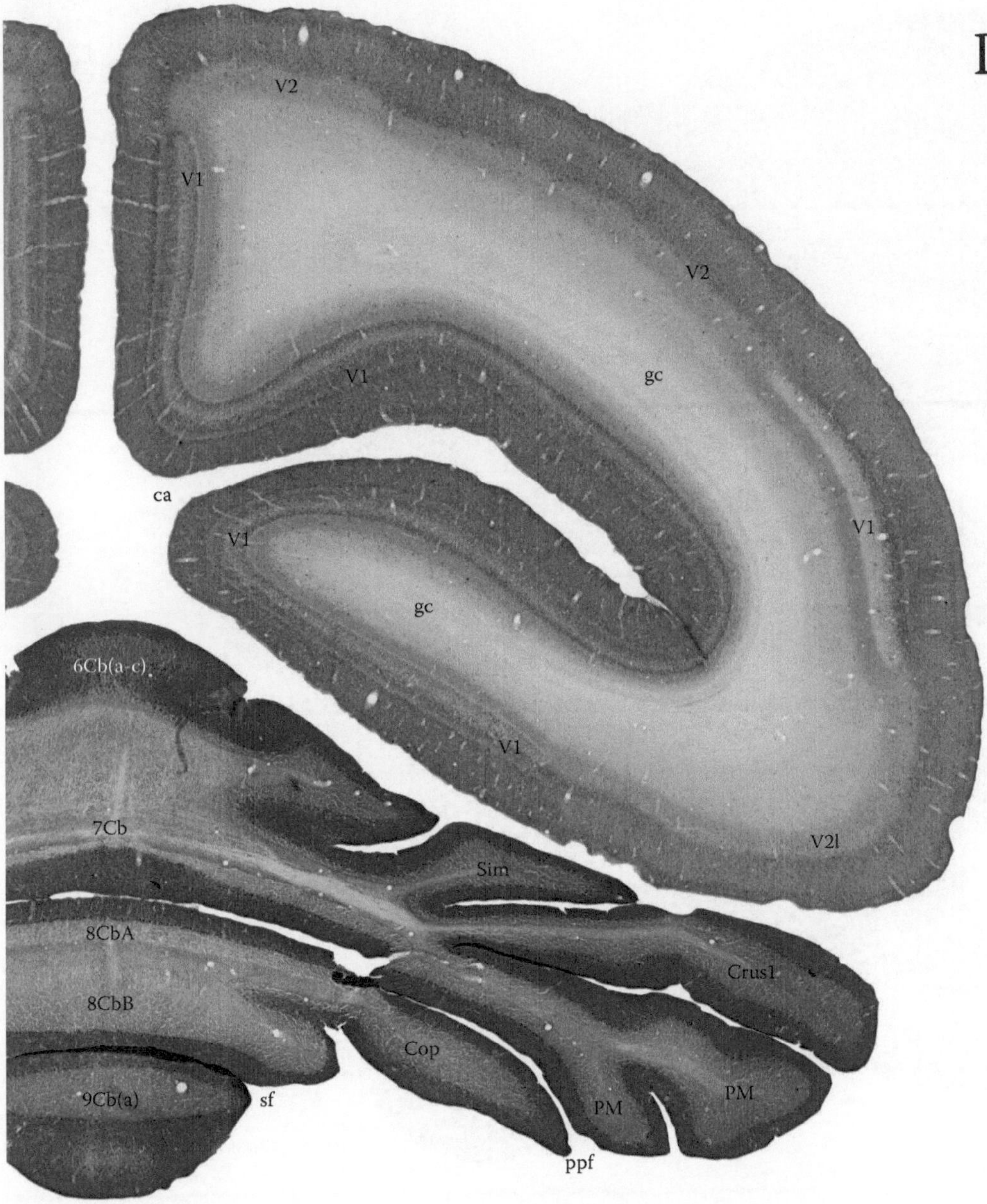


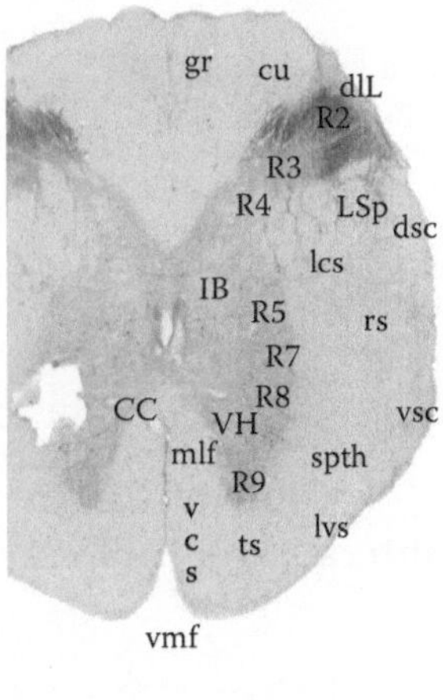


**Calbindin**
Interaural -07.50 mm

# Plate 53b

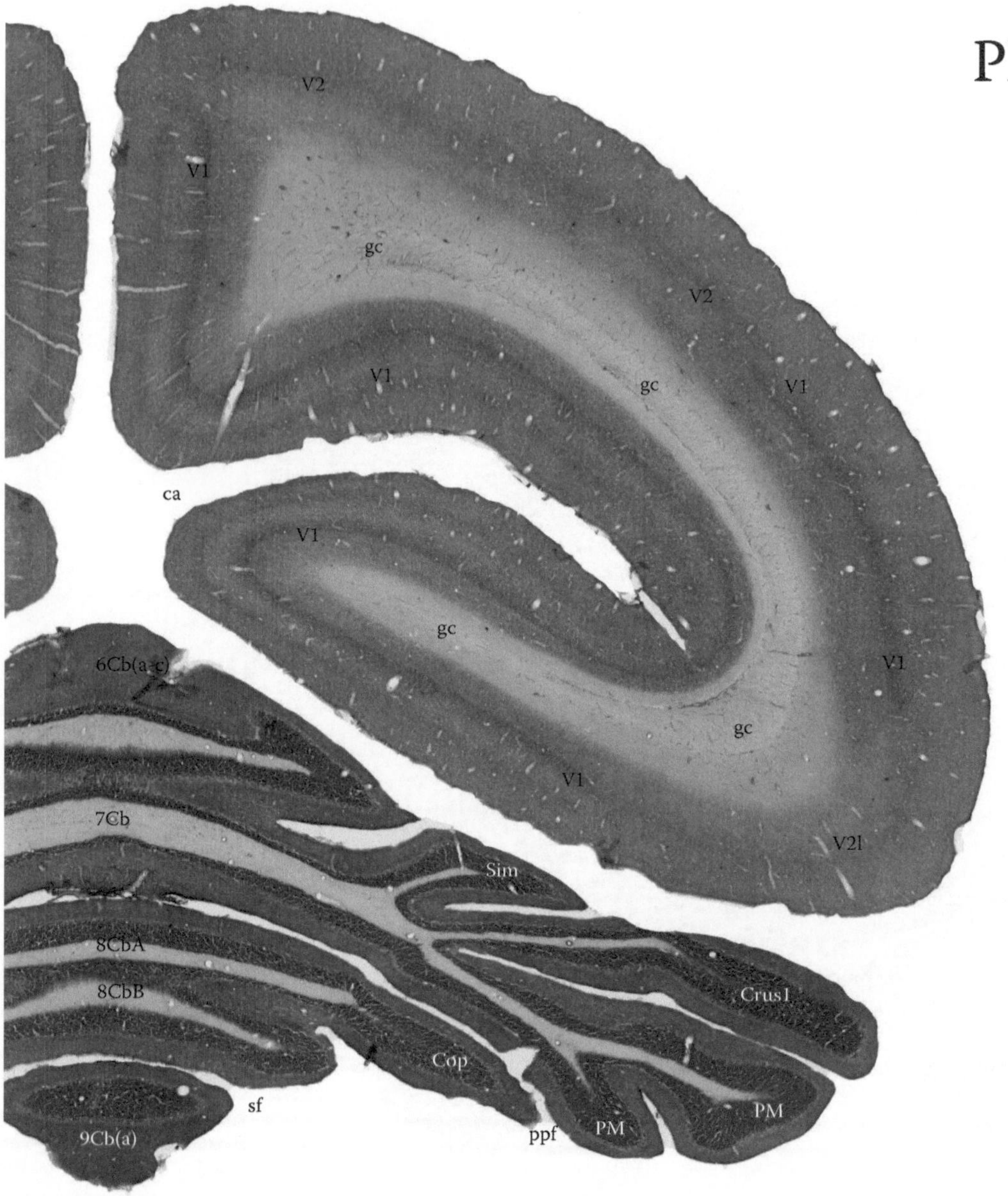


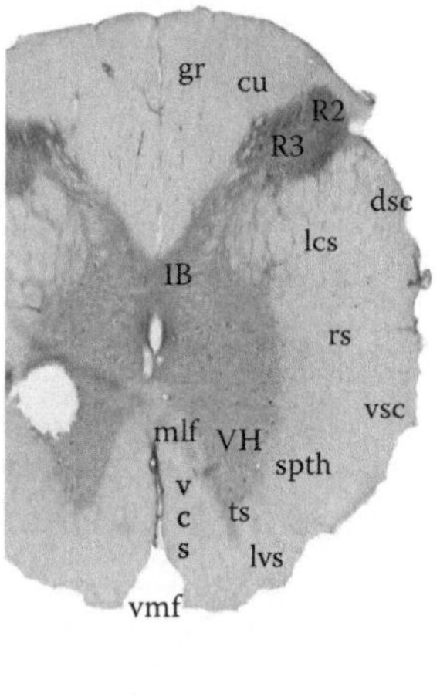


**Neurofilament**
Interaural -07.65 mm

# Plate 53c

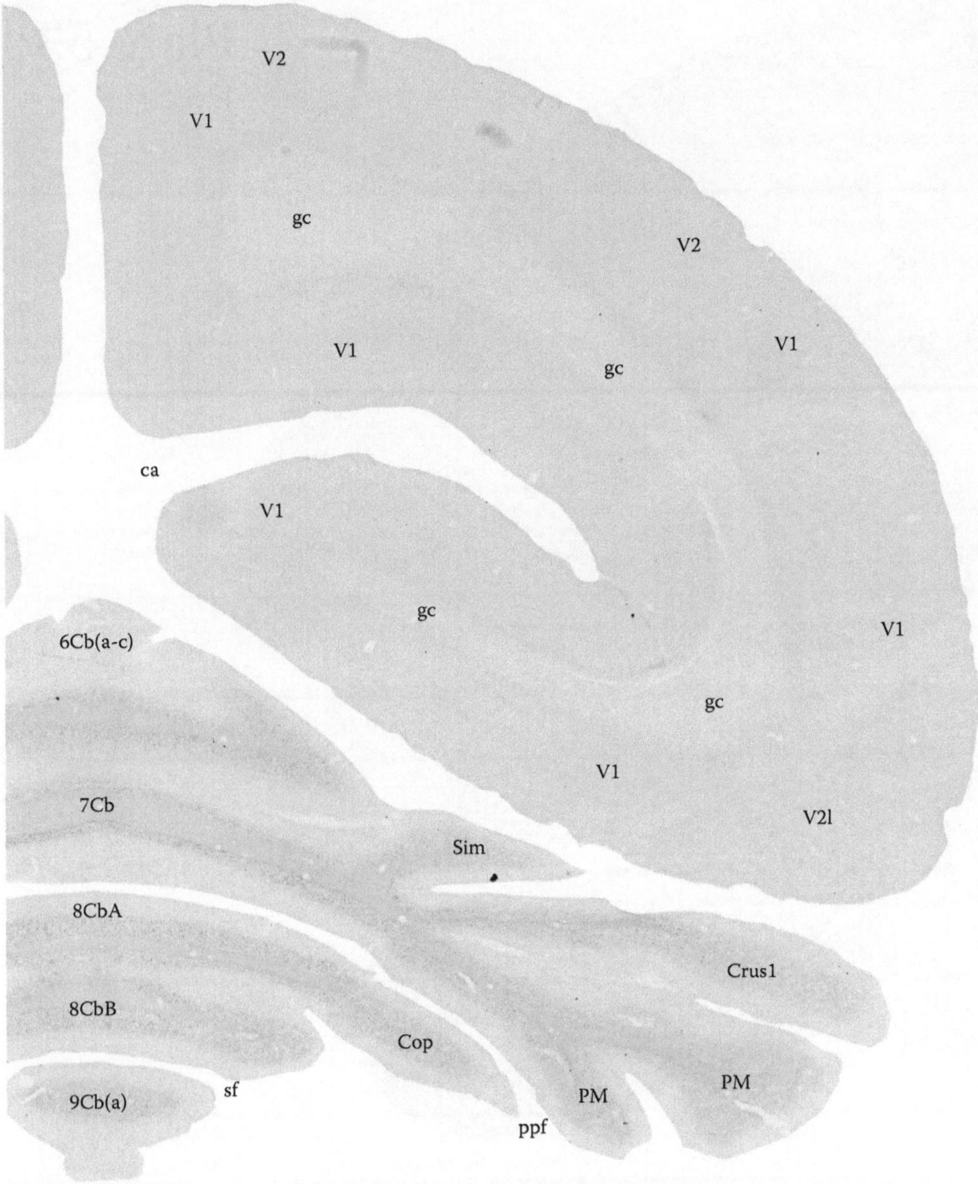


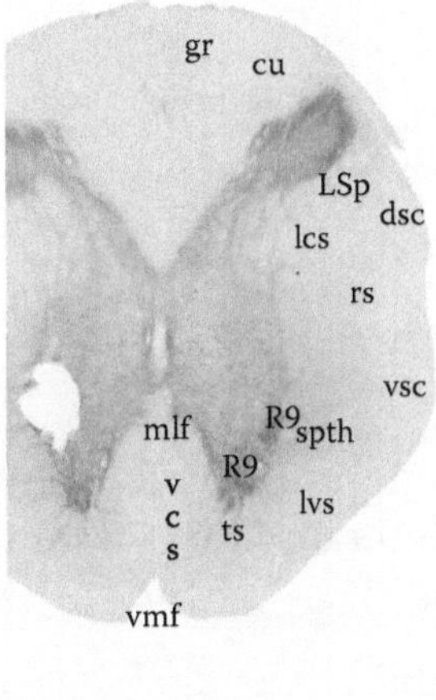


**Acetylcholinesterase**
Interaural -07.70 mm

# Plate 53d

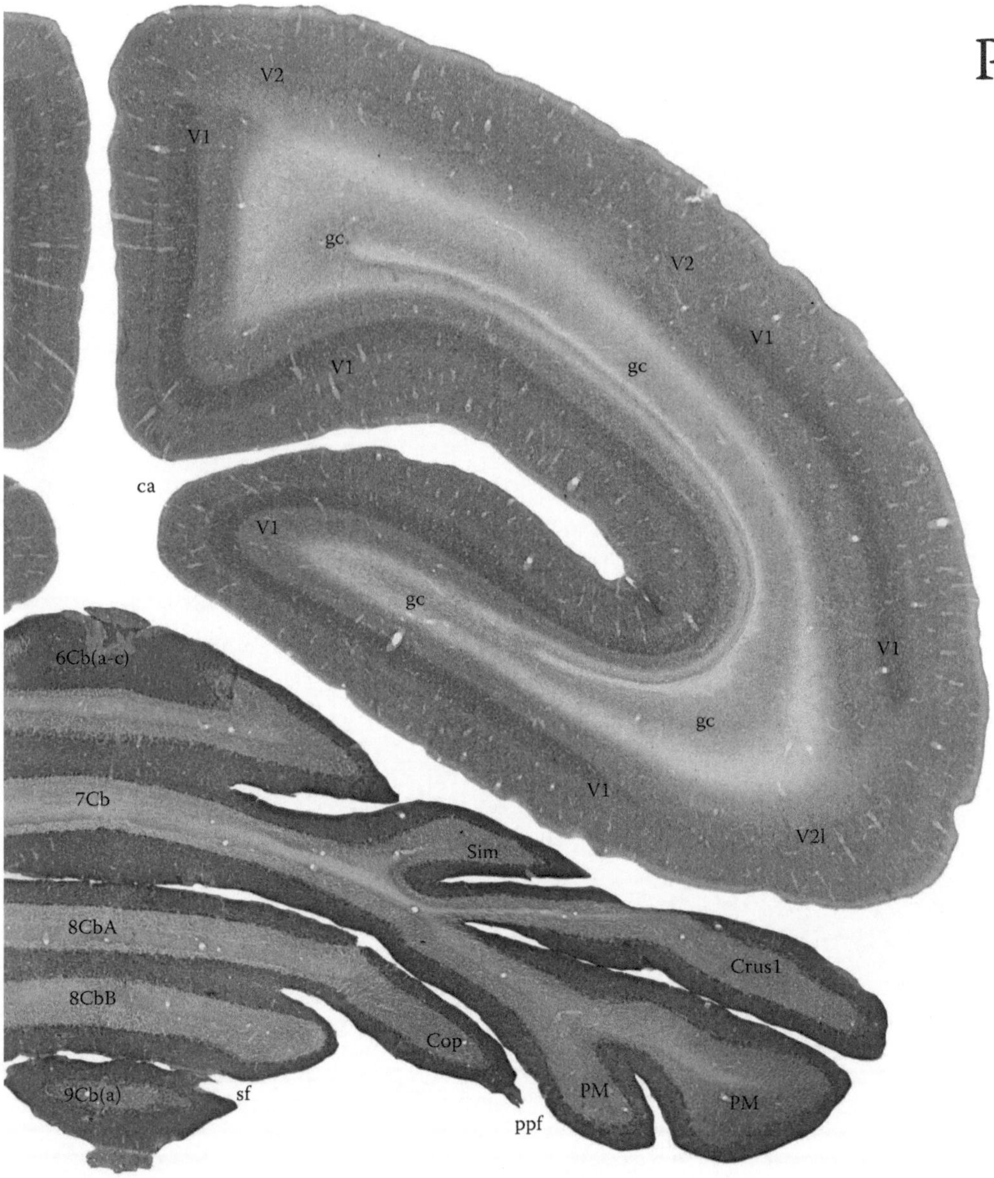


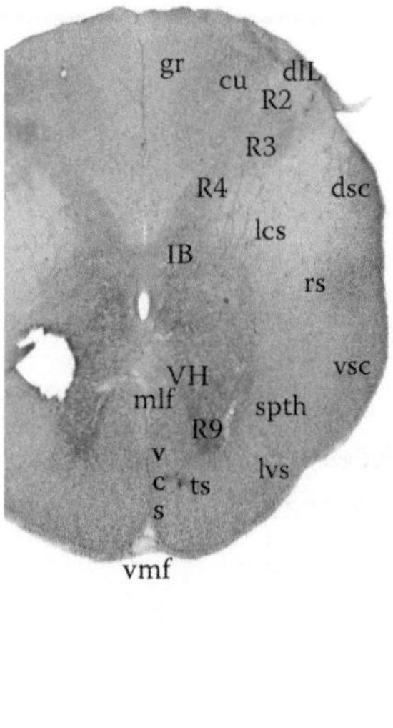


**Parvalbumin**
Interaural -07.75 mm

# Plate 53e

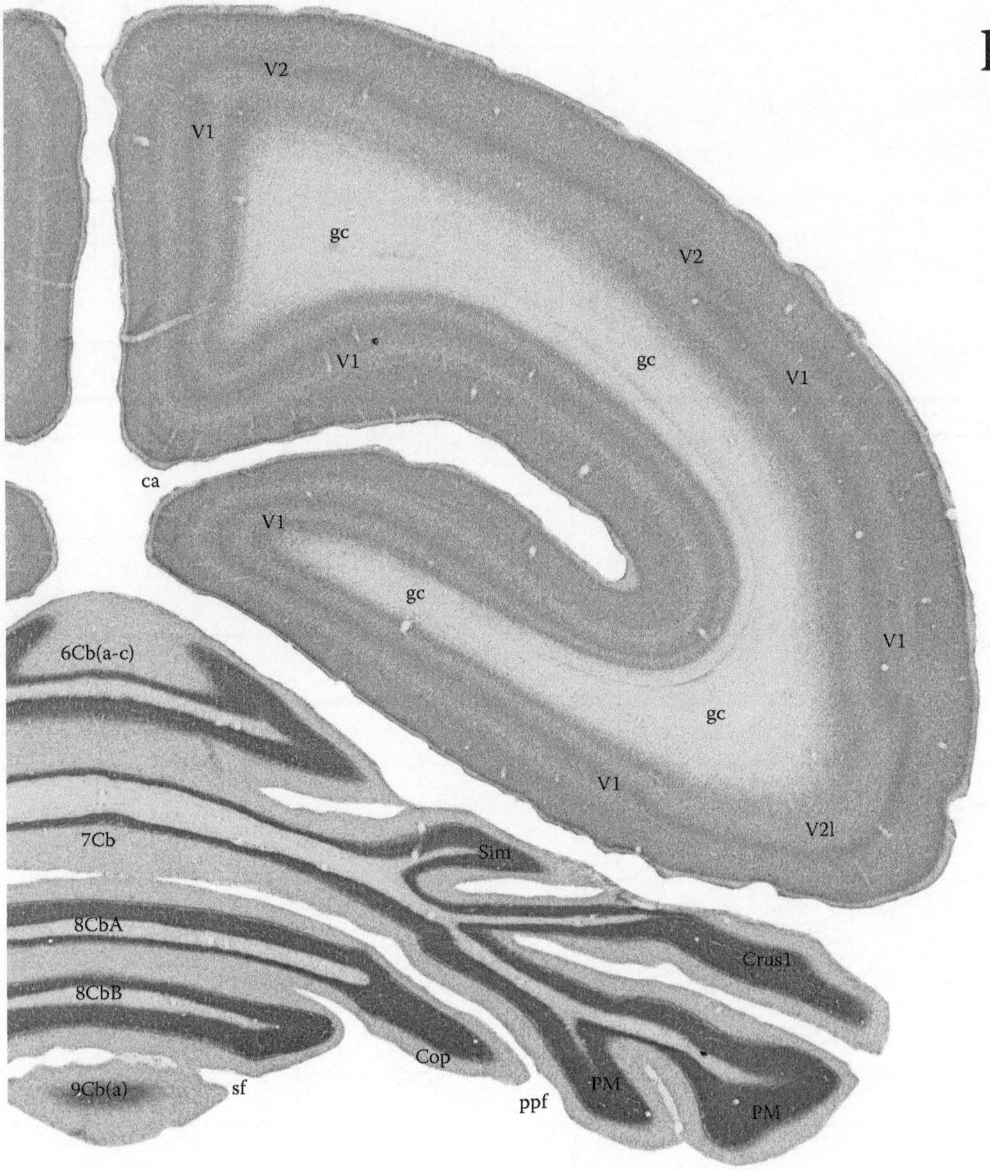


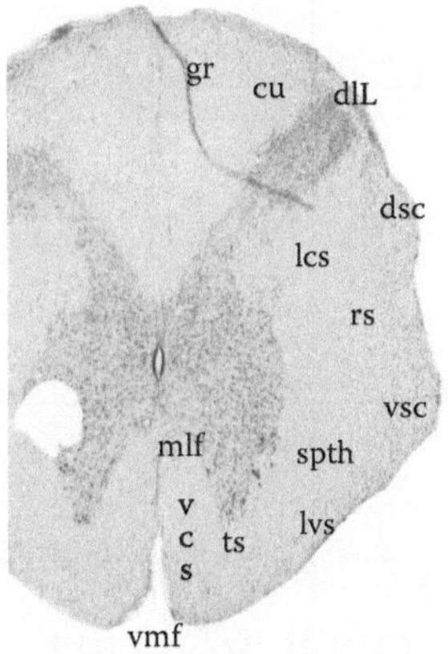


**Cresyl Violet**
Interaural -07.80 mm

# Plate 53

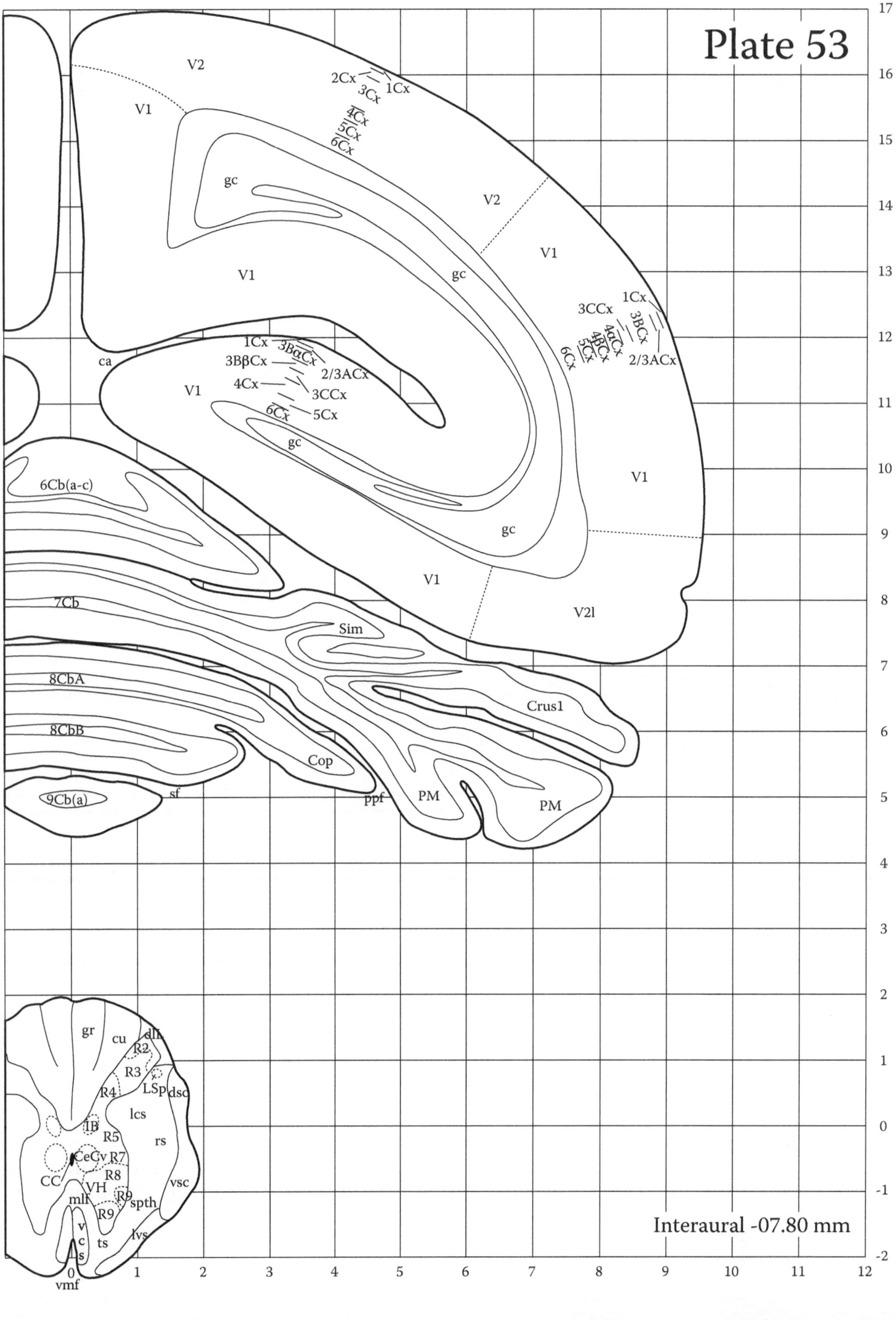


Interaural -07.80 mm

| | |
|---|---|
| 6Cb(a-c) | cerebell. lobule 6(a-c) |
| 7Cb | cerebell. lobule 7 |
| 8CbA | cerebell. lobule 8A |
| 8CbB | cerebell. lobule 8B |
| 9Cb(a) | cerebell. lobule 9(a) |
| ca | calcarine sulcus |
| CC | central canal |
| CeCv | central cervical nu. |
| Cop | copula |
| Crus1 | crus 1, ansiform lob. |
| cu | cuneate fasciculus |
| dlL | dorsolat. tr. Lissauer |
| dsc | dors. spinocerebell. tr |
| gc | geniculocalcarine tr. |
| gr | gracile fasciculus |
| IB | intern. basal nu. |
| lcs | lat. corticospinal tr. |
| LSp | lat. spinal nu. |
| lvs | lat. vestibulosp. tr. |
| mlf | med. longit. fasc. |
| PM | paramedian lobule Cb |
| ppf | prepyramidal fiss. |
| R2 | Rexed 2, spinal cord |
| R3 | Rexed 3, spinal cord |
| R4 | Rexed 4, spinal cord |
| R5 | Rexed 5, spinal cord |
| R7 | Rexed 7, spinal cord |
| R8 | Rexed 8, spinal cord |
| R9 | Rexed 9, spinal cord |
| rs | rubrospinal tract |
| sf | secondary fissure Cb |
| Sim | simplex lobule Cb |
| spth | spinothalamic tr. |
| ts | tectospinal tract |
| V1 | V1 Cx |
| V2 | V2 Cx |
| V2l | V2 (lat.) Cx |
| vcs | vent. corticospinal tr. |
| VH | ventral horn |
| vmf | vent. median fiss. |
| vsc | vent. spinocereb. tr. |

# Plate 54

Interaural -08.15 mm

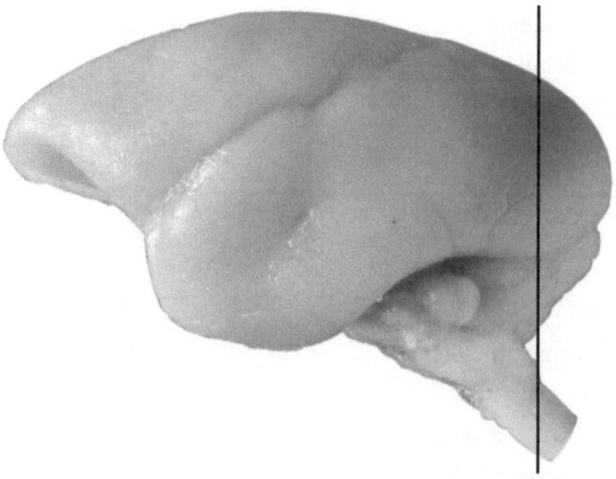

| | |
|---|---|
| 6Cb(a-c) | cerebell. lobule 6(a-c) |
| 7CbA | cerebell. lobule 7A |
| 7CbB | cerebell. lobule 7B |
| 8CbA | cerebell. lobule 8A |
| 8CbB | cerebell. lobule 8B |
| ca | calcarine sulcus |
| gc | geniculocalcarine tr. |
| PM | paramedian lobule Cb |
| ppf | prepyramidal fiss. |
| Sim | simplex lobule Cb |
| V1 | V1 Cx |
| V2 | V2 Cx |

# Plate 54a

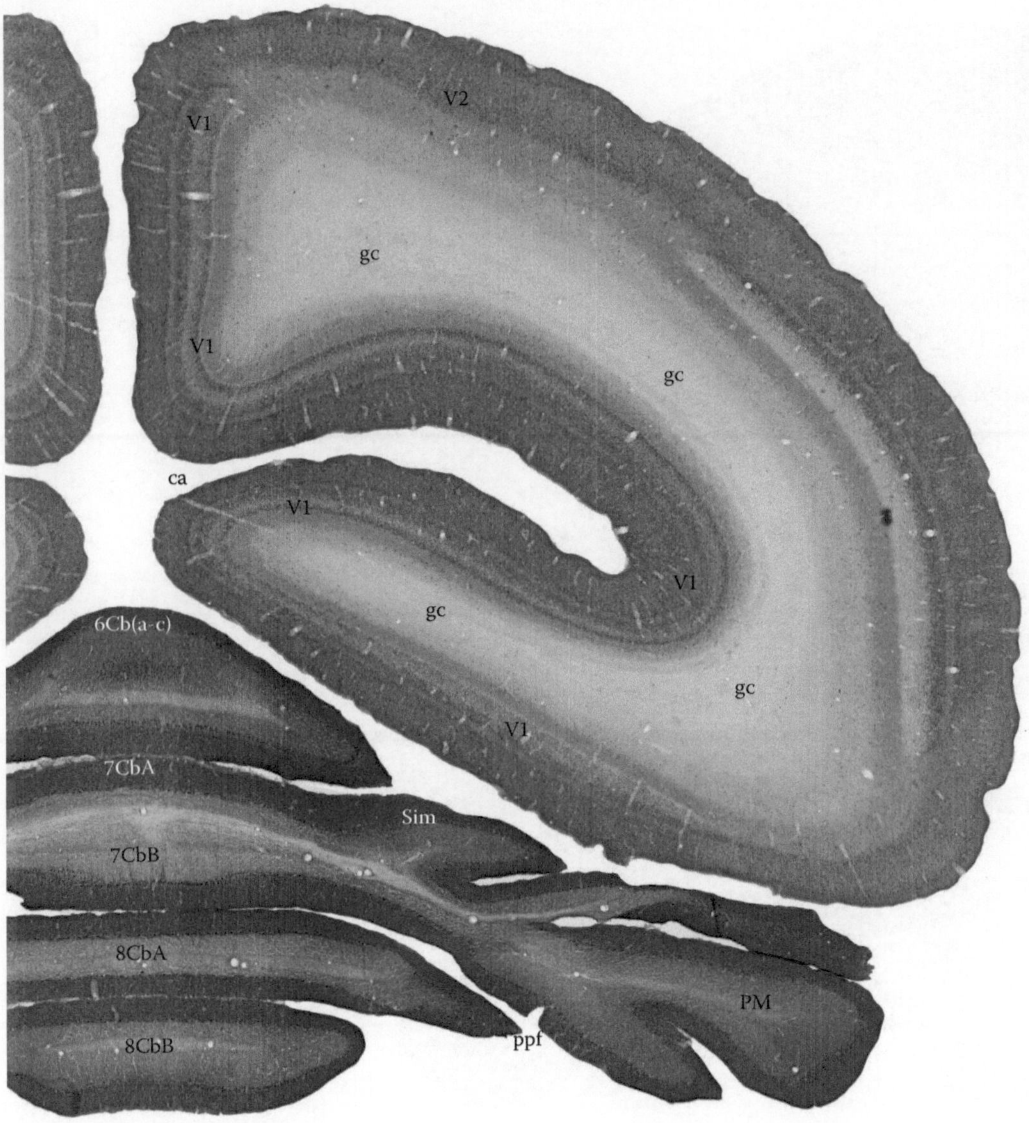


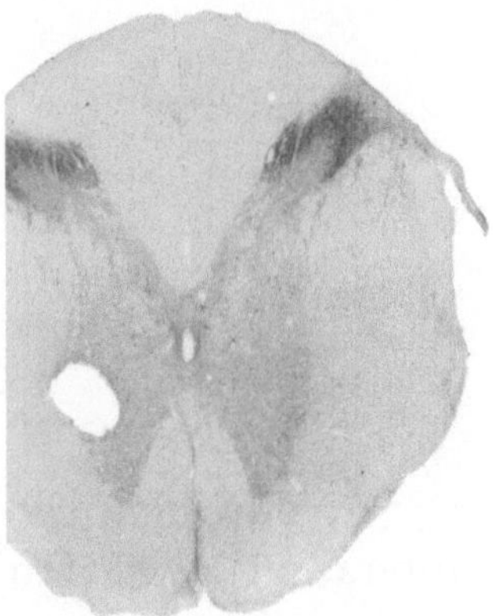

**Calbindin**
Interaural -08.00 mm

# Plate 54b

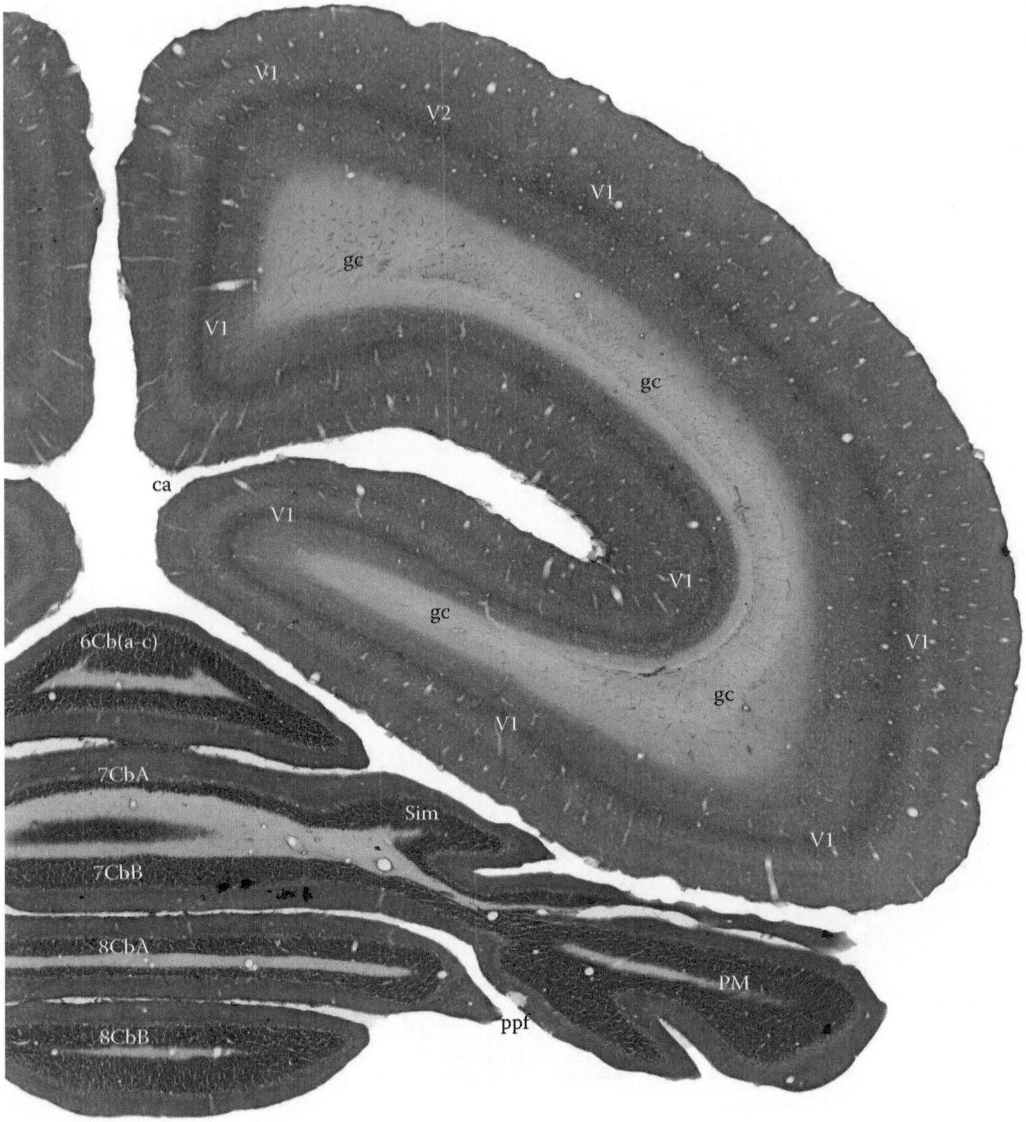


**Neurofilament**
Interaural -08.15 mm

# Plate 54c

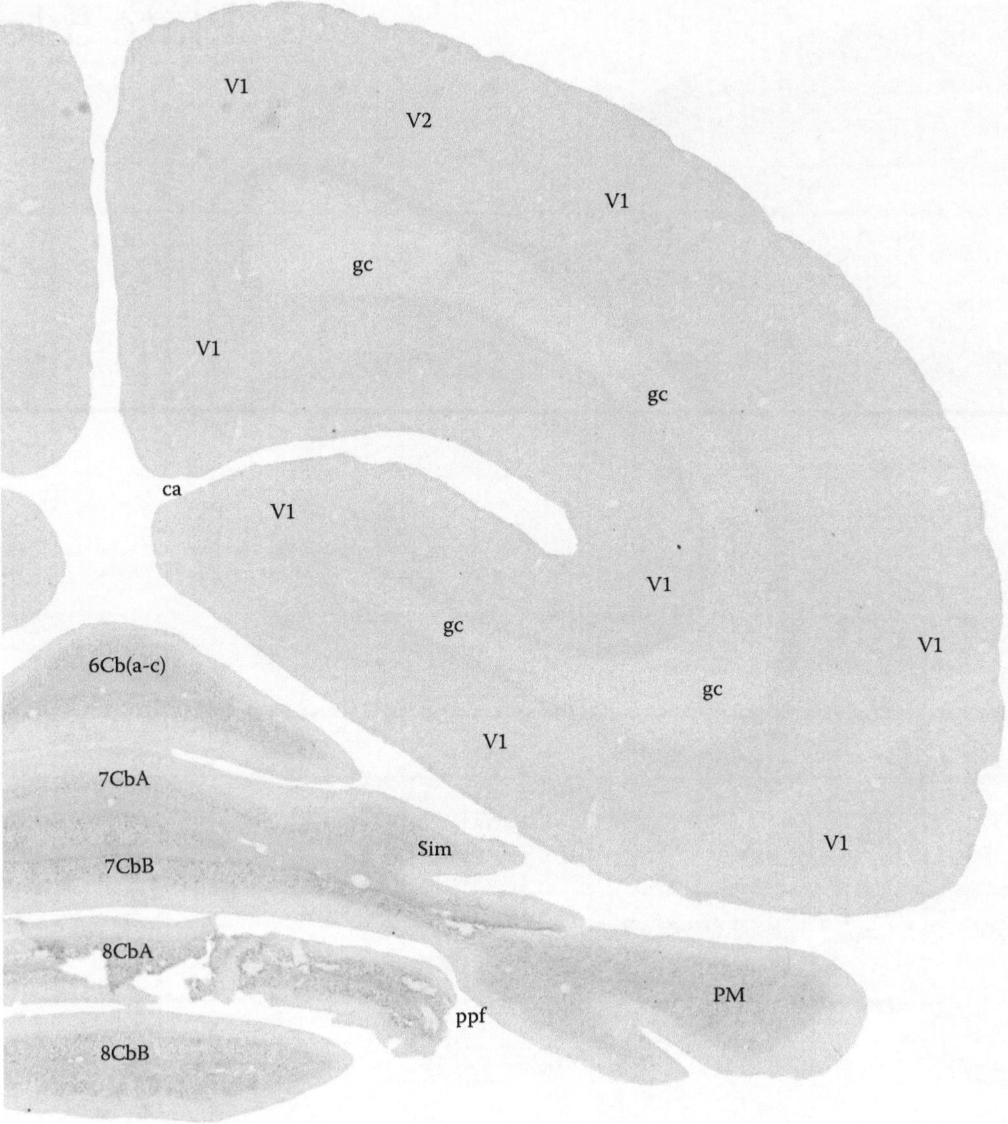


**Acetylcholinesterase**
Interaural -08.20 mm

# Plate 54d

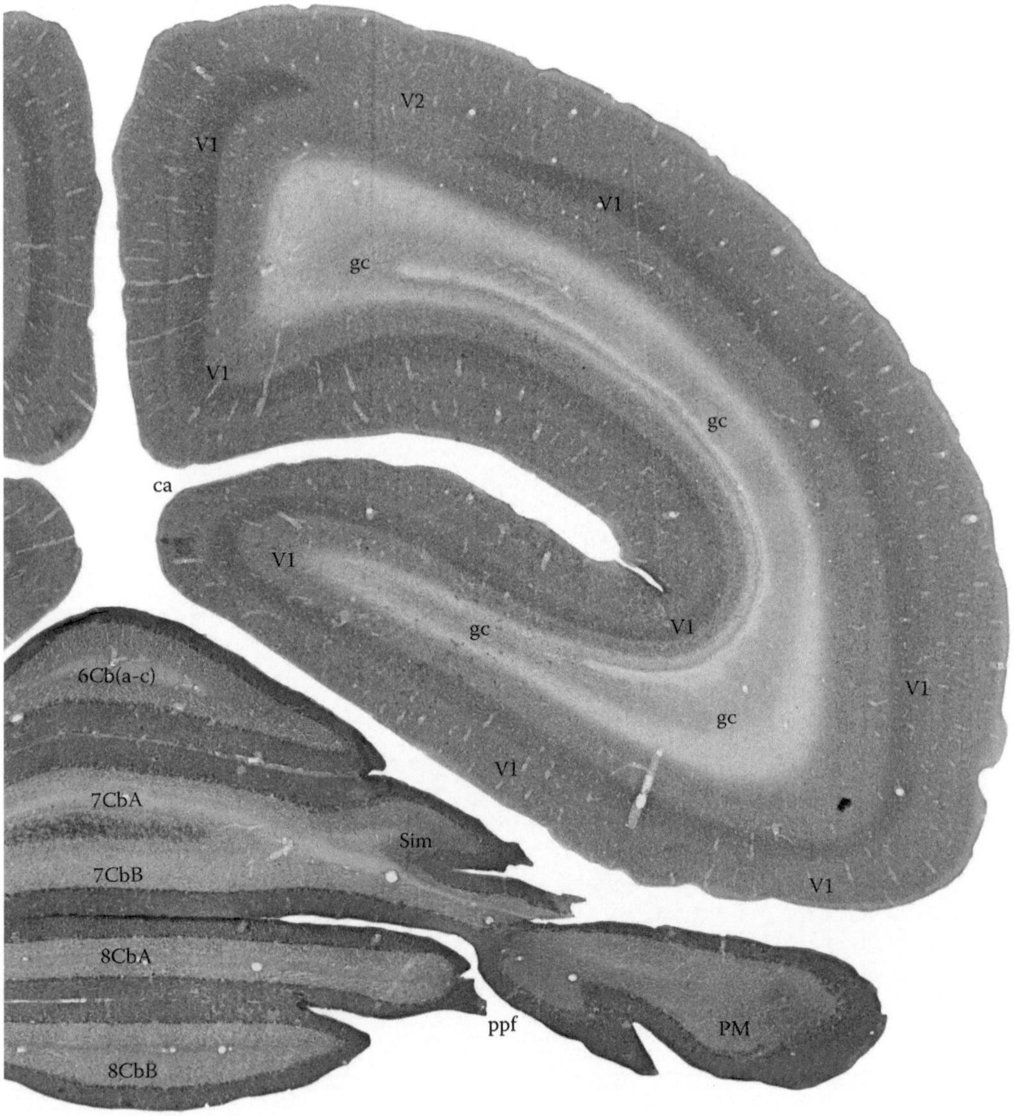


**Parvalbumin**
Interaural -08.25 mm

# Plate 54e

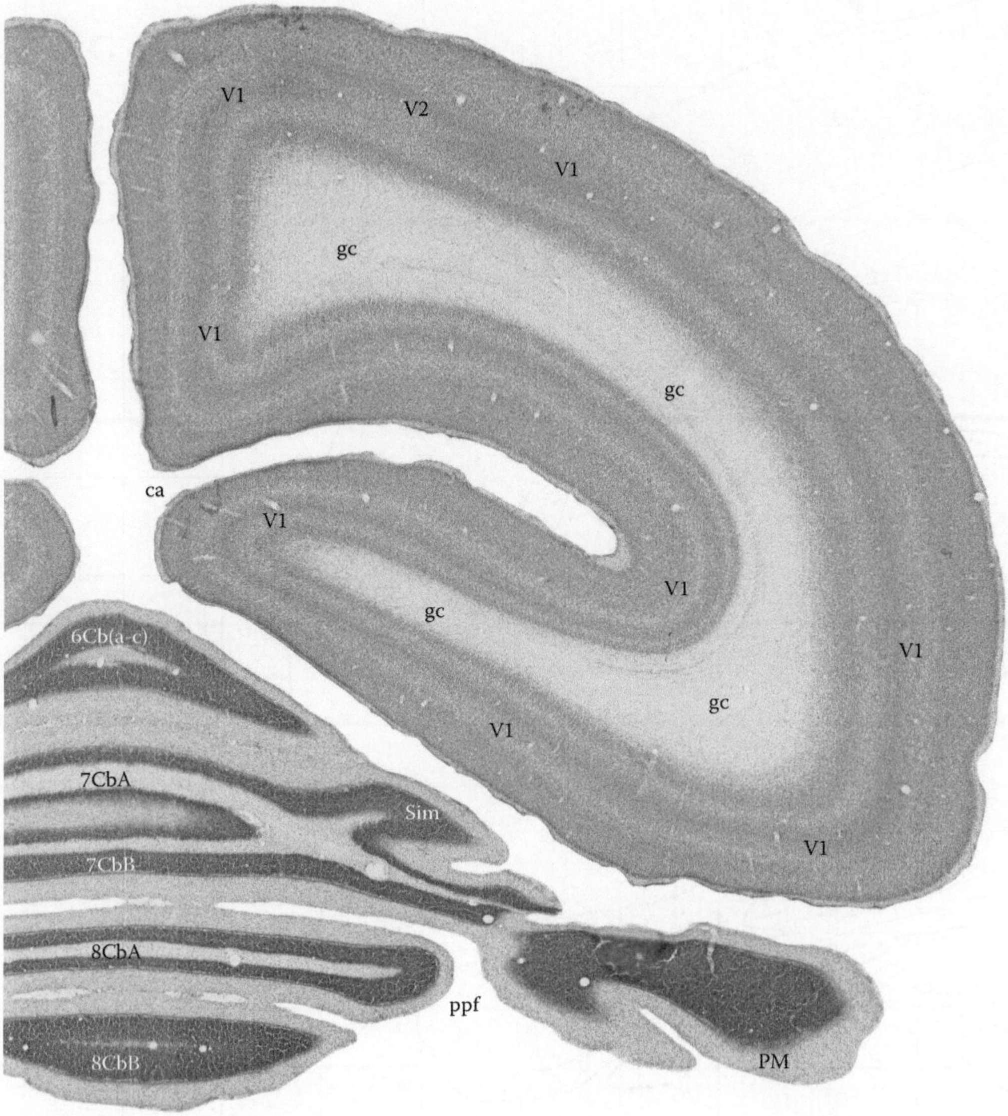


**Cresyl Violet**
Interaural -08.30 mm

# Plate 54

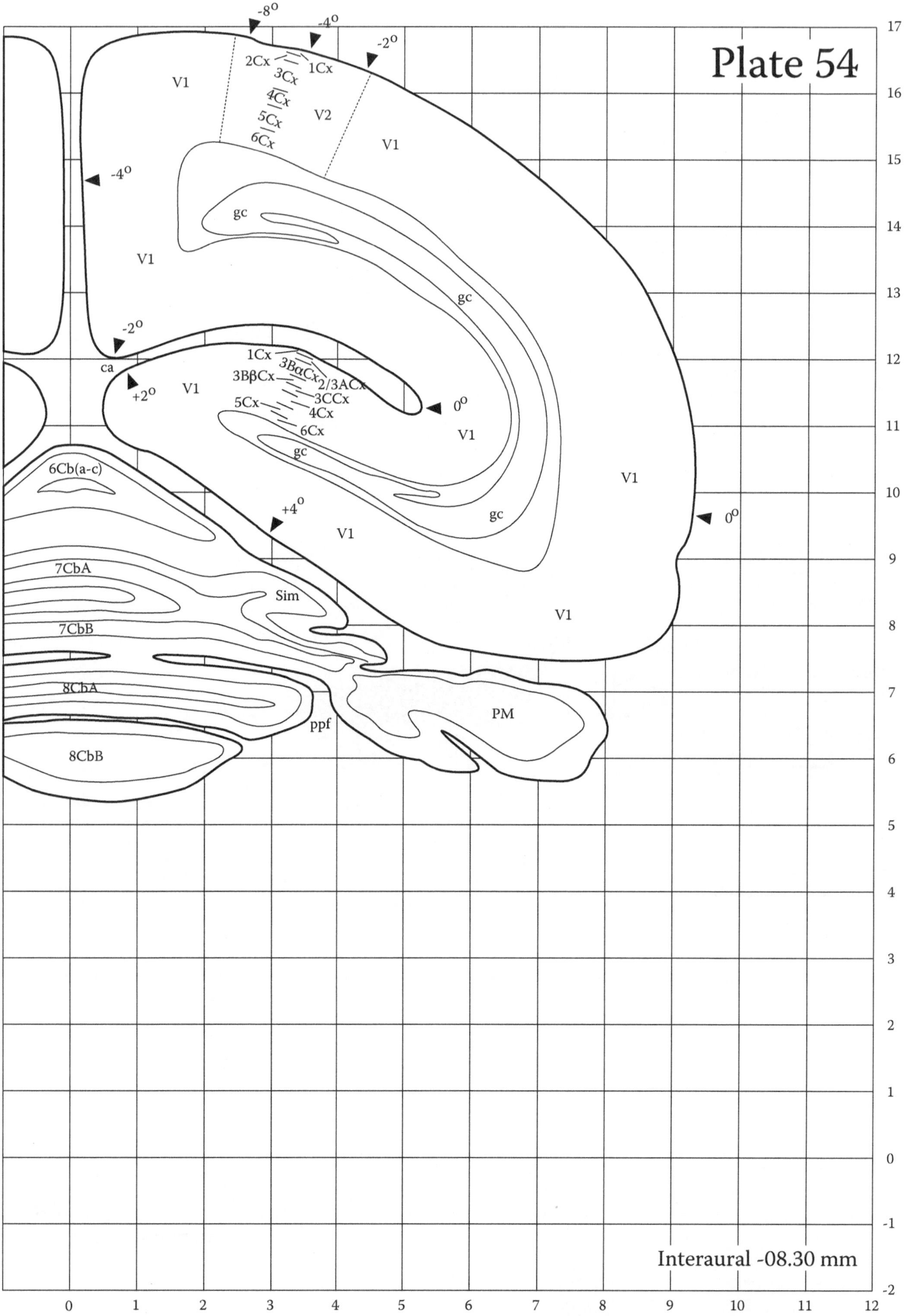


Interaural -08.30 mm

| | |
|---|---|
| 6Cb(a-c) | cerebell. lobule 6(a-c) |
| 7CbA | cerebell. lobule 7A |
| 7CbB | cerebell. lobule 7B |
| 8CbA | cerebell. lobule 8A |
| 8CbB | cerebell. lobule 8B |
| ca | calcarine sulcus |
| gc | geniculocalcarine tr. |
| PM | paramedian lobule Cb |
| ppf | prepyramidal fiss. |
| Sim | simplex lobule Cb |
| V1 | V1 Cx |
| V2 | V2 Cx |

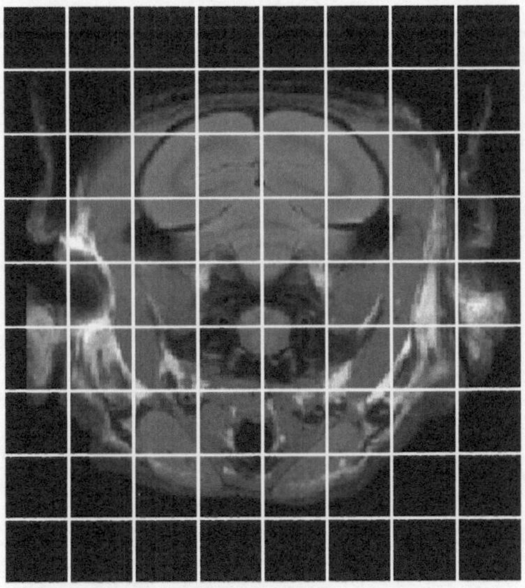

Interaural -08.35 mm

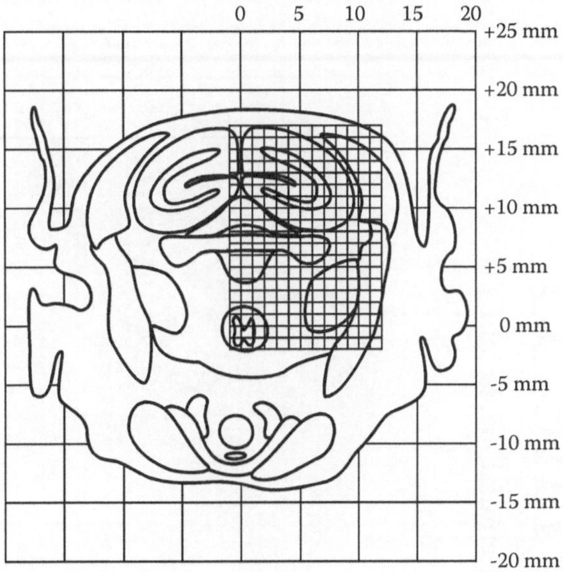

# Plate 55

Interaural -08.80 mm

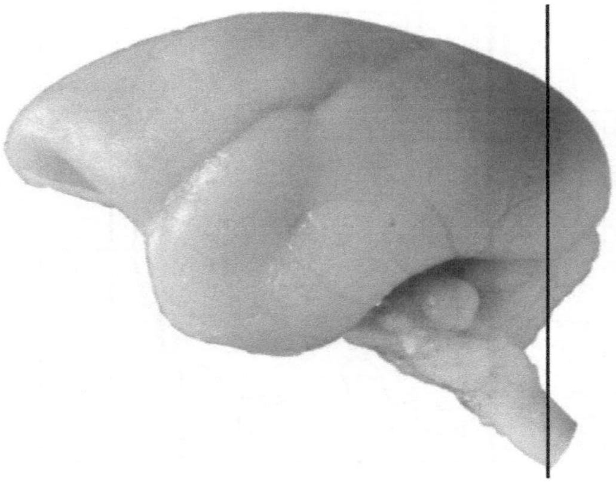

| | |
|---|---|
| 6Cb(a-c) | cerebell. lobule 6(a-c) |
| 7CbA | cerebell. lobule 7A |
| 7CbB | cerebell. lobule 7B |
| 8CbA | cerebell. lobule 8A |
| 8CbB | cerebell. lobule 8B |
| ca | calcarine sulcus |
| gc | geniculocalcarine tr. |
| ppf | prepyramidal fiss. |
| V1 | V1 Cx |

# Plate 55a

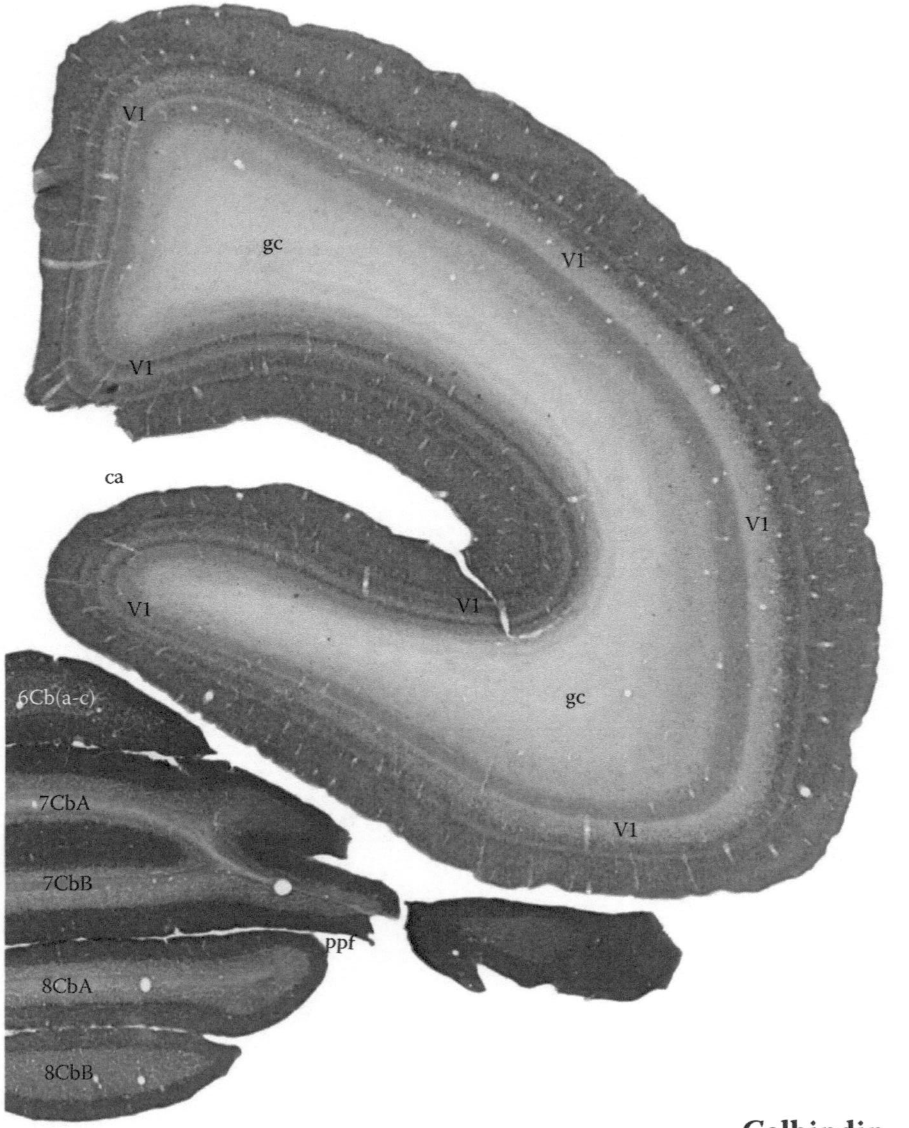


**Calbindin**

Interaural -08.50 mm

# Plate 55b

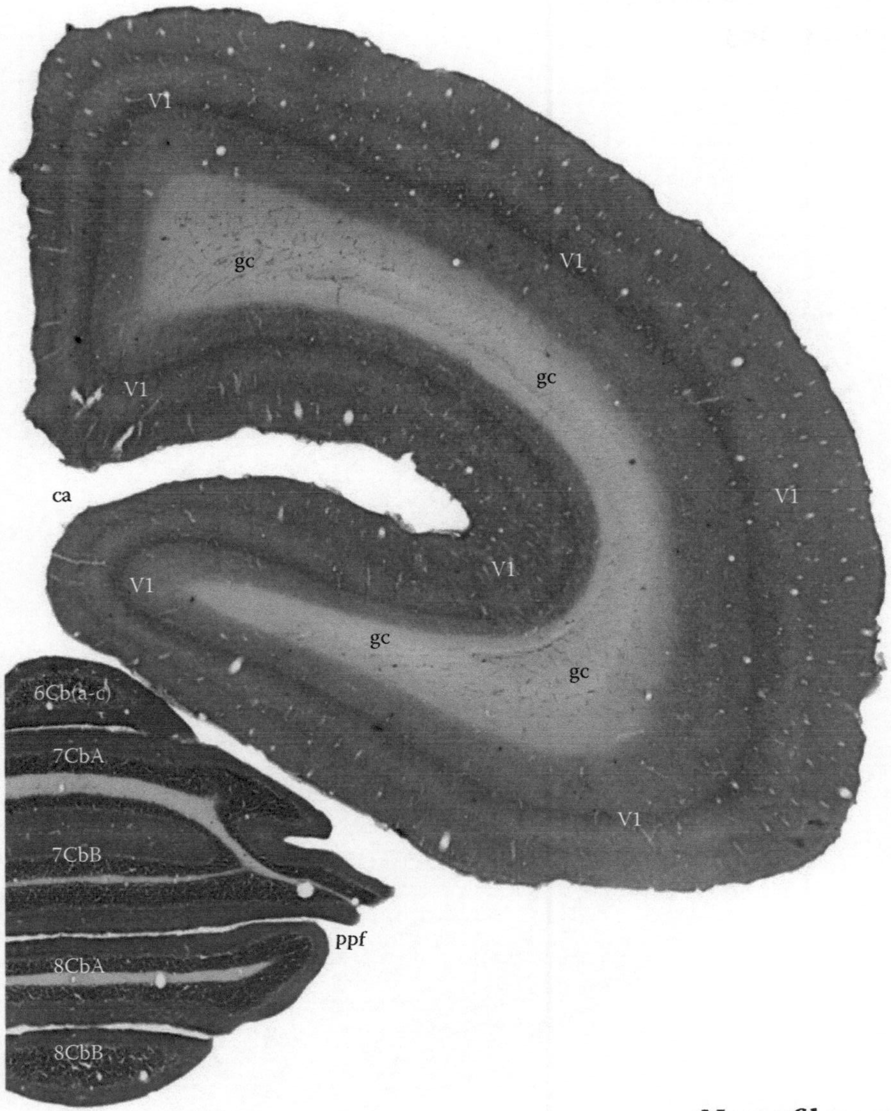


**Neurofilament**
Interaural -08.65 mm

# Plate 55c

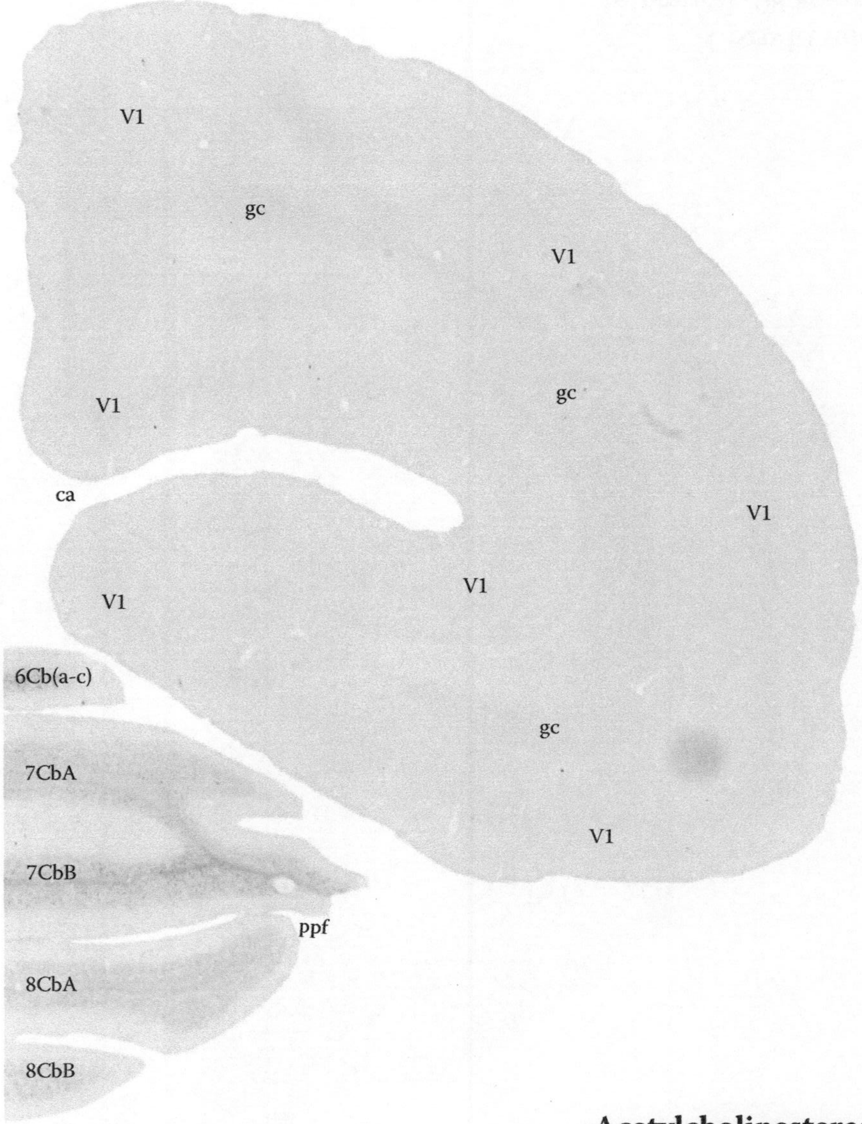


**Acetylcholinesterase**
Interaural -08.70 mm

# Plate 55d

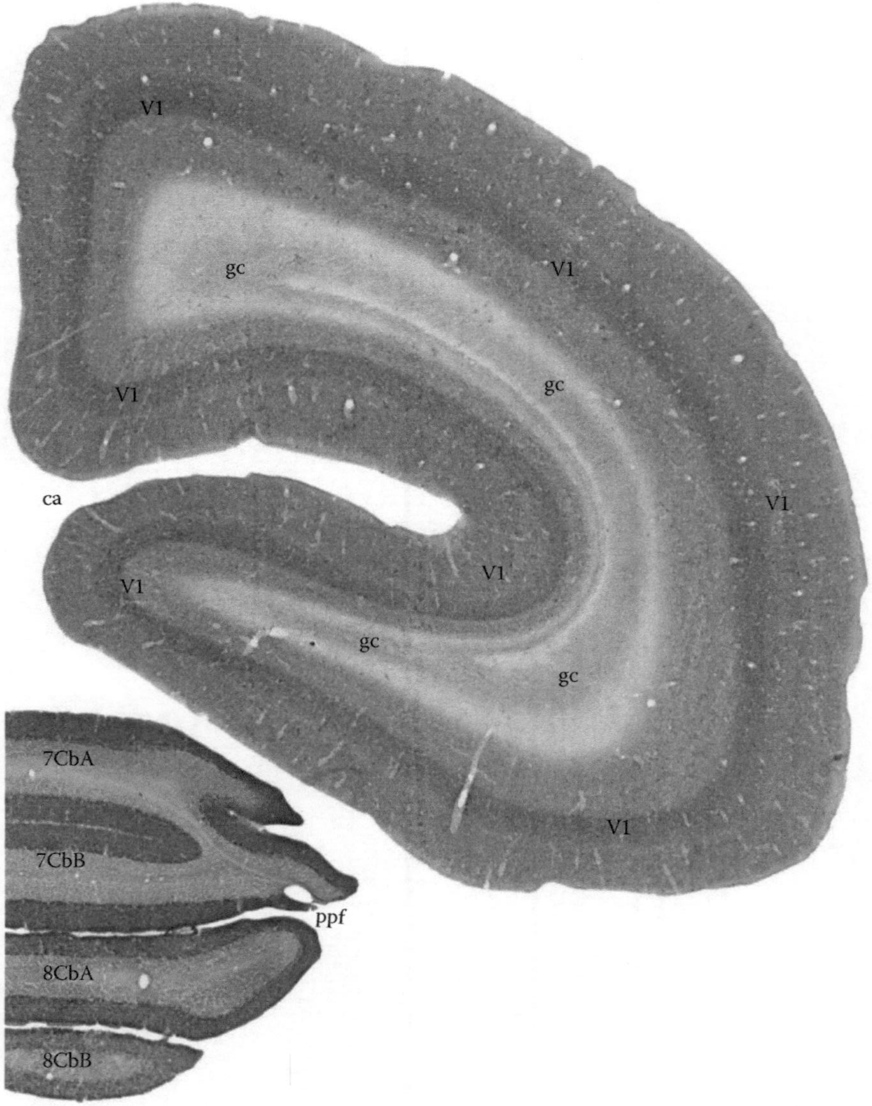


**Parvalbumin**
Interaural -08.75 mm

# Plate 55e

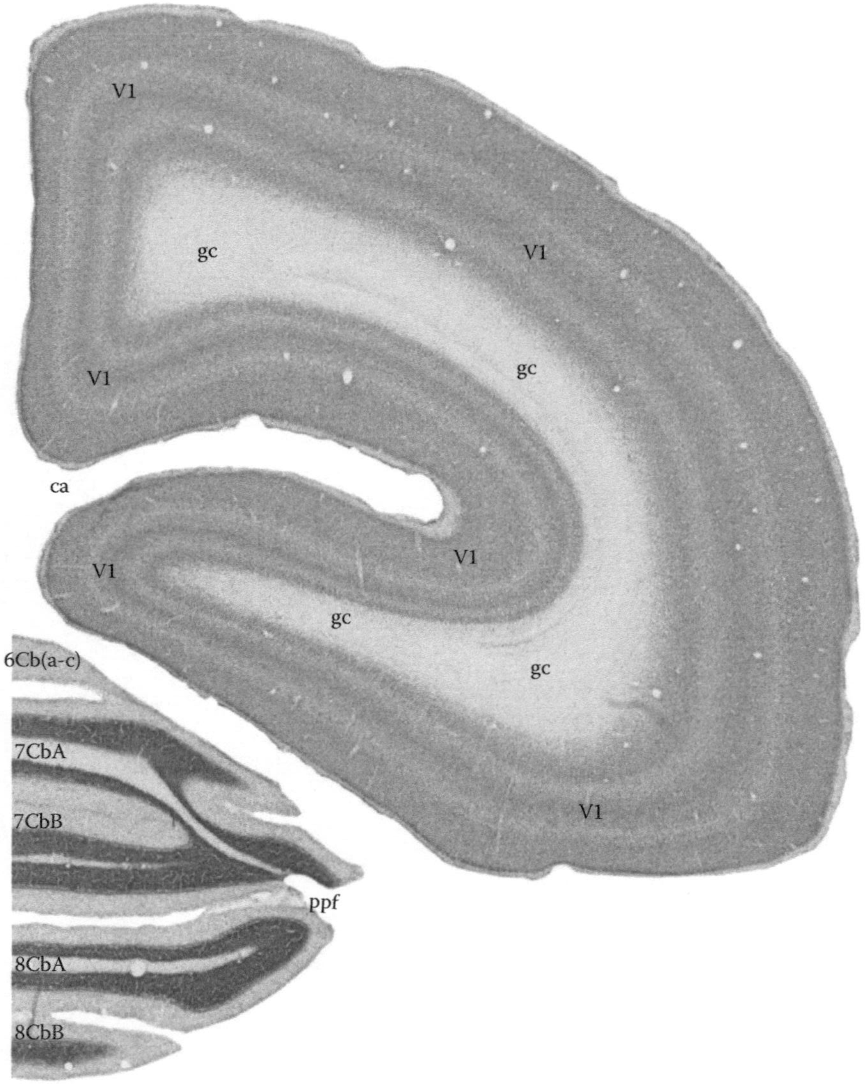


**Cresyl Violet**
Interaural -08.80 mm

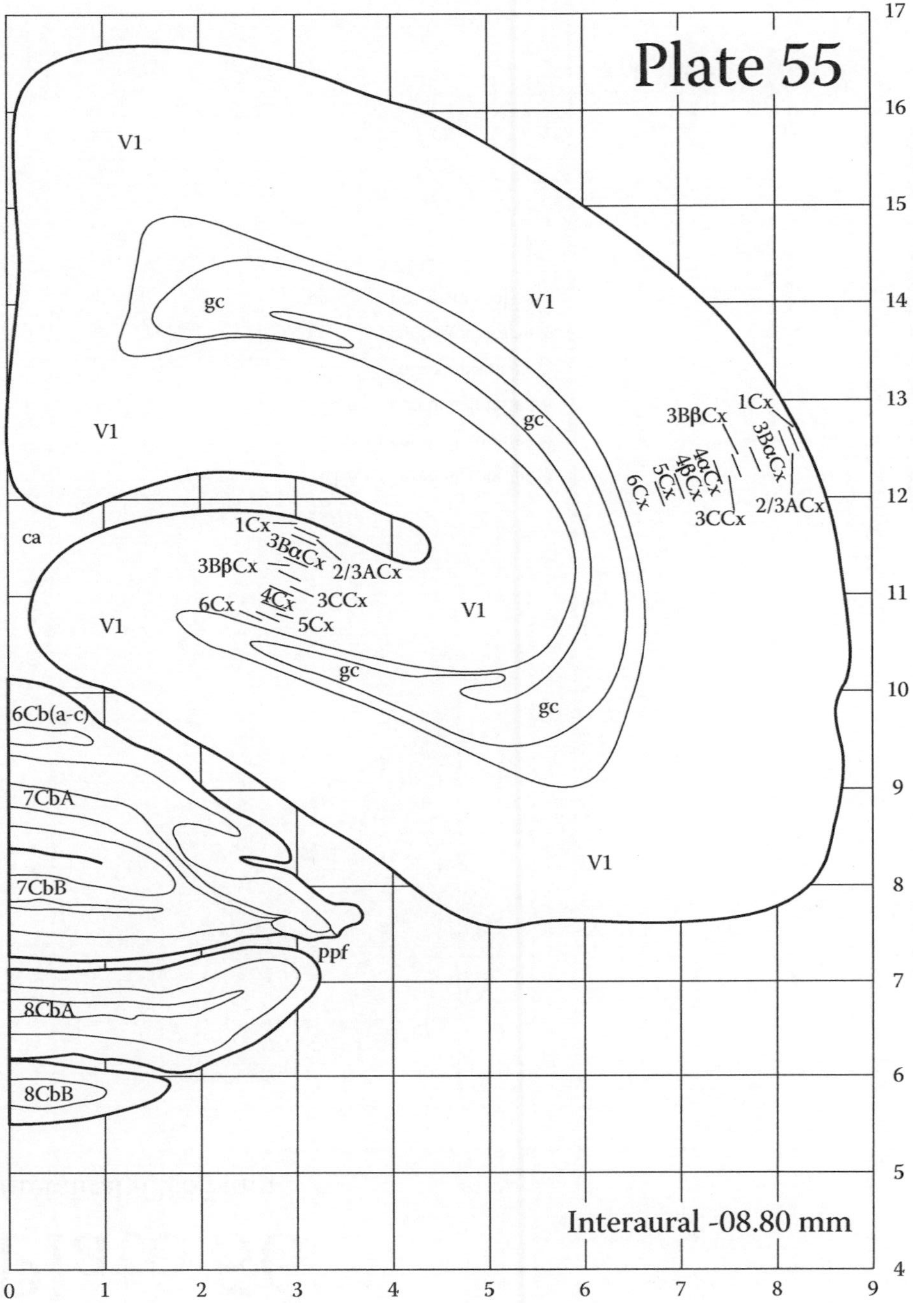


| | |
|---|---|
| 6Cb(a-c) | cerebell. lobule 6(a-c) |
| 7CbA | cerebell. lobule 7A |
| 7CbB | cerebell. lobule 7B |
| 8CbA | cerebell. lobule 8A |
| 8CbB | cerebell. lobule 8B |
| ca | calcarine sulcus |
| gc | geniculocalcarine tr. |
| ppf | prepyramidal fiss. |
| V1 | V1 Cx |

# Plate 56

Interaural -09.35 mm

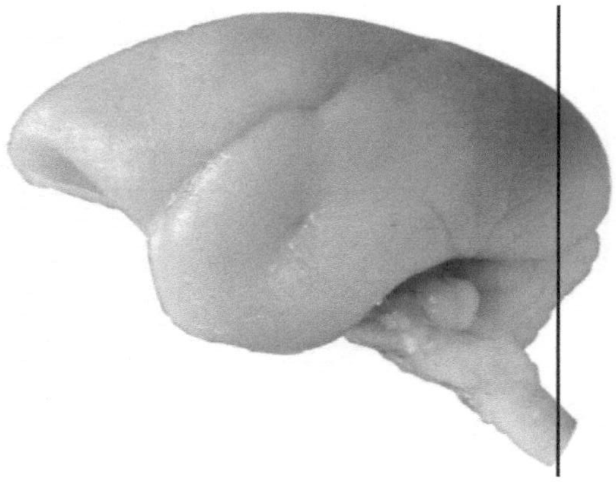

| | |
|---|---|
| 7CbA | cerebell. lobule 7A |
| 7CbB | cerebell. lobule 7B |
| 8CbA | cerebell. lobule 8A |
| ca | calcarine sulcus |
| gc | geniculocalcarine tr. |
| ppf | prepyramidal fiss. |
| V1 | V1 Cx |

# Plate 56a

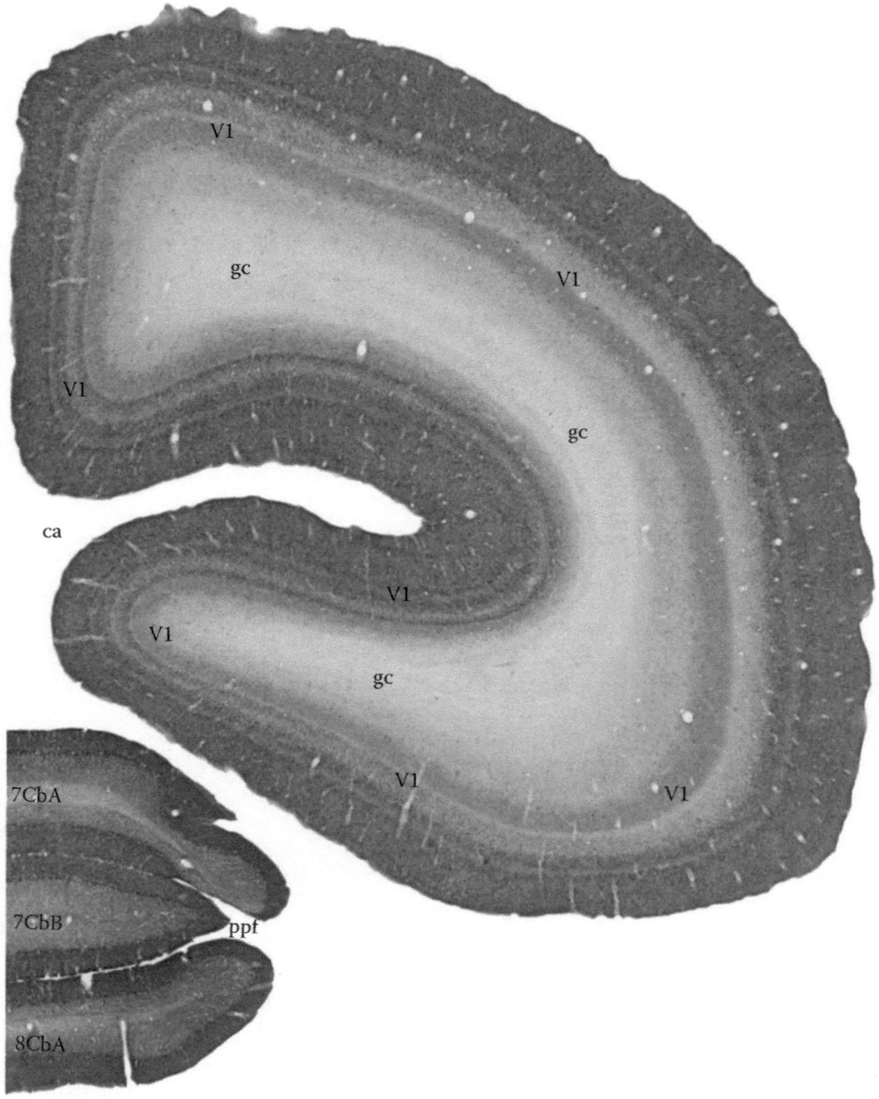


**Calbindin**

Interaural -09.05 mm

# Plate 56b

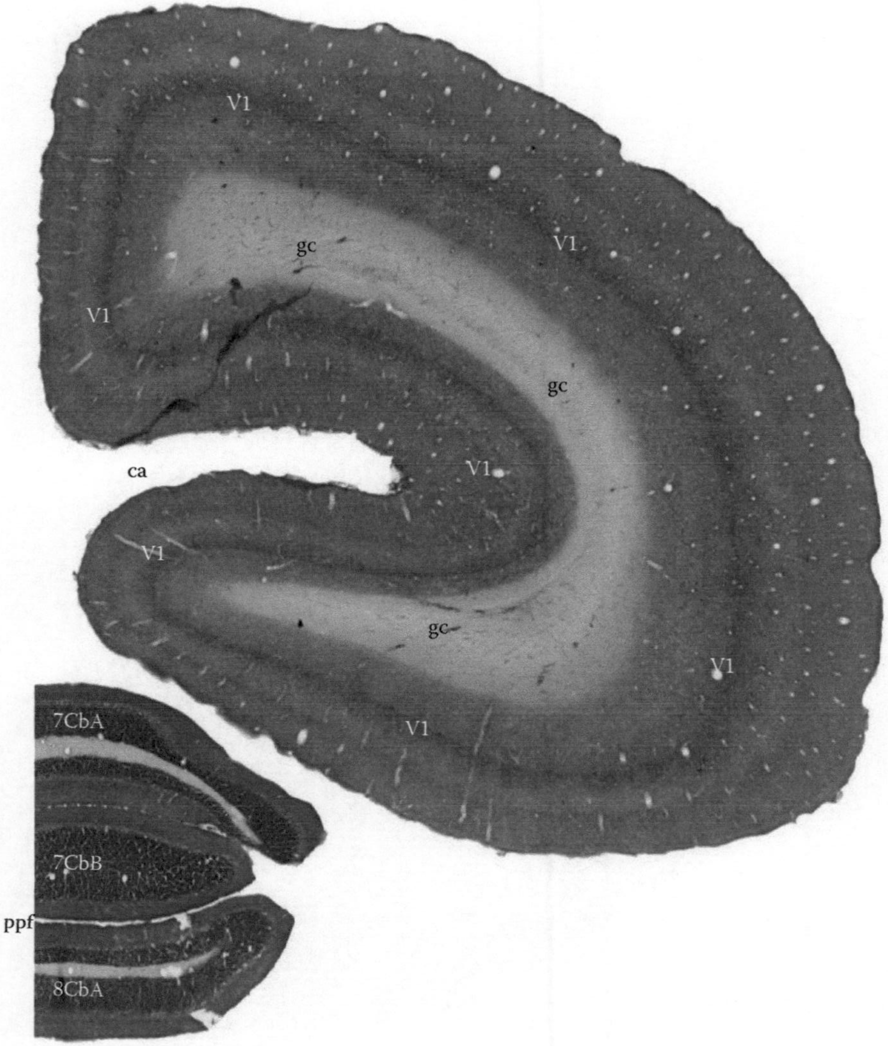


**Neurofilament**
Interaural -09.20 mm

# Plate 56c

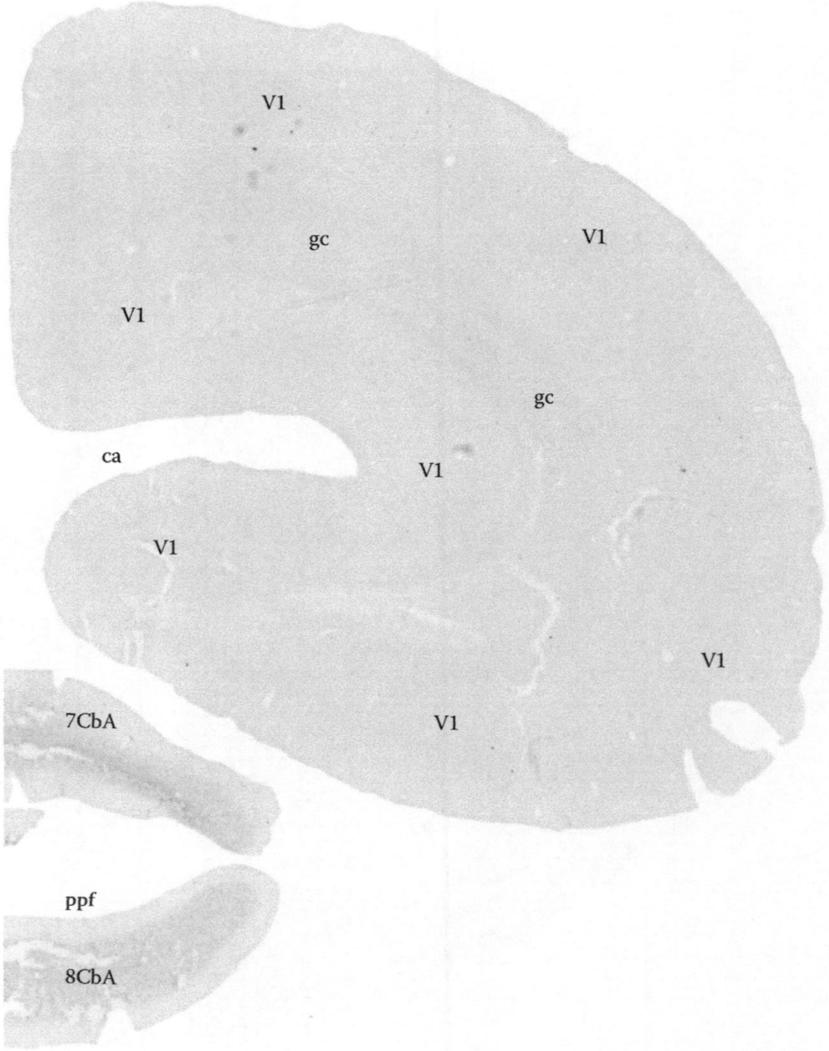


**Acetylcholinesterase**
Interaural -09.25 mm

# Plate 56d

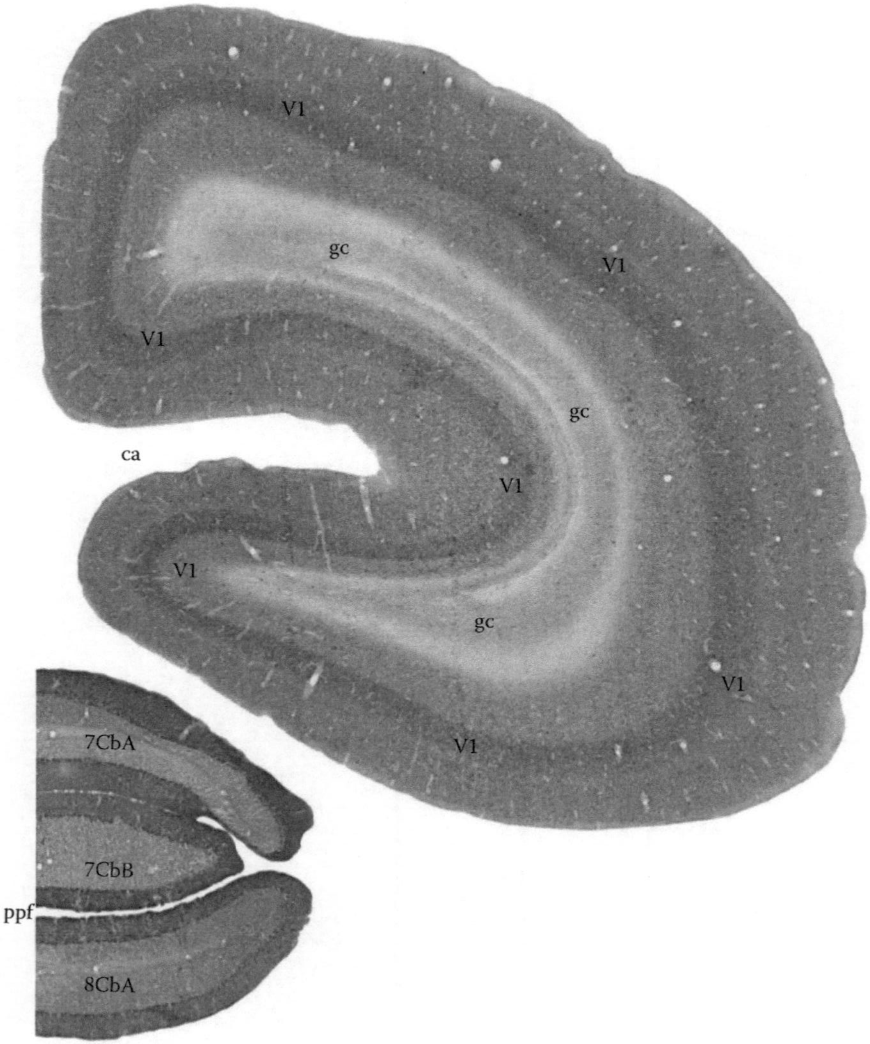


**Parvalbumin**
Interaural -09.30 mm

# Plate 56e

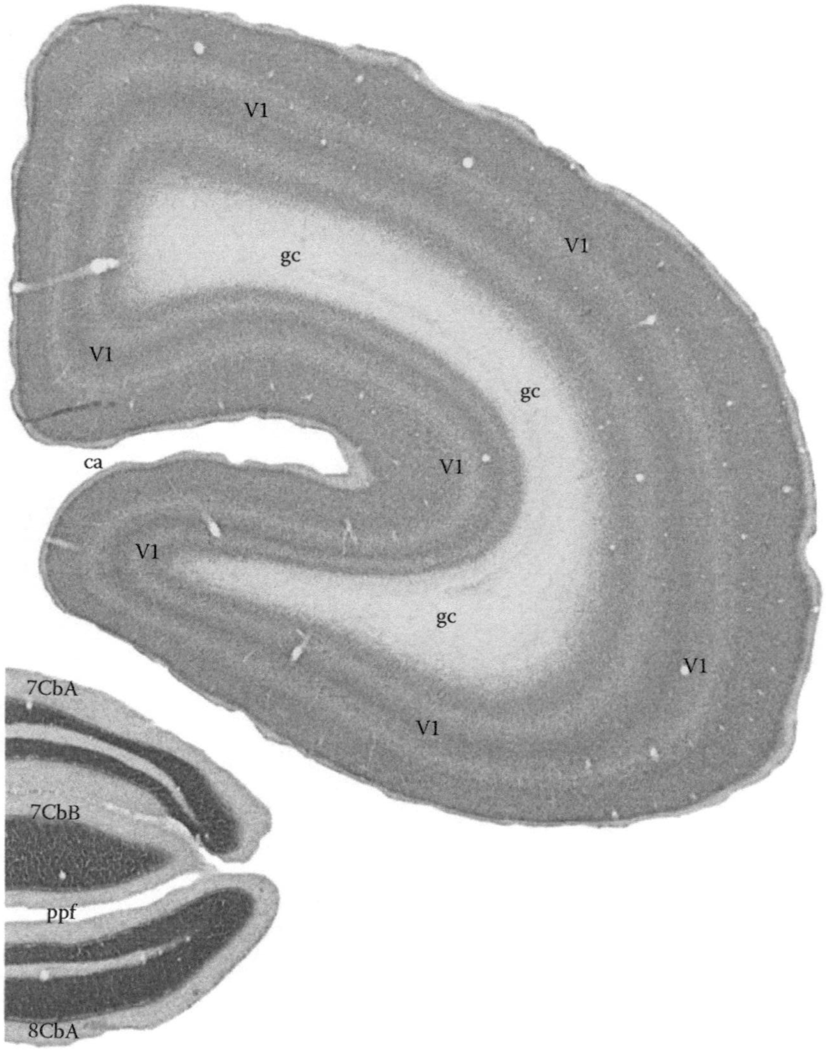


**Cresyl Violet**
Interaural -09.35 mm

# Plate 56

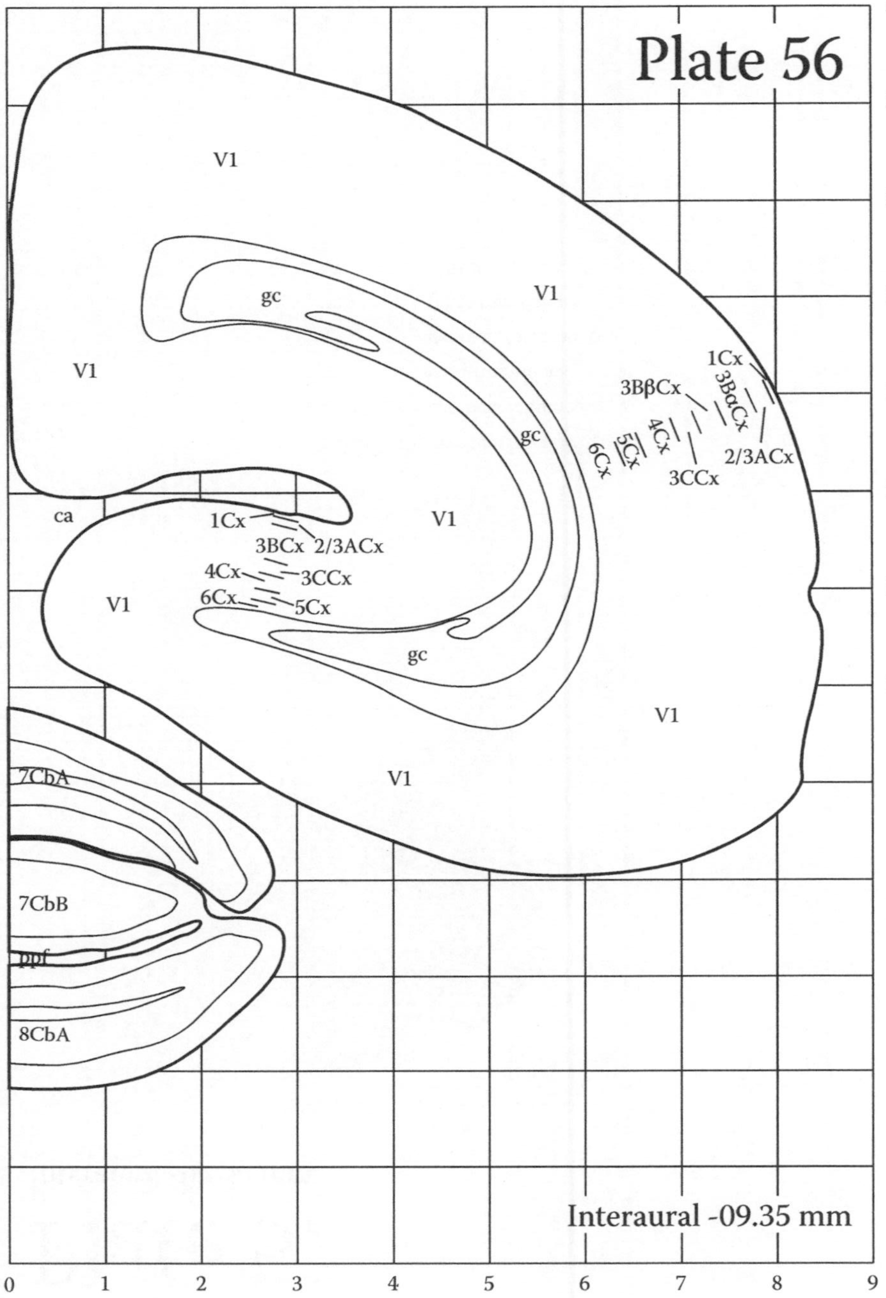


| | |
|---|---|
| 7CbA | cerebell. lobule 7A |
| 7CbB | cerebell. lobule 7B |
| 8CbA | cerebell. lobule 8A |
| ca | calcarine sulcus |
| gc | geniculocalcarine tr. |
| ppf | prepyramidal fiss. |
| V1 | V1 Cx |

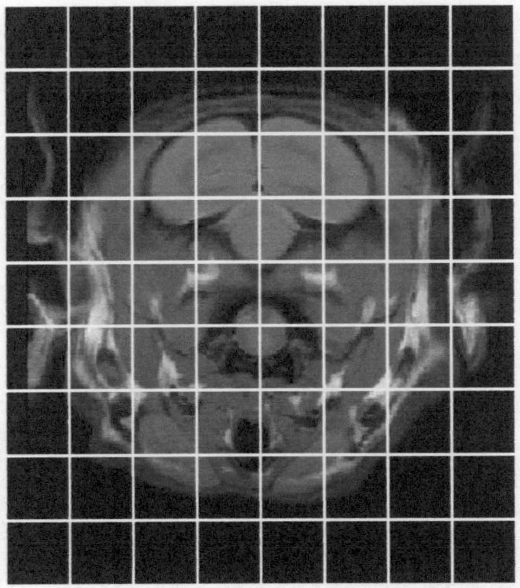

Interaural -09.30 mm

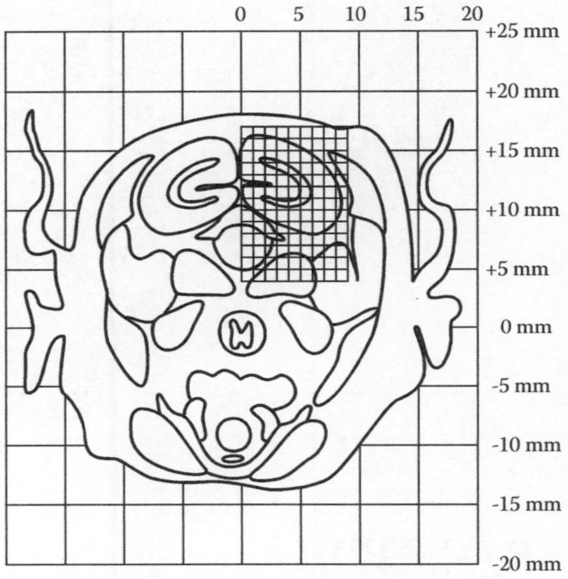

# Plate 57

Interaural -09.85 mm

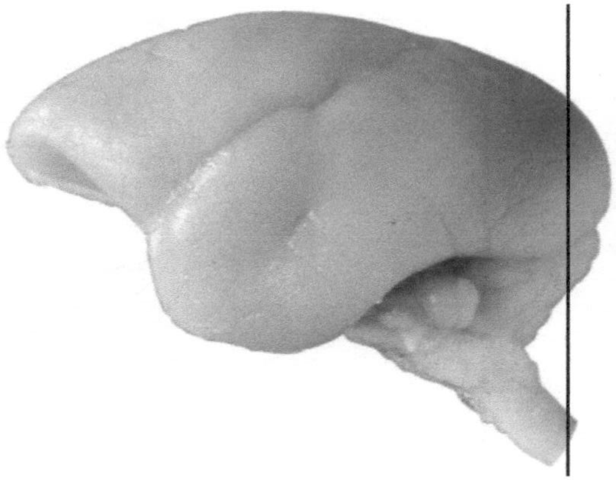

| | |
|---|---|
| 7CbA(b-d) | cerebell. lobule 7A(b-d) |
| 8CbA | cerebell. lobule 8A |
| ca | calcarine sulcus |
| gc | geniculocalcarine tr. |
| ppf | prepyramidal fiss. |
| V1 | V1 Cx |

# Plate 57a

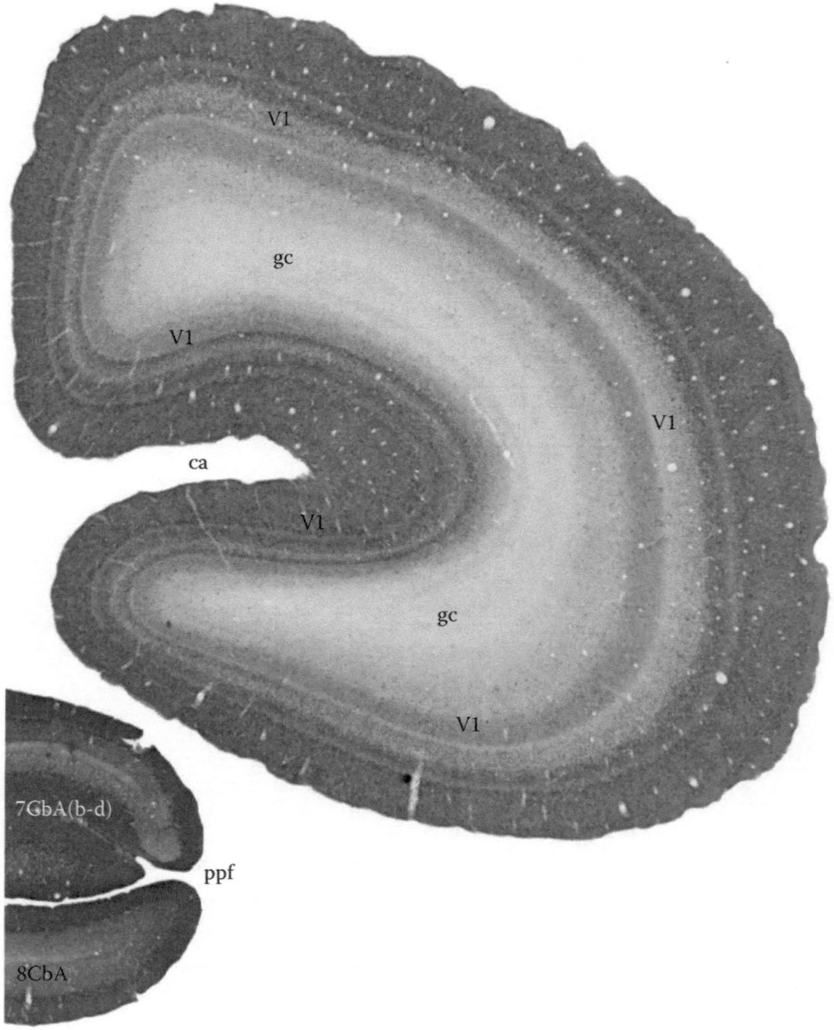


**Calbindin**

Interaural -09.55 mm

# Plate 57b

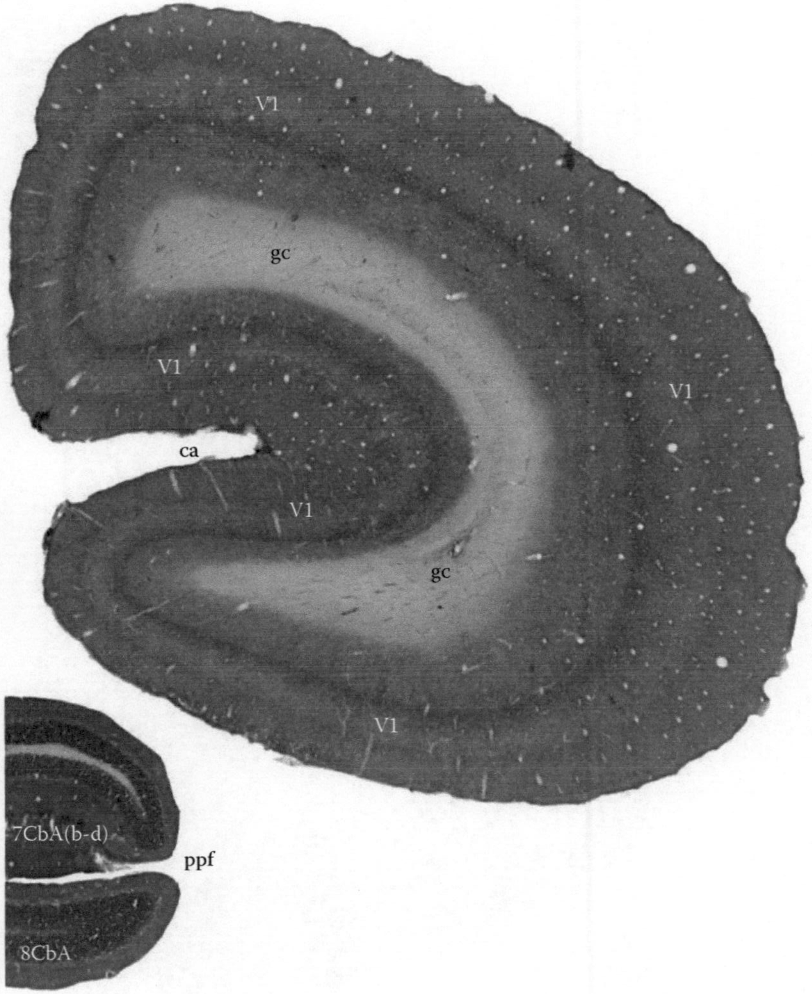


**Neurofilament**
Interaural -09.70 mm

# Plate 57c

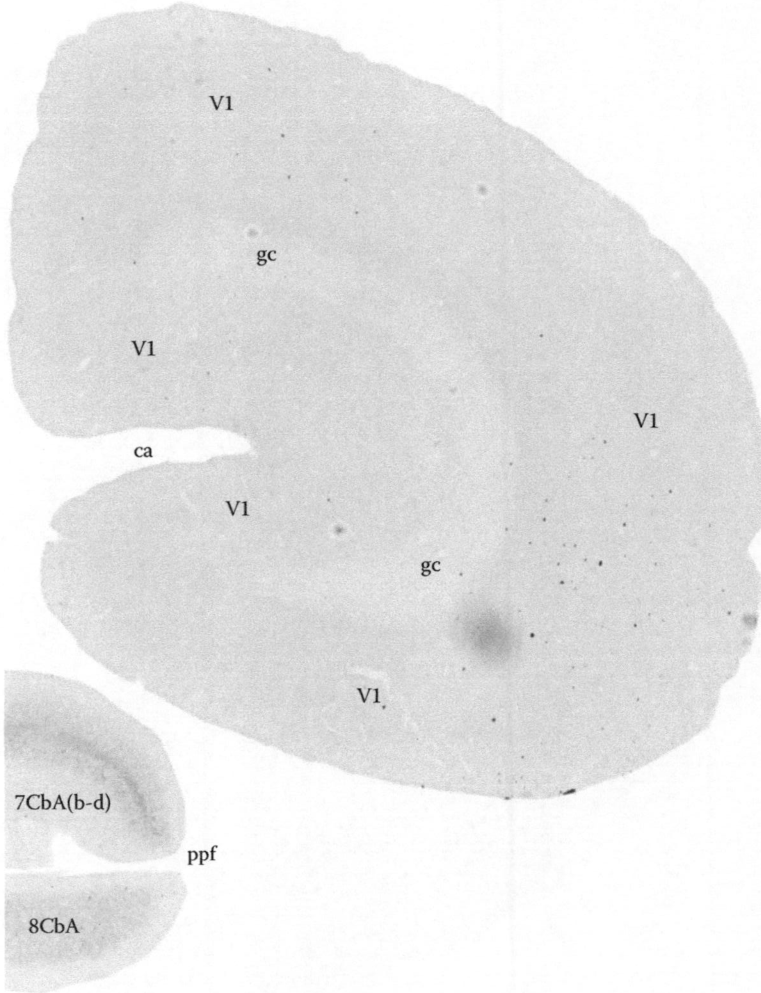


**Acetylcholinesterase**
Interaural -09.75 mm

# Plate 57d

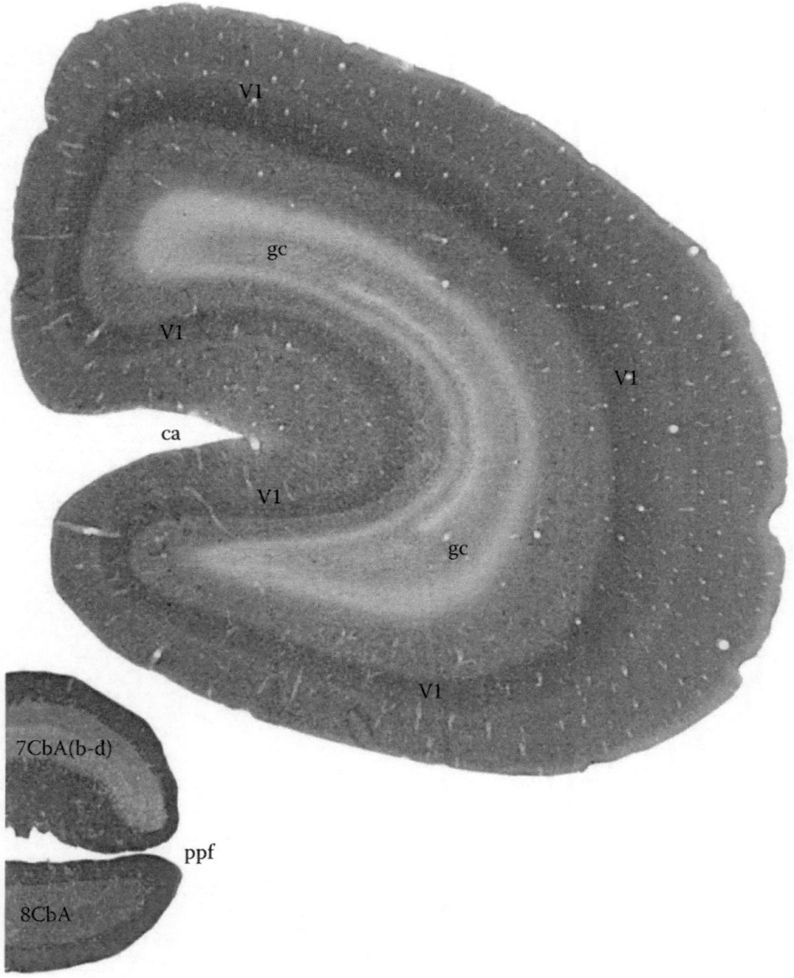


**Parvalbumin**
Interaural -09.80 mm

# Plate 57e

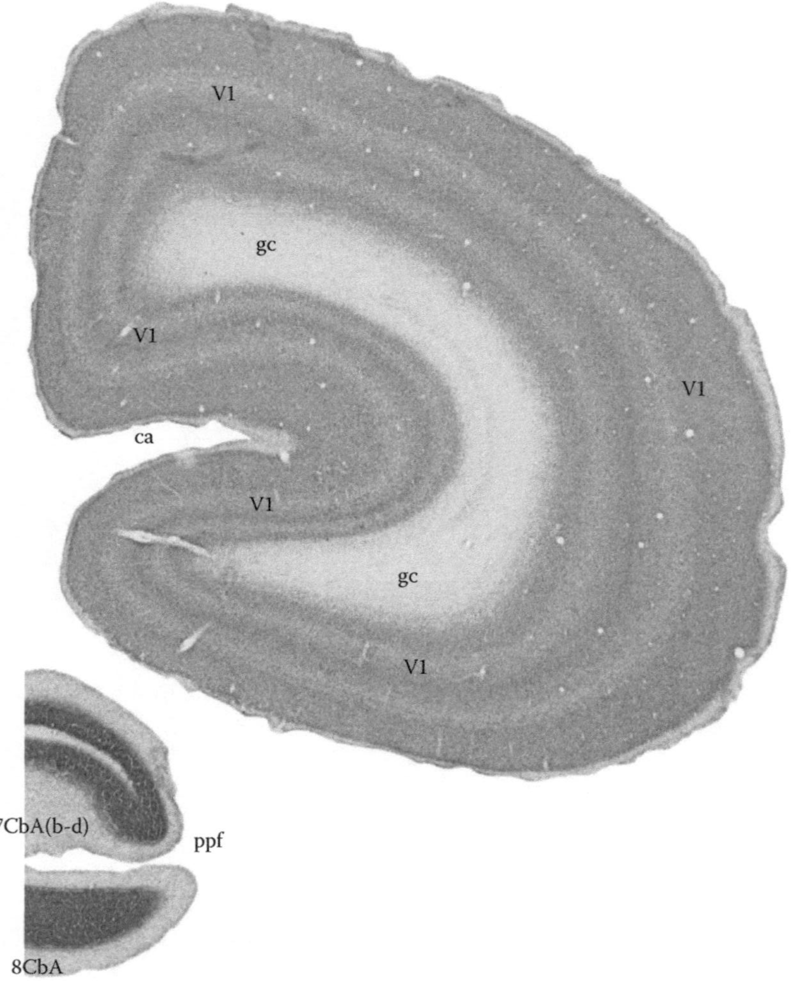


**Cresyl Violet**
Interaural -09.85 mm

# Plate 57

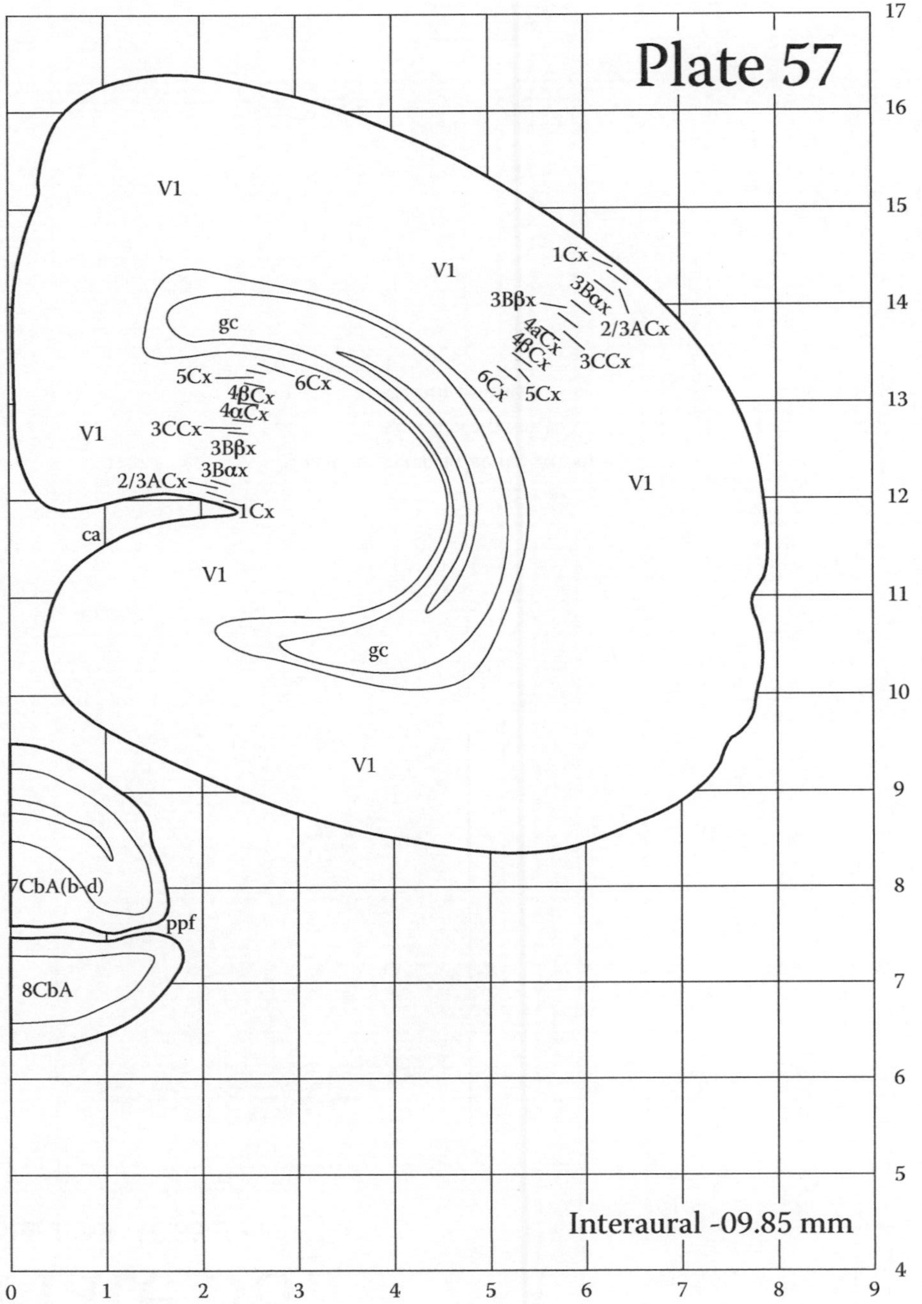


| | |
|---|---|
| 7CbA(b-d) | cerebell. lobule 7A(b-d) |
| 8CbA | cerebell. lobule 8A |
| ca | calcarine sulcus |
| gc | geniculocalcarine tr. |
| ppf | prepyramidal fiss. |
| V1 | V1 Cx |

# Plate 58

Interaural -10.35 mm

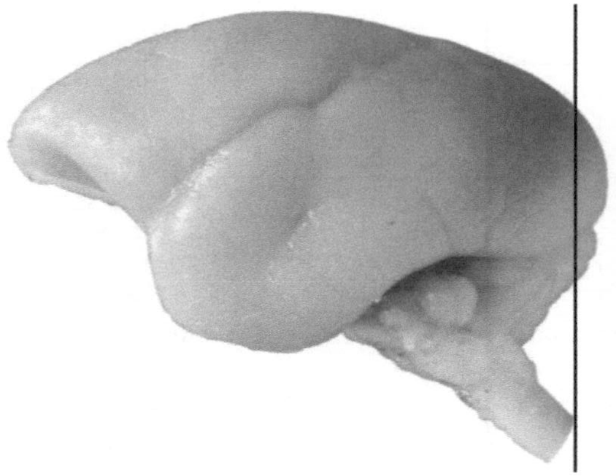

| | |
|---|---|
| 7CbA(b-d) | cerebell. lobule 7A(b-d) |
| gc | geniculocalcarine tr. |
| ca | calcarine sulcus |
| V1 | V1 Cx |

# Plate 58a

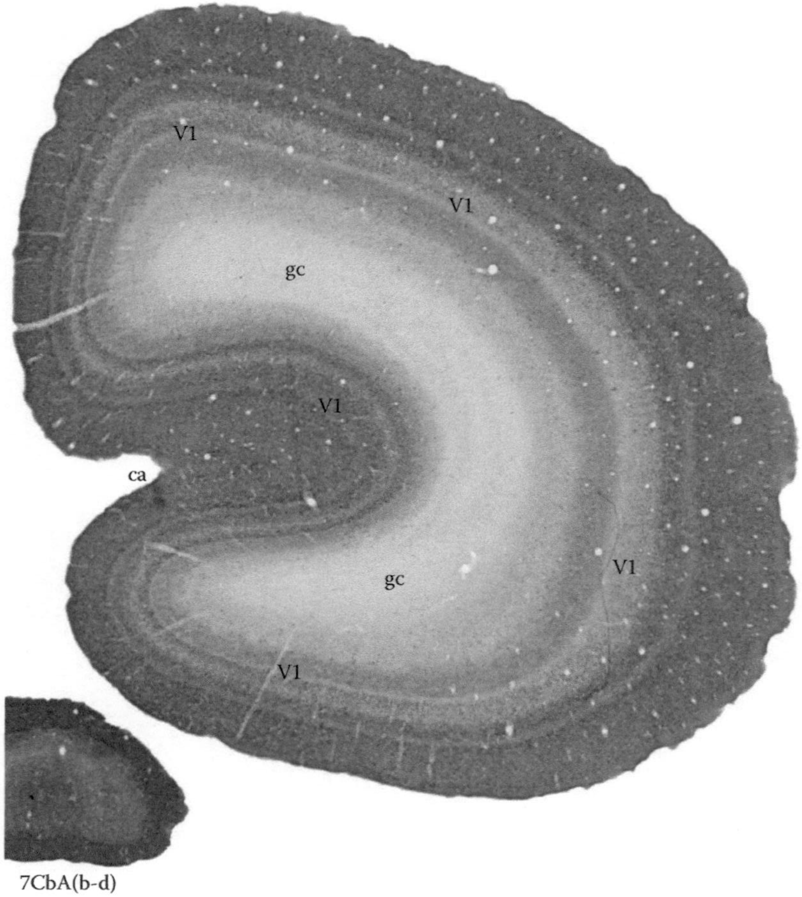


**Calbindin**
Interaural -10.05 mm

# Plate 58b

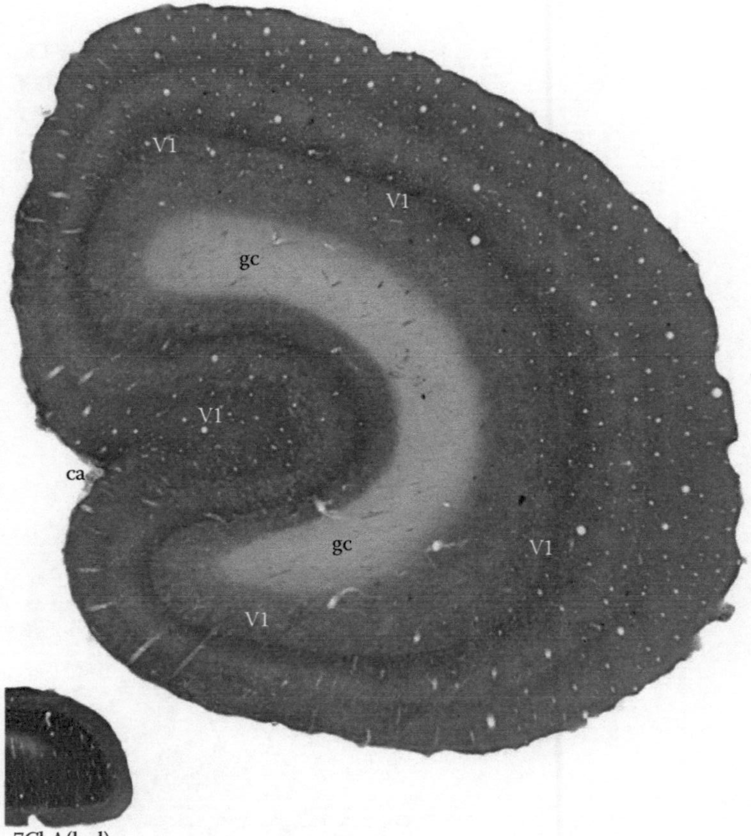


**Neurofilament**
Interaural -10.20 mm

# Plate 58c

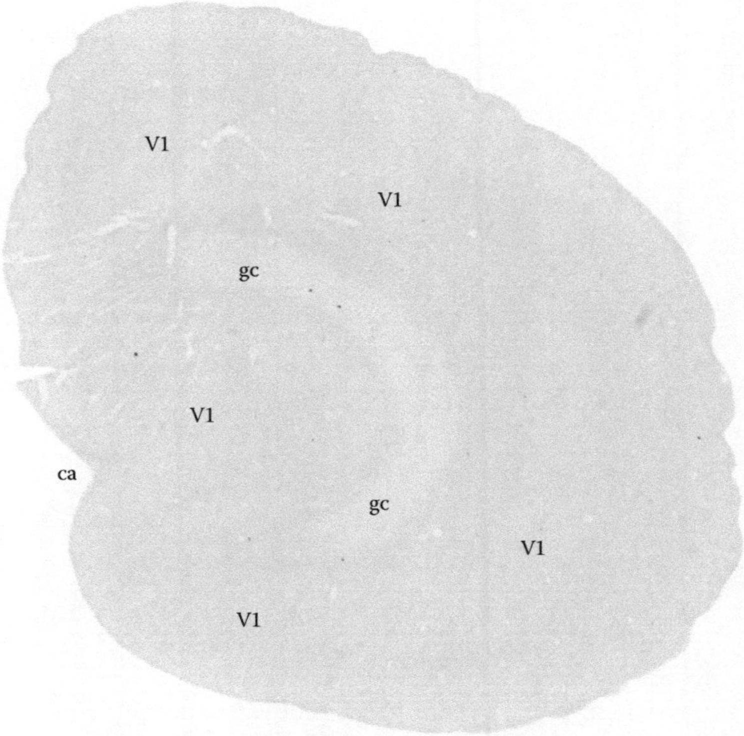


**Acetylcholinesterase**
Interaural -10.25 mm

# Plate 58d

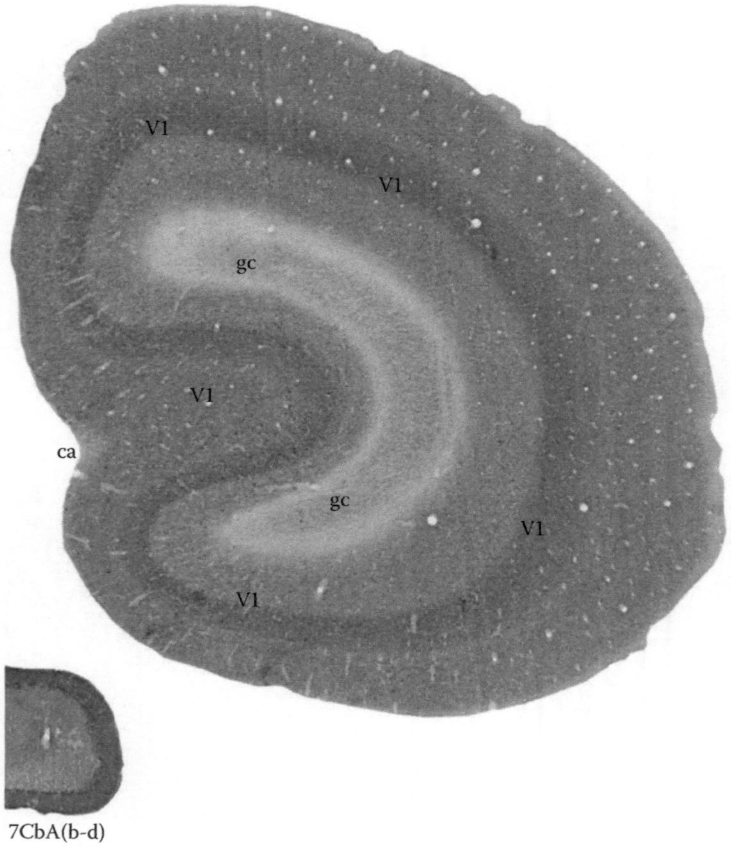


**Parvalbumin**
Interaural -10.30 mm

# Plate 58e

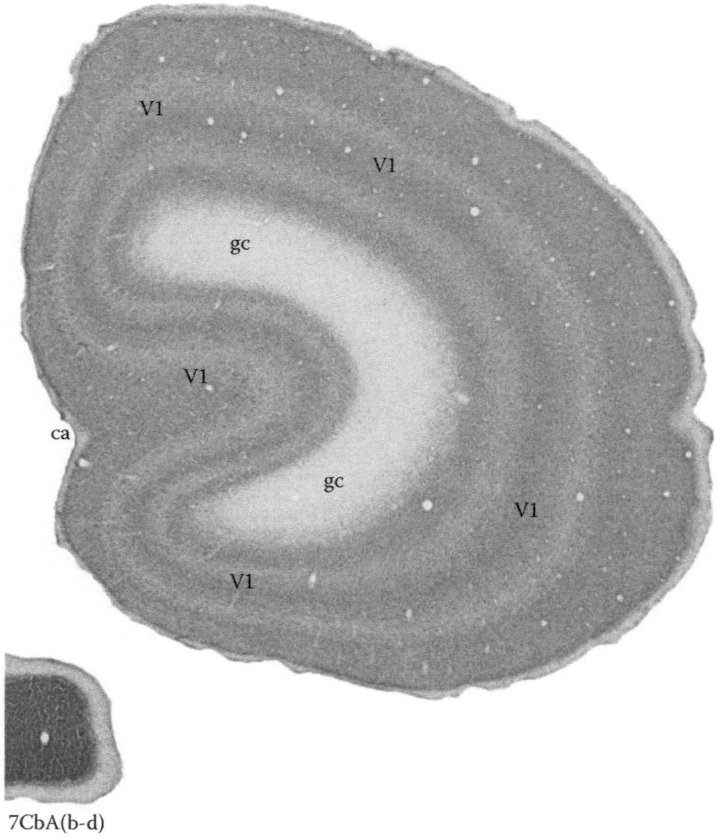


**Cresyl Violet**
Interaural -10.35 mm

# Plate 58

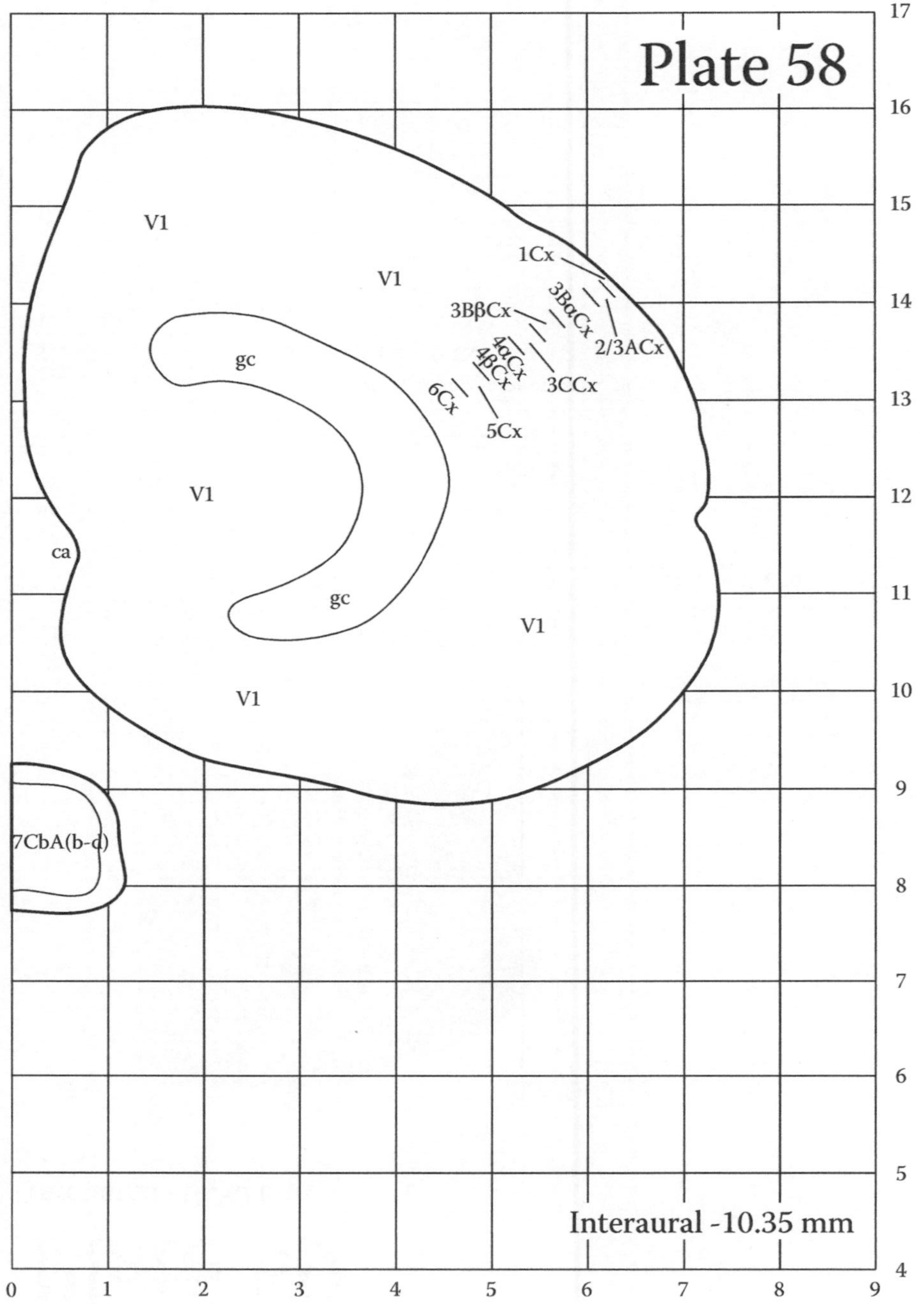


| | |
|---|---|
| 7CbA(b-d) | cerebell. lobule 7A(b-d) |
| gc | geniculocalcarine tr. |
| ca | calcarine sulcus |
| V1 | V1 Cx |

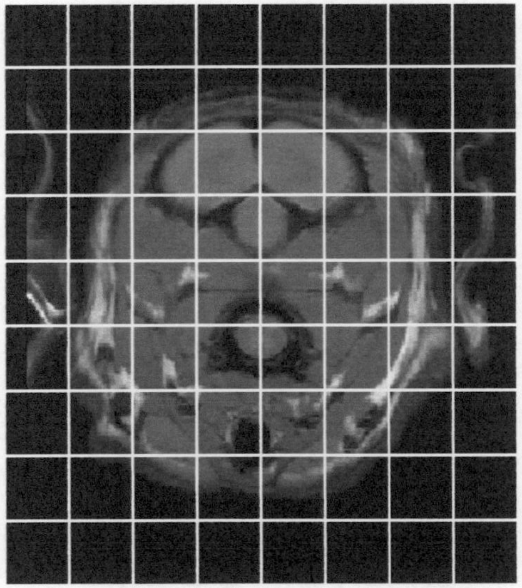

Interaural -10.25 mm

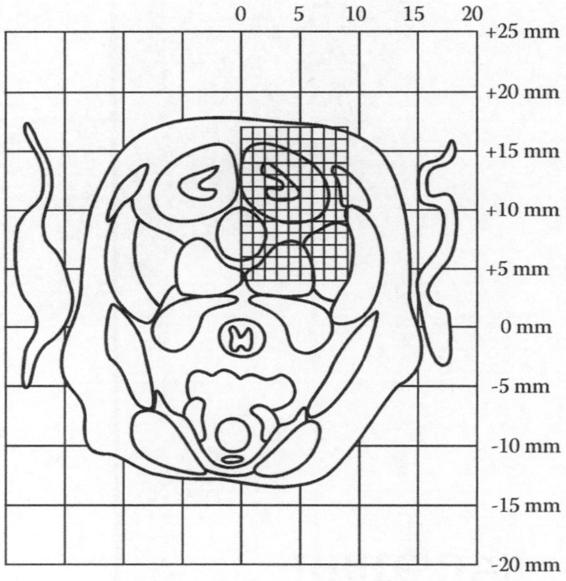

# Plate 59

Interaural -10.90 mm

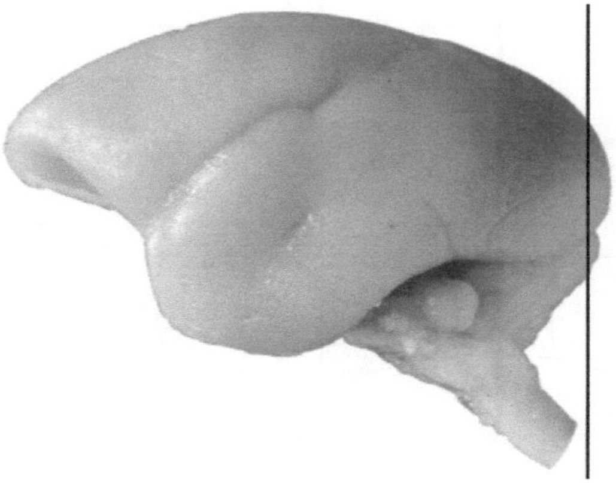

| | |
|---|---|
| gc | geniculocalcarine tr. |
| V1 | V1 Cx |

# Plate 59a

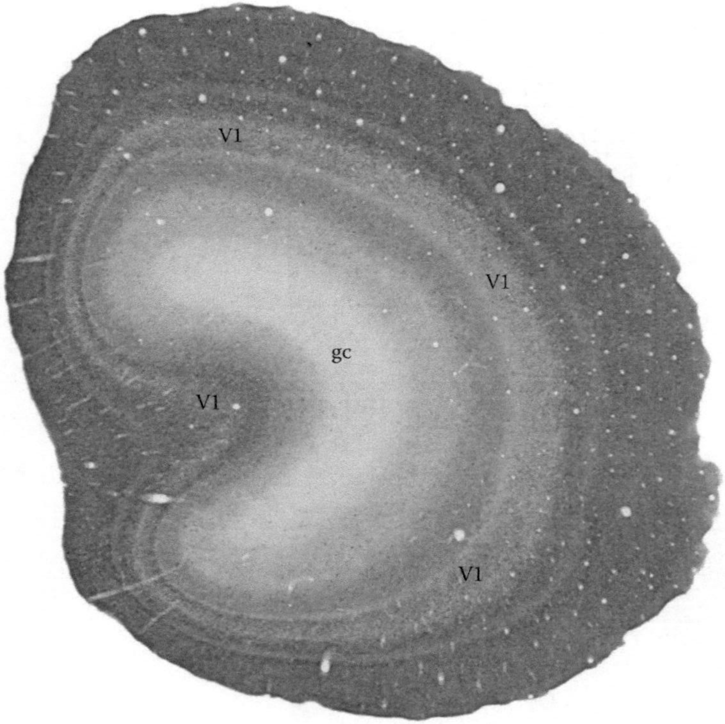


**Calbindin**
Interaural -10.60 mm

# Plate 59b

**Neurofilament**
Interaural -10.75 mm

# Plate 59c

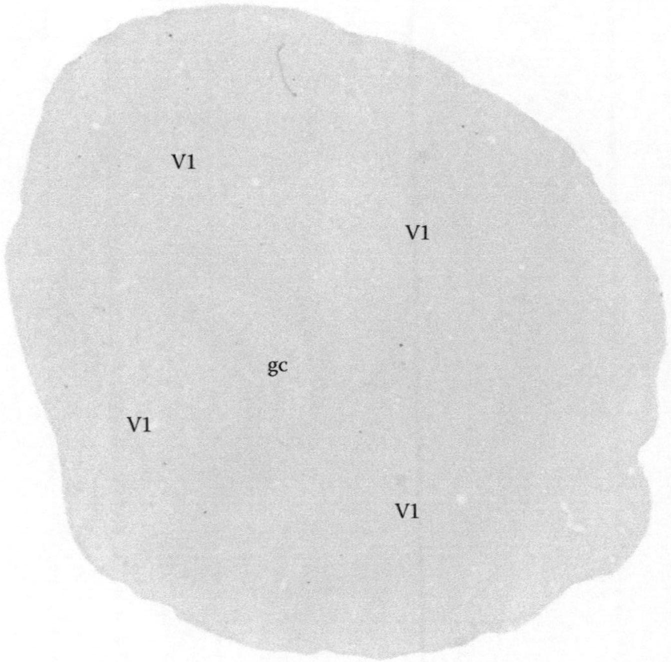


**Acetylcholinesterase**
Interaural -10.80 mm

# Plate 59d

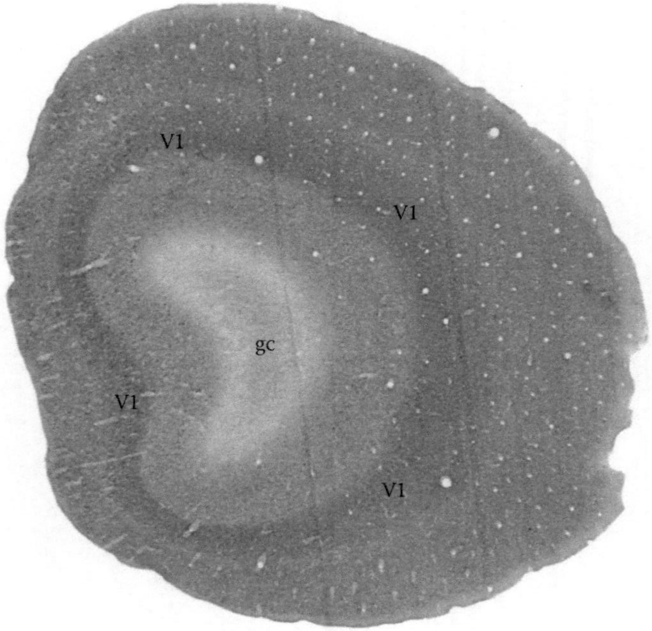


**Parvalbumin**
Interaural -10.85 mm

# Plate 59e

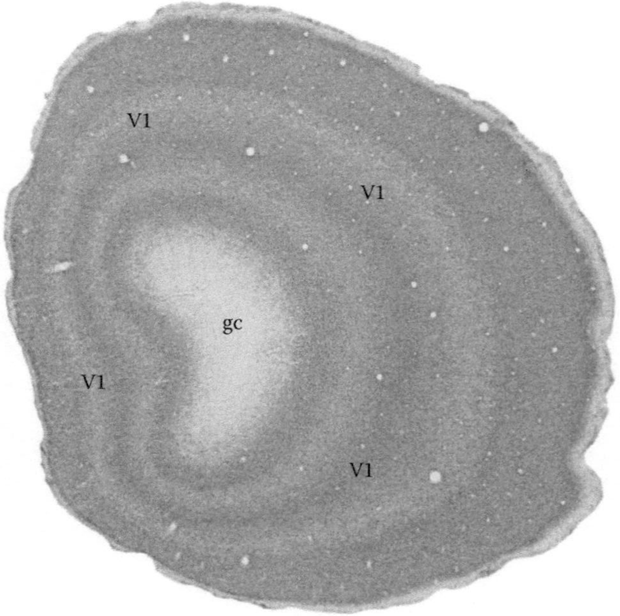


**Cresyl Violet**
Interaural -10.90 mm

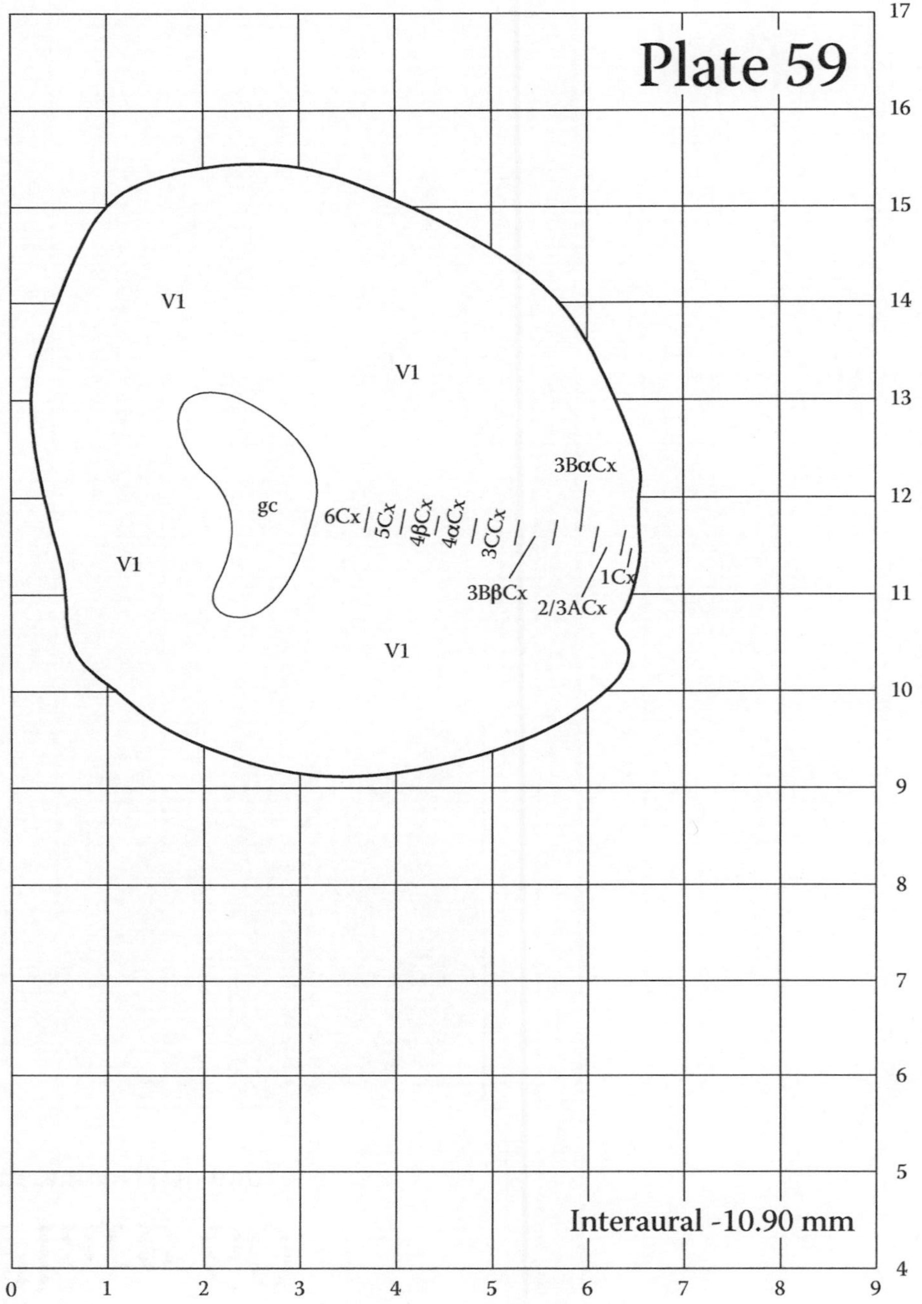


| | |
|---|---|
| gc | geniculocalcarine tr. |
| V1 | V1 Cx |

# Plate 60

Interaural -11.40 mm

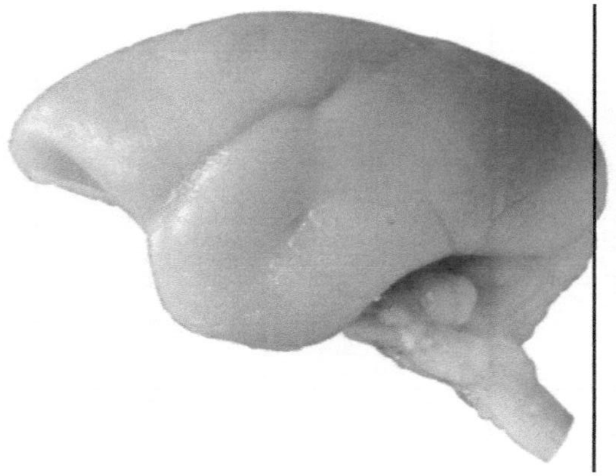

V1 V1 Cx

# Plate 60a

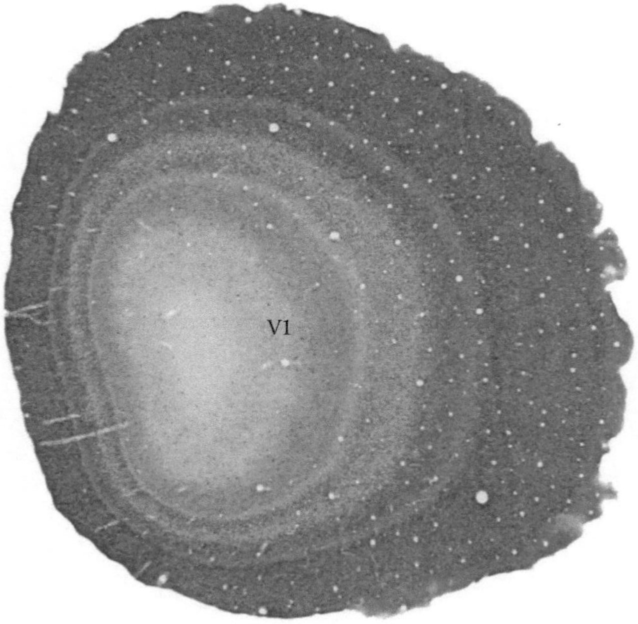


**Calbindin**
Interaural -10.10 mm

# Plate 60b

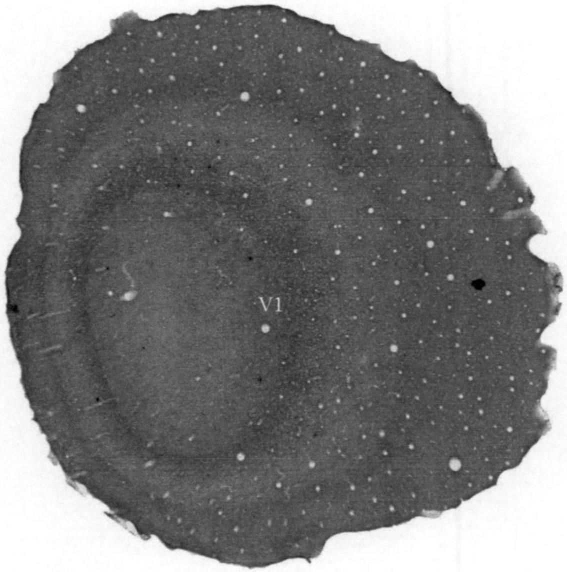


**Neurofilament**
Interaural -11.25 mm

# Plate 60c

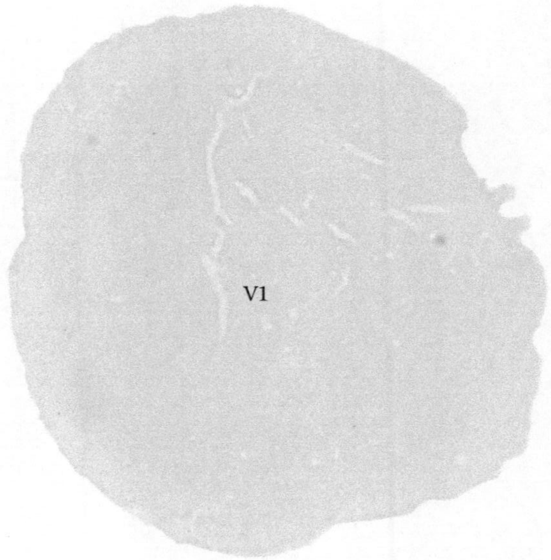


**Acetylcholinesterase**
Interaural -11.30 mm

# Plate 60d

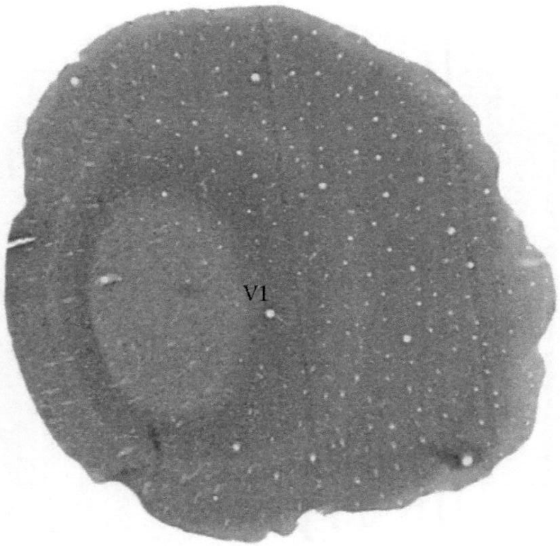


**Parvalbumin**
Interaural -11.35 mm

# Plate 60e

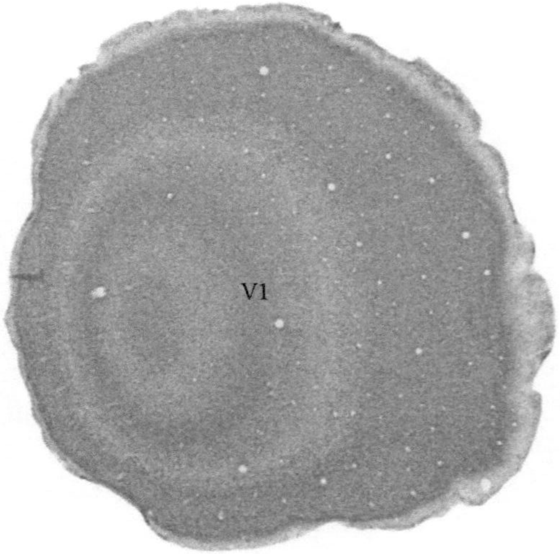


**Cresyl Violet**
Interaural -11.40 mm

# Plate 60

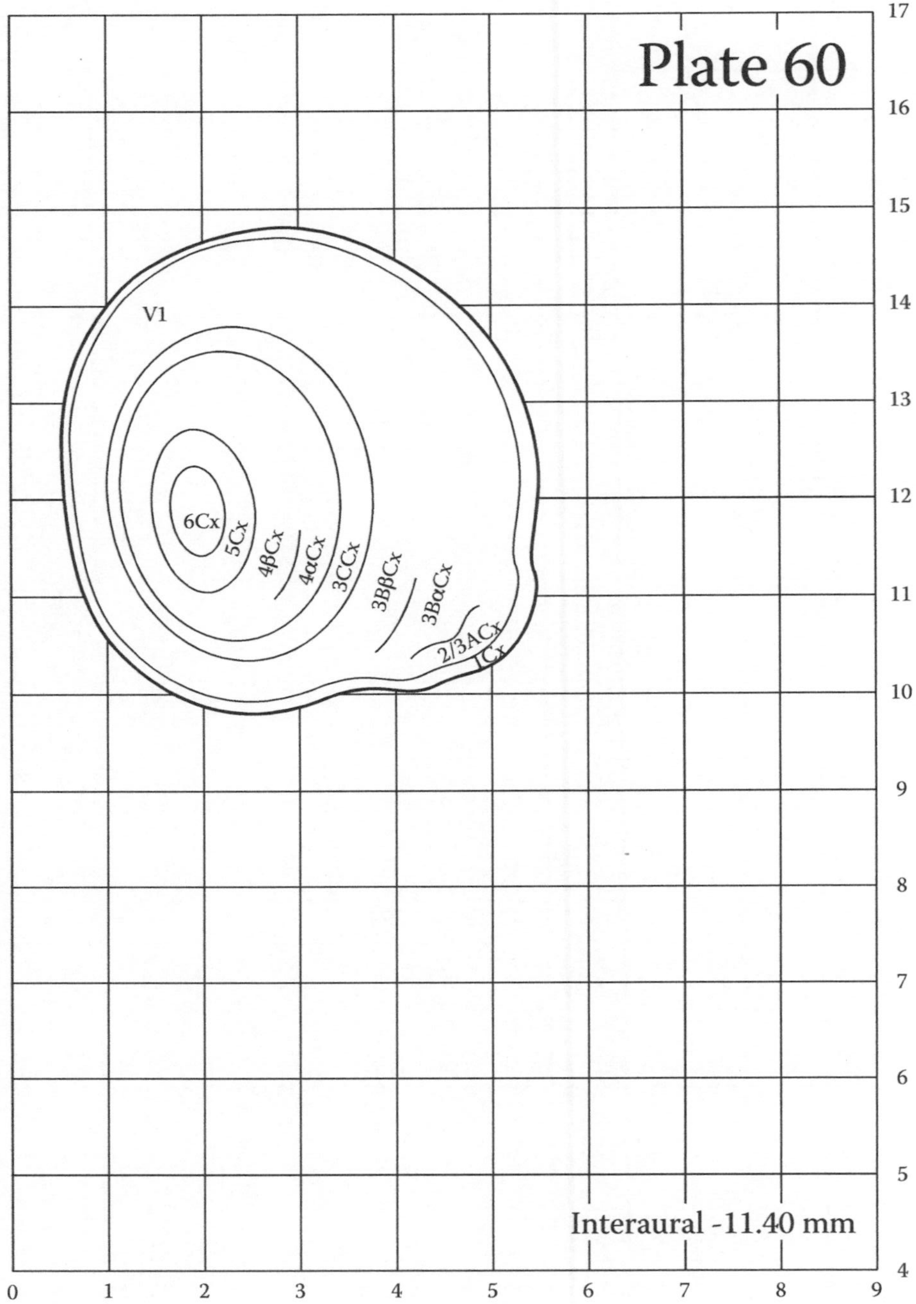


Interaural -11.40 mm

V1 V1 Cx

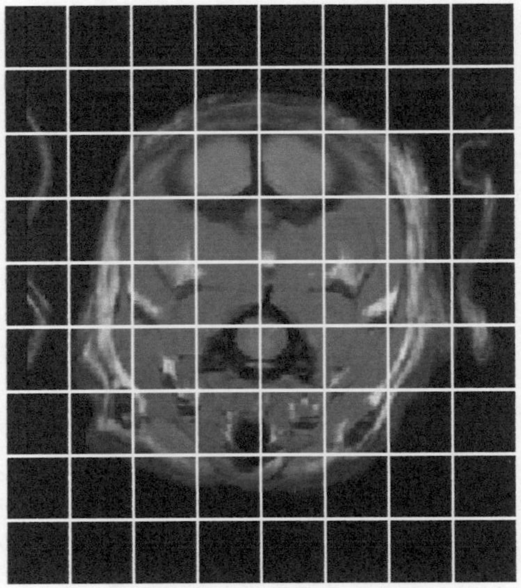

Interaural -11.20 mm

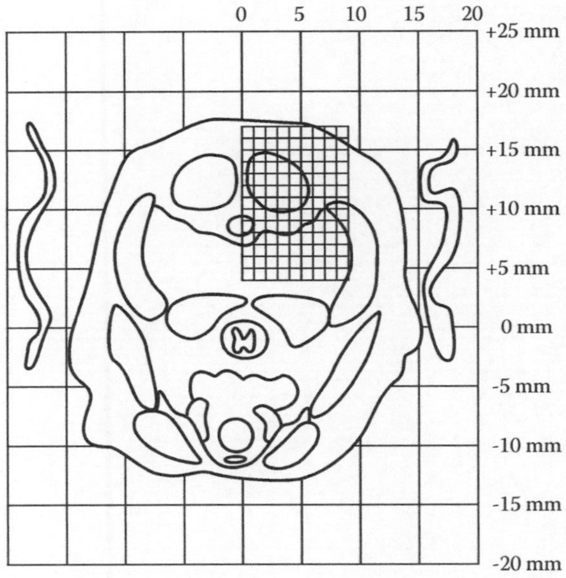

# Plate 61

Interaural -11.90 mm

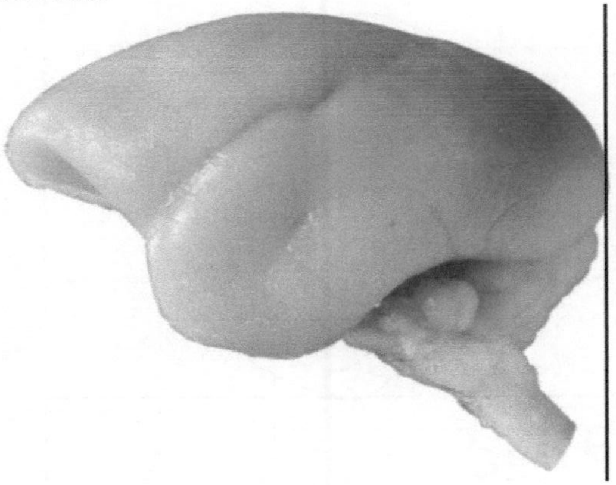

# Plate 61a

**Calbindin**
Interaural -11.60 mm

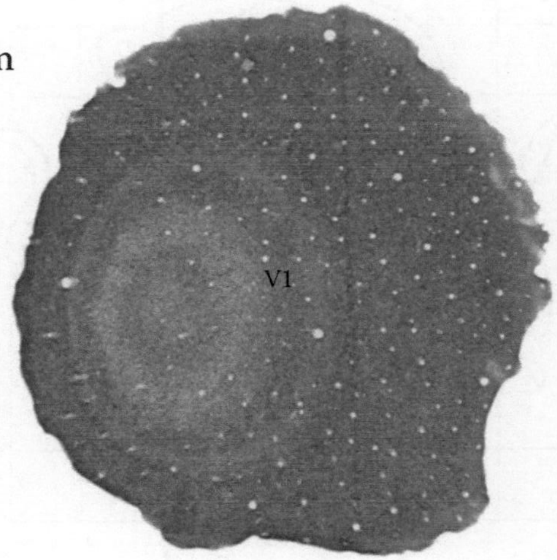


# Plate 61b

**Neurofilament**
Interaural -11.75 mm

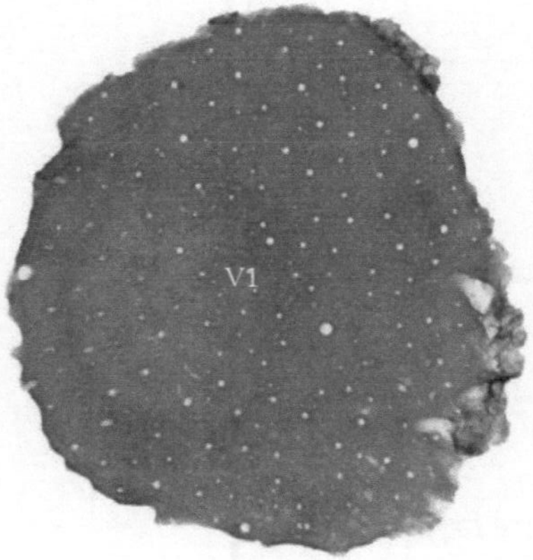


# Plate 61c

**Acetylcholinesterase**
Interaural -11.80 mm

# Plate 61d

**Parvalbumin**
Interaural -11.85 mm

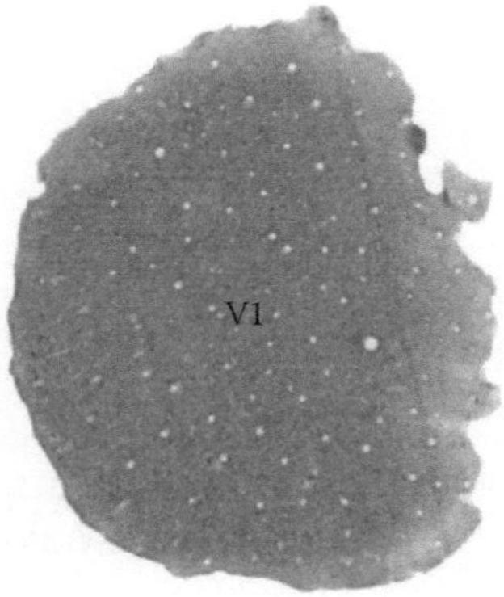


# Plate 61e

**Cresyl Violet**
Interaural -11.90 mm

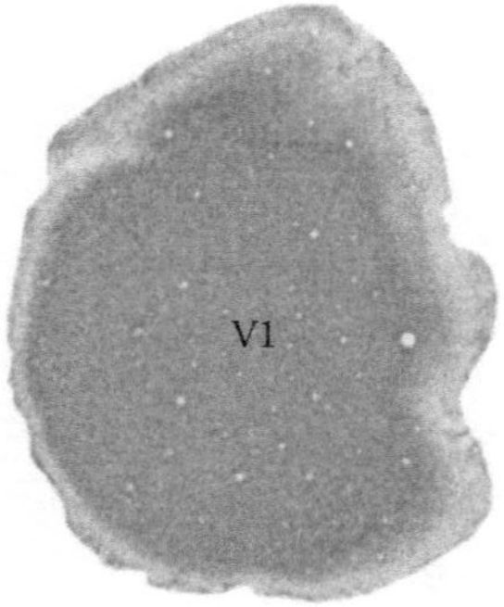


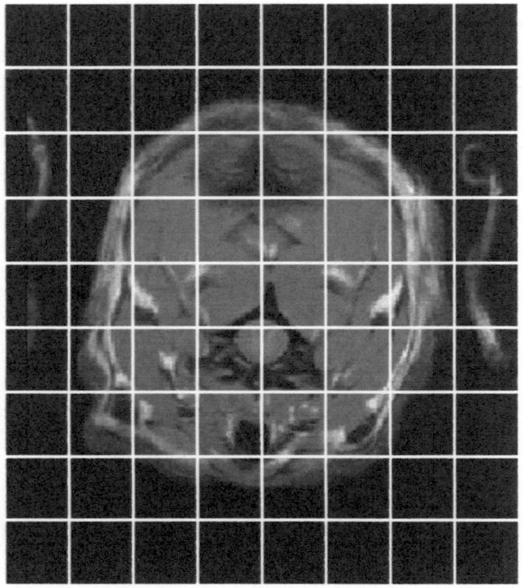

Interaural -12.10 mm

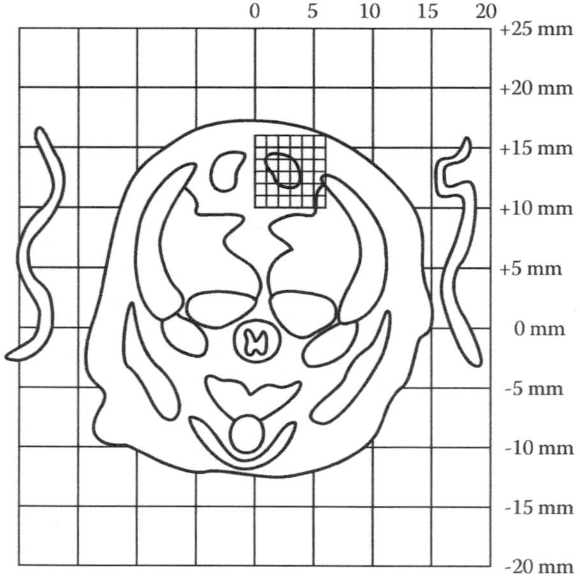


1Cx layer 1 cortex
2Cx layer 2 cortex
V1 V1 Cx

# Plate 61

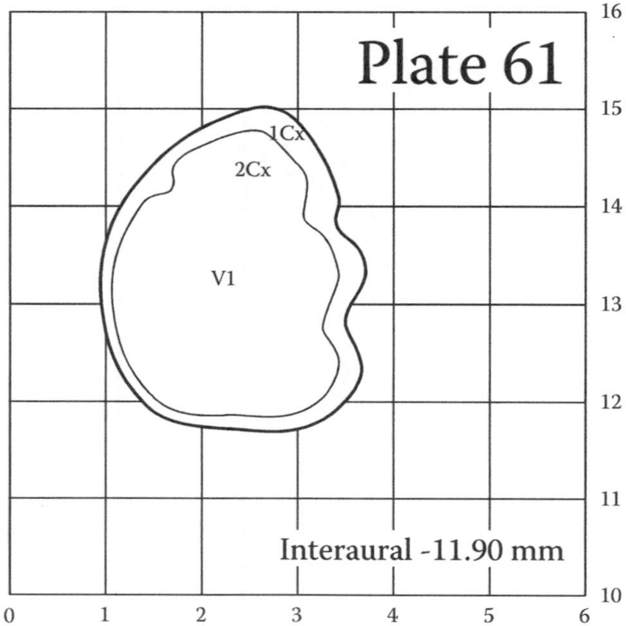

# Sagittal Plates

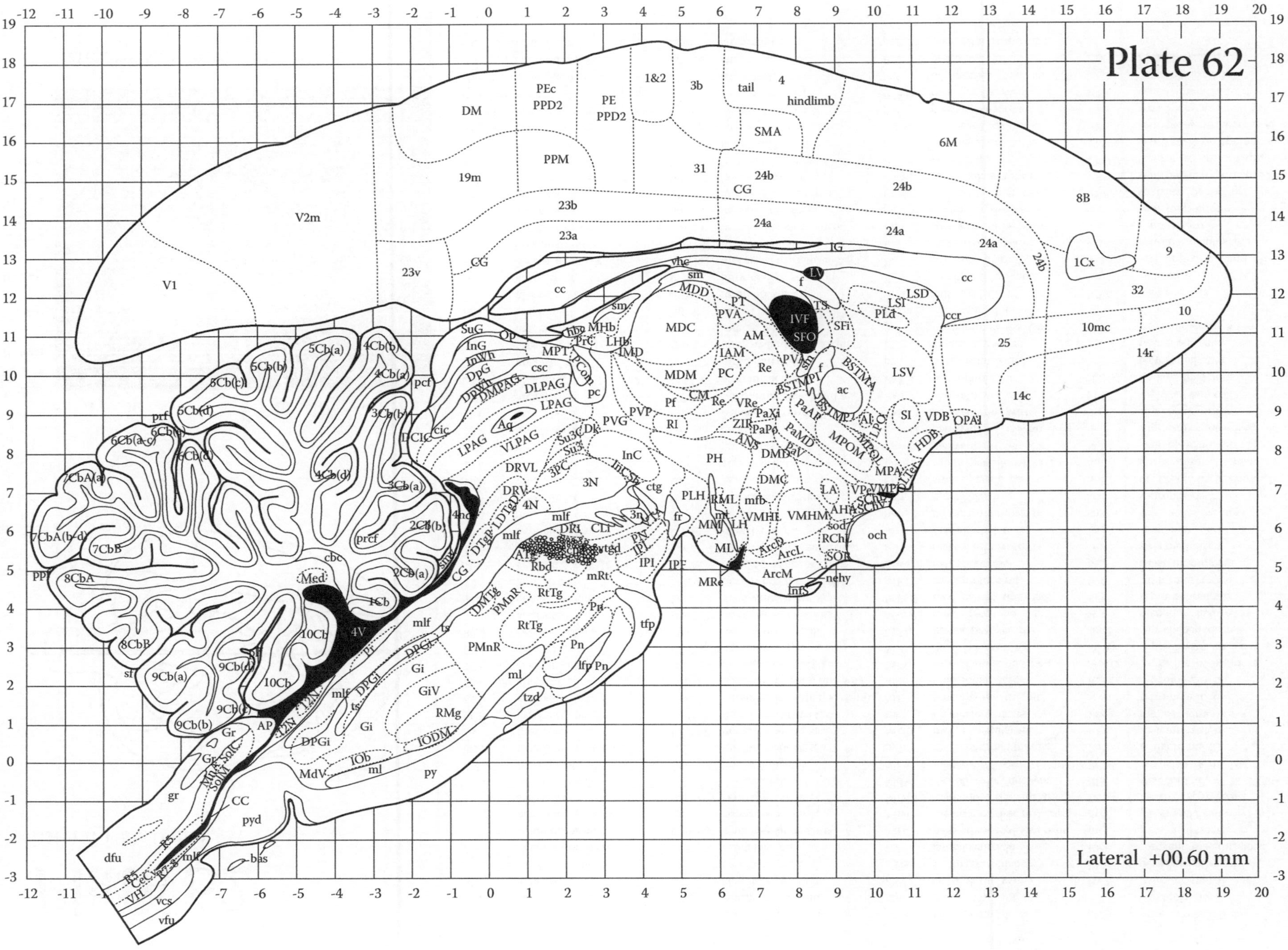
Plate 62
Lateral +00.60 mm
DM
PEc
PPD2
PE
PPD2
1&2
3b
tail
4
hindlimb
SMA
6M
PPM
19m
31
24b
CG
24b
8B
V2m
23b
24a
24a
24a
23a
IG
24b
1Cx
9
V1
23v
CG
cc
vhc
sm
LV
cc
32
f
MDD
PT
PVA
IVF
SFO
TS
LSI
LSD
PLd
ccr
10
MDC
AM
SFi
25
10mc
14r
IAM
BSTMA
LSV
MDM
PC
Re
ac
14c
CM
Pf
Re
VRe
SI
VDB
OPAl
PVG
PVP
RI
PaXi
ZIR
PaPo
HDB
PaMD
MPOM
InC
PH
DMD
PaV
MPA
DMC
LA
3N
PLH
RML
mfb
VMHM
och
VMHL
LH
MM
ML
ArcL
ArcM
MRe
InfS
nehy
SOR
RChL
sod
AHA
SuG
InG
InWh
DpG
DpWh
DMPAG
csc
DLPAG
LPAG
MPT
PCom
pc
hbc
MHb
LHb
PrC
IMD
Op
Aq
VLPAG
LPAG
DRVL
DRV
4N
3PC
Su3
Su3C
Dk
InCSh
ctg
fr
3n
mlf
DRI
CLi
IPL
IPI
IPF
Rbd
mRt
RtTg
Pn
tfp
lfp
PMnR
DMTg
PMnR
RtTg
ml
tzd
4V
mlf
Gi
GiV
RMg
IODM
IOb
ml
py
MdV
DPGi
Gi
AP
Gr
CC
pyd
gr
dfu
bas
vcs
vfu
VH
5Cb(a)
4Cb(b)
5Cb(b)
4Cb(a)
5Cb(c)
pcf
prf
6Cb(d)
3Cb(b)
6Cb(c)
6Cb(a-c)
6Cb(b)
DCIC
cic
7CbA(a)
4Cb(d)
3Cb(a)
4nc
7CbA(b-d)
7CbB
prcf
2Cb(b)
cbc
8CbA
Med
2Cb(a)
1Cb
ppf
8CbB
10Cb
9Cb(d)
9Cb(a)
10Cb
sf
9Cb(c)
9Cb(b)

# Plate 62

Lateral +00.60 mm

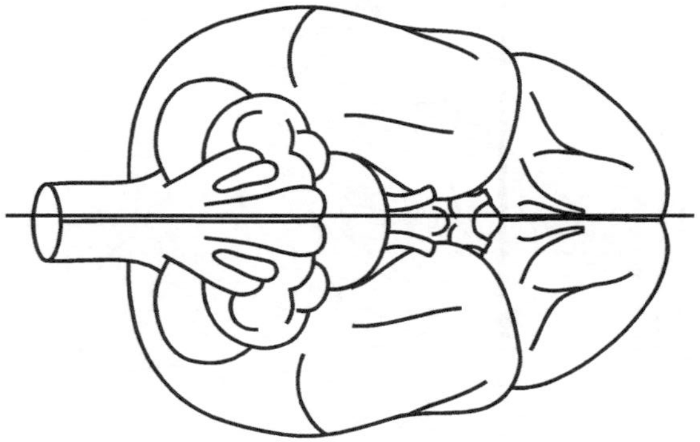

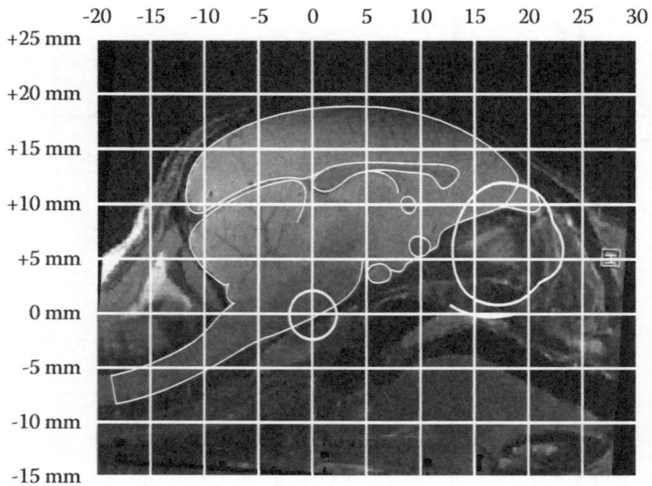


Lateral +00.50 mm

| | |
|---|---|
| 1&2 | areas 1/2 Cx |
| 2Cb | cerebell. lobule 2 |
| 3b | area 3b Cx |
| 3Cb | cerebell. lobule 3 |
| 3n | oculomotor nerve |
| 3N | oculomotor nu. |
| 3PC | oculomotor nu., parv. |
| 4 | area 4 Cx |
| 4Cb | cerebell. lobule 4 |
| 4N | trochlear nu. |
| 4nd | trochlear nerve decn |
| 4V | fourth ventricle |
| 5Cb | cerebell. lobule 5 |
| 6Cb | cerebell. lobule 6 |
| 6M | area 6M Cx |
| 7Cb | cerebell. lobule 7 |
| 8B | area 8B Cx |
| 8Cb | cerebell. lobule 8 |
| 9 | area 9 Cx |
| 9Cb | cerebell. lobule 9 |
| 10 | area 10 Cx |
| 10Cb | cerebell. lobule 10 |
| 10mc | area 10mc Cx |
| 12N | hypoglossal nu. |
| 14c | area 14c Cx |
| 14r | area 14r Cx |
| 19m | area 19m Cx |
| 23a | area 23a Cx |
| 23b | area 23b Cx |
| 23v | area 23v Cx |
| 24a | area 24a Cx |
| 24b | area 24b Cx |
| 25 | area 25 Cx |
| 31 | area 31 Cx |
| 32 | area 32 Cx |
| ac | ant. comm. |
| AHA | ant. hypoth. area, ant. |
| Al | alar nu. |
| AM | anteromed. thal. nu. |
| ANS | access. neurosec. nu. |
| AP | area postrema |
| Aq | cerebral aqueduct |
| ArcD | arc. hypoth. nu., dors. |
| ArcL | arc. hypoth. nu., lat. |
| ArcM | arc. hypoth. nu., med. |
| ATg | ant. tegmental nu. |
| bas | basilar artery |
| BSTMA | bed nu. st, med. ant. divn |
| BSTMP | bed nu. st, med. post. divn |
| BSTMPI | bed nu. st, med. post. int. |
| cbc | cerebellar commiss. |
| cc | corp. callosum |
| CC | central canal |
| ccr | corp. call. rostrum |
| CeCv | central cervical nu. |
| CG | cingulate gyrus Cx |
| cic | comm. inf. coll. |
| CLi | caudal lin. nu. |
| CM | cent. medial thal. nu. |
| csc | comm. sup. coll. |
| ctg | central tegm. tract |
| DCIC | dors. Cx inf. coll. |
| dfu | dorsal funiculus |
| Dk | nu. Darkschewitsch |
| DLPAG | dorsolat. periaq. gr. |
| DM | dorsomed. visual Cx |
| DMC | dorsmed. hypoth. nu., comp. |
| DMD | dorsmed. hypoth. nu., dors. |
| DMPAG | dorsomed. periaq. gr. |
| DMTg | dorsomed. tegm. area |
| DpG | deep gray sup. coll. |
| DPGi | dors. paragigant. nu. |
| DpWh | deep white sup. coll. |
| DRI | dors. raphe nu., interf. |
| DRV | dors. raphe nu., vent. |
| DRVL | dors. raphe nu., ventrolat. |
| DTgP | dors. tegm. nu., pericent. |
| f | fornix |
| fr | fasciculus retroflexus |
| Gi | gigantocell. retic. nu. |
| GiV | gigantocell. retic. nu., vent. |
| gr | gracile fasciculus |
| Gr | gracile nu. |
| hbc | habenular comm. |
| HDB | horiz. limb diag. band nu. |
| IAM | interantmed. nu. |
| IG | induseum gris. |
| IMD | intermediodors. thal. nu. |
| InC | interst. nu. Cajal |
| InCSh | interst. nu. Cajal, shell |
| InfS | infundib. stalk |
| InG | intermed. gr. sup. coll. |
| InWh | intermed. wh. sup. coll. |
| IOβ | inf. oliv. nu., subn. β |
| IODM | inf. oliv. nu., dorsomed. |
| IPF | interpedunc. fossa |
| IPI | interped. nu., interfasc. |
| IPL | interped. nu., lat. |
| IVF | intervent. foramen |
| LA | lateroant. hypoth. nu. |
| LDTgD | laterod. tegm nu., dors. |
| lfp | longit. fibers pons |
| LH | lat. hypothal. area |
| LHb | lat. habenular nu. |
| LPAG | lat. periaqued. gr. |
| LPO | lat. preopt. area |
| LSD | lat. sept. nu., dors. |
| LSI | lat. sept. nu., int. |
| LSV | lat. sept. nu., vent. |
| LTer | lamina terminalis |
| LV | lateral ventricle |
| MDC | mediodors. thal. nu., cent. |
| MDD | mediodors. thal. nu., dors. |
| MDM | mediodors. thal. nu., med. |
| MdV | medull. retic., nu. vent. |
| Med | med. deep Cb nu. |
| mfb | med. forebrain bundle |
| MHb | med. habenular nu. |
| ml | medial lemniscus |
| ML | med. mamill. nu., lat. |
| mlf | med. longit. fasc. |
| MM | med. mamill. nu., med. |
| MnA | median access. nu. |
| MPA | medial preopt. area |
| MPOL | med. preopt. nu., lat. |
| MPOM | med. preopt. nu., med. |
| MPT | med. pretect. nu. |
| MRe | mamill. recess |
| mRt | mesen. retic. formn |
| mt | mamillothal. tr. |
| nehy | neurohypophyseal tr. |
| och | optic chiasm |
| Op | strat. opt. sup. coll. |
| OPAI | orbital periallocortex |
| PaAP | paravent. nu., ant. parv. |
| PaMD | paravent. nu., magno. dors. |
| PaPo | paravent. nu., post. |
| PaV | paravent. nu., vent. |
| PaXi | paraxiphoid nu. |
| pc | post. commissure |
| PC | paracent. thal. nu. |
| pcf | preculminate fiss. Cb. |
| PCom | paracomm. nu. |
| PE | par. field E Cx |
| PEc | par. field E Cx, caud. |
| Pf | parafascicular nu. |
| PH | post. hypothal. area |
| PLd | paralambdoid septum |
| plf | posterolat. fissure |
| PLH | pedunc. lat. hypothal. |
| PMnR | paramedian raphe nu. |
| Pn | pontine nu. |
| PN | paranigral nu. |
| PPD2 | post. par. Cx, dors. 2 |
| ppf | prepyramidal fiss. |
| PPM | post. par. Cx, medial |
| Pr | prepositus hypogl. nu. |
| PrC | precomm. nu. |
| prcf | precentral fissure Cb |
| prf | prim. cerebell. fiss. |
| PT | paratenial nu. |
| PVA | paravent. thal. nu., ant. |
| PVG | periventricular gray |
| PVP | paravent. thal. nu., post. |
| py | pyramidal tract |
| pyd | pyramidal decn |
| R5 | Rexed 5, spinal cord |
| Rbd | rhabdoid nu. |
| RChL | retrochiasm. nu, lat. |
| Re | reuniens nu. |
| RI | rost. interstit. nu. |
| RMg | raphe magnus nu. |
| RML | retromamill. nu., lat. |
| RtTg | reticulotegm. nu. |
| SChD | suprachiasm. nu., dors. |
| SChV | suprachiasm. nu., vent. |
| scpd | sup. cb. ped. decn |
| sf | secondary fissure Cb |
| SFi | septofimbrial nu. |
| SFO | subfornical organ |
| SI | subst. innominata |
| sm | stria medullaris |
| SMA | supp. motor area Cx |
| smv | sup. medull. velum |
| sod | suproptic decn |
| SolC | nu. of solitary tr., comm. |
| SolM | nu. of solitary tr., med. |
| SOR | supraoptic nu, retroch. |
| Su3 | supraoculomotor PAG |
| Su3C | supraoculomotor cap |
| SuG | sup. gray sup. coll. |
| tfp | transv. fibers pons |
| ts | tectospinal tract |
| TS | triang. sept. nu. |
| tzd | trapezoid decussation |
| V1 | V1 Cx |
| V2m | V2 (med.) Cx |
| vcs | vent. corticospinal tr. |
| VDB | vert. limb diag. band. nu. |
| vfu | ventral funiculus |
| VH | ventral horn |
| vhc | vent. hipp. comm. |
| VLPAG | ventrolat. periaq. gr. |
| VMHL | ventromed. hypoth. nu., lat. |
| VMHM | ventromed. hypoth. nu., med. |
| VMPO | ventromed. preopt. nu. |
| VPe | vent. periventricular zone |
| VRe | vent. reuniens nu. |
| VTA | vent. tegment. area |
| vtgd | vent. tegm. decn |
| ZIR | zona incerta, rost. |

# Plate 62a

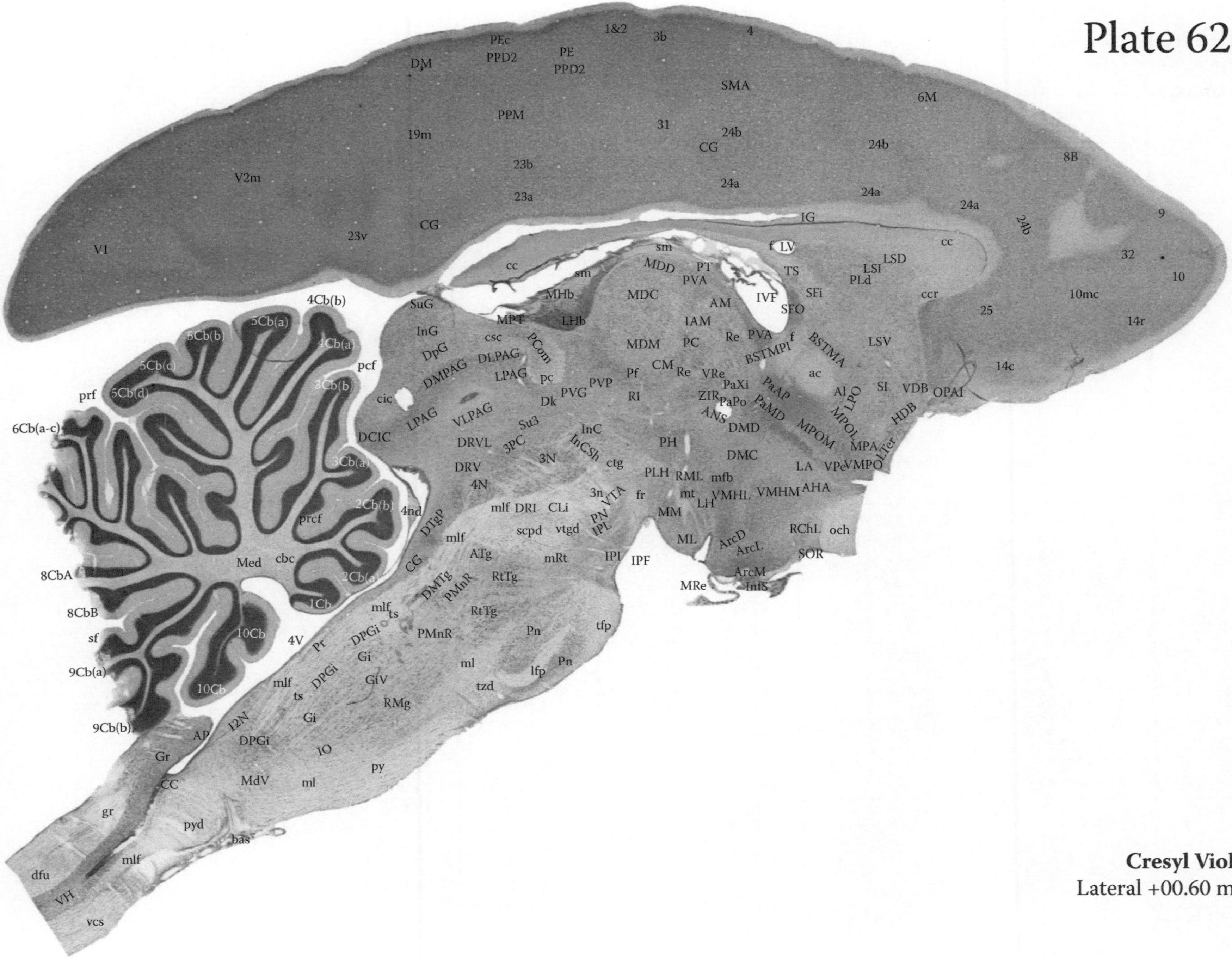


**Cresyl Violet**
Lateral +00.60 mm

# Plate 62b

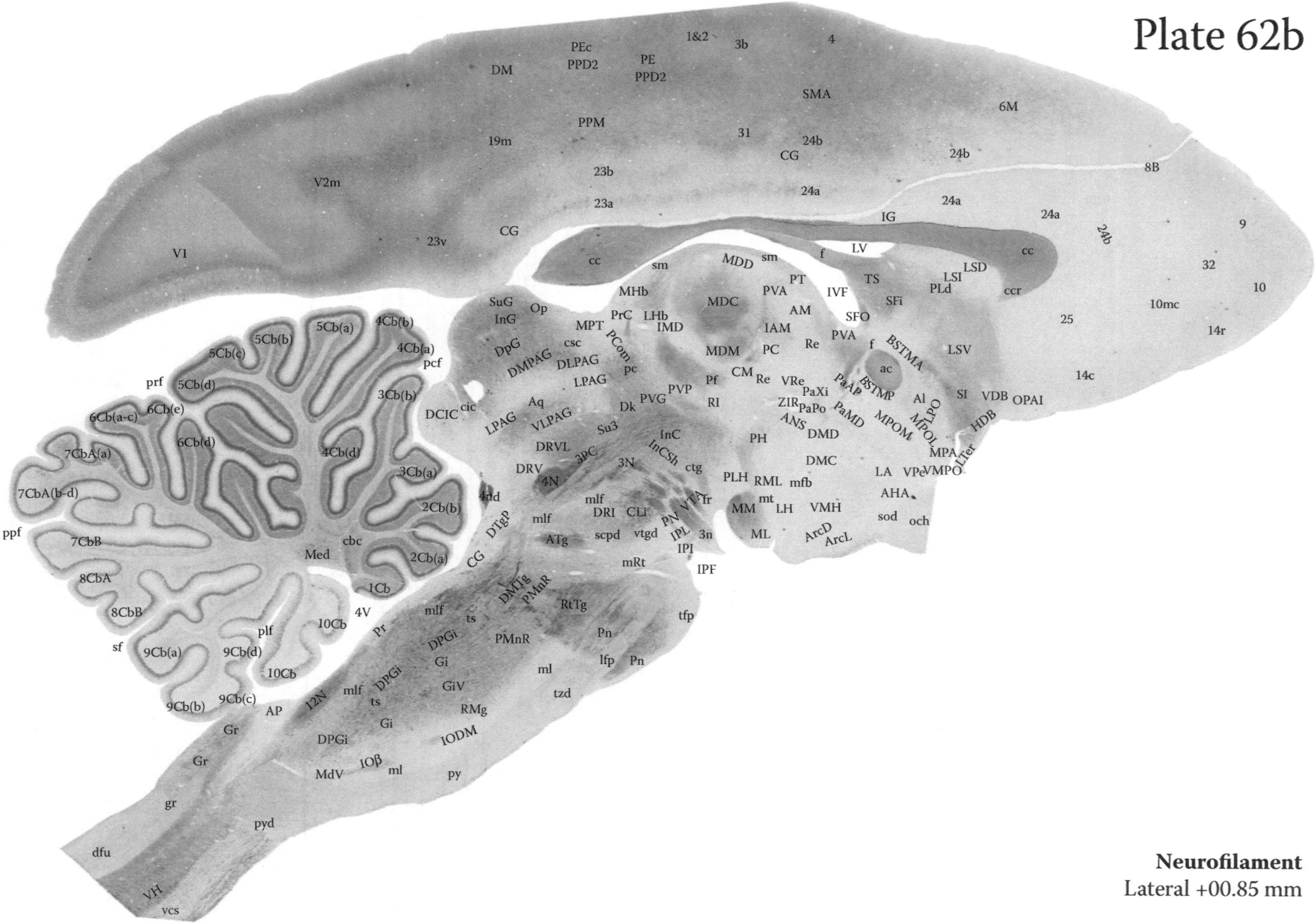


**Neurofilament**
Lateral +00.85 mm

# Plate 63

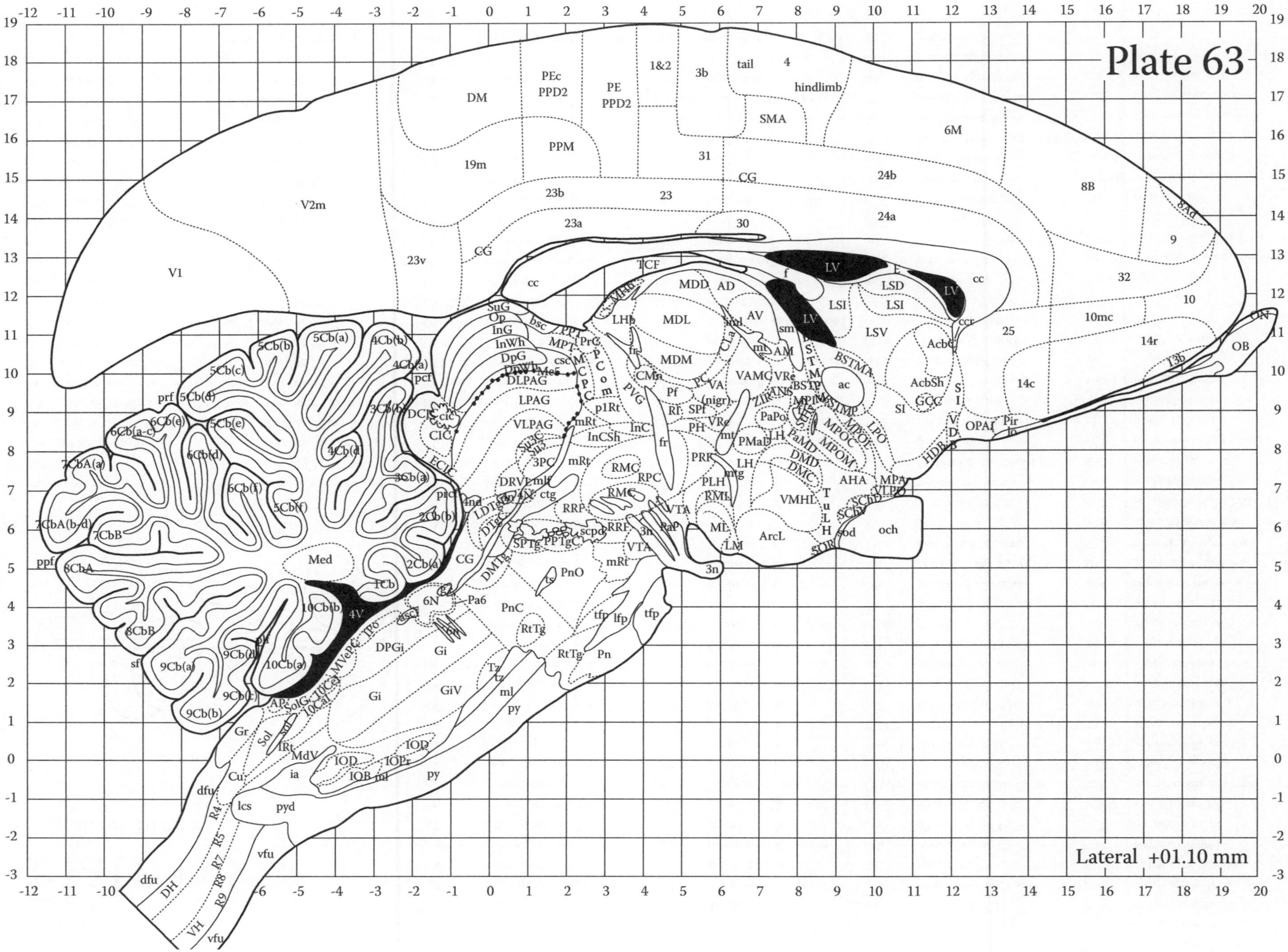


Lateral +01.10 mm

# Plate 63

## Lateral +01.10 mm

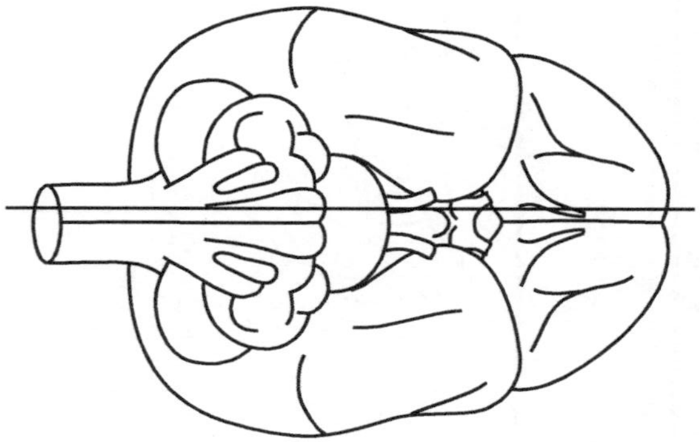

| | |
|---|---|
| 1&2 | areas 1/2 Cx |
| 1Cb | cerebell. lobule 1 |
| 2Cb | cerebell. lobule 2 |
| 3b | area 3b Cx |
| 3Cb | cerebell. lobule 3 |
| 3n | oculomotor nerve |
| 3PC | oculomotor nu., parv. |
| 4 | area 4 Cx |
| 4Cb | cerebell. lobule 4 |
| 4n | trochlear nerve |
| 4N | trochlear nu. |
| 4nd | trochlear nerve decn |
| 4V | fourth ventricle |
| 5Cb | cerebell. lobule 5 |
| 6Cb | cerebell. lobule 6 |
| 6M | area 6M Cx |
| 6n | abducens n. |
| 6N | abducens nu. |
| 7Cb | cerebell. lobule 7 |
| 8Ad | area 8Ad Cx |
| 8B | area 8B Cx |
| 8Cb | cerebell. lobule 8 |
| 9 | area 9 Cx |

| | |
|---|---|
| 9Cb | cerebell. lobule 9 |
| 10 | area 10 Cx |
| 10CaI | dors. nu. 10, caud. int. |
| 10Cb | cerebell. lobule 10 |
| 10CeI | dors. nu. 10, cent. int. |
| 10mc | area 10mc Cx |
| 13b | area 13b Cx |
| 14c | area 14c Cx |
| 14r | area 14r Cx |
| 19m | area 19m Cx |
| 23 | area 23 Cx |
| 23a | area 23a Cx |
| 23b | area 23b Cx |
| 23v | area 23v Cx |
| 24a | area 24a Cx |
| 24b | area 24b Cx |
| 25 | area 25 Cx |
| 30 | area 30 Cx |
| 31 | area 31 Cx |
| 32 | area 32 Cx |
| ac | ant. comm. |
| AcbC | accumb. nu. core |
| AcbSh | accumb. nu. shell |

| | |
|---|---|
| AD | anterodors. thal. nu. |
| AHA | ant. hypoth. area, ant. |
| AM | anteromed. thal. nu. |
| ANS | access. neurosec. nu. |
| AP | area postrema |
| ArcL | arc. hypoth. nu., lat. |
| asc7 | ascending 7n |
| AV | anterovent. thal. nu. |
| bsc | brach. sup. coll. |
| BSTMA | bed nu. st, med. ant. divn |
| BSTMP | bed nu. st, med. post. divn |
| BSTMPI | bed nu. st, med. post. int. |
| BSTMPM | bed nu. st, med. post. med. |
| cc | corp. callosum |
| ccr | corp. call. rostrum |
| CG | cingulate gyrus Cx |
| cic | comm. inf. coll. |
| CIC | cent. nu. inf. coll. |
| CLa | centrolat. thal. nu., ant. |
| CMn | centromedian thal. nu. |
| csc | comm. sup. coll. |
| ctg | central tegm. tract |
| Cu | cuneate nu. |
| DCIC | dors. Cx inf. coll. |
| dfu | dorsal funiculus |
| DH | dorsal horn |
| DLPAG | dorsolat. periaq. gr. |
| DM | dorsomed. visual Cx |
| DMC | dorsmed. hypoth. nu., comp. |
| DMD | dorsmed. hypoth. nu., dors. |
| DMTg | dorsomed. tegm. area |
| DpG | deep gray sup. coll. |
| DPGi | dors. paragigant. nu. |
| DpWh | deep white sup. coll. |
| DRVL | dors. raphe nu., ventrolat. |
| DTgC | dors. tegm. nu., cent. |
| E | ependyma |
| ECIC | extern. Cx inf. coll. |
| f | fornix |
| fr | fasciculus retroflexus |
| g7 | genu of facial n. |
| GCC | gran. cell clust. |
| Gi | gigantocell. retic. nu. |
| GiV | gigantocell. retic. nu., vent. |

| | |
|---|---|
| Gr | gracile nu. |
| HDB | horiz. limb diag. band nu. |
| ia | intern. arcuate fibers |
| iml | intern. medull. lamina |
| InC | interst. nu. Cajal |
| InCSh | interst. nu. Cajal, shell |
| InG | intermed. gr. sup. coll. |
| InWh | intermed. wh. sup. coll. |
| IOB | inf. oliv. nu., subn. B |
| IOD | inf. oliv. nu., dors. |
| IOPr | inf. oliv. nu., princ. |
| IPo | interpositus nu. |
| IRt | intermed. retic. formn |
| lcs | lat. corticospinal tr. |
| LDTgD | laterod. tegm nu., dors. |
| lfp | longit. fibers pons |
| LH | lat. hypothal. area |
| LHb | lat. habenular nu. |
| LM | lat. mamill. nu. |
| lo | lat. olf. tract |
| LPAG | lat. periaqued. gr. |
| LPO | lat. preopt. area |
| LSD | lat. sept. nu., dors. |
| LSI | lat. sept. nu., int. |
| LSV | lat. sept. nu., vent. |
| LV | lateral ventricle |
| MCPC | magnocell. nu. pc |
| MDD | mediodors. thal. nu., dors. |
| MDL | mediodors. thal. nu., lat. |
| MDM | mediodors. thal. nu., med. |
| MdV | medull. retic., nu. vent. |
| Me5 | mesen. trigem. nu. |
| Med | med. deep Cb nu. |
| MHb | med. habenular nu. |
| ml | medial lemniscus |
| ML | med. mamill. nu., lat. |
| mlf | med. longit. fasc. |
| MPA | medial preopt. area |
| MPOC | med. preopt. nu., comp. |
| MPOL | med. preopt. nu., lat. |
| MPOM | med. preopt. nu., med. |
| MPT | med. pretect. nu. |
| mRt | mesen. retic. formn |
| mt | mamillothal. tr. |

| | |
|---|---|
| mtg | mamillotegm. tr. |
| MVePC | med. vestib. nu., parv. |
| OB | olfactory bulb |
| och | optic chiasm |
| ON | olf. n. layer, olf. bulb |
| Op | strat. opt. sup. coll. |
| OPAI | orbital periallocortex |
| p1Rt | pros. 1 retic. formn |
| Pa6 | parabducens nu. |
| PaMD | paravent. nu., magno. dors. |
| PaP | parapeduncular nu. |
| PaPo | paravent. nu., post. |
| PC | paracent. thal. nu. |
| pcf | preculminate fiss. Cb. |
| PCom | paracomm. nu. |
| PE | par. field E Cx |
| PEc | par. field E Cx, caud. |
| Pf | parafascicular nu. |
| PH | post. hypothal. area |
| Pir | piriform Cx |
| plf | posterolat. fissure |
| PLH | pedunc. lat. hypothal. |
| PMaD | premamill. nu., dorsal |
| Pn | pontine nu. |
| PnC | pontine caudal nu. |
| PnO | pontine oral nu. |
| PPD2 | post. par. Cx, dors. 2 |
| ppf | prepyramidal fiss. |
| PPM | post. par. Cx, medial |
| PPT | post. pretectum |
| PPTgC | peduncpont. tegm. nu., com |
| PrC | precomm. nu. |
| prcf | precentral fissure Cb |
| prf | prim. cerebell. fiss. |
| PRF | prerubral field |
| PVG | periventricular gray |
| py | pyramidal tract |
| pyd | pyramidal decn |
| R4 | Rexed 4, spinal cord |
| R5 | Rexed 5, spinal cord |
| R7 | Rexed 7, spinal cord |
| R8 | Rexed 8, spinal cord |
| R9 | Rexed 9, spinal cord |
| RI | rost. interstit. nu. |

| | |
|---|---|
| RMC | red nu., magnocell. |
| RML | retromamill. nu., lat. |
| RPC | red nu., parvicellular |
| RRF | retrorubral field |
| RtTg | reticulotegm. nu. |
| SChD | suprachiasm. nu., dors. |
| SChV | suprachiasm. nu., vent. |
| scpd | sup. cb. ped. decn |
| sf | secondary fissure Cb |
| SI | subst. innominata |
| sm | stria medullaris |
| SMA | supp. motor area Cx |
| sod | suproptic decn |
| sol | solitary tract |
| Sol | nu. of solitary tr. |
| SolG | nu. of solitary tr., gelat. |
| SOR | supraoptic nu, retroch. |
| SPf | subparafasc. thal. nu. |
| SPTg | subpedunc. tegm. nu. |
| StHy | striatohypothal. nu. |
| Su3 | supraoculomotor PAG |
| Su3C | supraoculomotor cap |
| SuG | sup. gray sup. coll. |
| TCF | transverse cerebral fiss. |
| tfp | transv. fibers pons |
| ts | tectospinal tract |
| TuLH | tub. lat. hypothal. |
| tz | trapezoid body |
| Tz | nu. of trapezoid body |
| V1 | V1 Cx |
| V2m | V2 (med.) Cx |
| VA | vent. ant. thal. nu. |
| VAMC | vent. ant. thal. nu., medioce. |
| VDB | vert. limb diag. band. nu. |
| vfu | ventral funiculus |
| VH | ventral horn |
| VLPAG | ventrolat. periaq. gr. |
| VLPO | ventrolat. preopt. nu. |
| VMHL | ventromed. hypoth. nu., lat. |
| VRe | vent. reuniens nu. |
| VTA | vent. tegment. area |
| ZIR | zona incerta, rost. |

# Plate 63a

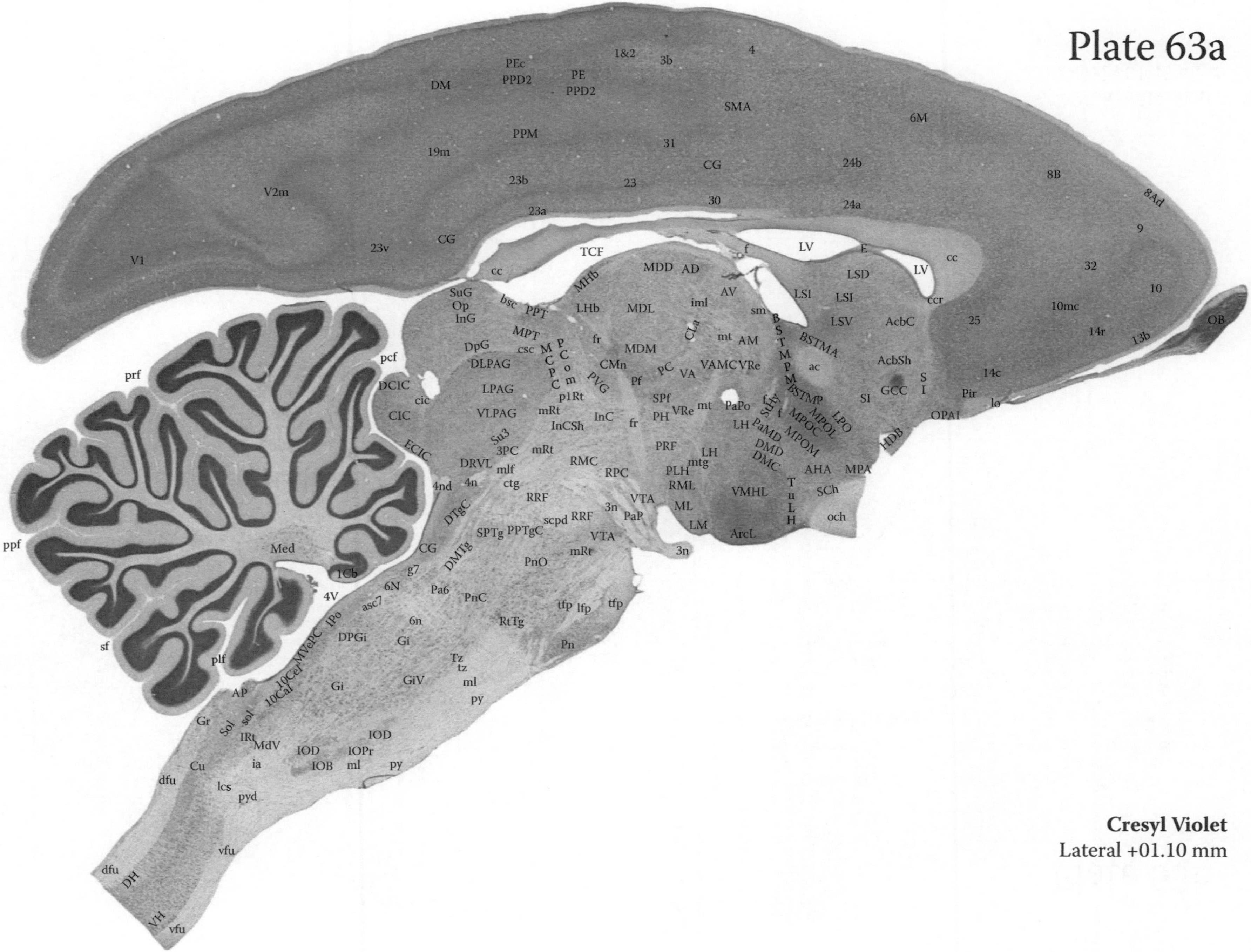


**Cresyl Violet**
Lateral +01.10 mm

# Plate 63b

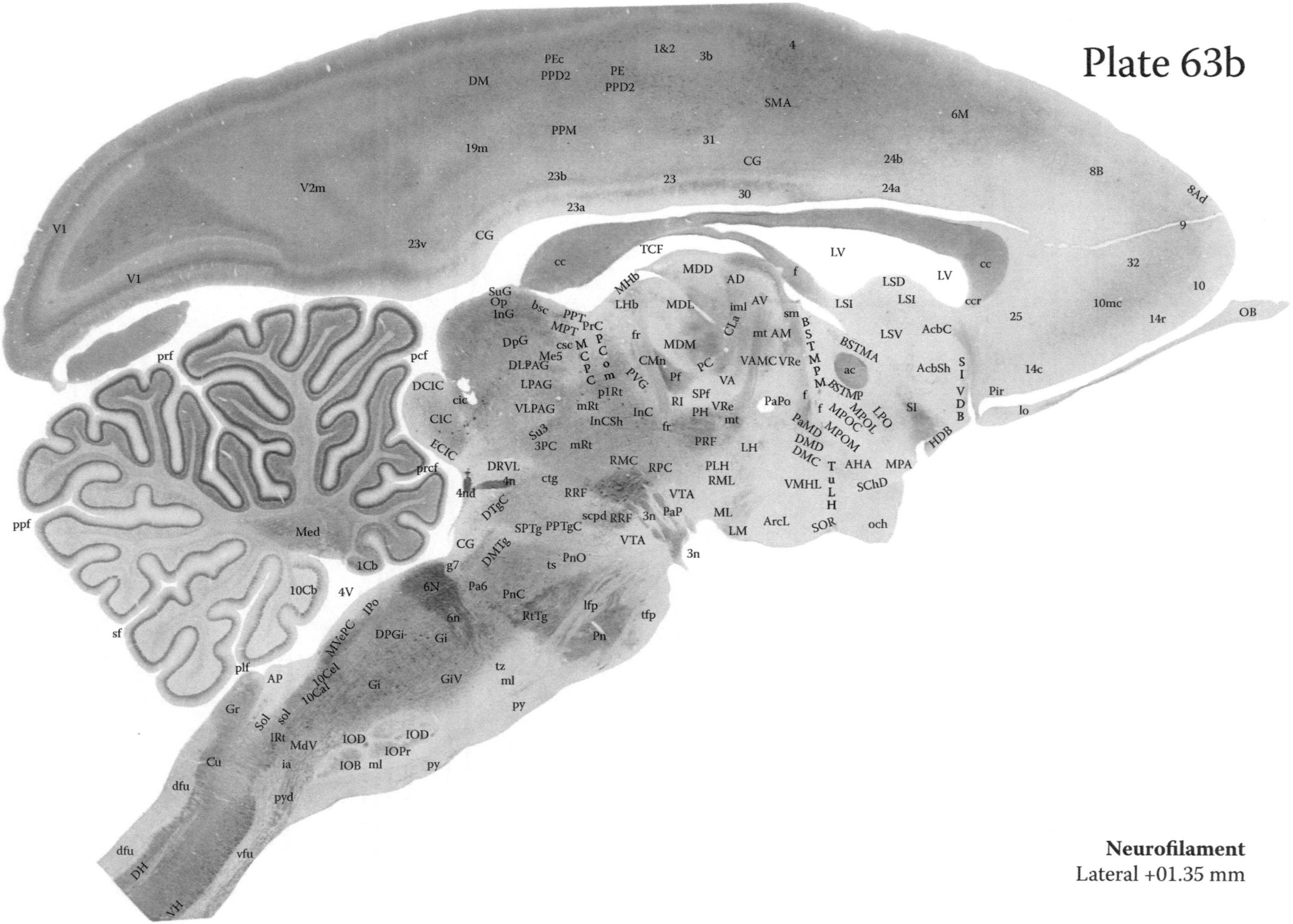


**Neurofilament**
Lateral +01.35 mm

# Plate 64

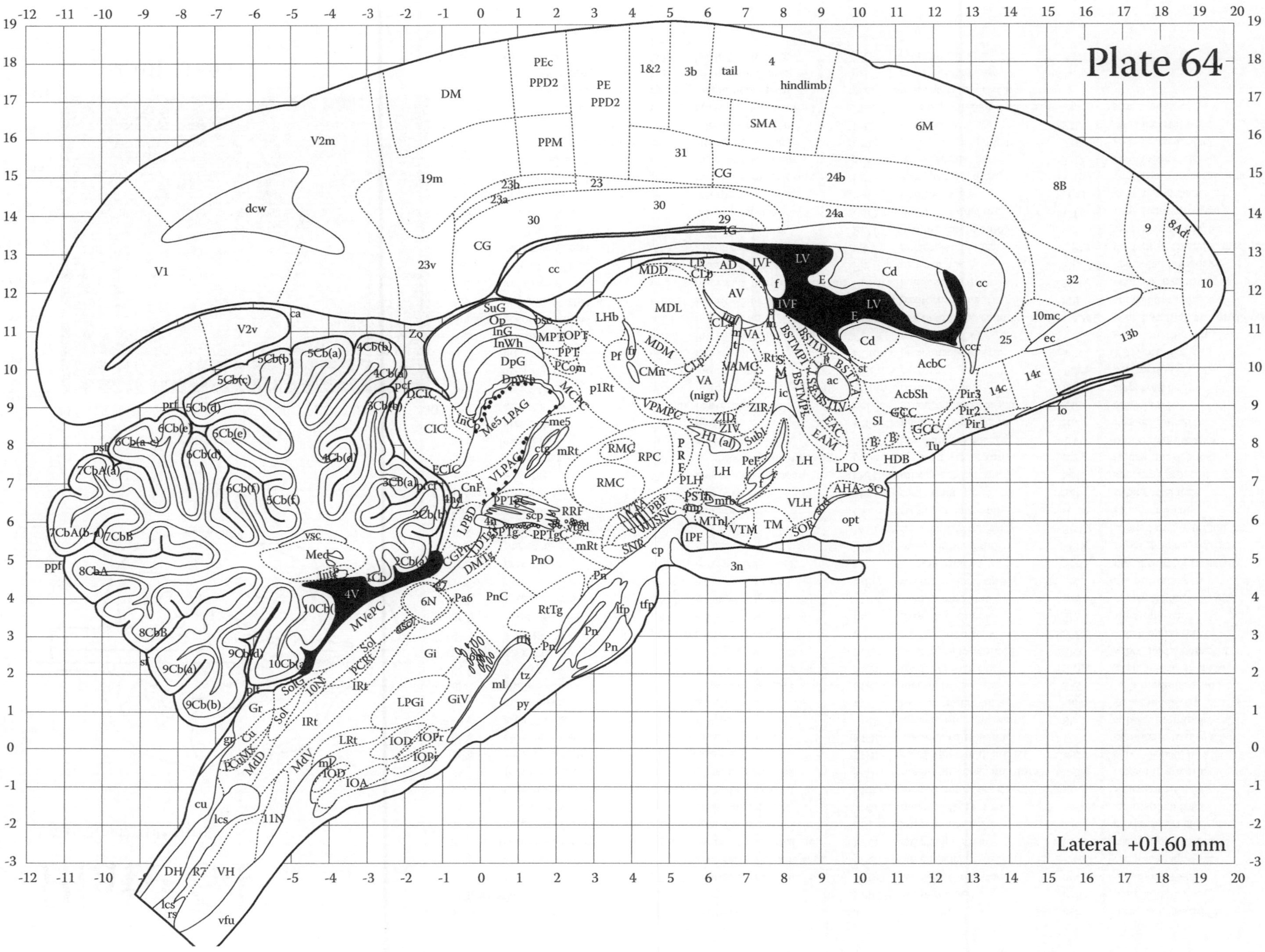

Lateral +01.60 mm

# Plate 64

Lateral +01.60 mm

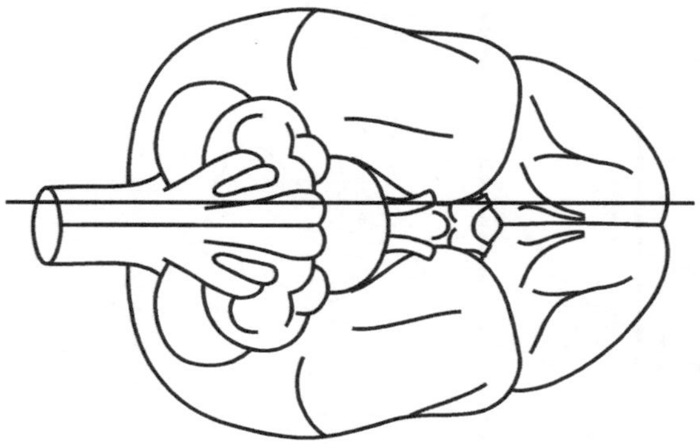

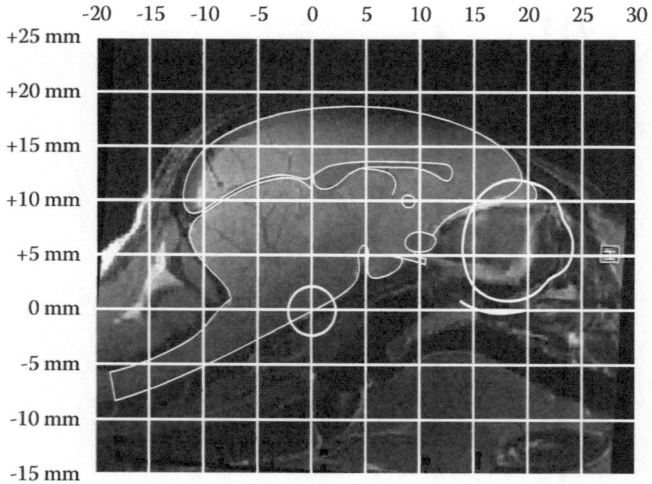


Lateral +01.45 mm

| | |
|---|---|
| 1Cb | cerebell. lobule 1 |
| 2Cb | cerebell. lobule 2 |
| 3b | area 3b Cx |
| 3Cb | cerebell. lobule 3 |
| 3n | oculomotor nerve |
| 4Cb | cerebell. lobule 4 |
| 4n | trochlear nerve |
| 4nd | trochlear nerve decn |
| 4V | fourth ventricle |
| 5Cb | cerebell. lobule 5 |
| 6Cb | cerebell. lobule 6 |
| 6M | area 6M Cx |
| 6n | abducens n. |
| 6N | abducens nu. |
| 7Cb | cerebell. lobule 7 |
| 8Ad | area 8Ad Cx |
| 8B | area 8B Cx |
| 8Cb | cerebell. lobule 8 |
| 9Cb | cerebell. lobule 9 |
| 10Cb | cerebell. lobule 10 |
| 10mc | area 10mc Cx |
| 10N | dorsal nu. 10n |
| 11N | accessory motor nu. |
| 13b | area 13b Cx |
| 14c | area 14c Cx |
| 14r | area 14r Cx |
| 19m | area 19m Cx |
| 23a | area 23a Cx |
| 23b | area 23b Cx |
| 23v | area 23v Cx |
| 24a | area 24a Cx |
| 24b | area 24b Cx |
| ac | ant. comm. |
| AcbC | accumb. nu. core |
| AcbSh | accumb. nu. shell |
| AD | anterodors. thal. nu. |
| AHA | ant. hypoth. area, ant. |
| asc7 | ascending 7n |
| AV | anterovent. thal. nu. |
| B | basal nu. (Meynert) |
| bsc | brach. sup. coll. |
| BSTLA | bed nu. st, lat. ant. divn |
| BSTLD | bed nu. st, lat. dors. divn |
| BSTLP | bed nu. st, lat. post. divn |
| BSTLV | bed nu. st, lat. vent. divn |
| BSTMPI | bed nu. st, med. post. int. |
| BSTMPL | bed nu. st, med. post. lat |
| ca | calcarine sulcus |
| ccr | corp. call. rostrum |
| Cd | caudate nu. |
| CGPn | central gray pons |
| CIC | cent. nu. inf. coll. |
| CLa | centrolat. thal. nu., ant. |
| CLp | centrolat. thal. nu., post. |
| CMn | centromedian thal. nu. |
| CnF | cuneiform nu. |
| cp | cerebral peduncle |
| ctg | central tegm. tract |
| cu | cuneate fasciculus |
| Cu | cuneate nu. |
| DCIC | dors. Cx inf. coll. |
| dcw | deep cort. white |
| DH | dorsal horn |
| DM | dorsomed. visual Cx |
| DMTg | dorsomed. tegm. area |
| DpG | deep gray sup. coll. |
| DpWh | deep white sup. coll. |
| EAC | extend amyg. central |
| EAM | extend amyg. medial |
| ec | external capsule |
| ECIC | extern. Cx inf. coll. |
| f | fornix |
| fr | fasciculus retroflexus |
| g7 | genu of facial n. |
| GCC | gran. cell clust. |
| Gi | gigantocell. retic. nu. |
| GiV | gigantocell. retic. nu., vent. |
| gr | gracile fasciculus |
| Gr | gracile nu. |
| H1 (al) | H prerubral field |
| HDB | horiz. limb diag. band nu. |
| ic | intern. caps. |
| IG | induseum gris. |
| iml | intern. medull. lamina |
| InCo | intercoll. nu. |
| InG | intermed. gr. sup. coll. |
| IntP | intermed. post. deep Cb nu. |
| InWh | intermed. wh. sup. coll. |
| IOA | inf. oliv. nu., subn. A |
| IOD | inf. oliv. nu., dors. |
| IOPr | inf. oliv. nu., princ. |
| IPF | interpedunc. fossa |
| IRt | intermed. retic. formn |
| IVF | intervent. foramen |
| lcs | lat. corticospinal tr. |
| LD | laterodors. thal. nu. |
| LDTgV | laterod. tegm nu., vent. |
| lfp | longit. fibers pons |
| LH | lat. hypothal. area |
| LHb | lat. habenular nu. |
| lo | lat. olf. tract |
| LPAG | lat. periaqued. gr. |
| LPBD | lat. parabr. nu., dors. |
| LPGi | lat. paragigant. nu. |
| LPO | lat. preopt. area |
| LRt | lat. reticular nu. |
| MCPC | magnocell. nu. pc |
| MdD | medull. retic., nu. dors. |
| MDD | mediodors. thal. nu., dors. |
| MDL | mediodors. thal. nu., lat. |
| MDM | mediodors. thal. nu., med. |
| MdV | medull. retic., nu. vent. |
| me5 | mesen. trigem. tr. |
| Me5 | mesen. trigem. nu. |
| Med | med. deep Cb nu. |
| mfb | med. forebrain bundle |
| ml | medial lemniscus |
| mp | mamill. peduncle |
| MPT | med. pretect. nu. |
| mRt | mesen. retic. formn |
| mt | mamillothal. tr. |
| MTnl | med. terminal nu. |
| MVePC | med. vestib. nu., parv. |
| Op | strat. opt. sup. coll. |
| opt | optic tract |
| OPT | olivary pretect. nu. |
| p1Rt | pros. 1 retic. formn |
| Pa6 | parabducens nu. |
| PBP | parabrach. pigm. nu. |
| pcf | preculminate fiss. Cb. |
| PCom | paracomm. nu. |
| PCRt | parv. retic. nu. |
| PCuMx | pericuneate matrix |
| PE | par. field E Cx |
| PEc | par. field E Cx, caud. |
| PeF | perifornical nu. |
| Pf | parafascicular nu. |
| Pir | piriform Cx |
| plf | posterolat. fissure |
| PLH | pedunc. lat. hypothal. |
| PnC | pontine caudal nu. |
| PnO | pontine oral nu. |
| PPD2 | post. par. Cx, dors. 2 |
| ppf | prepyramidal fiss. |
| PPM | post. par. Cx, medial |
| PPT | post. pretectum |
| PPTgC | peduncpont. tegm. nu., com. |
| prcf | precentral fissure Cb |
| prf | prim. cerebell. fiss. |
| PRF | prerubral field |
| psf | post. sup. fissure Cb |
| PSTh | parasubthal. nu. |
| py | pyramidal tract |
| RMC | red nu., magnocell. |
| RPC | red nu., parvicellular |
| RRF | retrorubral field |
| rs | rubrospinal tract |
| Rt | reticular nu. |
| RtTg | reticulotegm. nu. |
| scp | sup. cerebell. ped. |
| sf | secondary fissure Cb |
| SI | subst. innominata |
| sm | stria medullaris |
| SM | stria medullaris nu. |
| SMA | supp. motor area Cx |
| SNC | subst. nigra, compact |
| SNR | subst. nigra, retic. |
| SO | supraoptic nu. |
| sod | suproptic decn |
| Sol | nu. of solitary tr. |
| SolG | nu. of solitary tr., gelat. |
| SOR | supraoptic nu, retroch. |
| SPTg | subpedunc. tegm. nu. |
| st | stria terminalis |
| SubI | subincertal nu. |
| SuG | sup. gray sup. coll. |
| tfp | transv. fibers pons |
| TM | tuberomamill. area |
| tth | trigeminothal. tr. |
| Tu | olf. tubercle |
| tz | trapezoid body |
| V2m | V2 (med.) Cx |
| V2v | V2 (vent.) Cx |
| VA | vent. ant. thal. nu. |
| VAMC | vent. ant. thal. nu., mediocent. |
| vfu | ventral funiculus |
| VLH | vent. lat. hypothal. |
| VLPAG | ventrolat. periaq. gr. |
| VPMPC | vent. post. thal. nu., med. parv. |
| vsc | vent. spinocereb. tr. |
| VTA | vent. tegment. area |
| vtgd | vent. tegm. decn |
| VTM | vent. tuberomamill. area |
| ZID | zona incerta, dors. |
| ZIR | zona incerta, rost. |
| ZIV | zona incerta, vent. |
| Zo | strat. zonale SC |

# Plate 64a

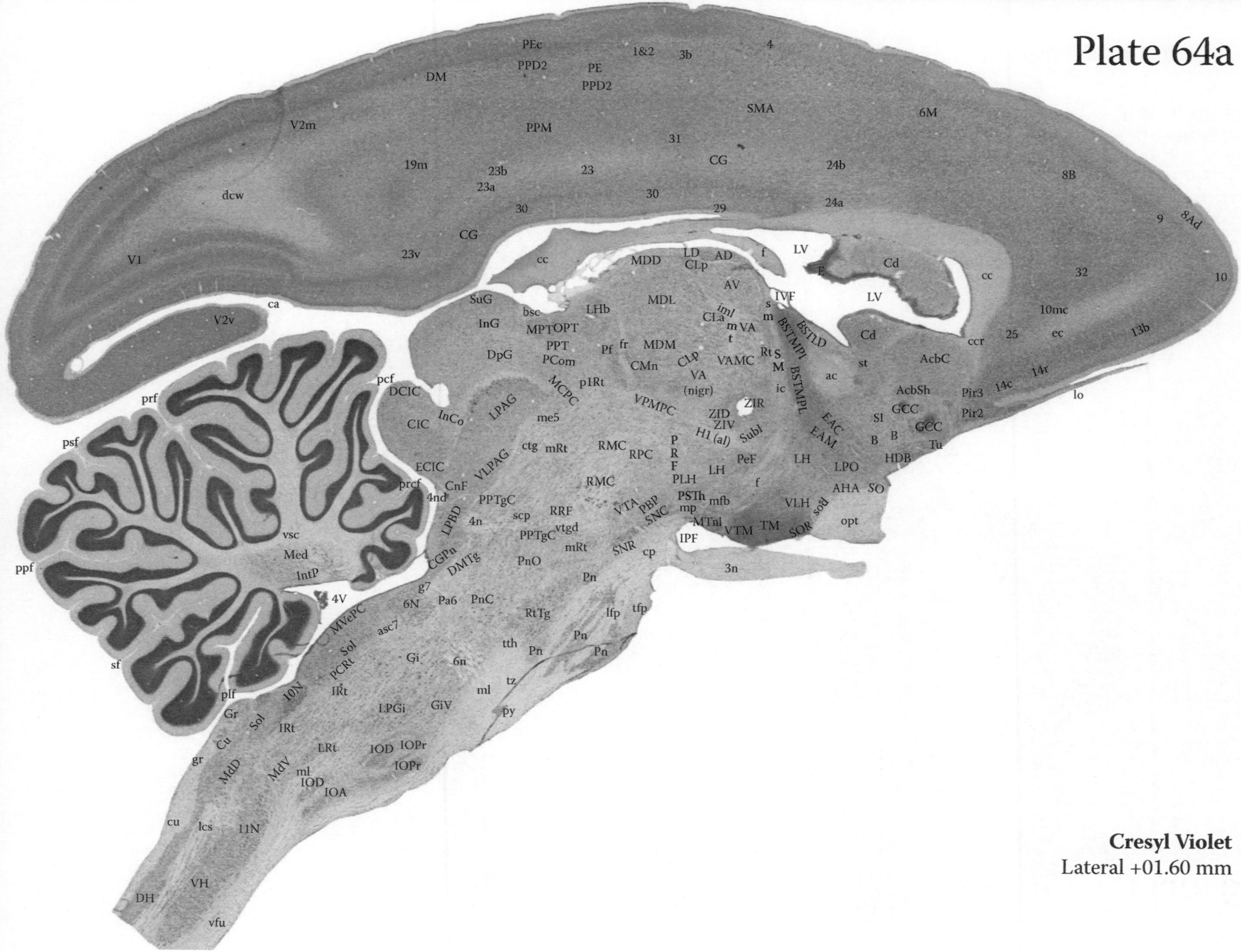


**Cresyl Violet**
Lateral +01.60 mm

# Plate 64b

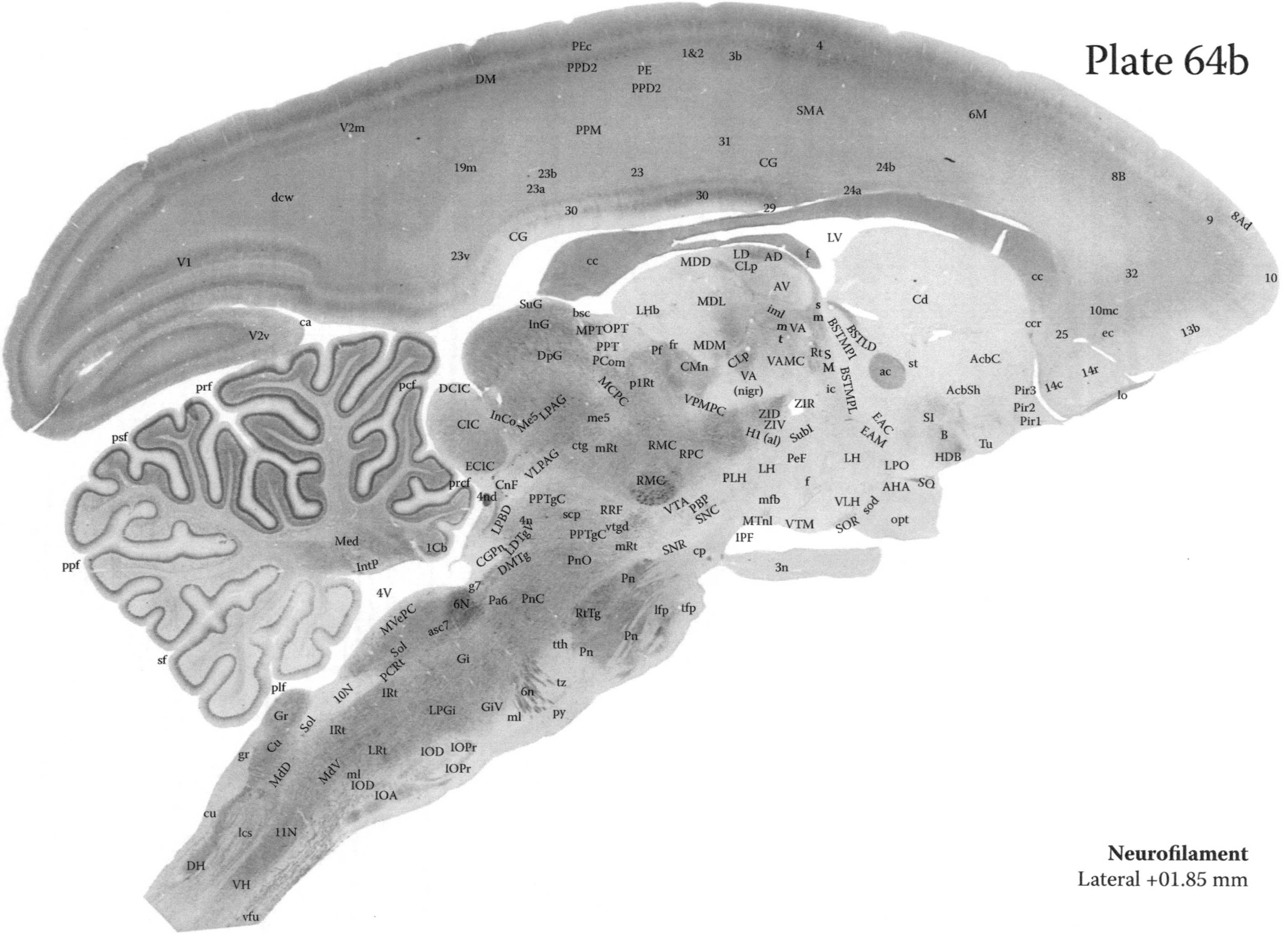


**Neurofilament**
Lateral +01.85 mm

# Plate 65

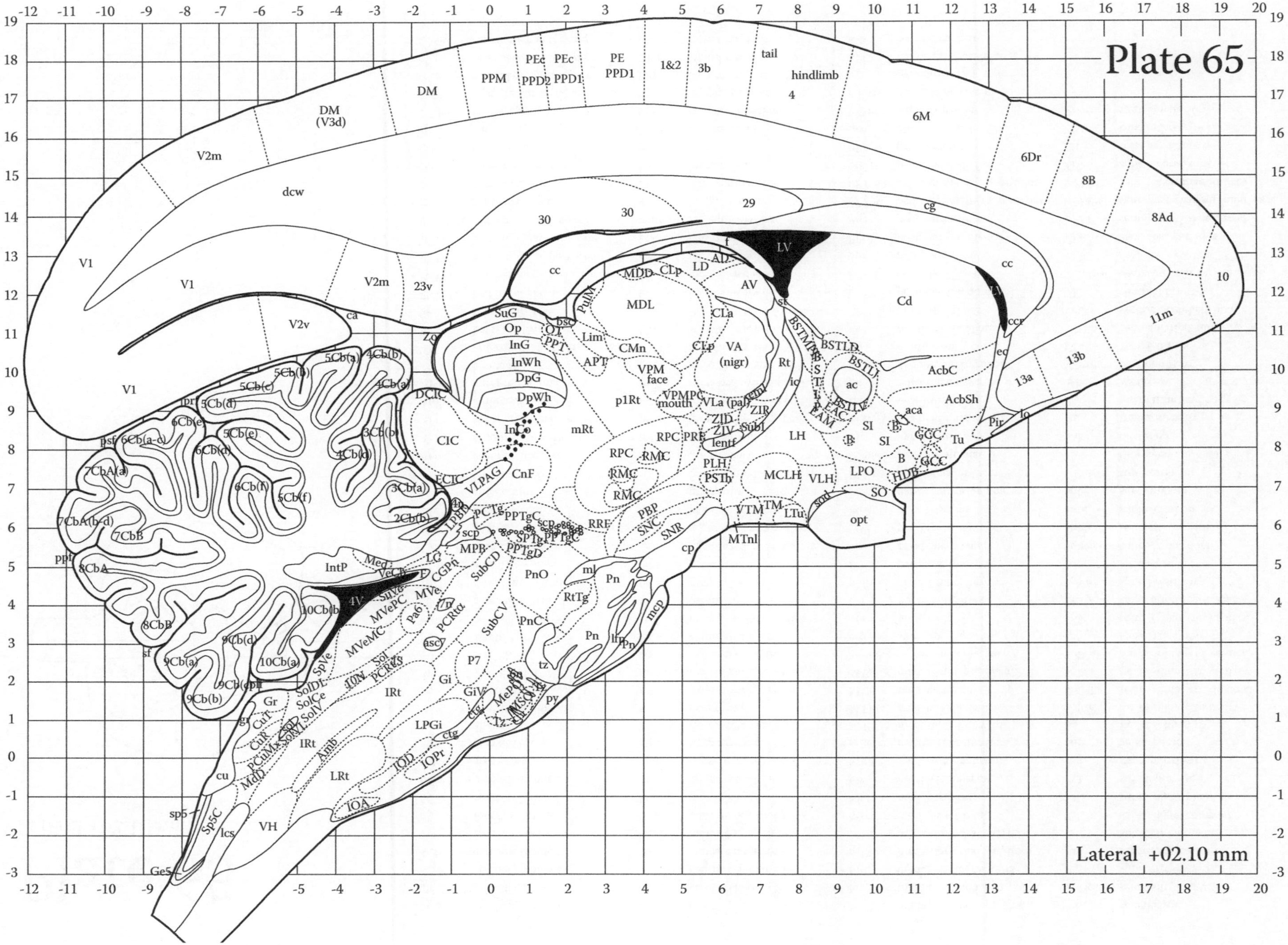
PPM
PEc PPD2
PEc PPD1
PE PPD1
1&2
3b
tail
hindlimb 4
DM
DM (V3d)
6M
6Dr
8B
8Ad
V2m
dcw
30
29
cg
V1
V2v
23v
ca
cc
LV
AD
MDD
CLp
LD
AV
MDL
PulM
Lim
CMn
APT
bsc
OT
PPT
SuG
Op
InG
InWh
DpG
DpWh
InCo
Zo
DCIC
CIC
ECIC
VLPAG
CnF
mRt
p1Rt
VPM face
VPMPC mouth
VLa (pal)
VA (nigr)
CLa
Rt
ic
st
BSTMPM
BSTLD
BSTLJ
BSTLV
EAC
EAM
ac
aca
Cd
AcbC
AcbSh
ccr
ec
lo
Pir
Tu
GCC
SI
B
LH
LPO
HDB
SO
opt
sod
VLH
MCLH
LTu
TM
VTM
MTnl
PLH
PSTh
lentf
ZIR
ZID
ZIV
Sub1
PRE
RPC
RMC
RRE
PBP
SNC
SNR
cp
mcp
ml
Pn
RtTg
PnO
PnC
tz
py
lfp
PPTgC
PPTgD
SPTg
scp
MPB
LPBD
PCTg
LC
CGPn
SubCD
SubCV
P7
Gi
GiV
ctg
MePO
MSO
Tz
Med
VeCb
IntP
4V
MVe
MVePC
MVeMC
Pa6
7n
PCRtα
PCRtA
asc7
Sol
10N
SpVe
SolDL
SolCe
SolIM
SolV
IRt
LPGi
IOD
IOPr
IOA
Amb
LRt
Gr
gr
Cu
CuR
PCuMx
MdD
cu
sp5
Sp5C
lcs
VH
Ge5
5Cb(a)
4Cb(b)
4Cb(a)
5Cb(b)
5Cb(c)
5Cb(d)
5Cb(e)
5Cb(f)
3Cb(b)
3Cb(a)
4Cb(d)
2Cb(b)
6Cb(a-c)
6Cb(d)
6Cb(e)
6Cb(f)
7CbA(a)
7CbA(b-d)
7CbB
8CbA
8CbB
9Cb(a)
9Cb(b)
9Cb(d)
10Cb(a)
10Cb(b)
ppf
psf
prf
sf
Lateral +02.10 mm

# Plate 65

Lateral +02.10 mm

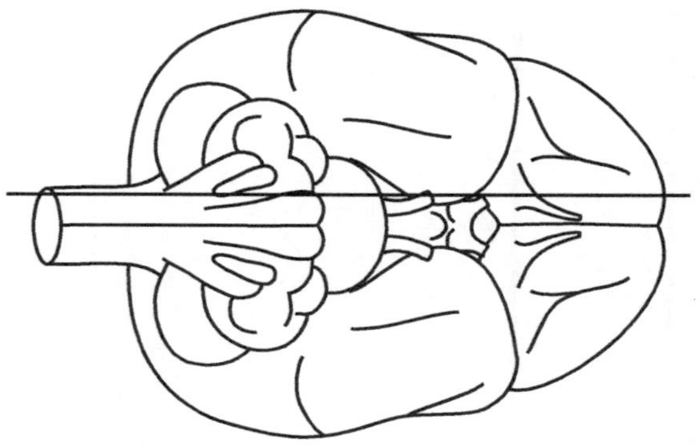

| | |
|---|---|
| 2Cb | cerebell. lobule 2 |
| 3b | area 3b Cx |
| 3Cb | cerebell. lobule 3 |
| 4Cb | cerebell. lobule 4 |
| 4n | trochlear nerve |
| 4V | fourth ventricle |
| 5Cb | cerebell. lobule 5 |
| 6Cb | cerebell. lobule 6 |
| 6Dr | area 6Dr Cx |
| 6M | area 6M Cx |
| 6n | abducens n. |
| 7Cb | cerebell. lobule 7 |
| 7n | facial nerve |
| 8Ad | area 8Ad Cx |
| 8B | area 8B Cx |
| 8Cb | cerebell. lobule 8 |
| 9Cb | cerebell. lobule 9 |
| 10Cb | cerebell. lobule 10 |
| 10N | dorsal nu. 10n |
| 11m | area 11m Cx |
| 13a | area 13a Cx |
| 13b | area 13b Cx |
| 23v | area 23v Cx |
| ac | ant. comm. |
| aca | ant. comm. ant. |
| AcbC | accumb. nu. core |
| AcbSh | accumb. nu. shell |
| AD | anterodors. thal. nu. |
| Amb | ambiguus nu. |
| APT | ant. pretect. nu. |
| asc7 | ascending 7n |
| AV | anterovent. thal. nu. |
| B | basal nu. (Meynert) |
| bsc | brach. sup. coll. |
| BSTLD | bed nu. st, lat. dors. divn |
| BSTLJ | bed nu. st, lat. juxtacap. divn |
| BSTLP | bed nu. st, lat. post. divn |
| BSTLV | bed nu. st, lat. vent. divn |
| BSTMPL | bed nu. st, med. post. lat |
| ca | calcarine sulcus |
| cc | corp. callosum |
| ccr | corp. call. rostrum |
| Cd | caudate nu. |
| cg | cingulum |
| CGPn | central gray pons |
| CIC | cent. nu. inf. coll. |
| CLa | centrolat. thal. nu., ant. |
| CLp | centrolat. thal. nu., post. |
| CMn | centromedian thal. nu. |
| CnF | cuneiform nu. |
| cp | cerebral peduncle |
| ctg | central tegm. tract |
| cu | cuneate fasciculus |
| CuR | cuneate nu., rotunda |
| CuT | cuneate nu., triang. |
| DCIC | dors. Cx inf. coll. |
| dcw | deep cort. white |
| DM | dorsomed. visual Cx |
| DpG | deep gray sup. coll. |
| DpWh | deep white sup. coll. |
| E | ependyma |
| EAC | extend amyg. central |
| EAM | extend amyg. medial |
| ec | external capsule |
| ECIC | extern. Cx inf. coll. |
| eml | extern. medull. lamina |
| f | fornix |
| GCC | gran. cell clust. |
| Ge5 | gelat. layer Sp5C |
| Gi | gigantocell. retic. nu. |
| GiV | gigantocell. retic. nu., vent. |
| gr | gracile fasciculus |
| Gr | gracile nu. |
| HDB | horiz. limb diag. band nu. |
| ic | intern. caps. |
| InCo | intercoll. nu. |
| InG | intermed. gr. sup. coll. |
| IntP | intermed. post. deep Cb nu. |
| InWh | intermed. wh. sup. coll. |
| IOA | inf. oliv. nu., subn. A |
| IOD | inf. oliv. nu., dors. |
| IOPr | inf. oliv. nu., princ. |
| IRt | intermed. retic. formn |
| IS | inf. salivatory nu. |
| LC | locus coeruleus |
| lcs | lat. corticospinal tr. |
| LD | laterodors. thal. nu. |
| lentf | H2 lenticular fasc. |
| lfp | longit. fibers pons |
| LH | lat. hypothal. area |
| Lim | limitans nu. |
| lo | lat. olf. tract |
| LPBD | lat. parabr. nu., dors. |
| LPGi | lat. paragigant. nu. |
| LPO | lat. preopt. area |
| LRt | lat. reticular nu. |
| LTu | lat. tuberal nu. |
| LV | lateral ventricle |
| MCLH | magno. nu. lat. hypothal. |
| mcp | middle cerebell. ped. |
| MdD | medull. retic., nu. dors. |
| MDD | mediodors. thal. nu., dors. |
| MDL | mediodors. thal. nu., lat. |
| Med | med. deep Cb nu. |
| MePO | med. periolivary nu. |
| ml | medial lemniscus |
| MPB | med. parabrach. nu. |
| mRt | mesen. retic. formn |
| MSO | med. sup. olivary nu. |
| MTnl | med. terminal nu. |
| MVe | med. vestib. nu. |
| MVeMC | med. vestib. nu., magn. |
| MVePC | med. vestib. nu., parv. |
| Op | strat. opt. sup. coll. |
| opt | optic tract |
| OT | nu. of optic tract |
| p1Rt | pros. 1 retic. formn |
| P7 | perifacial nu. |
| Pa6 | parabducens nu. |
| PBP | parabrach. pigm. nu. |
| PCRt | parv. retic. nu. |
| PCRtα | parv. retic. nu., alpha |
| PCTg | paracollicular tegm. |
| PCuMx | pericuneate matrix |
| PE | par. field E Cx |
| PEc | par. field E Cx, caud. |
| Pir | piriform Cx |
| plf | posterolat. fissure |
| PLH | pedunc. lat. hypothal. |
| Pn | pontine nu. |
| PnC | pontine caudal nu. |
| PnO | pontine oral nu. |
| PPD1 | post. par. Cx, dors. 1 |
| PPD2 | post. par. Cx, dors. 2 |
| ppf | prepyramidal fiss. |
| PPM | post. par. Cx, medial |
| PPT | post. pretectum |
| PPTgC | peduncpont. tegm. nu., comp. |
| PPTgD | peduncpont. tegm. nu., diff. |
| prf | prim. cerebell. fiss. |
| PRF | prerubral field |
| psf | post. sup. fissure Cb |
| PSTh | parasubthal. nu. |
| PulM | pulvinar nu., med. |
| py | pyramidal tract |
| RPC | red nu., parvicellular |
| RRF | retrorubral field |
| Rt | reticular nu. |
| RtTg | reticulotegm. nu. |
| scp | sup. cerebell. ped. |
| sf | secondary fissure Cb |
| SI | subst. innominata |
| SNC | subst. nigra, compact |
| SNR | subst. nigra, retic. |
| SO | supraoptic nu. |
| sod | suproptic decn |
| sol | solitary tract |
| Sol | nu. of solitary tr. |
| SolCe | nu. of solitary tr., cent. |
| SolDL | nu. of solitary tr., dorsolat. |
| SolV | nu. of solitary tr., vent. |
| SolVL | nu. of solitary tr., ventlat. |
| sp5 | spinal trigeminal tr. |
| Sp5C | spin. trigem. nu., caud. |
| SPTg | subpedunc. tegm. nu. |
| SpVe | spinal vestib. nu. |
| st | stria terminalis |
| SubCD | subcoer. nu., dors. |
| SubCV | subcoer. nu., vent. |
| SubI | subincertal nu. |
| SuG | sup. gray sup. coll. |
| SuVe | sup. vestib. nu. |
| TM | tuberomamill. area |
| Tu | olf. tubercle |
| tz | trapezoid body |
| Tz | nu. of trapezoid body |
| V1 | V1 Cx |
| V2m | V2 (med.) Cx |
| V2v | V2 (vent.) Cx |
| VA | vent. ant. thal. nu. |
| VeCb | vestibulocerebell. nu. |
| VH | ventral horn |
| VLa (pal) | ventrolat. thal. nu., ant. |
| VLH | vent. lat. hypothal. |
| VLPAG | ventrolat. periaq. gr. |
| VPM | vent. post. thal. nu., med. |
| VPMPC | vent. post. thal. nu., med. parv. |
| VTM | vent. tuberomamill. area |
| ZID | zona incerta, dors. |
| ZIR | zona incerta, rost. |
| ZIV | zona incerta, vent. |
| Zo | strat. zonale SC |

# Plate 65a

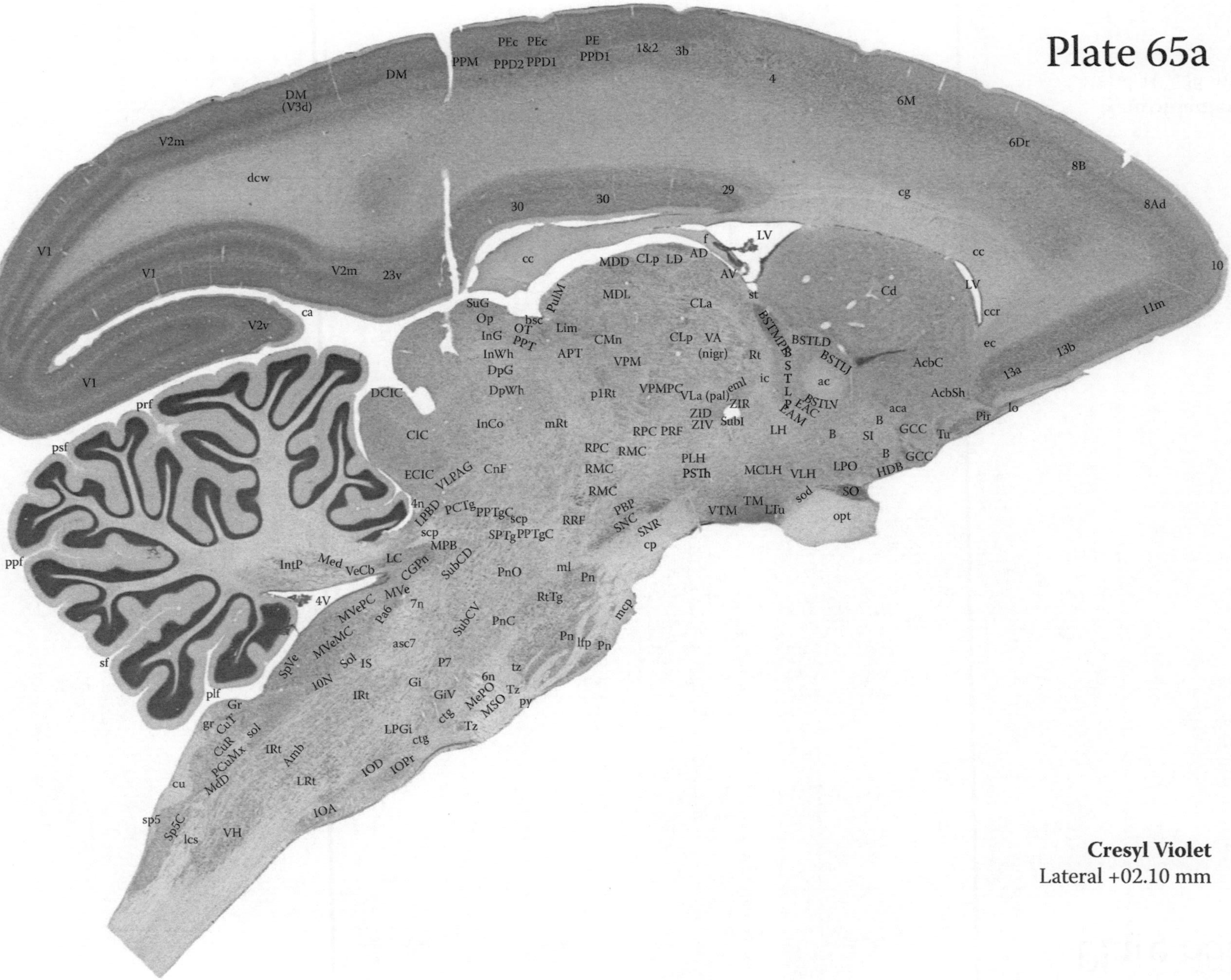


**Cresyl Violet**
Lateral +02.10 mm

# Plate 65b

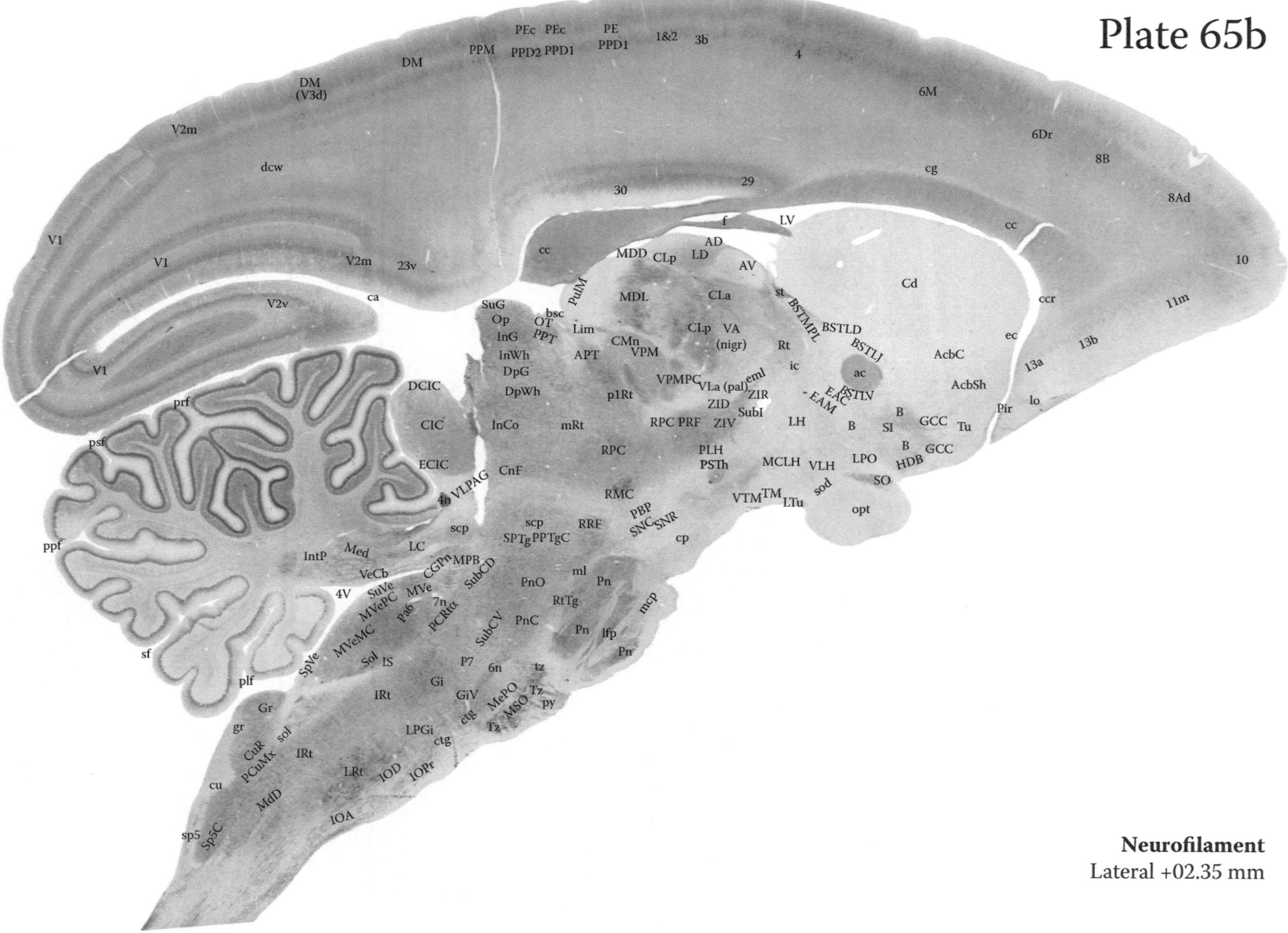


**Neurofilament**
Lateral +02.35 mm

# Plate 66

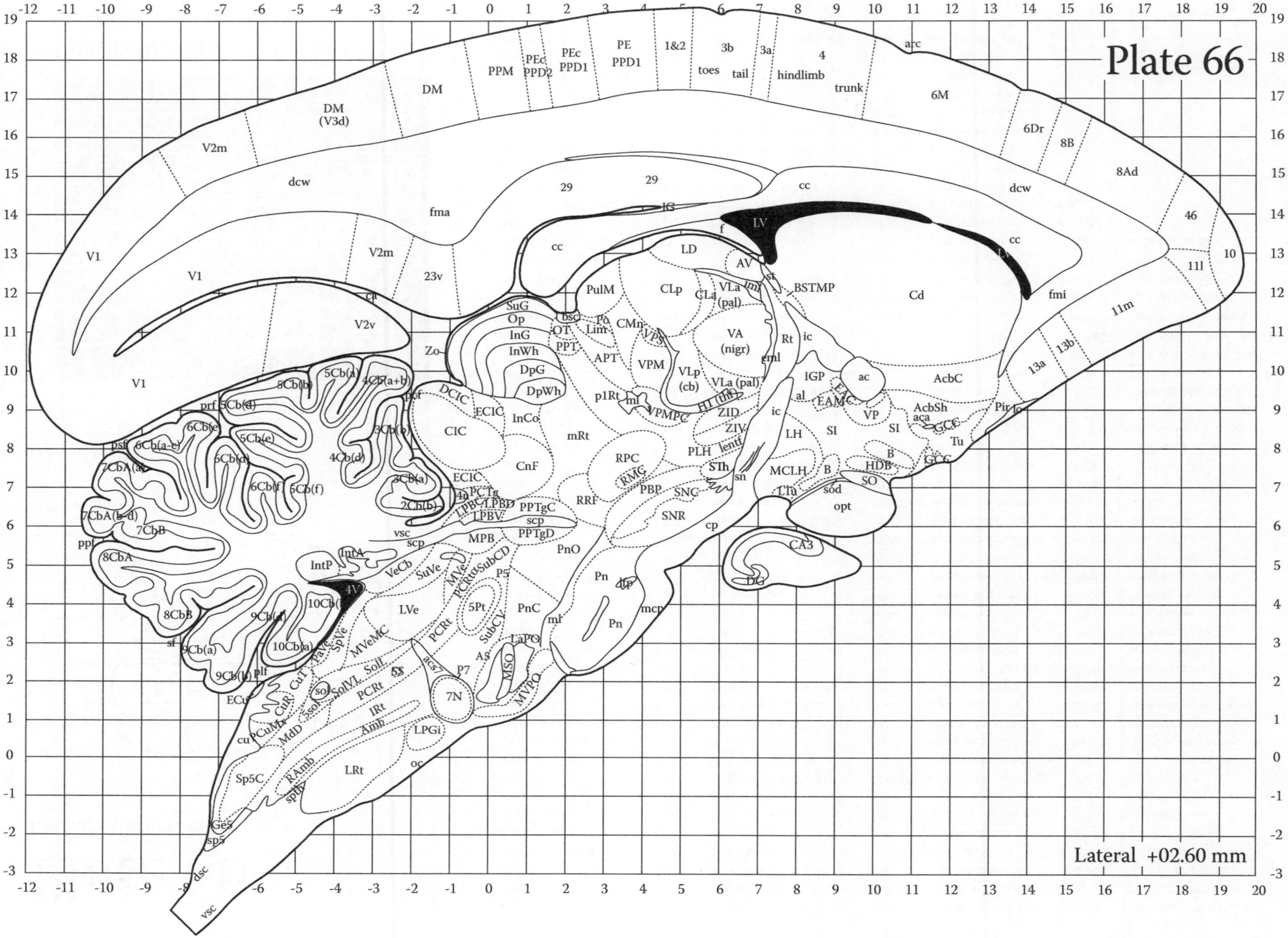

Plate 66
Lateral +02.60 mm
V1
V2m
V2v
dcw
DM
DM (V3d)
PPM
PEc PPD2
PEc PPD1
PE PPD1
1&2
3b
3a
toes
tail
4
hindlimb
trunk
arc
6M
6Dr
8B
8Ad
46
10
11l
11m
13a
13b
fma
29
IG
cc
LV
f
23v
cg
Zo
SuG
Op
InG
InWh
DpG
DpWh
DCIC
ECIC
CIC
InCo
CnF
LD
AV
st
sm
BSTMP
PulM
CLp
CLa
VLa (pal)
VA (nigr)
VLp (cb)
VPM
VPS
APT
Po
Lim
CMn
bsc
OT
PPT
p1Rt
ml
VPMPC
Rt
ic
eml
ZID
ZIV
lentf
PLH
STh
sn
mRt
RPC
RMC
PBP
SNC
SNR
RRF
PPTgC
PPTgD
scp
MPB
LPBC
LPBD
LPBV
PCTg
4n
PnO
PnC
Pn
mcp
tfp
cp
Cd
fmi
AcbC
AcbSh
aca
GCC
Tu
Pir
lo
ac
IGP
al
EA
EAM
VP
SI
LH
MCLH
B
HDB
SO
sod
Ltu
opt
CA3
DG
IntA
IntP
VeCb
SuVe
MVe
LVe
4V
PCRtA
SubCD
SubCV
P5
5Pt
LaPO
MSO
MVPO
As
P7
acs7
7N
PCRt
MVeMC
SpVe
PaVe
SolL
SolVL
sol
5sol
CuT
CuR
PCuMx
cu
ECu
MdD
IRt
Amb
RAmb
LPGi
LRt
oc
Sp5C
sp5
Ge5
spth
dsc
vsc
psf
prf
pcf
ppf
plf
sf
5Cb(a)
5Cb(b)
5Cb(d)
5Cb(e)
5Cb(f)
4Cb(a+b)
4Cb(d)
3Cb(a)
3Cb(b)
2Cb(b)
6Cb(a-c)
6Cb(d)
6Cb(e)
6Cb(f)
7CbA(a)
7CbA(b-d)
7CbB
8CbA
8CbB
9Cb(a)
9Cb(b)
9Cb(d)
10Cb(a)
10Cb(c)

# Plate 66

Lateral +02.60 mm

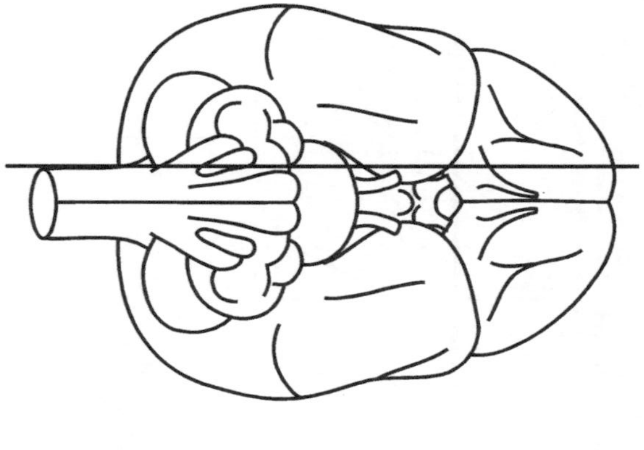

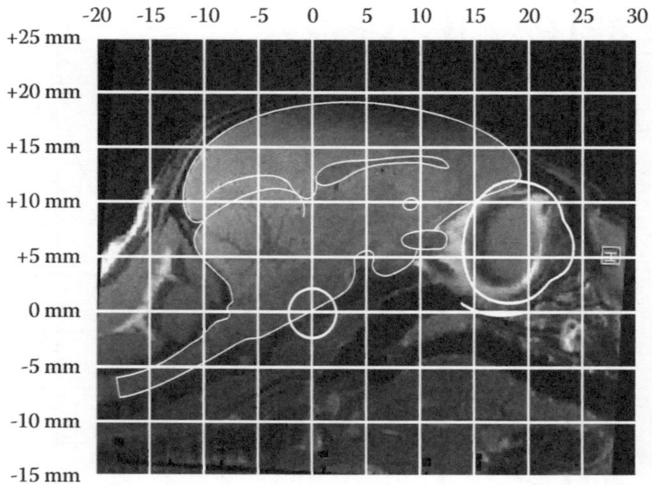


Lateral +02.40 mm

| | |
|---|---|
| 2Cb | cerebell. lobule 2 |
| 3a | area 3a Cx |
| 3b | area 3b Cx |
| 3Cb | cerebell. lobule 3 |
| 4Cb | cerebell. lobule 4 |
| 4n | trochlear nerve |
| 4V | fourth ventricle |
| 5Cb | cerebell. lobule 5 |
| 5Pt | trigem. nu., pteryg. |
| 5Sol | trigemsol. transn zone |
| 6Cb | cerebell. lobule 6 |
| 6Dr | area 6Dr Cx |
| 6M | area 6M Cx |
| 7Cb | cerebell. lobule 7 |
| 7N | facial (motor) nu. |
| 8Ad | area 8Ad Cx |
| 8B | area 8B Cx |
| 8Cb | cerebell. lobule 8 |
| 9Cb | cerebell. lobule 9 |
| 10Cb | cerebell. lobule 10 |
| 11l | area 11l Cx |
| 11m | area 11m Cx |
| 13a | area 13a Cx |
| 13b | area 13b Cx |
| 23v | area 23v Cx |
| A5 | A5 noradr. cells |
| ac | ant. comm. |
| aca | ant. comm. ant. |
| AcbC | accumb. nu. core |
| AcbSh | accumb. nu. shell |
| acs7 | accessory facial nu. |
| al | ansa lenticularis |
| Amb | ambiguus nu. |
| APT | ant. pretect. nu. |
| arc | arcuate sulcus |
| AV | anterovent. thal. nu. |
| B | basal nu. (Meynert) |
| bsc | brach. sup. coll. |
| BSTMP | bed nu. st, med. post. divn |
| ca | calcarine sulcus |
| CA3 | cornu ammonis 3 |
| cc | corp. callosum |
| Cd | caudate nu. |
| CIC | cent. nu. inf. coll. |
| CLa | centrolat. thal. nu., ant. |
| CLp | centrolat. thal. nu., post. |
| CMn | centromedian thal. nu. |
| CnF | cuneiform nu. |
| cp | cerebral peduncle |
| cu | cuneate fasciculus |
| CuR | cuneate nu., rotunda |
| CuT | cuneate nu., triang. |
| DCIC | dors. Cx inf. coll. |
| dcw | deep cort. white |
| DG | dentate gyrus |
| DM | dorsomed. visual Cx |
| DpG | deep gray sup. coll. |
| DpWh | deep white sup. coll. |
| dsc | dors. spinocerebell. tr |
| EAC | extend amyg. central |
| EAM | extend amyg. medial |
| ECIC | extern. Cx inf. coll. |
| ECu | ext. cuneate nu. |
| eml | extern. medull. lamina |
| f | fornix |
| fmi | forceps minor |
| GCC | gran. cell clust. |
| Ge5 | gelat. layer Sp5C |
| H1 (thf) | H1 thalamic fasc. |
| HDB | horiz. limb diag. band nu. |
| ic | intern. caps. |
| IG | induseum gris. |
| IGP | int. globus pallidus |
| iml | intern. medull. lamina |
| InCo | intercoll. nu. |
| InG | intermed. gr. sup. coll. |
| IntA | intermed. ant. deep Cb nu. |
| IntP | intermed. post. deep Cb nu. |
| InWh | intermed. wh. sup. coll. |
| IRt | intermed. retic. formn |
| LaPO | lat. periolivary nu. |
| LD | laterodors. thal. nu. |
| lentf | H2 lenticular fasc. |
| lfp | longit. fibers pons |
| LH | lat. hypothal. area |
| Lim | limitans nu. |
| lo | lat. olf. tract |
| LPBC | lat. parabr. nu., cent. |
| LPBD | lat. parabr. nu., dors. |
| LPBV | lat. parabr. nu., vent. |
| LPGi | lat. paragigant. nu. |
| LRt | lat. reticular nu. |
| LTu | lat. tuberal nu. |
| LV | lateral ventricle |
| LVe | lat. vestib. nu. |
| MCLH | magno. nu. lat. hypothal. |
| mcp | middle cerebell. ped. |
| MdD | medull. retic., nu. dors. |
| ml | medial lemniscus |
| MPB | med. parabrach. nu. |
| mRt | mesen. retic. formn |
| MSO | med. sup. olivary nu. |
| MVe | med. vestib. nu. |
| MVeMC | med. vestib. nu., magn. |
| MVPO | mediovent. perioliv. nu. |
| oc | olivocerebellar tr. |
| Op | strat. opt. sup. coll. |
| opt | optic tract |
| OT | nu. of optic tract |
| p1Rt | pros. 1 retic. formn |
| P5 | peritrigeminal nu. |
| P7 | perifacial nu. |
| PaVe | paravestibular nu. |
| PBP | parabrach. pigm. nu. |
| pcf | preculminate fiss. Cb. |
| PCRt | parv. retic. nu. |
| PCRtα | parv. retic. nu., alpha |
| PCTg | paracollicular tegm. |
| PCuMx | pericuneate matrix |
| PE | par. field E Cx |
| PEc | par. field E Cx, caud. |
| Pir | piriform Cx |
| plf | posterolat. fissure |
| PLH | pedunc. lat. hypothal. |
| Pn | pontine nu. |
| PnC | pontine caudal nu. |
| PnO | pontine oral nu. |
| Po | post. thal. nu. |
| PPD1 | post. par. Cx, dors. 1 |
| PPD2 | post. par. Cx, dors. 2 |
| ppf | prepyramidal fiss. |
| PPM | post. par. Cx, medial |
| PPT | post. pretectum |
| PPTgC | peduncpont. tegm. nu., compact |
| PPTgD | peduncpont. tegm. nu., diff. |
| prf | prim. cerebell. fiss. |
| psf | post. sup. fissure Cb |
| PulM | pulvinar nu., med. |
| RAmb | retroambiguus nu. |
| RMC | red nu., magnocell. |
| RPC | red nu., parvicellular |
| RRF | retrorubral field |
| Rt | reticular nu. |
| scp | sup. cerebell. ped. |
| sf | secondary fissure Cb |
| SI | subst. innominata |
| sn | striatonigral fibers |
| SNC | subst. nigra, compact |
| SNR | subst. nigra, retic. |
| SO | supraoptic nu. |
| sod | suproptic decn |
| sol | solitary tract |
| SolL | nu. of solitary tr., lat. |
| SolVL | nu. of solitary tr., ventlat. |
| sp5 | spinal trigeminal tr. |
| Sp5C | spin. trigem. nu., caud. |
| spth | spinothalamic tr. |
| SpVe | spinal vestib. nu. |
| SS | sup. salivatory nu. |
| st | stria terminalis |
| STh | subthalamic nu. |
| SubCD | subcoer. nu., dors. |
| SubCV | subcoer. nu., vent. |
| SuG | sup. gray sup. coll. |
| SuVe | sup. vestib. nu. |
| Tu | olf. tubercle |
| V1 | V1 Cx |
| V2m | V2 (med.) Cx |
| V2v | V2 (vent.) Cx |
| VA | vent. ant. thal. nu. |
| VeCb | vestibulocerebell. nu. |
| VLa | ventrolat. thal. nu., ant. |
| VLp | ventrolat. thal. nu., post. |
| VP | vent. pallidum |
| VPM | vent. post. thal. nu., med. |
| VPMPC | vent. post. thal. nu., med. parv. |
| VPS | vent. post. thal. nu., sup. |
| vsc | vent. spinocereb. tr. |
| ZID | zona incerta, dors. |
| ZIV | zona incerta, rost. |
| Zo | strat. zonale SC |

# Plate 66a

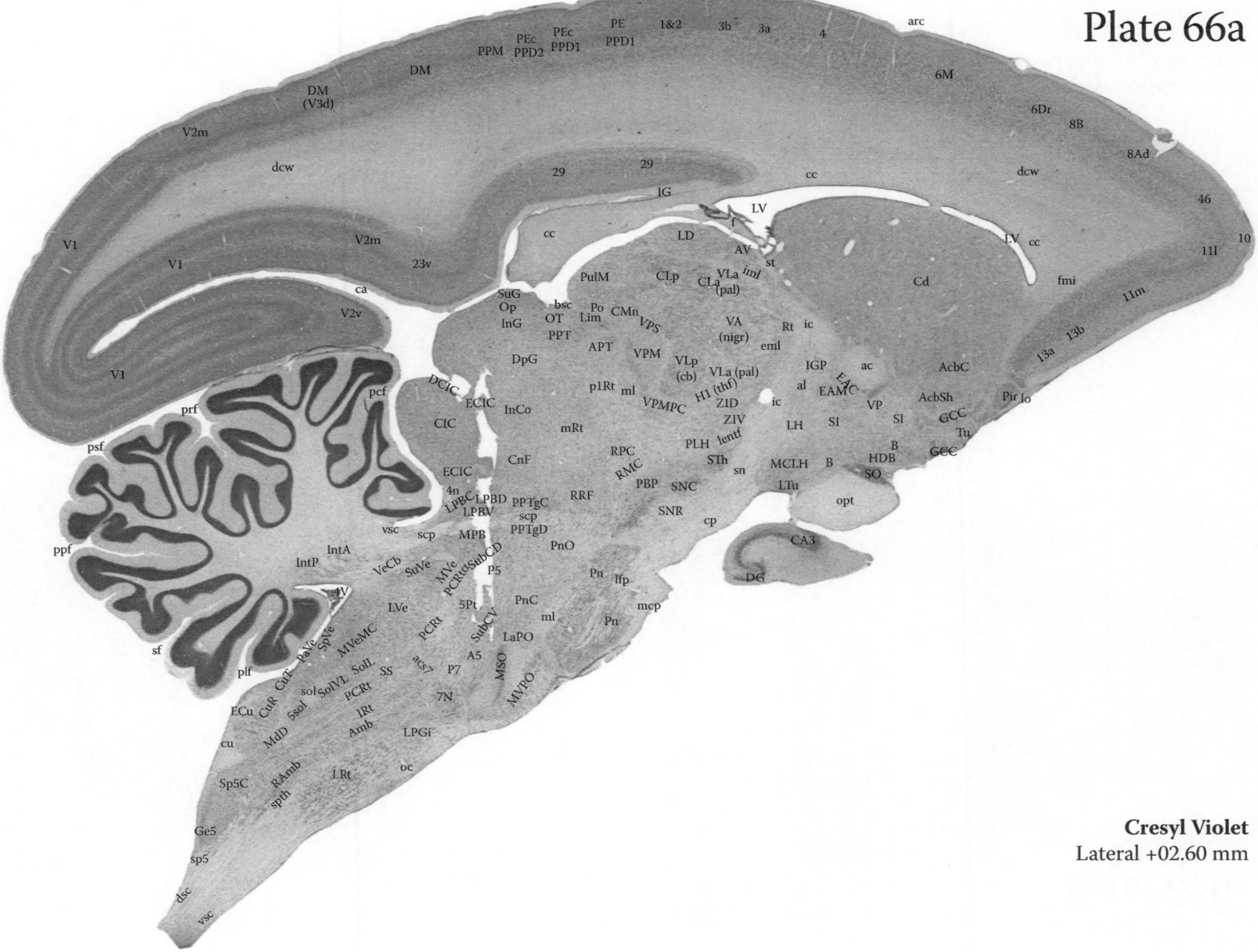


**Cresyl Violet**
Lateral +02.60 mm

# Plate 66b

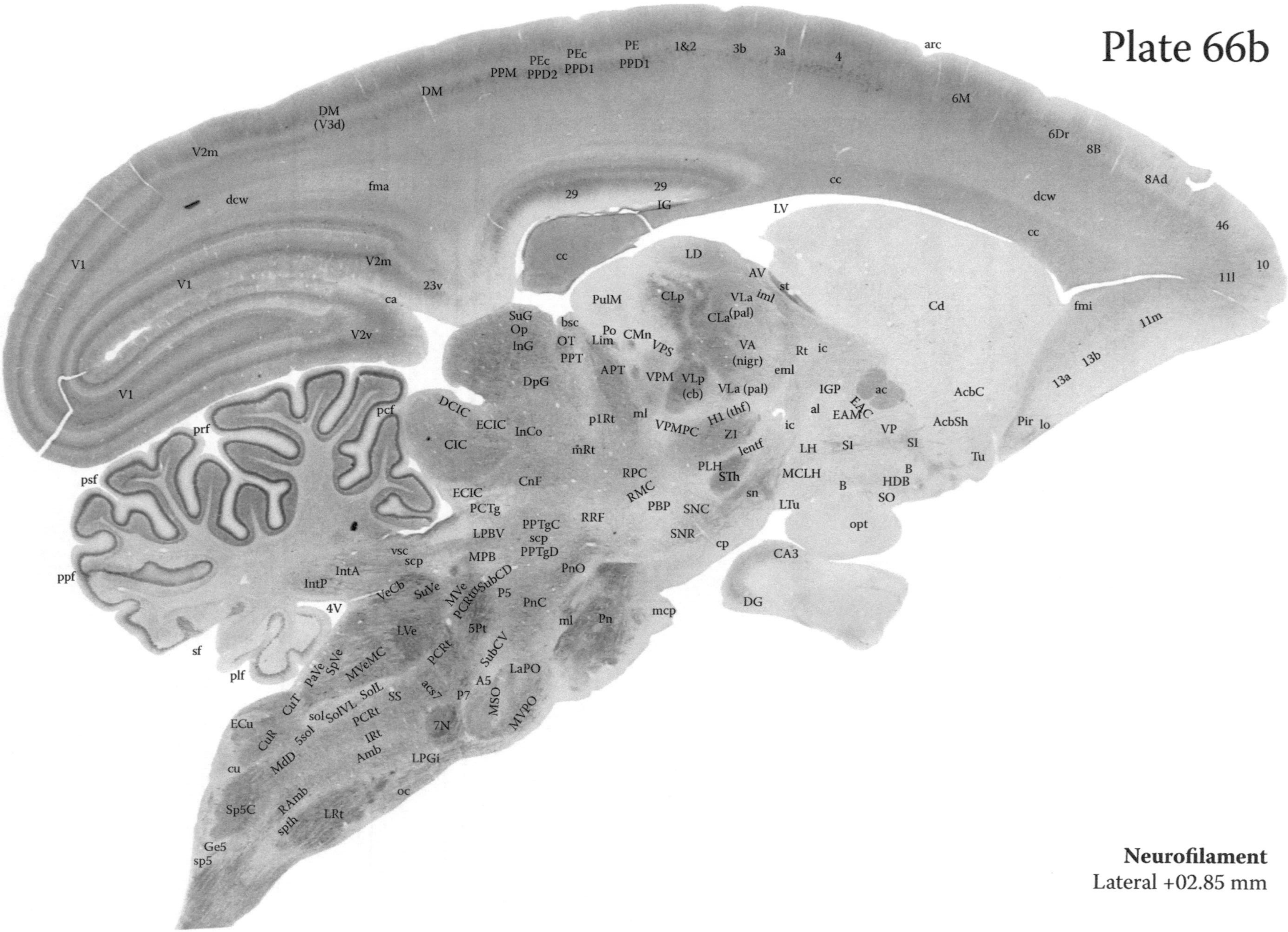


**Neurofilament**
Lateral +02.85 mm

# Plate 67

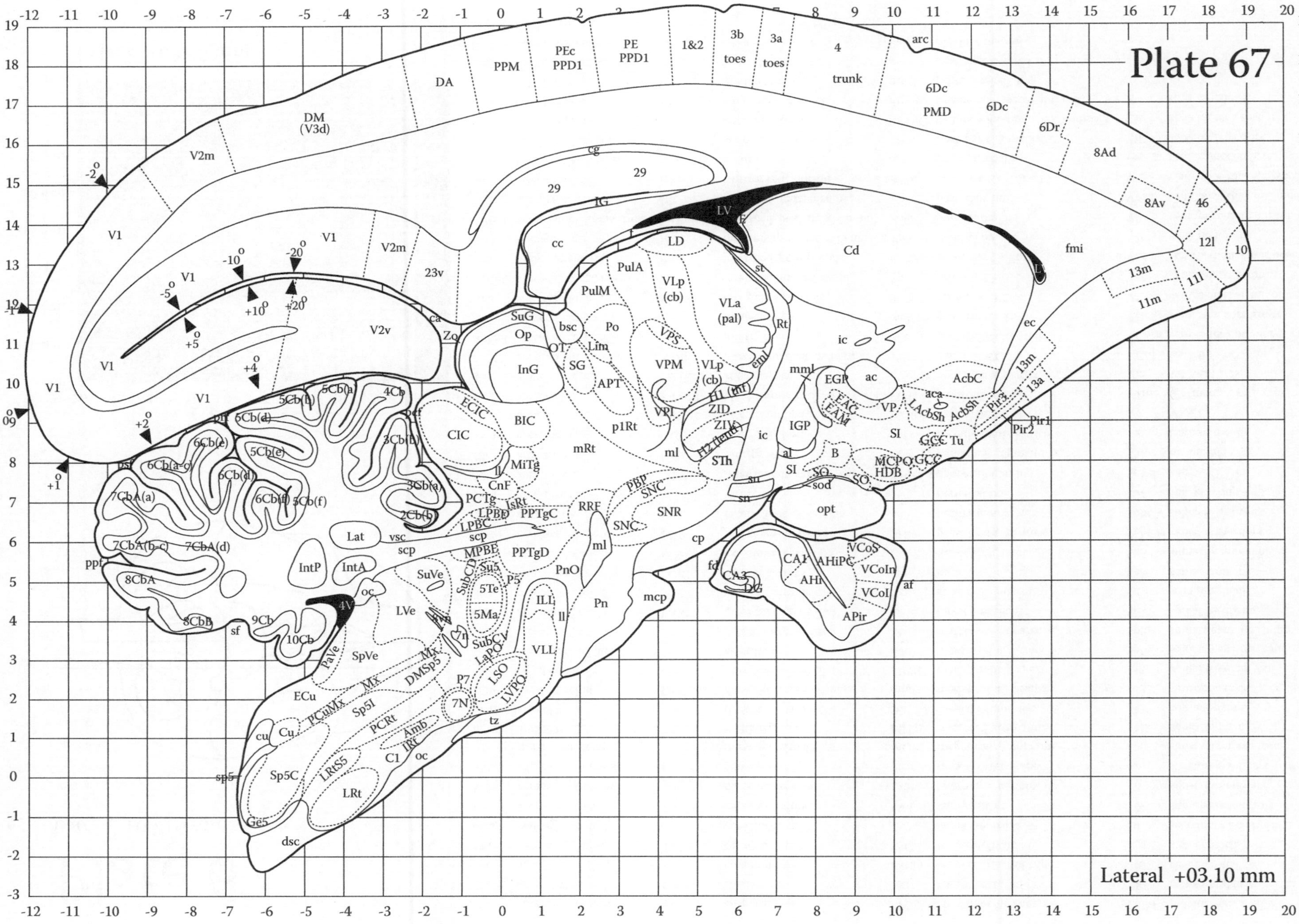

DA
PPM
PEc PPD1
PE PPD1
1&2
3b toes
3a toes
4 trunk
arc
6Dc PMD
6Dc
6Dr
8Ad
8Av
46
12l
10
13m
11l
11m
13a
DM (V3d)
V2m
V1
V2v
23v
29
cg
IG
cc
LV
LD
Cd
fmi
ec
st
PulA
PulM
VLp (cb)
VLa (pal)
Rt
ic
mml
EGP
ac
AcbC
aca
LAcbSh
AcbSh
Pir1
Pir2
VP
SI
GCC
Tu
MCPO
HDB
IGP
al
B
SO
sod
opt
SuG
Op
bsc
Po
OT
Lim
SG
InG
APT
VPS
VPM
VPl
eml
ZID
ZIV
STh
sn
ml
p1Rt
mRt
BIC
ECIC
CIC
MiTg
CnF
PCTg
LPBD
LPBC
IsRt
PPTgC
PPTgD
RRF
PBP
SNC
SNR
cp
mcp
Pn
PnO
MPBE
Su5
5Te
5Ma
P5
ILL
ll
VLL
SubCV
LaPO
LSO
LVPO
7N
P7
7n
tz
SuVe
LVe
SpVe
PaVe
Mx
DMSp5
PCuMx
Sp5I
PCRt
Amb
IRt
C1
oc
LRt
LRtS5
Sp5C
Cu
cu
ECu
sp5
Ge5
dsc
4V
Lat
IntA
IntP
vsc
scp
9Cb
10Cb
8CbB
8CbA
sf
ppf
psf
pcf
4Cb
5Cb(a)
5Cb(b)
5Cb(d)
5Cb(e)
5Cb(f)
6Cb(a-c)
6Cb(d)
6Cb(e)
6Cb(f)
7CbA(a)
7CbA(b-c)
7CbA(d)
3Cb(a)
3Cb(b)
2Cb(b)
ca
Zo
fd
CA1
CA3
DG
AHiPC
AHi
VCoS
VCoIn
VCoI
APir
af
Lateral +03.10 mm

# Plate 67

Lateral +03.10 mm

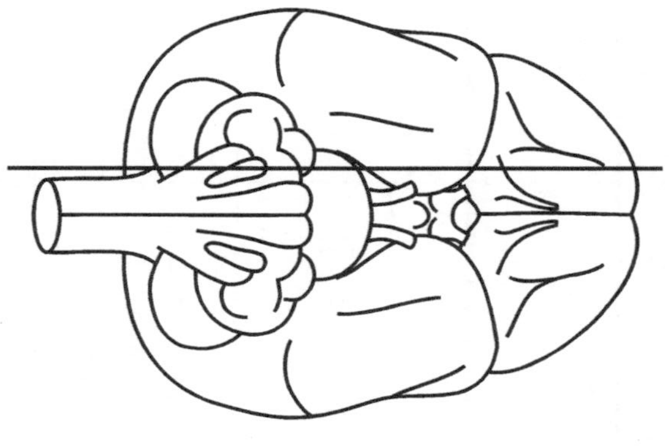

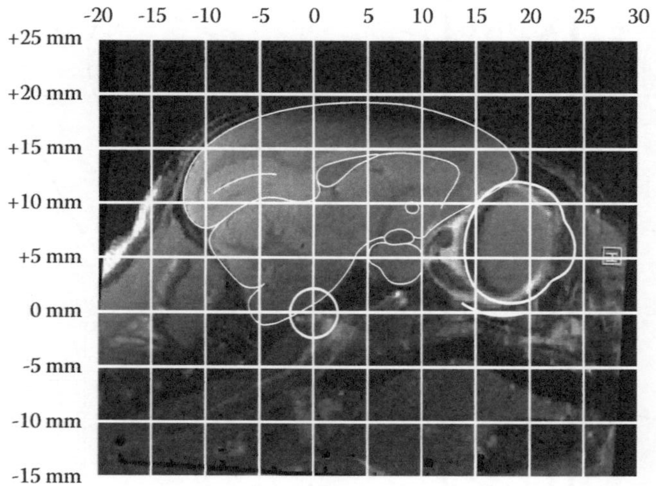


Lateral +03.35 mm

| | |
|---|---|
| 2Cb | cerebell. lobule 2 |
| 3a | area 3a Cx |
| 3b | area 3b Cx |
| 3Cb | cerebell. lobule 3 |
| 4Cb | cerebell. lobule 4 |
| 4V | fourth ventricle |
| 5Cb | cerebell. lobule 5 |
| 5Ma | trigem. nu., masset. |
| 5Te | trigem. nu., temp. |
| 6Dc | area 6Dc Cx |
| 6Dr | area 6Dr Cx |
| 7Cb | cerebell. lobule 7 |
| 7n | facial nerve |
| 7N | facial (motor) nu. |
| 8Ad | area 8Ad Cx |
| 8Av | area 8Av Cx |
| 8Cb | cerebell. lobule 8 |
| 8vn | vestibular divn 8n |
| 9Cb | cerebell. lobule 9 |
| 10Cb | cerebell. lobule 10 |
| 11l | area 11l Cx |
| 11m | area 11m Cx |
| 12l | area 12l Cx |
| 13a | area 13a Cx |
| 13m | area 13m Cx |
| 23v | area 23v Cx |
| ac | ant. comm. |
| aca | ant. comm. ant. |
| AcbC | accumb. nu. core |
| AcbSh | accumb. nu. shell |
| af | amygdaloid fiss. |
| AHi | amygdalohipp. area |
| AHiPC | amygdalohipp., magn. |
| al | ansa lenticularis |
| Amb | ambiguus nu. |
| APir | amygdalopir. trans. area |
| APT | ant. pretect. nu. |
| arc | arcuate sulcus |
| B | basal nu. (Meynert) |
| BIC | brach. inf. coll. nu. |
| bsc | brach. sup. coll. |
| C1 | C1 adrenaline cells |
| ca | calcarine sulcus |
| CA1 | cornu ammonis 1 |
| CA3 | cornu ammonis 3 |
| cc | corp. callosum |
| Cd | caudate nu. |
| cg | cingulum |
| CIC | cent. nu. inf. coll. |
| CnF | cuneiform nu. |
| cp | cerebral peduncle |
| cu | cuneate fasciculus |
| Cu | cuneate nu. |
| DA | dorsoant. visual Cx |
| DG | dentate gyrus |
| DM | dorsomed. visual Cx |
| DMSp5 | dorsomed. spin. 5 nu. |
| dsc | dors. spinocerebell. tr |
| E | ependyma |
| EAC | extend amyg. central |
| EAM | extend amyg. medial |
| ec | external capsule |
| ECIC | extern. Cx inf. coll. |
| ECu | ext. cuneate nu. |
| EGP | ext. globus pallidus |
| eml | extern. medull. lamina |
| f | fornix |
| fd | fimbriodent. sulcus |
| fmi | forceps minor |
| GCC | gran. cell clust. |
| Ge5 | gelat. layer Sp5C |
| H1 (thf) | H1 thalamic fasc. |
| H2 (lentf) | H2 lenticular fasc. |
| HDB | horiz. limb diag. band nu. |
| ic | intern. caps. |
| IG | induseum gris. |
| IGP | int. globus pallidus |
| ILL | intermed. nu. lat. lemn. |
| InG | intermed. gr. sup. coll. |
| IntA | intermed. ant. deep Cb nu. |
| IntP | intermed. post. deep Cb nu. |
| IRt | intermed. retic. formn |
| IsRt | isthmic retic. formn |
| LAcbSh | accumb. shell lat. |
| LaPO | lat. periolivary nu. |
| Lat | lat. deep Cb nu. |
| LD | laterodors. thal. nu. |
| Lim | limitans nu. |
| ll | lateral lemniscus |
| LPBC | lat. parabr. nu., cent. |
| LPBD | lat. parabr. nu., dors. |
| LRt | lat. reticular nu. |
| LRtS5 | lat. retic. nu., subtrigem. |
| LSO | lat. sup. olivary nu. |
| LV | lateral ventricle |
| LVe | lat. vestib. nu. |
| LVPO | laterovent. perioliv. nu. |
| mcp | middle cerebell. ped. |
| MCPO | magnocell. preopt. nu. |
| MiTg | microcell. tegm. nu. |
| ml | medial lemniscus |
| mml | med. medull. lamina |
| MPBE | med. parabr. nu., ext. |
| mRt | mesen. retic. formn |
| Mx | matrix sensory area |
| oc | olivocerebellar tr. |
| Op | strat. opt. sup. coll. |
| opt | optic tract |
| OT | nu. of optic tract |
| p1Rt | pros. 1 retic. formn |
| P5 | peritrigeminal nu. |
| P7 | perifacial nu. |
| PaVe | paravestibular nu. |
| PBP | parabrach. pigm. nu. |
| pcf | preculminate fiss. Cb. |
| PCRt | parv. retic. nu. |
| PCTg | paracollicular tegm. |
| PCuMx | pericuneate matrix |
| PE | par. field E Cx |
| PEc | par. field E Cx, caud. |
| Pir1 | piriform Cx 1 |
| Pir2 | piriform Cx 2 |
| Pir3 | piriform Cx 3 |
| PMD | premotor Cx dors. |
| Pn | pontine nu. |
| PnO | pontine oral nu. |
| Po | post. thal. nu. |
| PPD1 | post. par. Cx, dors. 1 |
| ppf | prepyramidal fiss. |
| PPM | post. par. Cx, medial |
| PPTgC | peduncpont. tegm. nu., compact |
| PPTgD | peduncpont. tegm. nu., diff. |
| prf | prim. cerebell. fiss. |
| psf | post. sup. fissure Cb |
| PulA | pulvinar nu., ant. |
| PulM | pulvinar nu., med. |
| RRF | retrorubral field |
| Rt | reticular nu. |
| scp | sup. cerebell. ped. |
| sf | secondary fissure Cb |
| SG | supragenic. thal. nu. |
| SI | subst. innominata |
| sn | striatonigral fibers |
| SNC | subst. nigra, compact |
| SNR | subst. nigra, retic. |
| SO | supraoptic nu. |
| sod | suproptic decn |
| sp5 | spinal trigeminal tr. |
| Sp5C | spin. trigem. nu., caud. |
| Sp5I | spin. trigem. nu., interpol. |
| SpVe | spinal vestib. nu. |
| st | stria terminalis |
| STh | subthalamic nu. |
| Su5 | supratrigeminal nu. |
| SubCD | subcoer. nu., dors. |
| SubCV | subcoer. nu., vent. |
| SuG | sup. gray sup. coll. |
| SuVe | sup. vestib. nu. |
| Tu | olf. tubercle |
| tz | trapezoid body |
| V1 | V1 Cx |
| V2m | V2 (med.) Cx |
| V2v | V2 (vent.) Cx |
| VCoI | vent. inf. cort. amyg. |
| VCoIn | vent. inter. cort. amyg. |
| VCoS | vent. sup. cort. amyg. |
| VLa | ventrolat. thal. nu., ant. |
| VLL | vent. nu. lat. lemn. |
| VLp | ventrolat. thal. nu., post. |
| VP | vent. pallidum |
| VPI | vent. post. thal. nu., inf. |
| VPM | vent. post. thal. nu., med. |
| VPS | vent. post. thal. nu., sup. |
| vsc | vent. spinocereb. tr. |
| ZID | zona incerta, dors. |
| ZIV | zona incerta, rost. |
| Zo | strat. zonale SC |

# Plate 67a

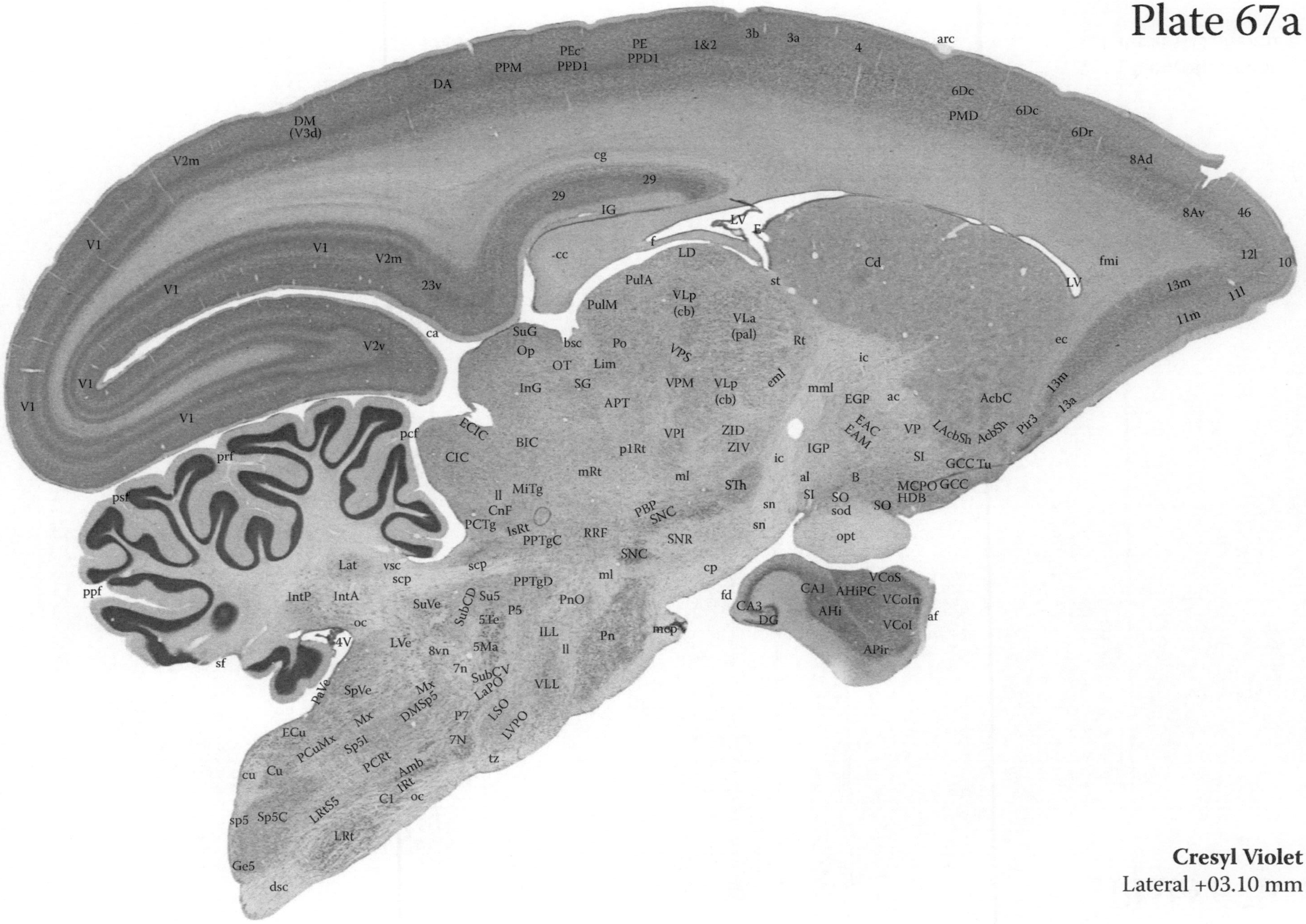


**Cresyl Violet**
Lateral +03.10 mm

# Plate 67b

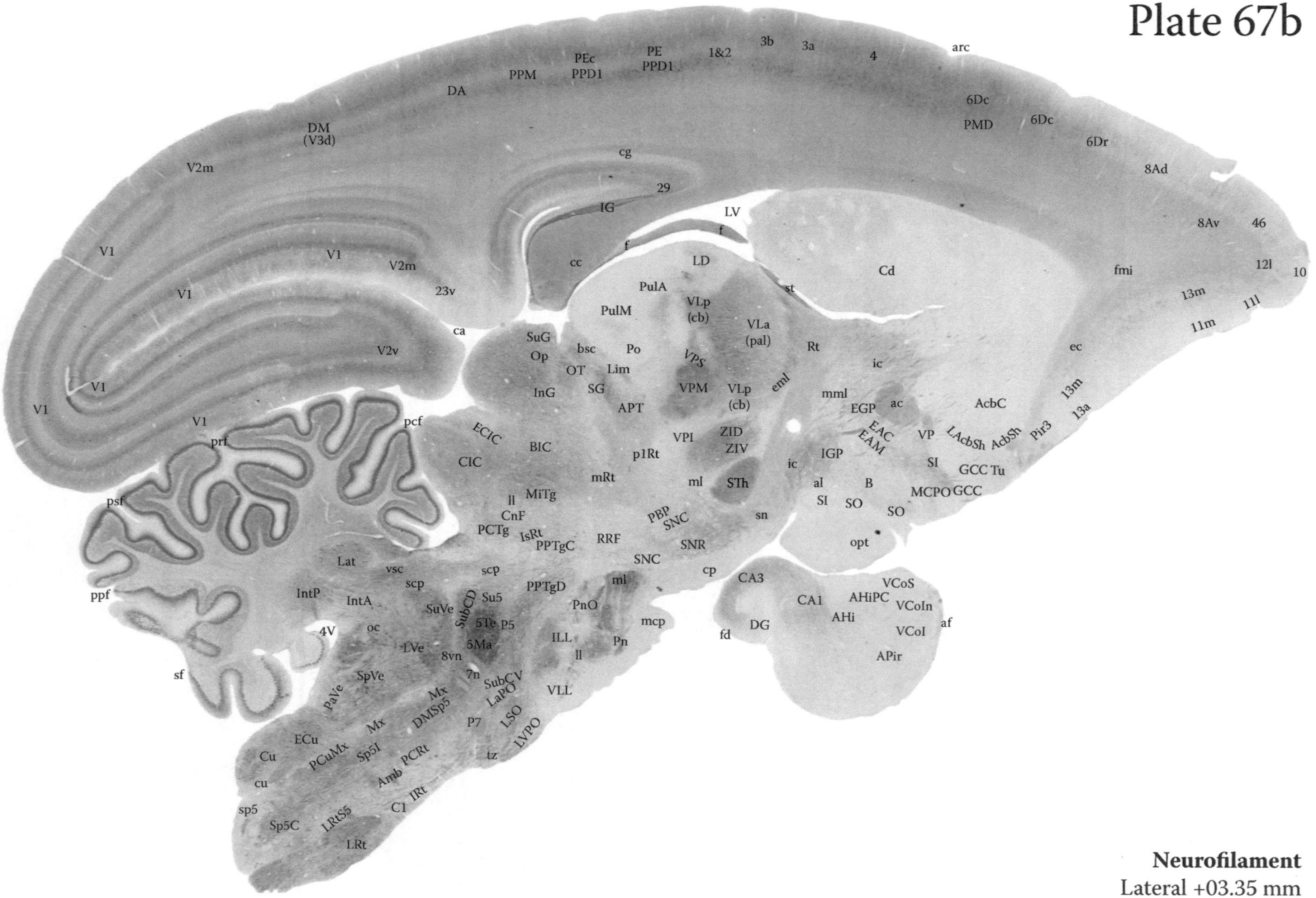


**Neurofilament**
Lateral +03.35 mm

# Plate 68

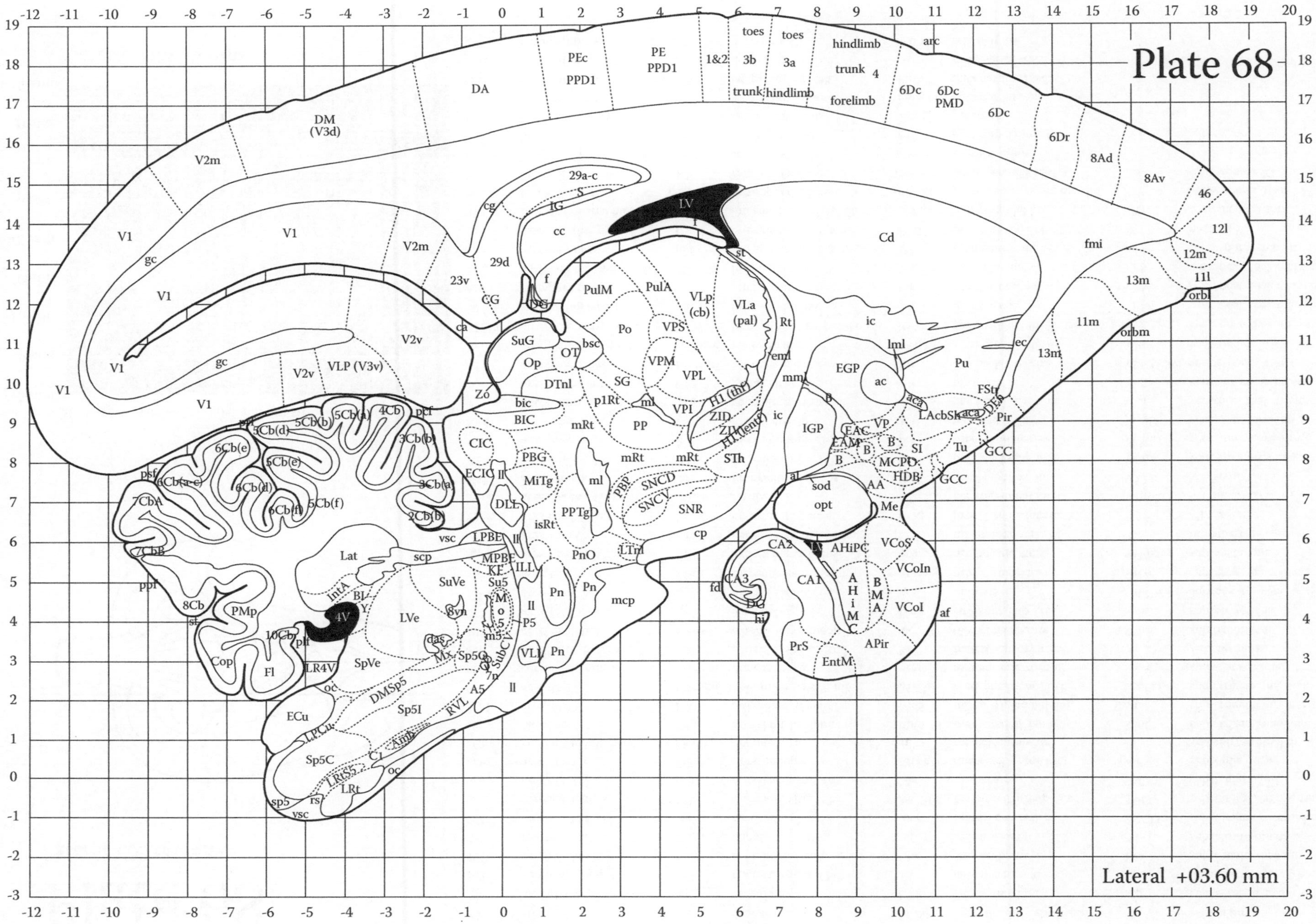
toes
toes
hindlimb
arc
PEc
PPD1
PE
PPD1
1&2
3b
3a
trunk
4
DA
trunk
hindlimb
forelimb
6Dc
6Dc
PMD
6Dc
DM
(V3d)
6Dr
8Ad
V2m
8Av
46
29a-c
S
LV
12l
cg
IG
V1
V1
cc
Cd
fmi
12m
V2m
gc
29d
23v
11l
f
13m
V1
GG
PulM
PulA
VLp
(cb)
VLa
(pal)
orbl
ic
Rt
11m
V2v
ca
SuG
bsc
Po
VPS
orbm
V1
gc
V2v
VLP (V3v)
Op
OT
VPM
eml
lml
13m
V1
EGP
Pu
Zo
DTnl
SG
VPL
ac
mml
FStr
V1
bic
p1Rt
ml
H1 (thf)
aca
BIC
PP
VPI
ZID
ic
LAcbSh
aca
Pir
mRt
VP
4Cb
pcf
IGP
EAC
5Cb(b)
5Cb(a)
3Cb(d)
3Cb(b)
CIC
ZIV
H1(lenf)
B
6Cb(e)
PBG
mRt
mRt
STh
EAM
B
SI
Tu
GCC
5Cb(e)
B
MCPO
psf
ECIC
ml
8Cb(a-e)
MiTg
PBP
SNCD
HDB
GCC
6Cb(d)
3Cb(a)
sod
AA
7CbA
DLL
SNCV
5Cb(f)
PPTgD
SNR
opt
Me
6Cb(f)
2Cb(b)
isRt
LPBE
vsc
cp
CA2
LV
AHiPC
VCoS
7CbB
Lat
scp
MPBE
PnO
LTm
ILL
KF
VCoIn
SuVe
CA1
Su5
CA3
fd
ll
Pn
Pn
mcp
B
M
A
A
H
i
M
C
ppf
IntA
BI
Y
8Cb
4V
DG
VCoI
af
8vn
M
o
5
P5
hi
LVe
PMp
st
das
m5
10Cb
APir
plf
Sp5O
VLl
Pn
Cop
Mx
PrS
SubCV
EntM
Fl
LR4V
SpVe
7n
A5
ll
oc
DMSp5
RVL
ECu
Sp5I
LPCu
Amb
C1
Sp5C
LRt55
oc
LRt
sp5
rs
vsc
Lateral +03.60 mm

# Plate 68

Lateral +03.60 mm

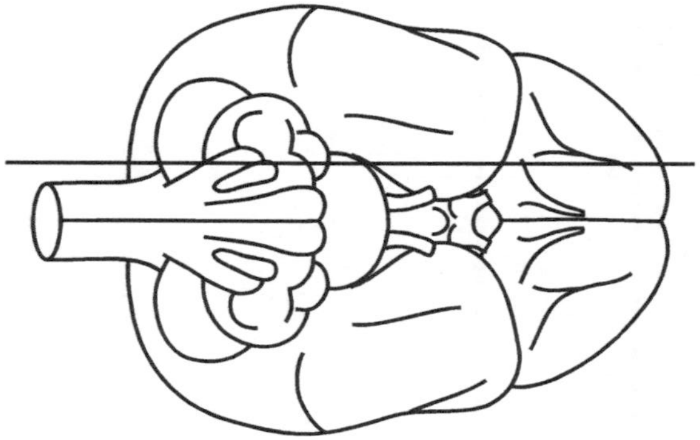

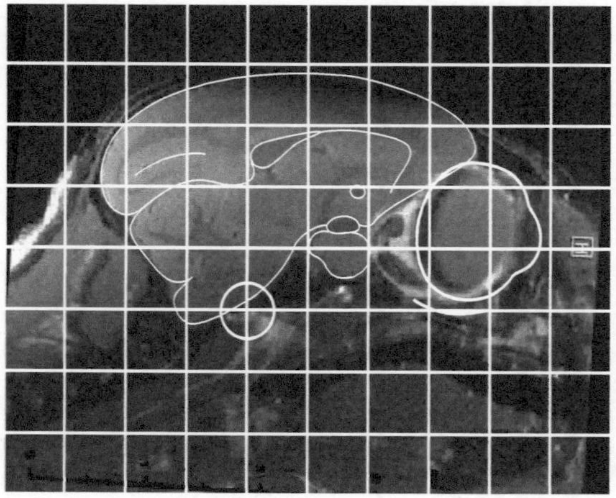


Lateral +03.35 mm

| | |
|---|---|
| 2Cb | cerebell. lobule 2 |
| 3a | area 3a Cx |
| 3b | area 3b Cx |
| 3Cb | cerebell. lobule 3 |
| 4Cb | cerebell. lobule 4 |
| 4V | fourth ventricle |
| 5Cb | cerebell. lobule 5 |
| 6Cb | cerebell. lobule 6 |
| 6Dc | area 6Dc Cx |
| 6Dr | area 6Dr Cx |
| 7Cb | cerebell. lobule 7 |
| 7n | facial nerve |
| 8Ad | area 8Ad Cx |
| 8Av | area 8Av Cx |
| 8Cb | cerebell. lobule 8 |
| 8vn | vestibular divn 8n |
| 10Cb | cerebell. lobule 10 |
| 11l | area 11l Cx |
| 11m | area 11m Cx |
| 12l | area 12l Cx |
| 12m | area 12m Cx |
| 13m | area 13m Cx |
| 23v | area 23v Cx |
| 29a-c | area 29a to c Cx |
| 29d | area 29d Cx |
| A5 | A5 noradr. cells |
| AA | ant. amyg. area |
| AAD | ant. amyg. dors. |
| ac | ant. comm. |
| aca | ant. comm. ant. |
| af | amygdaloid fiss. |
| AHiMC | amygdalohipp., parv. |
| AHiPC | amygdalohipp., magn. |
| al | ansa lenticularis |
| Amb | ambiguus nu. |
| APir | amygdalopir. trans. area |
| arc | arcuate sulcus |
| B | basal nu. (Meynert) |
| BI | basal interstitial nu. |
| bic | brach. inf. coll. |
| BIC | brach. inf. coll. nu. |
| BMA | basomed. amyg. nu., ant. |
| bsc | brach. sup. coll. |
| C1 | C1 adrenaline cells |
| ca | calcarine sulcus |
| CA1 | cornu ammonis 1 |
| CA2 | cornu ammonis 2 |
| CA3 | cornu ammonis 3 |
| cc | corp. callosum |
| Cd | caudate nu. |
| cg | cingulum |
| CG | cingulate gyrus Cx |
| CIC | cent. nu. inf. coll. |
| Cop | copula |
| cp | cerebral peduncle |
| DA | dorsoant. visual Cx |
| das | dorsal acoustic stria |
| DEn | dors. endopir. nu. |
| DG | dentate gyrus |
| DLL | dorsal nu. lat. lemn. |
| DM | dorsomed. visual Cx |
| DMSp5 | dorsomed. spin. 5 nu. |
| DTnl | dors. term. nu. |
| EAC | extend amyg. central |
| EAM | extend amyg. medial |
| ec | external capsule |
| ECIC | extern. Cx inf. coll. |
| ECu | ext. cuneate nu. |
| EGP | ext. globus pallidus |
| eml | extern. medull. lamina |
| EntM | entorhin. Cx, med. |
| f | fornix |
| fd | fimbriodent. sulcus |
| Fl | flocculus |
| fmi | forceps minor |
| FStr | fundus striatum |
| gc | geniculocalcarine tr. |
| GCC | gran. cell clust. |
| H1 (thf) | H1 thalamic fasc. |
| H2 (lentf) | H2 lenticular fasc. |
| HDB | horiz. limb diag. band nu. |
| hi | hippocamp. fiss. |
| ic | intern. caps. |
| IG | induseum gris. |
| IGP | int. globus pallidus |
| ILL | intermed. nu. lat. lemn. |
| IntA | intermed. ant. deep Cb nu. |
| isRt | isthmic retic. formn |
| KF | Kölliker-Fuse nu. |
| LAcbSh | accumb. shell lat. |
| Lat | lat. deep Cb nu. |
| ll | lateral lemniscus |
| lml | lat. medull. lamina |
| LPBE | lat. parabr. nu., ext. |
| LPCu | lat. pericuneate nu. |
| LR4V | lat. recess 4V |
| LRt | lat. reticular nu. |
| LRtS5 | lat. retic. nu., subtrigem. |
| LTnl | lat. terminal nu. |
| LV | lateral ventricle |
| LVe | lat. vestib. nu. |
| m5 | motor root 5n |
| mcp | middle cerebell. ped. |
| MCPO | magnocell. preopt. nu. |
| Me | medial amyg. nu. |
| MiTg | microcell. tegm. nu. |
| ml | medial lemniscus |
| mml | med. medull. lamina |
| Mo5 | motor trigeminal nu. |
| MPBE | med. parabr. nu., ext. |
| mRt | mesen. retic. formn |
| Mx | matrix sensory area |
| oc | olivocerebellar tr. |
| Op | strat. opt. sup. coll. |
| opt | optic tract |
| orbl | orbital sulc. lat. |
| orbm | orbital sulc. med. |
| OT | nu. of optic tract |
| p1Rt | pros. 1 retic. formn |
| P5 | peritrigeminal nu. |
| PBG | parabigeminal nu. |
| PBP | parabrach. pigm. nu. |
| pcf | preculminate fiss. Cb. |
| PE | par. field E Cx |
| PEc | par. field E Cx, caud. |
| Pir | piriform Cx |
| plf | posterolat. fissure |
| PMD | premotor Cx dors. |
| PMp | paramedian lob. Cb, post. |
| Pn | pontine nu. |
| PnO | pontine oral nu. |
| Po | post. thal. nu. |
| PP | peripedunc. nu. |
| PPD1 | post. par. Cx, dors. 1 |
| ppf | prepyramidal fiss. |
| PPTgD | peduncpont. tegm. nu., diff. |
| prf | prim. cerebell. fiss. |
| PrS | presubiculum |
| psf | post. sup. fissure Cb |
| Pu | putamen |
| PulA | pulvinar nu., ant. |
| PulM | pulvinar nu., med. |
| rs | rubrospinal tract |
| Rt | reticular nu. |
| RVL | rost. ventrolat. area |
| S | subiculum |
| scp | sup. cerebell. ped. |
| sf | secondary fissure Cb |
| SG | supragenic. thal. nu. |
| SI | subst. innominata |
| SNCD | subst. nigra, comp., dors. |
| SNCV | subst. nigra, comp., vent. |
| SNR | subst. nigra, retic. |
| sod | suproptic decn |
| sp5 | spinal trigeminal tr. |
| Sp5C | spin. trigem. nu., caud. |
| Sp5I | spin. trigem. nu., interpol. |
| Sp5O | spin. trigem. nu., oral. |
| SpVe | spinal vestib. nu. |
| st | stria terminalis |
| STh | subthalamic nu. |
| Su5 | supratrigeminal nu. |
| SubCV | subcoer. nu., vent. |
| SuG | sup. gray sup. coll. |
| SuVe | sup. vestib. nu. |
| Tu | olf. tubercle |
| V1 | V1 Cx |
| V2m | V2 (med.) Cx |
| V2v | V2 (vent.) Cx |
| VCoI | vent. inf. cort. amyg. |
| VCoIn | vent. inter. cort. amyg. |
| VCoS | vent. sup. cort. amyg. |
| VLa | ventrolat. thal. nu., ant. |
| VLL | vent. nu. lat. lemn. |
| VLP (V3v) | ventlatpost. (V3v) Cx |
| VP | vent. pallidum |
| VPI | vent. post. thal. nu., inf. |
| VPL | vent. post. thal. nu., lat. |
| VPM | vent. post. thal. nu., med. |
| VPS | vent. post. thal. nu., sup. |
| vsc | vent. spinocereb. tr. |
| Y | nucleus Y |
| ZID | zona incerta, dors. |
| ZIV | zona incerta, rost. |
| Zo | strat. zonale SC |

# Plate 68a

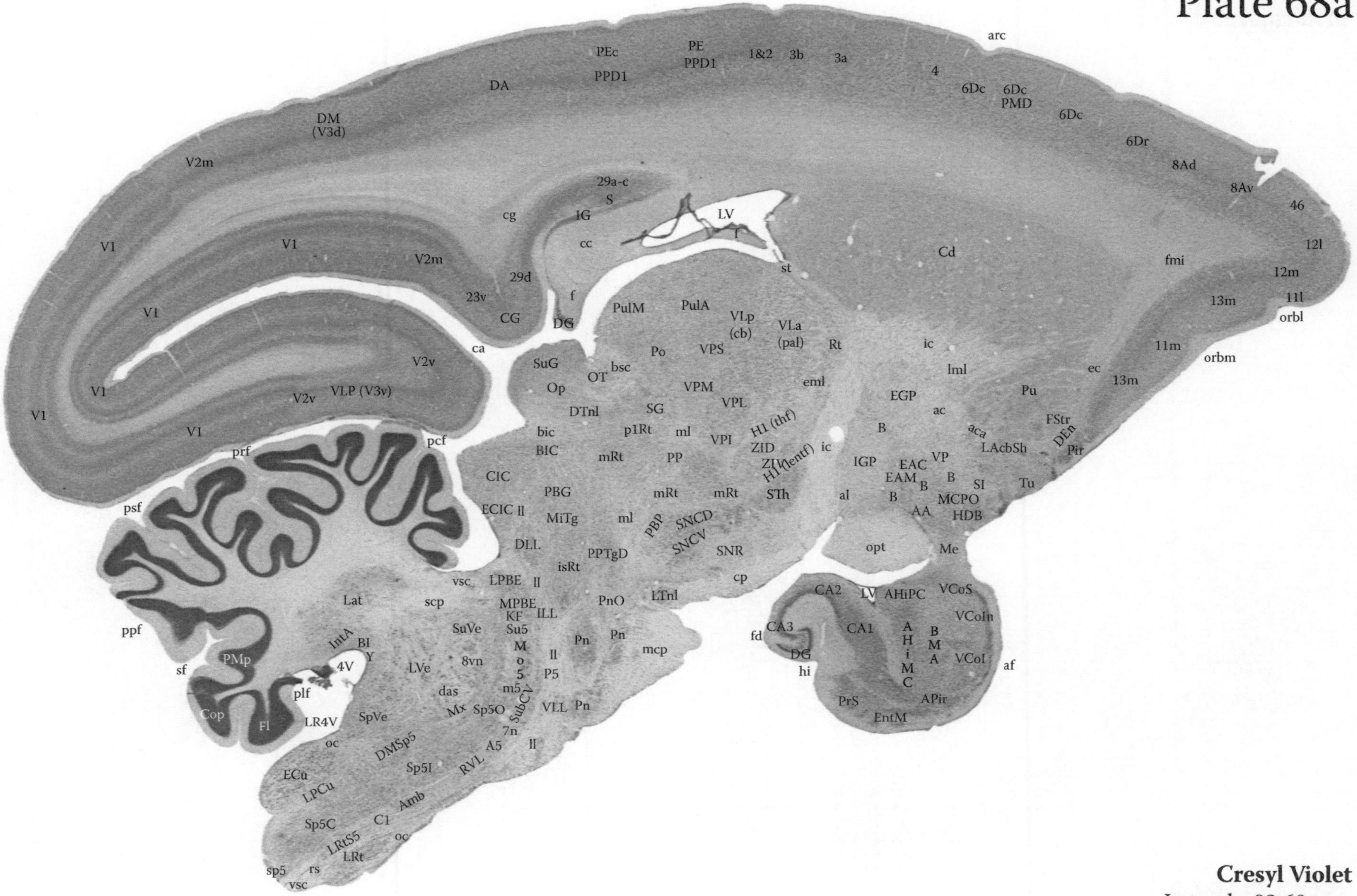


**Cresyl Violet**
Lateral +03.60 mm

# Plate 68b

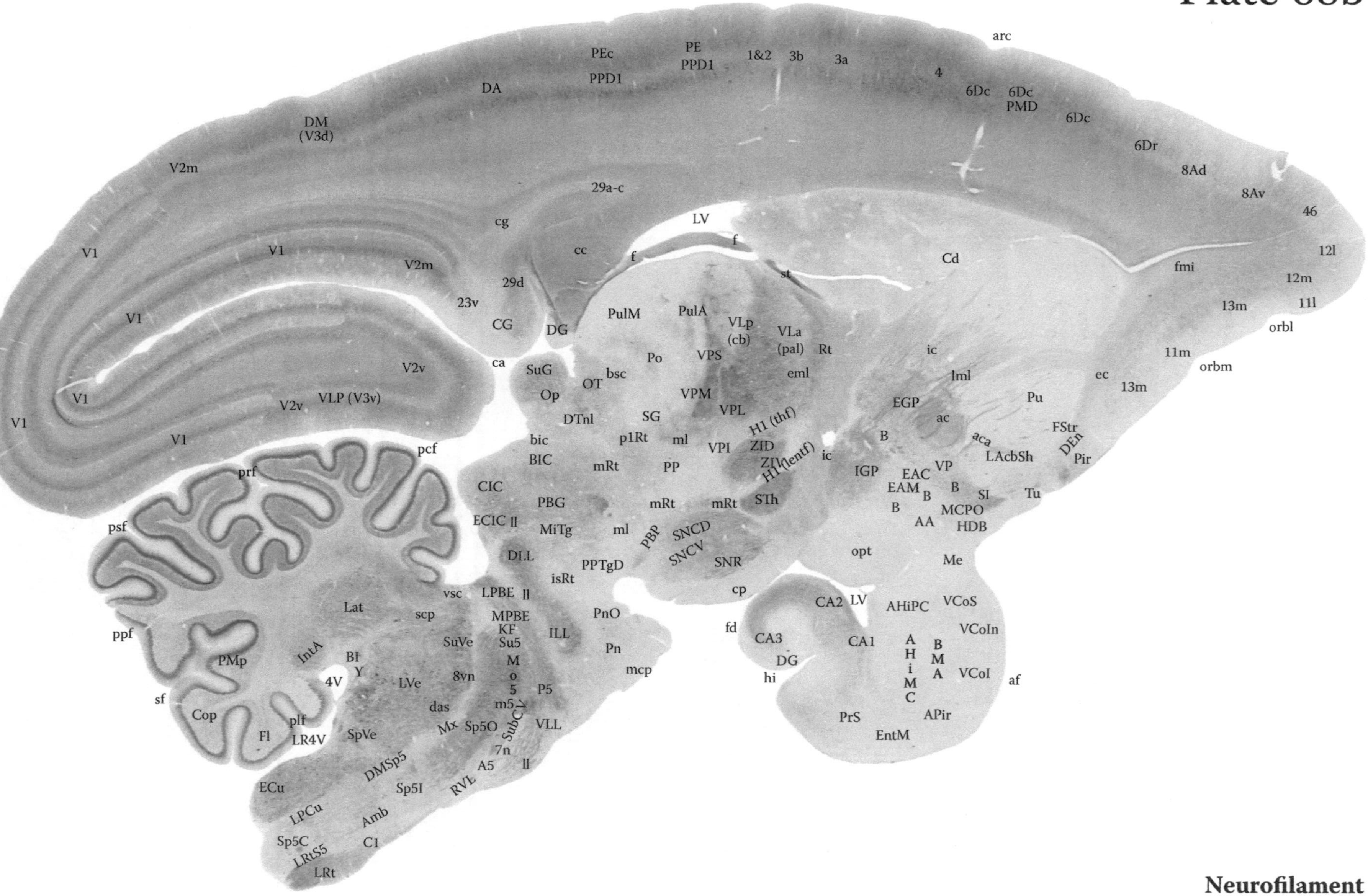


**Neurofilament**
Lateral +03.35 mm

# Plate 69

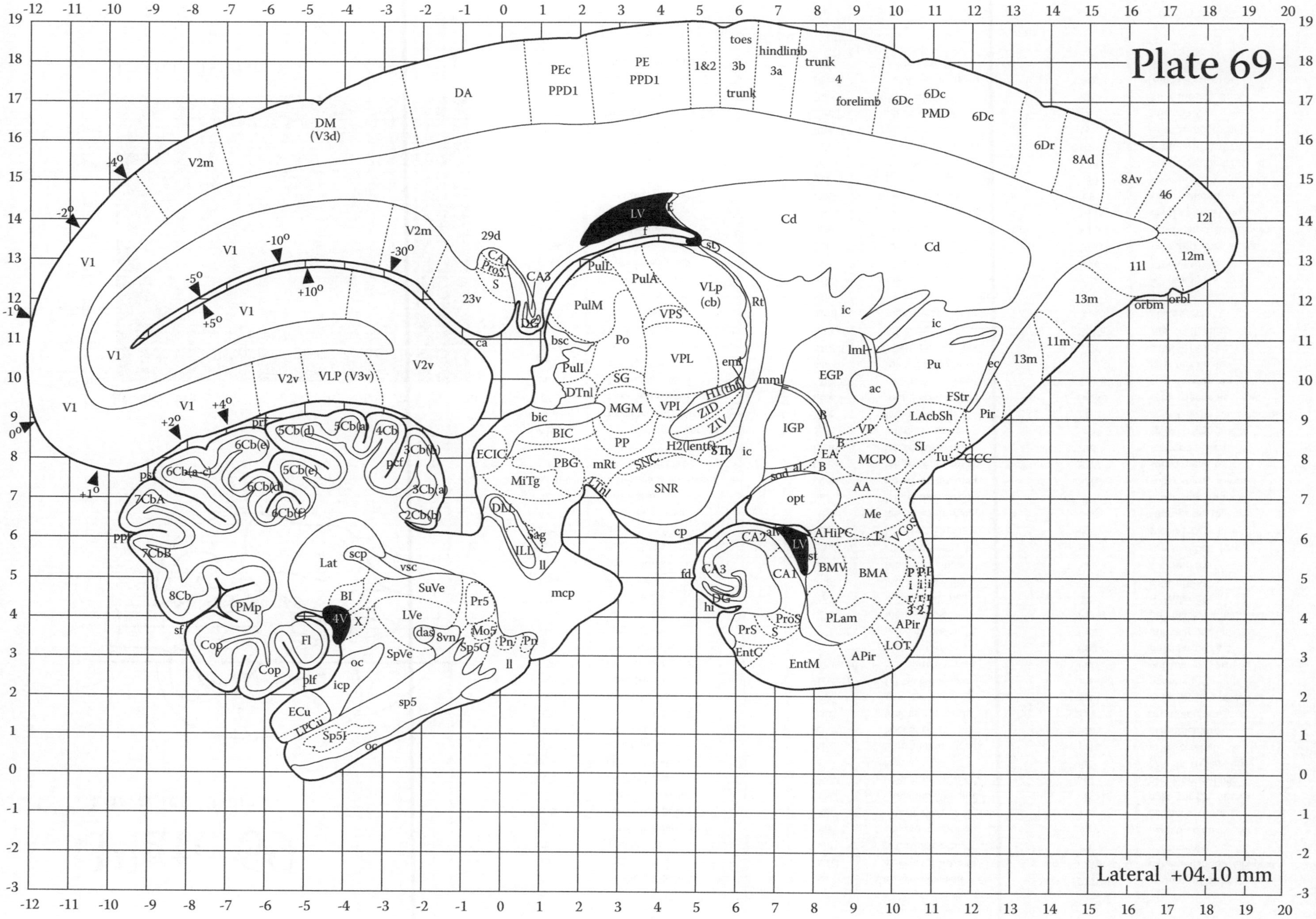


Lateral +04.10 mm

# Plate 69

Lateral +04.10 mm

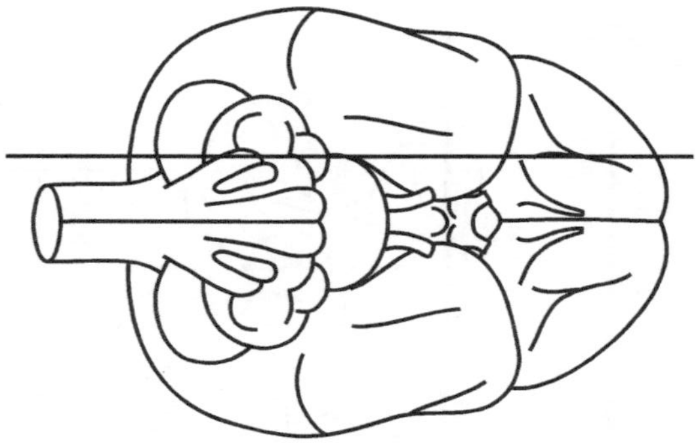

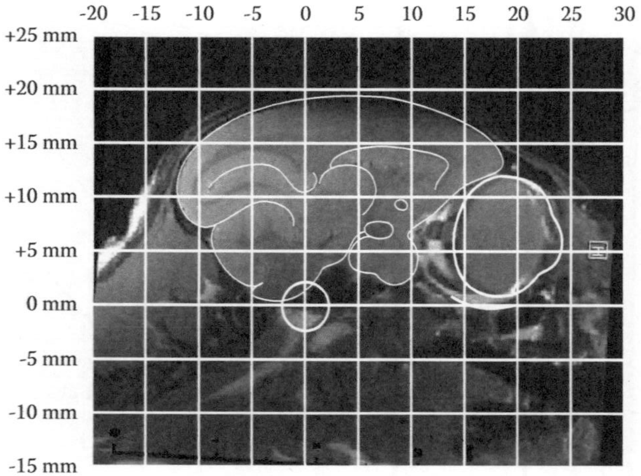


Lateral +04.30 mm

| | |
|---|---|
| 2Cb | cerebell. lobule 2 |
| 3a | area 3a Cx |
| 3b | area 3b Cx |
| 3Cb | cerebell. lobule 3 |
| 4Cb | cerebell. lobule 4 |
| 4V | fourth ventricle |
| 5Cb | cerebell. lobule 5 |
| 6Cb | cerebell. lobule 6 |
| 6Dc | area 6Dc Cx |
| 6Dr | area 6Dr Cx |
| 7Cb | cerebell. lobule 7 |
| 8Ad | area 8Ad Cx |
| 8Av | area 8Av Cx |
| 8Cb | cerebell. lobule 8 |
| 8vn | vestibular divn 8n |
| 11l | area 11l Cx |
| 11m | area 11m Cx |
| 12l | area 12l Cx |
| 12m | area 12m Cx |
| 13m | area 13m Cx |
| 23v | area 23v Cx |
| 29d | area 29d Cx |
| AA | ant. amyg. area |
| ac | ant. comm. |
| AHiPC | amygdalohipp., magn. |
| al | ansa lenticularis |
| alv | alveus |
| APir | amygdalopir. trans. area |
| B | basal nu. (Meynert) |
| BI | basal interstitial nu. |
| bic | brach. inf. coll. |
| BIC | brach. inf. coll. nu. |
| BMA | basomed. amyg. nu., ant. |
| BMV | basomed. amyg. nu., vent. |
| bsc | brach. sup. coll. |
| ca | calcarine sulcus |
| CA1 | cornu ammonis 1 |
| CA2 | cornu ammonis 2 |
| CA3 | cornu ammonis 3 |
| Cd | caudate nu. |
| Cop | copula |
| cp | cerebral peduncle |
| DA | dorsoant. visual Cx |
| das | dorsal acoustic stria |
| DG | dentate gyrus |
| DLL | dorsal nu. lat. lemn. |
| DM | dorsomed. visual Cx |
| DTnl | dors. term. nu. |
| E | ependyma |
| EA | extend amygdala |
| ec | external capsule |
| ECIC | extern. Cx inf. coll. |
| ECu | ext. cuneate nu. |
| EGP | ext. globus pallidus |
| eml | extern. medull. lamina |
| EntC | entorhin. Cx, caud. |
| EntM | entorhin. Cx, med. |
| f | fornix |
| fd | fimbriodent. sulcus |
| Fl | flocculus |
| FStr | fundus striatum |
| GCC | gran. cell clust. |
| H1(thf) | H1 thalamic fasc. |
| H2(lentf) | H2 lenticular fasc. |
| hi | hippocamp. fiss. |
| I | intercal. nu. amyg. |
| ic | intern. caps. |
| icp | inf. cerebell. ped. |
| IGP | int. globus pallidus |
| ILL | intermed. nu. lat. lemn. |
| LAcbSh | accumb. shell lat. |
| Lat | lat. deep Cb nu. |
| ll | lateral lemniscus |
| lml | lat. medull. lamina |
| LOT | nu. of lat. olf. tract |
| LPCu | lat. pericuneate nu. |
| LTnl | lat. terminal nu. |
| LV | lateral ventricle |
| LVe | lat. vestib. nu. |
| mcp | middle cerebell. ped. |
| MCPO | magnocell. preopt. nu. |
| Me | medial amyg. nu. |
| MGM | med. gen. nu., med. |
| MiTg | microcell. tegm. nu. |
| mml | med. medull. lamina |
| Mo5 | motor trigeminal nu. |
| mRt | mesen. retic. formn |
| oc | olivocerebellar tr. |
| opt | optic tract |
| orbl | orbital sulc. lat. |
| orbm | orbital sulc. med. |
| PBG | parabigeminal nu. |
| pcf | preculminate fiss. Cb. |
| PE | par. field E Cx |
| PEc | par. field E Cx, caud. |
| Pir1 | piriform Cx 1 |
| Pir2 | piriform Cx 2 |
| Pir3 | piriform Cx 3 |
| PLam | paralam. amyg. nu. |
| plf | posterolat. fissure |
| PMD | premotor Cx dors. |
| PMp | paramedian lob. Cb, post. |
| Pn | pontine nu. |
| Po | post. thal. nu. |
| PP | peripedunc. nu. |
| PPD1 | post. par. Cx, dors. 1 |
| ppf | prepyramidal fiss. |
| Pr5 | princip. trigem. nu. |
| prf | prim. cerebell. fiss. |
| ProS | prosubiculum |
| PrS | presubiculum |
| psf | post. sup. fissure Cb |
| Pu | putamen |
| PulA | pulvinar nu., ant. |
| PulI | pulvinar nu., inf. |
| PulL | pulvinar nu., lat. |
| PulM | pulvinar nu., med. |
| Rt | reticular nu. |
| S | subiculum |
| Sag | sagulum nu. |
| scp | sup. cerebell. ped. |
| sf | secondary fissure Cb |
| SG | supragenic. thal. nu. |
| SI | subst. innominata |
| SNC | subst. nigra, compact |
| SNR | subst. nigra, retic. |
| sod | suproptic decn |
| sp5 | spinal trigeminal tr. |
| Sp5I | spin. trigem. nu., interpol. |
| Sp5O | spin. trigem. nu., oral. |
| SpVe | spinal vestib. nu. |
| st | stria terminalis |
| STh | subthalamic nu. |
| SuVe | sup. vestib. nu. |
| Tu | olf. tubercle |
| V1 | V1 Cx |
| V2m | V2 (med.) Cx |
| V2v | V2 (vent.) Cx |
| VCoS | vent. sup. cort. amyg. |
| VLp | ventrolat. thal. nu., post. |
| VLP (V3v) | ventlatpost. (V3v) Cx |
| VP | vent. pallidum |
| VPI | vent. post. thal. nu., inf. |
| VPL | vent. post. thal. nu., lat. |
| VPS | vent. post. thal. nu., sup. |
| vsc | vent. spinocereb. tr. |
| X | nucleus X |
| ZID | zona incerta, dors. |
| ZIV | zona incerta, rost. |

# Plate 69a

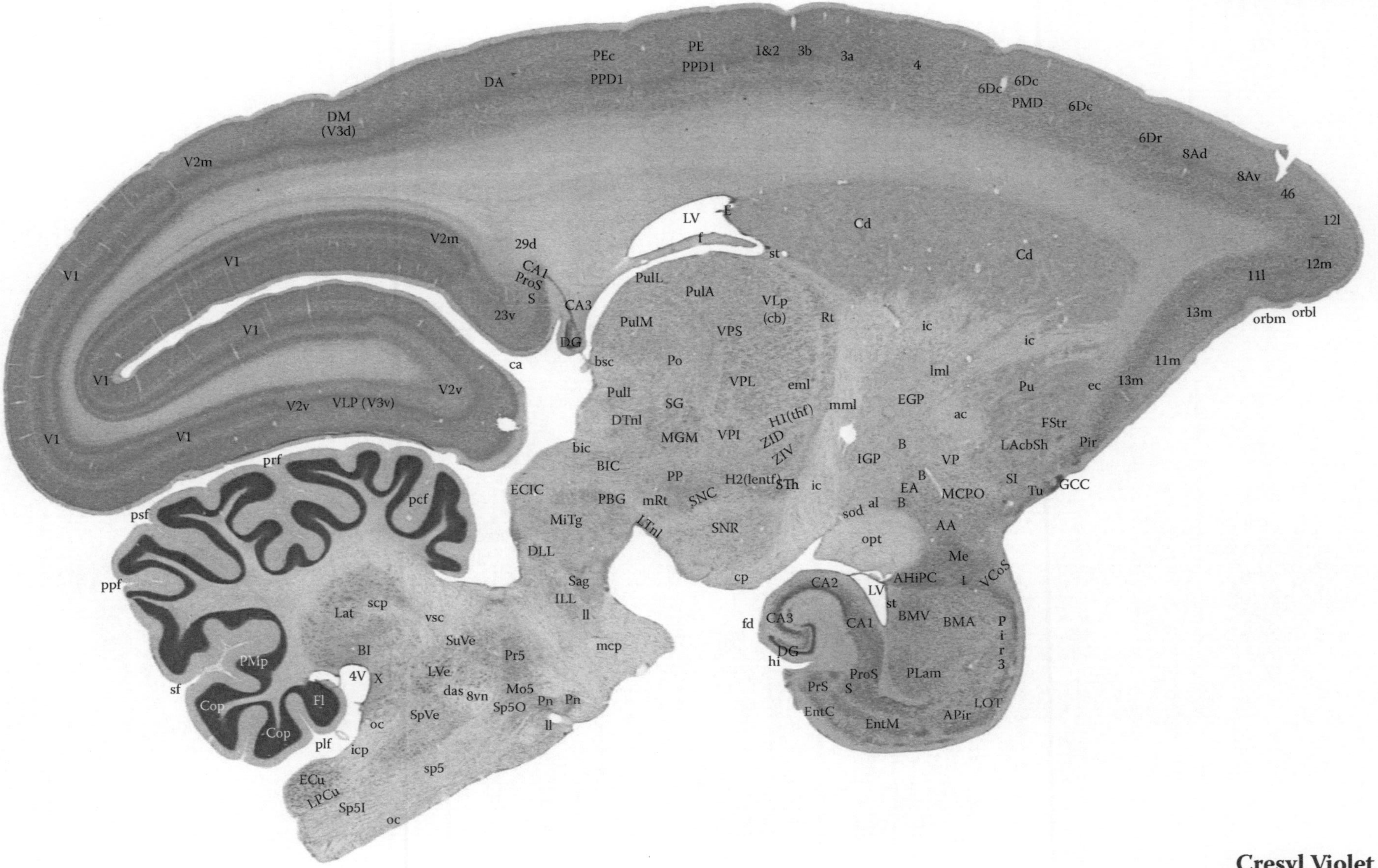


**Cresyl Violet**
Lateral +04.10 mm

# Plate 69b

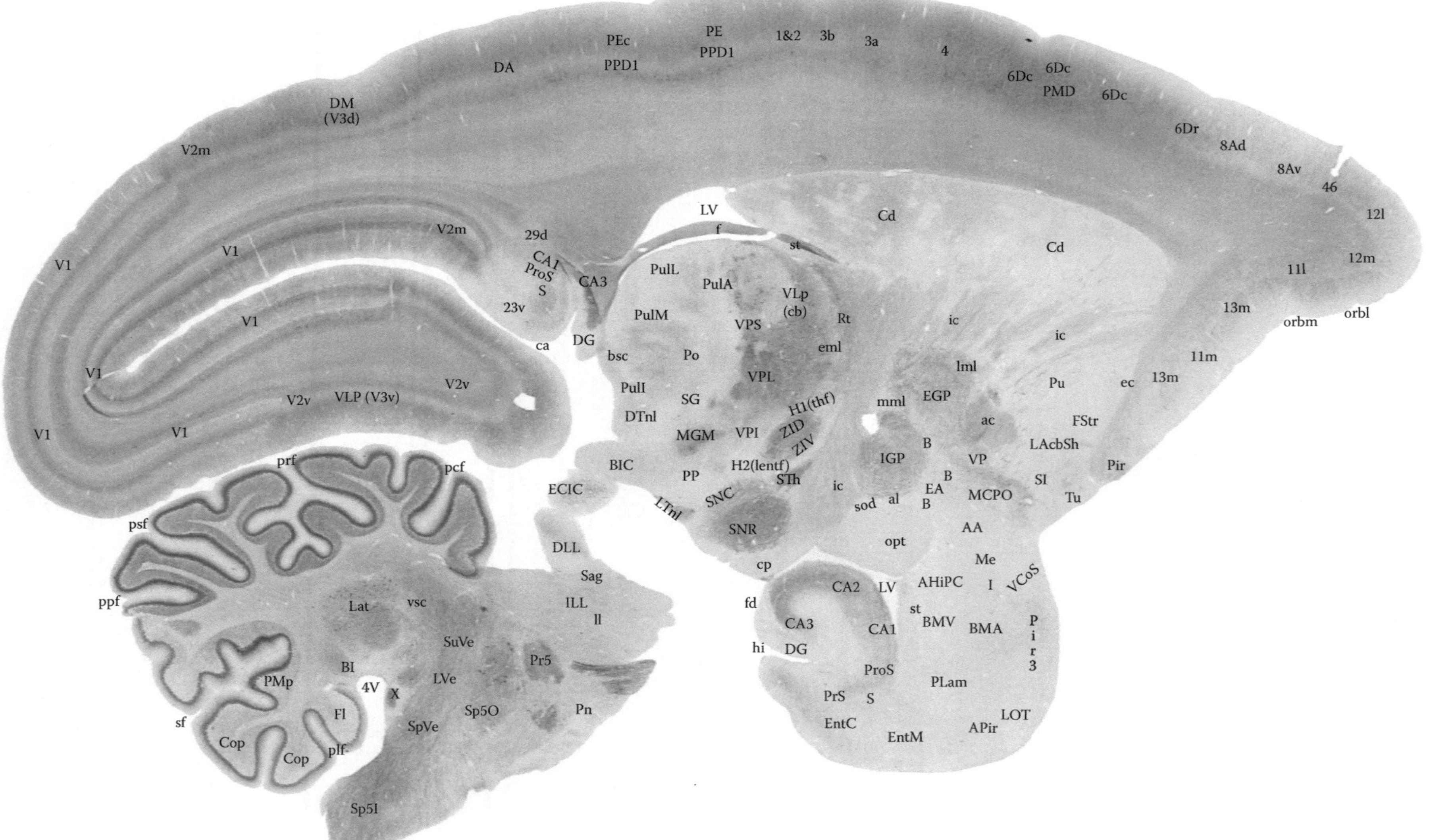


**Neurofilament**
Lateral +04.35 mm

# Plate 70

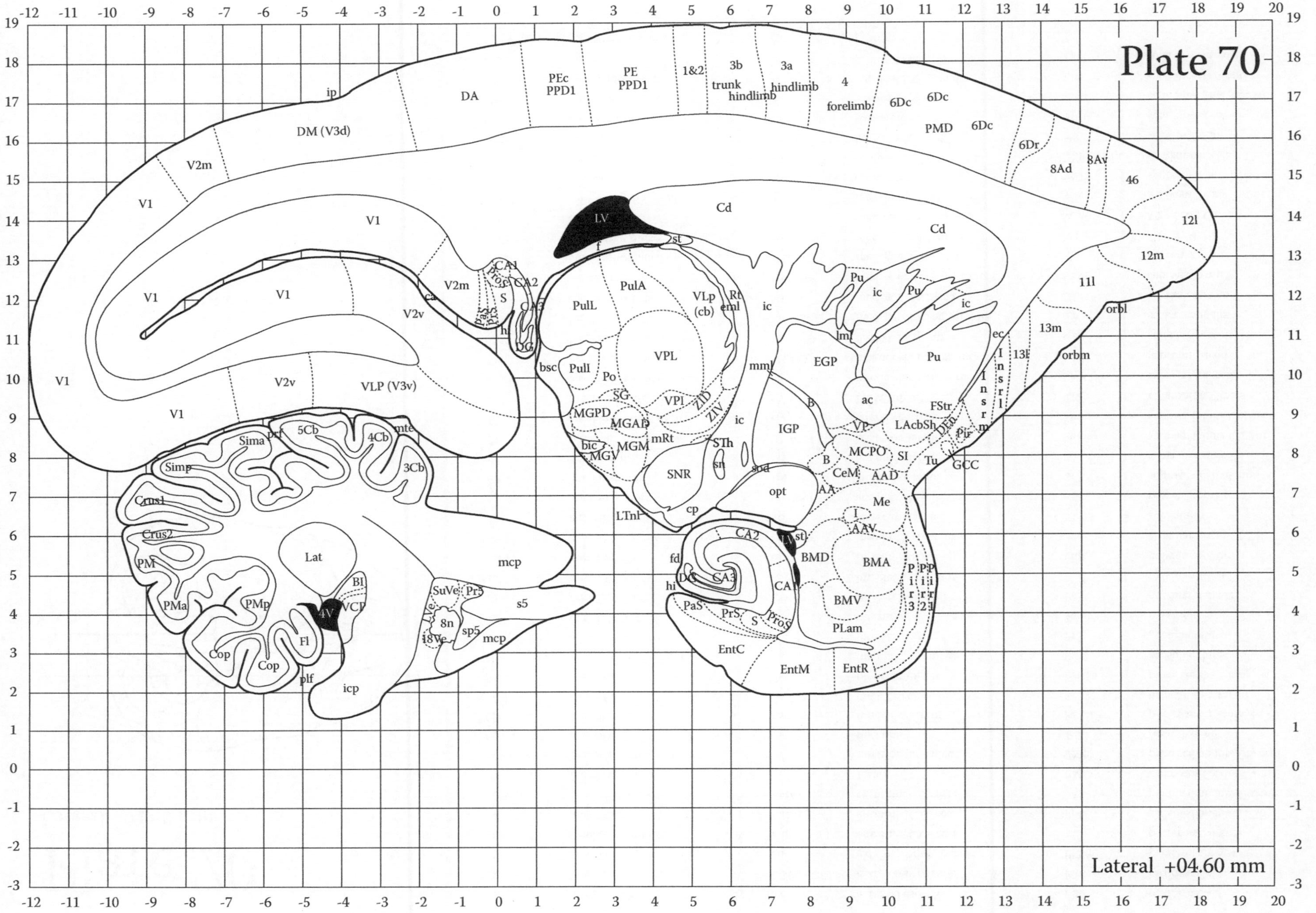


Lateral +04.60 mm

# Plate 70

Lateral +04.60 mm

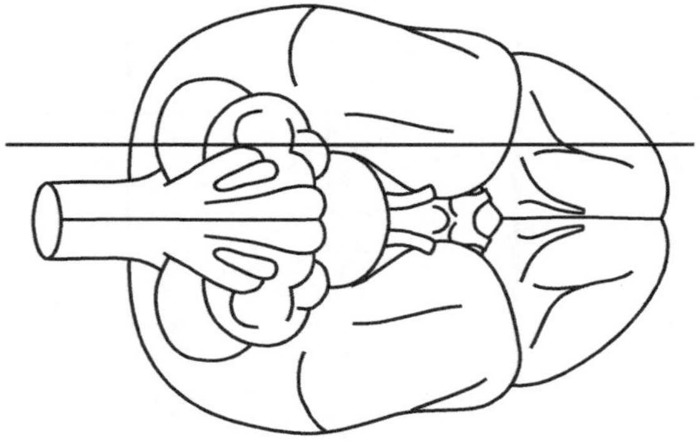

| | |
|---|---|
| 1&2 | areas 1/2 Cx |
| 3a | area 3a Cx |
| 3b | area 3b Cx |
| 3Cb | cerebell. lobule 3 |
| 4 | area 4 Cx |
| 4Cb | cerebell. lobule 4 |
| 4V | fourth ventricle |
| 5Cb | cerebell. lobule 5 |
| 6Dc | area 6Dc Cx |
| 6Dr | area 6Dr Cx |
| 8Ad | area 8Ad Cx |
| 8Av | area 8Av Cx |
| 8n | vestibulocochlear nerve |
| 11l | area 11l Cx |
| 12l | area 12l Cx |
| 12m | area 12m Cx |
| 13l | area 13l Cx |
| 13m | area 13m Cx |
| 46 | area 46 Cx |
| AA | ant. amyg. area |
| AAD | ant. amyg. dors. |
| AAV | ant. amyg. vent. |
| ac | ant. comm. |
| B | basal nu. (Meynert) |
| BI | basal interstitial nu. |
| bic | brach. inf. coll. |
| BMA | basomed. amyg. nu., ant. |
| BMD | basomed. amyg. nu., dors. |
| BMV | basomed. amyg. nu., vent. |
| bsc | brach. sup. coll. |
| ca | calcarine sulcus |
| CA1 | cornu ammonis 1 |
| CA2 | cornu ammonis 2 |
| CA3 | cornu ammonis 3 |
| Cd | caudate nu. |
| CeM | cent. amyg. nu., med. |
| Cop | copula |
| cp | cerebral peduncle |
| Crus1 | crus 1, ansiform lob. |
| Crus2 | crus 2, ansiform lob. |
| DA | dorsoant. visual Cx |
| DEn | dors. endopir. nu. |
| DG | dentate gyrus |
| DM (V3d) | dorsomed. visual Cx |
| ec | external capsule |
| EGP | ext. globus pallidus |
| eml | extern. medull. lamina |
| EntC | entorhin. Cx, caud. |
| EntM | entorhin. Cx, med. |
| EntR | entorhin. Cx, rost. |
| f | fornix |
| fd | fimbriodent. sulcus |
| Fl | flocculus |
| FStr | fundus striatum |
| GCC | gran. cell clust. |
| hi | hippocamp. fiss. |
| I | intercal. nu. amyg. |
| I8Ve | interst. nu. 8vn |
| ic | intern. caps. |
| icp | inf. cerebell. ped. |
| IGP | int. globus pallidus |
| Insrl | insular Cx, rost. lat. |
| Insrm | insular Cx, rost. med. |
| ip | intraparietal sulc. |
| LAcbSh | accumb. shell lat. |
| Lat | lat. deep Cb nu. |
| lml | lat. medull. lamina |
| LTnl | lat. terminal nu. |
| LV | lateral ventricle |
| LVe | lat. vestib. nu. |
| mcp | middle cerebell. ped. |
| MCPO | magnocell. preopt. nu. |
| Me | medial amyg. nu. |
| MGAD | med. gen. nu., anterodors. |
| MGM | med. gen. nu., med. |
| MGPD | med. gen. nu., posterodors. |
| MGV | med. gen. nu., vent. |
| mml | med. medull. lamina |
| mRt | mesen. retic. formn |
| mte | med. temp. sulc. |
| opt | optic tract |
| orbl | orbital sulc. lat. |
| orbm | orbital sulc. med. |
| PaS | parasubiculum |
| PE | par. field E Cx |
| PEc | par. field E Cx, caud. |
| Pir | piriform Cx |
| Pir1 | piriform Cx 1 |
| Pir2 | piriform Cx 2 |
| Pir3 | piriform Cx 3 |
| PLam | paralam. amyg. nu. |
| plf | posterolat. fissure |
| PM | paramedian lobule Cb |
| PMa | paramedian lob. Cb, ant. |
| PMD | premotor Cx dors. |
| PMp | paramedian lob. Cb, post. |
| Po | post. thal. nu. |
| PPD1 | post. par. Cx, dors. 1 |
| Pr5 | princip. trigem. nu. |
| prf | prim. cerebell. fiss. |
| ProS | prosubiculum |
| PrS | presubiculum |
| Pu | putamen |
| PulA | pulvinar nu., ant. |
| PulI | pulvinar nu., inf. |
| PulL | pulvinar nu., lat. |
| Rt | reticular nu. |
| S | subiculum |
| s5 | sensory root 5n |
| SG | supragenic. thal. nu. |
| SI | subst. innominata |
| Sima | simplex lobule Cb, ant. |
| Simp | simplex lobule Cb, post. |
| sn | striatonigral fibers |
| SNR | subst. nigra, retic. |
| sod | suproptic decn |
| sp5 | spinal trigeminal tr. |
| st | stria terminalis |
| STh | subthalamic nu. |
| SuVe | sup. vestib. nu. |
| Tu | olf. tubercle |
| V1 | V1 Cx |
| V2m | V2 (med.) Cx |
| V2v | V2 (vent.) Cx |
| VCP | vent. coch. nu., post. |
| VLP (V3v) | ventlatpost. (V3v) Cx |
| VP | vent. pallidum |
| VPI | vent. post. thal. nu., inf. |
| VPL | vent. post. thal. nu., lat. |
| ZID | zona incerta, dors. |
| ZIV | zona incerta, rost. |

# Plate 70a

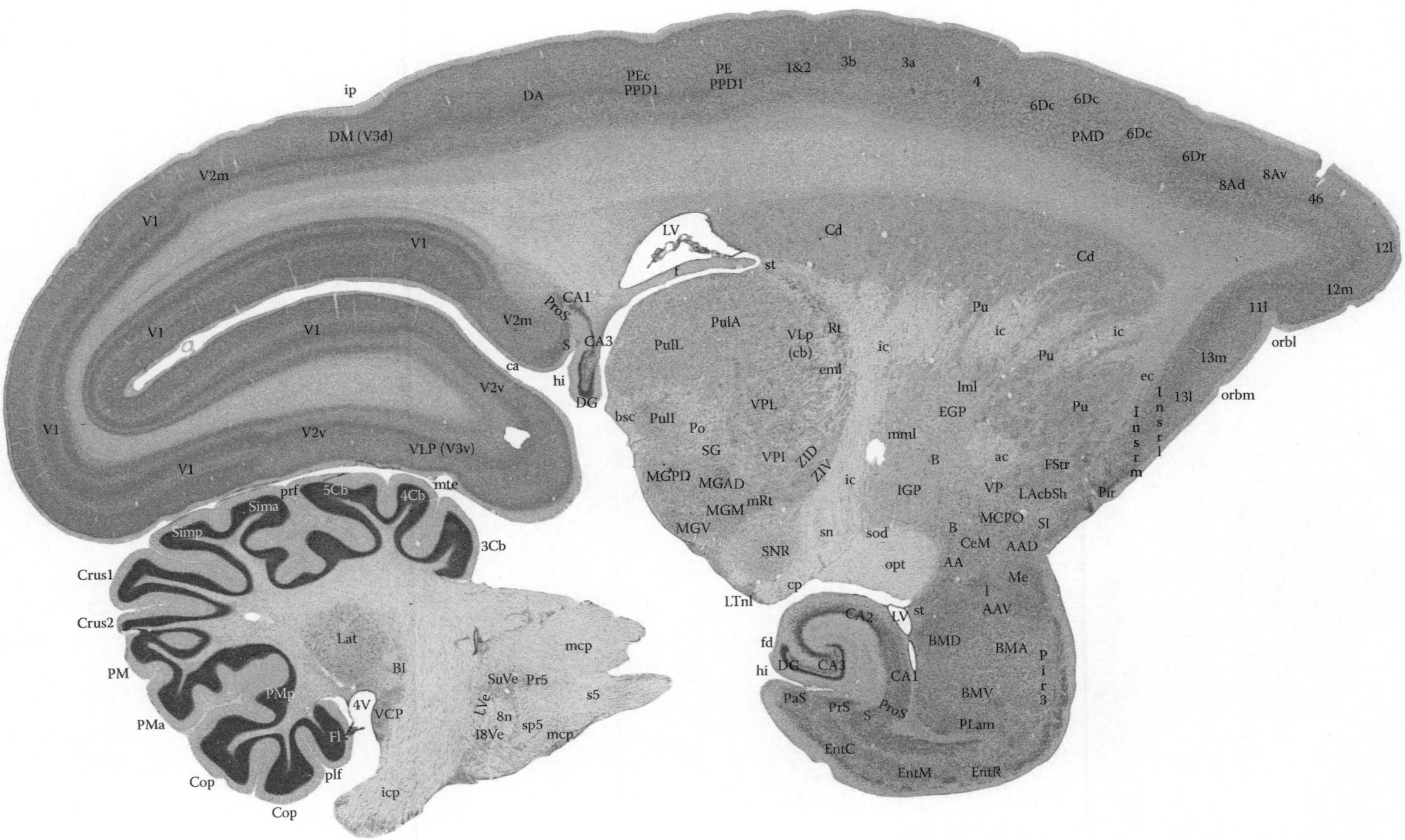


Cresyl Violet
Lateral +04.60 mm

# Plate 70b

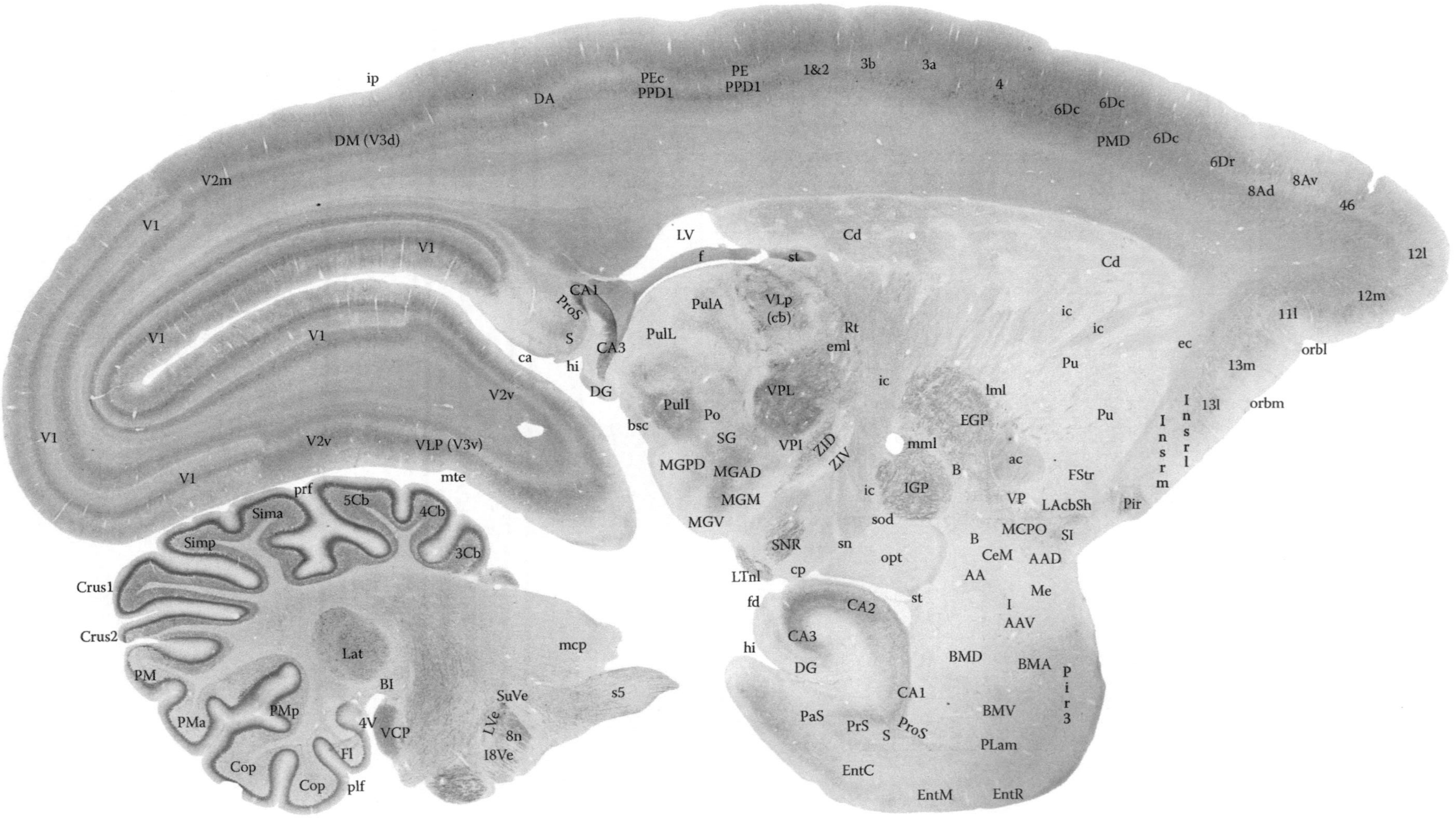


**Neurofilament**
Lateral +04.85 mm

# Plate 71

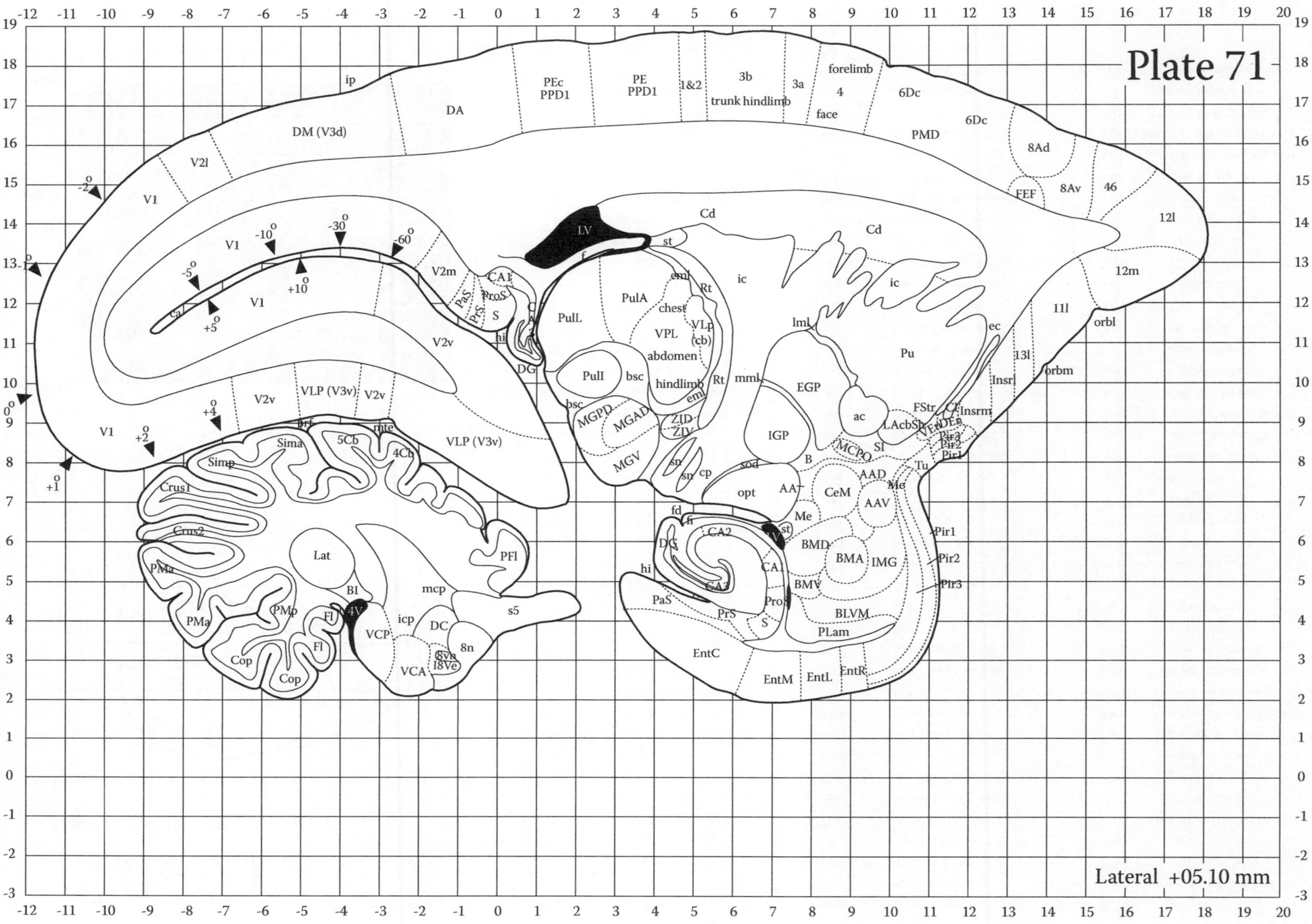


Lateral +05.10 mm

# Plate 71

Lateral +05.10 mm

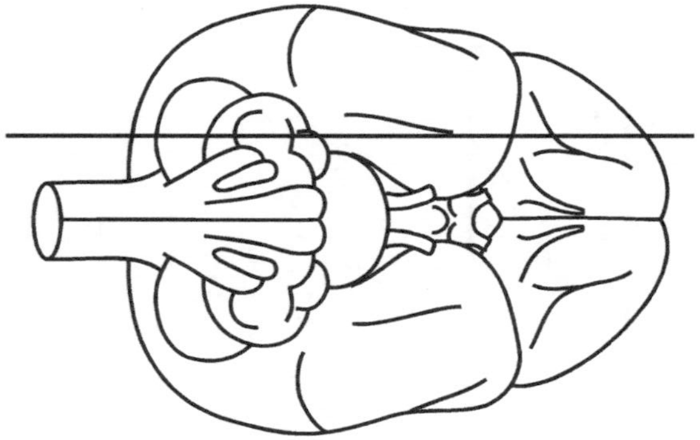

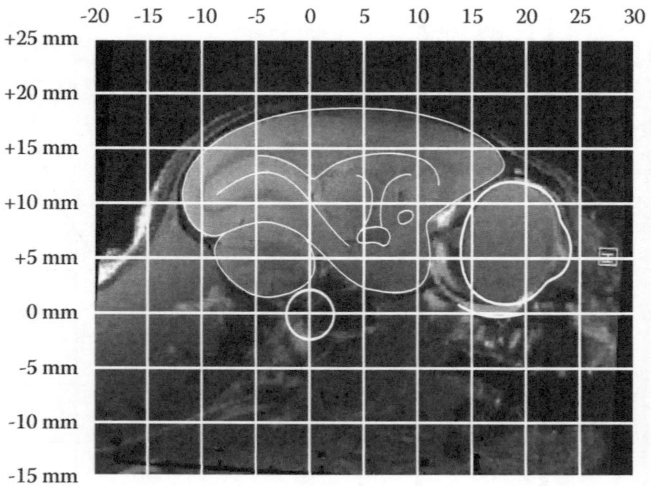


Lateral +05.25 mm

| | |
|---|---|
| 1&2 | areas 1/2 Cx |
| 3a | area 3a Cx |
| 3b | area 3b Cx |
| 4 | area 4 Cx |
| 4Cb | cerebell. lobule 4 |
| 4V | fourth ventricle |
| 5Cb | cerebell. lobule 5 |
| 6Dc | area 6Dc Cx |
| 8Ad | area 8Ad Cx |
| 8Av | area 8Av Cx |
| 8n | vestibulocochlear nerve |
| 8vn | vestibular divn 8n |
| 11l | area 11l Cx |
| 12l | area 12l Cx |
| 12m | area 12m Cx |
| 13l | area 13l Cx |
| 46 | area 46 Cx |
| AA | ant. amyg. area |
| AAD | ant. amyg. dors. |
| AAV | ant. amyg. vent. |
| ac | ant. comm. |
| B | basal nu. (Meynert) |
| BI | basal interstitial nu. |
| BLVM | basolat. amyg. nu., vent. med. |
| BMA | basomed. amyg. nu., ant. |
| BMD | basomed. amyg. nu., dors. |
| BMV | basomed. amyg. nu., vent. |
| bsc | brach. sup. coll. |
| ca | calcarine sulcus |
| CA1 | cornu ammonis 1 |
| CA2 | cornu ammonis 2 |
| CA3 | cornu ammonis 3 |
| Cd | caudate nu. |
| CeM | cent. amyg. nu., med. |
| Cl | claustrum |
| Cop | copula |
| cp | cerebral peduncle |
| Crus1 | crus 1, ansiform lob. |
| Crus2 | crus 2, ansiform lob. |
| DA | dorsoant. visual Cx |
| DC | dorsal cochlear nu. |
| DEn | dors. endopir. nu. |
| DG | dentate gyrus |
| DM (V3d) | dorsomed. visual Cx |
| ec | external capsule |
| EGP | ext. globus pallidus |
| eml | extern. medull. lamina |
| EntC | entorhin. Cx, caud. |
| EntL | entorhin. Cx, lat. |
| EntM | entorhin. Cx, med. |
| EntR | entorhin. Cx, rost. |
| f | fornix |
| fd | fimbriodent. sulcus |
| FEF | frontal eye field Cx |
| fi | fimbria |
| Fl | flocculus |
| FStr | fundus striatum |
| hi | hippocamp. fiss. |
| I8Ve | interst. nu. 8vn |
| ic | intern. caps. |
| icp | inf. cerebell. ped. |
| IGP | int. globus pallidus |
| IMG | intramedull. gr. amyg. |
| Insrl | insular Cx, rost. lat. |
| Insrm | insular Cx, rost. med. |
| ip | intraparietal sulc. |
| LAcbSh | accumb. shell lat. |
| Lat | lat. deep Cb nu. |
| lml | lat. medull. lamina |
| LV | lateral ventricle |
| mcp | middle cerebell. ped. |
| MCPO | magnocell. preopt. nu. |
| Me | medial amyg. nu. |
| MGAD | med. gen. nu., anterodors. |
| MGPD | med. gen. nu., posterodors. |
| MGV | med. gen. nu., vent. |
| mml | med. medull. lamina |
| mte | med. temp. sulc. |
| opt | optic tract |
| orbl | orbital sulc. lat. |
| orbm | orbital sulc. med. |
| PaS | parasubiculum |
| PE | par. field E Cx |
| PEc | par. field E Cx, caud. |
| PFl | paraflocculus |
| Pir1 | piriform Cx 1 |
| Pir2 | piriform Cx 2 |
| Pir3 | piriform Cx 3 |
| PLam | paralam. amyg. nu. |
| PMa | paramedian lob. Cb, ant. |
| PMD | premotor Cx dors. |
| PMp | paramedian lob. Cb, post. |
| PPD1 | post. par. Cx, dors. 1 |
| prf | prim. cerebell. fiss. |
| ProS | prosubiculum |
| PrS | presubiculum |
| Pu | putamen |
| PulA | pulvinar nu., ant. |
| PulI | pulvinar nu., inf. |
| PulL | pulvinar nu., lat. |
| Rt | reticular nu. |
| S | subiculum |
| s5 | sensory root 5n |
| SI | subst. innominata |
| Sima | simplex lobule Cb, ant. |
| Simp | simplex lobule Cb, post. |
| sn | striatonigral fibers |
| sod | suproptic decn |
| st | stria terminalis |
| Tu | olf. tubercle |
| V1 | V1 Cx |
| V2l | V2 (lat.) Cx |
| V2m | V2 (med.) Cx |
| V2v | V2 (vent.) Cx |
| VCA | vent. coch. nu., ant. |
| VCP | vent. coch. nu., post. |
| VEn | vent. endopir. nu. |
| VLp | ventrolat. thal. nu., post. |
| VLP (V3v) | ventlatpost. (V3v) Cx |
| VPL | vent. post. thal. nu., lat. |
| ZID | zona incerta, dors. |
| ZIV | zona incerta, rost. |

# Plate 71a

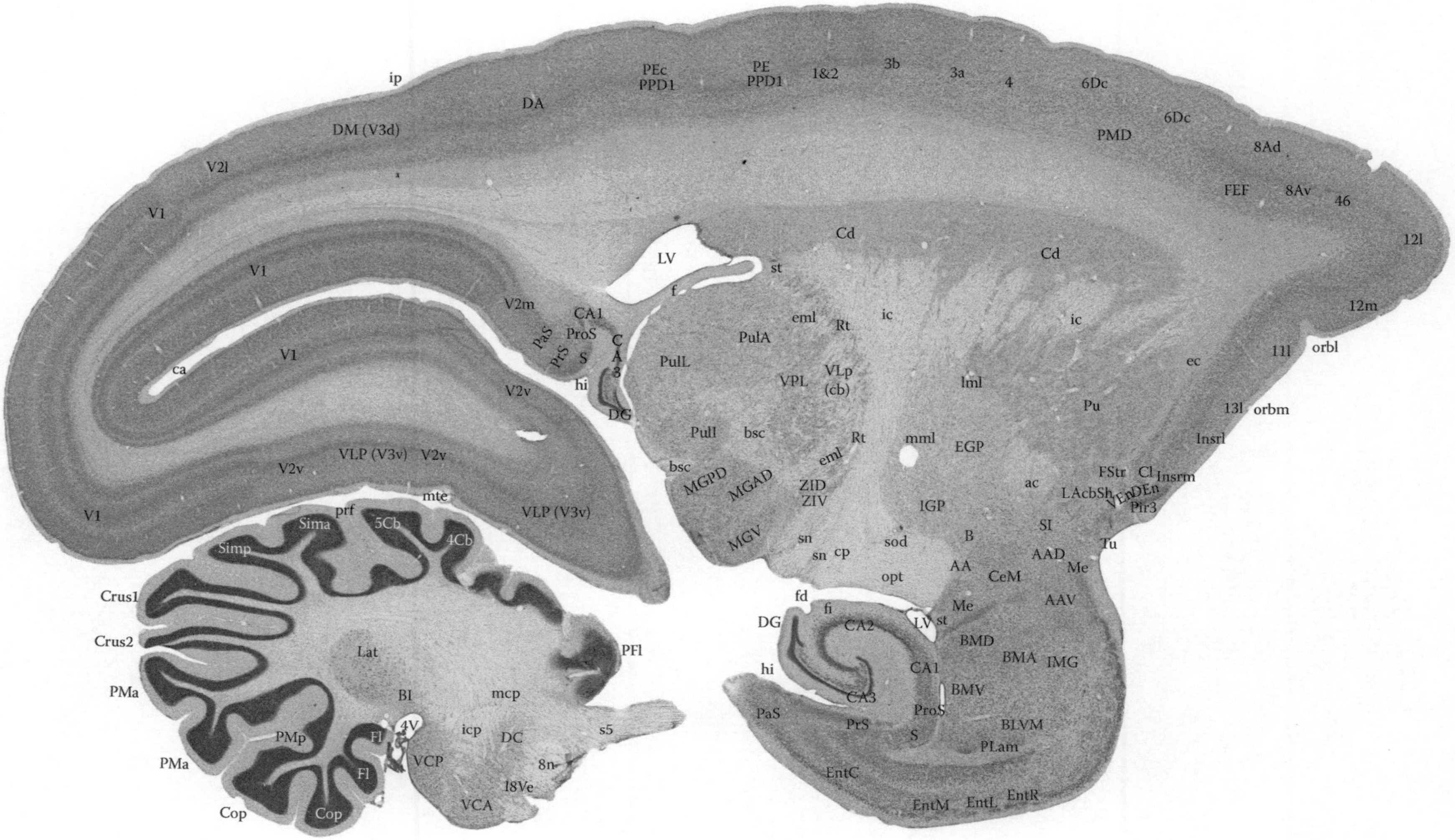


**Cresyl Violet**
Lateral +05.10 mm

# Plate 71b

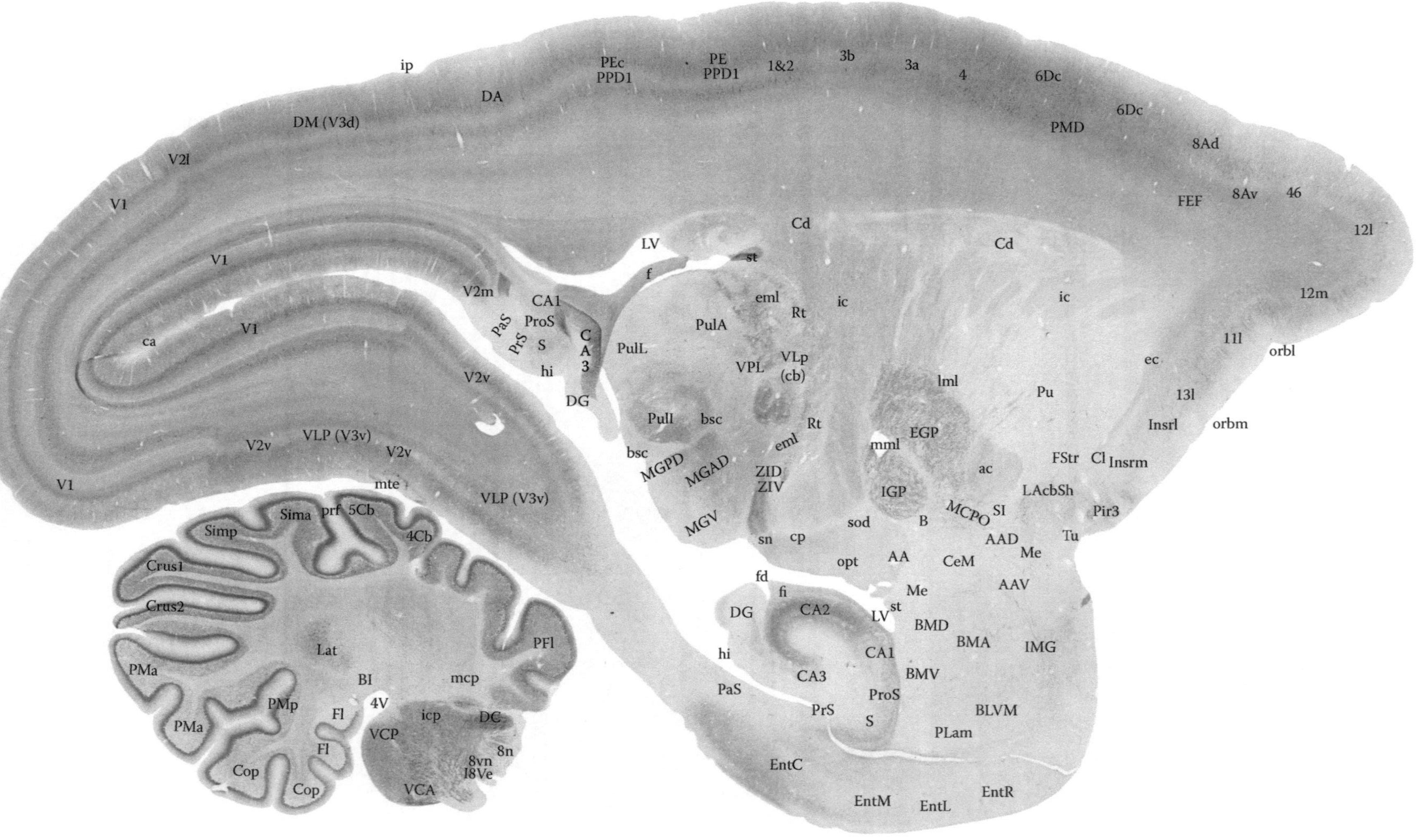


**Neurofilament**
Lateral +05.35 mm

# Plate 72

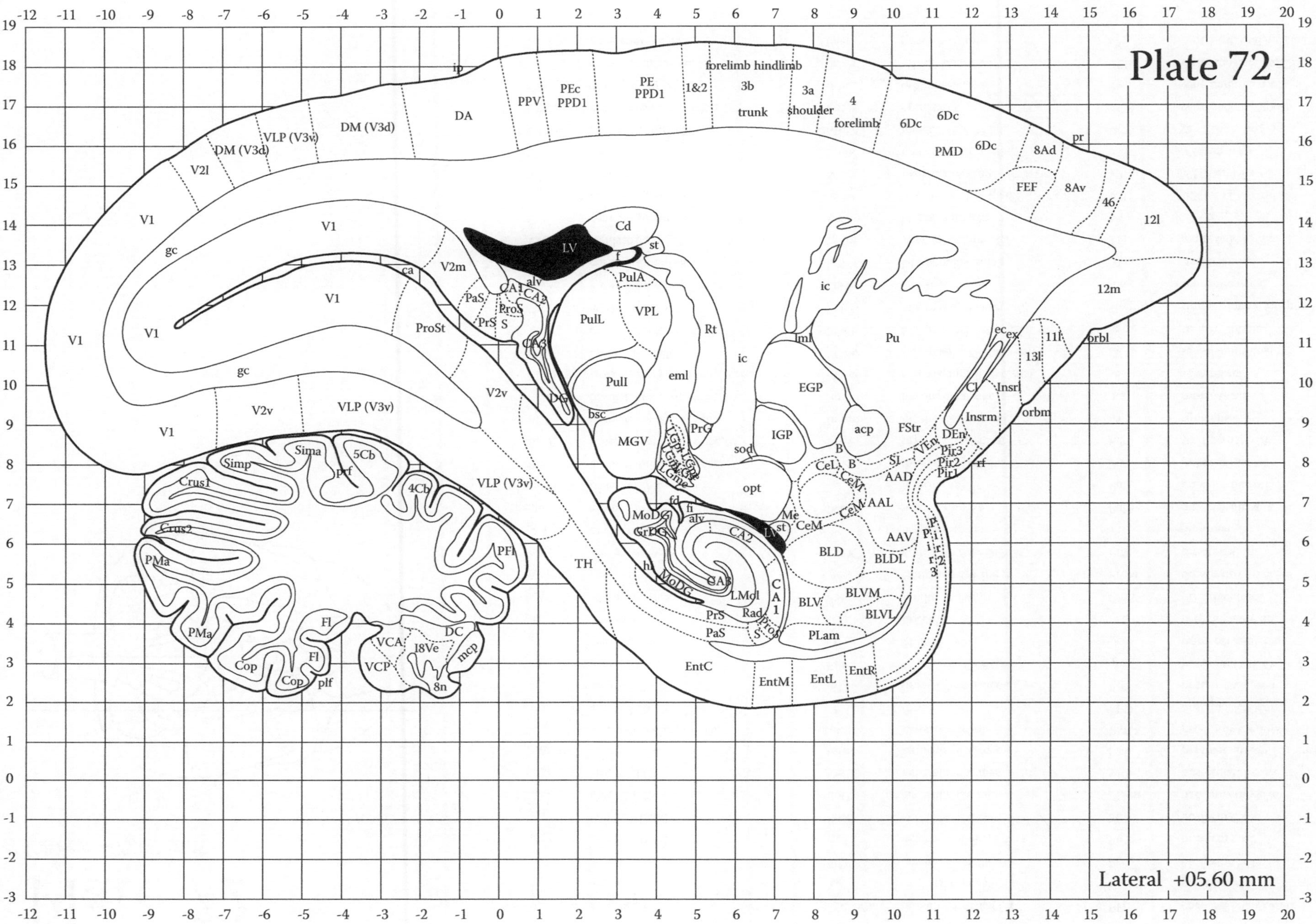


Lateral +05.60 mm

# Plate 72

Lateral +05.60 mm

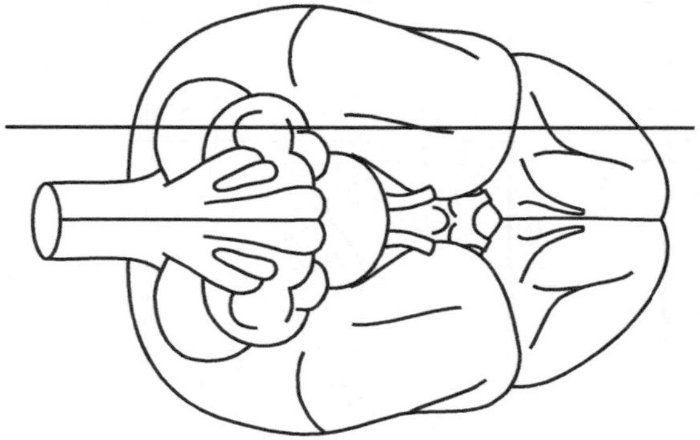

| | | | | | |
|---|---|---|---|---|---|
| 1&2 | areas 1/2 Cx | DG | dentate gyrus | PE | par. field E Cx |
| 3a | area 3a Cx | DM (V3d) | dorsomed. visual Cx | PEc | par. field E Cx, caud. |
| 3b | area 3b Cx | ec | external capsule | PFl | paraflocculus |
| 4 | area 4 Cx | EGP | ext. globus pallidus | Pir1 | piriform Cx 1 |
| 4Cb | cerebell. lobule 4 | eml | extern. medull. lamina | Pir2 | piriform Cx 2 |
| 5Cb | cerebell. lobule 5 | EntC | entorhin. Cx, caud. | Pir3 | piriform Cx 3 |
| 6Dc | area 6Dc Cx | EntL | entorhin. Cx, lat. | PLam | paralam. amyg. nu. |
| 8Ad | area 8Ad Cx | EntM | entorhin. Cx, med. | plf | posterolat. fissure |
| 8Av | area 8Av Cx | EntR | entorhin. Cx, rost. | PMa | paramedian lob. Cb, ant. |
| 8n | vestibulocochlear nerve | ex | extreme capsule | PMD | premotor Cx dors. |
| 11l | area 11l Cx | f | fornix | PPD1 | post. par. Cx, dors. 1 |
| 12l | area 12l Cx | fd | fimbriodent. sulcus | PPV | post. par. Cx, vent. |
| 12m | area 12m Cx | FEF | frontal eye field Cx | pr | prefrontal sulcus |
| 13l | area 13l Cx | fi | fimbria | prf | prim. cerebell. fiss. |
| 46 | area 46 Cx | Fl | flocculus | PrG | pregeniculate nu. |
| AAD | ant. amyg. dors. | FStr | fundus striatum | ProS | prosubiculum |
| AAL | ant. amyg. lat. | gc | geniculocalcarine tr. | ProSt | prostriate cortex |
| AAV | ant. amyg. vent. | GrDG | gran. cell layer DG | PrS | presubiculum |
| acp | ant. comm. post. | hi | hippocamp. fiss. | Pu | putamen |
| alv | alveus | I8Ve | interst. nu. 8vn | PulA | pulvinar nu., ant. |
| B | basal nu. (Meynert) | ic | intern. caps. | PulI | pulvinar nu., inf. |
| BLD | basolat. amyg. nu., dors. | IGP | int. globus pallidus | PulL | pulvinar nu., lat. |
| BLDL | basolat. amyg. nu., dors. lat. | Insrl | insular Cx, rost. lat. | Rad | strat. radiat. CA |
| BLV | basolat. amyg. nu., vent. | Insrm | insular Cx, rost. med. | rf | rhinal fiss. |
| BLVL | basolat. amyg. nu., vent. lat. | ip | intraparietal sulc. | Rt | reticular nu. |
| BLVM | basolat. amyg. nu., vent. med. | LGme | LG, magnocell. external | S | subiculum |
| bsc | brach. sup. coll. | LGmi | LG, magnocell. internal | SI | subst. innominata |
| ca | calcarine sulcus | LGpe | LG, parvicell. external | Sima | simplex lobule Cb, ant. |
| CA1 | cornu ammonis 1 | LGpi | LG, parvicell. internal | Simp | simplex lobule Cb, post. |
| CA2 | cornu ammonis 2 | LGs | LG, superficial | sod | suproptic decn |
| CA3 | cornu ammonis 3 | lml | lat. medull. lamina | st | stria terminalis |
| Cd | caudate nu. | LMol | strat. lacun. mol. CA | TH | temp. field H Cx |
| CeL | cent. amyg. nu., lat. | LV | lateral ventricle | V1 | V1 Cx |
| CeM | cent. amyg. nu., med. | mcp | middle cerebell. ped. | V2l | V2 (lat.) Cx |
| Cl | claustrum | Me | medial amyg. nu. | V2m | V2 (med.) Cx |
| Cop | copula | MGV | med. gen. nu., vent. | V2v | V2 (vent.) Cx |
| Crus1 | crus 1, ansiform lob. | MoDG | molecular layer DG | VCA | vent. coch. nu., ant. |
| Crus2 | crus 2, ansiform lob. | opt | optic tract | VCP | vent. coch. nu., post. |
| DA | dorsoant. visual Cx | orbl | orbital sulc. lat. | VEn | vent. endopir. nu. |
| DC | dorsal cochlear nu. | orbm | orbital sulc. med. | VLP (V3v) | ventlatpost. (V3v) Cx |
| DEn | dors. endopir. nu. | PaS | parasubiculum | VPL | vent. post. thal. nu., lat. |

# Plate 72a

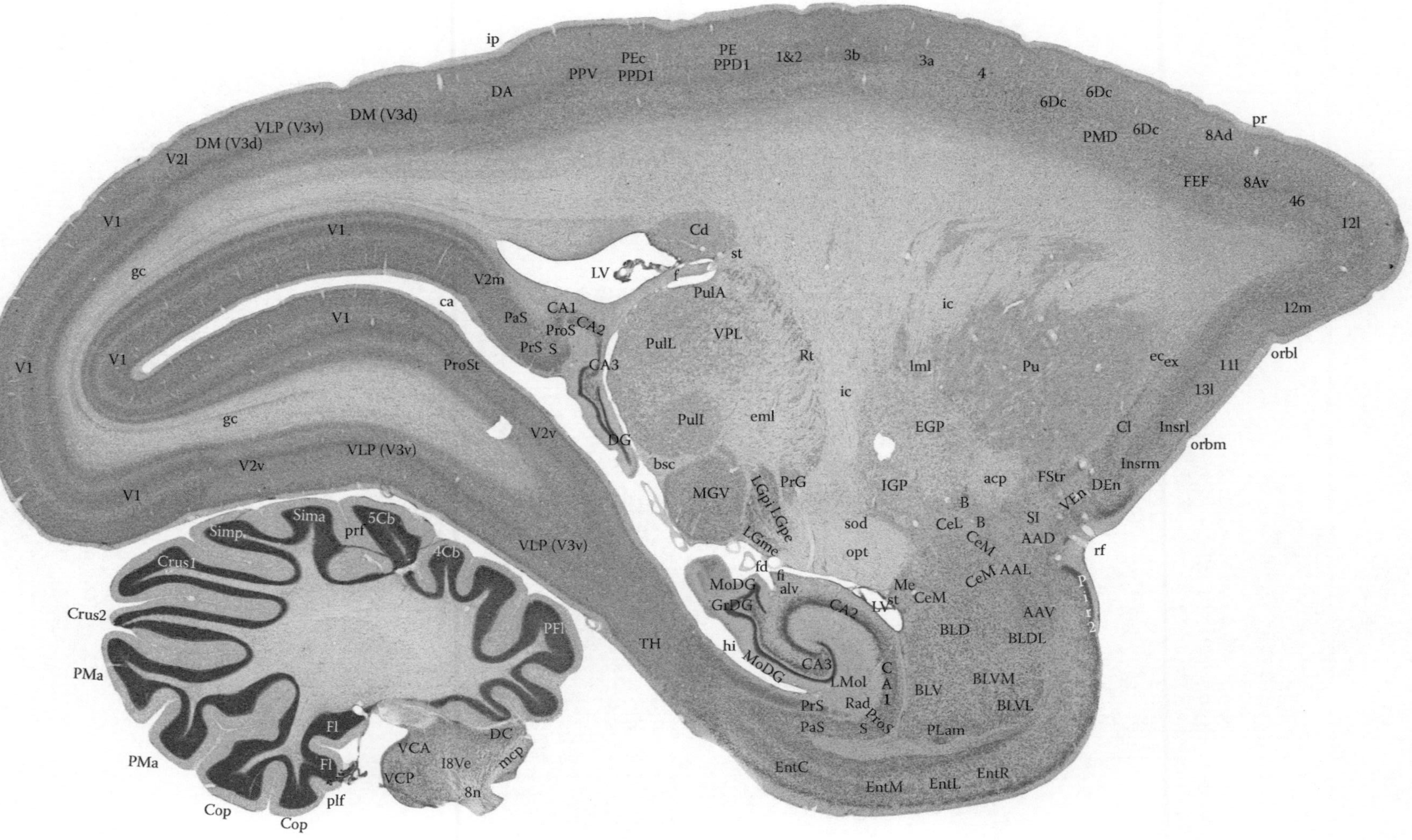


**Cresyl Violet**
Lateral +05.60 mm

# Plate 72b

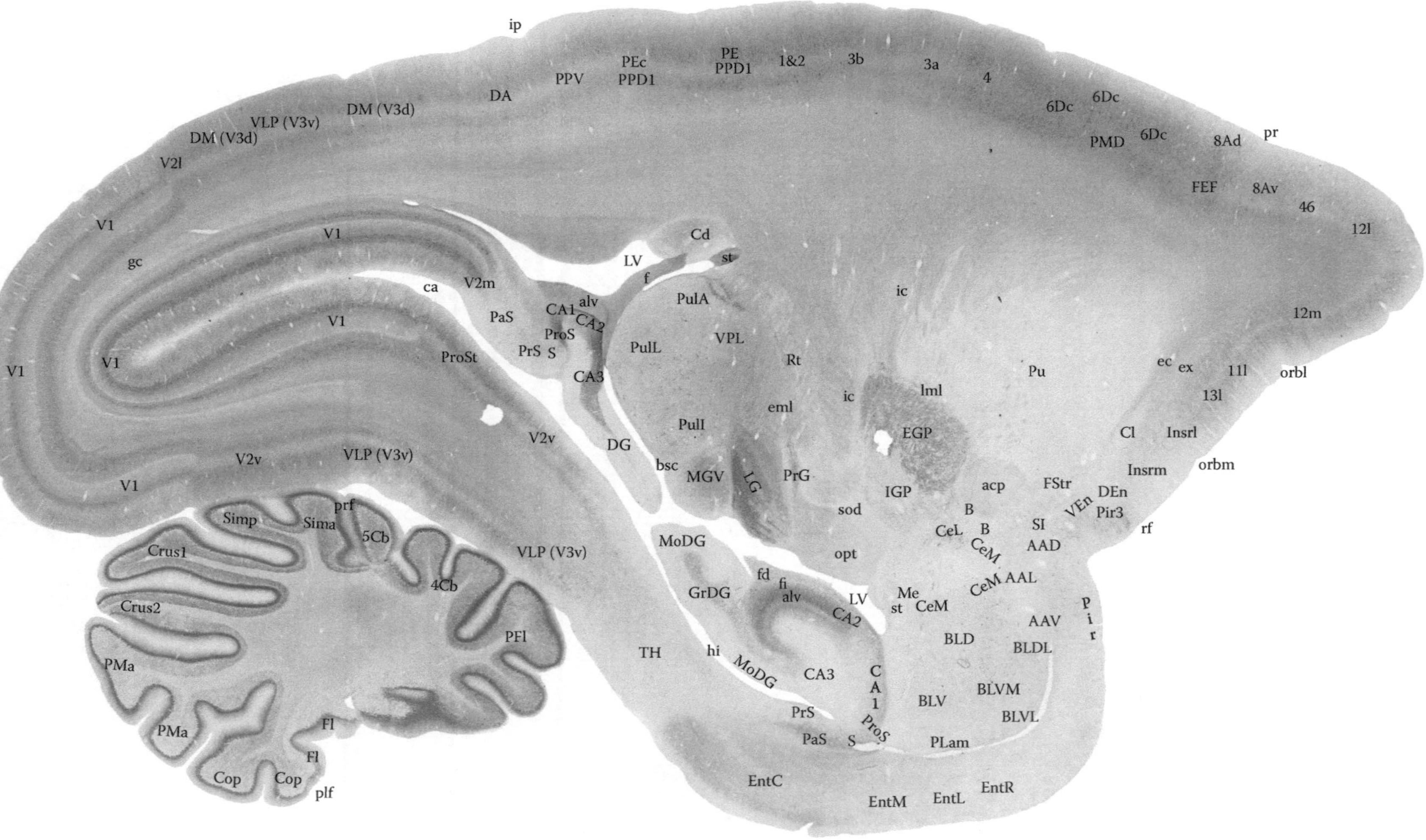


**Neurofilament**
Lateral +05.85 mm

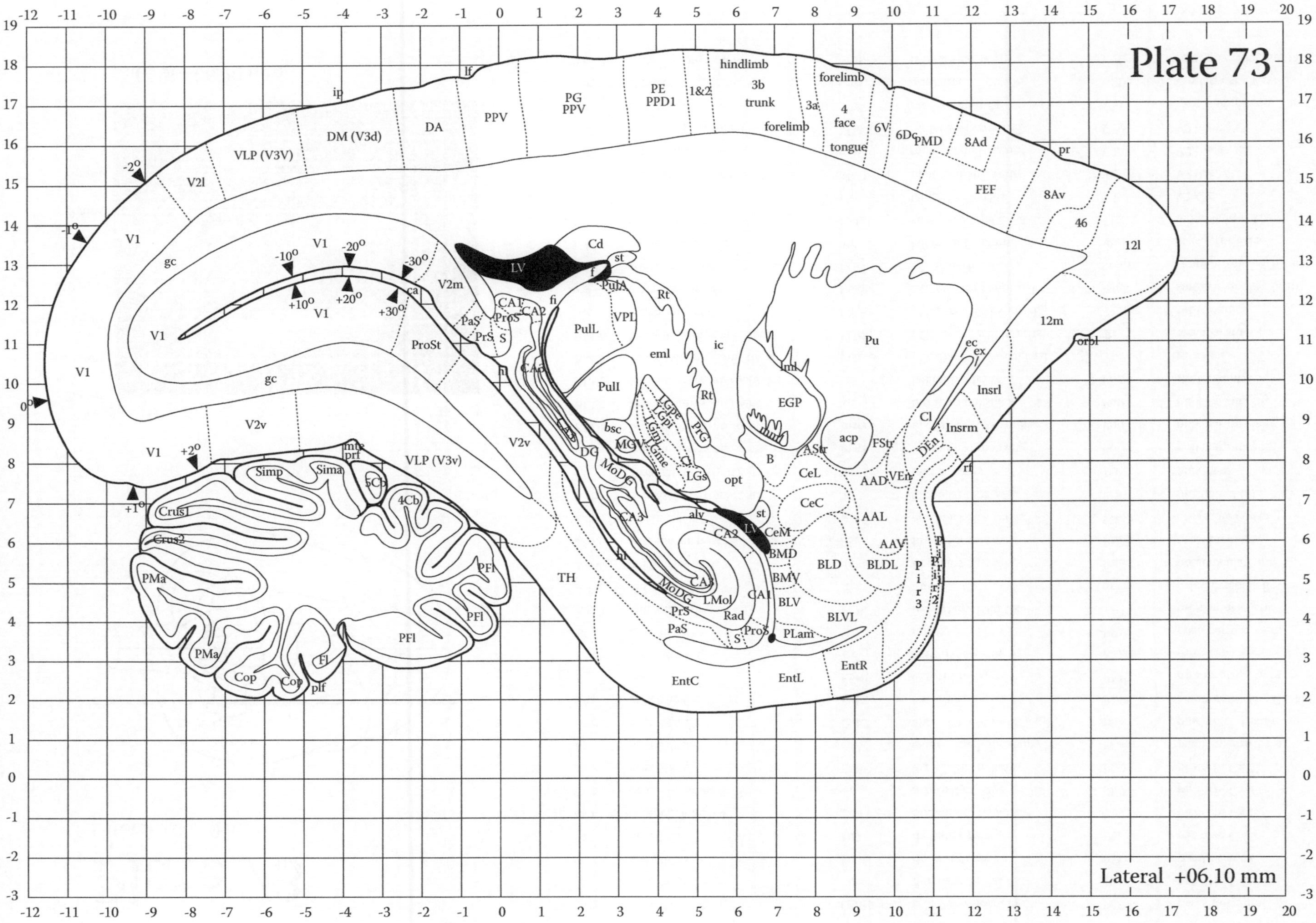
Plate 73
Lateral +06.10 mm
hindlimb
forelimb
3b
trunk
3a
4
face
tongue
forelimb
PE
PPD1
1&2
PG
PPV
PPV
DA
DM (V3d)
VLP (V3V)
V2l
ip
lf
6V
6Dc
PMD
8Ad
pr
FEF
8Av
46
12l
12m
orbl
V1
gc
V1
V1
V1
V2m
ca
ProSt
V2v
V2v
VLP (V3v)
-1o
-2o
0o
+1o
+2o
-10o
-20o
-30o
+10o
+20o
+30o
LV
Cd
st
f
PulA
fi
CA1
CA2
ProS
PaS
PrS
S
CA3
PulL
VPL
Rt
eml
ic
PulI
LGpe
LGpi
LGmi
LGme
Rt
PrG
bsc
MGV
DG
MoDG
CA3
CL
LGs
opt
alv
CA2
LV
st
CeM
BMD
BMV
CA1
LMol
Rad
ProS
S
PrS
PaS
MoDG
CA3
hf
Pu
lml
EGP
mml
B
AStr
acp
FStr
CeL
CeC
AAD
VEn
AAL
AAV
BLD
BLDL
BLV
BLVL
PLam
Cl
DEn
ec
ex
Insrl
Insrm
rf
Pir3
Pir2
TH
EntC
EntL
EntR
Simp
Sima
mfs
prf
5Cb
4Cb
Crus1
Crus2
PMa
PMa
PFl
PFl
PFl
Fl
Cop
Cop
plf

# Plate 73

Lateral +06.10 mm

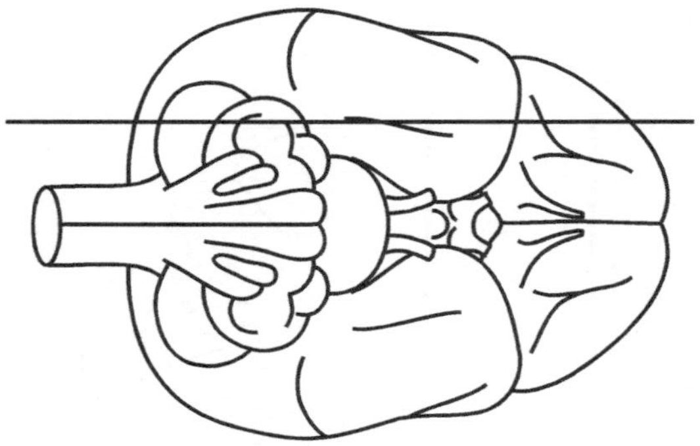

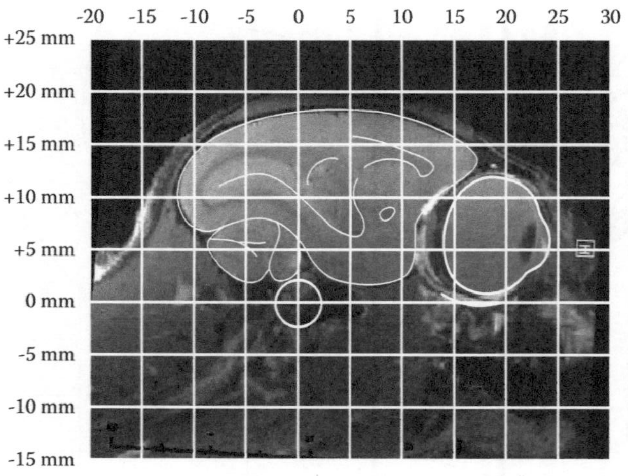


Lateral +06.20 mm

| | | | | | |
|---|---|---|---|---|---|
| 1&2 | areas 1/2 Cx | Crus2 | crus 2, ansiform lob. | PaS | parasubiculum |
| 3a | area 3a Cx | DA | dorsoant. visual Cx | PE | par. field E Cx |
| 3b | area 3b Cx | DEn | dors. endopir. nu. | PFl | paraflocculus |
| 4 | area 4 Cx | DG | dentate gyrus | PG | par. field G Cx |
| 4Cb | cerebell. lobule 4 | DM (V3d) | dorsomed. visual Cx | Pir1 | piriform Cx 1 |
| 5Cb | cerebell. lobule 5 | ec | external capsule | Pir2 | piriform Cx 2 |
| 6Dc | area 6Dc Cx | EGP | ext. globus pallidus | Pir3 | piriform Cx 3 |
| 6V | area 6V Cx | eml | extern. medull. lamina | PLam | paralam. amyg. nu. |
| 8Ad | area 8Ad Cx | EntC | entorhin. Cx, caud. | plf | posterolat. fissure |
| 8Av | area 8Av Cx | EntL | entorhin. Cx, lat. | PMa | paramedian lob. Cb, ant. |
| 12l | area 12l Cx | EntR | entorhin. Cx, rost. | PMD | premotor Cx dors. |
| 12m | area 12m Cx | ex | extreme capsule | PPD1 | post. par. Cx, dors. 1 |
| 46 | area 46 Cx | f | fornix | PPV | post. par. Cx, vent. |
| AAD | ant. amyg. dors. | FEF | frontal eye field Cx | pr | prefrontal sulcus |
| AAL | ant. amyg. lat. | fi | fimbria | prf | prim. cerebell. fiss. |
| AAV | ant. amyg. vent. | Fl | flocculus | PrG | pregeniculate nu. |
| acp | ant. comm. post. | FStr | fundus striatum | ProS | prosubiculum |
| alv | alveus | gc | geniculocalcarine tr. | ProSt | prostriate cortex |
| AStr | amygdalostr. trans. zone | hi | hippocamp. fiss. | PrS | presubiculum |
| B | basal nu. (Meynert) | ic | intern. caps. | Pu | putamen |
| BLD | basolat. amyg. nu., dors. | Insrl | insular Cx, rost. lat. | PulA | pulvinar nu., ant. |
| BLDL | basolat. amyg. nu., dors. lat. | Insrm | insular Cx, rost. med. | PulI | pulvinar nu., inf. |
| BLV | basolat. amyg. nu., vent. | ip | intraparietal sulc. | PulL | pulvinar nu., lat. |
| BLVL | basolat. amyg. nu., vent. lat. | lf | lateral fissure | Rad | strat. radiat. CA |
| BMD | basomed. amyg. nu., dors. | LGme | LG, magnocell. external | rf | rhinal fiss. |
| BMV | basomed. amyg. nu., vent. | LGmi | LG, magnocell. internal | Rt | reticular nu. |
| bsc | brach. sup. coll. | LGpe | LG, parvicell. external | S | subiculum |
| ca | calcarine sulcus | LGpi | LG, parvicell. internal | Sima | simplex lobule Cb, ant. |
| CA1 | cornu ammonis 1 | LGs | LG, superficial | Simp | simplex lobule Cb, post. |
| CA2 | cornu ammonis 2 | lml | lat. medull. lamina | st | stria terminalis |
| CA3 | cornu ammonis 3 | LMol | strat. lacun. mol. CA | TH | temp. field H Cx |
| Cd | caudate nu. | LV | lateral ventricle | V1 | V1 Cx |
| CeC | cent. amyg. nu., cent. | MGV | med. gen. nu., vent. | V2l | V2 (lat.) Cx |
| CeL | cent. amyg. nu., lat. | mml | med. medull. lamina | V2m | V2 (med.) Cx |
| CeM | cent. amyg. nu., med. | MoDG | molecular layer DG | V2v | V2 (vent.) Cx |
| Cl | claustrum | mte | med. temp. sulc. | VEn | vent. endopir. nu. |
| Cop | copula | opt | optic tract | VLP (V3V) | ventlatpost. (V3v) Cx |
| Crus1 | crus 1, ansiform lob. | orbl | orbital sulc. lat. | VPL | vent. post. thal. nu., lat. |

# Plate 73a

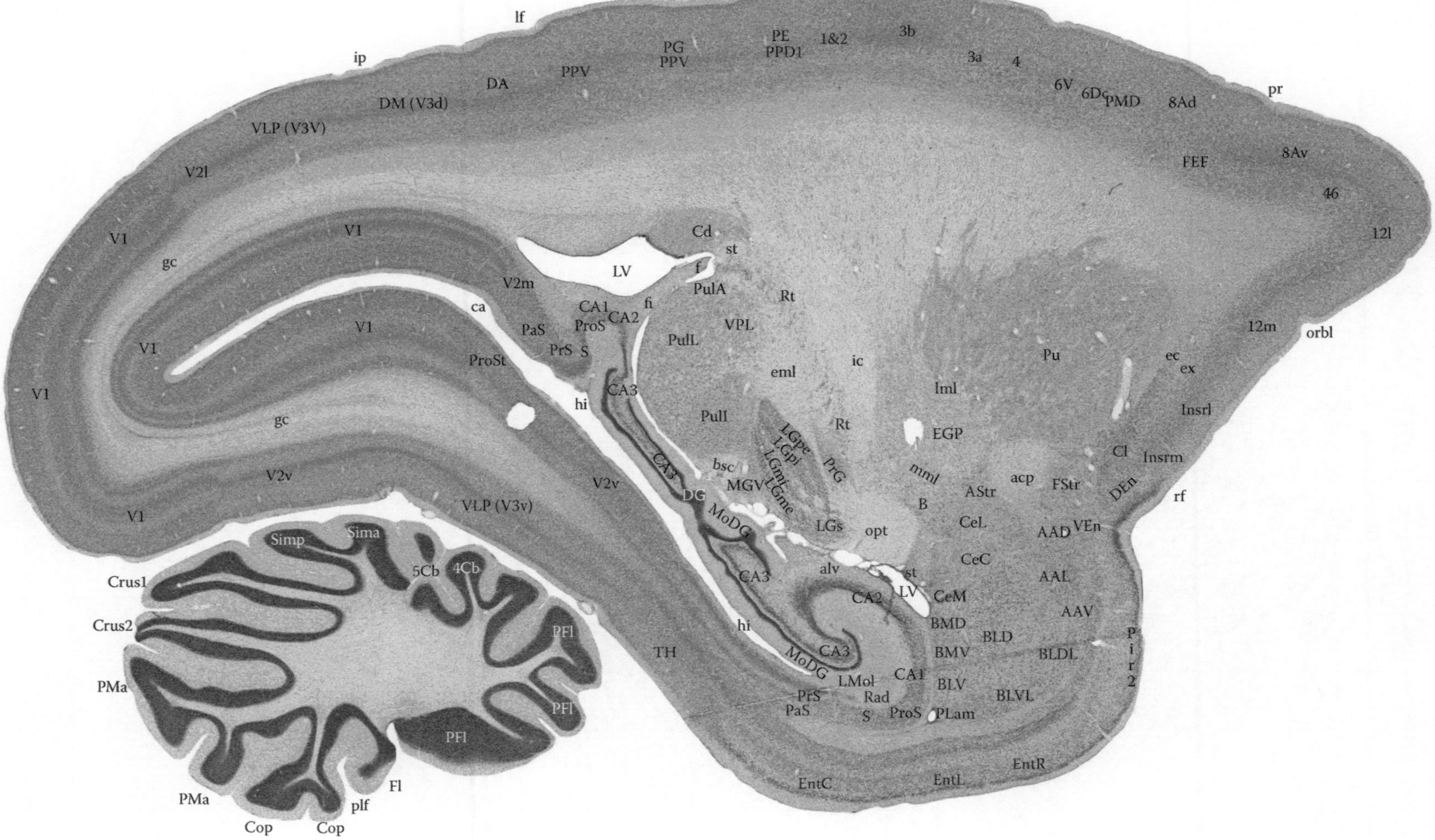


**Cresyl Violet**
Lateral +06.10 mm

# Plate 73b

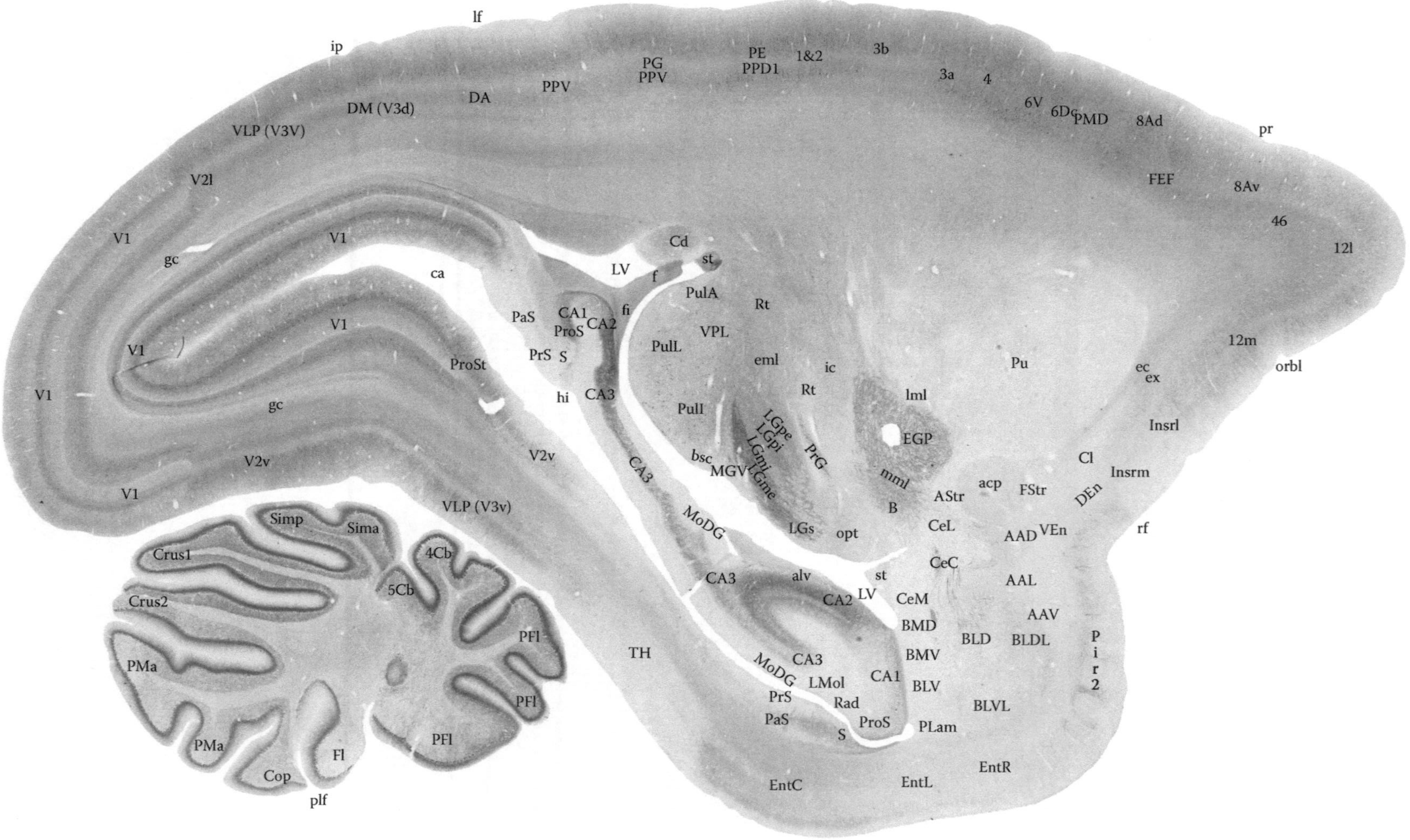


**Neurofilament**
Lateral +06.35 mm

# Plate 74

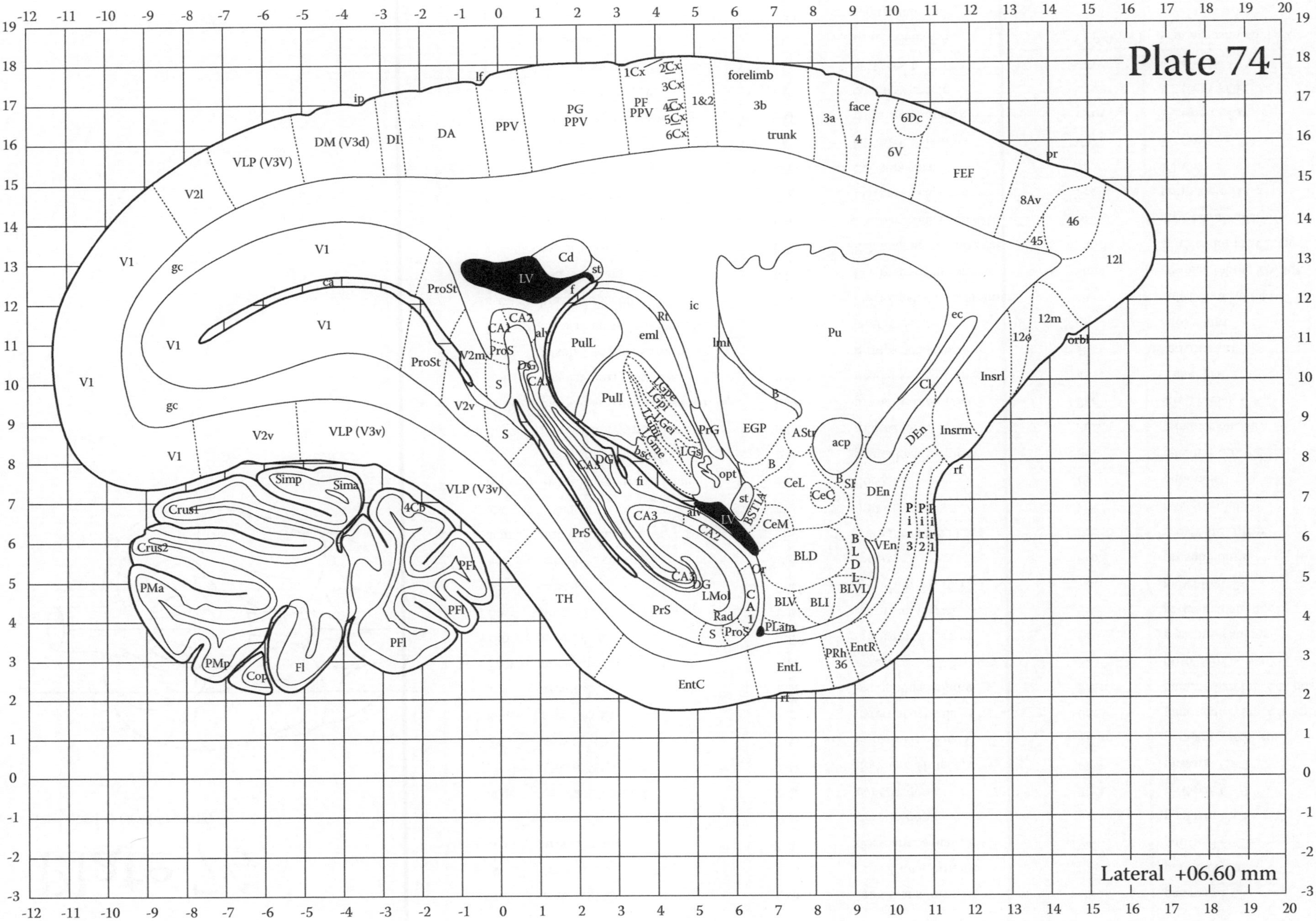


Lateral +06.60 mm

# Plate 74

Lateral +06.60 mm

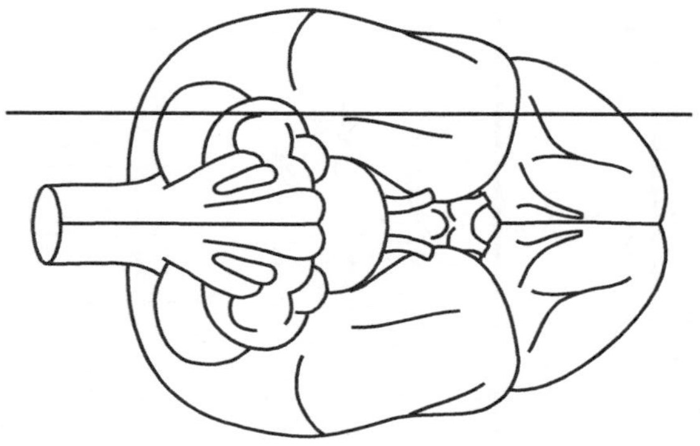

| | |
|---|---|
| 1&2 | areas 1/2 Cx |
| 3a | area 3a Cx |
| 3b | area 3b Cx |
| 4 | area 4 Cx |
| 4Cb | cerebell. lobule 4 |
| 6Dc | area 6Dc Cx |
| 6V | area 6V Cx |
| 8Av | area 8Av Cx |
| 12l | area 12l Cx |
| 12m | area 12m Cx |
| 12o | area 12o Cx |
| 36 | area 36 Cx |
| 45 | area 45 Cx |
| 46 | area 46 Cx |
| acp | ant. comm. post. |
| alv | alveus |
| AStr | amygdalostr. trans. zone |
| B | basal nu. (Meynert) |
| BLD | basolat. amyg. nu., dors. |
| BLDL | basolat. amyg. nu., dors. lat. |
| BLI | basolat. amyg. nu., int. |
| BLV | basolat. amyg. nu., vent. |
| BLVL | basolat. amyg. nu., vent. lat. |
| bsc | brach. sup. coll. |
| BSTIA | bed nu. st, intraamyg. |
| ca | calcarine sulcus |
| CA1 | cornu ammonis 1 |
| CA2 | cornu ammonis 2 |
| CA3 | cornu ammonis 3 |
| Cd | caudate nu. |
| CeC | cent. amyg. nu., cent. |
| CeL | cent. amyg. nu., lat. |
| CeM | cent. amyg. nu., med. |
| Cl | claustrum |
| Cop | copula |
| Crus1 | crus 1, ansiform lob. |
| Crus2 | crus 2, ansiform lob. |
| DA | dorsoant. visual Cx |
| DEn | dors. endopir. nu. |
| DG | dentate gyrus |
| DI | dorsoint. visual Cx |
| DM (V3d) | dorsomed. visual Cx |
| ec | external capsule |
| EGP | ext. globus pallidus |
| eml | extern. medull. lamina |
| EntC | entorhin. Cx, caud. |
| EntL | entorhin. Cx, lat. |
| EntR | entorhin. Cx, rost. |
| f | fornix |
| FEF | frontal eye field Cx |
| fi | fimbria |
| Fl | flocculus |
| gc | geniculocalcarine tr. |
| ic | intern. caps. |
| Insrl | insular Cx, rost. lat. |
| Insrm | insular Cx, rost. med. |
| ip | intraparietal sulc. |
| lf | lateral fissure |
| LGei | LG, external interlaminar |
| LGme | LG, magnocell. external |
| LGmi | LG, magnocell. internal |
| LGpe | LG, parvicell. external |
| LGpi | LG, parvicell. internal |
| LGs | LG, superficial |
| lml | lat. medull. lamina |
| LMol | strat. lacun. mol. CA |
| LV | lateral ventricle |
| opt | optic tract |
| Or | strat. oriens CA |
| orbl | orbital sulc. lat. |
| PF | par. field F Cx |
| PFl | paraflocculus |
| PG | par. field G Cx |
| Pir1 | piriform Cx 1 |
| Pir2 | piriform Cx 2 |
| Pir3 | piriform Cx 3 |
| PLam | paralam. amyg. nu. |
| PMa | paramedian lob. Cb, ant. |
| PMp | paramedian lob. Cb, post. |
| PPV | post. par. Cx, vent. |
| pr | prefrontal sulcus |
| PrG | pregeniculate nu. |
| PRh | perirhinal Cx |
| ProS | prosubiculum |
| ProSt | prostriate cortex |
| PrS | presubiculum |
| Pu | putamen |
| PulI | pulvinar nu., inf. |
| PulL | pulvinar nu., lat. |
| Rad | strat. radiat. CA |
| rf | rhinal fiss. |
| Rt | reticular nu. |
| S | subiculum |
| SI | subst. innominata |
| Sima | simplex lobule Cb, ant. |
| Simp | simplex lobule Cb, post. |
| st | stria terminalis |
| TH | temp. field H Cx |
| V1 | V1 Cx |
| V2l | V2 (lat.) Cx |
| V2m | V2 (med.) Cx |
| V2v | V2 (vent.) Cx |
| VEn | vent. endopir. nu. |
| VLP (V3V) | ventlatpost. (V3v) Cx |

# Plate 74a

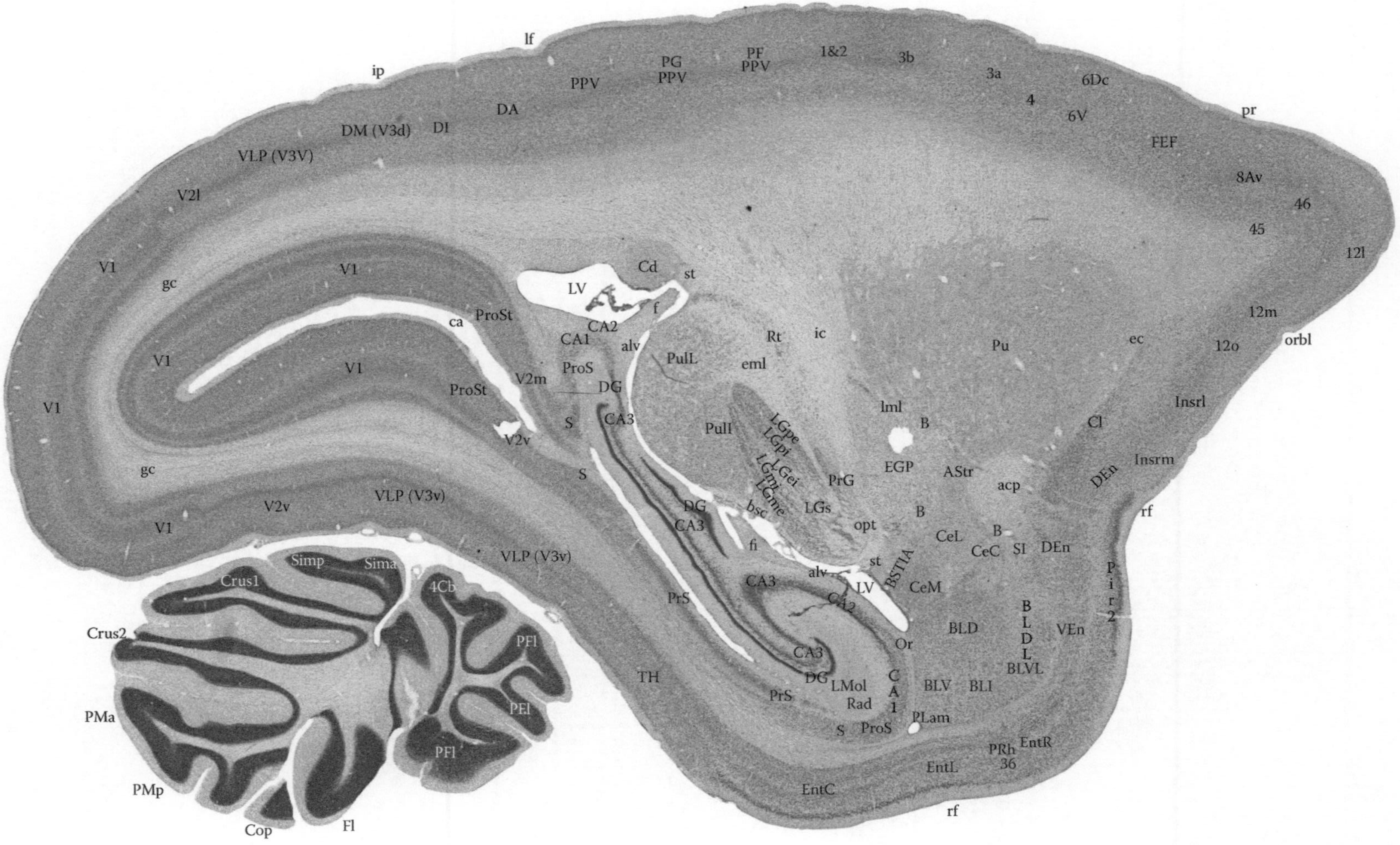


**Cresyl Violet**
Lateral +06.60 mm

# Plate 74b

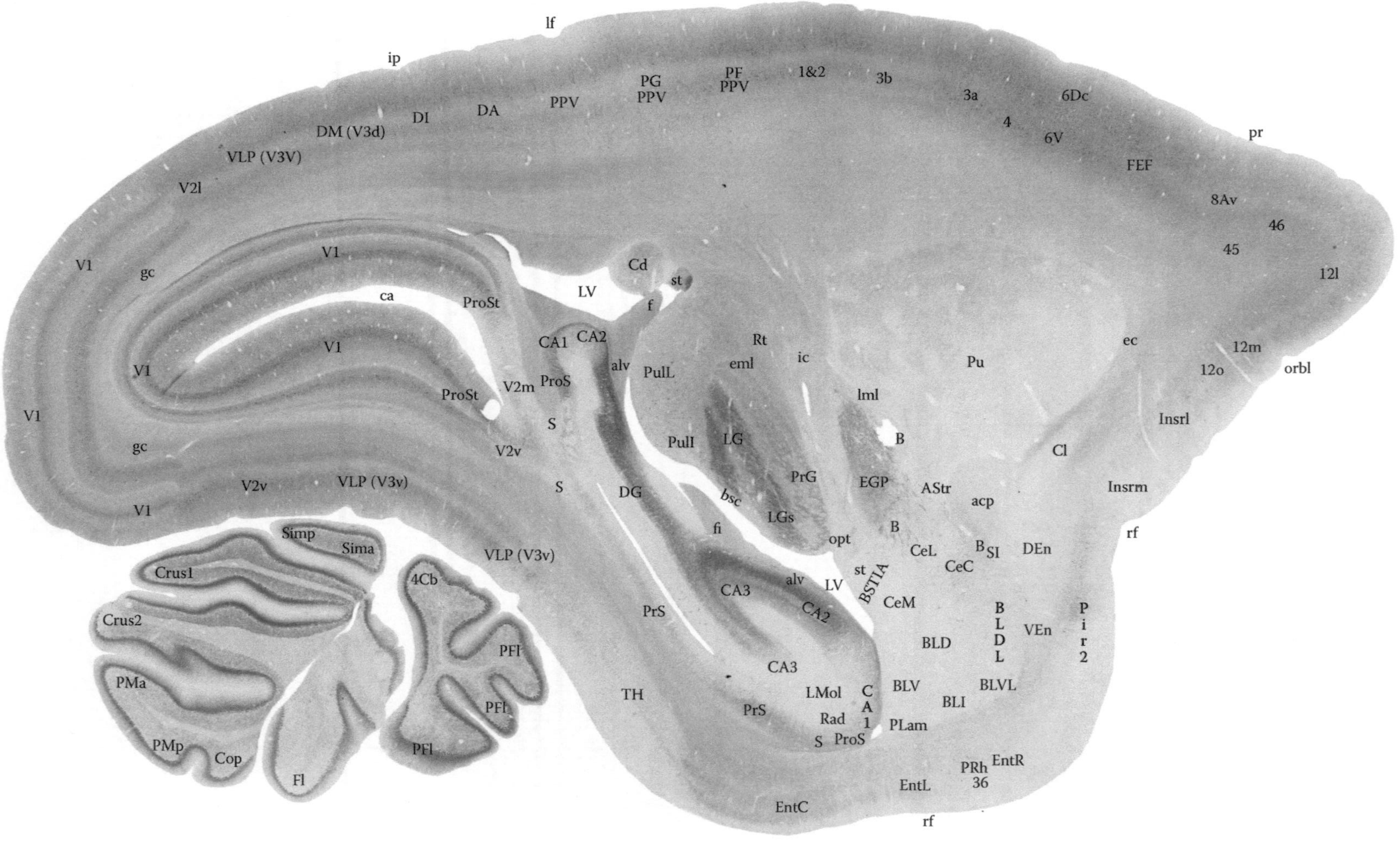


**Neurofilament**
Lateral +06.85 mm

# Plate 75

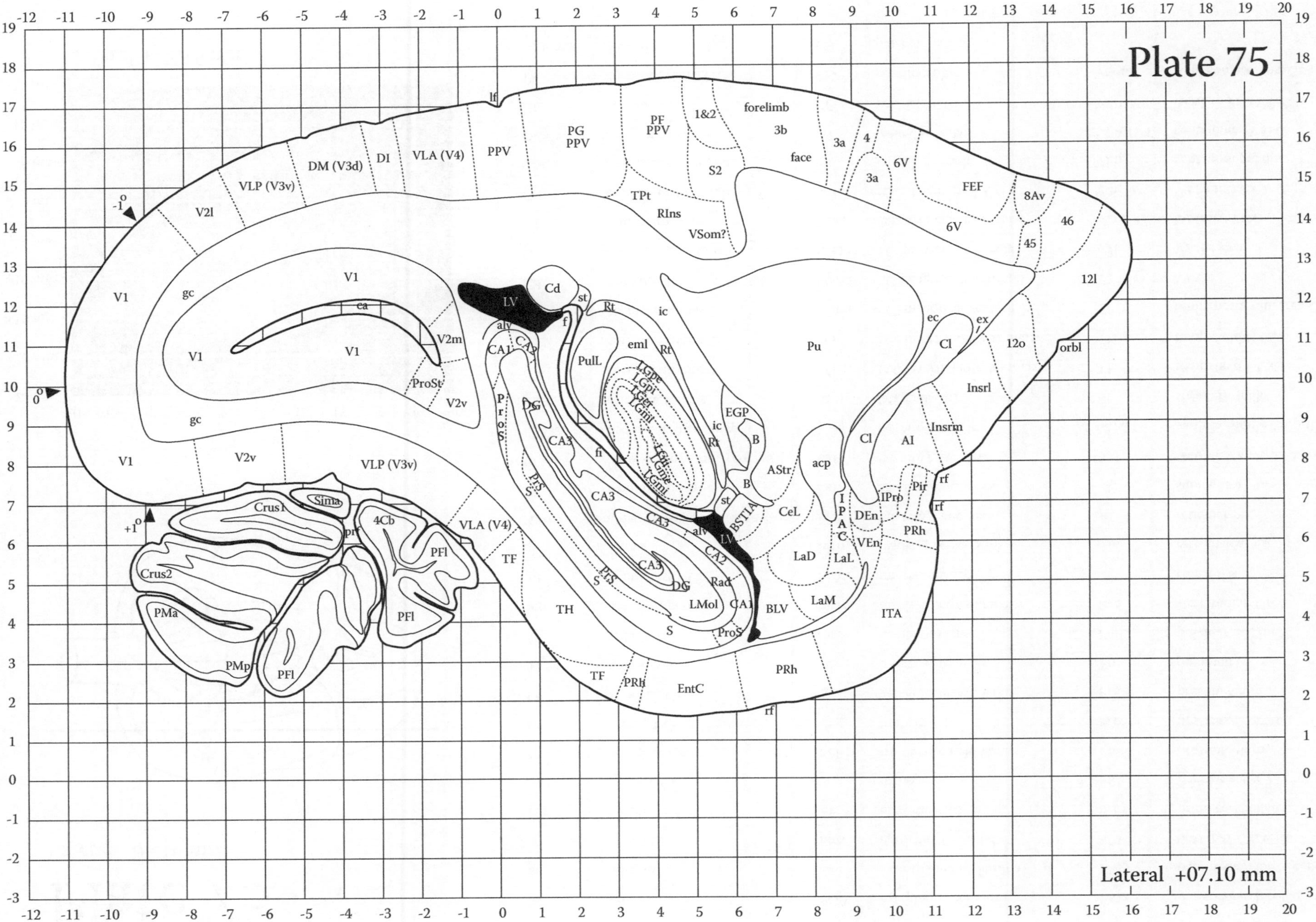
lf
PPV
PG
PPV
PF
PPV
1&2
forelimb
3b
face
S2
3a
4
3a
6V
FEF
8Av
6V
46
45
12l
orbl
12o
ex
ec
Cl
Insrl
Insrm
AI
Cl
TPt
RIns
VSom?
VLA (V4)
DI
DM (V3d)
VLP (V3v)
V2l
V1
gc
V1
ca
V1
V1
gc
V1
V2v
VLP (V3v)
V2m
ProSt
V2v
VLA (V4)
TF
TH
TF
PRh
EntC
PRh
rf
ITA
Cd
st
LV
alv
f
CA1
CA3
Pro
S
DG
CA3
fi
PrS
S
CA3
CA3
CA3
CA2
alv
LV
DG
Rad
LMol
CA1
S
ProS
PrS
S
Rt
ic
eml
Rt
PulL
LGpe
LGpi
LGei
LGmi
LGii
LGme
LGmi
Pu
EGP
ic
Rt
B
B
st
BSTIA
AStr
acp
I
P
A
C
CeL
LaD
LaL
LaM
BLV
DEn
VEn
IPro
Pir
rf
rf
PRh
Crus1
Sima
prf
4Cb
PFl
PFl
Crus2
PMa
PMp
PFl
-1°
0°
+1°
Lateral +07.10 mm

# Plate 75

Lateral +07.10 mm

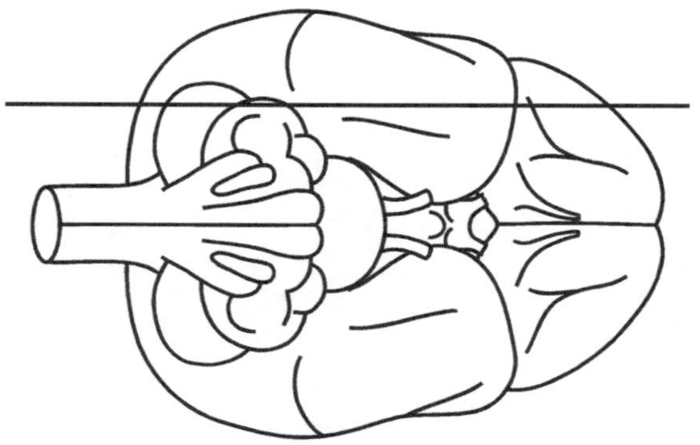

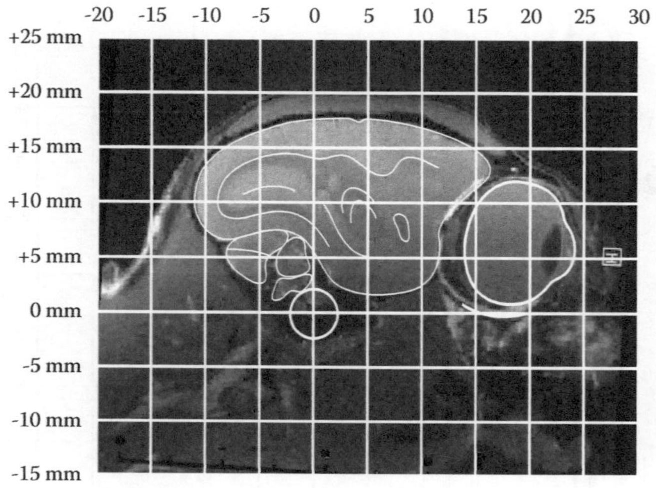


Lateral +07.15 mm

| | |
|---|---|
| 1&2 | areas 1/2 Cx |
| 3a | area 3a Cx |
| 3b | area 3b Cx |
| 4 | area 4 Cx |
| 4Cb | cerebell. lobule 4 |
| 6V | area 6V Cx |
| 8Av | area 8Av Cx |
| 12l | area 12l Cx |
| 12o | area 12o Cx |
| 45 | area 45 Cx |
| 46 | area 46 Cx |
| acp | ant. comm. post. |
| AI | agran. insul. Cx |
| alv | alveus |
| AStr | amygdalostr. trans. zone |
| B | basal nu. (Meynert) |
| BLV | basolat. amyg. nu., vent. |
| BSTIA | bed nu. st, intraamyg. |
| ca | calcarine sulcus |
| CA1 | cornu ammonis 1 |
| CA2 | cornu ammonis 2 |
| CA3 | cornu ammonis 3 |
| Cd | caudate nu. |
| CeL | cent. amyg. nu., lat. |
| Cl | claustrum |
| Crus1 | crus 1, ansiform lob. |
| Crus2 | crus 2, ansiform lob. |
| DEn | dors. endopir. nu. |
| DG | dentate gyrus |
| DI | dorsoint. visual Cx |
| DM (V3d) | dorsomed. visual Cx |
| ec | external capsule |
| EGP | ext. globus pallidus |
| eml | extern. medull. lamina |
| EntC | entorhin. Cx, caud. |
| ex | extreme capsule |
| f | fornix |
| FEF | frontal eye field Cx |
| fi | fimbria |
| gc | geniculocalcarine tr. |
| ic | intern. caps. |
| Insrl | insular Cx, rost. lat. |
| Insrm | insular Cx, rost. med. |
| IPAC | interst. nu. acp |
| IPro | insular proisocortex |
| ITA | inf. temp. ant. Cx |
| LaD | lat. amyg. nu., dors. |
| LaL | lat. amyg. nu., lat. |
| LaM | lat. amyg. nu., med. |
| lf | lateral fissure |
| LGei | LG, external interlaminar |
| LGii | LG, internal interlaminar |
| LGme | LG, magnocell. external |
| LGmi | LG, magnocell. internal |
| LGpe | LG, parvicell. external |
| LGpi | LG, parvicell. internal |
| LMol | strat. lacun. mol. CA |
| LV | lateral ventricle |
| orbl | orbital sulc. lat. |
| PF | par. field F Cx |
| PFl | paraflocculus |
| PG | par. field G Cx |
| Pir | piriform Cx |
| PMa | paramedian lob. Cb, ant. |
| PMp | paramedian lob. Cb, post. |
| PPV | post. par. Cx, vent. |
| prf | prim. cerebell. fiss. |
| PRh | perirhinal Cx |
| ProS | prosubiculum |
| ProSt | prostriate cortex |
| PrS | presubiculum |
| Pu | putamen |
| PulL | pulvinar nu., inf. |
| Rad | strat. radiat. CA |
| rf | rhinal fiss. |
| RIns | retroinsular Cx |
| Rt | reticular nu. |
| S | subiculum |
| S2 | second. somat. area Cx |
| Sima | simplex lobule Cb, ant. |
| st | stria terminalis |
| TF | temp. field F Cx |
| TH | temp. field H Cx |
| TPt | temp. paritemp. area Cx |
| V1 | V1 Cx |
| V2l | V2 (lat.) Cx |
| V2m | V2 (med.) Cx |
| V2v | V2 (vent.) Cx |
| VEn | vent. endopir. nu. |
| VLA (V4) | ventlatant. (V4) Cx |
| VLP (V3v) | ventlatpost. (V3v) Cx |
| VSom? | vent. somatosens. Cx |

# Plate 75a

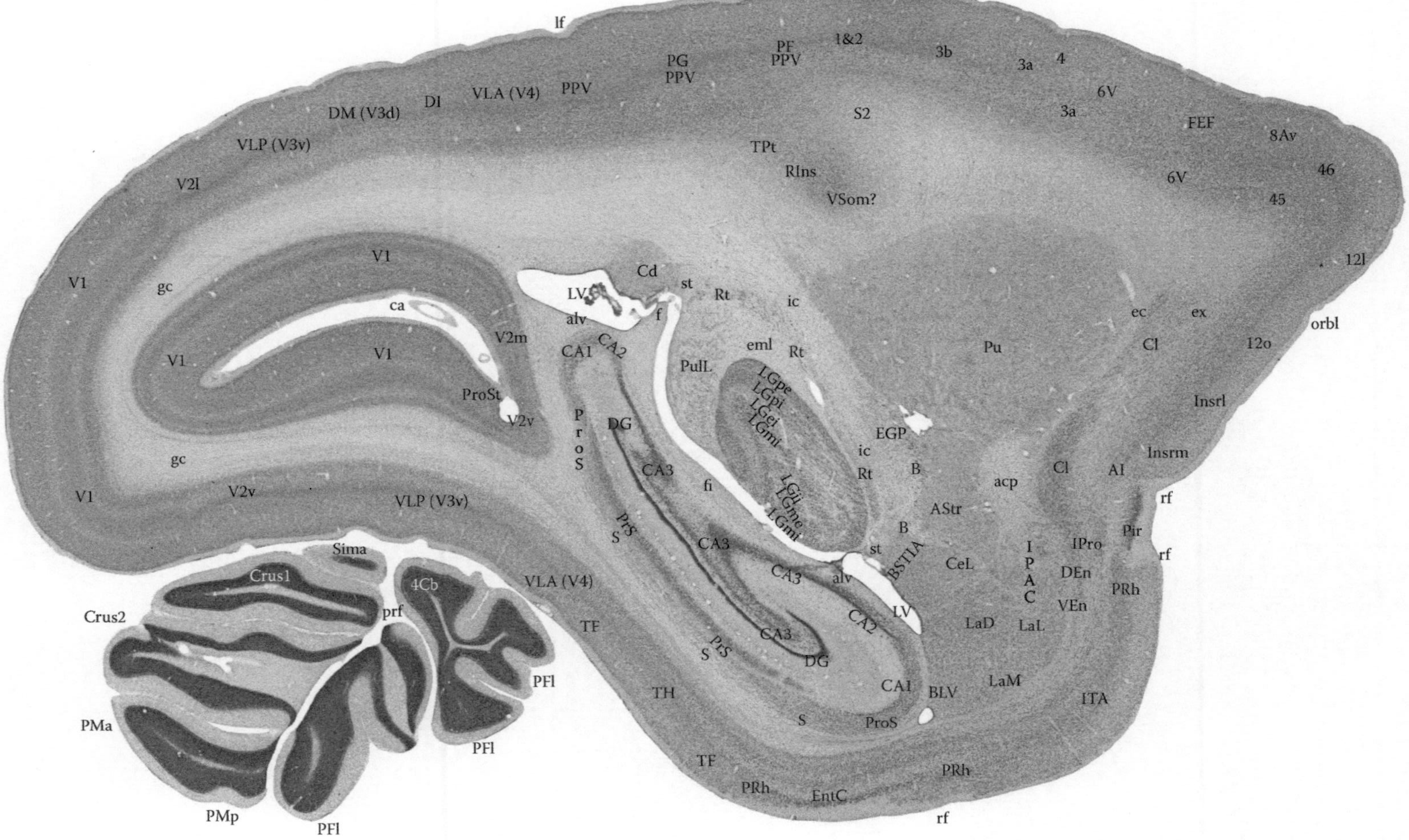


**Cresyl Violet**
Lateral +07.10 mm

# Plate 75b

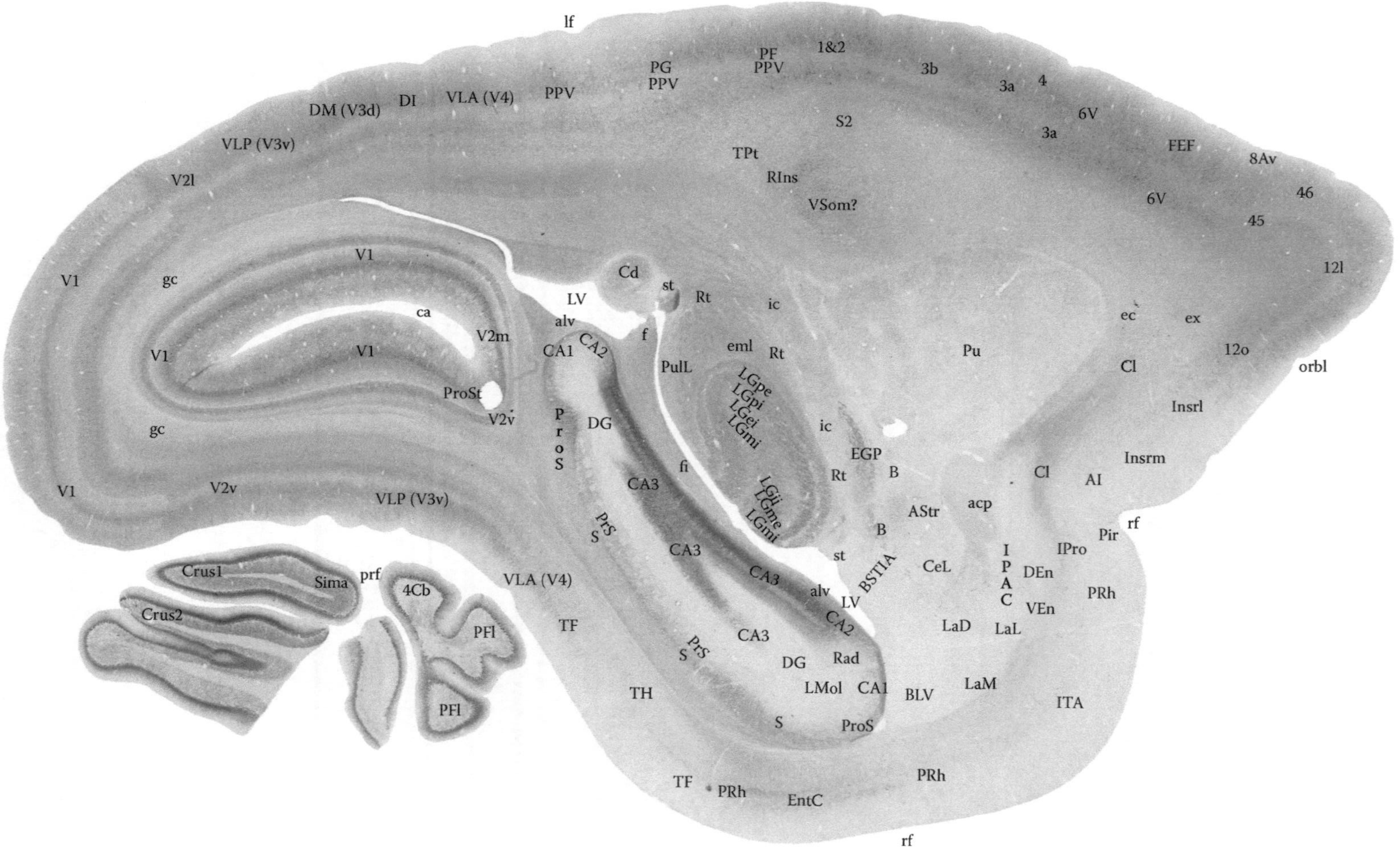


**Neurofilament**
Lateral +07.35 mm